HANDBUCH DER ASTROPHYSIK

HERAUSGEGEBEN VON

G. EBERHARD · A. KOHLSCHÜTTER
H. LUDENDORFF

BAND V / ERSTE HÄLFTE

DAS STERNSYSTEM

ERSTER TEIL

BERLIN
VERLAG VON JULIUS SPRINGER
1932

DAS STERNSYSTEM

ERSTER TEIL

I

BEARBEITET VON

FR. BECKER · A. BRILL

R. H. CURTISS † · K. LUNDMARK

MIT 173 ABBILDUNGEN

BERLIN

VERLAG VON JULIUS SPRINGER

1932

Softcover reprint of the hardcover 1st edition 1932

ISBN-13: 978-3-642-88847-2 e-ISBN-13: 978-3-642-90702-9
DOI: 10.1007/978-3-642-90702-9

Inhaltsverzeichnis.

Chapter 1.

Classification and Description of Stellar Spectra.

By Prof. R. H. CURTISS †, Ann Arbor.

(With 14 illustrations.)

Kapitel 2.

Zur Statistik der Spektraltypen.

Von Dr. Fr. BECKER, Bonn.

(Mit 7 Abbildungen.)

Kapitel 3.

Die Temperaturen der Fixsterne.

Von Prof. Dr. A. BRILL, Neubabelsberg.

(Mit 5 Abbildungen.)

Chapter 4.

Luminosities, Colours, Diameters, Densities, Masses of the Stars.

By Prof. Dr. KNUT LUNDMARK, Lund.

(With 147 illustrations.)

Inhalt der zweiten Hälfte.

Chapter 1.

Classification and Description of Stellar Spectra.

By

R. H. CURTISS[1] †-Ann Arbor.

With 14 illustrations.

a) The Beginning of Stellar Spectrum Analysis.

1. FRAUNHOFER. To JOSEPH VON FRAUNHOFER we are indebted for the first analyses of the spectra of fixed stars. Though he was concerned with the study of spectra primarily from the standpoint of an optician and maker of achromatic objectives, FRAUNHOFER's truly scientific interest led him instinctively to record and describe new phenomena which came under his observation. Thus he counted at least 576 lines and bands in the solar spectrum and mapped 324 of them, whereas a very few of the number were useful to him in his own work. In the same way FRAUNHOFER has given us excellent descriptions of stellar spectra as a by-product of investigations whose primary purpose was a comparison of the refrangibility of the light received from various celestial sources.

FRAUNHOFER's first stellar spectroscope was adapted from the instrument with which he mapped the lines of the solar spectrum[2]. This instrument employed a slit about a half millimeter in width (WOLLASTON had used a slit twice as wide in his discovery of seven of the stronger solar lines), through which sunlight passed to a 60° prism of flint glass distant 7,6 meters from the slit and large enough to cover the 29 millimeter objective of a small theodolite. Suspecting that the dark lines and bands in the solar spectrum seen with this spectroscope might be connected in some way with diffraction at the slit, FRAUNHOFER modified the experiment by substituting for the slit a small circular orifice of the same diameter as the slit width. A cylindrical lens was then inserted between the prism and the theodolite objective to widen the spectrum. This arrangement

[1] This article was completed shortly before the death of the author, December 25, 1929. Much of it, however, was written in the spring of 1926 during a visit to the Mt. Wilson Observatory. In accordance with the expressed wish of Professor CURTISS, grateful acknowledgement is here made of the many courtesies shown him by Director WALTER S. ADAMS and the staff of the Mt. Wilson Obervatory, by Director ROBERT G. AITKEN and the staff of the Lick Observatory, by Director EDWIN B. FROST and the staff of the Yerkes Observatory, and by Director HARLOW SHAPLEY and the staff of Harvard Observatory. Professor CURTISS would undoubtedly also have expressed his indebtedness to his colleague, Dr. W. C. RUFUS, who discussed with him the scope of this article, and who furnished valuable suggestions at later stages of the work. Other members of the staff of the Detroit Observatory assisted in various ways; Professor D. B. MCLAUGHLIN was especially helpful.

[2] Denkschr d kgl Akad d Wiss zu München 5, p. 193 (1817).

proved so successful that FRAUNHOFER decided to apply it, without the small orifice, to the study of the planet Venus, and of some of the brighter stars. Thus the objective prism became the earliest form of star spectroscope.

FRAUNHOFER's description of his first observations of stellar spectra with his small objective prism was published in 1817[1] and was copied widely in scientific publications[2]. Following an account of observations of the spectrum of Venus he wrote:

"With the same arrangement of apparatus I have made studies of the light of several fixed stars of first magnitude. Since the light of these stars is much fainter than that of Venus the brightness of the spectrum is of course much less. Nevertheless I have seen without question in the spectrum of the light of Sirius three broad bands which appear to show no resemblance to those in sunlight. Further in the spectra of the light of other fixed stars of the first magnitude we discern streaks; but in respect to such features these stars appear to differ among themselves."

FRAUNHOFER was not satisfied with these observations. After presenting them he announced his intention to repeat them with suitable modifications including the use of a greater objective. By so doing he hoped to arouse interest in these studies among skilled scientific observers, and also at the same time to effect an accurate comparison of the refrangibility of starlight and sunlight—a matter of great interest to FRAUNHOFER and one prominent in current scientific discussion at that time.

FRAUNHOFER published his promised observations of stellar spectra with improved apparatus in 1823[3]. He employed the objective prism spectroscope again but this time with a telescope of approximately 11 centimeters aperture before which was mounted a flint glass prism of 37° 40′ refracting angle of the same aperture as the telescope objective. Of his observations with this relatively powerful instrument FRAUNHOFER wrote:

"The spectra of the light of Mars and Venus contain the same fixed lines as that of sunlight and these lines lie exactly in the same positions. At least this is true of the lines, D, E, b, and F, whose relative positions could be accurately determined. In the spectrum of the light of Sirius I was not able to discern any lines in the orange and red regions, but a very strong band could be recognized in the green and two extraordinarily strong bands in the blue, which did not appear to resemble any of the lines of planetary spectra. We determined the positions of these bands with the micrometer. Castor gives a spectrum like that of Sirius. Notwithstanding the faintness of the light the band in the green was bright enough for measurement and I found it exactly in the same place as for Sirius. I could recognize the bands in the blue but the light was not strong enough for determinations of their positions. In the spectrum of Pollux I saw numerous though faint lines which appeared like those of Venus. I saw the D line very clearly, and it was exactly in the same position as in planetary spectra. Capella gives a spectrum in which the same fixed lines appear at D and b as in sunlight. The spectrum of Betelgeuse contains numerous fixed lines which are sharply defined when the air is steady. Although this spectrum seems at first sight to show no resemblance to that of Venus, yet I find in it lines similar

---

[1] Denkschr d kgl Akad d Wiss zu München 5, p. 193 (1817).

[2] Gilberts Ann d Phys 56, p. 264 (1817); Zschokke's „Überlieferungen zur Geschichte unserer Zeit“ (1817), p. 442; Bibliothèque universelle Genève 6 (1817); Astr Abhandl von Schumacher 2, p. 13 (1823); Edinburgh Phil J 9 (1823); 10 (1824).

[3] Gilberts Ann d Phys 74, p. 337 (1823); Edinburgh J of Science 7 (1827); 8 (1828); Astr Abhandl von Schumacher (Altona) 2, p. 40 (1823).

to D and b in sunlight and exactly in their positions in the solar spectrum. In the spectrum of Procyon a few lines could be recognized but with difficulty and not distinctly enough for reliable determinations of position. I think I saw a line in the orange at the position of D."

These observations, though limited in scope, did bring out some of the bolder resemblances and differences among the spectra of hydrogen-, solar, and red stars and of the light of the sun viewed either direct, or reflected from bodies in the solar system. They were sufficient to quiet argument respecting exclusively telluric origin of the spectral features of celestial objects. They revealed a fertile field for research which was not re-entered successfully until four decades had passed. FRAUNHOFER in presenting the results quoted above stated that his investigations in this direction were far from being completed. However in the three years of his life that followed he was occupied with other researches and it remained for JOHANN VON LAMONT to continue the work. LAMONT[1] repeated the observations with FRAUNHOFER's instrument apparently without extending them further. He also used in this connection the large Munich refractor combining a prism with the ocular of this telescope, but he secured no satisfactory results, as he states, because of atmospheric disturbances.

b) Early Classifications of Stellar Spectra.

2. DONATI. The next contribution to our knowledge of stellar spectra was made by GIOVANNI BATTISTA DONATI[2] in a memoir dated 1860 and published two years later. To gather and focus stellar light, DONATI used a nonachromatic lens of 0,41 meters aperture and 1,58 meters focal length mounted equatorially. He used a cylindrical lens to perform some of the functions of a slit with some economy of light. And in addition he placed a diaphragm at the focus of the telescope in order to fix the position of his star image. After passage through the diaphragm and cylindrical lens the light fell in turn upon an achromatic collimator lens, a flint prism of $60° 53'$ angle mounted on a graduated circle, and a view telescope of 24 millimeters aperture and 170 millimeters focal length. A micrometer of the movable bar type with an eyepiece magnifying twelvefold completed the equipment. AMICI assisted with the construction of this instrument which served as a model for others. DONATI's practice was to set his micrometer bar on a solar line during the day and to measure the positions of stellar lines with reference to this setting.

DONATI gives us diagrams showing the relative positions of stellar lines and the apparent ends of the spectra of fifteen stars with positions in the solar spectrum of the lettered FRAUNHOFER lines from B to H and K supplied for comparison. He gives also brief descriptions of the dark lines with the measured angular separations referred usually to the solar F. The stars observed were: Sirius, Vega, Procyon, Regulus, Fomalhaut, Spica, and Rigel, classed as white stars; Castor, Altair, and Capella, classed as yellow stars; Arcturus and Pollux, classed as orange stars; Aldebaran, Betelgeuse, and Antares, classed as red stars.

DONATI was puzzled by differences between his results and those of FRAUNHOFER, which seemed to him very great. Thus FRAUNHOFER saw F in the green whereas DONATI saw it in the blue of the spectrum. FRAUNHOFER also saw the D line in the spectra of Pollux, Capella, Betelgeuse, and Procyon, whereas DONATI was not able to see this line in any stellar spectrum. The explanations of these

[1] Jahresber d Sternw bei München (1838), p. 90.

[2] B. DONATI, Intorno alle Strie degli Spettri Stellari. Nuovo Cimento 15, 292; Annali del Museo Fiorentino (1862). M N 23, p. 100 (1863).

differences seem very simple, for color estimates of spectra, especially when faint, may vary with different observers and with the direction in which one is working along the spectrum. Further, since the red end of the stellar spectra observed by DONATI fell near or short of the *D* line, it seems not strange that he experienced difficulty in recognizing it. He suggested in explanation that for stars of different colors the spectrum lines may change. However he remarked that almost all of his fifteen stars had a line differing very little in position from the solar *F*. He therefore presumed that all these *F*-like striae were relatively constant, but with a greater or less refraction due to a difference in the refractive quality of the light of the particular star.

DONATI's observations deserved a better discussion. His measures and descriptions indicate that he observed $H\beta$, $H\gamma$, and $H\delta$ in Sirius, Vega, and Procyon, that he observed $H\beta$ and $H\gamma$ also in Regulus, Castor, Altair, and possibly Capella and Fomalhaut, and $H\beta$ possibly in some stars of later type. He measured *b* of magnesium in Arcturus, Pollux, Aldebaran, Betelgeuse, and Antares, and bands of titanium oxide near $H\beta$ in Betelgeuse and Antares. Also he saw $H\beta$ in Spica, Rigel, and Arcturus, $H\gamma$ or *G* in Arcturus and Pollux, and a titanium oxide band at λ 5500 in Aldebaran, Betelgeuse, and Antares. DONATI's most important deduction contains the germ of the idea of classification and thus in its earliest form may be quoted here. "I have grouped separately the stars of different colors, giving in each group the first place to the most conspicuous striae, and so in order; classifying them from HUMBOLDT's Cosmos, and SCHMIDT, in Astronomische Nachrichten. The Table indicates that the stellar lines have a certain relation to the star's color. Thus the white stars seem to have a family likeness; and likewise the yellow, orange, and red ones." DONATI's classification is found in Table 1, p. 15.

3. RUTHERFURD. In a letter to the editors of the American Journal of Science and Arts, dated December 4, 1862, L. M. RUTHERFURD[1] submitted the results of his observations of stellar spectra for the previous year. RUTHERFURD used a FITZ equatorial telescope with an aperture of 28,6 centimeters and a focal distance of 4,27 meters. To this he attached a slit spectroscope with collimator, 60° flint glass prism, view telescope with five-power eyepiece, and a scale telescope showing bright graduations on a dark background. At first the slit was placed extrafocally in order to broaden the spectrum, but this wasteful arrangement was discarded later when a cylindrical lens was placed between the collimator lens and the prism and the slit placed at the focus of the large objective.

RUTHERFURD identified the bands in stellar spectra as such and proposed a mode of classification of stellar spectra based on observations of twenty-three stars, for seventeen of which he gives his results in diagram and also in words. His principal results are contained in the following quotation.

"In the spectrum of Jupiter are found two bands in the red and orange, between *C* and *D*, which are not found in the solar spectrum. It may be that these bands, as well as those so remarkable in α Orionis, Aldebaran, and β Pegasi, are absorption bands due to the action of the atmosphere of these bodies; still it is possible that the application of sufficient optical power will resolve them into lines.

The star spectra present such varieties that it is difficult to point out any mode of classification. For the present I divide them into three groups; first, those having many lines and bands and most nearly resembling the sun, viz., Capella, β Geminorum, α Orionis, Aldebaran, γ Leonis, Arcturus, and β Pegasi.

---

[1] Amer J of Sci and Arts (2) 35, p. 71 (dated 1862).

These are all reddish or golden stars. The second group, of which Sirius is the type, present spectra wholly unlike that of the sun, and are white stars. The third group, comprising α Virginis, Rigel, and others, are also white stars, but show no lines; perhaps they contain no mineral substances or are incandescent without flame."

Later RUTHERFURD improved his spectroscope mounting and introduced a comparison spectrum which was visible during observations of a star. Under date of March 31, 1863, he reported that he had established the existence in the spectrum of Arcturus of the lines, *D*, *E*, *b*, and *G*, and that he had found almost with certainty that each line in the spectrum of the star has its counterpart in the solar spectrum[1]. But RUTHERFURD appears to have made no further changes in his scheme of classification. As it stood however it seems to have deserved more recognition then is usually accorded it. In effect RUTHERFURD's three classes correspond respectively to Harvard classes G to M, A to F, and B. SECCHI's prominent differentiation between solar and banded spectra is not made. But from some considerations, such as the stellar temperature scale, something can be said for RUTHERFURD's scheme.

4. CARPENTER. During the year 1862 (see Table 1 below) the classic researches of Father ANGELO SECCHI began. Before turning to them, however, passing mention should be made of the work of J. CARPENTER[2] at Greenwich, which was published in April, 1863. The spectroscope employed was attached to the then great equatorial telescope. According to AIRY's description of the instrument[3], the pencil of light from the object glass converged to form the star image in a moderately small hole in the end of a small tube. Diverging from this focus the light fell on a considerable area of the prism, after emergence from which it was received on a combination of lenses, which caused the pencils for the different colors to converge. By observing at the principal focal line the spectrum of a star was given the desired width. In order to establish the relative positions of stellar and solar lines an eye-piece with cross thread was mounted at the upper end of the spectroscope in such a way that it could be inserted immediately behind the focus of the large objective. By this means the star image at night could be placed exactly at the center of a small hole through which light was admitted for solar observations. Thus by micrometer measures the positions of star lines could be referred to and compared with solar lines observed on the previous or following day.

With this device CARPENTER observed the spectra of nineteen stars. He saw the bands in four or five red stars and noted their diffuse edges on the side of greater wave length. He observed the hydrogen lines in a number of white stars and various lines including *D* in solar stars. A most significant feature of his report is the arrangement of stars in his sketch of the lines in the spectra of the objects observed. CARPENTER places four Type III stars in one group, nine Type I stars in another, and at the bottom a miscellaneous group in which solar stars predominate. In this arrangement there is clearly evident a scheme of classification by resemblances and differences that approximated that which SECCHI has evolved in 1866 (see Table 1, p. 15).

5. SECCHI. SECCHI was led to attempt the study of stellar spectra by receiving intelligence simultaneously of DONATI's investigations described above and of a new compact direct vision spectroscope by HOFFMAN, AMICI, and JANSSEN[4]. By using the new spectroscope with the MERZ refractor at Rome he hoped to secure results superior to those of DONATI. The new spectroscope, though ordered

[1] Amer J of Sci and Arts (2) 35, p. 407 (1863).

[2] M N 23, p. 190 (1863). [3] M N 23, p. 188 (1863). [4] C R 55, p. 576 (1862).

in May, 1862, did not arrive until the following December. In the meantime SECCHI and JANSSEN worked jointly, using a small spectroscope which the latter had brought to Rome. As the result of their efforts, which were reported to the Paris Academy, they detected sodium and other metallic lines in celestial spectra[1]. SECCHI's reports of his own investigation began to appear in 1863[2].

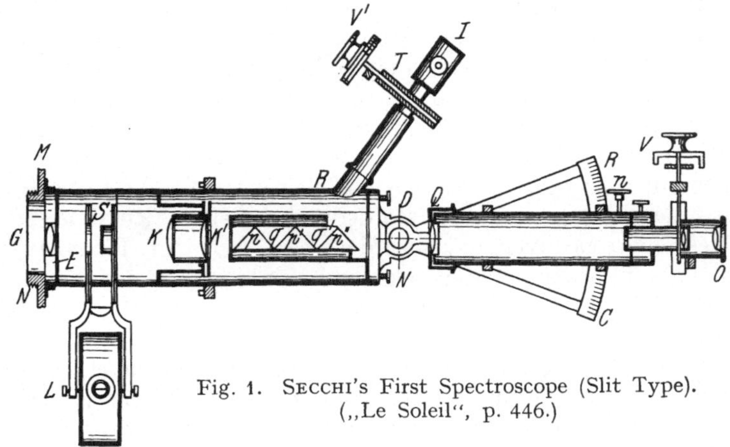

Fig. 1. SECCHI's First Spectroscope (Slit Type). („Le Soleil", p. 446.)

In the course of his study and survey of stellar spectra, SECCHI made use of three principal instruments[3]. The first was the direct vision slit spectroscope and spectrometer, received in December, 1862, with a scale telescope added by SECCHI, the light from which was reflected into the eye-piece from the last prism surface. This instrument is well shown in the view of Figure 1. This illustration is so complete that no further description is necessary though it may be stated that the letter E marks a cylindrical lens. In order to economize light SECCHI dispensed with the view telescope, observing the light as it came from the prism system directly with his eye or with an eye-piece of low power and finally, for fainter spectra, without the cylindrical lens. The general appearance of faint spectra to the seventh magnitude, though not the position of the lines, could be observed in this way for purposes of classification.

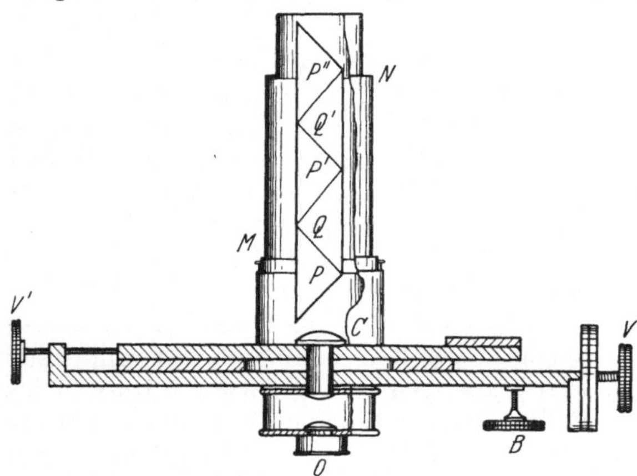

Fig. 2. SECCHI's Ocular Spectroscope. („Die Sterne", p. 82.)

SECCHI's second type of instrument in its improved form (1867) was a much simplified small spectroscope including a special ocular. This ingenious device is shown in Figure 2. It consisted of a direct vision compound prism, P, Q, P', Q', P'', 8 centimeters long, below which was mounted a cylindrical lens, C, whose axis was normal to the plane of dispersion of the prism. The cylindrical lens was placed close to the prism and two or three centimeters above the common focus

[1] Bull meteor del Coll Romano 2, No. 1 (1863).
[2] Bull meteor del Coll Romano 2, No. 14, p. 108 (1863).
[3] Die Sterne p. 70 et seq. (1878); Le Soleil, seconde Partie, p. 443 (1877).

of the objective and ocular. The linear dispersion could be varied by altering the distance of the prism from the ocular. Within the ocular were two fine metallic pointers which could be moved along the spectrum by micrometer screws. To secure maximum light transmission the ocular lenses were also of cylindrical type, giving a power of 20. In order to establish a reference point for the measurement of positions of lines the prism could be displaced slightly to allow a small part of the direct rays of the star to come to a focus. For less exact measurements the finder of the telescope could be used to fix the position of the star observed. Or when the principal lines had been determined with the slit spectroscope other features were measured differentially with the smaller instrument. This instrument in combination with the MERZ refractor yielded measurable spectra of stars down to the ninth magnitude.

A third instrument employed by SECCHI was of the objective-prism type in which a large prism of 162 millimeters aperture and 12° angle, by MERZ, was mounted over either the MERZ object glass of 4,43 meters focal length or a 160 millimeter objective of half this focal length. These instruments yielded excellent spectra of relatively great dispersion. Indeed when used with the MERZ refractor this dispersion was too great for observations of the fainter stars. Beautiful diagrams of spectra observed with the small telescope were published by SECCHI in 1872[1] (see reproduction below).

SECCHI presented to the Paris Academy on July 15, 1863, a report of his observations, with his first spectroscope, of thirty-five of the brighter stars[2]. In this communication he described the observed stellar spectra in two groups which represent his first attempt at a classification of stars according to their spectra. He used the word, "class" and not "type" in describing his groups.

The first class included yellow or red ("colored") stars whose spectra were characterized in general by the presence of many dark bands, especially in the less refrangible region. The spectra were so discontinuous that they could be compared to those of the electric spark in the apparatus of RUHMKORFF. These bands were usually badly defined at the edges and resembled those in the spectrum of the atmosphere of the earth and of the planets. The positions of the bands in the spectra of the stars of this class agreed in general with the strong rays of the solar spectrum which were in the regions *C*, *D*, *E*, *F* of FRAUNHOFER; but there were groups which did not have the same strength as in the sun, although among the many solar rays some lines of the second order of intensity could always be found which coincided with those of the stars. The *D* line was commonly found in these stars. In this class SECCHI placed Antares, Betelgeuse, Aldebaran, Algol (evidently a slip, possibly ϱ Persei was intended), β Pegasi, Arcturus, and β Ursae Minoris.

The second class included the white stars. These stars had in general a small number of discontinuities in their spectra, usually in the less refrangible part, and these bands were well defined at their borders. The dark lines of these stars were oftentimes not in agreement with those of the sun, especially the rays *G* and *H*. However the line *F* was common to all of the stars observed although not always the strongest line, as for example in Arcturus, Spica, and Rigel. The *D* line was lacking or very difficult to see in these stars except in α Lyrae and α Virginis. As stars of class 2, SECCHI cited Sirius, Rigel, β Scorpii, Castor, ζ and ε Ursae Majoris, α Lyrae, δ Orionis; α Lyrae was notable for the presence of some faint bands in the more refrangible region.

[1] Atti dell' Accademia Pontificia de' Nuovi Lincei 25, p. 177 (1872).
[2] C R 57, p. 71 (1863).

In August, 1866, SECCHI communicated his second scheme of stellar classification to the Paris Academy of Sciences[1]. He was using at that time, in addition to his first spectroscope, a simplified form in which a cylindrical lens, ten centimeters in front of focus, together with a small direct vision prism, between the cylindrical lens and focus, produced a widened spectrum which was observed with the micrometer eye-piece of the MERZ refractor. With this equipment SECCHI could recognize the hydrogen lines in stars of the seventh magnitude. SECCHI'S second classification included three classes as follows:

1. The colored stars, which are of the type of α Orionis, α Scorpii, β Pegasi, etc., which have spectra with wide bands.

2. The white stars, feebly colored, which have spectra crossed by fine lines; Arcturus, α Ursae Majoris, β Aquilae, Capella, Procyon, etc.

3. The blue stars, of which Sirius, Vega, α Aquilae, etc., are typical. These are the most numerous and are characterized by a large band in the blue at the place of *F*, another large band in the inner violet, and also occasionally a third in the extreme violet, with very sharp lines which are visible only in the spectra of brighter stars.

In the above report of 1866, SECCHI announced the preparation of a catalog of stellar spectra. The first of its kind, this catalog of 220 fixed stars classified on the basis of their spectra was published in October, 1866[2]. The classifications in this catalog were called types, the descriptions of which follow.

I. Type of the white and blue stars. Stars of this predominant type include α Lyrae, α Aquilae, Sirius, etc. They are characterized by a strong dark band in the green-blue at the position of the solar *F*, and a second band in the violet a little closer to *F* than the *G* band, and for the brighter stars a third in the extreme violet. The stars γ Cassiopeiae and β Lyrae are mentioned as exceptions under this type since they show a bright *F* line. However, in this classification they are designated as of the third type. The Orion stars are regarded as a modification of this type with lines usually narrow and a more or less visible *F*, but with bands very difficult to see in the violet region and no wide bands.

II. Type with large zones. This type, remarkable for its bands, includes red or orange colored stars. The type includes α Orionis, α Tauri, Antares, β Pegasi, and α Herculis, the most singular of the family. The spectrum of this star resembles a series of columns illuminated from the side. (A drawing of this spectrum appears in Figure 3.)

III. Type of the yellow stars with narrow lines or very weak bands. Arcturus, Capella, Pollux, etc., are examples. These stars may be classed in the solar type, for their spectra are exactly like that of the sun with fine lines similarly placed. We find in these stars the principal solar lines, *B, D, b, E, F, G, H*, of FRAUNHOFER and a great number of secondary lines. The banded spectra of Type II merge by slight gradations into the solar type.

As a result of this first survey of stellar spectra SECCHI made some deductions which are especially interesting as the earliest by-products of classification of stellar spectra. He observes that the stars variable in irregular period (α Orionis, α Herculis, etc.) have spectra with many zones. This spectral constitution, indicative of vast absorbing atmospheres, led to the thought that their variability is due to disturbances in the atmospheric masses surrounding them. SECCHI was not able to find any variability of spectral type in Algol during minimum. He

[1] C R 63, pp. 324, 364 (1866).

[2] C R 63, p. 621 (1866); 64, p. 345 (1867); A N 73, p. 129 (1867); Rep Brit Assoc 1868, p. 166.

found that in certain regions of the sky special types predominated and that in his whole list the two principal types were nearly in numerical balance. The distribution of the 220 stars among the types was as follows: Type I (White stars), 95; Orion stars, 11; Type II (Banded spectra), 20; Type III (Solar stars), 94.

At this time, August, 1866, SECCHI had modified his simple ocular spectroscope used earlier in the year by placing the cylindrical lens very close to the eye-piece, and a direct vision prism twelve centimeters long near the lens. The eye-pieces gave powers ranging from 400 to 800 and were movable over the spectrum.

With the extension of the number of stars to 316 this catalog was published in 1867[1] in a memoir by SECCHI under the title, "Sugli Spettri Prismatici dei Corpi Celesti". The scheme of classification was essentially unchanged as compared with that of August, 1866. The three types were called respectively, "type of α Lyrae", "type of α Herculis" and "solar type". The presence of narrow metallic lines of sodium and magnesium in Type I was noted in addition to wide $H\alpha = C$, $H\beta = F$, and $H\gamma = V$ identified with hydrogen by HUGGINS; also a line designated as W which SECCHI later identified with $H\delta$ of hydrogen[2]. The Orion stars were placed in Type I though they were recognized as a family by themselves. The spectrum of the Orion type was described as a special modification of the first type, in which the lines of the first type were present but notably narrow and in which also there were many sharp lines scattered over the entire spectrum. The color was predominantly green. The emission line stars β Lyrae and γ Cassiopeiae were counted under Type I. Type II was described as that of large and strong bright zones, six or seven in number, separated by dark lines and indistinct and nebulous intervals. The principal features were sensibly in the same position in different stars but there were notable differences in the widths of the zones and in the intensity of the various dark regions. In many cases Type II spectra approached to solar type in appearance, as in Aldebaran. The need for more accurate measures of the narrow solar lines was emphasized. Criteria for distinction between Type II and Type III were found in the b region and in the F line which is easily seen in Type III but is lacking in Type II. Cases of "fusion" between Type I and solar type were noted and Polaris was cited as an example. The distribution of stars among the types in this extended catalog was 164, 12, and 140 for Types I, II, and III respectively, preserving closely the balance between hydrogen- and solar stars. The majority of stars examined belonged to the first type, which was considered by SECCHI probably true of the unexplored regions of the sky.

With the completion of this catalog SECCHI was able to make several generalizations with reference to the spectra of the brighter stars. He was struck with the great uniformity characterizing these spectra and the scarcity of types. He observed the tendency for different types to dominate in certain regions of the sky. Thus stars of Type I predominate in Lyra, Ursa Major, and Taurus, whereas the solar type stars prevail in Cetus, Eridanus, Cepheus, and Draco. He noted especially the presence of a modification of the first type in the constellation of Orion, partaking in color of the nature of the great nebula. Such a coincidence he attributed to the distribution of primitive matter in space. Variation in width and diffuseness he noted in the hydrogen lines in different stars of Type I and suggested that differences in the temperature (probably high) and concentration of the hydrogen in stellar atmospheres might be responsible for such variety. Stars of solar type were notably constant in spectral structure though showing

---

[1] C R 64, p. 774 (1867); Memorie della Soc Italiana delle Scienze (3) 1, p. 67 (1867).
[2] C R 67, p. 373 (1868).

more variety than Type I. Variation in the *b* group was noted. In spectra of zonal or second type he found some of the strong lines of the solar type such as $H\alpha$ and $H\beta$ of hydrogen, but the fundamental lines of Type I were emissive or of secondary character. He noted that the lucid zones were more or less resolvable. He could predict that a red star would show a spectrum of zones, but aside from the conclusion that the existence of these zones indicated low temperatures his discussion of them was not significant at this time.

Preceding a reprint[1] of the description of individual spectra in this catalog, dated 1867, SECCHI refers to his spectral types as follows:

"Le stelle per brevità sono referite a tre tipi; 1° quello delle stelle bianchi come α Lira; 2° delle stelle gialle come Arcturo (α Boote); 3° delle stelle colorate in rosso o arancio, come α Ercole e α Orione."

This new numbering of the types was maintained throughout the rest of SECCHI's work.

In his second memoir[2], which was presented to the Società Italiana in November, 1868, the fourth type of stellar spectrum was described and seventeen stars of this type were cataloged, the brightest of which was assigned a magnitude of 5,5. Twenty-five stars of Type III were listed also in this memoir. The existence of a fourth type had been suspected in January, 1868[3] and confirmed in August, 1868[4]. At this time[5] SECCHI was using the very efficient prismatic ocular described above as of this date and illustrated in Figure 2. The description of the fourth type of stellar spectrum follows.

Type IV. The typical star is 152 Schjellerup in Ursa Major. This spectrum consists of only three principal luminous bands, the brightest one in the green, a faint one in the blue, and one quite bright in the red which is often subdivided into smaller zones. The luminous bands are separated by dark intervals. This type differs in its essentials from the third not alone in the distribution of the zones or luminous bands, which are of double width as compared with Type III, but also in that the light in these zones falls off abruptly on the edge of shorter wave length and fades gradually to darkness on the edge of greater wave length, whereas in the third type this distribution is reversed. In the spectrum of Type III not alone are the flutings double in number in a given space, but the maximum brightness is on the red side and the minimum on the edge toward the violet. If we compare the third type spectrum to a series of flutings, the fourth would be represented by cavities, supposing the direction of illumination unchanged. Bright lines like those of the metals are visible in the brighter edges of the colored zones. This spectrum has analogies with those of a gas and notably with that of carbon though reversed. In many of these stars the bright green zone stands out strongly, the blue and red regions being relatively faint. Dark lines in the green occur very near those of magnesium *b* but may be due to carbon. Their great width leads to the belief that they are not metallic. The spectra of the fourth type exhibit greater variety than those of the other three.

SECCHI explained his difficulty in observing the $H\alpha$ line in some stars as due to limitations of his spectroscope. He restated the descriptions of his types in English late in 1868[6] but added nothing remarkable except the statement that

---

[1] Memorie della Soc Italiana delle Scienze (3), 1, p. 105 (1867).

[2] Memorie della Soc Italiana delle Scienze (3), 2, p. 73 (1868). — [The first memoir and the catalog were published also in a quarto volume under the title "Sugli Spettri Prismatici delle Stelle Fisse", and both memoirs and the catalog were published in an octavo volume under the title "Sugli Spettri Prismatici dei Corpi Celesti" (1868).]

[3] C R 66, p. 124 (1868). [4] C R 67, p. 373 (1868). [5] C R 65, p. 979 (1867).

[6] Chem News 18, p. 168 (1868); Amer. Reprint 3, p. 304.

the second and third type spectra differ from one another, not in the metallic lines but in the nebulous bands. He made no significant changes in his classification after this date unless a somewhat less prominent mention of the Orion stars and the creation of a fifth type for γ Cassiopeiae and β Lyrae be so considered[1]. However, he continued to give attention to problems in stellar spectroscopy bearing upon the meaning of his classification. Thus he found that the spectrum of sunspots showed resemblances to spectra later than the solar type, namely those of Arcturus, Aldebaran, and Betelgeuse, and was led to suggest that red stars might owe their variability to sunspot phenomena[2].

He endeavored particularly to identify the substances producing the bands of third and fourth type spectra. In 1867 he noted in the flame issuing from a BESSEMER converter a spectrum with bands corresponding to those of α Herculis but reversed[3]. In 1868 he associated some of the bands with those of water vapor[4]. In July, 1869, a study of the spectrum of Antares confirmed this conclusion[5]. A little earlier the same year he concluded that the bands of fourth type stars are produced by absorption of a carbon compound. To reach this conclusion he passed a spark from platinum terminals through benzine vapor[6]. In August, 1869, he came to the decision that the black bands of the fourth type are found in the third, showing that carbon is present in both[7]. The same year he examined three bright line stars of WOLF and considered that they belonged to Type IV[8]. At this time, if not earlier, the fact that the bands in Type III and Type IV spectra were dark, as emphasized by VOGEL in 1874, must have been realized by SECCHI.

The powerful objective-prism spectroscope described above now became available and special studies were made of the hydrogen lines in Type I spectra and the structure of the bands in Types III and IV. A pressure of perhaps three atmospheres was indicated in the absorbing layers of Vega and Sirius[9]. In 1871 the principal bands of α Herculis were not resolved into lines[10] but finally in 1872 the spectrum of α Herculis was found decomposible into colonnades and lines[11].

An important contribution by SECCHI in 1872[12] includes excellent maps of the spectra of α Orionis and Sirius which are shown somewhat reduced in Figure 3.

There followed then a period of diminished activity after which SECCHI's excellent résumé of the methods and results of his work in classification of stellar spectra appeared in his well known volumes, "Le Soleil", second edition in 1877, and "Die Sterne"[13], published in 1878, the year of his death. In these volumes SECCHI gives an account of his studies in stellar spectroscopy. In the course of this discussion he restates his scheme of classification of stellar spectra, adding comments based on his later observations with the objective-prism instrument and a discussion of the composition and physical condition of stellar atmospheres. In this final statement of the classification the Orion stars are not mentioned. But in the discussion SECCHI states that these objects exhibit generally the first type spectrum though possessing narrow lines which could place them in the second type. He asks whether their green color could be due to the nebulous material through which we observe them.

[1] Die Sterne, p. 70 et seq. (1878); Le Soleil, seconde Partie, p. 443 (1877).
[2] C R 68, p. 959 (1869). [3] C R 65, p. 389 (1867), p. 562 (1867).
[4] Memorie della Soc Italiana delle Scienze (3) 1, p. 100 (1867).
[5] C R 69, p. 163 (1869).
[6] C R 68, p. 1086 (1869). [7] C R 69, p. 549 (1869).
[8] C R 69, pp. 39, 163 (1869). [9] C R 69, p. 1053 (1869).
[10] C R 71, p. 252 (1870); Pogg. Ann. 131, p. 156 (1867).
[11] C R 75, p. 655 (1872).
[12] Atti dell' Accademia Pontificia de' Nuovi Lincei 25, p. 177 (1872).
[13] Die Sterne, p. 70 et seq. (1878). — Le Soleil, seconde Partie, p. 443 (1877).

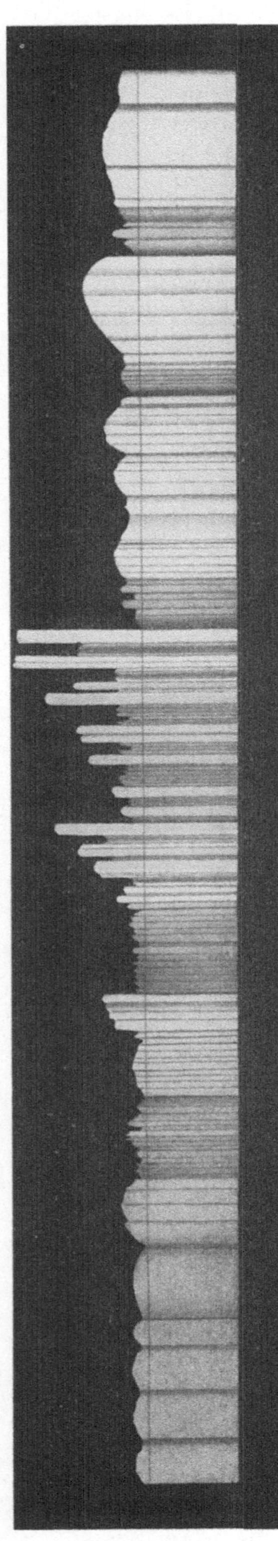

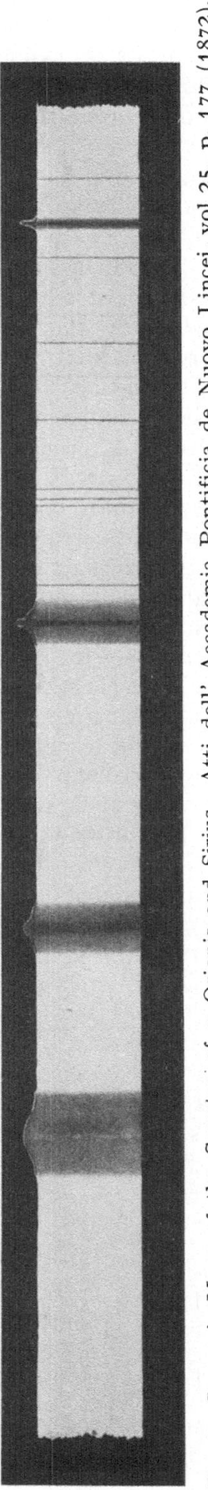

Fig. 3. SECCHI's Maps of the Spectra of α Orionis and Sirius. Atti dell' Accademia Pontificia de Nuovo Lincei, vol 25, p. 177 (1872).

Again in this account, though the Orion stars are relegated to an obscure niche, the emission stars, γ Cassiopeiae and β Lyrae, are raised to the level of a new type numbered five.

In the descriptions of the types given below, references to variations are omitted. These changes in the spectrum of a given star held SECCHI's attention in his later studies and were considered important since they showed that with the same composition a star might exhibit wide variation of spectral type. However, his observed variations of the spectra of Vega, Sirius, Arcturus, Aldebaran and α Hydrae appear not to be substantiated, nor those of α Orionis and α Herculis. Cases are known however that illustrate the point. SECCHI's final statement of his types, condensed from the descriptions in "Die Sterne" and "Le Soleil", is given here.

I. The first type is that of the white or blue stars such as Sirius, α Lyrae, β, γ, δ, ε, ζ, η Ursae Majoris, Castor, Markab, α Ophiuchi, and the like. The spectrum of these stars is relatively strong in the blue and violet regions and nearly continuous, being interrupted by four strong dark lines, belonging to hydrogen. In the cases of the brightest stars of this type all four lines are visible but for the fainter stars as a rule only $H\beta$. This line is moreover generally very broad with diffuse edges as in the case of Sirius. This broadened and diffuse structure shows that the hydrogen atmosphere of this star is of high temperature and density. Also traces of other lines (e.g. of sodium in the yellow and magnesium and iron in the green) are noted but only under very good observing conditions, as they are extraordinarily weak.

This class is very numerous and includes almost one half of the visible stars.

It is to be noted that in some stars of this type, as for example, Procyon (F5), α Aquilae (A5), α Virginis (B2), and others, many narrow lines are observed at points where hardly a trace is found in the other spectra. These stars furnish a transition to the following type. Thus the separation between the several types is not sharp.

SECCHI's drawing of the spectrum of Sirius is found in Figure 3.

II. The second type is that of the yellow stars. Their spectrum conforms precisely to that of the sun. They have many fine lines which differ in width as in Arcturus as compared with Pollux. The hydrogen lines are present but they are very narrow and always less marked than in stars of the first type. The lines of sodium, hydrogen, iron, and magnesium are observed more easily than the others. (Calcium is mentioned.) Capella, Pollux, α Ceti, and α Ursae Majoris and many other stars belong to this type. Aldebaran may be considered as furnishing a transition from second to third type. The stars of the second type are very numerous and embrace somewhat less than one half of the stars.

III. The third type includes the orange and red stars. The spectrum consists of dark and bright lines associated with or in bands or zones. Regarded as luminous, these bands are brightest on the edge of greater wave length and fade toward the violet, thus appearing like a series of fluted columns illuminated from the side of greater wave length. If the spectrum is complete, nine such column shaped bands appear always in the same position and are probably due to oxides. Remarkable for their spectra of this type are α Orionis, α Scorpii, o Ceti, β Pegasi, α Herculis, and many other stars.

These spectra must be regarded as composed of two parts which are superposed. The one consists of broad dark bands which produce the appearance of shadows in a row of columns; the other consists of dark metallic lines, which are fundamental and the same as those of the second type. A spectrum of the first kind, with broad bands, is shown by α Herculis, in which the flutings are usually not resolved even with the great objective prism. Only at times, namely, when the star had a vivid red color, did SECCHI succeed in recognizing the dark lines. In α Orionis the dark lines are strong and the zones weak. Hydrogen is present in these stars but difficult to observe because the lines of this element are partly due to emission.

The complicated structure and the wealth of lines of these spectra can be better observed from the enlarged drawing of the spectrum of α Orionis in Figure 3 taken from Volume 25 of the "Atti dell' Accademia dei Nuovi Lincei". Here the relative intensity of the bright zones is indicated by varying heights.

IV. The fourth type includes very remarkable stars, for the most part of a blood red color, none of which in SECCHI's list exceeds the fifth magnitude in brightness. Spectra of this type contain three fundamental luminous zones, colored yellow, green, and blue, and in some a trace of a red zone appears. Some of the luminous zones coincide closely in their limits with those of the third type and have almost doubled width as compared with those of third type. Thus if we eliminate alternate bands, 2, 4, and 6, from those of Type III, we have approximately the arrangement in Type IV, but with the fundamental difference that the flutings in the two types are graduated in the opposite sense, for the luminous zones in Type IV are brightest on the side toward the violet whereas those of Type III are brightest on the side toward the red. The one appears to be the negative of the other. In some spectra of Type IV the middle zone is divided into several bands while in others it is quite continuous. In others there are bright lines, generally very rare and weak and resolvable into several components but not into true metallic lines. The dark line in the yellow does not coincide with the sodium line although at the position of the *D* line there is a trace of a dark band. Finally, this type is extraordinarily variable in its details.

V. There are some few stars belonging to none of these four types, as they possess isolated simple lines and show bright lines instead of dark ones. They constitute a fifth type — that of the bright lines.

The outstanding example of these exceptional stars is γ Cassiopeiae in whose spectrum the hydrogen lines appear not dark by absorption but bright. β Lyrae shows bright hydrogen lines also at times but not always, since it is variable.

SECCHI's lists of stars in his volume "Die Sterne" (1878) include 168 objects of Type I, 154 of Type II, 25 of Type III, 17 of Type IV, and 2 of Type V; a total of 366 objects. However, he examined the spectra of nearly all of the principal stars and enough others in their neighborhood to bring the total observed to at least four thousand. The number of which detailed descriptions had been made by him up to April, 1867, was 400, and at that time 500 had been examined.

In his analysis and interpretation of studies of stellar spectra SECCHI apparently had isolated states of color and spectrum prominently in mind and did not apply in any important degree the principles of evolution or development. He did conclude that stars of the first type are surrounded by a denser and hotter hydrogen atmosphere than those of the second. The third type differs from the second only in the predominance of different substances. When a star varies between second and third type it may be due to eruptions bringing different substances into the atmosphere. Sunspot phenomena on the one hand and stellar light variability on the other fell in with this idea. The conclusions which he stressed particularly are found in the statement, "The star spectra of the first and second types show absorption lines caused, like those of the sun, by metallic vapors; those of the third and fourth types show, in addition to the metallic lines, those of other gases most probably produced by carbon in the form of the oxide and other combinations. On this basis the stars of the third and fourth types have lower temperatures than the stars of the Types I and II".

The termination of SECCHI's researches marked the end of the early years of stellar spectral classification. The rapid progress and development made during the years from 1860 to 1868 is brought out in Table 1. The general agreement among observers working independently is quite striking as showing the natural character of the division into types. However, RUTHERFURD was the only one of the early observers who assigned the helium stars to a separate group. SECCHI regarded them as a family in Type I with lines sharp enough to place them in Type II. SECCHI's first classification followed RUTHERFURD in grouping solar and banded spectra in one group while his later researches resulted mainly in subdividing this group. His final types recognized broad fundamental differences, and some intermediate gradations, which have found place in all subsequent classifications.

6. PICKERING. In 1891[1], PICKERING proposed a second Type V which has come to be associated so commonly with SECCHI's classification that it is properly referred to here. This type included the WOLF-RAYET stars and the planetary nebulae, "whose spectra resemble each other closely and consist mainly of bright lines and bands". In this paper WOLF-RAYET stars were divided into three classes. In the first and second classes the emission line at λ 469 was the most

[1] A N 127, p. 1 (1891).

conspicuous; in the third class, that at λ 464. To designate these classes Mrs. FLEMING[1] appears to have used the symbols OI, OII, and OIII, which in three specific cases corresponded respectively to Ob, Oc, and Oapec by Miss CANNON.

7. Early Classifications of Stellar Spectra.

Table 1.

DONATI 1860—1862	RUTHERFURD 1862	CARPENTER April 1863	SECCHI July 1863	SECCHI Aug. 1866	SECCHI Oct. 1866	SECCHI 1867	SECCHI 1867—1868
	Group 3 Like Spica and Rigel						
White	Group 2 Type of Sirius	Group 2	Class 2 White stars (Sirius, etc.)	Class 3 Wide hydrogen lines Sirius	Type I Vega	Type I Vega	Type I Vega
Yellow Orange		(Group 3)		Class 2 Narrow Lines Arcturus	Type III Sun	Type III Sun	Type II Arcturus
	Group 1 Lines and bands		Class 1 Dark bands				
Red		Group 1		Class 1 Wide bands Betelgeuse	Type II Betelgeuse	Type II α Herculis	Type III α Herculis
							Type IV 152 Schjellerup

8. D'ARREST. SECCHI's survey of stellar spectra was continued by D'ARREST at Copenhagen during the years 1871 to 1875[2]. Examination of the spectra of eleven thousand stars led to the discovery of only eighty of the third type and five of the fourth. D'ARREST succeeded in resolving bands in Type III which SECCHI found difficult. He felt also that there existed many exceptions to the relations between color and spectral type, so strongly emphasized by SECCHI. DUNÉR[3] suggested that the fault was with D'ARREST and might have been due to a lack of sensitivity of the latter's eyes to red light.

c) Later Classifications of Stellar Spectra.

9. VOGEL. In 1874[4] HERMANN CARL VOGEL proposed a classification of stellar spectra based upon many observations of visual spectra at BOTHKAMP with a direct vision slit spectroscope attached to a 293,5 mm refractor, both by SCHROEDER. Fundamentally VOGEL's classification differed from that of SECCHI in that the guiding principle was that of stellar development rather than formal character but the result adhered to SECCHI's scheme in a general way. VOGEL regarded SECCHI's third and fourth types as parallel divisions under one, his third class. He also made three divisions of the first type, thus separating out as SECCHI did either formally or by discussion the Orion stars and bright line

[1] Harv Ann 56, p. 225 (1912).
[2] A N 84, p. 263 (1874); 84, p. 369 (1874); 85, p. 249 (1875); 86, p. 53 (1875).
[3] Kgl Svenska Vet Akad Handl 21, No. 2 (1884).
[4] A N 84, p. 113 (1874); Publ Astroph Obs Potsdam 3, No. 3 (1883).

objects like γ Cassiopeiae. He placed WOLF-RAYET stars with others in a subdivision under Class II, the solar stars.

VOGEL's first classification is stated below following closely the translation by FROST[1].

"A rational classification of the stars according to their spectra is probably only to be obtained by proceeding from the standpoint that the phase of development of the particular body is in general mirrored in its spectrum. Three clearly distinguished classes may then be recognized, as follows:

1. Stars in such a high state of incandescence that the metallic vapors contained in their atmospheres can exert an exceedingly slight absorption, so that either none or only very faint lines can be perceived in the spectrum. (The white stars are included here.)

2. Yellow stars, the metallic vapors in whose atmospheres reveal themselves by strong lines of absorption in the spectrum, as in the case of our sun; and finally

3. Red stars, whose heat of incandescence is so far diminished that the substances in their atmospheres can enter into chemical combinations, which modern researches have shown to be characterized by a greater or smaller number of broad absorption bands.

If we examine the spectra of the stars of SECCHI's third and fourth types we shall see that they both fall under the above-mentioned third class, being only distinguished by the different arrangement of the dark bands, or in other words by the different composition of the atmospheres about the incandescent bodies. For this reason it seems to me advisable not to consider SECCHI's fourth type as distinct from the third, although readily distinguished from it in appearance.

I beg to propose the following classification, which seems to correspond to our present knowledge of stellar spectra:

Class I. Spectra in which the metallic lines appear very faint and fine, or are entirely invisible, and whose more refrangible portions—the blue and violet—are conspicuous by their intensity.

a) Spectra in which, in addition to the faint metallic lines, the hydrogen lines are present and conspicuous for their breadth and intensity. (Most of the white stars, as Sirius and Vega, are included here.)

b) Spectra in which the metallic lines are few in number, and very faint or entirely imperceptible, and in which the hydrogen lines are lacking[2] (β, γ, δ, and ε Orionis).

c) Spectra in which the hydrogen- and D_3 emission lines are visible. (At present β Lyrae and γ Cassiopeiae are the best known examples of this class.)

Class II. Spectra in which the metallic lines are very prominent. The more refrangible portions of the spectrum are much fainter than in the preceding class, and weak bands occasionally occur in the less refrangible portions.

a) Spectra with very numerous metallic lines, which are especially intense in the yellow and green. The hydrogen lines are generally strong but never so broad as in Class Ia; in a few stars, however, they are faint, and then faint bands composed of numerous close lines can be ordinarily recognized in the less refrangible portions. (Capella, Arcturus, Aldebaran.)

b) Spectra in which a number of emission lines appear in addition to the dark lines and the faint bands. (Here would be included T Coronae, and in all

[1] Scheiner's Astronomical Spectroscopy, p. 236 (1894).

[2] In 1888 (A N 119, p. 97), on the basis of photographic spectra secured at Potsdam, VOGEL altered the last clause to read, "and the strong hydrogen lines of Class Ia are lacking".

probability also the stars in Cygnus observed by WOLF and RAYET, as well as the variable R Geminorum, although the faintness of this star has not allowed the detection of dark lines, but only of certain dark bands in the red and yellow.)

Class III. Spectra throughout which, in addition to dark lines, numerous dark bands are distributed; the more refrangible portions of the spectrum are strikingly faint.

a) In addition to the dark lines, bands are present, the most conspicuous of them being dark and sharply defined toward the violet, but faint and indistinct toward the red. (α Herculis, α Orionis, β Pegasi.)

b) Spectra in which dark and very broad bands may be seen, with the direction of their increasing intensity the reverse of that for the previous sub-division, so that the most prominent bands are sharply defined and darkest on the red side and gradually shade off toward the violet. (The stars of this group are all faint. Examples are Nos. 78, 152, 273, etc., of SCHJELLERUP's catalog of red stars.)"

VOGEL restated his classification in 1895[1], availing himself of the data gained by photography and particularly of the identification, as due to helium, of many significant lines in stellar spectra. He modified his first classification by differentiating farther the spectra of Class I and by redefining Ib. Classes II and III were not changed. Superb spectra in any division or subdivision were denoted by double exclamation marks following the designation, especially good examples by single exclamation points.

The revision of VOGEL's Class I follows:

Class I. These are continuous stellar spectra whose more refrangible portions, blue and violet, are especially strong. These spectra are interrupted by the complete series of hydrogen lines, which appear as dark, broad, diffuse, seldom sharply delimited bands, or again as narrower absorption lines, and in general exceed considerably in strength the metallic lines showing in the spectrum. Quite frequently the lines of hydrogen and other substances appear as emission lines on a continuous background.

Ia1. Spectra in which only broad, strong lines of hydrogen appear and no other lines are visible. (Probably none are known today.)

Ia2. Spectra in which metallic lines, including calcium, magnesium and sodium, appear in addition to hydrogen, but no lines of helium. The *K* line of calcium is sharp, not approaching those of hydrogen in breadth. The other metallic lines are faint and difficult to detect with small dispersion.

Ia3. Spectra in which the *K* line appears of nearly equal intensity with the hydrogen lines, and in a few cases is sharply defined at the edges or, broader and stronger than the hydrogen lines, forms a conspicuous companion for the blend of calcium *H* and $H\varepsilon$. In the spectra of this subdivision the lines of helium are not visible; instead numerous and strong lines of the metals always appear, especially those of iron; but the hydrogen lines always predominate. $H\delta$ stands out prominently among the lines, and $H\gamma$ is stronger than *G*. This subdivision forms the direct transition to Class II, in which the hydrogen lines no longer play a dominating role over the lines of the metals, while the calcium lines *K* and *H* stand unrivalled for their extraordinary breadth and diffuseness and the *G* group appears broad and strong in comparison with $H\gamma$.

[1] Publ Astrophys Obs Potsdam 12, pt. 1, No. 39 (1899); see review by Frost Ap J 10, p. 362 (1898); Sitzgsber d Kgl Preuß Akad d Wiss 1895, p. 947.

Ib. Spectra in which helium lines appear in addition to those of hydrogen th ugh the latter always predominate. The more conspicuous helium lines are those of wave lengths, 3965, 4026, 4472, 5016 and 5876 (D_3). (λ 3889 is so close to $H\zeta$ that it appeared unavailable for the recognition of helium in stellar spectra). Also more or less numerous lines of calcium, magnesium, sodium, and iron are found in spectra of this division.

Ic1. Spectra with hydrogen emission lines.

Ic2. Spectra in which emission lines of calcium, magnesium, and other metals are present in addition to emission lines of hydrogen. (The fainter component of β Lyrae.)

VOGEL noted that in the spectra of Classes II and III, though the regions beyond $H\gamma$ were often similar, exact studies brought out differences. Thus in Class III the lines were stronger and broader and the fading of the spectrum toward the violet was more pronounced. To distinguish between divisions IIIa and IIIb in the photographic region the greater or smaller extension toward the violet provided a criterion, though PICKERING[1] found the photographic spectra of VOGEL's divisions IIIa and IIIb so different from each other that he was not able to write them in one class.

In his attempt to devise an orderly classification representing successive stages of development VOGEL met with limited success. Divisions Ib and IIb were clearly out of place and the position of Ic was in doubt. In his second classification VOGEL felt that some of the distinctions might be formal.

A comparison of VOGEL's final classification with SECCHI's and certain later ones is found in Table 5.

SCHEINER adopted VOGEL's classification as the basis of extensive studies and discussions[2] but at the same time criticized these classifications more or less constructively. He characterized the first division Ib as "spectra in which the hydrogen lines and the few metallic lines all appear to be of equal breadth and definition". He pointed out that dark lines are present in division Ic spectra and suggested that this bright line group should stand at the top of the sequence. He found the subdivisions of division Ia superfluous in many cases and not applicable in the visual region. He devised a classification of the hydrogen lines in spectra of VOGEL's first type, thus introducing a three-dimensional concept. In this classification the first division included lines of various degrees of darkness, in the second division central emission was evident but fainter than the adjacent continuous spectrum, and in the third division the emission exceeded the neighboring continuous spectrum in brightness. He explained the sequence as due to increasing extent of atmosphere relative to the photosphere of the stars.

10. HUGGINS. WILLIAM HUGGINS facilitated the work of classifiers of stellar spectra by his intsnsive studies of about one hundred stars, and himself proposed independently a equence of eleven stars in 1879[3] following in some degree the order of the classfications of VOGEL. His chief contributions in this connection were (some in association with Mrs. HUGGINS or W. A. MILLER): Beginning[3, 4] in 1863, the discovery of the identity of certain spectra with those of the sun, the identification of prominent lines in the spectra of Sirius and Vega with those of hydrogen, the detection of five lines including D and b in the spectra of these two stars, the detection of many other solar lines in stellar spectra, and the determination of the chemical origins of many lines in celestial spectra. In 1864 he

[1] A N 127, p. 1 (1891).

[2] Scheiner, Spektralanalyse der Gestirne, Chap. 5 (1890); Populäre Astrophysik, p. 588 (1908); Publ Astrophys Obs Potsdam 7, pt. II, p. 222 (1895).

[3] Phil Trans 171, p. 669 (1880). [4] Phil Trans 154, p. 413 (1864).

discovered the bright nebular lines[1]. In 1879 he discovered ultra-violet lines of hydrogen in the so-called HUGGINS (BALMER) series[2,3]. He made unsuccessful attempts to photograph dark lines in stellar spectra in 1863[4] but achieved marked success, of note particularly in the ultra-violet region, when he resumed the project in 1875[5]. In the meantime, however, he had been anticipated by HENRY DRAPER[6] in the production of the first successful stellar spectrogram.

HUGGINS' stellar sequence based on the behavior of hydrogen and calcium *K* in the photographic region was intended to follow on the whole the order of change beginning with the more simple spectra. This sequence with the corresponding symbols in the VOGEL and Harvard classifications is given below:

HUGGINS SEQUENCE	VOGEL	HARVARD
{ Sirius	Ia 2	A0
{ Vega	Ia 2!!	A0
η Ursae Majoris	Ib!	B3
α Virginis	Ib!	B2
α Aquilae	Ia 3	A5
β Orionis	Ib!	B8p
α Cygni	Ia 2	A2p
Capella	IIa!!	G0
Arcturus	IIa	K0
Aldebaran	IIa	K5
Betelgeuze	IIIa	Ma

11. DUNÉR. Following an extensive survey, N. C. DUNÉR[7] published, in 1884, a new catalog of Class III stars employing VOGEL's divisions IIIa and IIIb. In this catalog the total of IIIa stars reached 297, and of IIIb stars, 55. As a result of this survey and study of stellar spectra he offered some corrections affecting the descriptions of SECCHI's spectral types. Thus he questioned the relative faintness of the red zone in Type IV, the existence of bright lines terminating the luminous zones on the violet edge, the interpretation of the bands as emission features, and the reality of short period changes in spectra of Types III and IV. He found differences in the relative intensities of the zones in different stars. He questioned PECHÜLE's suggestion to the effect that IIIa stars are transformed into the IIIb class by a catastrophe and that IIIb corresponds to a later stage of development than IIIa. DUNÉR felt that IIIa and IIIb were coordinate divisions and that the discovery of transition spectra between IIa and IIIb could be hoped for and cited the star BD $+ 38°$ 3957 as a possible example of such an intermediate type. He pointed out that such stars would be rare, and hard to find and classify. In this connection he suggested that the line spectrum rather than the bands was the best indication of the development of a star.

12. LOCKYER. J. NORMAN LOCKYER approached the problems of stellar spectral classification with a background of ten years of research in solar physics. He also had founded a laboratory in which celestial spectra could be interpreted by comparison with radiation from various terrestrial sources. From the first his efforts in the direction of stellar spectral classification were influenced by a dissociation theory according to which elements as well as compounds assumed simpler or "finer" forms with increased temperatures. Later a rising and falling temperature sequence of stellar states with increasing condensation was associated

[1] Phil Trans 154, p. 437 (1864). [2] Phil Trans 171, p. 669 (1880).
[3] London R S Proc 48, p. 216 (1890); Sid Messenger 9, p. 318 (1890).
[4] Phil Trans 154, p. 413 (1864). [5] London R S Proc 25, p. 445 (1876).
[6] Amer J of Sci (3) 18, p. 419; Nature 21, p. 83 (1879); Amer J of Sci (3) 13, p. 95 (1877); Nature 15, p. 218 (1876); Phil Mag (5) 3, p. 238.
[7] Kgl Svenska Vet Akad Handl 21, No. 2 (1884).

with the spectral divisions; and a "meteoritic hypothesis" influenced his classification of spectra, particularly at the beginning of stellar existence.

LOCKYER's first classification, in 1874[1], was a restatement from a new point of view of SECCHI's findings. He suggested that the first named classes belonged to stars that were hotter and, therefore, possessed of atmospheres made simpler by dissociation. The scheme follows:

Class α: These spectra are simpler than, but of the same kind as, the sun's. The blue end of the spectrum is open. Lines of hydrogen are present; lines of metals are exceedingly thin.

Class β: These spectra are represented by the sun.

Class γ: These spectra are more complex and of a different kind, being composed of channeled spaces and bands. The blue end is closed. Lines of hydrogen are absent and metallic lines are reduced in thickness and intensity.

In 1888[2] LOCKYER published his well known arched curve representing the rise and fall of stellar temperatures with increasing condensation. In this early form it was associated with VOGEL's spectral classes. In its later form in association with LOCKYER's own spectral types it is shown in Figure 4, and discussed below. But even at this time, before his own stellar observations had begun, LOCKYER felt the need of a new classification of stellar spectra, which would represent a sequence in time, following the rise and fall of the temperature curve, and present in order the spectral phenomena attending the process of condensation on the meteoritic hypothesis. Based on all available visual observations this second classification divided stars into six spectral groups with several subdivisions.

A restatement and development of this classification, given below, followed in 1892[3]. LOCKYER's own photographic observations of stellar spectra had begun in 1890 with two six-inch objective prism cameras and a slit spectrograph attached to a thirty-inch reflector. Spectrograms of 171 stars were already available. These photographic spectra were classified in three main tabular groups based on the amount of general absorption at the blue end with a fourth group for bright line stars. No stars of SECCHI's Type IV had been photographed by LOCKYER up to this time. The three main groups were: *A*, stars in which there is no remarkable continuous absorption either in the ultraviolet or violet, and the hydrogen lines are broad; *B*, stars in which there is a continuous absorption in the ultra-violet, extending to about *K*, and the hydrogen lines are relatively thin; *C*, stars in which there is a continuous absorption in the ultra-violet or violet, extending to about $H\gamma$, and the hydrogen lines are narrow. These groups were further subdivided on the basis of the behavior of the lines into seventeen sub-groups. Here as when dealing with visual spectra it was "not possible to place all the stars in one line of temperature". Thus there were spectra in which the hydrogen lines were of the same average width, while the remaining lines were almost entirely different. Further than this, distinctions were evident as stated in the classification given below.

The meteoritic hypothesis at this time occupied perhaps its strongest position and influenced materially the descriptions of the groups at the beginning and on the ascending branch of the sequence. This hypothesis assumed that "all celestial bodies are or have been, swarms of meteorites, the difference between

[1] Bakerian Lecture, Phil Trans 164, p. 492 (1874).

[2] London R S Proc 44, p. 1 (1888).

[3] Phil Trans 184, p. 675 (1893); London R S Proc 52, p. 326 (1892); Nature 47, p. 261 (1893).

them being due to different stages of condensation. The life of a star is characterized first by an increase of temperature due to collisions in a condensing nebula; then, after the apex of temperature has been reached, the gradual cooling down of a mass of gas and vapor." In stars of Group I below, the meteorites are assumed to be so far apart that atmospheric radiation predominates, whereas with greater condensation in Groups II and III atmospheric absorption increases in prominence and becomes simpler at higher temperatures through more or less complete dissociation. In Group I the nebulae were thought to betray their meteoritic character by the presence of a "magnesium fluting" known as the chief nebular line at λ 5007, an identification shown to be incorrect by KEELER, and a "carbon fluting" at λ 4686, known to be due to ionized helium, and other lines erroneously identified with low temperature iron, calcium, and magnesium. These and other faulty identifications and conceptions, such as those relating to the bands of Group II, misled LOCKYER at this time in his attempt to classify spectra on the branches of his temperature curve and in the selection of the "hottest star".

LOCKYER's classification of 1892 follows:

Group I. Emission lines and emission flutings predominant.
- Sub-group α*.—Nebulae.
- Sub-group β.—Bright-line stars.

Group II. Mixed fluting radiation and absorption predominant. Much continuous absorption in the violet.
- Sub-group α.—Dark flutings probably extending from the red end to the region more refrangible than G (λ 4341). None of the stars of this sub-group have yet been photographed at KENSINGTON, but their existence is indicated by the discussion of DUNÉR's observations, and on the hypothesis that there must be an intermediate stage between the bright-line stars and stars like Mira Ceti.
- Sub-group β.—Dark flutings extending as far as G. E.g., Mira Ceti.
- Sub-group γ.—Stars in which the most refrangible dark fluting is at λ 4585. E.g., α Herculis.
- Sub-group δ.—Stars in which the most refrangible dark fluting is at λ 4763. In addition to the flutings there is a large number of dark lines. E.g., α Orionis.

Group III. Line absorption predominant, with increasing temperature. Less continuous absorption at the violet end.
- Sub-group α.—Stars with line spectra resembling those of Group II, Sub-group δ, but with only a single fluting in the red remaining. E.g., α Tauri.
- Sub-group β.—Continuous absorption in violet less than in sub-group α. The calcium lines are less intense, while lines at λ 4172, 4233, and 4177 have their intensity increased. E.g., γ Cygni.
- Sub-group γ.—Stars with spectra consisting of a relatively small number of dark lines. The hydrogen lines are of only moderate breadth, and among the additional lines are some which are seen bright in the solar chromosphere. In this subgroup the continuous absorption in the violet is almost a minimum. E.g., β Orionis.

* The Greek letters have been adopted to avoid confusion with the symbols employed in VOGEL's classification.

Group IV. Simplest line absorption predominant, the hydrogen lines being very broad.

Sub-group α.—The spectra are marked by the presence of fine lines at wave lengths 4024, 4471, 4481, the two latter being of almost equal intensity. E.g., β Persei.

Sub-group β.—Highest temperature. The spectra show additional faint lines, and λ 4471 almost disappears. E.g., α Andromedae.

Sub-group γ.—The lines of iron make their appearance, but the line of calcium at λ 4226 is not yet distinct. E.g., Sirius.

Group V. Line absorption predominant, with decreasing temperature.

Sub-group α.—The lines of hydrogen are still broad, and the line of calcium at λ 4226 is clearly visible. The grouping of lines about G, which is so characteristic of the solar spectrum, is not visible. E.g., β Arietis.

Sub-group β.—All the solar lines are now clearly visible, but the hydrogen lines are broader than in the solar spectrum. The grouping at G is only partially developed. E.g., Procyon.

Sub-group γ.—The spectra very closely resemble the solar spectrum, the characteristic grouping of lines about G being fully developed. Carbon absorption commencing in the ultra-violet. E.g., Capella.

Group VI. Carbon absorption predominant.

In 1899[1] and 1902 LOCKYER proposed a new (third) classification, based on chemical differences in which the arched temperature-condensation curve, the theory of dissociation of elements with increased temperature, and the meteoritic hypothesis all survived. His previous classifications had been based on criteria for the most part of unknown origin but for his new classification identifications of a majority of the previously unknown lines were available. The following account of LOCKYER's third classification is drawn from "The Catalogue of 470 of the Brighter Stars" (1902). The names of the groups are based upon those of typical stars in the same way that the names of the geological formations are based upon those of typical strata. Some of these groups were later divided into species. Thus the Alnitamian group was divided into four species[2] of which the typical stars were ι Orionis (Oe5), ζ Orionis (B0), ε Orionis (B0) and $\varkappa$ Orionis (B0). The subdivisions which were retained are referred to below. Proto-hydrogen is used here to designate ionized helium. Proto-metals are metallic vapors in such a high temperature state that they emit only enhanced lines. They are "finer" forms of the metals than those which produce arc lines. The substance called asterium is assumed to emit the lines of three helium series of which the "chief lines" used by LOCKYER are at λ 4009, 4388[3].

LOCKYER's third classification of stellar spectra follows.

Definitions of stellar genera.

Argonian (γ Argus).

Predominant.—Hydrogen and proto-hydrogen.

Fainter.—Helium, unknown gas (λ 4451, 4457), proto-magnesium, proto-calcium, asterium.

[1] London R S Proc 65, p. 186 (1899). [2] London R S Proc 84, p. 426 (1910).

[3] Catalogue of 470 of the Brighter Stars, Solar Phys Committee (1902); Inorganic Evolution p. 61 (1900).

Alnitamian (ε Orionis).

Predominant.—Hydrogen, helium, unknown gas (λ 4649.2), silicium (IV). Fainter.—Asterium, silicium (III), proto-hydrogen, proto-magnesium, proto-calcium, oxygen, nitrogen, carbon, silicium (II).

Proto-metallic lines relatively thick, hydrogen relatively thin.

Crucian (α Crucis).

Predominant.—Hydrogen, helium, asterium, oxygen, nitrogen, carbon.

Fainter.—Proto-magnesium, proto-calcium, silicium (III), unknown gas (λ 4649.2), silicium (II), silicium (IV).

Taurian (ζ Tauri).

Predominant.—Hydrogen, helium, proto-magnesium, asterium.

Fainter.—Proto-calcium, silicium (II), proto-iron, proto-titanium, proto-chromium, nitrogen, carbon, oxygen.

Rigelian (β Orionis).

Predominant.—Hydrogen, proto-calcium, proto-magnesium, helium, silicium (II).

Fainter.—Asterium, proto-iron, carbon, nitrogen, proto-titanium, proto-chromium, oxygen, silicium (III).

Cygnian (α Cygni).

Predominant.—Hydrogen, proto-calcium, proto-magnesium, proto-iron, silicium (II), proto-titanium, proto-chromium.

Fainter.—Proto-nickel, silicium (I), proto-vanadium, proto-manganese, proto-strontium, iron (arc), helium, silicium (III), asterium.

Polarian (α Ursae Minoris).

Predominant.—Proto-calcium, proto-titanium, hydrogen, proto-magnesium, proto-iron, and arc lines of calcium, iron, manganese, silicium (I).

Fainter.—The other proto-metals and metals occurring in the Sirian genus.

Aldebarian (α Tauri).

Predominant.—Proto-calcium, arc lines of iron, calcium, manganese, proto-strontium, hydrogen, silicium (I).

Fainter.—Proto-iron and proto-titanium.

Proto-metallic lines relatively thin, hydrogen relatively thick.

Achernian (α Eridani).

Same as Crucian.

Algolian (β Persei).

Predominant.—Hydrogen, proto-magnesium, proto-calcium, helium, silicium (II).

Fainter.—Proto-iron, asterium, carbon, proto-titanium, proto-manganese, proto-nickel.

Markabian (α Pegasi).

Predominant.—Hydrogen, proto-calcium, proto-magnesium, silicium (II).

Fainter.—Proto-iron, helium, asterium, proto-titanium, proto-manganese, proto-nickel, proto-chromium.

Sirian (α Canis Majoris).

Predominant.—Hydrogen, proto-calcium, proto-magnesium, proto-iron, silicium (II).

Fainter.—The lines of the other proto-metals and the arc lines of iron, calcium, manganese, silicium (I).

Procyonian (α Canis Minoris).

Same as Polarian.

Arcturian (α Bootis).

Same as Aldebarian.

Stars with fluted spectra.

Antarian (α Scorpionis).

Predominant.—Flutings of manganese.
Fainter.—Arc lines of metallic elements.

Piscian (19 Piscium).

Predominant.—Flutings of carbon.
Fainter.—Arc lines of metallic elements.

Figure 4 reproduces LOCKYER's arched temperature-condensation curve as published in 1915[1], showing the positions of the spectral genera along the sequence, the related Harvard classes, and the important chemical substances occurring

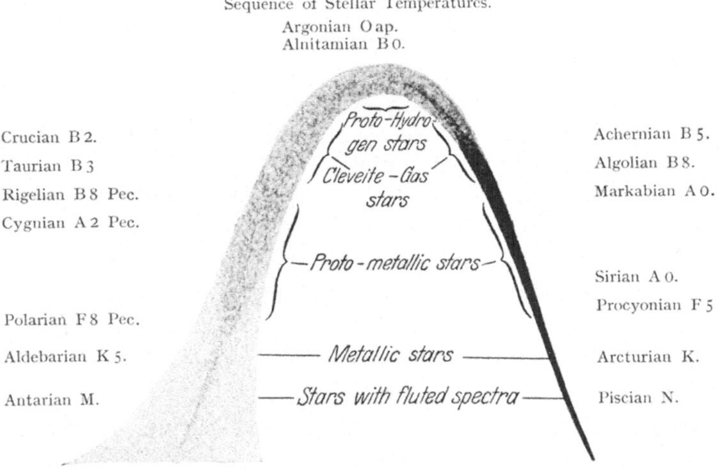

Fig. 4. LOCKYER's Temperature-Condensation Curve. Hill Observatory Bull Nr. 2, p 1.

at different temperature levels. The gaps below the Cygnian and above the Sirian types were left for two other possible classes, but these classes were omitted in 1915[2]. Cygnian and Sirian stars were then placed in parallel, but later in the same year[3] the Sirian, Procyonian and Aldebarian groups were split in two and designated by adding Roman numerals. The groups, thus paired on the two branches of the temperature curves, were then: Cygnian and Sirian I, Polarian and Procyonian I, Aldebarian II and Procyonian II (Solar type), Aldebarian I and Arcturian. Sirian II was left without a parallel of increasing temperature. The letters, *h*, *c* and *a*, with the group name, indicated respectively increasing, cooling, and hottest temperatures; suffix *p* denoted spectral peculiarity. Thus modified, LOCKYER's third classification took its final form. On this basis 641 of the fainter stars were classified in the years 1913 to 1916[3, 4].

It is within the province of this chapter to discuss the significance of LOCKYER's temperature-condensation curve (Figure 4) in so far as it affects the physical basis of his classification. We may note therefore that this arched

[1] Bull Hill Obs 4 (Dec. 1915).
[2] Bull Hill Obs 2 (Feb. 1915).
[3] Bull Hill Obs 3 (Oct. 1915).
[4] Bull Hill Obs 5 (Nov. 1916).

curve indicates on its two branches a difference in physical state which is primarily one of density—a difference connected by LOCKYER with his meteoritic hypothesis. The wide dotted curve at the bottom of the ascending arm represents nebulous conditions whereas the firm dark line at the bottom of the descending branch represents condensed stars. At the Algolian stage where the curve becomes black we are dealing no longer with swarms but with bodies having discs and therefore photospheres approaching the solar condition. The rise of temperature with the rising curve is assumed to be due to collisions in a condensing nebula. The fall of temperature on the descending branch is due to cooling of a mass of gas and vapor. The symmetrical form of the curve is not intended to denote equal or similar time intervals for the life process symbolized by the two branches. On the contrary the scarcity of stars on the ascending branch might indicate a much reduced time scale for the rising temperature phase as compared with the cooling stages[1].

The attempt to allocate stars to this curve on grounds of spectra and theory was too difficult to permit of complete success. From the added standpoint of densities the Piscian stars should be coördinated with the Antarian at the foot of the ascending branch and the dwarf M stars under a new name should replace the Piscian group at the end. Also on the descending branch stars more typical of dwarf densities and more surely of falling temperature could be found than Arcturus, for example, but LOCKYER's preference for conspicuous type stars and the weakness of his criteria at some points undoubtedly influenced his choice. On the ascending branch stars with Miss MAURY's *c* characteristics predominate[2] and all are more happily chosen, but, at least in a measure, accidentally, since LOCKYER's criteria for discrimination between spectra on the two branches were weak at and near the bottom of the curve where distinctions were more significant, and since selections were influenced by the meteoritic hypothesis and by the tendency to select bright stars. Lacking knowledge of densities and masses LOCKYER's selections are nothing short of remarkable.

It is undoubtedly true that LOCKYER's classification as such has been judged too harshly; and this is largely because of its association with strange nomenclature and untenable theories. Reference to Figure 4 makes clear that LOCKYER's genera are in vertical sequence with the Harvard classes as rearranged by Miss CANNON. LOCKYER reached that order through his search for a temperature sequence based on studies of laboratory and solar atmospheric spectra. Harvard observers came to it as the result of an attempt to classify spectra, as SECCHI had done, according to colors and distinctive characteristics, beginning with simpler spectra. It may, therefore, be considered accidental that the Harvard order proved to be a temperature sequence of valuable physical significance. If it had not, some modification of LOCKYER's classification would probably have been in use today. To LOCKYER and his associates we owe a distinct debt for their development and application of temperature criteria for the classification of stellar spectra.

13. Mc Clean. In 1898, FRANK McCLEAN[3] proposed a classification of stellar spectra in which SECCHI's first type was separated into three divisions and SECCHI's second, third and fourth types corresponded to divisions IV, V and VI, respectively. The classification accompanied a photographic study of 160 spectra of northern stars, made with an objective prism of twelve inches aperture and twenty degrees angle, fitted to a telescope of equal aperture and 11,25 feet focal

[1] Bull Hill Obs 6 (Aug. 1919). [2] Bull Hill Obs 2 (Feb. 1915).

[3] Phil Trans 191, p. 127 (1898); Spectra of Southern Stars, London (1898); Ap J 10, p. 367 (1899).

length. McCLEAN assumed throughout that his successive divisions were merely the manifestations of the successive physical states through which every star passes in the course of its career after evolving from a luminous nebula at the top of the sequence. The order conforms also to a satisfactory formal succession of types such as the Harvard sequence, as shown in Table 5. McCLEAN's classification follows.

Division I. The hydrogen-helium type. All stars whose spectra are characterized by the lines of hydrogen and helium.

a. Certain additional lines are present, possibly due to oxygen (stars closely connected with gaseous nebulae as indicated by comparisons of spectra).

b. Certain additional lines are present which have been attributed to calcium, barium, and magnesium. *K* of calcium first appears in this subdivision.

II. The hydrogen type. In this division the hydrogen spectrum attains its fullest development both in the strength of the lines and in the extent of the spectrum, which includes the ultra-violet series. *K* is narrower and is more sharply defined than in the divisions which precede and follow. Fine, delicate absorption lines, appearing to be due to calcium and titanium rather than iron, are present.

III. The hydrogen-iron type. The hydrogen lines remain very strong, and the calcium *K* and *H* are generally subordinate to them. In its more advanced examples the iron spectrum is fully developed. The brightness of the violet end and the obscurity of the red end of the spectrum remain the same as in the preceding divisions but this type is more closely allied to the solar type, Division IV, than to the hydrogen and helium types.

IV. The solar type. Its characteristics are elucidated by our intimate knowledge of the solar spectrum.

V. The first of the banded types. SECCHI's Type III. These spectra are closely allied to those of Division IV.

VI. SECCHI's Type IV. These spectra contain hydrocarbon bands and a line spectrum similar to that of α Tauri.

14. Miss CLERKE. Authors of treatises on Astronomy have edited the statements of the classifications of stellar spectra and in a few cases have devised systems of their own. The most notable of the latter is the classification of Miss AGNES M. CLERKE[1]. This is an exhaustive classification based upon broad and unmistakable distinctions. No hypothesis of development or affinity pervades the sequence. The constituents of the first four classes show absorption spectra only, while the four last are marked by emission as well. Miss CLERKE's classes are arranged in parallel with the divisions of other classifications in Table 5. The classes may be briefly described in their essentials.

Class I. Helium stars. Absorption of hydrogen and helium predominates. In some spectra the ζ PUPPIS series is represented while in others lines of oxygen, nitrogen, and silicon are found. Metallic lines (sodium, iron, calcium, and magnesium) are faint and scarce. The comparative prominence of λ 4481 is especially remarkable among metallic lines.

II. Hydrogen stars. Intense hydrogen absorption is distinctive. No ζ Puppis lines are present. Neutral helium lines are weak or absent. Calcium *H* and *K* are thin but distinct. Feeble iron lines are numerous.

---

[1] Problems in Astrophysics, p. 181 (1903).

III. Solar stars. The solar spectrum is shared by members of this class, with slight variations. The leading features are *H* and *K* of calcium. Other metallic lines are innumerable but mostly sharp and thin. Hydrogen is still present (four lines).

IV. Stars with fluted spectra. A linear absorption system, somewhat modified from the solar type, has superposed upon it a set of dusky flutings, about ten in number. None of these bands occur in the more refrangible part of the spectrum.

V. Carbon stars. These have banded spectra totally different in stamp from those of Class IV. Three particularly conspicuous shadings testify to strong absorption by carbon vapor. Subordinate bands are also present. With many variations, all are designed on the same pattern. Many dark lines, notably *D* and *E*, are present. HALE discovered photographically in 1898 several unfamiliar bright lines superposed on the dark shadings.

VI. Stars with fluted spectra showing bright hydrogen lines. All are pronounced variables with a possible exception. The flutings, essentially the same as those of Class IV, tend to deepen and widen with declining light. They overlie a metallic spectrum similar to Class IV. Vivid hydrogen rays come into view with the approach of each maximum and fade after it has passed. Helium D_3 shines in some members of the class.

VII. Helium stars with bright lines. These objects give spectra of Class I, variously emblazoned with rays of hydrogen, helium, and a few other substances. In the P Cygni variety bright and dark lines are ranged side by side; in the variety like γ Cassiopeiae they are superposed. The lowest terms in each series show the strongest emission.

VIII. WOLF-RAYET stars. The leading spectroscopic distinction is the display of the ζ Puppis lines of helium, which may be bright or dark. Emission bands in the blue are always visible. Two at least are simultaneously or alternately visible, λ 4688 of HeII and λ 4652. Neutral helium lines are both bright and dark. D_3 may be bright and λ 4472 dark. $H\alpha$ may be bright, $H\beta$ neutral, $H\gamma$ and $H\delta$ dark. No metallic lines, bright or dark, have been recognized.

15. SALET. P. SALET[1] suggested a classification of stars according to their spectra into six groups as follows:

Symbol	Group
H′	ζ Puppis stars (VOGEL's IIb).
He	Helium stars (Ib, Ic).
H	Hydrogen stars (Ia).
Fe	Solar stars (iron spectrum) (IIa).
Ti	Stars with titanium bands (IIIa).
C	Stars with carbon bands (IIIb).

d) The Harvard Classifications.

16. PICKERING. In connection with the classification of stellar spectra for the DRAPER Catalog (1890) EDWARD C. PICKERING[2] originated a simple system of designation of the stellar spectral classes by capital letters. Initially in this work the classification of SECCHI had been used, with letters to indicate smaller distinctions. Later a more detailed classification was adopted as described below, and

[1] P. SALET, Spectroscopie Astronomique, p. 385 (1909).
[2] Harv Ann 26, pt. I (1891).

to this scheme the earlier observations were made to conform. On this basis the first general classification of photographic spectra was made by Mrs. W. P. FLEMING in the DRAPER Catalog contained in Volume 27 of the Harvard Observatory Annals. The spectra classified in this catalog were of small dispersion not permitting detection of less prominent differences. Hence no very detailed classification was practicable.

These spectra were photographed with an objective-prism instrument of 20 cm aperture and 115 cm focus combined with a 13° prism. The first spectrogram was made in 1885. The catalog, giving spectral classifications for 10351 stars, was published in 1890.

The PICKERING-FLEMING classification of the DRAPER Catalog of 1890 may be stated as follows:

Description.

Class A. Only the hydrogen series and generally K are visible.

Class B. In addition to the lines of Class A other lines, found in many spectra of stars in Orion and Canis Major, are seen. Of these lines, λ 4026 and λ 4471 are most commonly observed.

Class C. Spectra of Type I (Classes A and B) in which $H\gamma$ and $H\delta$ appear double.

Class D. Spectra of Type I containing emission bands.

Class E. Only $H\beta$, H and K are visible and the continuous spectrum undergoes no sudden change of intensity.

Class F. In addition to $H\beta$, H and K, other hydrogen lines are visible. K is much stronger, and $H\gamma$ and $H\delta$ are rather faint as compared with Type I.

Class G. Lines in addition to those observed in Class F are visible.

Class H. This spectrum resembles Class F except that the continuous spectrum suddenly diminishes in intensity when the wave length becomes less than 4310.

Class I. Like Class H except that additional lines are seen.

Class K. Bright bands are visible which may be portions of the continuous spectrum between dark bands.

Class L. Other varieties of this form of spectrum.

Class M. SECCHI's third type. The intensity of the spectrum changes suddenly at λ 4762. Many such spectra are otherwise like Class K.

Class N. SECCHI's fourth type.

Class O. Spectra consisting mainly of bright lines.

Class P. Planetary nebulae.

Class Q. All spectra not included in the above classes.

17. PICKERING-FLEMING. In classifications of spectra in clusters, published in 1897[1], Mrs. FLEMING used the capital letters, with some omissions, with the following meanings.

Description.

Class A. Spectra of the first type in which the hydrogen lines alone are conspicuous.

Class B. In addition to the hydrogen lines, those characteristic of stars in the constellation of Orion are also present, especially those whose wave lengths are 4026 and 4472.

Class F. Spectra intermediate between first and second type in which H and K are equally intense and the hydrogen lines are fainter.

[1] Harv Ann 26, pt. II (1897).

Class E. Spectra of the second type in which H and K are strong, the hydrogen lines faint and no other lines visible.
Class G. Spectra like those of Class E but with other lines present.
Class H. Spectra like Class E but with the continuous spectrum more intense for rays of greater wave length than 4300 than for those whose wave lengths are less then 4300.
Class K. Spectra like those of Class G but with the continuous spectrum as in Class H.
Class M. Spectra of Secchi's Type III.
Class N. Spectra of Secchi's Type IV.
Class O. Spectra of Pickering's Type V.

Thus the Draper classification was brought closer to the form which it finally assumed. Of the letters dropped, C appears to have characterized defective plates. Class D, denoting the presence of emission, was less useful than B or A, denoting the class of spectrum, followed by a symbol to indicate emission. Class I spectra were advantageously distributed over other classes; and L appears not to have been really defined. The transposition of E and F and the redefinition of K were other significant changes.

18. Miss Maury. One of the first investigations undertaken at Harvard as a part of the Henry Draper Memorial was the detailed study of the spectra of the bright stars north of declination $-30°$. Miss Antonia C. Maury assumed this task in the year 1888. The results were published in 1897 in Harvard Annals Volume 28, Part I. Among other valuable contributions by Miss Maury in this volume there is found a classification of stellar spectra[1] which is without question the most complete, thorough, and comprehensive achievement of its kind by any one investigator in this field. That it did not find wider application was due to its appearance of excessive detail, and to the greater acceptability of the more flexible Draper Classification as revised and developed by Miss Cannon.

Like McClean, whose classification appeared in the same year, but in much more complete detail than McClean, Miss Maury arranged the main sequence of stellar spectra in its recognized formal order. Thus her comparison of her own groups with the classes of Pickering and Fleming led to the rejection of all but the essential letters in the Draper Classification and placed these essential letters in the "mystic order" of the Harvard sequence. Further she pointed out that her Group I (Class Oe5) stars were those which appeared to be most closely related to her Group XXII (Wolf-Rayet stars) and through them to the planetary nebulae.

Miss Maury's working material consisted of about 4800 photographs containing the spectra of 681 selected bright stars. The instrument employed was a photographic refractor of 280 mm aperture with a focal length of 389 cm. Four fifteen degree prisms could be used singly or in combination before the objective. The length of the spectrum from $H\varepsilon$ to $H\beta$ was 8 cm with four prisms. Only the brighter stars were photographed with four prisms; and of such brighter stars a number selected as typical were photographed also with two and one prisms in order that the detailed spectra of higher dispersion could be compared more accurately with low dispersion spectra made with one prism. The classification of Miss Maury depends chiefly on the comparative study of spectra made with one prism, taking into account the appearance as well as the position of spectral features.

This classification comprised a main sequence of nineteen groups denoted by Roman numerals and arranged vertically in a "progressive series" through

[1] Harv Ann 28 pt. I, p. 13 (1897).

which ran an uninterrupted graded succession of spectral characteristics, varying chiefly with temperature. These nineteen groups ranged from Class Oe5 to Class M3 of Miss CANNON's classification. Three additional groups at the end included long period variables with spectra of Group XIX, carbon stars, and WOLF-RAYET stars. Six transition groups were also recognized. The classification provided also for five horizontal divisions which might occur in each group. These divisions were denoted by small letters each belonging to a collateral series of which the main or "*a*" series extended from Groups II to XX. The other collateral series designated by letters, *c*, *ac*, *ab*, and *b* (with *p* for peculiar spectra) ran parallel with more or less of the main sequence. Of the 110 subdivisions (excluding transition groups) thus provided for normal spectra Miss MAURY used only forty-eight. But in addition there were peculiar stars in eight groups. Further, outside of the sequence, the letter C denoted composite spectra and L, spectra of "Orion" type with emission lines. The transitional group-divisions, as IV—V and XI—XII, added to the rest made a total of fifty-five effective compartments in the classification not including *p* stars.

It is not practicable to reprint the detailed description of fifty-five subdivisions, given in thirty-five quarto pages of the Harvard Annals. The reader will derive an adequate knowledge of the classification through the information given below. An explanation of the meaning of the horizontal divisions will be given, followed by a tabulation with condensed descriptions of groups in regard to type and a complete tabulation of the main vertical or progressive series of groups (the *a* series). Comparisons with the DRAPER and other classifications are also given in Table 5, p. 58.

The horizontal divisions into which any group of Miss MAURY's progressive or vertical series might be divided were defined as follows. The chief division *a* (including 355 of the 681 stars examined) contained stars whose spectral lines were well contrasted with the neighboring continuous spectrum (i. e. not hazy) and of average width. Division *b* comprised stars whose spectral lines were relatively wide and hazy though preserving about the same relative intensities as in *a*. The fainter lines thus widened become imperceptible. Division *c* included stars in whose spectra the Orion lines (chiefly of He I, He II, O II, Si II, Si III, Si IV, N III, C III, and Mg II), when present, and the hydrogen lines were narrow and well defined; the system of metallic lines differed from those in the solar spectrum in that some were in new positions, many were unusually intense, and the relative intensities were not solar; the *H* and *K* lines were more intense and the hydrogen lines less intense than in corresponding divisions of the same group. In general, Division *c* was distinguished by the strongly defined character of its lines, but Miss CANNON found that the difference between *c* and other spectra was more in the peculiar intensity of numerous prominent lines than in the actual width of the lines. Spectra intermediate between divisions *b* and *a* were designated by the combination *ba*; between *a* and *c* by *ac*. The combination *bc* could not occur. Typical stars of Division *a* were γ Orionis, α Lyrae, α Canis Minoris, the Sun, and α Boötis. Typical stars in Division *b* were δ Orionis, α Leonis, and α Aquilae. Typical stars in Division *c* were χ^2 Orionis, η Canis Majoris, β Orionis, α Cygni, ε Aurigae, and δ Canis Majoris. The collateral series *b* paralleled the main (*a*) sequence, among Miss MAURY's 681 stars, from Group I to Group X (Oe5 to A5). The *c* series paralleled the main (*a*) series from Group III to Group XIII (B2 to F8). Division *ac* occured in Group XIV (solar type) with the typical star, the Cepheid variable η Aquilae, but below this group only *a* stars were found. These lettered divisions though based on formal spectral characteristics have been found to represent important physical differences to be discussed below

The following tabulation shows the relation of Miss MAURY's groups to the types of SECCHI and PICKERING, and provides condensed descriptions of the main or bolder features. The spectral region involved is the photographic. The hydrogen lines are of course of the BALMER series. The Orion lines are mainly those of He I, Si II, Si III, Si IV, O II, N III, C III, and Mg II. The solar lines are those characteristic of the solar spectrum excluding the lines of hydrogen and calcium.

Tabulation of groups in regard to type.

Groups I—V. Stars of Orion Type.

Orion lines strong, numerous, declining in the later groups. Hydrogen lines of moderate intensity, increasing toward the first Type. Solar lines absent. *K* line faint.

Group VI. Stars intermediate between the Orion and first Types.

Groups VII—XI. Stars of SECCHI's first Type.

Orion lines in general absent. Hydrogen lines at maximum intensity, declining in the last two groups. Solar and calcium lines increasing toward the second Type.

Group XII. Stars intermediate between first and second Types.

Groups XIII—XVI. Stars of SECCHI's second Type.

Hydrogen lines not very strong and growing fainter. Solar lines extremeld numerous and increasing. Calcium lines seen as bands, which are strong any gradually increasing in strength.

Groups XVII—XX. Stars of SECCHI's third Type.

Hydrogen lines weak, bright in Group XX. Solar lines slightly increasing, then declining. Calcium lines strong, slightly decreasing. Bands and flutings replacing lines. The former fading in the direction of greater wave length.

Group XXI. Stars of SECCHI's fourth Type.

Two absorption bands present. Line absorption slight. Blue portion of the spectrum generally very faint.

Group XXII. Stars of fifth Type.

Wide bright bands superposed on a faint continuous spectrum, the strongest one of these probably coincident with a bright band in the spectrum of the gaseous nebulae, and most of the others probably coincident with hydrogen lines and prominent Orion lines.

Remaining Classes of spectra:

1. Group C. Composite spectra, probably of double stars.
2. Group L. Stars of Orion type having bright lines.

Table 2 contains a complete tabulation of the photographic criteria and descriptions of Miss MAURY in the main or *a* series (Group I contained only *b* stars) from Group I to XXp with Group XXII placed in position at the beginning of the sequence. Group XXI, in which only band 4862 is mentioned in the photographic region, and also a few references to the visual region, is omitted. Among the criteria, hydrogen-, Orion-, and solar lines are used in the sense described in the last paragraph above. Identifications of single line criteria for which wave lengths are used will be found below in connection with the DRAPER Classification. Spectral regions used as criteria are denoted by wave lengths of their limits connected by hyphens. The mathematical symbols for "is greater than" and "is less than" are used here in the sense, "is stronger than" and "is weaker than" respectively. The other mathematical symbols will be readily understood. Roman numerals always refer to spectral groups. Comparisons, such as weaker or stronger, always refer to the adjacent group to the left unless otherwise specified. The intensity of the hydrogen lines is expressed numerically in terms of that in the spectrum of Sirius. Prom. is used for prominent, and faint is used interchangeably with weak. V. signifies "very". Criteria that carry over are repeated on each page so that the full description of each group can be readily

Table 2. Miss MAURY's Classification of Stellar Spectra.

Criteria	XXII	I	II	III	IV	V	VI	VII
λ 4686	Strong E Band	0,5 × 4026						
$\frac{4200}{4026} = \frac{4543}{4026}$		About 0,5						
4089		Very strong	= 4026, > 4116	Rather narrow	Difficult	Not seen		
(1) 4097		Strong	About same	Rather narrow	Difficult	Not seen		
4116		Weak	> 1, 2, 3, 4	Rather narrow	Absent			
(2) 4070, 4073, 4076		Weak	> I	Rather narrow	Narrow	Not seen		
(3) 4121		Weak	> I	At least 2×II	Conspicuous	< 4128 + 4131		
(4) 4144			> I	At least 2×II	2×4120	> 4128 + 4131		
4472			V. prominent				Well marked	Trace
4026			= 4472				Well marked	In some spectra
4650		Much < II	V. prom. Oft. wide as $H\beta$	0,5×II				
4650/4472			> Unity					
4650/4026			> Unity	0,6 ±				
4388/4026				> 0,5				
4367					Stands out			
4481			V. weak	V. weak	> III		Moderate intensity	Strongest solar line
4481/4472						0,5	= or > 1	
4128, 4131						Blend > 4121	Mod. int. Each = or > 4144	Trace
Orion Lines	Wide E bands (Prom.)	Many lines	Seventy present	Eighty present	Like II and III	Much weaker	Fewer, weaker	Trace of some
Hydrogen Lines	Wide E bands	Strong, V. wide	0,2× Sirius	0,3—0,4 ×Sirius	0,4—0,5 ×Sirius	0,9 × Sirius	= or < Sirius	Maximum
Solar Lines		None	None	None			Present 25 to 45	Numerous but faint
K		V. weak	V. weak	Faint	> III	> IV	> 4026 < Sirius	
$K/H\delta$								Slightly > 0,1

Table 2. Miss Maury's Classification of Stellar Spectra. (Continued.)

Criteria	VIII	IX	X	XI	XI—XII	XII	XIII
Hydrogen	About = Sirius	0,95 × Sirius	0,9 × Sirius	0,7 × Sirius	0,4 × Sirius	0,25 × Sirius	0,2 × Sirius
Solar Lines	More and stronger	More and stronger	> IX More	Slightly > X	> XI < XII	Number 3,0 × VII. Strength 4,3 × VII	Increasing toward sun. Like sun
K						or H = 0,8 × sun = 0,7 × α Boö.	As in sun
$K/H\delta$	0,2	0,4—0,8	0,9—1,1	1,5—2,5	3,5		
$\frac{H+H\varepsilon}{K}$			> 1,0	Usually < 1,0	Slightly < 1,0	0,70—0,80	
G			Not continuous	Not continuous	Trace of band	Faint band much < XIV	Increasing, as in sun at end of group
Lines 4299—4315			Visible in G	Visible in G. Much < XII		Stronger	
4326				Much < XII		Not strong	

Table 2. Miss Maury's Classification of Stellar Spectra. (Continued.)

Criteria	XIV	XV	XVI	XVII	XVIII	XIX	XX	XXp
Hydrogen Lines	0,15 × Sirius	< 0,10 × Sirius	0,08—0,07 × Sirius	0,08—0,07 × Sirius	0,07—0,06 × Sirius	Like XVIII	Emiss.	Emiss.
Solar Lines	As in sun	About as in sun	Like XV	> XVI < XVIII	1,2 × sun > XV	Fewer, weaker.	Fewer, weaker	
K	Nearly at max[1]	Maximum 1,2 × sun	Maximum 1,2 × sun	1,2 × sun H = 1,4 × sun	Probably narrower	Like XVIII		
$\frac{H+H\varepsilon}{K}$		0,8						
Band G 4299—4315	Continuous 2 prism	Like sun, some lines stronger	Not cont. Most metallic l's stronger	Not cont. Two parts	Not cont. Stronger except 4308 and 4313	Less conspicuous. Two parts	Weaker, two parts	
$\frac{H\gamma + 4338}{4324 + 4326}$	> 1, usually = 2							
4326/Hγ	< 1	Nearly 2						
$\frac{\text{Region} > 4307}{\text{Region} < 4307}$	> 1	Greater ratio	Maximum marked	Like XVI	Less marked	Nearly 1,0		
4315—4368, 4470—4525	Faintly seen emission	Visible	V. strong	Like XVI	Weaker	Almost invis.		

[1] 2 or 3 times $H\delta$ in Sirius.

Table 2. Miss MAURY's Classification of Stellar Spectra. (Concluded.)

Criteria	XIV	XV	XVI	XVII	XVIII	XIX	XX	XXp
4614—4648	Faintly visible emission	Visible	V. strong	Like XVI	> 4556—4586	Narrower	Present	Bright maximum
$\frac{4326}{H\gamma + 4338}$		2,0						
$\frac{4227}{4384 + 4385}$		1,5—2,0	3,0—4,0					
4078—4096		Emission distinct	Visible		Present			
4144—4216, 4055—4078		Absorption						
3889 to Violet		Absorption						
g 4227		Much < *G*	Note[1]	Nearly as in XVI[1]	As in XVII Note[1]	Almost same as XVIII[1]	About same as XIX[1]	
H or *K*/4227			> 1					
4556—4586			Present emission	Clearly seen	> 4470—4525	> 4614—4648	Narrow, inconspic.	Bright maximum
Region < 3970			Faint		Faint			
Bands TiO_2 4762, 4954				Clearly shown	Strong	Stronger	Clear, strong	
4307—Hε					Nearly uniform			
Dark band 4586						Appears	Probably present	
4408—4423							Weak, bright	
Violet Abs.			> XV					> XX

[1] Most conspicuous except *H* and *K*.

obtained by glancing down the column below it. Much space with no sacrifice of completeness is saved by such a tabulation of a detailed spectral classification.

Comparisons of Miss MAURY's Groups with the divisions of other classifications are found in Table 5, p. 58.

19. MISS CANNON. The DRAPER classification of PICKERING and Mrs. FLEMING was greatly modified and extended by Miss ANNIE J. CANNON during her study of the spectra of bright southern stars, the results of which were published in 1901 in Harvard Annals, Volume 28, Part II.

Miss CANNON's working material in Harvard Annals, Volume 28, consisted of 5961 photographs on which 1122 stars were classified. These plates were made in the years 1891—1899. The instrument employed had a diametrical aperture of 33 cm and a focal length of 192 cm. One, two, or three prisms were used in front of the objective, giving a dispersion of 7,43 cm between *Hε* and *Hβ* with three prisms. Of the 1122 stars classified, 41 had been photographed with three prisms, 268 with two prisms, and 813 with one only. The plates taken with one prism were found most useful in making the general classification; those with two and three prisms were employed for detailed study of peculiarities and intensities of lines.

In 1897[1], Miss MAURY had selected the pertinent DRAPER letters and had arranged the symbols, B, A, F, G, K, M, N, in their recognized order in the Harvard sequence. At the same time she had placed S Monocerotis, a star of Group I and Class Oe5, at the beginning of her sequence, and had pointed out that this group appeared to be most closely related to Group XXII and through it to the planetary nebulae. Further she had redefined the significance of the DRAPER letters in greater detail by placing them in parallel with her own groups.

In Harvard Annals, Volume 28, Miss CANNON subdivided the DRAPER classes by affixing numbers from naught to nine to each DRAPER letter from B to K inclusive, thus providing symbols at decimal intervals of a class. However, not all such symbols were used, the actual number of compartments defined and employed by her being twenty-eight, the same number used by Miss MAURY in her main series including six transition groups. Further Miss CANNON placed Class O at the beginning of the sequence and subdivided it, as in Class M at the end, by affixing small letters, a, b, c, etc. to denote divisions not necessarily in a sequence. Class Oe5 (Miss MAURY's Group I) included stars preceding B0. Mrs. FLEMING[2] had apparently used the symbols, O I, O II, and O III in specific cases to designate spectra of Classes Ob, Oc, and Oap, previous to 1891. Miss CANNON adopted Class P for planetary nebulae and Class Q for peculiar spectra with bright lines.

Miss CANNON described the classes from Oa to Mb in Volume 28. In so doing she employed some forty-eight separate criteria in the sequence from Oe5 to Mb, of which thirty-five had been used by Miss MAURY. Eleven others not used by Miss MAURY were single lines or ratios which were used only once. The other two were the ζ Puppis lines and the ratio of $H\beta$ to λ 4762. Miss MAURY had employed forty-nine criteria in her descriptions of her main sequence, of which fourteen of relatively small importance were not used by Miss CANNON. Thus, the basic resemblance between the two classifications is evident. Miss MAURY isolated distinguishing as against merely descriptive features whereas Miss CANNON in general did not. However, in Harvard Annals, Volume 56, pp. 66—69, Miss CANNON published brief descriptions of each of her spectral divisions extracted in abridged form from Volume 28, and in the last version of the classification this concise, abbreviated form was maintained. In all her extensive classification she found it inexpedient to use the horizontal divisions *a*, *b*, and *c* of Miss MAURY, preferring to give the facts relating to the widths of lines in the remarks concerning individual stars.

Pecularities manifested themselves in five different ways. First, the widths of lines were found to be greater or less than normal. Second, notable departures from the relative line intensities of the typical star were present. A striking instance of this is found in the Class A spectra with abnormally strong silicon lines of wave lengths 4128 and 4131. Third, bright lines were present as in Class Md or in Classes O, B, and A. Fourth, lines were found periodically double. Finally, spectra were found to be composite as in ε Carinae.

e) The Development of the DRAPER Classification from 1901 to 1924.

20. General Changes. The development of the classification from the rather comprehensive descriptions of Harvard Annals, Volume 28, part I, 1901, through the very concise definitions of Volume 56, 1912, to the final statement in Volume 99, 1924, involved changes in content in the definitions of classes and the addition

[1] Harv Ann 28, pt. I, p. 10, Table I (1897). [2] Harv Ann 56, p. 225 (1912).

of some new classes and several new divisions. The sixteen pages devoted to the description of the Classes Oa to Mb in Volume 28 were reduced to three in Volume 56, and increased to four in Volume 99. Class N was added in Volume 56. Descriptions of Classes Pa, Pb, Pc, Pd, Pe, Pf, Mc, Md, S, R0, R3, R5, R8, Na, Nb, and Nc are found in Volume 99, having been added as described below. Pec. was applied to spectra of novae and some variable stars; Con., to spectra apparently continuous; while Q was dropped.

The changes in the descriptions tended away from references to individual plates and typical stars toward class characterizations. In general the criteria that survived in the definitions must have been those that proved useful in actual application. The lack of continuous numerical criteria in the final description may be seen at a glance in Table 3. The classification is evidently based on general appearances rather than quantitative criteria.

Notable changes in the description of the classes from Volume 28 to Volume 99 included omission of reference to bright lines of neutral helium in Class O. Several intensity estimates of the hydrogen series were omitted, leaving insufficient data to construct a continuous curve of hydrogen intensity, and all reference to hydrogen dark lines was omitted after Class K0. All references to the K line were omitted before Class B5. Certain small changes in numerical ratios probably resulted largely from the application of the description to classes rather than to typical stars.

21. Class P. Miss CANNON employed Class P in Volume 28, p. 163, without affix. In Harvard Annals Vol. 76, p. 20, Class P was subdivided by Miss CANNON into Pa, Pb, Pc, Pd, Pe, and Pf, of which Pd and Pe included most of the gaseous nebulae. The divisions were described and a fundamental criterion in arranging the order of the sequence at the end was the band 4686 of He II which is found in Class O. In the final statement of the HENRY DRAPER Catalog the bright band at 4650 was identified as the most conspicuous feature of Class Pf and not 4686 as in Volume 76. Otherwise the criteria in Class P were essentially unchanged. Also the typical nebulae were all retained.

22. Class M. In 1897, Classes Ma, Mb and Md were placed in parallel by Miss MAURY with her Groups XVII, XVIII, XIX, and XX (Harvard Annals 28, p. 10). Classes Mc and Md were used by Mrs. FLEMING[1] previous to 1900 in her studies of spectra of the third type. Class Md was used also in Harvard Catalogues of variable stars[2] in 1903 and 1907. Class Mc included spectra with band absorption stronger and continuous spectrum of shorter wave length relatively weaker than in Class Mb. Class Md was subdivided in detail by Mrs. FLEMING[3] and symbols were formed by adding numbers from one to ten, including 1,5, to the letters Md. Though no complete description of the subdivisions was ever published, a number of long period variables were classified on this basis in Table IX, p. 197, of the Harvard Annals, Volume 56. Class Md1, of which R Lyncis (Class Se) was the typical star, showed a spectrum resembling α Orionis (M0) but having $H\beta$ and $H\gamma$ strongly and nearly equally bright, with $H\delta$ an emission line barely visible. R Leonis, the typical star of Class Md10, had an absorption spectrum Mc (M 6,5—8,0) or later with $H\beta$ emission not seen, $H\gamma$ barely visible and $H\delta$ strongly marked. The other classes formed a nearly continuous sequence between these extremes. Miss CANNON employed Class Md in Harvard Annals 28, Part II, 1901, and described Classes Mc and Md in the HENRY DRAPER Catalog.

---

[1] Sci 8, p. 455 (1898); Publ Amer Astr Soc 1, p. 48; Harv Ann 56, No. 6 (1912).
[2] Harv Ann 48, No. 3 (1903); 55, pt. I (1907).
[3] Sci 8, p. 455 (1898); Publ Amer Astr Soc 1, p. 48; Harv Ann 56, No. 6 (1912).

23. Class R. Class R, or Type VI, was proposed by E. C. PICKERING in 1908[1]. Spectra of this kind were described in 1896[2] by Mrs. FLEMING as containing rays of much shorter wave length than ordinary fourth type stars. Later it was stated in Harvard publications that the blue end of these spectra was as extensive as in Class K, that *H* and *K* were well shown and that two well marked absorption bands were present, the one centered near 4227 and the other extending from 4640 to 4750. In Harvard Annals, Volume 56, p. 220, 1912, it is suggested as probable that stars can be found forming a continuous sequence from Class N to Class R, like that connecting Class B and Class M, but the relation of Class R to the Harvard sequence is not stated. The use of the designation N5R in Volume 56 of the Harvard Annals, p. 218, might indicate for Class R a position later in the sequence than Class N, on the basis of Miss CANNON's notation, but in a private letter, dated 1915, E. C. PICKERING stated, "it seems more probable that Class R should fall between Classes M and N".

DUNÉR[3] felt that Types III and IV of SECCHI were coördinate divisions each connected by intermediate spectra with Type II. HALE, ELLERMANN, and PARKHURST[4], after a thorough study and discussion of stars of SECCHI's fourth type, reached the conclusion that "stars of the third and fourth types should be classed together as coördinate branches leading back to stars like the sun." They summarized the characteristics common to stars of Class M and Class N as (1) red color, (2) tendency to variability, (3) resemblance of dark lines, (4) presence of lines widened in sunspots, (5) similar physical conditions revealed by spectra, (6) presence of bright lines, (7) dark bands of which those of cyanogen were common to both, and (8) connection of both types of spectra with spectra of solar stars. PARKHURST[5] found the color index of some Class R stars to parallel that of Class K.

W. C. RUFUS[6] and R. H. CURTISS[7] established the fact that Class R stars form the connecting link between Class N and the Harvard sequence. Class R and Class N form a branch joining the main sequence near Class K0 and running parallel with Class K and M. In the parallel relation, Classes K and R were found by RUFUS to resemble each other in (1) color index, (2) small tendency to variability, (3) the character and identity of absorption lines, (4) the shift in wave length of certain lines, (5) absence or weakness of bright lines, (6) percentage in the galactic region, and (7) residual radial velocity. Classes R—R9 differed from Class K—K9 in the significant presence of carbon bands in the former and the presence of weak titanium oxide bands in the later divisions of the latter. Furthermore the same characteristics which placed Class R in parallel with Class K progressed continuously through Class R connecting Class G with Class N but with the distinctive difference that the carbon bands, and no titanium bands, were always present in Classes R and N whereas titanium bands without carbon bands characterized Class M and the later divisions of Class K. The conclusion is now generally accepted that the spectral sequence divides, Classes K and M forming one branch and Classes R and N the other. A graphic demonstration of the reality of this R—N sequence is found in Plates G, H, I, and J published and discussed by Curtiss in Volume 2 of the Publications of the Detroit Observatory and elsewhere. Miss CANNON defined the divisions, R0, R3, R5, and R8 in the HENRY DRAPER Catalog and placed them before Class N.

[1] Harv Circ 145 (1908). [2] Harv Circ 9 (1896).

[3] Kgl Svenska Vet Akad Handl 21, No. 2 (1884).

[4] Yerkes Obs Publ 2, p. 385 (1903). [5] Ap J 35, p. 132 (1912).

[6] University of Michigan (Detroit) Obs Publ 2, p. 103 (1915).

[7] University of Michigan (Detroit) Obs Publ 2, p. 182 (1915); Pop Astr 25, p. 279 (1917); Sci Amer 83, p. 344 (1917).

24. Class N. The symbol Na was placed in parallel with her Group XXI by Miss MAURY in 1897. It was used in publications at Harvard as early as 1906[1], Nb as early as 1921[2]. Classes Na and Nb were described in Harvard Annals, Volume 91, 1918, the first volume of the HENRY DRAPER Catalog. Class Nc was introduced and described in 1923 in the eighth volume of the HENRY DRAPER Catalog. In volumes of the Catalog prior to this such spectra were described individually in the sections devoted to remarks.

25. Class S. Class S was adopted and described in the eighth volume (1923) of the HENRY DRAPER Catalog. In the volumes of the catalog prior to this, stars of this class were designated as Pec., and the nature of the peculiarities was described under Class Md. Stars of this class were studied by ESPIN and WRIGHT[3] and MERRILL[4]. The latter established the fact that important pecularities of these spectra were the presence of bands of zirconium oxide[5] and the strength of zirconium lines[6, 7]. The spectra of these stars probably form an ill-defined branch of the Harvard sequence running parallel with Class M but merging with Class M and possibly joining it in the later divisions.

f) Presentation of the DRAPER Classification.

26. Methods of Presentations. The best interpretation of the DRAPER classes is undoubtedly found in the actual 225300 classifications of the HENRY DRAPER Catalog; the best representation, in the spectra of the stars chosen as types for each division. Least expressive are the verbal descriptions appearing first in relatively permanent form in Harvard Annals, Volume 28, Part II, and in most recent restatement in the ninth volume (Harvard Annals, Volume 99) of the HENRY DRAPER Catalog. Such verbal descriptions are not used in general to classify spectra. They are intended to enable non-classifiers to comprehend the character of the sequence. At Harvard spectra are classified on the basis of their general appearance undoubtedly with definite type spectra well in mind or in view. Elsewhere the tendency has been to use line ratios as classification criteria or to select typical spectra at short intervals along the sequence and using these as standards to grade other spectra in relation to them.

In the presentation of the DRAPER Classification we shall employ five methods of exposition.

1. Photographic reproduction of typical spectra from Class P to Md of the main sequence and from G to N of the carbon branch.
2. The verbal descriptions of the classes.
3. A tabular description of the classes.
4. The variation curves of the main criteria, whether single lines, groups or series of lines.
5. Ratios of line intensities.
6. Comparisons with other classifications. (Section g.)

27. Photographic Reproductions. The photographic reproductions of typical stars of the DRAPER Classification are found in Figures 5, 6, 7, and 8. The originals were prepared by RUFUS and CURTISS[8] from spectrograms made with the one-prism spectrograph of the Detroit Observatory. The originals were enlarged and widened by a clepsydra device[9] designed by CURTISS, employing a technique

[1] Harv Circ 111, p. 2 (1906). [2] Harv Circ 224 (1921). [3] M N 72, pp. 546, 548 (1912).

[4] Ap J 56, p. 457 (1922) = Mt Wilson Contr 252; Ap J 58, p. 215 (1923) = Mt Wilson Contr 264.

[5] Publ A S P 35, p. 217 (1923). [6] Ap J 63, p. 13 (1926) = Mt Wilson Contr No. 306.

[7] Trans Int Ast Union 2, p. 121 (1925).

[8] University of Michigan (Detroit) Obs Publ 3, p. 258 (1923).

[9] University of Michigan (Detroit) Obs Publ 2, p. 182 (1916).

developed by him. The typical objects are named in the right hand margins. The classes are given in the left hand margins. Important features are designated in the upper and lower margins. Comments on the plates connecting them with the descriptions of the classes will now be given.

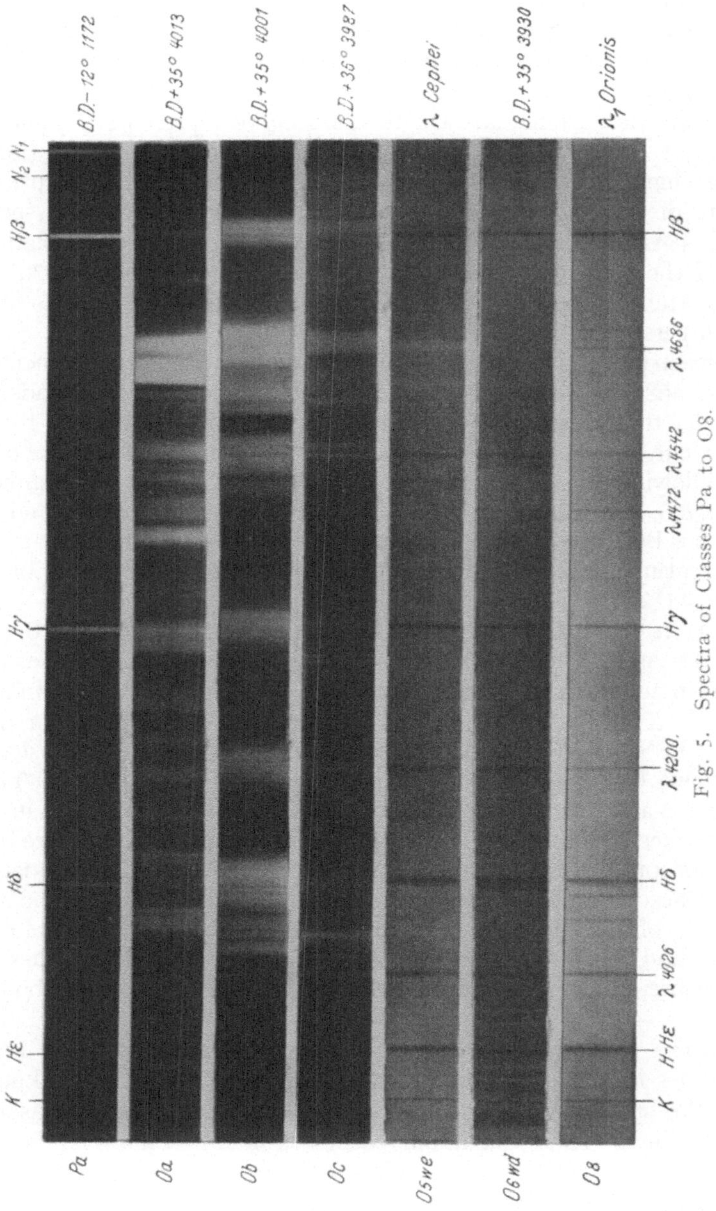

Fig. 5. Spectra of Classes Pa to O8.

Remarks concerning Fig. 5. In addition to the hydrogen bright lines in the spectrum of the planetary nebula B.D. −12° 1172 the bright lines N_1 and N_2 of ionized oxygen are also conspicuous. Numerous bright lines and bands characterize Classes Oa, Ob and Oc. The ordinary lines of the elements hydrogen and helium are present, also the ζ Puppis series and 4686 of ionized helium, and numerous unidentified lines and bands. Especially strong in Class Oa are the wide bands with center at approximately 4650, due to carbon, and 4686 previously mentioned. The line 4441 is also strong in Class Oa but is not found in the other classes. In Class Ob, 4686 is the characteristic feature, while 4650 is lacking. The hydrogen bright lines are stronger than in the preceding class. The ζ Puppis lines, 4542 and 4200, are prominent. Class Oc has a comparatively strong continuous spectrum which renders contrast difficult. The bright lines are weaker and narrower than in Classes Oa and Ob; 4686 is the strongest band; 4059 is strong and quite sharply defined.

The transition from Class Oc to Class O5we represents a change from a typical bright line to a typical absorption line spectrum. Indications of the impending change are suggested, however, in Class Oc by the absorption lines at the centers of some of the bright bands, especially the ζ Puppis series, 4026, 4200, and 4542. A similar feature is seen in the region corresponding to the position of the band *G* of the solar spectrum. The reversal of 4026, He I and II, may be traced through Classes Ob and Oa, while bright borders of this line and others persist in Class O5we.

The spectrum of λ Cephei, Class O6wd, shows a slight strengthening of the spectrum at 4686 and 4638. The background of the reproduction is made relatively dark in order that these emission features may be brought out, but they are not clearly evident. Hydrogen dark lines are the most prominent features of Class O5 and the following Classes O6 and O8. The ζ Puppis series, represented best by 4542 and 4200, is strongest in Class O5 but continues in Class O8 and is faintly seen in Class B0. The ordinary helium lines are also present but are not the strongest features, increasing in intensity in Classes O5 to B0, as brought out clearly by 4472.

Remarks concerning Fig. 6. The increase in the intensity of the hydrogen lines to Class A0 (A2) and the decrease through the following classes constitute one of the main features. The increase of calcium absorption is indicated by the lines *H*, *K* and 4227. The difference between the behavior of the *H*, $H\varepsilon$ blend and the single *K* band is notable. Helium decreases in intensity from B0 (B2) as shown by 4472 and 4026 and is evanescent in A0. Line 4649 of carbon and oxygen is also at maximum intensity in Class B0 but is absent in B5. Line 4481 of magnesium increases in strength from B0 to A0 (A2) where it becomes the most conspicuous line excepting the hydrogen series and *K*. In later classes 4481 is replaced by a low temperature line in the same position. The increase in the intensity of solar lines may be illustrated in numerous cases, for example 4326, 4384 and 4668. The increase in the number of metallic lines is also a special feature. The development of the band *G* of hydrocarbon may be traced.

Remarks concerning Fig. 7. The hydrogen series is no longer prominent. Lines *H* and *K* are broad and strong, but due to the weakness of this region of the spectrum it is difficult to observe their intensity as compared with other lines. Their maximum occurs near K3. Line 4227 is prominent and increases in intensity through Classes G, K, and probably M. The band *G* is conspicuous and appears almost continuous in Classes G5 and K0 with weaker absorption at the center; but becomes weaker and loses continuity in the following classes. Iron lines showing large change in intensity with spectral class from F to K or M are 4405, 4326 between $H\gamma$ and *G*, 4352 blended with magnesium, seen just at the right of $H\gamma$, and 4383 at the left of 4405, making with it a conspicuous pair in Class K5, both components of which are strengthened by blending with close lines. These four iron lines and the calcium line 4227 were used by ADAMS and KOHLSCHÜTTER in comparison with $H\gamma$ to form relative intensity curves in their quantitative method of spectral classification. Lines 4872 and 4958 compared with $H\beta$ were also used in the same way. The close pair, 4251 and 4254, which is well separated in these spectra, furnishes an interesting illustration of change in relative intensity, as 4254 increases to the end of the sequence while 4251 passes its maximum near K5 and is very faint below

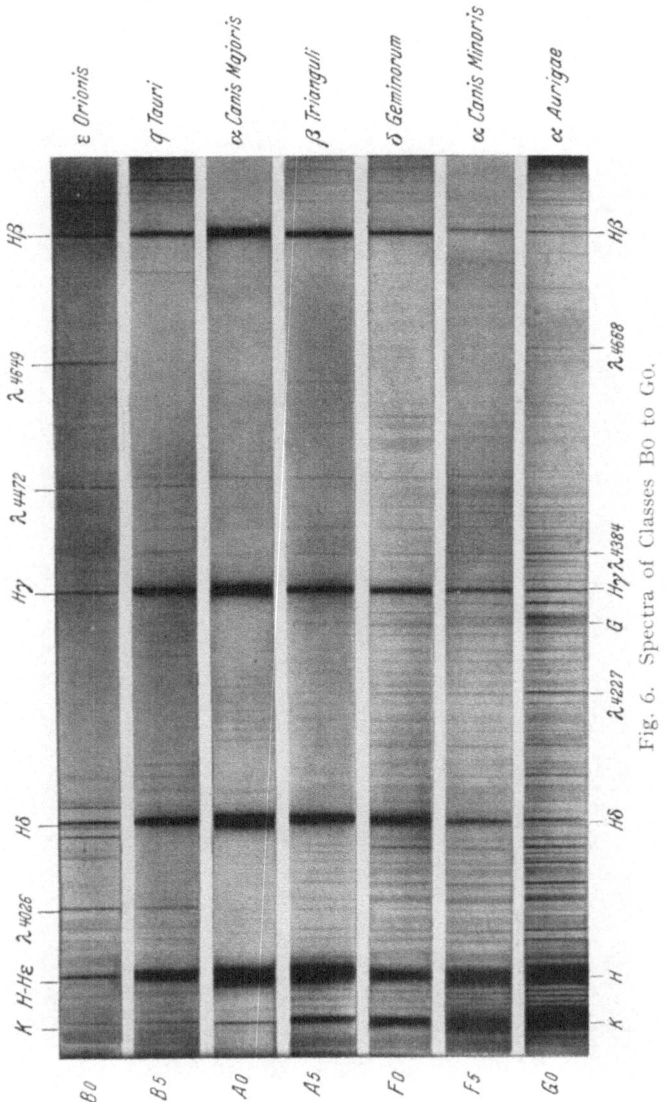

Fig. 6. Spectra of Classes B0 to G0.

M3. The lines strengthened in high luminosity stars, 4216, 4395, and 4408, may be noted in the spectrum of α Orionis.

The banded appearance typical of all Class M spectra begins to show in Class K5 and is most strongly marked in the last subdivision Md. The spectrum of *o* Ceti is enriched by the presence of bright hydrogen lines, as well as the characteristic titanium oxide bands.

Remarks concerning Fig. 8. The strong absorption band with head at 4737 usually attributed to carbon monoxide is one of the chief characteristics of Class R and N. In general the band *G* is strong on plates showing this region of the spectrum of Class R, although it is lacking in the spectrum of B.D. $-10°$ 5057, which was selected at Harvard as the type of Class R0. The

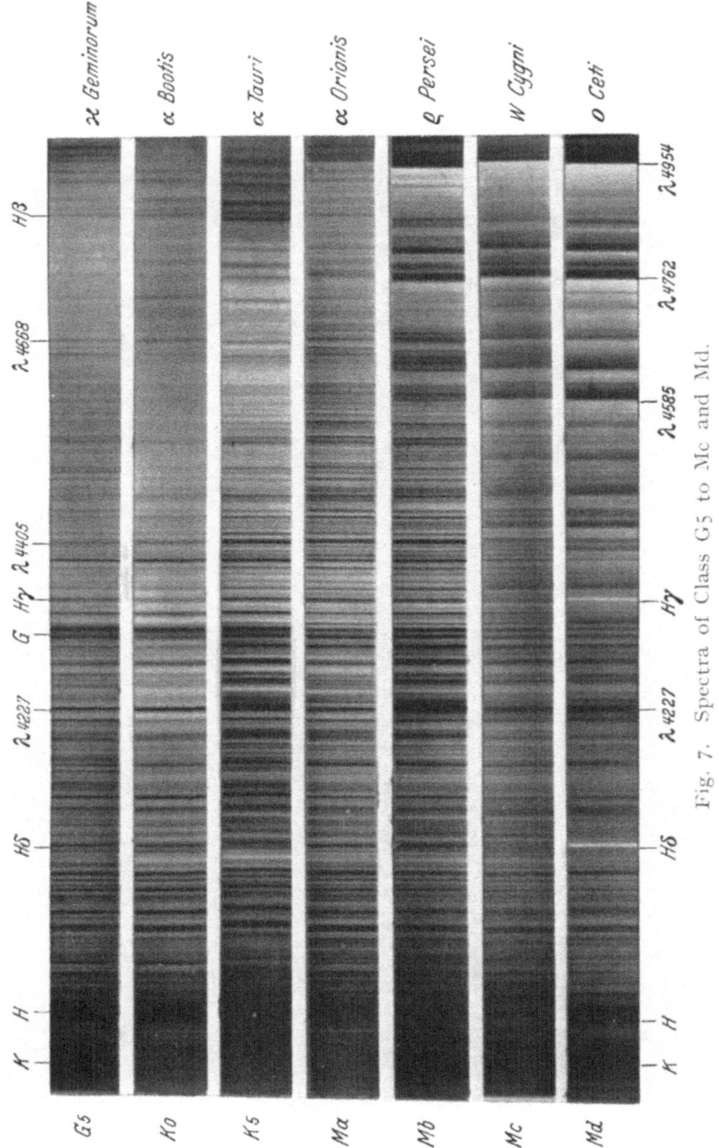

Fig. 7. Spectra of Class G5 to Mc and Md.

reproduction of the spectrum of that star is preceded here by another, B.D. $+42°$ 2811, which shows a closer resemblance to Class G than the selected type star, with regard to strength of the violet light, intensity of the hydrogen lines, and weakness of carbon absorption. Line 4227 is well marked in all Class R spectra photographed in that region but weaker than its counterpart in similar divisions of Class K. The subordinate lines of iron are apparently declining in strength through Classes R and N. This fact, together with the presence of a number of strong lines of scandium, vanadium, titanium and calcium in the red region of Class N spectra adds evidence to the conclusion clearly drawn from color indices that these are low temperature stars. A drop in the intensity of the continuous spectrum at 4216 due to cyanogen and another at

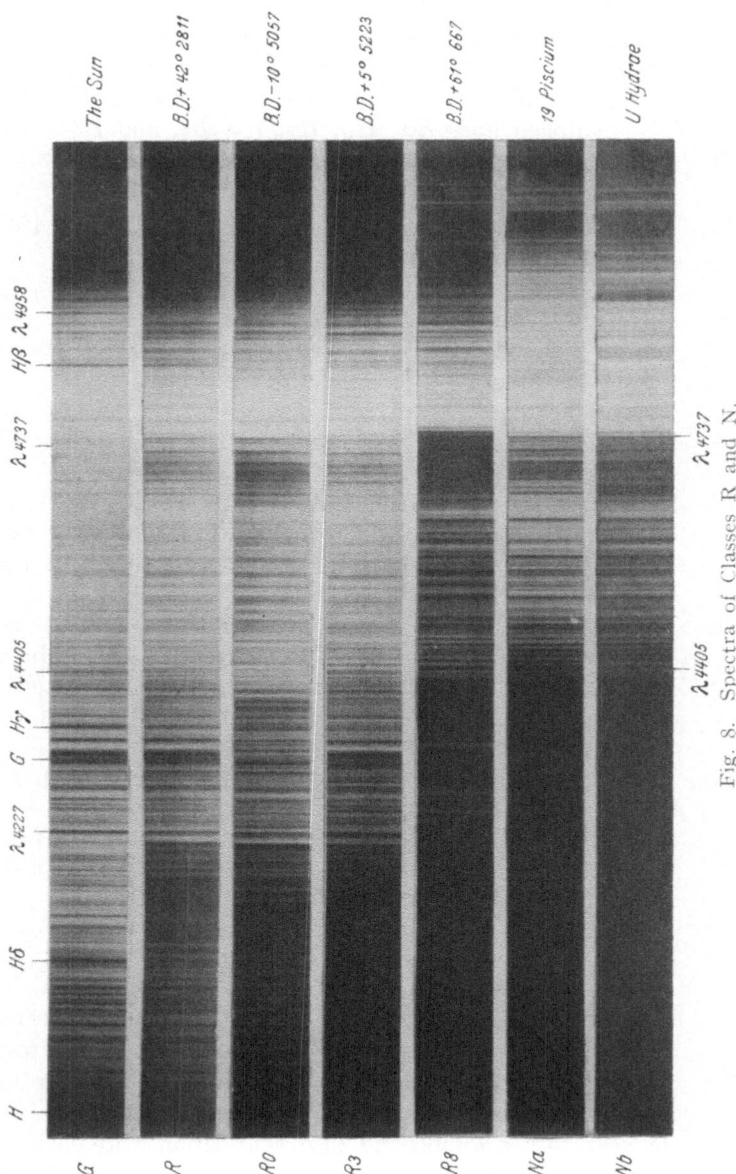

Fig. 8. Spectra of Classes R and N.

4396, possibly due to Group V of the Swan spectrum, are noticeable, Class N spectra being very weak on the violet side of the latter. Bright lines are present in some of the spectra of Class R stars and in Class N stars, but are not conspicuous. In Class Nc, which is not shown, the spectra contain no observed light of wave length shorter than $H\beta$.

28. Verbal Descriptions of DRAPER Classes. Verbal descriptions of the divisions of the DRAPER Classification are given below. In general they follow the wording of Miss CANNON in Harvard Annals, Volume 99, with slight modifications except for Class S, which has been rewritten on the basis of MERRILL's researches. The descriptions of other classes, such as K0, Na, and Nb, have been somewhat recast. The reader should look below for rediscussions of Classes P and O of the Draper Classification by PLASKETT and PAYNE.

Class Pa. Typical nebula, I.C.418, R.A. 5^h 22,8^m, Dec. — 12° 46′. (All positions referred to Equinox of 1900,0). The double line, 3726, 3729, is more conspicuous than the chief nebular lines, 5007,0 and 4959,0. The hydrogen lines $H\alpha$, $H\beta$, $H\gamma$, $H\delta$, $H\varepsilon$, and $H\zeta$ are bright.

Class Pb. Typical nebula, the Great Nebula of Orion. Lines 5007,0 and 4959,0 are more intense than in Class Pa.

Class Pc. Typical nebula, I.C.4997, R.A. 20^h 15,6^m, Dec. + 16°25′. Line 4363,4 is the most conspicuous.

Class Pd. Typical nebulae, N.G.C.6826, R.A. 19^h42,1^m, Dec. + 50°17′, and N.G.C. 6326, R.A. 17^h12,9^m, Dec. — 51°40′. The chief nebular line, 5007,0, is the strongest line. The greater number of gaseous nebulae belong to this and the following class.

Class Pe. Typical nebulae, N.G.C.7662, R.A. 23^h21,1^m, Dec. + 41°59′, and N.G.C.7009, R.A. 20^h58,7^m, Dec. — 11°46′. This class differs from class Pd in having line He II 4685,7 also present.

Class Pf. Typical nebula, N.G.C.40, R.A. 0^h7,6^m, Dec. + 71°32′. A bright band centered at 4650, probably due to carbon, is the most conspicuous feature of this spectrum and appears to ally it with spectra of Class O.

Class Oa. Typical stars, B.D.+35° 4013, R.A. 20^h 8,2^m, Dec. + 35°54′, and C.P.D.—60° 2578, R.A. 11^h5,8^m, Dec. — 60°26′. A broad, bright band, centered at 4650, is the most conspicuous feature. $H\gamma$ and $H\delta$ are bright, and several other bright bands are seen, but not the typical nebular lines such as 4363 or 5007.

Class Ob. Typical stars, B.D.+35° 4001, R.A. 20^h6,5^m, Dec. + 35°53′, and C.D.M.—23° 4553, R.A. 6^h50,0^m, Dec. — 23°48′. A wide, bright band, centered at 4686, is the most characteristic feature. The hydrogen lines $H\beta$, $H\gamma$ and $H\delta$ are bright, and also those of the ζ Puppis series due to ionized helium.

Class Oc. Typical stars, B.D.+36° 3987, R.A. 20^h13,3^m, Dec. + 37°7′ and C.D.M.—41° 10972, R.A. 16^h45,3^m, Dec. — 41°41′. The bands are narrower than in Classes Oa and Ob, and two well separated lines are seen at 4686 and 4638, the former due to ionized helium being twice as bright as the latter due to doubly ionized nitrogen (4634, 4642). The hydrogen lines and the ζ Puppis lines are bright. No dark lines are seen.

Class Od. Typical stars, ζ Puppis and λ Cephei. All lines are dark except 4686 and 4638, which are bright. Seven dark lines of the ζ Puppis series (5412, 4542, 4200, 4026, 3924, 3858, 3815, all of He II), have been photographed. The neutral helium line, 4471,6, is present but very faint in ζ Puppis. Several faint, dark lines between $H\beta$ and $H\gamma$ are seen in the spectrum of λ Cephei, but not in that of ζ Puppis. From here to Class A7 the dark hydrogen lines are paramount.

Class Oe. Typical star, 29 Canis Majoris, R.A. 7^h 14,5^m, Dec. – 24°23′. The spectrum resembles that of ζ Puppis. All lines are dark except He II 4686 and N III 4634, 4642, the two latter lines being the probable source of a bright band at 4638. Hydrogen lines and numerous helium and other dark lines are present. Lines N III 4097,5 and Si IV 4089,0 are at their maximum intensity, the latter continuing at maximum through Class B0. This represents a transition stage from PICKERING's Type V to SECCHI's Type I.

Class Oe5. Typical star, τ Canis Majoris, R.A. 7^h14,5^m, Dec. – 24°47′. All the lines are dark; no bright bands are seen. This spectrum is clearly intermediate between those of Classes Oe and B0. It resembles that of Class Oe in the presence and intensity of the ζ Puppis series, and that of Class B0 with respect to the helium lines, of which He 4026,3 and 4471,6 are strong. The lines 4649,3 (probably C III 4647, 4650) and He II 4685,9 are strong.

Class B0. Typical star, ε Orionis (Alnitam). The hydrogen lines strengthen and are 0,3 as intense as in the spectrum of α Canis Majoris. The ζ Puppis series is present, but much fainter than in Class Oe5. Oxygen lines are strong. The triplet, O II 4070,0, 4072,5, and 4076,1, is well marked. Line 4649,3, now probably a blend of oxygen and carbon, is slightly more intense than the helium lines 4026,3 and 4471,6, which are equally strong. Lines 4649,3, 4116,3, and 4089,0, reach their greatest intensity in this class and decrease very rapidly in succeeding classes of spectra.

Class B1. Typical stars, β Canis Majoris (Mirzam) and β Centauri. The BALMER series of hydrogen lines is seen from $H\beta$ to $H\tau$. The ζ Puppis series of ionized helium is not distinctly seen. The lines of helium are more intense while the trebly ionized silicon and doubly ionized carbon and singly ionized oxygen lines are fainter than in Class B0. Line He I 4471,6 exceeds O II 4649,3, while He I 4121,0 exceeds Si IV 4116,3 in intensity. The lines of Si III 4552, 4568, 4574 are reaching maximum strength.

Class B2. Typical stars, γ Orionis (Bellatrix) and α Lupi. The lines of helium are at their maximum intensity in this and the following class. Line Si IV 4116,3 is not seen, and lines Si IV 4089,0 and O II 4649,3 are faint and much weaker than in Class B1.

Class B3. Typical stars, π^4 Orionis and α Pavonis. The hydrogen lines strengthen appreciably and are about 0,5 as intense as in α Canis Majoris. The helium lines, while not stronger than in Class B2, are more prominent, due to the disappearance or extreme faintness of the lines 4070,0, 4072,5, 4076,1, 4089,0, 4116,3, and 4649,3. Helium lines having the greatest intensities are 3819,8, 4009,4, 4026,3, 4143,9, 4388,1, 4471,6, and 4922,1.

Class B5. Typical stars, q Tauri and φ Velorum. These spectra show an advance towards Class A0 in the increased intensity of the calcium line, K, and of the double silicon line 4128,1, 4131,1, which is stronger than the helium lines 4121,0, and fainter than 4143,9. Line Mg II 4481,3 is 0,7 as intense as 4471,6.

Class B8. Typical stars, β Persei and γ Gruis. The helium lines 4026,3 and 4471,6 are present, together with several lines of ionized metals prominent in the spectra of Class A0. Lines 4471,6 and 4481,3 are approximately equal. Line K is less intense than 4026,3.

Class B9. Typical stars, λ Aquilae and λ Centauri. The spectrum closely resembles that of Class A0, except that 4026,3 is seen and the line K is somewhat fainter than in Class A0.

Class A0. Typical star, α Canis Majoris (Sirius). The hydrogen lines are at their maximum intensity, and line K is 0,1 as intense as $H\delta$, or less. The line H of ionized calcium at 3968,6, when separated from $H\varepsilon$, 3970,3, is nearly as intense as line K. Line 4481,3 is the strongest except the hydrogen lines and line K. The lines Si II 4128, 4131 are at maximum strength. Many faint metallic lines may be present in spectra of this division.

Class A2. Typical stars, δ Ursae Majoris and ι Centauri. The line K is 0,3 to 0,5 as intense as $H\delta$. Solar lines are well marked, especially lines Mg II 4481,3, Ca 4226,9, and Fe II 4233,8. The two latter form a nearly equal pair. No helium lines are seen in this, or any following class.

Class A3. Typical stars, α Piscis Austrini (Fomalhaut) and τ^3 Eridani. The line K is more than 0,5 as intense as the compound line H and $H\varepsilon$, and is 0,8 as intense as $H\delta$. The metallic lines are more numerous and more intense than in Class A2, while the hydrogen lines are slightly fainter.

Class A5. Typical stars, β Trianguli and α Pictoris. The line K is 0,9 as intense as the compound line H and $H\varepsilon$, and more intense than $H\delta$. Line Mg II 4481,3 is no longer the most conspicuous among the solar lines. Lines Fe I 4299,4, Ti II 4300,7, and Fe I 4302,7 are well marked.

Class F0. Typical stars, δ Geminorum (Wasat) and α Carinae (Canopus). The hydrogen lines are now about 0,5 as intense as in α Canis Majoris. The line K is as strong as the blended lines H and $H\varepsilon$ combined, and about 3,0 as intense as $H\delta$. The lines Fe 4305,6, Ti 4308,0, and Ca 4309,5 and other lines which form the absorption band G are faint and inconspicuous.

Class F2. Typical star, π Sagittarii. This spectrum resembles Class F0, except that there is more appearance of continuity in the band G, due to increased strength of lines 4305,6 to 4315,2.

Class F5. Typical stars, α Canis Minoris (Procyon) and ϱ Puppis. The hydrogen lines are still predominant and are 2,0 as intense as in Class G0, while the metallic lines are fainter and less numerous than in Class G0. Line Fe 4325,9 is about 0,1 as strong as $H\gamma$. On plates with small dispersion, the FRAUNHOFER band G appears to be nearly continuous from 4299,4 to 4315,2. The compound line 4308,0 and 4309,5 is more intense than 4315,2. Line Ca 4226,9 is well marked among the numerous lines, but is not 0,5 as strong as $H\gamma$.

Class F8. Typical stars, β Virginis and α Fornacis (Zavijava). The spectrum resembles that of Class G0, except that the hydrogen lines are stronger, and a few of the metallic lines are fainter.

Class G0. Typical stars, α Aurigae (Capella) and β Hydri. The spectrum resembles closely that of the sun. The hydrogen lines are no longer conspicuous as a series. $H\gamma$ is 1,5 as intense as Fe 4325,9 and 3,0 as intense as the adjacent line, 4337,7, when the dispersion is sufficient to show the two lines separately. The lines Fe, Sr II, 4076,8 to 4077,9, $H\delta$, and Ca 4226,9 are nearly equal in intensity. The band G is continuous on photographs taken with one or two prisms. The bands H and K are very conspicuous. The continuous spectrum shows no very marked changes in the distribution of light, from $H\beta$ to $H\varepsilon$, although there is a slight gradual decrease from $H\gamma$ to $H\varepsilon$.

Class G5. Typical stars, $\varkappa$ Geminorum and α Reticuli. The hydrogen lines are slightly fainter than in Class G0. Line $H\gamma$ when combined with 4337,7 is equal to 4325,9; when separated, $H\gamma$ is fainter than 4325,9. Si 3905, 4103 are at maximum strength. Several spaces appear brighter than adjacent portions, and in the distribution of light there is a decided advance toward Class K0.

Class K0. Typical stars, α Bootis (Arcturus) and α Phoenicis (Nair al Zaurak). The sunspot spectrum. The hydrogen lines are fainter than in Class G5, $H\gamma$ being about 0,5 as strong as Fe 4325,9. Line Ca 4226,9 is 3,0 as intense as in Class G0, 2,0 as intense as the compound line 4172, and nearly 3,0 as intense as lines 4383 to 4385. The band G, extending from 4299 to 4315, is continuous and is more conspicuous than line 4226,9. Bands H and K reach their greatest intensity. The continuous spectrum shows a decided decrease from $H\gamma$ to $H\varepsilon$. Several portions appear brighter than adjacent parts, such as from 4077,9 to $H\delta$, 4215,7 to 4226,9, 4470 to 4525, and 4614 to 4648, approximately.

Class K2. Typical stars, β Cancri and v Librae. The spectrum resembles Class K5 in the increased intensities of several lines, as 4226,9, and a general faintness of the continuous spectrum toward the end of shorter wave length. The band G is still continuous.

Class K5. Typical star, α Tauri (Aldebaran). The bands H and K and line Ca 4226,9 are the most conspicuous features. The band G is no longer continuous, owing to the disappearance of several of the fainter lines. The double lines Fe 4383 to 4385 and Fe, Va 4405 to 4408, form a conspicuous pair, of which the former is somewhat stronger. Faint breaks in the continuous spectrum are seen at the wave lengths 4762, 4954, and 5168, which are the beginning of the absorption bands of Class M. There is also a sudden diminution in light at $H\beta$, which is nearly as well marked as the similar change at 4762. This class constitutes the transition stage from Type II to Type III of SECCHI.

Class Ma or M0—M2. Typical stars, α Orionis (Betelgeuse) and γ Hydri. The spectrum is banded. The bands, extending from 4762 to 4954 and from 5168 to 5445, are well marked. The change in light at $H\beta$ is much less conspicuous than at 4762. Several (relatively) bright spaces are seen, such as from 4556 to 4586 and from 4657 to 4668. The lines of the G band are well separated, and line Fe 4315,2 is very faint. Line Ca 4226,9 is the most conspicuous absorption line. The spectrum is faint toward the end of shorter wave length, so that bands H and K are generally barely visible[1].

Class Mb or M3—M5. Typical stars, ϱ Persei and γ Crucis. The edges of the absorption bands, at wave lengths 4762, 4954, 5168, and 5445, are strong and appear somewhat like bright bands. These apparently bright bands fade gradually toward the edge of shorter wave length.

[1] The bands are due to titanium oxide. The strongest absorption bands in Class M0 have heads at 4762, 4954, 5168, 5445. These bands are sharply bounded and more intense on the edge of smaller wave length, fading out toward the red.

Conspicuous bright bands of equal intensity are seen from 4556 to 4586 and from 4614 to 4626. Line 4226,9 is very wide and in some spectra appears to be as intense as $H\delta$ in the spectrum of α Canis Majoris (Class A0). Of the band G, lines 4299,4, 4300,7, and the compound line 4305,6, 4308,0, and 4309,5 are the only well marked lines remaining. (On isochromatic plates, absorption bands are also seen having edges at the wave lengths 5763, 5816, and 5857, approximately.)

Class Mc or M6—M10. Typical stars, W Cygni and RX Aquarii. The continuous spectrum is fainter, and the bright edged bands are stronger, than in Classes M0 and M3, so that the spectrum appears to be of a fluted character, and on plates of small dispersion many of the dark lines seem to have disappeared[1].

Class Md (M0e, M1e—M10e). Typical stars, χ Cygni and o Ceti. This class includes spectra of any division of Class M, in which at least one hydrogen line is bright. The spectra differ widely. Either $H\beta$, $H\gamma$, or $H\delta$, but seldom the first, may be the strongest bright line, while the underlying spectrum may belong to Classes M0, M3, M0—M8, or M6,5. The greater portion of the variable stars of long period have this class of spectrum. Indeed, any star having a spectrum of Class Me with giant characteristics is as a rule a long period variable.

Class S. Typical stars, π^1 Gruis and R Geminorum. A characteristic feature is a region of complicated structure from 4620 to 4660, containing several lines of absorption and possibly of emission, and also usually if not always one or more band heads (4620, 4637) in absorption due to zirconium oxide. The absorption lines 4536 (Ti—Zr?) and Ba II 4554 are very strong. So far as observed the region from 4450 toward the violet resembles that of Class M more closely than at greater wave lengths, and well marked absorption is present at Ca 4227, H, and K. As compared with Class M0, the continuous spectrum becomes increasingly fainter toward the violet from about 4500, and more particularly from 4450. Most of the spectra of this class have bright hydrogen lines of which the less refrangible are stronger in the photographic region.

Class R. This letter was assigned in 1908 to a few spectra which on photographs of small dispersion resemble those of Class N between $H\beta$ and $H\gamma$, but which contain so much blue light that the spectrum is visible as far as the calcium bands, H and K. A list of spectra assigned at that time to Class R is given in Harvard Circular 145. A careful study of these spectra shows that they may be subdivided into at least four classes, which are described below.

Class R0. Typical star, S.D.—10° 5057, ptm. magn. 7,04, R.A. $19^h 17{,}7^m$, Dec. — 10°53′. The distribution of light resembles that in Class G5 or K0, and the absorption bands K and H are well seen. The dark carbon band centered at 4700 is wide and strong, and the dark band at 4395 (probably of titanium, vanadium and iron) is about equal in strength to the G band. Lines Ca 4226,9, Fe 4233,8, 4236,1, and 4239,0 are well marked, and on photographs having small dispersion the appearance in this region is that of a wide, continuous band of

[1] Variables of Classes M0 to M6 at maximum light may at minimal phases emit spectra later than M8 as shown by strength of the titanium oxide bands.

absorption. This and the following classes belong to SECCHI's Type IV[1].

Class R3. Typical star, B.D. + 5° 5223, ptm. magn. 8,8, R.A. $23^h44{,}0^m$, Dec. + 5° 50′. The H and K bands of calcium are visible, but they are fainter than in Class R0, and the continuous spectrum between these bands and $H\gamma$ is not more than 0,5 as intense as in Class R0.

Class R5. Typical star, S.D. — 3° 1685, ptm. magn. 7,5, R. A. $6^h56{,}1^m$, Dec. — 3° 6′. In the region of shorter wave length than 4240, the continuous spectrum is barely visible on plates of normal exposure. When the dispersion is small, the spectrum appears to consist of three wide, bright bands, whose centers are at the approximate wave lengths 4300 (4220—4400), 4500 (4400—4700), 4840 (4700—5100), and whose intensities are estimated to be 3, 6, and 10, respectively.

Class R8. Typical star, B.D. + 61°667, ptm. magn. 7,92, R.A. $3^h57{,}2^m$, Dec. + 61°31′. The spectrum is very faint from 4240 to the violet, so that on photographs of high dispersion it is difficult to distinguish between this class and Class Na or N0.

Class Na = N0. Typical star, 19 Piscium, B.D. + 2°4709, var., R.A. $23^h41{,}3^m$, Dec. + 2°56′. The spectrum is visible as far toward the violet as the bands H and K, but the portion of the spectrum between 4240 and K is even fainter than in Class R8. The three bright spectral regions described in Class R5, which for that class have intensities 3, 6, and 10, have for Class N0 intensities 0, 8, and 10 respectively. These bright regions are best observed in this connection under low dispersion. Thus when the dispersion is small, the dark band 4700 appears to separate the spectrum into two wide bright bands, the portion from 4400 to 4700 being estimated as 0,8 as intense as that from 4700 to 5100.

Class Nb = N3. Typical star, B.D. + 67°350, ptm. magn. 7,39, R.A. $4^h40{,}8^m$, Dec. + 67°59′. The three bright spectral regions described in Class R5 have the relative estimated intensities 0, 6, 10 in Class N3.

Class Nc. Typical star, S Cephei, var. R.A. $21^h36{,}5^m$, Dec. + 78°10′. The spectrum contains little, or no, light of shorter wave length than $H\beta$. The most brilliant portion is from 5900 to 6800.

Class Pec. All spectra which cannot be assigned to any known class, considering their principal characteristics. This includes the spectra of Novae, and some variable stars.

Class Con. Spectra apparently continuous. This includes the spectra of nebulae without bright lines, or of clusters which resemble such nebulae with the dispersion employed. As these objects appear as surfaces, and since objective prisms are used, dark lines are not visible.

29. Tabular Descriptions of DRAPER Classes. The descriptions of the Harvard Classification of 1923 are given in full in Table 3. Notation is employed similar to that used in the tabulation of Miss MAURY's classification. Abbreviations, stated in the order of occurrence, have the following meanings: E, emission; Max., maximum; >, is stronger than; <, is weaker than; v f, very faint; inc. int., increasing intensity; num., numerous; well seen is used for well marked in some descriptions.

[1] The dark carbon bands in contrast with those of titanium oxide are sharply bounded and more intense on the edge of greater wave length, fading out toward the violet.

Table 3. Miss CANNON's Classification of Stellar Spectra. Oa—A0.

	Oa	Ob	Oc	Od	Oe	Oe5	B0	B1	B2	B3	B5	B8	B9	A0
4686	Wide E band	Wide E band[1]	E band $<$ Ob	E	E	Strong, dark								
4650	Wide E band[1]					Strong, dark	Max.	$<$ B0	f, $<$ B1	vf	Not seen			
4638			E band $<$ Ob	E	E									
4686/4638			2,0											
ζ Puppis Lines		E	E $<$ Ob	7 dark lines	Many, dark	Like Oe	Much fainter	Indistinct						
Neutral Helium				Faint 4472 dark	Many, dark	Like B0	Strong	Stronger	Max.	Max. More prom.	Present	Present	4472, 4026 present	4026 not seen
Hydrogen	E	E	E $<$ Ob	Dark	Dark	Note[1]	0,3 A0[1]	$H\beta$—$H\iota$[1]	Note[1]	0,5 A0[1]	Note[1]	Note[1]	Note[1]	Max.[1]
4097					Max.				Not seen					
$\frac{4650}{4026} = \frac{4650}{4472}$							Slightly $>$ 1	$<$ 1						
4121/4116								$>$ 1						
4116							Max.	$<$ B0	Not seen	Not seen				
4089							Max.	$<$ B0	f, $<$ B1	Not seen				
4070, 4073, 4076							Well marked	$<$ B0		vf				
K											Inc. int.		$<$ A0	$>$ 4481
$\frac{4128 + 4131}{4121}$											$>$ 1			
$\frac{4128 + 4131}{4144}$											$<$ 1			
4481/4472											0,7	1 ±		
Solar Lines												Present[2]	Like A0	$<$ 4481
K/4026												$<$ 1		
$K/H\delta$													Like A0	Not $>$ 0,1

[1] Most conspicuous feature.

[2] Fe 4174, 4179, 4233, 4384, 4585.

Table 3. Miss CANNON's Classification of Stellar Spectra. A0–K0.

	A0	A2	A3	A5	F0	F2	F5	F8	G0	G5	K0
Helium	4026 not seen	Not seen									
Hydrogen	Max.		< A2		0,5 A0		2,0 G0	> G0	Note[1]	< G0	< G5
K/H	Not > 0,1	0,3–0,5	0,8	> 1,0	3,0 ±						
$K/(H+H\varepsilon)$			0,5	0,9	= or > 1,0						
H and K	K > 4481								Note[2]		Max.
Solar Lines	< 4481	Well seen	> A2 more num.	> 4481			< G0	Some < G0			
4481		Well seen									
g 4227		Well seen					Well seen				3 G0
4227/4234		1 ±									
4299, 4301, 4303				Well seen							
Band G 4299–4315					Comp. lines faint	More continuity	Near continuous		Continuous 2 prisms		Continuous > 4227
$4326/H\gamma$							0,1		0,67	> 1	2,0
4309/4315							> 1,0				
$4227/H\gamma$							< 0,5				
$H\gamma/4338$									3,0		
$\frac{4077}{H\delta} = \frac{H\delta}{4227}$									1 ±		
Deg. $H\beta - H\gamma$ / Deg. $H\gamma - H\varepsilon$									> 1	Toward K0	Much > G5
$\frac{H\gamma + 4338}{4326}$										1,0	
4227/4172											2,0
4227/4384											3,0

Table 3. Miss CANNON's Classification of Stellar Spectra. K0–Md, S.

	K0	K2	K5	M0	M3	M6,5	Md	S
Hydrogen	< G5						Some emiss.	
H and K	Max.		Note[3]	Note[4]				Well seen
Solar Lines						Less well seen		
g 4227	3 G0	> K0	Note[3]	Note[3]	V wide			Well seen

[1] Not conspicuous as a series. [2] Very conspicuous. [3] Most conspicuous features.
[4] Generally barely visible because of faintness of violet region.

Table 3. Miss CANNON's Classification of Stellar Spectra. K0—Md, S. (Continued.)

	K0	K2	K5	M0	M3	M6,5	Md	S
G Band	Continuous > 4227	Continuous	Not continuous	Well separated	Five lines well seen.			
$4326/H\gamma$	2,0							
$\frac{\text{Reg. } H\beta - H\gamma}{\text{Reg. } H\gamma - H\varepsilon}$	Much $>$ G5							
4227/4172	2,0							
4227/4384	3,0							
4384/4406			$>$ 1,0					
Region short λ		Gen. faintness		Faint				
4078 — Hδ, 4216—4227, 4470—4525, 4614—4648		Bright spaces						
TiO_2 Bands			Appear	Well marked				
$H\beta/4762$			$<$ 1,0	Much $<$ 1,0				
4315 Fe				v f				
4556—4586, 4657—4668				Bright spaces				
$\frac{4556-4586}{4614-4626}$					= 1,0, bright conspic.			
Bright edges TiO_2 Bands					Strong	$>$M3		
Continuous spectrum						$<$ M3 Fluted.	M0, M3, or M6,5	Beyond 4450 $<$ M
ZrO Bands 4619, 4637								Present
Ba 4554								Strong

Table 3. Miss CANNON's Classification of Stellar Spectra. G5—Nc.

	G5	R0	R3	R5	R8	N0	N3	Nc
Continuous spectrum	Like R0	Like G5 or K0			Like N0	Like R8		
H and *K*	Near Max.	Well seen	$<$ R0					
Carbon Band	Not seen	Wide, strong						
$4395/G$		1 ±						
4227, 4234, 4236, 4239	4227 strong	Well marked						

Table 3. Miss CANNON's Classification of Stellar Spectra. G5—Nc. (Continued.)

	G5	R0	R3	R5	R8	N0	N3	Nc
Region $H\gamma - K$			Not $>$ 0,5 R0					
Region beyond 4240				Barely visible	Very faint	$<$ R8		
Reg. 4220—4400 / Reg. 4400—4700				0,5		0,0	0,0	
Reg. 4400—4700 / Reg. 4700—5100				0,6		0,8	0,6	
Region beyond $H\beta$								Little or no light
Reg. 5900 to 6800								Most brilliant region

30. Variation Curves of Spectral Criteria with DRAPER Class. Curves representing the variation of numerical intensities of the more important criteria are given in Figures 9 and 10. Numerical data in this connection are taken from the description of Miss MAURY and Miss CANNON in Harvard Annals, Volume 28, and from studies of H. H. PLASKETT[1], of MENZEL[2], and Miss PAYNE[3,4]. The divisions in Class O proposed by H. H. PLASKETT are adopted as described below.

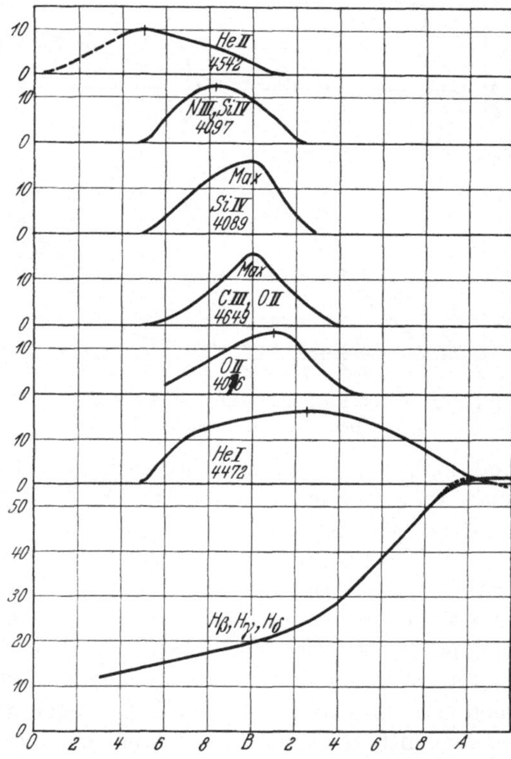

Fig. 9. Curves of Variation of Intensity of Lines in Spectra of Classes O and B.

In Fig. 9 curves are drawn showing the variation of intensity of selected spectral criteria with spectral class from O0 to A2. The intensity curve for He II 4542 was derived from Miss CANNON's estimates and from intensities derived from the hydrogen and 4472 curves through PLASKETT's ratios. The curves for Si IV (4089), N III, Si IV (4097), C III, O II (4649), O II (4076) and He I (4472) are based mainly on intensities by Miss PAYNE[3,4], modified by estimates of Miss CANNON reduced to the range of the former. The helium curve was strengthened by combining PLASKETT's ratios for $H\gamma$ and 4472 with Miss PAYNE's intensities for the former line, and the curve of N III, Si IV (4097) by combining Miss PAYNE's curve of Si IV (4089) with H. H. PLASKETT's ratio of 4089 to 4097.

[1] Publ Dom Astrophys Obs Victoria 1, pp. 360, 366 (1922). [2] Harv Circ 258 (1924).
[3] Harv Circ 252, 256 and 263 (1924); Stellar Atmospheres, p. 121 et seq., Append. (1925).
[4] Harv Bull 855 (1928).

The slopes of all these curves on the side of lower temperatures were drawn with the descriptions of Miss MAURY, Miss CANNON, and H. H. PLASKETT and the estimates of Miss CANNON as guides. The curve of hydrogen from B8 to A2 is based on the intensities by MENZEL for $H\beta$, $H\gamma$, and $H\delta$, adjusted by the use of relative intensities of Miss MAURY and Miss CANNON. It is extented into Class O on the basis of Miss CANNON's estimates, Miss PAYNE's intensities of $H\beta$ and $H\gamma$, and by adjustment with the curves for 4472, 4649, and 4542, using PLASKETT's ratios. The maximum of hydrogen indicated by the broken line curve is statistical. The approximate relative scale of the different curves was checked by ratios when these were available. Nevertheless the curves of Fig. 9 must be regarded as qualitative in a large degree since the scales of intensity employed are not well standardized and the variation in sharpness of lines in stars of the same spectral class leads to important uncertainties. The curve of He I 4472 might be drawn somewhat higher according to data in PAYNE's Stellar Atmospheres, page 121; and the curve of O II 4076 should be somewhat lower according to estimates of Miss CANNON. The scale of Miss PAYNE, which influences especially the shape of the curves other than those of He II and hydrogen, is possibly logarithmic. The unit of intensity used in Fig. 9 is such that the maximum of the hydrogen curve is numerically equal to four tenths of MENZEL's maximum for hydrogen as shown in Fig. 10.

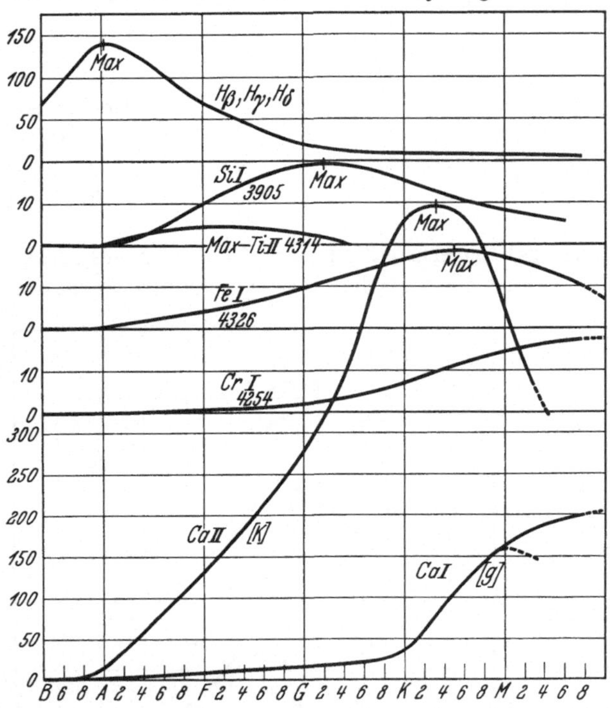

Fig. 10. Curves of Variation of Intensity of Lines in Spectra of Classes B5 to M8.

In Fig. 10, curves are drawn showing the variation of intensity of important criteria from Class B5 to Class M10. The observations are due to MENZEL[1] except in the case of the hydrogen curve which was strengthened by estimates of Miss MAURY and Miss CANNON, and in the case of Si I 3905 which is based on intensities by Miss PAYNE reduced to MENZEL's scale by a curve based on common estimates of Fe 4326. The hydrogen and calcium curves have been drawn on one fifth the scale of the other curves in Fig. 10 as the numbers along the vertical margin will show, except that MENZEL's intensities for K after Class F5 were reduced one-third in view of Miss CANNON's ratios of K and $H\delta$. The curves of Fig. 10 must be qualitative in some degree, especially the low temperature branch of the curve for the K band which is in a relatively faint region in Class M spectra. Also the maximum for K, which is shown here at Class K, varies greatly with absolute magnitude. To construct the curves of Fig. 10 on an accurate quantitative basis distinctions of absolute magnitude must be made and the intensity

[1] Harv Circ 258 (1924).

scale must be carefully standardized. It should be added that the stars selected by MENZEL were nearly all giants, and that he endeavored to make his scale linear.

The spectral classes at the maxima and at the limits of the curves in Figures 9 and 10 are given in Table 4. The limits are obviously uncertain. Miss PAYNE gives the maximum for Fe 4326 at K2 instead of K5, and for Ti (blend) 4314 at F5. In general, however, the agreement with her maxima in Stellar Atmospheres, p. 127, is good. YOUNG and HARPER assign a maximum to CaI *g* at K9 or M0.

Table 4. Spectral Class of the Maxima and Limits of Intensity Curves of Important Spectral Criteria.

Feature	High Temp. Limit	Max.	Low Temp. Limit
He II, 4200, 4542	O0	O5	B1,5
N III, Si IV, 4097	O5 (?)	O8+	B2,5
Si IV, 4089	O5 (?)	B0	B3
C III, O II, 4649	O5 (?)	B0	B4
O II, 4076	O5 (?)	B1	B5
He, 4472	O5 (?)	B2,5	A1,5
$H\beta$, $H\gamma$, $H\delta$	—	A0—A3	—
Ti (blend), 4314	B9	F2	G5 (?)
Si, 3905	A0	G1	—
Fe, 4326	B9	K5	—
Ca, *K*	B5	K0—K7	—
Ca, *g*	A0	None M0	—
Cr, 4254	A4	None	—
TiO_2 Bands	K5	None	—
Carbon Bands	R0	None (?)	—

A summary of known maxima of spectral lines with the kind of evidence on which the determination is based is here given, following Miss PAYNE[1].

Atom	Maximum	Evidence	Atom	Maximum	Evidence
H I	A0	Contour	CaII	K2	Contour
He I	B1,5	Estimate	TiII	F5	Line Depth
O II	B3	Estimate	FeI	None	Line Depth
Mg II	A3	Estimate	FeII	F5	Line Depth
Si I	G5	Estimate	SrII	K2	Line Depth
Ca (4435)	None	Estimate	BaII	M(?)	Estimate

31. Line Intensity Ratios for Classification Purposes. Line intensity ratios covering all stellar classes from B0 to M 5 are given below. Ratios in Class O due to H. H. PLASKETT are given in Table 8. The ratios given here are due to Miss MAURY, Miss CANNON, and MENZEL[2], whose publications are cited above. As stated above the stars whose spectra were studied by MENZEL were nearly all giants. When his observations were supplemented by those of Miss MAURY and Miss CANNON the agreement was always close. The Mount Wilson curves[3] could not be used directly since they are difference and not ratio curves. They apply to dwarf stars and in general diverge from curves based on MENZEL's intensities.

The ratios which may be used to advantage in classifying spectra within specified ranges are as follows: Class O5 to B0, $H\gamma/4542$; O6 to B0, $H\gamma/4542$, 4089/4097, $H\gamma/4472$, $H\gamma/4388$; O6 to B2, $H\beta/4649$; B1 to A0, 4472/4481; B2 to B8,4026/$H\delta$; A1 to F5, $H\delta/K$; F2 to K0, $H\gamma/g$; F5 to K0, $H\gamma/4352$; F5 to K5, $H\gamma/4384$, $H\gamma/4326$; F5 to M7, 4290/g; G0 to K5, $H\gamma/4405$; G8 to M0, 4272/g; K0 to M7, 4272/4275. In addition, in the range G0 to Ma, ADAMS and KOHLSCHÜTTER have used the combinations, $H\beta/4872$ and $H\beta/4957$. By the use of the above ratios spectra in the main spectral sequence may be classified.

[1] Harv Bull 867, p. 6 (1929). [2] Harv Circ 258 (1924).
[3] Mt Wilson Comm 23, p. 113 (1916).

Emphasis should be laid on the fact that the above intensity curves and intensity ratios, especially those concerning hydrogen and calcium lines, may not apply to super-giants or dwarfs, at least in the range from F0 to M10. Even in Class O9, a possible dependence of the ratio, 4089/4097, on luminosity is pointed out. However, in the earlier classes distinctions based on luminosity introduce less complication. The curves and ratios by MENZEL, supported well by estimates due to Miss MAURY and Miss CANNON, were derived essentially from spectra of giant stars and, especially in the later classes, should not be applied to accurate classification of stars of low or very high luminosity without further study. The problem of determining line intensity and intensity ratio curves based on different luminosities as well as spectral classes is an observational one now demanding attention.

Classification Ratios in Class B:

	B0	1	2	3	4	5	6	7	8	9	A0
4472/4481		9,0	7,3	5,5	3,0	1,8	1,5	1,3	1,0	0,67	0,25
4026/$H\delta$	0,60	0,57	0,53	0,44	0,32	0,20	0,15	0,12	0,10		

Classification Ratios in Class A:

	A0	1	2	3	4	5	6	7	8	9	F0	1	2	3	4	5
$H\delta/K$	9,0	6,0	3,5	2,0	1,3	1,0	0,82	0,67	0,56	0,49	0,42	0,36	0,31	0,27	0,24	0,21

Classification Ratios in Classes F0 to M5.

	$H\gamma/g$	$H\gamma/4352$	$H\gamma/4384$	$H\gamma/4326$	$H\gamma/4405$	$g/4272$	$g/4290$	4272/4275
F0	10,4							
1	8,7							
2	7,4							
3	6,3							
4	5,2							
5	4,3	8,2	7,2	6,7			1,5	
6	3,5	6,8	5,9	5,2			1,6	
7	2,9	5,3	4,5	3,8			1,7	
8	2,4	4,1	3,3	2,8			1,8	
9	1,9	3,2	2,2	2,0			1,9	
G0	1,5	2,5	1,3	1,6	6,5		2,0	
1	1,1	2,1	1,0	1,3	5,7		2,1	
2	0,82	1,9	0,86	1,1	5,0		2,2	
3	0,63	1,7	0,75	0,91	4,4		2,3	
4	0,50	1,6	0,66	0,89	3,8		2,4	
5	0,44	1,5	0,59	0,71	3,2		2,5	
6	0,39	1,4	0,54	0,63	2,7		2,6	
7	0,35	1,3	0,50	0,58	2,3		2,7	
8	0,31	1,2	0,47	0,54	1,8	3,0	3,0	
9	0,28	1,1	0,44	0,50	1,4	3,4	3,3	
K0	0,25	1,0	0,42	0,47	1,0	3,8	3,7	1,8
1			0,39	0,44	0,77	4,3	4,2	1,7
2			0,36	0,42	0,66	4,8	4,9	1,6
3			0,34	0,40	0,59	5,3	5,6	1,5
4			0,32	0,38	0,55	5,8	6,5	1,4
5			0,30	0,36	0,53	6,4	7,0	1,3
6						7,1	7,6	1,2
7						7,9	8,2	1,1
8						8,8	8,7	1,0
9						9,8	9,2	1,0
M0						11,0	9,7	0,91
1							10,2	0,83
2							10,8	0,77
3							11,2	0,72
4							11,6	0,67
5							12,0	0,62

Dating back to the spectral classifications of LOCKYER and Miss MAURY the existence of separable vertical sequences in the stellar spectral order has been recognized. That these are associated clearly with differences in luminosity is now well understood. Also it is well known that color index and effective wave length are related to absolute magnitude in later spectral types. Recently GERASIMOVIČ[1] has emphasized the well recognized importance of constructing curves of varying intensity of a given spectral line for stars of a given luminosity in the different DRAPER classes. If the method of classification by line intensity curves or by ratios is to be developed successfully, curves for super-giants, giants, intermediates, dwarfs and faint dwarfs, and possibly finer divisions may be necessary. The effect of varying active mass may thus be practically excluded.

g) Comparison of the Principal Stellar Spectral Classifications.

32. Comparison of Spectral Classifications. In Table 5, the divisions of the various classifications of stellar spectra are placed in parallel columns. The descriptions of spectral types are not always definite enough to permit the avoidance of every uncertainty in such a table. Thus there is not agreement in every detail with the less comprehensive table of the same nature by DE GRAMONT[2]. In the regions intermediate between well recognized types these uncertainties become most troublesome but the parallels in Table 5 will suffice for all practical purposes. The classification of Procyon, of division F5, as intermediate between Types I and II is well substantiated in SECCHI's later works[3].

In Table 5, column 1, type (V) is that of PICKERING whereas type V is due to SECCHI. I *O* refers to SECCHI's Orion family. VOGEL's second classification is referred to in column 2. MC CLEAN's division (I a) as described by him would include Clase O but there is some question as to whether he had such spectra in mind. LOCKYER's genera from the two branches of his curve are placed in one column here. There is some question as to whether the divisions Ma, Mb, Md, and the symbol Na should be assigned to the original DRAPER Classification of E. C. PICKERING and Mrs. FLEMING. This was done by Miss MAURY in 1897 probably on the basis of work by Mrs. FLEMING then unpublished. The author has added the symbol Mc in column 6. There is room for objection to the extension of divisions in early classifications to express parallelism with Harvard divisions in which only faint stars are represented. It seems wise to do this for Classes R and N but not for Class S. It is interesting to note that the conspicuous gaps in Miss CANNON's classification at A5 to F0, G0 to G5, G5 to K0, and K5 to M0, are not subdivided in other classifications, though they have been filled in partly by Miss CANNON herself in the HENRY DRAPER Extension and by classifiers employing finer distinctions in absolute magnitude studies. It cannot be doubted that there are among stellar spectra continuous sequences exhibiting every stage in the transition, by almost imperceptible degrees, from one type to the next throughout the whole spectral range. It is equally certain that there are horizontal sequences which cut across the spectral classes and lead to parallel temperature sequences. The horizontal sequence associated with absolute magnitude is established within many classes. Excitation sequences cutting across temperature sequences in Classes O to B and in Class M are recognized. Other horizontal sequences are probable. These complicate seriously the problem of stellar spectral classification.

[1] Harv Circ 311 (1927).
[2] Ann d Bur d Long 1909, p. 238.
[3] Pop Astr 36, p. 576 (1928).

Table 5. Comparison of the Principal Stellar Spectral Classifications.

SECCHI Type	VOGEL Class	McCLEAN Division	CLERKE Class	LOCKYER Genus	PICKERING[1] Class	MAURY Group	CANNON[2] Class
(V)	—	—	—	—	P	—	P
(V)	IIb	(Ia)	VIII	Argonian	O	XXII	Oa
(V)	IIb	(Ia)	VIII		O	XXII	Ob
(V)	IIb	(Ia)	VIII		O	XXII	Oc
(V)—I*O*	IIb	(Ia)	VIII		O	XXII	Od
(V)—I*O*	IIb	(Ia)	VIII		O	XXII	Oe
I*O*[2]	Ib	Ia	I		B	I	Oe5
I*O*	Ib	Ia	I	Alnitamian	B	II	B0
I*O*	Ib	Ia	I		B	III	B1
I*O*	Ib	Ia	I	Crucian	B	IV	B2
I*O*	Ib	Ia	I	Taurian	B	IV	B3
I*O*	Ib	Ib	I		BA	V	B5
I*O*—I	Ib	Ib	I	Algolian	BA	VI	B8
I*O*—I	Ib	Ib	I	Rigelian	BA	VI	B9
V	Ic1, Ic2	—	VII	Crucian	D	L	Oe5p—B9p
I	Ia2	II	II	Markabian	A	VII	A0
I	Ia2	II	II	Sirian	A	VIII	A0
I	Ia2	II	II	Cygnian	A	VIII	A2
I	Ia2	II	II		AF	IX	A2
I	Ia3	III	II		AF	IX—X	A3
I	Ia3	III	II		AF	X	A5
I	Ia3	III	II—III		F	XI	F0
I	Ia3	III	II—III		F	XI—XII	F2
I—II	Ia3—IIa	III	II—III	Procyonian	FG	XII	F5
II	IIa	IV	III	Polarian	G	XIII	F8
II	IIa	IV	III		G	XIV	G0
II	IIa	IV	III		GK	XIV—XV	G5
II	IIa	IV	III	Arcturian	K	XV	K0
II	IIa	IV	III		K	XV—XVI	K2
II—III	IIa—IIIa	IV—V	III—IV	Aldebarian	KM	XVI	K5
III	IIIa	V	IV	Antarian	Ma	XVII	M0 (Ma)
III	IIIa	V	IV		Ma	XVIII	M0 (Ma)
III	IIIa	V	IV		Mb	XIX	M3 (Mb)
III	IIIa	V	IV		(Mc)	—	M6,5 (Mc)
III	IIIa	V	VI		Md	XX	Md
—	—	—	—		—	—	S
IV	IIIb	VI	V			XXI (?)	R0
IV	IIIb	VI	V			XXI (?)	R3
IV	IIIb	VI	V			XXI (?)	R5
IV	IIIb	VI	V			XXI (?)	R8
IV	IIIb	VI	V	Piscian	Na	XXI	N0 (Na)
IV	IIIb	VI	V			XXI	N3 (Nb)
IV	IIIb	VI	V			XXI	Nc
—	—	—	—	—	—	—	Pec
—	—	—	—	—	—	—	Con

h) The DRAPER Classification. Additions and Modifications.

33. The International Union for Coöperation in Solar Research. In 1913, this Union adopted a resolution to the effect that, pending the establishment of a system that can be recommended for permanent and universal adoption, "the use of the DRAPER Classification be recommended in the form described in Volume 56, page 66, of the Annals of the Harvard College Observatory: except

[1] In collaboration with Mrs. FLEMING. [2] The DRAPER Classification.

that hereafter in accurate classification, a zero be added to the letters not followed by other numerals, and that the absence of any numeral be taken to indicate only a rough classification". The additions and modifications that were adopted in the final statement of the DRAPER Classification in Harvard Annals, Volumes 98 and 99, 1923 and 1924, have been described above. It remains to discuss further additions and modifications which have been adopted by the International Astronomical Union or have been suggested on good bases by individual investigators though not formally accepted.

34. The International Astronomical Union. The recommendations which have been made to the International Astronomical Union[1] by the committee on stellar spectral classification fall into three categories, (1) general underlying principles and methods, (2) additions and redefinitions of classes or divisions, (3) additions of new symbols of general significance. The first and third of these categories will be discussed here, the second under the discussion of individual classes which follows.

1. The classification should describe and be based solely on what can be seen in the spectrum of a given star suitably observed with appropriate instruments.
2. Each class should ultimately be precisely defined but not until the relations of the lines to physical and chemical conditions in the source are better understood.
3. The DRAPER Classification should be the sole basis for further extensions, and in such extensions symbols now a part of this classification should be retained with the old meaning or abandoned utterly.
4. The decimal system of classification should be used as the "main symbol" in all cases in which a continuous spectral sequence exists.
5. The notation by means of small letters appended to the capital letters should be retained in cases in which it has not been clearly established that a continuous spectral sequence exists.
6. Permissive (not mandatory) symbols, to denote peculiarities, are desirable affixes to spectral letters.
7. Additional distinctions should be based on the line and band absorption and emission.
8. Additional notation should be devised to describe as many as may be convenient of those spectral characteristics which are known to be common to any considerable number of stars. Such notation should be simple to print and convey as much information as may be practicable in a small compass.
9. The capital letters B, A, F, etc., when standing alone, indicate only the general character of the spectrum and should be so understood and used.
10. In cases of great uncertainty, SECCHI's types may be employed. These were described above, and compared with the DRAPER Classification in Table 5. The parallels which may be adopted are:

SECCHI's Types	Harvard Classes
I Orion I *O*	Od–B5
I	B8–F2
I–II	F5
II	F8–K5
III	M
IV	R, N
V (PICKERING)	Oa, Ob, Oc, Op

[1] Trans Int Astr Union 1 (1922); 2 (1925); 3 (1928).

11. Composite spectra should be denoted by the plus sign connecting the two superimposed classes or by printing the two classes vertically. Eclipsing variables showing two spectra would be included here.
12. The spectra of variable stars should be recorded normally as at maximum brightness. When the spectrum varies continuously, as in Cepheid variables, it may be recorded, for example, as F7 to G4; when discontinuously as in eclipsing variables, like a composite spectrum. Thus A0 + K or A0, K.
13. The spectrum of a nebula and its nucleus should be classified separately if possible.
14. In the selection of standard or typical spectra, it is recommended that the fundamental stars should be giant stars, preferably of absolute magnitude about 0 or + 1 (except in Class B where they must necessarily be brighter). Stars showing either the c-star or dwarf characteristics should be selected as auxiliary standards.
15. The terms earlier and later may be applied to spectra to denote their position in the sequence. Hotter and colder, whiter or redder, may be applied to stars.

Notation supplementing the main symbol to designate additional characteristics.

Characteristics connected with absolute magnitude and mass.

Prefix c. This denotes spectra of very bright stars or super-giants in which all lines normally are relatively narrow and sharp though not all stars with sharp lines are c-stars. In spectra later than A0, the hydrogen lines and the enhanced lines are abnormally strong for the general spectral class. The calcium *g* 4227 is abnormally weak as compared with $H\gamma$ or 4215 of ionized strontium. These spectral characteristics are found in the spectra of Miss MAURY's c-stars, of the Cepheid variables, the Pseudo-Cepheids, and of practically all other stars of exceptionally great luminosity, like ζ^1 Scorpii and β Orionis, of Class B. For spectra earlier than B0 this distinction disappears, so far as known, all the stars being bright.

Prefix g. This denotes the spectra of bright or giant stars. Enhanced lines are fairly strong, Calcium *g* 4227 has a moderate intensity for the spectral type. The low temperature lines such as 4435 Ca and 4454 Ca are relatively weak. The hydrogen lines are strong. In Class F, 4077 Sr II, 4215 Sr II, and 4290 Ti II are strong. In Classes G, K and M, 4077 and 4215 are strong. The differences between ordinary giant and dwarf stars do not become prominent until spectra later than F0 are reached. The prefix g will not be used in general for spectra earlier than F0.

Prefix d. This denotes spectra of faint or dwarf stars. In these spectra Ca *g* 4227 is strong for the class and 4435, 4454, and 4535 are strong. In Class F, 4077, 4215, and 4290 are weak. In Class M, 4607 Sr is relatively strong while the hydrogen lines are weak.

Width of Lines.

Suffix n. This denotes spectra showing all the lines unusually wide and diffuse (nebulous).

Suffix s. This may be used to qualify spectra in which the lines are sharp but in which the other super-giant "c" characteristics are not present.

Stationary Lines.

Suffix k. This denotes the presence of "stationary" *H* and *K* lines and also the *D* and possibly other lines. So far they are isolated only in stars of Classes O, B, and Q (Novae).

Emission Lines.

Suffix e. This denotes the presence of emission lines except in cases where emission is a normal feature as in Classes Oa, P, and Q. This symbol is useful more particularly in Classes O, B, A, M, N, R, and S. In Class M it displaces the old symbol, Md. In spectra of Classes O, B, and A, "e" may be followed by the Greek letter of the last hydrogen line in which emission is visible. If the emission has disappeared as in Pleione the "e" may be enclosed in parentheses.

Suffix er. This denotes cases in which the emission lines are conspicuously reversed.

Suffix eq. This denotes the presence of emission lines with absorption bounding the emission on the side of shorter wave length as in certain stages of the spectra of novae.

Suffix ep. This denotes the presence of emission lines differing considerably in character from those of a recognized group.

Suffix em. This denotes the presence of fairly conspicuous metallic emission lines in addition to those of hydrogen, e.g. γ Cassiopeiae, B0em; o Ceti, Mvemv + Bpemp.

Suffix Fe. This denotes the distinctive presence of iron lines as explained below under suffix p.

Variable Spectra.

These may be denoted by giving the limiting classes of the spectral range over which the star varies, or a suffix "v" may be used.

Suffix v. This denotes variation in a stellar spectrum other than changing line displacement due to orbital motion or pulsation. If the variability is peculiar to emission lines in the spectrum the suffix is written "ev".

Peculiarities.

Suffix p. This denotes miscellaneous peculiarities not sufficiently frequent or important to justify individual designation. Thus abnormal strength of the silicon lines, 4128, 4131, or of the strontium line, 4077, may be designated in this way. When the peculiarities are related to particular sets of lines the symbol of the element most affected may be added in parentheses, unidentified lines being designated by Un. and ionization by Roman numerals, e.g. α Andromedae, A0p (Mn II, Un), 61^1 Cygni, dK8(CaIIe). Lines arising from transitions involving metastable states in the atom such as the oxygen lines in gaseous nebulae and certain iron lines in η Carinae may be denoted by the symbol of the element in square brackets, e.g. Boss 5650, gM2ep [FeIIe].

Suffix pq. This denotes pecularities of a character suggestive of the spectrum of novae.

Exclamation point. This may be used as a modifier to indicate very marked occurence of the phenomenon designated by the preceding symbol.

Each suffix letter should follow immediately the symbol it qualifies.

i) Additions and Redefinitions Affecting Classes and Divisions of the Draper Classification.

35. Comparisons of Classifications from Objective Prism and Slit Spectra. Since the Draper classification is based upon objective prism spectra of low dispersion some difficulty has been anticipated in applying it to slit spectra, especially when the latter are of great dispersion.

When the attempt was made to classify slit spectra according to the DRAPER system it was found that the relative narrowness and greater resolution of these spectra, as compared with those made with the objective prism, affected the general impression of the relative intensity and position of lines. Thus the band G, which is described in the DRAPER classification with H and K as the most conspicuous feature in Class G0, is not conspicuous on slit spectrograms. Series of lines are also less conspicuous in the slit spectra. Thus Miss CANNON[1] found that the mental picture of each division by which she had classified objective prism spectra seemed out of proportion on first inspection of slit plates. After some study she was able to bring classifications of spectra made with the different types of spectrographs into good agreement. The average difference between estimates of spectral class by KOHLSCHÜTTER and Miss CANNON for 314 stars from O to M was 1,4 divisions or about one-seventh of an interval of one class. The former used Mt. Wilson single prism slit spectrograms, the latter the usual Harvard material. A majority of the classifications compared were in Classes B and A. More recent comparisons between classifications made with slit spectrograms and at Harvard show a similar agreement. Aside from the difference of appearance of slit and objective prism spectrograms, the use in the former of few or single criteria or of newly chosen typical stars has contributed to divergencies. Systems avoiding favored divisions and the use of absolute magnitude criteria in classification will depart in some measure from the Harvard results. It is to be expected that many of the additions and redefinitions affecting the DRAPER classifications would be derived from studies of slit spectra.

36. Class Q. The spectra of novae are described in detail in another chapter of the Handbuch. Only the more important classifications are given here.

Sir NORMAN LOCKYER[2] defined four distinct stages in the history of a nova as pictured in its spectrum.

1. A continuous spectrum stage in which in some cases dark lines are also observed as in the spectrum of a normal star.

2. The bright line or typical nova stage where the outstanding feature is a spectrum crossed by broad bright bands.

3. A stage marked by the presence of a line at about 4640.

4. A bright band or nebular stage where the continuous spectrum is absent or exceedingly faint and the bright lines first seen are replaced by others characteristics of the spectra of nebulae.

A classification of novae according to the entire spectral behavior of each was made by Miss CANNON[3]. Five classes of novae were proposed on the basis of their spectral history. The difficulty of devising a classification by stages to cover all cases is thus emphasized. However, our more complete modern studies suggest that there is a succession of stages through which a bright nova ordinarily passes.

F. J. M. STRATTON[4] has summarized the various stages through which Nova Geminorum II passed. He noted that Nova Persei II and Nova Aquilae III both followed the same general course. In briefest terms he stated that there is a gradual, if somewhat fluctuating, progression in spectral type in the order A, B, P, O as the star fades, and that an absorption spectrum emerges before the corresponding radiation spectrum in the cases of A and B spectra. He mentioned the following stages, but pointed out that intermediate stages of confusion where one type of spectrum is struggling with another are common.

---

[1] Harv Ann 56, p. 251 (1912).

[2] Publ Solar Phys Com, Phenomena of New Stars, p. 5 (1914).

[3] Harv Ann 76, p. 32 (1916). [4] Ann Cambridge Solar Phys Obs 4, pt. 1, p. 9 (1920).

1. An absorption spectrum of type A5 displaced, with weak radiations undisplaced.
2. An absorption spectrum of type A2p (α Cygni) displaced, with radiation spectrum undisplaced and with many absorptions doubled.
3. Superposed absorption spectra of types A2p (α Cygni) and B2 (γ Orionis) displaced by separate amounts, together with an α Cygni radiation spectrum undisplaced. The γ Orionis absorption spectrum increases in strength compared with the α Cygni absorption spectrum, and accompanying bright bands of γ Orionis type gradually appear and increase in strength.
4. α Cygni and γ Orionis radiation spectra (undisplaced).
5. γ Orionis and nebular radiation spectra (undisplaced).
6. Nebular radiation spectrum.
7. Nebular and WOLF-RAYET radiation spectra.

In 1922 the International Astronomical Union[1] retained the Harvard symbol Q to designate the spectra of new stars and devised the following provisional notation to distinguish successive stages through which a nova spectrum ordinarily passes. The International Astronomical Union proposed this scheme of classification as a basis for discussion. Bright bands due to hydrogen appear always to be present and are not referred to specifically. The range of subscript letters makes ample provision for the insertion of further divisions.

Qa. Absorption spectrum of faint lines. Bright bands inconspicuous.

Qb. Absportion spectrum of stronger lines, mainly enhanced metallic, many of which are double. Bright bands stronger.

Qc. Absorption spectrum of enhanced lines, oxygen, nitrogen, helium and associated elements. Bright lines of all of these elements.

Qu. Broad nebulous emission bands near 3480, 4515, and 4640, accompanied at times by one at 4379. The spectrum appears usually to occur in conjunction with other typical forms which it may modify through the extinction of some of their characteristic radiations, particularly 3445 and 4686.

Qx. Bright bands due to enhanced lines, and to oxygen, nitrogen, and helium, many of which are due to an ionized atom. Absorption lines faint.

Qy. Bright nebular bands in addition to the preceding.

Qz. Bright nebular bands. Weak WOLF-RAYET bands.

Qz5O. Both nebular and WOLF-RAYET emission bands are strong.

It was further provided that combinations of the small letters might be used to designate combinations of the above spectra. Thus Qbc would indicate a combination with Qb more prominent, and Qcb would show the reverse.

In 1925 NEWALL[2] suggested as modifications of the above scheme the following:

Qc. Absorption spectrum of metallic lines, oxygen, nitrogen, helium; enhanced lines predominating.

Qd. The lines of Qc present but with the gaseous lines predominating over the enhanced lines of the metals.

Qu. Omit the word "nebulous".

Probably there is no class of stellar spectra except Class P in which the divisions are more tentative than in Class Q. Many old records are confusing and modern studies must develop slowly with the occasional availability of bright novae. The order outlined by the above scheme is strongly indicated. But it must be kept in mind that Nova Centauri was totally unlike any other new star

[1] Trans Int Astr Union 1 (1922). [2] Trans Int Astr Union 2, p. 121 (1925).

yet photographed, that Nova Sagittarii II showed no nebular lines at the end of six months after its appearance, and that stars like η Carinae and P Cygni, whose spectra suggest a nova-like history, offer evidence indicating that the ordinary course of temporary stellar brightness may often be greatly modified.

37. Class P. This class was found among the original divisions of the DRAPER classification. It includes emission spectra of gaseous nebulae. It was subdivided by Miss CANNON[1] by adding subscript letters from a to f. This classification was descriptive like the rest of the DRAPER classification. However, in this case the alphabetical order of the subscript letters has proved to be essentially the reverse of that of ionization potential. The divisions of Class P by Miss CANNON appear with slight revision in the HENRY DRAPER Catalog[2]. The only essential change was the identification of 4650 as the most conspicuous feature of division Pf in the later version instead of 4686 He II as in Harvard Volume 76. In the later version all the typical nebulae of the earlier one were retained and others added. Miss CANNON's later version of the classification is found on another page. A comparison of her two versions follows.

Outline of Class P.

Division	Description		Typical Nebulae	
	Harvard Annals, 76	Harvard Annals, 99	Harvard Annals, 76	Harvard Annals, 99
Pa	3726, 3729 most characteristic feature 5007, 4959 faint $H\beta - H\zeta$ bright 3869 not seen	3726, 3729 more conspicuous than 5007, 4959 $H\beta - H\zeta$ bright	I.C. 418	Same
Pb	5007, 4959 much stronger than in Pa	5007, 4959 stronger than in Pa	Orion Neb.	Same
Pc	4363, $H\gamma$ characteristic	4363 most conspicuous	I.C. 4997	Same
Pd	5007 strongest line 4686 not seen	5007 strongest line	N.G.C. 6826	Also N.G.C. 6326
Pe	Like Pd but 4686 present	Like Pd but 4686 present	N.G.C. 7662	Also N.G.C. 7009
Pf	4686 strongest line	4650 most conspicuous	N.G.C. 40	Same

W. H. WRIGHT[3], on the basis of a thorough and extensive study with quartz and glass spectrographs, proposed a classification of nebular emission spectra which, though intended to be descriptive, separated these spectra into comparatively homogeneous groups. Complications were found to arise from the great extent of nebular surface. Further, the form of the object was found in many cases to be related to the spectrum. The Orion nebula was found to radiate a spectrum practically identical with that of a class of planetary nebulae.

WRIGHT was disposed to place nebulae with strong 4686 He II emission farther from the stellar stage in general. Thus he selected this line to characterize his nebular spectral Class I. The lines of shorter wave length than 3500 occur only when 4686 is present and serve to subdivide Class I. When 4686 is absent,

[1] Harv Ann 76, p. 20 (1916). [2] Harv Ann 91—99 (1918—1924).
[3] Lick Obs Publ 13, p. 260 (1918).

3869 has been adopted tentatively as a guide to characterize Class II. This line is of unknown origin but 3867 (unknown) and 4363 O III probably conform with 3869 in so far as intensity variations are concerned, the latter line in a general way. The subdivision of Class II on the basis of intensities of 3869 was not entirely an arbitrary one for it placed in the first subdivision those spectra which contained 4363 and excluded them from the second.

Of the prominent nebular lines WRIGHT found the significance of the pair at 3727 O II to be obscure in his scheme of classification and its use as a criterion hardly justifiable. 4363 O III was not used as a criterion because it was sometimes not visible through underexposure on spectrograms made with quartz spectrographs, or blended with $H\gamma$ in low dispersion or slitless instruments. 4959 O II and 5007 O II were apparently not critical in their behavior. Thus the classical nebular lines though prominent in other classifications were not used directly by WRIGHT.

Class III contains only three objects. It requires subdivision probably with respect to the occurrence of helium and the strength of the N lines.

WRIGHT's classification follows:

Class Ia. Typical Nebula, N.G.C. 7027. 4686 present. 3426 stronger than 3445.
Ib. Typical Nebula, N.G.C. 7009. 4686 present. 3445 stronger than 3426.
Ic. Typical Nebula, N.G.C. 6884. 4686 present. 3445, 3426 absent.
IIa. Typical Nebula, N.G.C. 6572. 4686 absent, 3869 present. Intensity of 3869 greater than 30 on WRIGHT's quartz spectrograph scale.
IIb. Typical Nebula, I.C. 2149. 4686 absent, 3869 present. Intensity of 3869 equal to or less than 30 on WRIGHT's quartz spectrograph scale.
III. Typical Nebulae, N.G.C.40, I.C. 418, B.D.+30°3639. 4686 and 3869 absent.

The identification by BOWEN[1] of many of the lines characteristic of nebular spectra with radiation from atoms of ionized oxygen and nitrogen together with the determination of the corresponding ionization and excitation potentials has provided a new basis for the classification of nebular spectra. BOWEN placed the nebulae, for which WRIGHT gave line intensities, in the order of the intensities with which 4959 and 5007 appeared in them. Since WRIGHT's intensities were referred to those of $H\beta$ and $H\gamma$ and since the lines 4959 and 5007 have an ionization potential about four times that of hydrogen, this arrangement of the nebulae in the order of intensity of 5007 and 4959 was an arrangement in order of intensity of ionization in these objects. BOWEN found that the O III lines (such as 4363, 4959, 5007) behave very much like He II lines (4200, 4541, 4686) as would be expected from the approximate equality of their ionization potentials which are respectively 54,8 and 54,18 volts. The O II lines (3726, 3729) showed characteristics similar to those of other elements of lower ionization potential. The only line of O IV to be expected in this range is probably WRIGHT's 3346 and the only strong N IV line is probably WRIGHT's 3426. These lines were observed by WRIGHT only when O III lines were very strong, which satisfied the condition that they are lines of high ionization potential. The line 3869 of WRIGHT's Class II appears to come from an unknown ion of low ionization potential.

With the help of this new material, Miss PAYNE[2] proposed a classification of spectra of gaseous nebulae to serve as a basis for discussion. This classification applies to the integrated light of the nebula and does not refer to its form or its nucleus. The spectra of diffuse and planetary nebulae appear to be similar enough to be classified on the same system.

[1] Ap J 67, p. 1 (1928). [2] Harv Bull 855, p. 1 (1928).

Miss PAYNE's suggested classification of nebular spectra follows the order suggested by BOWEN and described above and recalls the scheme suggested by WRIGHT with 4686 strong at the beginning of the series and 3727 at the end. However, it is notable that Miss PAYNE does not adopt any of WRIGHT's typical nebulae. The division P0 was suggested as a possibility by RUSSELL.

Miss PAYNE's classification follows:

Class P0. Typical nebulae, none. The nebular spectrum of highest excitation; it might be characterized by great strength of 3346 O IV (?) and 3426 N IV (?).

Class P1. N.G.C. 2440. 4959 O III and 5007 O III combined stronger than 4363 O III; 4686 He II strong.

Class P2. N.G.C. 6326. $(4959 + 5007) > 4363$; $(4959 + 5007)/H\beta > 20$.

Class P3. N.G.C. 1946. $(4959 + 5007) > 4363$; $20 > (4959 + 5007)/H\beta > 5$.

Class P4. N.G.C. 1714. $(4959 + 5007) > 4363$; $(4959 + 5007)/H\beta < 5$.

Class P5. N.G.C. 1722. $(4959 + 5007) = 4363$.

Class P6. N.G.C. 1748. $(4959 + 5007) < 4363$; 3729 weak

Class P7. I.C. 418. 3729 strong; 4959, 5007 weak.

Class P8. —3729 and hydrogen strong; 4959, 5007 absent.

Class P9. N.G.C. 1977. Continuous spectrum and hydrogen emission.

Class P10. et seq. Continuous spectrum of reflected star light, whose spectral class may be added in parenthesis, e.g., P10(B2).

Class P13. Merope Neb. Reflected spectrum Class B5—B8, which may be added in parenthesis, e.g., P13(B6).

The progressive variation of the lines used in Miss PAYNE's classification is brought out in table 6. The abbreviation S means a strong emission feature; M, medium; W, weak; a blank, absent.

Table 6. Line Variations in Class P (PAYNE).

Line	P1	P2	P3	P4	P5	P6	P7	P8	P9
3726, 3729 O II	M	M	M	M	M	M	S	S	S
$H\delta$, $H\gamma$, $H\beta$	W	W	W	M	M	M	M	S	S
4363 O III	W	W	M	M	M	S	M	M	W
4686 He II	M	—	—	—	—	—	—	—	—
4959, 5007 O III	S	S	S	M	M	W	W	—	—

A parallel between the classifications of Miss PAYNE, Miss CANNON, and WRIGHT is found in Table 7. The comparisons are necessarily rough, especially between the first and the last column, since the common criteria are few and not one of WRIGHT's typical nebulae was adopted by Miss PAYNE and only two by Miss CANNON. Also Miss CANNON and Miss PAYNE have only two typical nebulae in common. Generally speaking, however, Miss CANNON's classification is reversed in order by Miss PAYNE and WRIGHT's general order is maintained.

Table 7. Rough Comparison of Nebular Classifications.

PAYNE	CANNON		WRIGHT
	(1)	(2)	
P 0	—	—	
P 1	Pe, Pf	Pe	Ia, Ib, Ic
P 2	Pd	Pd	
P 3	Pd	Pd	
P 4	Pb	Pb	IIa, IIb
P 5	Pb	Pb	IIa, IIb
P 6	Pc	Pc	
P 7	Pa	Pa	III
P 8	Pa	Pa, Pf	III

Miss PAYNE considers as an approximation to the truth that nebular spectra of Classes P2 to P4 would be excited by stellar radiation of Classes O0 to O3; P5, by O4; P6, by O5; P7 by O6—O7; P9, by O8. No stars have been assigned to Classes O0, O1, and O2. Such stars might be difficult to observe because they emit light relatively weak in the observed photographic region but strong in the short waves which excite visible radiation in the surrounding nebulous gas. Class P10 is made to correspond to Class B1,5 approximately at the lower limit for excitation of emission lines. Later nebular spectra are due to reflected starlight.

Classes P2, P3 and P4 comprise the greater number of planetary nebulae. This crowding may suggest the need of a better subdivision of Class P or it may mean that nebular radiation in these classes is intrinsically brighter and more easily recognized, or that the corresponding stage is a persistent one. Another suggestion for the modification of Miss PAYNE's classification might favor the use of intensity ratios in divisions P7, P8 and P9. The lines $H\gamma$ and 4363, which are about equally strong in P7 but with the former growing relatively stronger in P8 and P9, might be used unless there are difficulties due to blending of these lines. Further, division P1 is set apart, perhaps arbitrarily, from the three succeeding divisions by the criterion that 4686 He II is strong. Under this criterion nebular spectra whose line intensity ratios in hydrogen and oxygen would place them in divisions P2, P3, and perhaps P4 are collected into P1. However, 4686 is excited in a volume more closely restricted about the nucleus and thus when strong in the integrated light of a gaseous nebula would indicate greater excitation conditions which might justify the creation of a spectral division. With quartz spectrograms better criteria for Class P1 might be found in the region near λ 3500.

A classification of gaseous nebulae by integrated light on a basis of effective temperatures of the nuclear star appears to have been well attained. Further studies will attempt to associate these classes with nebular forms and to classify radiation from different parts of the same nebula, especially in the cases of diffuse objects.

38. Class O. The subdivisions of Class O given above in the DRAPER Classification were described in 1901[1] and were of course used earlier at Harvard in the work of classification of stellar spectra. Previous to 1901 Mrs. FLEMING[2] appears to have used the symbols OI, OII, and OIII corresponding respectively in specific cases to Ob, Oc, and Oapec. That the spectral groups Oa to Oe5 stood so long is remarkable in view of the fact that it has been recognized for many years that they do not conform to a physical or a well marked descriptive sequence and that spectra in Class Oe5 share the dark line characteristics of those in Classes Od and Oe. The explanation of this slow progress is found in the facts that most stars of this class are faint and that the lines and bands found in their spectra, except those of hydrogen, are associated only with atomic states of high excitation which experiment and theory have described and interpreted only in recent years. Even now it cannot be said that the emission bands in WOLF-RAYET stars have been accounted for, though the suggestion is offered that they are emitted by stars, perhaps more massive than their fellows, in which matter is being expelled into space by radiation pressure under such conditions of velocity, temperature, etc., as to produce the quantum radiation bands observed.

By 1922, radiation knowledge and theory had reached such a stage that a classification of stars of Class O in which dark lines are well represented (Divi-

[1] Harv Ann 28, pt. II (1901). [2] Harv Ann 56, p. 225 (1912).

sions Od, Oe, and Oe5) could be successfully formulated. This was done by Doctor H. H. PLASKETT[1] on the basis of material secured with the 72-inch reflector of the Dominion Astrophysical Observatory. The result was an extension of the Harvard sequence in five graded steps from Class B0 over an upward range of 8000° K in temperature to a condition of excitation so marked that the dark lines of neutral helium were not visible (Class O5). Between Classes O5 and B0, divisions were numbered to conform to the Harvard sequence. Above this range by extrapolation an hypothetical class called O0 was set up by the criterion that the lines of ionized helium would be invisible. This left a space at the top for spectra earlier than any known.

H. H. PLASKETT's classification follows. The temperatures are those of FOWLER and MILNE[2]. Type stars suggested by J. S. PLASKETT[3] are included. Slight changes in the wording are made to aid clarity by maintaining a unidirectional concept of variation along the temperature-ionization sequence. Dark lines of ionized helium form the distinguishing feature of the class O absorption stars, with lines of triply ionized silicon, doubly ionized carbon, ionized oxygen and ionized nitrogen prominent. Of metallic lines only 4481 of ionized magnesium is present.

Class O0. Typical stars, none known. This class would be characterized by the absence of the ζ Puppis (He II) lines.

Class O5. Typical stars, B.D. 4° 1302, 35° 3930. Probable temperature, 30000° K. In this class the ordinary helium lines and enhanced nitrogen lines are not yet visible. In B.D. 35° 3930 there is a trace of ionized nitrogen emission at 4634 and 4641 and also a very characteristic line of unknown origin at 4603,8. The intensity ratios for the class are

$$\frac{4472}{4542} = 0{,}0; \quad \frac{4542}{H\gamma} = 0{,}6 .$$

$$\left[\frac{4388}{H\gamma} = 0{,}0; \quad \frac{4649}{H\beta} = 0{,}0\right] .$$

The ζ Puppis (He II) lines are at their strongest relative to the BALMER series of hydrogen. J. S. PLASKETT does not mention 4603,8 as characteristic.

Class O6. Typical stars, B.D. 56° 2617, 44° 3639. Temperature, 28000° K. In B.D. 44° 3639, Si IV 4089 is weak and Si IV 4116 is absent or very faint. There is no trace of the doubly enhanced carbon triplet at 4649 (4647, 4649, 4651). The intensity ratios which define the class are

$$\frac{4472}{4542} = 0{,}8; \quad \frac{4542}{H\gamma} = 0{,}5; \quad \frac{4089}{4097} = 0{,}6 .$$

Class O7. Typical stars, S Monocerotis, 9 Sagittae. Temperature, 26000° K. In 9 Sagittae, ionized nitrogen reaches its maximum strength. Ordinary helium is well marked and the enhanced carbon triplet, at 4649, is just beginning to appear.

The enhanced magnesium line 4481 is not yet visible though about to appear. The intensity ratios are

$$\frac{4472}{4542} = 1{,}4; \quad \frac{4542}{H\gamma} = 0{,}4; \quad \frac{4089}{4097} = 0{,}8;$$

$$\left[\frac{4388}{H\gamma} = 0{,}33; \quad \frac{4649}{H\beta} = 0{,}33\right] .$$

[1] Publ Dom Astrophys Obs Victoria 1, p. 366 (1922).
[2] M N 83, p. 403 (1923).
[3] Publ Dom Astrophys Obs Victoria 2, p. 294 (1924).

Class O8. Typical stars, λ_1 Orionis, A Cygni. Temperature, 24000° K. This spectrum is characterized by the decreased strength of the enhanced nitrogen lines, the strengthening of ordinary helium and of the enhanced carbon triplet at 4649. In λ_1 Orionis the fainter helium lines at 4121, 4144, 4388, 4713, are quite strong and 4481 is barely present. The intensity ratios are

$$\frac{4472}{4542} = 2{,}0; \qquad \frac{4542}{H\gamma} = 0{,}3; \qquad \frac{4089}{4097} = 1{,}0\,.$$

Class O9. Typical stars, 10 Lacertae, B.D. 34° 980. Temperature, 22000° K. The star 10 Lacertae is characterized by the strength and sharpness of its lines. Si IV, 4089, 4116 reach their maximum. Si III, 4552, 4568, are on the point of appearing and Mg II 4481 is clearly present. The enhanced nitrogen lines are present and the carbon triplet at 4649 is conspicuous. The O III lines, 3962 and 5592, are strong. The intensity ratios are

$$\frac{4472}{4542} = 2{,}7; \quad \frac{4542}{H\gamma} = 0{,}2; \quad \frac{4089}{4097} = 1{,}4; \quad \left[\frac{4388}{H\gamma} = 0{,}67; \quad \frac{4649}{H\beta} = 0{,}50\right].$$

The intensity ratios which characterize these classes are tabulated below with some extension into Class B.

Table 8. Line Intensity Ratios in Classes O5 to B8.

	$\frac{4472\ (\text{He I})}{4542\ (\text{He II})}$	$\frac{4542\ (\text{He II})}{H\gamma}$	$\frac{4089\ (\text{Si IV})}{4097\ (\text{N III})}$	$\frac{4388\ (\text{He I})}{H\gamma}$	$\frac{4649\ (\text{C III, O II})}{H\beta}$
	(1)	(2)	(3)	(4)	(5)
O 5	0,0	0,6	—	0,0	0,0
O 6	0,8	0,5	0,6	—	—
O 7	1,4	0,4	0,8	0,33	0,33
O 8	2,0	0,3	1,0	—	—
O 9	2,7	0,2	1,4	0,67	0,50
B 0	5,0	0,15	1,5	0,65	0,80
B 2	—	0,0	1,0	0,80	—
B 8	—	—	—	0,40	—

The first two of these ratios are most sensitive and are distinctive of decimal divisions from O6 to B0. Below this range these ratios are too great or too small for convenient use. The ratio, 4472 to 4542, is probably the best available. It shows no very considerable dispersion in any division though the ratios in successive divisions do overlap somewhat. The lines of the second ratio are rather far apart but it is the only ratio known that would extend above O5 where, however, it might not vary sufficiently to be useful for spectral classification. Indeed it is possible that, in Class O5, several stars are included that might be assigned to divisions from O0 to O4. The fourth ratio, as well as that of 4472 to $H\gamma$, which can be derived from ratios (1) and (2), is useful from O6 to O9 and again from B2 to B9. The fifth ratio also may be useful from O6 to O9 but should be determined carefully because of irregularities possible in the blend of carbon and oxygen. The last two ratios of Table 8 are not proposed as classifying criteria.

The ratio 4089 to 4097 is not available in the earlier divisions of the range O5 to B0, where these lines are diffuse. It does not change rapidly with spectral class. In Class O9 it shows wide dispersion partly due to the difficult character of these lines but possibly due also to relative sensitivity of these lines to conditions other than temperature and associated with mass or absolute magnitude. It is notable, though by no means conclusive of an absolute magnitude effect, that the numerical value of this ratio in PLASKETT's O9 stars increases consistently with the apparent visual magnitude. This is shown in Fig. 11. Also in O9—B0, the two well determined ratios conform to this rule though two roughly determined ratios for faint stars do not. In the other divisions of Class O PLASKETT does not give enough values of this ratio to bear on this point. The possibility of an absolute magnitude or mass effect should be kept in mind when this ratio is used.

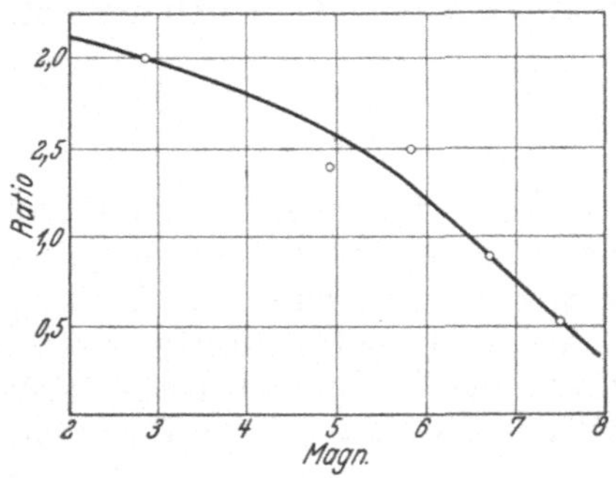

Fig. 11. Variation of the Line Intensity Ratio 4089/4097 with Apparent Magnitude in Class O9.

The descriptions in PLASKETT's classification are given in Table 9. The absence of or single reference to such features as the hydrogen and ζ Puppis lines is notable. This leaves the reader with a less complete picture of the spectrum than that which the Harvard descriptions convey. However, the ratios as given help to supply the details, and Fig. 12, which is also due to H. H. PLASKETT, gives a graphic picture of the spectral variations in Class O. Again in his direct references to the spectrum of the typical star, PLASKETT departs from the more recent practice at Harvard.

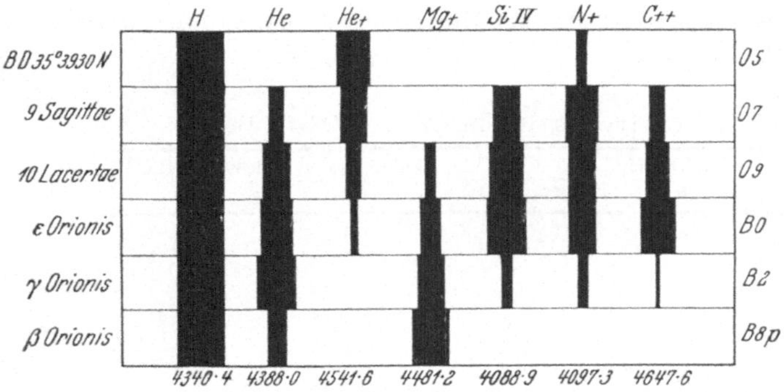

Fig. 12. Variation of Line Intensities in Class O. H. H. PLASKETT, Publ Dom Astrophys Obs Victoria 1, p. 356.

Table 9. The Descriptions of PLASKETT's Divisions of Class O.

Feature \ Class	O0	O5	O6	O7	O8	O9
ζ Puppis lines	absent					
Neutral helium		absent		well marked	stronger	
Enhanced nitrogen		absent		max.	less strong	present
4603,8		present				
4089			weak			max.
4116			weak or absent			max.
4649			no trace	appearing	stronger	conspicuous
4481				not visible	barely present	clearly present
4552, 4568						appearing
3962, 5592						strong

H. H. PLASKETT's classification of dark line stars in Class O is obviously fundamentally sound and should have been adopted earlier officially as it was by individuals. It distributed 45 classified O spectra as follows: O5, 5 stars; O6, 12; O7, 7; O8, 11; and O9, 10. Its only important limitation has been that it did not (and could not) place the emission features in any satisfactory sequence. Its author suggests as a working hypothesis that the WOLF-RAYET stars constitute a side chain with the O's, the former running in parallel with the dark line stars and possibly connecting at the top with the planetary nebulae. Physically they may be exceptionally massive stars which eventually may be "blown out" into planetary nebulae.

J. S. PLASKETT[1] has arranged fourteen WOLF-RAYET stars (Classes Oa, Ob, Oc, Op) in a sequence on the basis of intensity ratios of emission bands. As a preliminary step he provided that those Class O spectra in which relatively narrow emission lines are superposed on a characteristic O-type absorption spectrum should be designated by adding the suffix e. The WOLF-RAYET stars he limited to those objects which have strong broad emission bands standing out distinctly from the continuous spectrum. The elements found in WOLF-RAYET stars were practically identical with those in absorption O stars and the degree of excitation as shown by ionization in the emission stars was similar to that in the absorption class. In forming his sequence he used the first and second ratios of H. H. PLASKETT and three others as follows: 4511 N II to 4541 He II; 4861 $H\beta$ to 4686 He II; 4649 C III, O II to 4686 He II. He found that each and all of these five ratios in general varied progressively for one well defined arrangement of the fourteen stars. On this basis he divided tentatively the fourteen stars into four groups not conforming to the Harvard divisions in Class O, but did not propose symbols or characteristic descriptions. At the top of the sequence, emission bands of ionized helium were at their strongest relative to hydrogen, neutral helium, ionized nitrogen and the ionized carbon oxygen blend at 4649. Further down the sequence, the ionized helium lines weakened in comparison with the other lines. An apparent discontinuity in the ratio 4649 to 4686 might be ascribed to the uncertainty in the carbon-oxygen blend. However, the last division containing but one star might not be significant. A place for a further group between the third and fourth seemed to be left.

The sequence of emission O spectra, due to J. S. PLASKETT, becomes one of decreasing excitation if we assume similarity in the behavior of atoms in emission and absorption. He does not find, however that the sequence so determined corresponds to the temperature sequence determined from the absorption lines found in the same spectra. Thus the classification based on emission is in no sense continuous or accordant with that based on absorption. As J. S. PLASKETT states, "The absorption lined continuous spectrum, if we are to judge by the transition examples observed, is gradually replaced, when conditions are favorable, by the banded WOLF-RAYET emission while the continuous spectrum apparently becomes weaker and almost disappears, the final result being the WOLF-RAYET star with strong broad emission bands and no absorption except H and K, which . . . is external to the star." The transition states in Class O between stars with absorption only and stars with emission only are undoubtedly gradual and form an excitation sequence cutting across the vertical order of temperatures. Indeed, it is quite probable that the evolutionary order in the excitation sequence may be independent of the temperature changes within Class O limits at least. Further, in a single star of Class O, excitation (or non-vertical) variations may be expected

[1] Publ Dom Astrophys Obs Victoria 2, p. 351 (1924).

as in the case of many Be spectra. The WOLF-RAYET spectra in which emission predominates differ so strongly from the spectra having the same lines in absorption that it has been convenient to refer to them as a side chain paralleling the absorption O spectra. It would be more satisfactory to define several such chains. It should be noted, however, that the logical conception of the system of Class O stars is at least two-dimensional, the absorption defining one dimension and the emission the other.

P. W. MERRILL, in an unpublished suggestion for the consideration of the International Astronomical Union, proposed in 1925 as a working basis for a few years certain suggestions, modifications, and additions in connection with the nomenclature in Class O, as follows:

1. For the absorption spectrum classify by H. H. PLASKETT's system, O5 to O9.
2. When hydrogen emission lines similar to those in class B are seen add the symbol e.
3. When bright WOLF-RAYET lines are superposed on the continuous spectrum with absorption lines, add the symbol w in case the WOLF-RAYET lines are relatively faint and inconspicuous; when they are outstanding add the symbol w!. The letters a, b, c, d, may follow w to indicate the DRAPER class corresponding to the emission lines.
4. When the WOLF-RAYET lines are outstanding and the absorption spectrum is absent or not classified, use the DRAPER classification, Oa, Ob, Oc.

As a basis for discussion Miss CECILIA H. PAYNE[1] proposed in 1928 certain modifications and additions affecting the classification of O and B stars. In the same paper a classification of nebulae is given which is discussed elsewhere in this chapter. The suggestions in Classes O and B are as follows:

1. Adopt the DRAPER criteria in Class B.
2. Incorporate H. H. PLASKETT's classification of absorption O spectra, using the ratio 4541/4340 both for emission and absorption[2].
3. When simple emission is present as in the spectrum of Merope add e as adopted by the International Astronomical Union.
4. When emission of ionized iron is seen, as in the spectrum of β Monocerotis, add fe.
5. When emission lines of ionized helium (4686) and doubly ionized carbon or ionized nitrogen at 4640 are seen, as in the spectrum of ζ Puppis, add he.
6. If WOLF-RAYET bands are present on an absorption spectrum, as for γ Velorum, add w.
7. For a pure emission spectrum with WOLF-RAYET bands, as for B.D. $+35°$ 4013, add w!.

This classification provides six columns into which very early stellar spectra may go. The e column parallels the absorption sequence from O5 to A2 (or perhaps earlier). The fe column extends from B0 to B5 and the he, w and w! columns from O3 to O9 or O7 with several vacant divisions. Thus there is more or less parallelism pervading the six different sequences of the six columns. However, there is a question whether a parallelism based on similar criteria for absorption lines and emission bands is anything more than formal.

Among O stars bright enough to be classified Miss PAYNE found 45 of Class O, 15 of Class Oe, 17 of Class Ohe, 9 of Class Ow, and 12 of Class Ow!. Among

[1] Harv Bull 855, p. 1 (1928).

[2] Values of this ratio in the divisions of Class O are: O3: 1,0 and greater; O4: 0,8; and the values above by PLASKETT from O5 to O9.

fainter O stars a large number of w and w! spectra were observed, due probably to a discovery factor favoring the detection of conspicuous bright line objects. The preponderance among classified O spectra of those having absorption lines with or without emission is thus evident.

At the 1928 meeting of the International Astronomical Union the following recommendations concerning Class O were made and adopted[1].

1. For O-type stars which show a definite absorption spectrum, whether or not accompanied by bright lines or bands, H. H. PLASKETT's decimal system of classification should be adopted. The classes for which criteria are given range from O5 to O9. The upper subdivisions are left for future discoveries, O0 denoting the theoretical limit of a star so hot that it shows no spectral lines at all. Such objects may perhaps be found among the nuclei of planetary nebulae.

2. Stars which show bright lines of hydrogen, and in some cases of metals, resembling those in Types B0e—B3e should be denoted by the suffix e.

Example: H.D. 60848, O8e.

3. Stars showing the characteristic WOLF-RAYET emission lines or bands, and in which the absorption spectrum can be classified as above, should be denoted by the suffix ew. When the bands are conspicuous they should be described as ew!.

Examples: 9 Sagittae, O7ew, H.D. 193793, O6ew!.

4. For stars in which the WOLF-RAYET bands are prominent and the absorption spectrum cannot be classified, the notation of the DRAPER Catalogue, Oa, Ob, Oc, Od, should be retained pending further investigation.

It may prove possible to classify these stars on the basis of the excitation potentials of the bright bands, but a detailed study of the correlation between the degrees of excitation indicated by emission and absorption when both are present will be necessary before these stars can be connected with the general sequence. J. S. PLASKETT's investigation[2] of this question has led him to the conclusion that any such correlation is very doubtful, and that WOLF-RAYET bands may make their appearance at any stage of the (Class O) absorption spectrum.

5. If it is desired to describe both the absorption and emission spectra, the letters a, b, c, d, being the notation of the DRAPER Catalogue, may be added to the suffix ew according to the appearance of the WOLF-RAYET bands.

Examples: ζ Puppis, O5ewd, γ Velorum, O6ew!a.

In recent unpublished studies, H. H. PLASKETT has divided Ow stars into two classes, OW I and OW II. In the former the intensity ratio 4686 (He II)/4649 (C III, O II) is greater than unity; in the latter, less than unity. This divides these spectra into two classes of decreasing excitation as determined by emission features.

39. Class B. RUTHERFURD (1862)[3] was the first to place stars like Spica (B2) and Rigel (B8p) in a group according to their spectra. He did this not because of the presence of Orion lines nor because of their blue color but because they showed no lines. SECCHI recognized in 1866[4] the Orion stars as a family in Type I with narrow spectrum lines. Later (1878)[5] though preserving their place in Type I he observed that the narrow lines could place them in Type II. VOGEL

---

[1] Trans Int Astr Union 3, p. 167 (1928).

[2] Publ Dom Astrophys Obs Victoria 2, p. 352 (1924).

[3] Amer J of Sci and Arts (2) 35, p. 71 (dated 1862).

[4] C R 63, p. 621 (1866); 64, p. 345 (1867); A N 73, p. 129 (1867); Rep Brit Assoc 1868, p. 166.

[5] Die Sterne, p. 70 et seq. (1878); Le Soleil, seconde Partie, p. 443 (1877).

(1874)[1], with phase of development prominently in mind, placed Orion stars in a subdivision of his Class I. LATER (1899)[2] when the identity of helium had been established, he used the helium lines as a criterion for his class Ib following hydrogen stars. LOCKYER (1899)[3] first placed the helium stars in subgroups IIIγ and IVα on his rising scale of temperatures near and overlapping the maximum. LATER (1910)[4] he distributed helium stars over six or seven genera at the top of his temperature curve. Miss MAURY (1897) placed the helium stars in six divisions at the beginning of her sequence and arranged the letters of the DRAPER Classification in their recognized order. McCLEAN (1897)[5] placed the hydrogen-helium stars in Division I at the top of his sequence of spectra corresponding to the successive physical states through which a star passed. Miss MAURY and McCLEAN appear to have been the first to give helium stars a position in the spectral sequence above the hydrogen stars, an order which has been recognized ever since.

The lines of neutral helium are distinctive of Class B. However, no helium line certainly reaches the intensity of the strongest hydrogen lines in the photographic spectrum except possibly for certain stars with spectra in Divisions B0—B2 or thereabouts. The helium maximum occurs near B2 or B3.

R. H. CURTISS[6] finds that the helium stars which emit visible emission lines are of relatively great average luminosity in their class. Such stars are known to be characterized by effective temperatures relatively low for their class. Their atmospheres are undoubtedly relatively extensive and relatively tenuous.

Considerably more than two hundred spectra of Class B are known to contain emission lines, to all of which hydrogen emission is common. Metallic lines and lines of other gases than hydrogen may also appear in emission. When emission lines of hydrogen only are present the letter e is added to the spectral symbol. According to a suggestion of Miss PAYNE[7] the symbol em may be added instead of e when metallic emission is fairly conspicuous in addition to that of hydrogen, and he when emission of ionized helium (4686) and doubly ionized carbon (4650) are present. By action of the International Astronomical Union the symbol Fe may be used to call attention to the distinctive presence of iron lines. When the emission is conspicuously reversed the symbol r follows e. When emission and absorption lie side by side (absorption to violet) in the same line, as in P Cygni, the symbol q follows e.

Class Be spectra were classified by MERRILL[8] according to the strength of the emission lines. The several groups were: the γ Cassiopeiae group with intensely bright $H\alpha$ and strong bright $H\beta$; the b^2 Cygni group with strong bright $H\alpha$ and bright reversed $H\beta$; the Electra group with weak bright $H\alpha$ and absorption $H\beta$; the φ Persei group with intensely bright $H\alpha$, bright reversed $H\beta$, emission lines variable. To these may be added the P Cygni group with emission and absorption side by side (absorption to violet).

In studies of stellar luminosity and parallax, B stars have been classified by several observers. D. L. EDWARDS[9] in his earlier work, measuring line intensity with an extinction wedge[10], classified B stars according to the intensity of 4026 He I and 4472 He I. Line 4481 Mg II showed a maximum near B5 which limited its usefulness. However, it was used as a check. EDWARDS found that his classification filled the Harvard gaps, B4, B6, B7, and that his curves of line intensity were continuous over the whole B interval. EDWARDS' mean classifications

[1] A N 84, p. 113 (1874); Publ Astroph Obs Potsdam 3, No. 3 (1883).
[2] Publ Astroph Obs Potsdam 12, pt. 1, No. 39 (1899).
[3] London R S Proc 65, p. 186 (1899).
[4] London R S Proc 84. p. 426 (1910). [5] Phil Trans 191, p. 127 (1898).
[6] J R Astron Soc of Canada 20, p. 19 (1926). [7] Harv Bull 855, p. 3 (1928).
[8] Lick Bull 7, p. 176 (1913). [9] M N 83, p. 47 (1922). [10] M N 82, p. 226 (1922).

from all three intensity curves were generally not very different from the Harvard classifications. For determination of absolute magnitude the ratios $4144/H\delta$ and $4388/H\gamma$ were used. The helium lines so used were of the diffuse singlet series whereas those employed in classification were of the doublet series. LATER[1], EDWARDS used the ratio $4472/H\gamma$ for absolute magnitude determination thus bringing out the fact that absolute magnitude varies with spectral division in Class B. The method was found to break down for stars with very diffuse lines, emission lines and other similar peculiarities and for nearly all stars of class later than B5[2].

EDWARDS, in 1927[3] proposed divisions to supply the gaps at B4, B6, and B7 in the Harvard classification. The relative intensities of 4121, 4144, and 4472 of He I, 4481 Mg II and 4128, 4131 Si II, as well as the general prominence or otherwise of hydrogen, helium and calcium lines, were used in this classification. The Harvard descriptions with some additions are used for divisions B3, B5 and B8. The divisions are applicable especially to objective prism spectrograms of low dispersion. A comparison of EDWARDS' classifications of individual stars with the mean of the Harvard values revealed the tendency to favor B3 and B8 in the latter. Within each of the divisions below EDWARDS found that luminosity increased with line sharpness on the average. EDWARDS' classification follows:

Class B3. Hydrogen lines about 0,5 as intense as in Sirius. Helium lines more prominent than in B2 owing to extreme faintness of some O, Si, and C lines. Line 4472 much stronger than 4481; 4121 stronger than the silicon double 4128, 4131.

Class B4. Helium lines less prominent than in B3; 4472 stronger than 4481. Si double approximately equal to 4121 and fainter than 4144.

Class B5. Increased intensity in K line and Si double which is stronger than 4121 but fainter than 4144. 4481 is 0,7 as intense as 4472. K usually just visible.

Class B6. Si double much stronger than 4121, and nearly equal to 4144; 4472 slightly stronger than 4481; hydrogen lines increased slightly in intensity; helium lines less prominent.

Class B7. Si double equal to 4144; 4121 scarcely visible; 4481 equal to 4472; 4026 and 4472 easily seen while other helium lines are faint or invisible; K much fainter than 4026.

Class B8. Lines 4026 and 4472 are present together with several lines prominent in class A0; K is less intense than 4026; 4472 is slightly fainter than 4481; Si double is stronger than 4144.

These divisions of Class B are summarized, so far as the chief lines are concerned, in the following tables. The symbols $>>$, $>$, and $\overset{=}{>}$ are used to indicate much stronger, stronger and slightly stronger, with corresponding meanings for reversed symbols.

Table 10. Principal Criteria in Class B.

Division	4472 and 4481	Si Double and Helium	K
B 3	He $>>$ Mg	4121 $>$ Si $<<$ 4144	Usually invisible
B 4	He $>$ Mg	4121 $=$ Si $<$ 4144	Faint or absent
B 5	He $>$ Mg	4121 $<$ Si $<$ 4144	Faint
B 6	He $\overset{=}{>}$ Mg	4121 $<<$ Si $\overset{=}{<}$ 4144	$<<$ 4026
B 7	He $=$ Mg	. . . Si $=$ 4144	$<<$ 4026
B 8	He $\overset{=}{<}$ Mg	. . . Si $>$ 4144	$<<$ 4026

Remarks: Hydrogen and ionized calcium increasing in prominence; helium decreasing in prominence from B3 to B8.

[1] M N 84, p. 366 (1924). [2] M N 85, p. 439 (1925); 87, p. 364 (1927); 88, p. 175 (1928).
[3] M N 87, p. 367 (1927).

Table 11. Principal Criteria in Class B.

Criteria	B3	B4	B5	B6	B7	B8
4472/4481	>> 1	> 1	> 1	⋝ 1	= 1	⋜ 1
4121/Si*	> 1	= 1	< 1	<< 1	—	—
4144/Si*	>> 1	> 1	> 1	⋝ 1	= 1	< 1
K	Usually absent	Faint or absent	Faint	> B5	> B6	> B7
K/4026	—	—	—	<< 1	<< 1	< 1

Hydrogen increasing; helium decreasing from B3 to B8.
* Si = 4128 + 4131.

ADAMS and JOY[1] followed the Harvard classification as closely as possible in Class B and used spectrograms of typical stars for reference purposes. They found that the spectra listed at Harvard as B5 could be classed advantageously as B7 and the other divisions in Class B revised to correspond. They, and EDWARDS subsequently, used the letters n and s to denote diffuse and sharp lines with combinations to denote intermediate and extreme widths.

LINBLAD and SCHALÉN[2] devised a preliminary system of classification of B and A spectra of very small dispersion (1,4 mm between $H\gamma$ and $H\varepsilon$) based on estimations of the width of the hydrogen lines and of color in objective prism spectra. After a preliminary classification in the Harvard sequence a further classification was made according to the extension of the wings of the hydrogen lines and according to a quantitative estimation of color by comparison of the regions $H\delta$—$H\varepsilon$ and 4400—$H\gamma$. The zero of color was defined by an average star of classes $\sigma+$ or σ. The new symbols were used as suffixes added to the Harvard letter and number. The special classes were defined as follows:

- τ. The hydrogen lines are very narrow, so that they are invisible or almost invisible with the dispersion used (see above).
- $\tau-$. The hydrogen lines are very narrow, but appear sufficiently clearly to be identified without difficulty.
- $\sigma+$. The hydrogen lines are somewhat wider, but do not increase materially in width from strong to weak images of a sequence.
- σ. The hydrogen lines increase somewhat in width toward faint images.
- $\sigma-$. The sharpness of the lines is less pronounced but is still noticeable. The color is always near the adopted zero.
- ϱ. The wings of the hydrogen lines are strong. The color is near zero.
- μ. The wings of the hydrogen lines are very strong and the color may be slightly more advanced.
- $\nu-$. The color is perceptibly more advanced than in preceding classes. The hydrogen lines are winged.
- ν. The color is considerably more advanced than in classes ϱ and μ. The absorption at λ 3906 due to arc lines of Si and Fe is often marked.

The symbols τ and $\tau-$ apply mainly to Class B spectra. This classification supplements and cuts across the Harvard scheme as does the c, a, and b notation of Miss MAURY. It cannot be used to advantage with high dispersions and must be adjusted to the instrumental conditions employed. It is useful in correlation of stellar luminosity and spectrum.

40. Class A. The classification of the A stars in the one-dimensional Harvard system depends in large measure on the increasing intensity of the *K* line of calcium in relation to the declining hydrogen lines. This method gives results that

[1] Ap J 57, p. 294=Mt Wilson Contr 262 (1923).
[2] Med fr Astron Obs Upsala No. 17=Ark Mat Astr Fys 20 A, No. 7 (1927).

are in general fairly consistent, but the divisions formed are not entirely homogeneous and other variables must be introduced in a satisfactory classification. The great variety in the intensity and character of the hydrogen and metallic lines in such stars makes further studies of their classification important. It is probable that no one line could be used with greater success than *K*. But the position of the *K* line well removed from the effective center of the photographic spectrum in many spectrographs makes desirable a method of classification based on other characteristics. The measurement of widths, intensities and contours of hydrogen and other lines with modern self registering equipment is contributing valuable material in this connection.

The spectra of Class A are dominated by the BALMER series of hydrogen. The lines of this series reach their maximum intensity in Class A0 according to the preface of the HENRY DRAPER Catalogue[1]. Miss MAURY's published intensity estimates would indicate that this maximum occurs fully as early as A0. Miss CANNON's estimates would place it very nearly at A0—possibly at A1. MENZEL's studies of individual stars most of which were typical giants indicated a maximum at A3, or on the average in Classes A2 to A4, but the mean intensity in this range was often exceeded in B8, A0 and A5 spectra. Apparently no significant maximum of the hydrogen lines can be derived from a limited number of estimates. The spectral class of this maximum as given in the HENRY DRAPER Catalogue is based on the examination of many spectra and is entitled to greater weight than any other on statistical grounds. Since the maximum of hydrogen is a characteristic adopted in the classification of A0 stars, only conflicting criteria, possibly due to a preponderance of peculiar or non-average spectra in any given group of stars, would be expected to displace this maximum. However, the non-homogeneity of the divisions in Class A includes wide variations in the widths of the hydrogen lines and makes difficult or impracticable the determination of any exact position of the hydrogen maximum under the present system. An adjustment of the criteria in the neighborhood of A0 on an extensive statistical basis after a separation into several sequences depending on line width will be needed if it is desired to set up maximal points for the hydrogen series.

Neutral helium is not seen normally after Class A0. Ionized magnesium (4481) reaches a maximum at A3, then declines and is replaced by a low temperature iron line in later classes. Lines *H* and *K* of ionized calcium increase rapidly in Class A and surpass the hydrogen lines individually after Division A5. So called metallic lines and *g* are well seen in Class A2 and become conspicuous in later divisions. The *G* band appears near the end of the class.

For some years the opinion has been current that some revision of Class A would be desirable. W. S. ADAMS[2] suggested some years ago that further distinction was needed among stars of this class. He doubted whether the range A to F was sufficient to include the vast number of stars having this general type of spectrum. Miss A. J. CANNON at the same time[3] expressed the opinion that some of the intervals in Class A were not exactly "true". She suggested that A5 be called A8 and that A3 be called A5. Some justification for this is found in more recent intensity and intensity ratio curves. E. B. FROST[4] felt that Class A in particular lacked subgroups. Under A and A2 he found many spectra included which with high dispersion were quite different. F. SCHLESINGER[5] thought that the use of objective prism spectra made without hand guiding would probably account, among other things, for the wide diversity of spectra classified under A. — PARKHURST[6] found under small dispersion, not permitting the use of VOGEL's

[1] Harv Ann 91, Preface (1918). [2] Ap J 33, p. 262 (1911). [3] Ap J 33, p. 268 (1911).
[4] Ap J 33, p. 277 (1911). [5] Ap J 33, p. 296 (1911). [6] Ap J 33, p. 289 (1911).

classification, that the DRAPER system could be applied satisfactorily. Between A and F spectra were classified with precision by comparing the widths of K against $H\delta$ and $H\varepsilon$. In the HENRY DRAPER Extension Miss CANNON uses the divisions A4, A7, and A8.

Miss MAURY[1] divided stars of Class A into several parallel series on the basis of line widths. Miss CANNON[2] prefered to recognize great sharpness or diffuseness of lines by appended remarks. In studies of stellar luminosities several observers have used a classification of A spectra in which line width is used as a second dimension supplementing the DRAPER classification. However, this sorting of spectra by line widths leaves the system still inadequate. The difficulty appears to be due largely to the presence of so-called abnormalities within Class A.

One of the notable abnormalities within Class A is the unusual strength of the silicon lines, 4128, 4131 or of the strontium lines, 4077, 4216. This peculiarity occurs chiefly in divisions A0, A2, and A3, where the silicon lines are normally at maximum intensity. However, the strontium lines have their normal maximum at Class K2 or K5 and thus present a more striking anomaly. Studies by LUYTEN[3] would seem to indicate that the absolute brightness of these silicon and strontium stars is about that of a normal A star. Miss A. V. DOUGLAS[4] found twenty-four strontium stars and fifteen silicon stars, all in Class A, to be 0,24 and 0,5 magn. brighter respectively than the average stars of the same spectral type. However, there are too few parallaxes, proper motions and radial velocities available for statistical treatment of these objects. Possible explanations are found in the assumption of an intervening strontium cloud or of abnormal abundance of the element, neither of which is entirely acceptable. The circumstances indicate rather that the problem of the silicon and strontium stars is one involving the atom and its energy supply.

Another peculiarity, found by WOODS[5] in several Class A spectra, is the extreme faintness of the K line for the spectral division indicated by the other features. The spectroscopic absolute magnitude of such stars averaged about one magnitude too faint.

Further peculiarities involving lines of a single element or a group of elements are the occurrence of enhanced lines of manganese, broadly winged, and of unusual strength in an A0p star, α Andromedae (LOCKYER and BAXANDALL[6]), and the presence of variable lines ascribed to the rare earths, notably europium and terbium, in the spectrum of the A2 star, α^2 Canum Venaticorum[7]. The spectrum of the super-giant α Cygni (A2p) is notable for its rich array of fine sharp lines[8], many of which though awaiting identification are possibly doubly enhanced lines which, like the observed occurrence of $H\alpha$ emission and helium absorption, might have been predicted in a star of Class A temperature with an unusually extensive atmospheric envelope. The star α Cygni is representative of a large class of stars but is the only one bright enough to be readily observed. Ionized iron and titanium are strong in α and γ Cygni; ionized chromium in ε Ursae Majoris (A0p).

---

[1] Harv Ann 28, pt. I (1897).
[2] Harv Ann 28, pt. II (1901).
[3] Harv Bull 797 (1924).
[4] Ap J 64, p. 262 (1926).
[5] M N 87, p. 389 (1927).
[6] London RS Proc 77 A, p. 550 (1906).
[7] LUDENDORFF, A N 173, p. 1 (1906); BELOPOLSKY, Bull Ac Imp St Pet 6, p. 12 (1913); Pulk Bull 6, No. 10 (1915); BAXANDALL, Obs 36, p. 440 (1913); STRATTON, Obs 36, p. 461 (1913); KIESS, Publ Detroit Obs Univ of Mich 3, p. 106 (1923).
[8] LOCKYER, Catalogue of 470 of the Brighter Stars, Bull Sol Phys Com (1902); WRIGHT, Lick Bull 10, p. 100 (1921).

An interesting peculiarity, observed more readily in Class A and B spectra with relatively few lines, is the presence of absorption bands mainly in the region between $H\delta$ and $H\gamma$ and especially in the neighborhood of λ 4200[1]. SHAPLEY has substantially abandoned his theory that the source of these bands is found in absorption of substances, notably cyanogen, brought near the star by comets and meteors[2] and has concluded that some of the bands observed at Harvard may be due to absorption in the optical train of the spectrograph[3]. Mrs. McLAUGHLIN[4] finds strong evidence of the identity of these bands with auroral and perhaps other telluric bands. Tests at Ann Arbor for the presence of these bands in continuous spectra of terrestrial sources give negative results.

In their studies of stellar luminosities, ADAMS and JOY[5], disregarding the K line because of its poor definition or absence, based their accurate classification of A spectra on the intensities of the metallic arc lines. Spectra of standard stars classified by KOHLSCHÜTTER and checked closely at Harvard were used in defining the spectral divisions. The divisions were as follows:

B8 4472 and 4481 equal.

B9 4026 and 4472 just visible.

A0 Helium lines not visible. Prominent arc lines, as 4326 and 4384, not present.

A1, A2, A3, etc. Increasing intensity of the arc lines.

Letters n and s were appended to the spectral symbols to denote nebulous and sharp lines, and c was prefixed to indicate characteristics, including exceptional sharpness of lines, found in spectra like that of α Cygni. Absolute brightness declined on the average with later spectral type in both the n and the s series; and the stars with s lines averaged considerably brighter in the earlier subdivisions and somewhat brighter in the later ones.

BERTIL LINDBLAD[6] found evidence of a systematic difference in the Mount Wilson type for n and s spectra which would appear to account for the absolute magnitude difference between n and s stars observed by ADAMS and JOY for the same spectral division. In Classes A0 to A3 at least, the Mt. Wilson system seemed to have the character of a luminosity classification. LINDBLAD found that a comparison of the density of the regions 3884—3907 and 3907—3935 gave a correlation with absolute magnitude for stars with spectra of Classes B8—A3. In 1924, O. STRUVE obtained relations between magnitude and the width of 4481 throughout the range of A stars[7].

Miss PRISCILLA FAIRFIELD[8] found a close agreement between line widths and Mount Wilson n and s line characters. She found that the Mount Wilson classification especially of n spectra did not parallel the Harvard classification and this with the well known change of intrinsic luminosity with Harvard spectral class appeared to account in the main for the difference found between the absolute magnitudes of n and s stars at Mt. Wilson.

Miss A. V. DOUGLAS[9] used a classification of A stars following very closely that of ADAMS and JOY. Correlations were found to exist between luminosity and the line intensity ratios, 4215/4227, 4233/4227, 4535/4481, 4549/4481. Correlations between line width and luminosity were found for 4481, $H\delta$ and K. In the case of $H\delta$ this correlation was well marked in stars with sharp line spectra

[1] SHAPLEY, Harv Bull 805 (1924); LOCKYER, M N 86, p. 486 (1926); Mrs. McLAUGHLIN, Pop Astr 36, p. 601 (1928).

[2] Harv Bull 857 (1928).

[3] Harv Bull 862 (1928).

[4] Pop Astr 36, p. 601 (1928).

[5] Mt Wilson Contr 11, p. 273 = Ap J 56, p. 242 (1922).

[6] Ap J 59, p. 305 (1924).

[7] Abstracts of Theses, Univ of Chicago Sci Serv 2, p. 57 (1926).

[8] Harv Circ 264 (1924).

[9] Ap J 64, p. 263 (1926).

but was absent in stars with wide lines. This ambiguity is shown also in the case of line ratios and indicates some general complexity in the atmospheres of these stars.

In his studies of stellar luminosities H. C. WOODS[1] classified stars in seven subdivisions from B9 to A5. Spectra showing 4026 and 4472 of helium faintly were classed in B9. The K line was not used on account of poor focus in that region. The intensity of the metallic lines as a whole was estimated for classification and 4481 of ionized magnesium was invariably seen. A second dimension was set up in each subdivision on the basis of the width and sharpness of the hydrogen lines. Seven grades of line character were used from the sharpest, s, through n s +, n s, + n s, n, n + to n n, the most nebulous. The grades of line character were connected with absolute magnitude in each subdivision, the sharper lines with greater absolute brightness.

The classification of LINDBLAD and SCHALÉN for B and A stars is described above under Class B.

ABETTI[2] classified spectra of divisions B9 to F5 in the course of studies of absolute magnitude. For the determination of spectral subclass he used spectra classified at Mount Wilson (see above) and constructed curves connecting spectral subdivisions with the width of K and with the difference in width of $H + H\varepsilon$ and K. For the interval A5 to F5 he employed also a number of metallic lines for classification purposes. The spectra were divided according to the width of the lines $H\zeta$ and $H + H\varepsilon$ into groups with diffuse (n), well defined (s), and average (sn) lines. The widths were measured to microns and standardized by measurement of the intensity of the continuous spectrum. For each group a curve was drawn connecting absolute magnitude and spectral class.

YOUNG and HARPER[3] in their classification of spectra earlier than F0 used the ratios $K/H\delta$ and $K/(H + H\varepsilon)$.

Examples of spectra listed in Transactions of the International Astronomical Union, Vol. 1, p. 102, include the c-stars: B1, ε Canis Majoris; B8, β Orionis; A0, η Leonis; and giant stars: B0, ε Orionis; B2, γ Orionis; B5, α Gruis; A0, α Lyrae; A5, β Leonis.

41. Classes F, G, K. Notable features of the spectral interval from Class F0 to Class M0 are the gaps from G0 to G5, G5 to K0 and K5 to M0. The first two of these large intervals are bridged over without subdivision in all classifications and the latter is a region of division in most if not all classifications. In the HENRY DRAPER Extension[4], Miss CANNON used the divisions K1, K3, K7 and K8, in addition to K2 and K5, the two former not being used frequently. Those engaged in absolute magnitude studies have closed in the gaps noted above.

The difficulties of classification in this interval raise the question as to whether the several classes, largely based as they are on temperature ionization, are spaced as advantageously in the later as in the earlier parts of the sequence. It is well known that the temperature intervals are much smaller between the later classes. On the other hand since color index varies quite uniformly with spectral class, it follows that the ratio of visual to photographic brightness varies more rapidly with temperature in the later classes. The distribution of stars among the spectral classes[5] reveals a marked condensation in Class K0 and one less marked in Class G5. Apparently the spectral range from G0 to K0 is not

[1] M N 87, p. 388 (1927).
[2] Pubbl della R Univ degli Studi di Firenze, Fasc 41 (1924); 42 (1925).
[3] Publ Dom Astrophys Obs Victoria 3, p. 14 (1924).
[4] Harv Ann 100 (1925—1928).
[5] Proc Am Ac Arts and Sci 59, p. 217 (1924).

too small for a class interval nor is that from K0 to M0. But the divisions, especially in Class K, appear to need readjustment. RUSSELL[1] found relatively less homogeneity among Class K spectra of fainter stars and suggested the desirability of closer definition of the limits especially toward Class M. In the dwarf star classification at Mount Wilson, K5 is much more nearly midway between K0 and M0 than in the Harvard classification of giant stars. BAXANDALL[2] found the step from K0 to K5 in relatively bright stars apparently shorter than from K5 to M0. The evidence of color indices suggests strongly that what is now called K5 is really much nearer to M0 than to K0. In the case of giant stars, but not of dwarfs, physical data should be correlated with K2 and K5 as if they were K5 and K8 respectively[3].

Spectra of Classes F, G, and K are characterized by the presence of many metallic and other lines. In spectra so complex, rapid classifications from low dispersion spectrograms must be based on bolder features such as the K, $H + H\varepsilon$, $H\delta$, $H\gamma$, $H\beta$, G, and g lines. A glance at Table 3 reveals the fact that this is true of the DRAPER classification. PARKHURST[4] suggested that the relative widths of G and g be used in classifying low dispersion spectra of Classes F, G, and K. NEWALL[5] suggested that the presence or absence of the G group due to H_nC_n might be made a point of distinction between Class F and Class G. Miss CANNON[6] pointed out that when $H\gamma$ and 4326 were of equal intensity the star was classed as G0.

Generally speaking the line criteria available for spectral classification do not present variations so conspicuous in the range G0 to M0 as in some earlier spectral intervals. The decline of hydrogen lines is retarded; H and K are wide and in early Class K reach a broad maximum; G is difficult to use in slit spectra. Metallic lines do not show large rapid variations. The g line varies rapidly in Class G but rounds over more or less in Class K. Hydrogen and calcium lines vary also with absolute brightness.

Little attention was paid in the DRAPER Classification to individual solar lines or line ratios, as Table 3 reveals. But intensive studies of individual solar lines in stars of different spectral divisions and particularly in the sun-spot- as compared with the normal solar spectrum led to the recognition of important variations in these fainter features. Such variations were seen to be associated in an important way with temperature and related conditions under which spectral features were characterized as furnace, arc, spark and super-spark lines in the scale of increasing excitation, or more simply as normal and enhanced. At the same time the occurrence of many enhanced lines in the chromospheric spectrum led to the idea that pressure might be a factor affecting the occurrence of spectral features. The theory of ionization and atomic energy states has defined the function of temperature and pressure in the production of any spectral line and has placed great emphasis on the study of individual features.

RUSSELL[7] wrote, in 1911, long before the significance of the behavior of spectral lines in different stellar spectra was well understood, "I would add the suggestion that a comparative study should be made of the spectra of stars of very different total luminosity but the same spectral class." This was done by HERTZSPRUNG and with most significant results by ADAMS and KOHLSCHÜTTER and later by many others. In the paper which furnished the basis for all the valuable studies of spectroscopic parallaxes which followed, ADAMS and KOHL-

[1] Ap J 33, p. 293 (1911).
[2] M N 86, p. 524 (1926).
[3] Trans Int Astr Un 1, p. 97 (1922).
[4] Ap J 33, p. 289 (1911).
[5] Trans Int Astr Un 2, p. 121 (1925).
[6] Trans Int Astr Un 2, p. 208 (1925).
[7] Ap J 33, p. 293 (1911).

SCHÜTTER[1] separated the spectra of 700 stars, of types ranging from A to M, into two groups in the first of which were included intrinsically faint stars of large proper motion and with measured parallaxes and in the second intrinsically bright stars of very small proper motion. Comparisons of spectra in these two luminosity groups brought out the facts that:

1. The continuous spectrum of intrinsically bright stars was relatively fainter in the violet as compared with the red end of the spectrum. This effect was small for A spectra, increased with later spectral class and was twice as marked for K as for F stars.
2. The hydrogen lines were abnormally strong in a considerable number of stars of small proper motion (presumably stars of high luminosity), especially in titanium oxide stars.
3. Certain lines were weak in large proper motion (intrinsically faint) stars and strong in the small proper motion (intrinsically bright) stars and conversely. The sensitive lines that were stronger in high luminosity stars were 4216, 4395, 4408; those weak in such stars were 4325, 4435, 4456, 4535.

Pairs whose relative intensity variation showed the greatest and most definite changes with type were:

F8 to G6; $g/H\gamma$, $4326/H\gamma$, $H\gamma/4352$, $H\gamma/4405$, $H\gamma/4384$;

G6 to K9; $4326/H\gamma$, $H\gamma/4352$, $H\gamma/4405$, $H\beta/4872$, $H\beta/4957$.

Pairs selected as a basis for investigation of individual stars for absolute magnitude were:

F8 to K9; 4216/4250, 4395/4415, 4408/4415, 4456/4462, 4456/4495.

MENZEL[2] suggested

F to M; 4375/4384, 4375/4371.

ADAMS[3] finally adopted before 1916 the following pairs of lines for classification of stellar spectra by relative intensities: F0 to G5, $H\gamma/g$, $H\gamma/4384$; F0 to M0, $H\gamma/4352$; F3 to M0, $H\gamma/4326$, $H\gamma/4405$; G0 to M0, $H\beta/4872$, $H\beta/4957$. The variation curves based on these pairs depended on differences and not ratios. Only stars of large proper motion (intrinsically faint) were used in deriving the curves.

ADAMS and JOY[4], also ADAMS, JOY, STRÖMBERG and BURWELL[5], used two methods of classification of spectra in the range F0 to K0, in 1921. In each case the aim was to conform as closely as possible to the DRAPER system. Method (1). The "measured class" was determined from reduction tables based upon standard spectra. These tables gave values of intensity differences for the pairs $4326/H\gamma$, $H\gamma/4352$, and $H\gamma/4405$, for the whole range F0 to M0; $g/H\gamma$ and $H\gamma/4384$ from F0 to G0; $H\beta/4872$ and $H\beta/4957$ from G0 on. The corresponding curves were published in 1916. Method (2). The estimated spectral class was determined by direct comparison with typical spectra selected from the Harvard list and placed side by side with the spectrum to be "estimated", in a HARTMANN spectrocomparator. The typical spectra were selected at intervals of five spectral divisions, namely, F0, F5, G0, G5, K0, K5. In estimating the spectral division especial consideration was given to Ca Ig and the chromium lines 4255 and 4275, which varied rapidly with spectral type and formed valuable criteria. It was thought probable that all three of these lines varied to some extent with absolute

[1] Ap J 40, p. 385 (1914) = Mt Wilson Contr 5, p. 67.

[2] Harv Circ 252, 256 and 263 (1924); Stellar Atmospheres, p. 121 et seq., Append. (1925).

[3] Wash Nat Ac Proc 2, p. 143 (1916).

[4] Ap J 46, p. 332 (1917) = Mt Wilson Contr 7, p. 318.

[5] Ap J 53, p. 32 (1921) = Mt Wilson Contr 9, p. 423.

magnitude, but this effect seemed to be slight except in extreme cases like those of the Cepheid variables. The stars whose spectra were regarded as typical are found in Table 14, p. 88. The pairs of lines used for the determination of absolute magnitude were: In Classes A to F7, 4077/4072, 4290/4271 and in Classes F8 to M (except M giants), 4215/4250, 4455/4462, 4455/4495. For M giants 4077 and 4215 were compared directly with the same lines in β Andromedae and μ Geminorum. For Cepheids the pairs 4077/4072, 4215/4250 and 4290/4271 were employed in conjunction with special tables.

Comparison between "estimated" and "measured" spectral classes showed but little difference for stars of absolute magnitude 4,0 or fainter and for stars brighter than 4,0 as a group in Classes F0 to K0 or K3. But for high luminosity stars of later K and of M classes the differences became large due to the fact that the hydrogen lines are abnormally strong, thus affecting the measures based on comparison of hydrogen with other lines.

Almost all recent classification at Mount Wilson has been based on comparison with plates of standard spectra of which a very complete list has been worked out (see Table 14). The "measured" spectral types have been given up because they do not conform to the DRAPER system among giant stars, since the intensity of the hydrogen lines, for the more complex spectra at least, is a criterion of absolute magnitude rather than spectral type. In α Orionis (M0) for example, the hydrogen lines are at least as strong as in the Sun (G0). "Measured" types can be used in absolute magnitude determination, and for dwarf stars the types may be "measured" for spectral classification with good success. Estimated classifications agree more closely with the Harvard results. Thus for 151 brighter stars the average difference between Harvard classifications and Mount Wilson estimates was 1,6 spectral divisions whereas between Harvard classifications and Mount Wilson measures the average difference was 2,2 spectral divisions. The systematic difference was not greater than 1,2 divisions in either case and was apparently smaller for estimated values.

YOUNG and HARPER[1] classified spectra at first by measures of intensities of pairs of lines but later discarded this method for the simpler procedure of eye estimates. Their typical stars are found in Table 14. Spectral types when measured were found to require correction for absolute magnitude and when estimated, the indefinite nature of the Harvard definitions of spectral classes gave difficulty. Multiple criteria in the DRAPER system were often found to be contradictory. Also since the hydrogen lines are relatively faint in dwarf stars and g is relatively strong, Class F spectra of intrinsically faint stars are classed too late in the DRAPER system.

For types earlier than Class F, YOUNG and HARPER used the classification ratio, $K/H\delta$ or $K/(H + H\varepsilon)$ as given by Harvard for the various divisions (see Table 3). For stars of Class F0 or later they used the ratio $H\gamma/g$, which was taken as unity at about Class F8 (see Section f, ciph. 31). For classes later than F8 up to and including K9 they used the absolute strength of the line g. They assumed that g reaches its maximum strength near K9 or M0 and in the late M types is much weaker. When their criteria indicated a spectrum as late as K5, they looked for the presence of titanium oxide bands and when these were present the star was placed in Class M. If the titanium oxide bands were not present the spectrum was placed in Class K, its exact place being fixed by the strength of CaI g.

A comparison between Victoria and Mount Wilson classifications of the same spectra shows that the agreement is good except for the early F and late K

[1] J Astr Soc Canada 18, p. 9 (1924); Publ Dom Astroph Obs Victoria 3, p. 1 (1924).

spectra. For any Mount Wilson division the dispersion in the Victoria estimates averages about three tenths of a class interval, for stars of the first twelve hours of right ascension. A comparison between Victoria and Harvard classifications shows that the former generally average later, divisions F5 to F8 being exceptions. The average dispersion in the Victoria estimates of spectra in each Harvard division is about eight tenths of a class interval. Part of the lack of agreement between Victoria and Harvard classifications is due to the fact that not all intermediate divisions are used in the latter. Part is due to divergence of the single criteria used at Victoria from those of the DRAPER system. Part is accidental. Since the Victoria observers needed a system agreeing only on the whole with the Harvard system no attempt was made to readjust their criteria.

Table 12. Lines Used for Absolute Magnitude.

Ratio	YOUNG	HARPER
4071,9/4077,9	A2 — M6	A2 — M6
4161,6/4167,6	A2 — K2	A2 — K9
4202,2/4207,0	G5 — M6	G4 — M6
4215,7/4250,6	A2 — K2	A2 — K2
4247,2/4250,6	A2 — K2	A2 — K9
4258,5/4260,6	F2 — M6	A5 — M6
4271,7/4290,1	A2 — G4	A2 — G4
4455,1/4462,0	F8 — K9	F8 — K9
4455,1/4494,7	G2 — K9	G0 — K8
4455,1/4482,4	G6 — M6	G6 — M6
4455,1/4489,4	G5 — M6	G6 — M6
4482,4/4494,7	M0 — M6	G8 — M6
4489,4/4494,7	G5 — M6	G8 — M6
4494,7/4496,5	G5 — M6	G8 — M6

The lines used by YOUNG and HARPER in absolute magnitude determination together with the spectral range over which they found them applicable are given in Table 12.

SHAPLEY and others, in absolute magnitude studies at Harvard[1] in the range G5 to K5, subdivided the classifications of the HENRY DRAPER Catalogue more closely in some cases and used prominently as a criterion of absolute magnitude the intensity ratio 4326/4215. The cyanogen absorption bands and the hydrogen, calcium and manganese lines as luminosity criteria were also considered.

LINDBLAD[2] developed a scheme for classification of spectra in the range G to M, using types by ADAMS and his collaborators as standards. The classification was based on the appearance and relative strength of the hydrogen lines, *K*, *g*, titanium oxide bands, and the general energy distribution in the spectrum. Since the decrease of the intensity of the hydrogen series and the increase of the intensity of *g* with advancing spectral type proceeded more rapidly in the dwarf series, while the increase in redness took place faster in the giant series, and since *g* had enormous strength in dwarfs of Classes K5 to M, these characteristics served in a great number of cases to determine the spectral type and to decide whether the star was a giant or a dwarf. A valuable absolute magnitude criterion was found in the cyanogen bands with first heads at λ 3590, 3883, and 4216, that at 3883 being the strongest. The development of these bands is at a maximum for giant stars between G5 and K5 and diminishes strongly toward the F and M types. In dwarf stars of all types the bands are only faintly developed. For absolute magnitude studies the intensities of the two spectral regions on either side of λ 3889 were compared by the method of exposure ratios. In the spectral range from G8 to K0 the absolute magnitude effect was a maximum and here the difference in exposure ratio between giants and dwarfs corresponded to nearly one magnitude, but the effect decreased rapidly from G5 to G0. In the same way, spectral regions 4144—4184 and 4227—4272 were compared, thus employing

[1] Harv Circ 228 (1921); 232 (1922); 243, 246 (1923). [2] Ap J 55, p. 85 (1922).

a second cyanogen absorption region; also the regions 4227—G and G—4383 were compared.

W. B. RIMMER[1] in absolute magnitude studies over the range F0 to M3 based his classifications on Mount Wilson determinations. Spectral class was taken from Mt. Wilson Contr. 199 or was estimated by eye comparisons with reference to typical spectra chosen from a list of 1646 Mount Wilson stars. Three or more type stars chosen for range in absolute magnitude, were selected in each reference division. Thus the grading of classes allowed at the same time an approximate estimate of absolute magnitude to be made. Pairs of lines used for absolute magnitude determination were:

4078/4072	F0 to G8	Double weight, F0 to F7
4227/4216	F0 to G4	
4216/4250	F0 to M3	Double weight, F8 to K5
4290/4272	F0 to F6	Used as check
4444/4455	F6 to Mb	
4455/4462	G5 to K6	

For the most luminous stars in Classes F and G the first and fifth of the above ratios were used and also the pairs 4173/4179, 4395/4384, 4400/4384, and 4418/4384. In highly luminous stars in Classes K and M, the lines stronger in the brighter stars were 4078 Sr II, $H\gamma$, 4375 Ti II and 4444 Ti II. Pairs used for absolute magnitude determination were 4326/$H\gamma$, $H\gamma$/4352, 4375/4384, and 4444/4455.

MACKLIN[2] in absolute magnitude studies for the range F8 to K3 used Mount Wilson classifications. Owing to the small dispersion employed, the close pairs used at Mount Wilson were not available. Lines λ 4227 Ca I and 4326 Fe I were found to decrease slightly in intensity with increasing absolute magnitude while λ 4174, 4215 Sr II, 4290 Ti II, 4353 V II, 4387 Ti II and 4444 Ti II increased notably with increase of absolute brightness. The pairs used to determine absolute magnitudes were 4227/4174, 4326/4290, 4326/4387 and 4326/4444.

BAXANDALL[3] studied the intensity variations of many lines in the region between λ 4326 and λ 4716 in five relatively bright stars ranging in spectral class from G0 to M0. He found that the majority of the lines showing considerable increase of intensity from G0 to M0 were relatively low temperature lines of iron, vanadium, chromium, and titanium. Those lines which weakened considerably were high temperature lines and components in the G band.

GERASIMOVIČ[4] has emphasized the well-recognized importance of constructing curves of varying intensity of a given spectral line for stars of a given luminosity in the different DRAPER Classes. The effect of varying active mass is thus practically excluded. For giants, as compared with dwarfs, not only does the decrease of pressure encourage ionization in any spectral division but the variation of level of a given line absorption with luminosity may lead to displacement of maxima toward the early spectral types thus opposing the "pressure effect". These and other factors complicate the problem and permit of a solution only from the observational side. Intensity curves for Sr II 4077, 4215 from F0 to K5 bring out the well known sensitivity of these lines to absolute luminosity. On the other hand the relative insensitivity of a group of Fe I lines to absolute luminosity in the range F0 to M0, is brought out. The intensity maximum for Fe I is found in the region K0 to K5 and is not clearly different for giants and dwarfs.

A full discussion of the observed changes in stellar line intensity associated with differences in absolute magnitude will be found in another part of the Handbuch. Table 13 due to Miss PAYNE[5] giving variations of line intensity

[1] Mem R A S 62 IV, p. 113 (1923); 64 I, p. 1 (1925).
[2] M N 85, p. 444 (1925).
[3] M N 86, p. 524 (1926).
[4] Harv Circ 311 (1927).
[5] PAYNE, Stellar Atmospheres, p. 149 (1925).

Table 13. Variation of Line Intensity with Absolute Magnitude.

Line	Element	Observed Effect						
		F0	F2	F5	G0	G5	K0	Ma
3933	Ca II	+ 1	..	..	..	..	0 ?	+
3944	Al I	+ 0	+	− 2	..	..	..	−
3953	Fe I	+ 0	+	− 2	..	..	..	..
3961	Al I	+ 0	+	− 2	..	..	..	−
3968	Ca II	+ 1	..	− 1	..	..	0 ?	+
3999	Ti I	+ 0	..	− 1	..	..	..	..
4005	Fe I	+ 0	..	0	..	..	..	..
4031	Mn I	− 2	− 1	− 1	..	..	0	..
4041	Mn I	− 1	− 1	− 2	..	..	− 0	..
4046	Fe I	− 1	− 1	0	− 1	− 1	− 2	..
4064	Fe I	− 1	− 2	..	+ 2	0	− 2	..
4068	FeMn I	− 1	..	..	+ 1	− 1	..	..
4072	Fe I, —	0	+ 4	..	..	− 1	− 2	..
4077	Sr II	0	0	+ 1	S	0	+ 3	+
4084	Fe I	− 1	+ 1	0	0	..	..	..
4101	H	− 1	− 1	+ 1	− 4	0	− 1	+
4132	Fe I	0	+ 2	+ 1	..	+ 0	− 1	..
4135	Fe I	− 1	..	0	..	+ 0	..	..
4144	Fe I	− 1	0	+ 2	0	+ 0	− 1	..
4167	?	− 1	..	− 1	− 1 W	0	..	..
4172	Fe II	+ 1	+ 1	+ 3	+ 1	+ 0	..	..
4177	Fe II	+ 1	..	+ 2	..	..	..	..
4215	Sr II	0	0	+ 2	+ 3	+ 0	+ 1	+
4227	Ca I	− 1	− 1	0	− 2	− 0	− 2	−
4247	Sc II	0	..	+ 1	..	+ 1	..	..
4250	Fe I	0	− 2	− 1	− 2	− 1	− 2	..
4254	Cr I	− 1	..	− 1	− 1	− 1	− 1	..
4260	Fe I	− 1	− 2	− 1	− 2	..	− 1	−
4272	Fe I	− 1	− 2	..	0	+ 0	− 1	..
4275	Cr I	− 1	..	..	0	+ 0	− 1	..
4290	Cr I	0	− 2	0	..	+ 0	− 1	..
4298	Ti I, Ca I	− 1	− 2	0	..	0	..	..
4308	Fe I	0	+ 1	0	..	..	− 1	..
4315	Fe II	0	+ 1	+ 2	..	..	..	..
4321	Sc II	+ 1	..	..	S	..	..	..
4326	Fe I	0	+ 1	+ 1	− 2 W	0	− 2	..
4340	H	− 2	− 7	− 1	− 4	− 0	− 1	+
4352	CrMg I	0	+ 2	0	+ 1	+ 0	− 3	..
4360	Cr I	0	..	+ 1	..	..	..	..
4370	Fe I	+ 1	..	0	..	+ 0	..	..
4376	Y II	− 1	+ 2	0	..	+ 1	+ 2	..
4383	Fe I	0	+ 3	+ 1	− 2	− 4	− 2	..
4405	Fe I	− 1	..	..	0	− 1	− 2	..
4415	Fe II	+ 1	0	+ 3	0	0	− 1	..
4435	Ca I	− 1	..	+ 1	..	0	− 2	−
4444	Ti II	0	0	+ 1	S	− 3	..	..
4455	Ca I	− 2	..	− 1	W	− 1	− 3	−
4476	Fe I	− 1	..	+ 1	− 1	..	..	..
4481	Mg II	− 1	0	0	0	+ 0	+ 1	..
4490	Fe I	0	..	..	+ 1	− 0	+ 2	..

with absolute magnitude was based on all available estimates. A positive number or positive sign means an increase in intensity from dwarf to giant star, such increase being larger for larger numbers; a negative number or negative sign, a decrease; and a zero, no change. The letters "W" and "S" refer to weakening or strengthening of lines from the solar spectrum to that of Capella as observed by BAXANDALL[1].

In the classification of the spectra of highly luminous stars such as Cepheids, criteria may prove conflicting. Between maximum and minimum light in Cepheid variables important classification criteria were found to vary in intensity by ADAMS and SHAPLEY[2] as follows:

Feature	At Max. Light	At Min. Light
$H\gamma$	Strong	Much weakened
Fe II, Ti II, Sr II and Cr II	Strong	Much weakened
λ 4481	Very strong	Much weakened
g	Strong	Strengthened
Ca, Fe, Ti and Cr (Low Temp.)	Weak	Strengthened
Continuous spectrum (ALBRECHT)	Strong in violet	Weakened in violet

Miss A. C. MAURY[3] before 1897 recognized the "c" characteristcs in spectra of Cepheid stars. S. ALBRECHT[4] discovered the displacement of the intensity maximum in some Cepheid spectra toward the violet with increasing light and toward the red with fading brightness. Miss A. J. CANNON[5] observed a change in spectral class of TT Aquilae roughly from G at maximum light to K at or near minimum light (later values, G5 to K2), including concordant variations in the continuous spectrum. Miss I. LEHMANN[6] found changes in the intensity of some lines with the change in total light. F. G. PEASE[7], referring especially to the hydrogen lines and Ca II K, interpreted changes in the spectrum of RS Boötis as variations in spectral class from B8 at maximum light to F0 at minimum.

ADAMS and SHAPLEY[8] and SHAPLEY[9] classified Cepheid spectra mainly by hydrogen lines, though calcium lines and G were used to some extent in F and G. The ratio $G/H\gamma$ was applied in Class F. This hydrogen-line classification was intimately connected with light variation and color index. The spectral class was invariably earlier at light maximum as compared with that at light minimum. Eleven Cepheids showed an average difference of 0,8 of a spectral interval from maximum to minimum light.

ADAMS and JOY[10] determined spectral classes of Cepheids by characteristics which seemed to be primarily a function of general spectral type, such as intensities of certain arc lines, and found an average difference of but one or two tenths of a spectral division between maximum and minimum light. The median value of the hydrogen-line classification[11] was about 0,3 of a spectral interval earlier than the estimates at Mount Wilson[12]. The estimated class agreed well with the hydrogen-line type at light minimum. Studies of line displacements by ALBRECHT[13], of spectral class by Miss CANNON[14], of line intensities by PANNEKOEK and REESINCK[15], and of photographic spectral intensities by GYLLENBERG[16] point toward variations of spectral class in Cepheids similar to those found by PEASE[7], ADAMS[8], and SHAPLEY[9]. Several other studies of relative intensities of lines compared at

[1] Researches on the Chemical Origin of Various Lines in Solar and Stellar Spectra. (Solar Physics Committee 1910.)

[2] Wash Nat Ac Proc 2, p. 136 (1916). [3] Harv Ann 28, pt. I (1897).

[4] Lick Bull 4, p. 131 (1907). [5] Harv Circ 129 (1907); Pop Astr 31, p. 188 (1923).

[6] Pulk Mitteil 5, p. 176 (1913). [7] Publ A S P 26, p. 256 (1914).

[8] Wash Nat Ac Proc 2, p. 136 (1916).

[9] Wash Nat Ac Proc 12, p. 208 (1916); Mt Wilson Contr 124 (1916); 159 (1919).

[10] Wash Nat Ac Proc 4, p. 129 (1918) = Mt Wilson Comm 53.

[11] Ap J 55, p. 206 (1921) = Mt Wilson Contr 226. [12] Mt Wilson Contr 199 (1921)

[13] Lick Bull 4, p. 131 (1907); Ap J 54, p. 161 (1921). [14] Pop Astr 31, p. 188 (1923)

[15] B A N 3, p. 47 (1926) [16] Lund Medd (2), No. 24 (1920).

Table 14. Typical Stars.

Class	Harvard	Mt. Wilson		Int. Ast. Un. 1, p. 102		Victoria	Class
		Giants	Dwarfs	Giants	Dwarfs		
F0	δ Geminorum α Carinae	ϱ Tauri		θ Scorpii			F0
2	π Sagittarii	20 Pegasi				ν Herculis	2
5	α Can. Min. ϱ Puppis	ε Ceti		ε Ceti	α Can. Min.		5
7						γ Serpentis	7
8	β Virginis α Fornacis	θ Draconis			β Virginis		8
G0	α Aurigae β Hydri	α Aurigae	8 Can. Ven.	τ Persei	The Sun		G0
1						ζ Herculis	1
5	κ Geminorum α Reticuli	η Andromedae	Boss 3075	η Piscium	μ Herculis	β Boötis	5
K0	α Boötis α Phoenicis	α Boötis	Boss 814	α Boötis	70 Ophiuchi Br.	α Boötis	K0
2	β Cancri υ Librae	ι Aurigae	70 Ophiuchi Br.				2
3						π Herculis	3
4					70 Ophiuchi Ft.		4
5	α Tauri	40 Lyncis	Lal. 31055	α Tauri		γ Draconis	5
8			61^1 Cygni		61^1 Cygni	δ Ophiuchi	8
M0	α Orionis γ Hydri	β Andromedae	Lal. 39866	δ Virginis		β Andromedac	M0
1		ν Virginis	A. Oe. 4961			α Orionis	1
2		α Ceti	Lal. 21185			η Geminorum	2
3	ϱ Persei γ Crucis	μ Geminorum	Krueger 60	β Pegasi	Lal. 21185	β Pegasi	3
4		ϱ Persei	A. Oe. 17415				4
5		α Herculis	Barnard's Star				5
6	W Cygni RX Aquarii	45 Arietis				α Herculis	6
7		R Aquarii*				R Lyrae	7
8		R Leonis*		45 Arietis		45 Arietis	8
9		o Ceti**					9

* Normal Maximum. ** Normal Minimum.

maximum and minimum light have revealed increased ionization near the former phase.

Stars selected as typical for spectral classification at Harvard, Mount Wilson, and Victoria, and also a list designated as giant and dwarf by the committee on stellar spectral classification of the International Astronomical Union, are found in Table 14. The Harvard type stars in the northern hemisphere are prevailingly giant for G0 and later classes but are dwarfs for Classes F5 and F8. The Harvard typical stars assigned in Table 14 to M0, M3, and M6 belong to the more general designations Ma, Mb, and Mc.

Emission line spectra, so numerous at the ends of the spectral sequence, are so rare in Classes F, G, and K that no characteristic group can be recognized.

42. Class M. The absorption bands of titanium oxide which are characteristic of Class M have constituted a criterion which has influenced all stellar spectral classifications dating back to SECCHI[1] and probably to CARPENTER[2] and RUTHERFURD[3]. The notation corresponding to Class M in the various classifications is found in Table 5. The Harvard notation, Ma, Mb, Mc and Md, undoubtedly arose initially in the impression that the separate groups so designated were quite unlike. Thus E. C. PICKERING[4] thought that these four groups differed from each other more than Classes G, K and Ma. Certainly spectra classified under Md (Me) differed widely and included cases in which the underlying spectrum was not of Class M at all. Miss CANNON[5] suggested in 1911 that Ma be called M0, and Mb, M5; and that Mc might be designated by another letter since there was as much difference between Mc and Ma as between Ma and K0. However, it was so clear that the groups Ma, Mb and Mc formed a sequence that the designations M0, M3, and M8 were respectively suggested for them in 1925 by the International Astronomical Union[6]. Probably the last symbol should be more commonly M6,5. The correspondence between the two systems as determined by usage is roughly: Ma = M0—M2, Mb = M3—M5, Mc = M6—M10. The spectra in Class Md, which are characterized by variable emission lines, are classified according to the absorption features, the presence of emission being denoted by the addition of the symbol "e".

Criteria used in the DRAPER Classification are not in general easy to follow from M0 to M8. The great width of g and the disintegration of G are used as far as M3. The increased faintness of the photographic continuous spectrum is referred to. But the criterion which appears to extend progressively through Class M is the conspicuousness of the titanium oxide bands. They emerge in K5 (K8), and are well marked in M0. In M3 and M6,5—8, the classifier's attention, like that of SECCHI, seems to be transferred to the bright margins of the titanium oxide bands which appear to increase in contrast with the neighboring absorption from M0 to M8. DUNÉR[7] suggested that the line spectrum rather than the bands was the best indication of the development of a star. HERTZSPRUNG[8] considered that the titanium oxide bands had no necessary connection with the temperature scale. However, they are of great significance in the Harvard Classification.

The Mount Wilson Classification[9] by intensity differences between a dark hydrogen line $H\gamma$ or $H\beta$ and neighboring metallic lines was found to be applicable to Class M in the dwarf branch of the sequence. But for M stars of high luminosity,

[1] C R 57, p. 71 (1863). [2] M N 23, p. 190 (1863).
[3] Amer J of Sci and Arts (2) 35, p. 71 (dated 1862).
[4] Ap J 33, p. 289 (1911). [5] Ap J 33, p. 268 (1911).
[6] Trans Int Astr Un 2, p. 119 (1925). [7] Kgl Svenska Vet Akad Handl 21, No. 2 (1884).
[8] Trans Int Astr Un 2, p. 208 (1925). [9] Wash Nat Ac Proc 2, p. 157 (1916).

though the presence of bands placed them definitely in Class M, the hydrogen lines were abnormally strong. The hydrogen classification would place α Orionis, for example, in division G2. Other features than hydrogen lines found to be strong in M giants as compared with M dwarfs were enhanced lines of iron, titanium, strontium and vanadium, whereas low temperature lines of calcium (including g), titanium, chromium and strontium were found strong in M dwarfs and weak in M giants.

The failure of the measured classification when applied to M giants led to the adoption of a method of estimates[1] based on general characteristics such as high and low temperature lines, the intensity of g, the prominence of bands and the appearance of many other features. For this purpose spectrograms of typical stars (see Table 14) were selected and the spectrum to be classified was compared with these in turn. Accordance between measured and estimated classes was close for absolute magnitudes of 4,0 and fainter. But for stars brighter than 4,0 the measured divisions in Class M especially, but also in Class K, were earlier than the estimated, due to increase in strength of the hydrogen lines with greater luminosity.

In Table 13 some of the lines affected by luminosity in Class M0 are listed. ADAMS and JOY[2] found useful for absolute magnitude determination the low temperature lines of iron, 4206, 4258, 4389, and 4489; also $H\gamma$, 4150 and the two strontium lines, 4077 and 4215, all of which show intensity increase with absolute magnitude. In M dwarfs Sr I 4607 showed a marked increase in intensity. The prominence of certain enhanced lines in the spectra of giants as compared with dwarfs of the same type is due to low atmospheric density which promotes ionization. On such lines the effect of a small change of temperature in the opposite direction is of slight consequence. But the sensitive low temperature lines are influenced directly by the reduction of temperature and thus appear strengthened in the spectra of giants. Differences in level and relative abundance are doubtless effective also.

RIMMER[3], using typical spectra as standards, found late M stars difficult to classify. His results showed strong divergences from those derived at Mount Wilson. His absolute magnitude pairs in Class M were $4326/H\gamma$, $H\gamma/4352$, 4375/4384 and 4444/4455.

Typical stellar spectra selected by ADAMS in Class M were α Ceti (giant) and 34 Groombridge Br (dwarf) in Ma, 51 Geminorum (giant) in Mb, and Boss 660 (giant) in Mc. A classification based upon the intensities of the titanium oxide bands exclusively and defined by typical spectra was proposed by ADAMS, JOY, and MERRILL[4]. The typical stars selected are found in Table 14. The star o Ceti must be used carefully because of its composite spectrum. Typical spectra in Class M adopted at Harvard and Victoria and some mentioned in the Transactions of the International Astronomical Union 1, p. 102 (1922), are found in Table 14.

In spectra of Class Me the hydrogen lines are bright and narrow, usually with no absorption components visible. $H\gamma$ and $H\delta$ are generally bright and conspicuous. $H\varepsilon$ is absent or extremely weak. $H\beta$ is weak or absent except when the underlying spectrum is of an early division of Class M, or possibly a late one of Class K, in which case $H\beta$, $H\gamma$ and $H\delta$ may have approximately equal intensities. Weaker emission lines, especially at λ 3905, 4138, 4178 and 4202 are not unusual. The above description applies especially to the descending branch of the light curve.

---

[1] Ap J 46, p. 313 (1917) = Mt Wilson Contr 142. [2] Publ A S P 35, p. 326 (1923).
[3] Mem R A S 64, 1, p. 1 (1925). [4] Trans Int Astr Un 2, p. 119 (1925).

Mrs. FLEMING[1], in 1898, subdivided Md spectra using divisions from one to ten, including 1,5. Though no complete description of these subdivisions has been published, Mrs. FLEMING's classifications, in Table 9, p. 197, Harvard Annals 56, reveal the nature of the scheme. Class Md1 of which R Lyncis (now classed as Se) was the typical star showed a spectrum resembling α Orionis (M1) but having $H\beta$ and $H\gamma$ strongly and nearly equally bright, with $H\delta$ an emission line barely visible. R Leonis, the typical star of Class Md10 had an absorption spectrum of Mc (M6—M8—M10) with $H\beta$ emission not seen, $H\gamma$ barely visible and $H\delta$ strongly marked. The other divisions formed a nearly continuous sequence between these extremes. On page 225 of Harvard Annals 56, Miss CANNON interpreted these divisions of Md spectra as follows:

Md1, continuous spectrum of Class Ma (M0), $H\beta$ and $H\gamma$ bright, the former the brighter.

Md4, $H\beta$ invisible as emission, $H\delta$ present as emission but fainter than $H\gamma$.

Md6, background spectrum Mb (M3); $H\delta$ emission as strong as that at $H\gamma$; $H\zeta$ also bright.

Md9, $H\delta$ emission stronger than that at $H\gamma$; $H\zeta$, $H\eta$, $H\theta$, and $H\iota$ present as emission.

Md10, background spectrum Mc (M6—M10); $H\delta$ emission much stronger than that at $H\gamma$.

In 1919, Miss CANNON[2] stated that it is the exception rather than the rule for $H\beta$ to be bright when the spectrum is of Class M, and, if bright, it is never the strongest emission line. Spectra of Class Md may be divided into two groups according to whether $H\gamma$ or $H\delta$ is the strongest emission line. When $H\gamma$ is strongest the spectrum is usually of an early division of Class M, and sometimes even of K5. When $H\delta$ is the strongest emission line the spectrum is of a later division of Class M and the bands of titanium oxide are strongly marked. During the light variation changes occur in the class of spectrum, the intensity of certain absorption lines, the relative intensity of the hydrogen emission lines and in the distribution of light in the continuous spectrum. The emission at $H\delta$ may appear first and greatly exceed $H\gamma$ until near maximum light when the latter may reach equality with the former. Further the relative intensities of these two lines may vary at different maxima of the same star. The spectral class may vary as much as a whole spectral interval during the light variation, and concomitant variations in Ca *g* and other absorption lines may occur.

Miss CANNON has found that Mrs. FLEMING's Classes Md2, Md3—Md10 correspond closely in characteristics of the absorption features to the divisions of Class M adopted by the International Astronomical Union. But in the HENRY DRAPER Catalogue Mrs. FLEMING's subdivisions of Class Md were not employed. Accordingly the designation Md was used without numeral, and additional facts such as the intensities of the hydrogen emission lines were given in the remarks. It was evident that no accurate subdivisions could be made until observations of Md spectra had been obtained at different points of the light curve, since changes in the relative intensity of the hydrogen lines and probably in the class of spectrum were connected with the light variation of any star.

MERRILL[3] has given more detailed descriptions of the intensity variations of emission lines in spectra of Class Me during the light cycle. Curves representing the variations of $H\delta$, $H\gamma$, Fe I 4202, Fe I 4308, and Mg I 4571 are reproduced in Band VI, S. 142 of this Handbuch. Briefly stated, his results show that $H\delta$ emission appears shortly after light minimum, followed by $H\gamma$ at about the

[1] Sci 8, p. 455 (1898); Publ American Astr Soc 1, p. 48; Harv Ann 56, No. 6 (1912).
[2] Pop Astr 27, p. 527 (1919). [3] Ap J 53, p. 185 (1921) = Mt Wilson Contr 200.

middle of the ascending branch, 4202 just before light maximum, and the other two lines (4308 and 4571) on the descending branch of the light curve. These lines come to a maximum in the same order along the descending branch of the light curve except that the maximum of 4202 seems to precede that of $H\gamma$ and to coincide nearly with that of $H\delta$. Further, these five lines disappear in the original order, $H\delta$ shortly before and 4571 shortly after minimum light. Shortly before light minimum 4571 is in some stars the strongest emission line in the photographic region. At light maximum, Si I 3905,5 is the strongest emission line next to those of hydrogen. The other three metallic lines mentioned above attain a maximum strength of about one-third that of the hydrogen lines. The identifications of the metallic lines above, originally due to STEBBINS[1], indicate the expected low temperature characteristics. These identifications of emission lines other than those of hydrogen have been questioned by BAXANDALL[2]. Later FROST and LOWATER[3] have mentioned the presence of Si I 4103,1 which was to be expected to occur with Si I 3905,5. The iron identifications including Fe I 4376 are accepted provisionally. MERRILL[4] has described the behavior of lines of H, He, Li, Na, K, Mg, Ca, Sr, Ba, Al, Ti, Zr, V, Cr, Mn, and Fe in stellar spectra of Classes M, N, and S. His characteristic intensities of the hydrogen emission lines at maximum light are found in Table 15.

Table 15. Characteristic Intensities of Bright Hydrogen Lines in Variable Stars at Maximum Light.

	Class M	Class S	Class N
$H\alpha$ . . .	2	15	10
$H\beta$ . . .	2	12	10
$H\gamma$ . . .	20	5	5
$H\delta$ . . .	30	3	2
$H\varepsilon$ . . .	2	—	—
$H\zeta$ . . .	15	1	Trace
$H\eta$ . . .	8	—	—
$H\theta$ . . .	5	—	—
$H\iota$ . . .	7	—	—
$H\varkappa$ . . .	1	—	—

In the spectrum of *o* Ceti at the exceptionally low maximum of February 1924 (magnitude 4,7), JOY[5] observed a number of previously unrecorded emission bands or lines flanked by absorption on the side of smaller wave length. Recently BAXANDALL[6] has called attention to the close agreement of position of these lines with heads of bands of aluminium oxide. The laboratory wave lengths of these bands are: 4537,5, 4648,0, 4672,0, 4694,6, 4842,3, 4866,4. Further study led BAXANDALL to conclude that aluminium oxide is represented also as absorption in the normal spectrum of *o* Ceti, the most definite coincidences occurring in the cases of the band heads 4715,5, 4736,0, 4842,4 and 5439,5.

43. Class S. Class S was adopted and described in the eighth volume (1923) of the HENRY DRAPER Catalog. In the volumes of this catalog prior to this, spectra of Class S were designated as Pec and the nature of the peculiarities was described under Class Md. The designation, Class S, was suggested for these spectra by action of the International Astronomical Union in 1922[7]. There such spectra were characterized as most complicated in the region λ 4500 to 4700, which appeared to consist of both absorption and emission lines, with absorption bands present at about λ 4650 and λ 6470. Most of the stars known to belong to this type were long period variables with spectra showing hydrogen emission.

Spectra of this class have been studied by ESPIN and WRIGHT[8] and more extensively by MERRILL[9, 10, 11]. The description of Class S spectra given under the

[1] Lick Bull 2, p. 92 (1903). [2] Obs 46, p. 82 (1923).
[3] Ap J 58, p. 278 (1923). [4] Ap J 63, p. 13 (1926) = Mt Wilson Contr 306.
[5] Ap J 63, p. 281 (1927) = Mt Wilson Contr 311. [6] M N 88, p. 679 (1928).
[7] Trans Int Astr Un 1, p. 98 (1922). [8] M N 72, pp. 546, 548 (1912).
[9] Ap J 56, p. 457 (1922) = Mt Wilson Contr 252; Ap J 58, p. 215 (1923) = Mt Wilson Contr 264.
[10] Publ A S P 35, p. 217 (1923). [11] Ap J 63, p. 13 (1926) = Mt Wilson Contr 306.

DRAPER Classification has been restated above by the present author on the basis of MERRILL's results. That description may now be amplified.

In addition to the probable bands of zirconium oxide with heads at 4620 and 4637, every one of the thirteen bands of this compound listed by EDER and VALENTA between 6229 and 6612, together probably with several additional band heads, have been observed as absorption by MERRILL[1] in spectra of Class S. These absorption bands were the most prominent features in the red region of Class S spectra except for the $H\alpha$ emission in Se variable stars. The band in this region extending from 6470 to greater wave lengths was particularly notable. Though spectra have been found by MERRILL which in other respects resemble those of Class S but in which the zirconium oxide bands are faint or absent, it is felt that these bands should be recognized as outstanding features of this class[2]. Like the titanium oxide absorption bands, those of zirconium oxide are sharply bounded and more intense at the edge of shorter wave length, fading out toward the red.

Zirconium appears also to be represented by a number of absorption lines in the blue region. The strength of the line Ba I 4554 makes it an outstanding feature, and Sr I 4607 is generally one-half or two-thirds as intense but varying with the total light if the latter is variable. These two lines are much stronger in Class S than in Classes K and M. A broad absorption space at 4644 is notable. Hydrogen absorption is of moderate intensity in spectra of non-variable stars of Class S. In the region of greater wave length than 4500, numerous emission lines and bands are found, notably those at 4511,4 and 4521,4 and three probably due to Fe II at 4584, 4924, and 5018. The whole region from 4500 to 4700 is remarkably complicated.

Almost without exception, variable stars in Class Se have exhibited very conspicuous hydrogen emission near maximum phase, and, conversely, those stars in Class S which have bright hydrogen lines are long period variables. The characteristic relative intensities of the bright hydrogen lines in Class Se are found in Table 15. When titanium oxide bands are present in the spectrum the decrease from $H\beta$ to $H\delta$ is less rapid. $H\delta$ may equal or even exceed $H\gamma$ when the bands are strong. Bright hydrogen lines are found in a greater proportion of the spectra in later divisions of Class S. The hydrogen emission in long period variables varies with the total light very much as do similar features in Class Me variables. In addition to the bright lines of hydrogen and probable enhanced iron, the following emission lines, found also in Me variables, are observed: λ 4138, 4178, 4202, 4225, 4308, 4352, 4373, 4376, 4427, 4461, 4511, 4521, 4571. Absorption lines and bands, as well as the emission features, vary with total light.

The general appearance of the typical Class S spectrum is that associated with high luminosity and low density. Indeed, the S-type stars may bear the same relation to those of Class M that Cepheids and Pseudo-Cepheids do to ordinary F and G stars. The typical Class S spectrum conforms more nearly in general appearance to early Class M, whereas the periods and velocities of the S stars are more nearly the same as those of divisions M7e or M8e. Class S is clearly more closely related to Class M than to Class N. Indeed, in some spectra of Class S, titanium oxide bands may be present with zirconium oxide bands as in χ Cygni, AA Cygni and H. D. 156957[3]; and other spectra appear to be intermediate between Class S and Class M both in the possession of bands of the two

---

[1] Publ A S P 35, p. 217 (1923). [2] Trans Int Ast Union 2, p. 121 (1925).

[3] Publ A S P 36, p. 353 (1924).

compounds and in the character of the remaining spectral features (e. g. H. D. 22649 and H. D. 35155).

Spectra of Class S are considered to form a third branch of the giant sequence similar to G5—K—M and G5—R—N. The S branch leaves the main giant sequence at about K5 or later and may rejoin it near M8 where temperatures are too low for zirconium oxide bands to appear but where titanium oxide bands may be present. The character of the main and branching sequences is shown in the following diagram:

```
                   R—N0—N3—N8—
                  /
O—B—A—F—G—K—M0—M3—M8— —
                   \          /
                    S————Se
```

44. Classes R and N. Spectra of Classes R and N are characterized by the presence of the SWAN bands of carbon as absorption and are thus identical with SECCHI's fourth type. These dark bands are sharp on the edge of greater wave length and are degraded toward the violet end of the spectrum. The groups of bands beginning with λ 5635 (Group II), 5165 (Group III), and 4737 (Group IV) are usually the most prominent, while that beginning at λ 4383 (Group V) is less conspicuous. SHANE[1] found the three band heads at λ 6191 (Group I), 6122 (Group I), and 6060 (Group I) to be relatively narrow features on spectrograms of the visual region. The SWAN bands appear to increase in strength on the average from R0 to R5 and from N0 to N4 in a roughly parallel manner[2]. From division R5 to N0 the data are scattered but are consistent with a decline of intensity. Data in divisions later than N5 are unreliable in the Group IV region. Representatives of Groups I to V of the SWAN bands were apparently measured by VOGEL[3] and DUNÉR[4] and more completely by HALE, ELLERMAN, and PARKHURST[5], RUFUS[2] and SHANE[1].

Absorption bands of cyanogen are conspicuous features of Class R spectra but decline in prominence in Class N. These bands have heads on the edge of greater wave length and are degraded toward the violet. The conspicuous groups of cyanogen bands in the photographic region of Class R spectra are those with heads of greatest wave length at λ 4216 (Group III), 3883 (Group IV), and 3590 (Group V). Group II of cyanogen absorption bands at λ 4503, 4515, 4533, 4555, 4578, and 4609, measured by HALE, ELLERMAN and PARKHURST[5], are relatively narrow, having very little extension toward the violet and, under low dispersion, resembling broad lines. RUFUS[6] found that the cyanogen bands of Group II were stronger in spectra of Class N. SHANE[7] confirms this increase of intensity of Group II bands to the end of Class R, beyond which they are nearly constant in divisions N0, N3, and N5. He establishes also the important fact that the cyanogen absorption in the bands of Groups III and IV, already very strong in division R0, increases rapidly to a maximum at R5, beyond which the decrease is even more rapid to an effective disappearance at division N3, where the bands of Group II of cyanogen, as stated above, continue very nearly at maximum intensity.

The *G* hydrocarbon band was found by BAXANDALL[8] to strengthen on the G—R branch of the spectral sequence. SHANE[9] found an increase in the intensity

[1] Lick Bull 10, p. 85 (1920).
[2] Publ Univ of Mich (Detroit) Obs 2, p. 140 (1916); Lick Bull 13, p. 126 (1928).
[3] Publ Astrophys Obs Potsdam 4, pt. 1 (1894).
[4] Ap J 9, p. 119 (1899). [5] Publ Yerkes Obs 2, p. 365 (1903).
[6] Publ Univ of Mich (Detroit) Obs 2, p. 139 (1916). [7] Lick Bull 13, p. 125 (1928).
[8] M N 86, p. 524 (1926). [9] Lick Bull 13, p. 127 (1928).

of this band to division R3 where it appears in unusual strength. Beyond R3 it decreases slowly at first and then more rapidly until in N3 it is nearly or quite absent. Some irregularities in this order were observed, as is evident in Fig. 8 of this chapter of the Handbuch where B. D. $-10°$ 5057 shows little or no *G* absorption in Class R0.

SHANE[1] has found unknown absorption bands in Class N spectra with heads toward the violet at λ 6789, 6807, 6826, 6846, 6865, 6892, and 6909, and SANFORD[2] two others at λ 4866 and 4976 in spectra later than N0 with possible bands at λ 4905, 4932, and 5035 when those at λ 4866 and 4976 are strong. SHANE[3] added three more bands at λ 4510, 4572, and 4642. Six of these bands fall into similar groups, with similarities in separation and relative intensities, as follows:

Group I	Group II	Description
4642	4976	Wide, strongly degraded toward red
4572	4905	Narrow
4540	4866	Moderate width, slightly degraded toward red

Most of these bands may be seen in the spectrum of U Hydrae in Fig. 8 of this chapter of the Handbuch. It is possible that the first band in each of these Groups I and II was seen by DUNÉR[4]. The intensities of these bands increase rapidly at division N5 proceeding down the sequence, and are useful in classifications at divisions N5 and N6.

HALE, ELLERMAN, and PARKHURST[5] identified in Class N stars the elements: carbon, hydrogen, vanadium, calcium, magnesium, sodium, iron, nickel and manganese. SHANE[6] added scandium, yttrium, titanium and chromium; MERRILL[7] ionized strontium and ionized barium. The presence of lanthanum, cadmium and zirconium is probable. RUFUS[8] found carbon, hydrogen, vanadium, calcium, magnesium, iron, nickel, titanium, chromium and possibly zinc and zirconium in Class R.

Bright lines in spectra of carbon stars have been observed by DUNÉR, KEELER, and CAMPBELL visually, and as prominent photographic features by HALE, ELLERMAN, PARKHURST, RUFUS, and SHANE. SHANE[9] noted close correspondence in Class N with lines of Ti, V, Fe, and Cr, but considered no identifications established except for hydrogen emission.

SHANE[10] found that the spectral variations in Class N variables were similar in their general character to those described above in Me variables[11].

Only a superficial examination of Fig. 8 of this chapter of the Handbuch der Astrophysik reveals a series of re-entrant steps proceeding down toward the violet termination of the typical carbon stellar spectra with abrupt edges at λ 4220, 4400, and 4740, the first heads of cyanogen Group III, and of SWAN Groups V and IV, respectively. The intensities or relative intensities of these roughly flat regions 4220 to 4400, 4400 to 4700, 4740 to 5100, and the region beyond 4220, are the main criteria used in the DRAPER Classification to allocate carbon spectra in the divisions of Classes R and N. Since the intensities in these regions are affected both by variations in the bands involved as well as by the decline, along the sequence, in the strength of the continuous spectrum of shorter

[1] Lick Bull 10, p. 79 (1920).
[2] Publ A S P 38, p. 177 (1926).
[3] Lick Bull 13, p. 123 (1928).
[4] Ap J 9, p. 119 (1899).
[5] Publ Yerkes Obs 2, p. 385 (1903).
[6] Lick Bull 10, p. 79 (1921).
[7] Ap J 63, p. 13 (1926) = Mt Wilson Contr 306.
[8] Publ Univ of Mich (Detroit) Obs 2, p. 103 (1916).
[9] Lick Bull 10, p. 79 (1920).
[10] Lick Bull 10, p. 91 (1920).
[11] Handb d Astrophys 6, p. 145 (1928).

wave length, the DRAPER sequence may not represent a temperature order throughout Classes R and N. The cyanogen band at 4220 declines in intensity along the sequence, R5 to N0, within which it would affect classifications. If, however, there are two roughly parallel increases in the strength of the SWAN bands of Groups IV and V in the ranges, R0 to R5 and N0 to N4, some departures from the temperature sequence may be expected in the classifications. SHANE[1] has pointed out that the Harvard classification in R and N follows well the order of increasing faintness in the more refrangible region. Probably the classifier avoided the regions most affected by band absorption and employed those parts of the continuous spectrum adjoining the band heads on the side of greater wave length. Nevertheless, since the band absorption does not follow unidirectional changes with temperature in all cases, it is probable that in the R—N branch the temperature order and dependence on temperature change is departed from in a greater degree than in other parts of the spectral sequence.

SHANE[2] has proposed a new decimal classification of R and N spectra in the photographic region. He has pointed out that classification criteria are of particular significance in this region and that in the spectral range N3 to N6 visual spectra exhibit no appreciable variation. He has chosen new criteria to follow the Harvard system. Thus the DRAPER order is not affected. The bands are brought more definitely into the classification and the decrease of spectral intensity toward the violet remains a criterion of prime significance. The proposed classification follows.

SHANE's Classification of R and N Spectra.

Class R0: Cyanogen bands of Groups III and IV strong. Spectral intensity distribution as in Class G or K. SWAN bands weak.

Class R3: Intermediate between R0 and R5.

Class R5: Cyanogen bands very strong. Considerably weaker in short wave lengths than Class R0. SWAN bands very strong.

Class R8: Cyanogen bands much weaker than in Class R5. SWAN bands show a great range of intensity in this class. Spectral intensity distribution intermediate between Class R5 and Class N0.

Class N0: Cyanogen bands very weak. Spectrum can be photographed as far as *H* and *K* only with difficulty. SWAN bands weak or moderate.

Class N3: Cyanogen bands missing. Spectrum extremely weak to the violet of 4383. SWAN bands strong.

Class N5: Spectrum very faint to the violet of the SWAN band at 4737 A. Unidentified bands at 4979 and 4686 strong.

Classes N6 and N7: Continue in greater strength the characteristics of Class N5.

The studies of HALE, ELLERMAN, and PARKHURST (1912)[3,4] by which the conclusion was reached, previously suggested by DUNÉR, that "stars of the third and fourth type should be classed together as coördinate branches leading back to the sun" have been discussed above in this section—also the studies of RUFUS[5] and CURTISS[6] (1916) by which stars with spectra of Class R were shown to form the link connecting stars having spectra of Class N with the main sequence near

[1] Lick Bull 10, p. 91 (1920).
[2] Lick Bull 13, p. 128 (1928).
[3] Yerkes Obs Publ 2, p. 385 (1903).
[4] Ap J 35, p. 132 (1912).
[5] Publ Univ of Mich (Detroit) Obs 2, p. 103 (1915).
[6] Publ Univ of Mich (Detroit) Obs 2, p. 182 (1915); Pop Astr 25, p. 279 (1917); Sci Amer 83, p. 344 (1917).

Class K0. A series of spectra exposed by RUFUS and enlarged by CURTISS to illustrate the sequence is shown in Fig. 8.

Apparently no spectrum has been observed that could be classed as intermediate between Classes M and N or between Classes K and R. The SWAN bands of carbon which characterize Classes R and N are not found in the same spectrum with bands of titanium oxide or zirconium oxide, which characterize spectra of Classes M and S and late Class K. This is true notwithstanding the fact that the remaining spectral features are much alike in the corresponding classes and that carbon and titanium are both found in the several branches of the sequence.

W. C. RUFUS[1] pointed out that the exclusive occurrence of titanium oxide and SWAN bands respectively in stars of Classes K and M and Classes R and N may be due to relative abundance combined with suppressive action by the compound having ascendancy over the other in that part of the star where the compounds in question form and absorb. It is well known[2] that a slight excess of oxygen in the source of light destroys the SWAN spectrum and produces the second band spectrum due to carbon dioxide. Thus the proportion of oxygen in the atmosphere of a star may be a prominent factor in the determination of the absorption bands. It is thought that the characteristic banded absorption in Class R and Class N spectra is due to an oxide of carbon and it is known that the characteristic banded absorption in Class M stars is due to titanium oxide.

R. H. CURTISS[3] offered the explanation that, when conditions thermal and otherwise favor the production of oxides, either carbon or titanium will combine with oxygen most actively, and that this most active element will monopolize in a given star the supply of oxygen in that portion of the atmosphere where bands are forming. In this event the bands of the oxide of the more active element will appear. If the supply of oxygen exceeds the amount necessary to exhaust the available supply of the more active element, the oxide of the less active element will form and suppress the absorption of the more active element. The absorption of the oxide of the less active element will thus appear to the exclusion of the other. Under this explanation, the occurrence of the bands of both carbon and titanium is still left open in extremely rare cases, an example of which may yet be found.

45. Classification of Stellar Spectra in the Visual Region. The early classifications of stellar spectra covered the visual region and extended into the photographic. Classifications subsequent to that of SECCHI were generally based on the photographic region. Several writers[4] have advocated modern studies of stellar spectra in the visual region for classification purposes. W. W. CAMPBELL[5] pointed out that $H\alpha$, D_1, D_2, D_3, and other lines could hardly fail to show significant variations from one spectral type to another. His visual observations in this region are notable. Miss CANNON[6] examined the portion from $H\beta$ to D_3 for six stars of Class B and found little that helped in the classification except the line He II 5413. J. S. PLASKETT[7] mentioned especially the lines $H\alpha$, $H\beta$, D_1, D_2, D_3, and b for classification in the visual region.

V. M. SLIPHER[8] published excellent reproductions of spectra of the stars Rigel (B8), Sirius (A0), Procyon (F5), the Sun (G0), Arcturus (K0), Betelgeuse (M1), and Mira Ceti (M), extending from λ 4300 to λ 7200 or further. These afforded a good record of spectral changes in the visual region from Class B8

[1] Publ Univ of Mich (Detroit) Obs 2, p. 142 (1916).
[2] BALY, Spectroscopy, p. 444 (1905). [3] Pop Astr 25, p. 279 (1917).
[4] Ap J 33, p. 267 (1911); PAYNE's Stellar Atmospheres, p. 200 (1925).
[5] Ap J 33, p. 267 (1911). [6] Ap J 33, p. 268 (1911). [7] Ap J 33, p. 292 (1911).
[8] Enc Brit, 11th Ed., 21, p. 717 (1911).

to late M. In Class B, D_3, and other lines of neutral helium, at λ 6678 and 7065, are distinctive. $H\alpha$ is wide and strong in Class A0. Lines $b_{1,2,3}$ and $D_{1,2}$ are well shown in Class F5 but increase through Class G0 to Class K0. In Class M, a strong band head effaces b while $D_{1,2}$ remains strong. Heavy dark bands of titanium oxide extended over the visual region of Class M.

MERRILL[1] using a large prism over the Harvard 24-inch reflector in combination with dicyanin dyed plates, extended the photography of stellar spectra in the most favorable case to λ 8700 ±. A titanium oxide band at λ 7590 and other features of importance were observed in the region of greater wave length than λ 7000.

In general the DRAPER Classification does not include reference to the visual region. In Class Mb, absorption bands, having edges at λ 5763 and 5857 ± are noted, and in Class Nc the most brilliant portion is from λ 5900 to λ 6800.

SHANE[2] found it possible to fall into serious error in classifying spectra of Class R and N if the visual region only were used.

Intensive studies of individual stars have partially prepared the way for a classification of stellar spectra extended to the visual region. Such studies include that of α Orionis by NEWALL and COOKSON[3], of α Cygni by WRIGHT[4], and of α Persei by DUNHAM[5].

j) Classifications of Stellar Spectra not based on Estimates of Line Intensities.

46. Classification of Spectra by Line Displacements. S. ALBRECHT[6] found progressive displacements of a number of spectral lines with changing stellar spectral class over the range from F to Mb. Spectrograms made with the three-prism MILLS spectrographs were used, having dispersions at $H\gamma$ of 10,5 and 12,5

Table 16. Suggested List of Lines to be included in the Measures for Radial Velocities. (The wave lengths given in part *b* were determined from the stellar spectra.)

a) Lines whose Wave Lengths vary		*b*) Lines whose Wave Lengths are Constant	
4246,9	F to Mb	4245,435	F to Mb
54,5	F to K2	50,293	F to Mb
67,9	F to Mb	50,951	F to Mb
74,9	F to K2	71,319	F to Mb
88,1	F to K5	94,273	F to Mb
93,2	F8 to K5	4313,041	F to Mb[7]
4315,1	F to Mb	18,867	F to Mb
21,0	F to K5	28,101	G to Mb
34,0	F to Mb	37,223	F to Mb
44,6	F to Mb	39,721	F to Mb
52,0	F to Mb	76,103	F to Mb
52,9	F to Mb	79,364	G to Mb
90,1	F to Mb	83,725	F to Mb
95,2	F to Mb	4401,611	F to Ma
4425,6	F to Ma	06,803	F8 to Mb
30,7	F to K5	07,851	F to Mb
35,2	F to Mb	08,572	F to Mb
64,7	F to K2	27,444	F to Mb
68,7	F to Mb	42,526	F to Mb
69,5	F to K5	47,911	F to Mb

[1] Bull Bur Stand 14, p. 487 (1919).
[2] Lick Bull 13, p. 128 (1928).
[3] M N 67, p. 482 (1907).
[4] Lick Bull 10, p. 100 (1921).
[5] Contr Princeton Obs No. 9 (1929).
[6] Ap J 24, p. 333 (1906).
[7] May vary slightly.

Ångströms per millimeter. The progressions in measured displacement were due to changing intensity of component lines in blends. For any particular line a smooth curve could be drawn through points plotted with displacements, $\triangle\lambda$, as ordinates and spectral class as abscissa.

The inverse of this process was employed by ALBRECHT[1] in the development of a method by which spectral class could be determined by comparison of measured wave lengths for any stellar spectrum with curves representing the variation of wave lengths as measured for a series of spectra ranging over the spectral classes involved (F to Mb). Variations in wave lengths ranging up to 0,35 Ångströms were found for individual lines from Class F to Class M. The lines suggested by ALBRECHT for use in classifying spectra with the dispersions of the three-prism MILLS spectrographs are given in Table 16.

The stars selected by ALBRECHT for the determination of his classification curves are given in Table 17. The adopted and measured spectral classes together with probable errors are found in the several columns of the table.

Table 17.

Star	Type		
	DRAPER Classification	ALBRECHT Classification	Probable Error
ι Carinae	F	F2,6	± 1,0
α Circini	F	F0,8	± 1,7
π Sagittarii	F2	F4,3	± 2,7
κ Reticuli	F5	F3,9	± 1,1
b Velorum	F5 pec.	F3,7 pec.	± 1,9
α Fornacis	F8	F6,7	± 0,7
Venus	G(0)	G0,3	± 0,6
α Triang. Austr.	K2	K2,0	± 0,5

This method when applied by ALBRECHT[2] gave variations of spectral class with phase in the spectrum of the Cepheid variable l Carinae, corresponding to those determined by classifications based on the hydrogen line intensities. Lines subject to the larger variations of wave length with phase in the spectrum of l Carinae were λ 4258,4, 4268,0, 4288,1, 4334,0, 4344,6, 4391,9, 4395,2, 4399,9, 4412,2, 4416,9, 4417,8, 4422,0, 4425,6, 4435,2, 4464,7, 4468,6, and 4469,4. Curves were determined for these and other lines. ALBRECHT[3] had found indications of such line shifts in η Aquilae but found no definite trace of them in low dispersion (one-prism) spectrograms of Y Ophiuchi and T Vulpeculae.

Doctor M. ALBERTA HAWES[4] applied the method of classification of stellar spectra by line displacements to one-prism spectrograms (dispersion 40 Ångströms per millimeter at $H\gamma$) made with the $37^1/_2$-inch reflector of the University of Michigan. The stars selected for this study are listed in Table 18. They were chosen by R. H. CURTISS on the basis of spectral class, absolute magnitude, and availability. All the stars adopted were intrinsically bright. The range of Harvard classes was from F0 to K0. Table 18 contains for each star the visual and absolute magnitudes, the Harvard, Mt. Wilson, and Victoria classes, the Michigan classes with their probable errors.

Ninety-nine individual lines, for which were derived well determined variation curves of wave length referred to spectral class, afforded abundant material for the determination of spectral division with great accuracy. The largest variations in wave length from Class F3 to Class G7 were 0,58 Ångströms. Lines showing

[1] Ap J 33, p. 130 (1911).
[2] Ap J 54, p. 161 (1921).
[3] Lick Bull 4, p. 131 (1907).
[4] Unpubl Thesis, Univ of Mich (1929).

Table 18.

Star	HARV. Vis. Mag.	Mt. W. Abs. Mag.	VICT. Abs. Mag.	HARV. Class	Mt. W. Est.	Mt. W. Meas.	VICT. Class	MICH. Class	p.e.
β Cor. Bor.	3,7	− 0,3	− 0,4 − 0,2	F0p	F2	F4	F2	F3,2	± 0,6
ν Herc.	4,5	− 0,2	+ 1,2 + 1,5	F0	F2p	F3	F3	F3,5	± 0,2
20 Can. Ven.	4,7	− 1,4	+ 0,7 + 1,0	F0	F4	F3	F3	F4,1	± 0,2
ε Aurigae	3,4	− 2,0	− 2,9 − 2,7	F5p	F5p	F9	F2	F3,4	± 0,4
β Drac.	3,0	− 3,5	− 2,1 − 1,5	G0	G0p	F8	F9	F9,0	± 0,3
29 Monoc.	4,4	− 4,2	− 2,1 − 2,5	G0	G3p	F9	G0	F8,5	± 0,3
ε Leonis	3,1	− 0,9	—	G0p	G0	G0	—	G0,1	± 0,2
ε Urs. Min.	4,4	− 1,4	—	G5	G2	G2	—	G2,6	± 0,4
o Urs. Maj.	4,5	− 0,8	—	G0	G2	G2	—	G2,5	± 0,4
η Herc.	3,6	+ 0,9	+ 1,1 + 1,0	K0	G5	G5	G6	G5,6	± 0,4
ε Virg.	3,0	0,0	+ 1,4 + 1,4	K0	G6	G6	G6	G5,0	± 0,3
β Herc.	2,8	− 1,0	+ 0,7 + 1,0	K0	G5	G6	G5	G5,1	± 0,4
ε Boötis	2,7	− 0,9	+ 0,5 + 1,0	K0	G8	G7	G6	G6,3	± 0,3

progressive displacements greater than 0,2 Ångströms were: 4041,3, 4048,7, 4052,0, 4101,8, 4195,3, 4222,2, 4270,2, 4289,7, 4298,0, 4344,4, 4385,6, 4395,2, 4398,0, 4442,3, 4450,4, 4464,5, 4468,5, 4470,9, 4476,1, 4482,2, 4501,3, 4629,5, and 4654,6.

This method of classification as applied involved considerable labor but yielded results of high accuracy. It was fully as successful with low dispersion as with high dispersion spectrograms. This was partly due in the cases discussed above to the selection of stars of similar luminosity for the low dispersion studies at Ann Arbor. With single-prism dispersion classifications more accurate than those "estimated" should be made with no great expenditure of time as a by-product of radial velocity studies.

47. Classification of Spectra by Measurement of Effective Wave-Length. SCHWARZSCHILD[1] was the first to determine effective wave-lengths of stellar spectra produced by an objective grating. The determinations were visual and were made in connection with interference measures of double stars. COMSTOCK[2] and LAU[3] subsequently published results of their determinations of effective wave-lengths by similar methods. COMSTOCK measured effective wave-lengths of fifty-four stars, LAU of sixty-eight stars, and both published a correlation of the same with spectral type. The table by LAU showing the relation between visual effective wave-length and spectral class follows:

Spectral Class	Effective Wave-Length COMSTOCK	LAU
B	565,3 μμ	565,3 μμ
A	564,0	565,9
FG	568,2	570,8
K	573,8	576,8
M	573,4	584,7

It was probably anticipated that much greater changes in effective wave-length with spectral class would be observed photographically.

[1] A N 139, p. 353 (1896). [2] Ap J 5, p. 26 (1897). [3] A N 173, p. 81 (1906).

The HENRY brothers[1] employed a coarse grating placed before an objective to study diffraction images photographically in connection with atmospheric dispersion. HERTZSPRUNG[2] employed a grating camera (aperture, 81 mm; focal length, 1230 mm; grating constant, 2,802 mm) on the same principle to measure the effective wave-length of diffraction images of stars, pointing out the importance of reducing measured effective wave-length to standard image intensity. He published a partial comparison of effective wave-length with Harvard spectral class. With results similar to HERTZSPRUNG'S, BERGSTRAND[3] measured effective wave-lengths of a few stars with the 33 cm Upsala refractor under a grating having a constant of 8,04 mm. Later BERGSTRAND[4] employing the Meudon reflector, derived a correlation for ordinary plates between effective wave-length and spectral class as shown in column 2 of Table 19. With orthochromatic plates he observed a wider range of effective wave-length. He noted that the stars could be divided sharply into "white" and "yellow" stars.

HERTZSPRUNG, using the STEINHEIL astrograph at Copenhagen, determined the effective wave-length (reduced to standard image intensity) of a number of stellar spectra, especially in clusters[5], reducing his results to the zenith. With a coarse grating (6 mm interval) in front of the Mount Wilson 60-inch reflector stopped down to 40 inches HERTZSPRUNG[6] observed the effective wave-length of a number of absolutely faint stars and of cluster stars. The correlation between spectral class and effective wave-length, obtained by plotting HERTZSPRUNG'S measures, is shown in column 3 of Table 19 for the cluster stars as far as Class F5 and for the intrinsically faint stars. The effective wave-lengths of the cluster stars, in parentheses, appear to depart from those of the intrinsically faint stars after Class F5.

Table 19. Correlation between Harvard Spectral Class and Effective Wave-Length.

Class	Effective Wave-Length				
	BERGSTRAND	HERTZSPRUNG	LINDBLAD	WOLF	Greenwich
B0	415 μμ	4190 ± A	408 μμ	416 μμ	4130 ± A
B5		4215		420	4176
A0	422	4238	413	425	4250
A5		4263		429	4269
F0	429	4287	420	433	4270
F5		4312		438	4278
G0	436	4350 (4395)	428	442	4315
G5		4392		447	4394
K0	443	4440 (4560)	433	452	4465
K5		4488		457	4539
M	450	4530	444	457	4539 ±
N	460—470			471	4572 ±

BERGSTRAND and LINDBLAD[7], using a twin ZEISS triplet (aperture, 15 cm; focal length, 150 cm) with a wire grating before one of the objectives (grating constant, 1,3422 mm) determined the effective wave-length of fixed stars. LINDBLAD[8], using the same instrument, derived a correlation between spectral class and effective wave-length similar to that given in column four of Table 19. Only

[1] Conférence Astrographique, Circ. 8 (1901).
[2] B A 25, p. 5 (1908).
[3] A N 177, p. 241 (1908).
[4] Nova Acta R S Scient Ups (4) 2, No. 4 (1909); C R 148, p. 1079 (1909).
[5] Publ Astrophys Obs Potsdam 22, No. 63 (1913).
[6] Ap J 42, pp. 92, 111 (1915).
[7] Ark Mat Astr Fys 11, No. 17 (1916).
[8] Ap J 46, p. 206 (1917).

the faintest measurable images were considered by him and the results were corrected for selective extinction in our atmosphere. With this correlation the spectral class of several nebulae and globular clusters was determined by LUNDMARK and LINDBLAD[1] employing the equation:

$$\text{Spectral Class} = \text{A0} + \frac{\lambda_{\text{eff}} - 415}{6}.$$

LINDBLAD[2] found effective wave-length to bear a close relation to spectral class, proving on the whole to be analogous to color index. This correlation between effective wave-length and spectral class for faint stellar images is given in column four of Table 19. For Classes G, K and M the values of effective wave-length are valid for the giant series. For dwarfs the values are smaller. For later spectral types the effective wave-length was closely related to absolute magnitude as KAPTEYN, VAN RHIJN, ADAMS, MONK and others had noted. The minimum wave-length visible was also observed by LINDBLAD as a function of spectral type and absolute magnitude. LINDBLAD'S[3] curves connecting spectral type, effective wave-length, and minimum wave-length are shown in Figures 13 and 14. LUNDMARK and LINDBLAD[4] continued the spectral classification of nebulae and clusters by measures of effective wave-length finding very short wave-length (< 416 $\mu\mu$) for Class P.

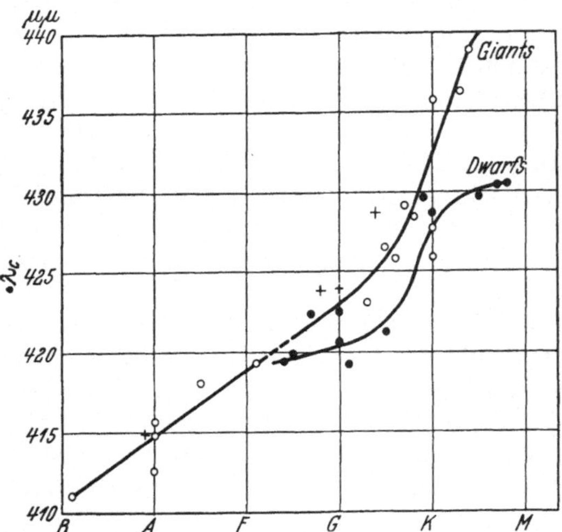

Fig. 13. Correlation of Effective Wave-Length and Spectral Class. (LINDBLAD, Ap J 49, p. 298.)

M. WOLF[5] measured effective wave-lengths of stellar spectra in the range B to M, using an objective grating with constant 2,292 mm over a reflector of 71 cm aperture and 281 cm focal length. Correcting for zenith distance and the exposure factor, WOLF determined effective wave-lengths for twenty stars of known DRAPER Class. The results when plotted yield the values in column 5 of Table 19. The stars used were for the most part of great absolute brightness. Those of Classes M0 and M3 yielded effective wave-lengths smaller than an extension of the data for earlier classes would indicate.

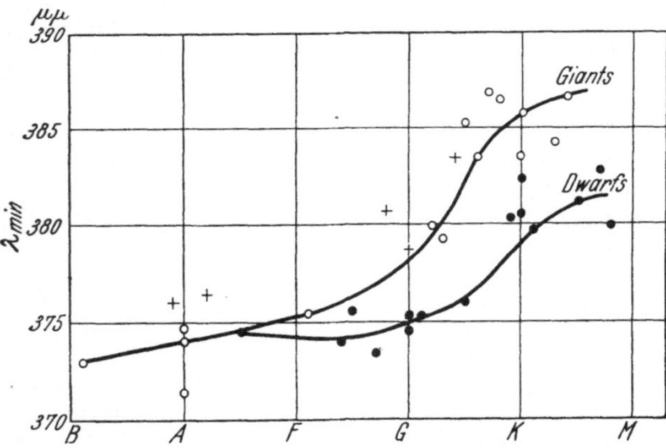

Fig. 14. Correlation of Minimum Wave-Length and Spectral Class. (LINDBLAD, Ap J 49, p. 299.)

[1] A N 205, p. 161 (1917).
[2] Ark Mat Astr Fys 13, No. 26 (1918); Ap J 49, p. 289 (1919).
[3] Ap J 49, p. 289 (1919). [4] Ap J 46, p. 206; 50, p. 376 (1919). [5] A N 213, p. 49 (1920).

Applied to the determination of spectral types of stars of considerable proper motion the results were markedly of too early class for K and M spectra.

Various investigators who have observed effective wave-lengths of stellar spectra have discussed the sources of error involved. Different results are obtained on the same star with different photographic emulsions and with different instruments having different chromatic properties (see BERGSTRAND[1]). Reflectors give results in better agreement than refractors. With the same instrument differences in focus (see ROSENBERG[2]), density of image, zenith distance and atmospheric disturbances lead to differences in measured effective wave-length. To study and compare these errors BERGSTRAND and ROSENBERG[3] proposed a list of twenty-five standard stars ranging from Class B2 to M and in color index from $-0{,}46$ to $+1{,}87$. These stars are circumpolars chosen from the Yerkes Actinometry in the photographic magnitude range from 4,22 to 6,97. A star of Class N and color index $+2{,}72$ was also recommended for observation.

MARTIN and DAVIDSON[4] used a 30-inch reflector of 3476 mm focus stopped down to 20 inches with a grating constant of 2,846 mm to observe effective wave-lengths. Their purpose was to test the usefulness of effective wave-lengths in the determination of spectral class and color index of faint stars. Studies of circumpolar stars including the standard stars of BERGSTRAND and ROSENBERG yielded a correlation between wave-length and spectral class similar to that found in the last column of Table 19. DYSON and MARTIN[5] studied effective wave-lengths with ordinary and stained plates with and without a screen as BERGSTRAND[6] had done. Their correlation between effective wave-length measured on Seed Gilt Edge plates and spectral class was very similar to the more complete Greenwich result[7] in the last column of Table 19. The effective wave-length changed rapidly with spectral type from B0 to A0 and from G0 to M, but from A0 to G0 the change was small. With a panchromatic plate and suitable screen they observed a very sensitive and approximately linear relationship between effective wave-length and spectral class over the range B8 to M. They did not find effective wave-length a sufficient criterion of luminosity in particular cases, but there was clear evidence that the redder stars are of greater luminosity than the bluer stars of the same spectral class.

The extent to which luminosity must be taken into account in the determination of spectral class at K0 by measures of effective wave-length is brought out by the work of BALANOWSKY[8] who found for the spectral range G5 to K3 that the relations between effective wave-length, reduced to K0, and spectroscopic absolute magnitude Ms could be expressed by the formulae:

$$\text{(Giants)}\quad \lambda_{\text{eff.}} = 429{,}2\,\mu\mu - 1{,}70\,\text{Ms}.$$

$$\text{(Dwarfs)}\quad \lambda_{\text{eff.}} = 423{,}8\,\mu\mu - 0{,}33\,(\text{Ms} - 5{,}0).$$

J. H. PETERSON[9] employing the apparatus and technique used by LINDBLAD studied the effective wave-lengths of photographic grating-spectra of 1020 stars in a Milky Way region in Aquila. The relation between effective wave-length and spectral type was quite similar to that of LINDBLAD. It appeared from this investigation that distinction between white and red stars could be made successfully by measures of effective wave-length but scarcely between the successive

[1] SEELIGER-Festschr p. 386 (1925).
[2] A N 213, p. 329 (1921).
[3] A N 215, p. 447 (1922).
[4] M N 82, p. 65 (1921); 84, p. 428 (1924).
[5] M N 86, p. 28 (1925).
[6] Ap J 33, p. 130 (1911).
[7] Determination of Effective Wave-Lengths of Stars at Greenwich. 1925.
[8] A N 226, p. 393 (1925).
[9] Ark Mat Astr Fys 19a, No. 3 (1925).

Harvard spectral classes, a conclusion which BERGSTRAND[1] had reached in a similar program with the same instrument.

G. EBERHARD[2] discussed the sources of the discrepancies between the effective wave-length scales of different observers. He pointed out that the color curve of the objective, focussing, field curvature, atmospheric disturbances, disturbances in the optical parts (especially in reflectors), measuring practice (whether bisecting the spectral image or setting on the center of density or otherwise) could not account for the discrepancies observed. He studied errors involved in the photographic process and concluded that the photographic plate exercised an important complicated influence on the measured effective wave-length.

The measures of effective wave-length in Table 19 were secured with reflecting telescopes except for those of LINDBLAD in column four. The agreement among reflector measures of effective wave-length is found to be good especially where the dispersion due to giant and dwarf branching is taken into account in spectra of later type; but with the reflector the usable field is relatively limited. With the refractor, measured effective wave-lengths are in a great degree characteristic of the chromatic properties of the objective and are sensitive to focal changes. With either type of telescope, valuable though rough determinations of spectral class of faint objects can be made.

48. Determination of Spectral Class by Measures of Color Index and Heat Index. The relations among stellar spectral class, color index, and heat index are considered elsewhere in this Handbuch. Applying these known relations, the determination of spectral class by measures of color index, though complicated by the uncertainties introduced in later classes by dependence upon absolute luminosity, can be effected advantageously in studies of faint objects. A well known example of this is found in SHAPLEY's studies of faint stars in globular clusters[3]. Following the suggestion of SEARES[4], SHAPLEY used the letters b, a, f, g, k, m to designate classes of color, defining the latter so that, under certain average conditions, stars of spectral Class B will have the color b, A corresponds to a, etc. The color classes were referred to as hypothetical spectra which for any given star might deviate from the mean, due to changes in spectral intensity dependent on absolute magnitude, extent of atmosphere, etc. For the mean of a large number of stars the relation between spectrum and color index for the Mount Wilson 60-inch reflector was taken as

$$\text{Color Index} = 0{,}4 \times \text{spectrum},$$

where the color index was expressed in magnitudes and the "spectrum" had the values -1, 0, $+1$, $+2$, etc., corresponding to B0, A0, F0, G0, etc. HERTZSPRUNG[5] derived a typical equation connecting color index (C. I. = Gött. ptg—Harv. vis. magn.), spectral class (Sp), and spectroscopic absolute magnitude (Ms) where Class G is taken as 2, K as 3, all in the range G6 to K7, as follows:

$$\text{C. I.} = +0{,}721\,(\text{Sp} - 3{,}05) - 0{,}0474\,(\text{Ms} + 2^{\text{m}}{,}655).$$

HERTZSPRUNG[6] discussed a series of objective grating plates of the Pleiades made with the Mount Wilson 60-inch reflector. These plates afforded 9972 effective wave-lengths of 1246 stars. This material was used for a more thorough determination of the reduction to standard image intensity and comparisons were made with the results of others. With the improved corrections for image intensity,

[1] A N (Jubiläumsnummer), p. 3 (1921).
[2] SEELIGER-Festschr., p. 115 (1925).
[3] Mt Wilson Contr Nos 115, 116 (1915); 117 (1916).
[4] Wash Nat Ac Proc 1, p. 481 (1915).
[5] A N 208, p. 271 (1918).
[6] Mem R Ac Copenhagen (8) 4, No. 4 (1925).

effective wave-lengths of 129 stars near the north pole[1] gave the following relation between color index (C. I.) in the Göttingen Actinometry and effective wave-length ($\lambda_{\text{eff.}}$) (see similar formula for Harvard color index[2]),

$$\text{C. I.} = \frac{1}{200}(\lambda_{\text{eff.}} - 4216\,\text{A}).$$

The color index of any star (photographic—visual magnitudes) depends for its particular value on many factors. For any combination of telescope, photographic plate, and light filter the results may differ widely from those secured under different conditions. But such results can be brought into close accord by application of the proper factors. Mean values of color index corresponding to the DRAPER Classes are:

Spectrum	C. I.	Spectrum	C. I.	Spectrum	C. I.
B0	− 0,33	gG0	+ 0,67	dG0	+ 0,57
B5	− 0,18	gG5	0,92	dG5	0,65
A0	0,00	gK0	1,12	dK0	0,78
A5	+ 0,20	gK5	1,57	dK5	0,98
F0	+ 0,33	gM0	1,73	dM0	1,45
F5	+ 0,47	N	2,7		

From B0 to gM0 these color indices were determined by E. S. KING[3]. The differences between dwarfs and giants are due to SEARES[4] but reduced approximately to the Harvard scale. The value for Class N was determined by PARKHURST[5].

In illustration of actual application of the relation between spectral class and color index to the determination of spectral class of faint stars, a recent study by C. J. KRIEGER[6] of color indices and spectral types in the Scutum cloud may be cited. KRIEGER's correlation between color index and spectral class was secured with a reflector employing CRAMER Instantaneous Isochromatic Plates and WRATTEN K2 and 35 D filters. The color indices are about 35 per cent larger than those of the I. A. U. color system[7]. KRIEGER's correlation follows:

Spectrum	C. I.	Spectrum	C. I.	Spectrum	C. I.
B5	− 0,50	gG0	+ 0,98	dG0	+ 0,75
A0	± 0,00	gG5	+ 1,40	dG5	+ 1,02
A5	+ 0,35	gK0	+ 1,91	dK0	+ 1,38
F0	+ 0,55	gK5	+ 2,42		
F5	+ 0,72	gM0	+ 2,92		

The correlation for giant stars between heat indices (visual—radiometric magnitudes) determined by PETTIT and NICHOLSON[8] is as follows:

Spectrum	H. I.	Spectrum	H. I.
A0	0,0	M0	1,8
F0	0,1	M2	2,2
G0	0,5	M4	2,9
K0	1,0	M6	4,2
K5	1,6	M8	5,0
S Me variables at min (M10)			8,5

These results refer to the zenith at Mount Wilson, and represent observations made with a thermocouple provided with a rock salt window and a telescope involving two reflections from fresh silver.

[1] Ap J 55, p. 370 (1922) = Mt Wilson Contr 231.
[2] Ap J 42, p. 96 (1915).
[3] Harv Ann 76, p. 125 (1915).
[4] Ap J 55, p. 198 (1922) = Mt Wilson Contr 226.
[5] Ap J 36, p. 218 (1912).
[6] Lick Bull 14, p. 95 (1929).
[7] Trans I A U 1, p. 69 (1922).
[8] Trans I A U 3, p. 150 (1928).

Well determined mean correlations between spectral classes and reciprocal temperatures are given by HERTZSPRUNG[1] and BRILL[2].

k) Catalogs of Stellar Spectra.

49. Bibliography. SECCHI issued the first catalog of stellar spectra in October, 1866[3]. This catalog contained 220 stars classified according to one of SECCHI's earlier systems described above. With the extension of the number of stars to 316 this catalog was published in 1867[4] in a memoir by SECCHI under the title, "Sugli Spettri Prismatici dei Corpi Celesti". Subsequently SECCHI adopted the numbering of his spectral types now accepted and in "Die Sterne", p. 90—96 (1878) catalogued 366 stars on this system.

D'ARREST continued SECCHI's survey of stellar spectra at Copenhagen during the years 1871—1875. The resulting classifications were published in the Astronomische Nachrichten, Volumes 84, 85 and 86.

VOGEL and MÜLLER, in 1893, published a catalog containing classifications of 4051 stars with magnitudes to 7,5 between declinations —1° and +20° in the Publikationen des Astrophysikalischen Observatoriums zu Potsdam, Band 3, S. 134. VOGEL's earlier classification was employed and estimated star colors were also tabulated.

DUNÉR, in 1884, published a catalog of stars of VOGEL's divisions III a and III b with totals of 297 and 55 stars respectively in the two divisions. This catalog appeared in the memoir, "Les étoiles à spectres de la troisième classe".

A catalog of 1217 stars of Types IIIa and IIIb was published by SCHEINER and FROST in Astronomical Spectroscopy, p. 400, 1894.

MCCLEAN published in 1898 a photographic catalog of 160 spectra of northern stars in the Philosophical Transactions of the Royal Society, 191, p. 127 and, in 1899, the "Spectra of Southern Stars".

VOGEL and WILSING, in 1899, published a catalog containing classifications of 528 stars in the Publikationen des Astrophysikalischen Observatoriums zu Potsdam, Band 12, S. 11. VOGEL's revised classification was employed.

LOCKYER issued in 1902 his Catalogue of 470 of the Brighter Stars classified in respect to his genera, with types by PICKERING, VOGEL, and MCCLEAN added for comparison. This catalog was extended in the Bulletins of the Hill Observatory, Nos. 3 (1915) and 5 (1916), to include 641 of the fainter stars.

The Harvard catalogs of stellar spectra began with the DRAPER Catalogue (Harvard Annals, Vol. 27) of 1890 which contained classifications of 10351 stars north of —25° declination by Mrs. W. P. FLEMING. This was supplemented in 1897 by classifications of 681 stars in clusters by Mrs. FLEMING. The DRAPER letters were used to characterize spectra. Miss ANTONIA C. MAURY's Catalogue of the Spectra of 681 Stars North of — 30°, classified on her own detailed system, appeared in Harvard Annals, Volume 28, Part I, p. 106. Miss ANNIE J. CANNON's several catalogs of stellar spectra include: (1) Spectra of Bright Southern Stars, in Harvard Annals, Volume 28, Part II, p. 111, containing classification of 1122 stars (1909); Classification of 1477 Stars

[1] Leiden Ann 14, pt. 1 (1922); B A N 1, p. 204 (1923).

[2] Berlin-Babelsberg Veröffentl 5, Heft 1 (1924); Z f Phys 31, p. 726 (1925).

[3] C R 63, p. 621 (1866); 64, p. 345 (Feb. 1867); A N 73, p. 129 (1867); Rep Brit Assoc 1868, p. 166.

[4] C R 64, p. 774 (Apr. 1867); Memorie della Soc Italiana delle Science (3) 1, p. 67 (1867).

by Means of their Photographic Spectra, Classification of 1688 Southern Stars by Means of their Spectra, and the Spectra of 745 Double Stars, in Harvard Annals, Volume 56, p. 71, 113, and 231 (1912).

The HENRY DRAPER Catalogue in Harvard Annals, Volumes 91 to 99, (1918 to 1924) is the final authoritative catalog of stellar spectra. It contains classifications by Miss CANNON of 225300 stars and nebulae on the DRAPER system together with photometric and photographic magnitudes. This catalog resulted from an attempt to include all stars that could be classified on the plates of the HENRY DRAPER Memorial. For southern stars whose spectra were photographed at Arequipa much fainter stars could be classified.

The HENRY DRAPER Catalogue is being extended by Miss CANNON in Harvard Annals, Volume 100, to fainter stars. The zone $+55°$ to $+60°$ (1855) and regions in the Milky Way have been selected for this extension. The total of classifications in the HENRY DRAPER Catalogue has been increased to 240325.

l) Conclusion.

Recent years have witnessed a rapid advance in the analysis of line and band spectra. Almost every line which is of importance to the astrophysicist is now physically intelligible, even to the principal nebular lines of ionized oxygen. Thus the empirical classification of the spectral lines is substantially complete. It remains to develop a rational theory of astrophysical spectra and of stellar spectra in particular. Such a theory should connect the spectrum of a star with fundamental stellar characteristics including effective temperatures, surface gravity and atmospheric composition. But we are still far from being able to do this.

Although a complete theory is still to be formulated, marked progress has been made toward an understanding of the general nature of the Harvard sequence of spectral classes. That this depends chiefly upon changes of ionization with temperature has been shown clearly; that the average pressure in stellar atmospheres is surprisingly low is well established; that the opacity of an ionized gas prevents radiation from reaching us when its source is in the higher pressure layers of a star's atmosphere throws further light on the nature of stellar spectra.

The outcome of all investigations to date has been to confirm the practical value of the empirical system of the DRAPER Catalogue. Recent changes have dealt with spectral types near the extremes of the sequence where the faintness of spectra in some cases has made analysis difficult and where the advance in our knowledge has enabled us to select and interpret criteria related to the temperature-ionization sequence.

The classification of the future will undoubtedly be based on physical principles in addition to temperature-ionization and will be expressed numerically in terms of definite parameters. Such a classification will be based upon the whole range of accessible knowledge and particularly upon detailed and accurate data. In simplified and abridged form it should conform to the requirements of the observer with smaller dispersion. This classification may well be confined to the spectral lines and bands, of whose intensities, structures, widths, contours, wave lengths, etc. more accurate studies should be made. With such material available we can surely apply to good advantage in stellar spectral classification our knowledge of the physical interpretation of spectral features according to energy level, series relationships, magnetic separation and similar phenomena. Further advance may be made by the introduction of additional dimensions or coordinates in the system of classification, such as width of lines, emission features, criteria for absolute magnitude, mass and density. But the color index and, more generally, the energy distribution in the continuous spectrum should be regarded as closely allied properties not normally included in the classification.

For the numerical expression of the behavior of spectral lines at least three parameters will be necessary. They include a quantity based on temperature, another depending on the abundance of neutral atoms, and a third depending on the abundance of ionized atoms. The two latter parameters will differ considerably for different elements. Other parameters will be involved. Indeed it is obvious that a complete numerical specification of the characteristics of a stellar spectrum will not be a simple matter. Our knowledge of the mechanism of radiation must be greatly extended before a complete physical classification of stellar spectra will be possible on a basis of absorption or emission lines and bands.

Kapitel 2.

Zur Statistik der Spektraltypen.

Von

FR. BECKER-Bonn.

Mit 7 Abbildungen.

a) Einleitung.

Die folgenden Paragraphen geben eine kurze Zusammenfassung unserer Kenntnisse über die Häufigkeit und die Verteilung der einzelnen Spektraltypen unter den Sternen. Allerdings muß die Behandlung eines Gebietes, das nicht so sehr der Astrophysik wie der Stellarstatistik angehört und aus diesem Zusammenhange gelöst wird, mehr oder weniger fragmentarisch bleiben; unsere Ausführungen sind daher lediglich als Ergänzung zu dem vorhergehenden Kapitel ,,Classification and Description of Stellar Spectra" zu betrachten.

1. Vorbemerkung. Das Klassifizierungssystem ist in allen hier besprochenen Untersuchungen das Harvard-System. Die Vergleichung der einzelnen Ergebnisse untereinander wird aber häufig dadurch erschwert, daß keine einheitliche Regel für die Gruppierung der Spektralklassen existiert. So ist z. B. mit der Sammelbezeichnung A in einigen Arbeiten der Spektralbereich von A0 bis A9 gemeint, in anderen dagegen der Bereich B8—A3, mit der Bezeichnung F das Intervall F0—F9 oder A5—F2 usf. Soweit der Modus aus den betreffenden Arbeiten ersichtlich ist, werden im folgenden die beiden Arten der Gruppierung durch die Ausdrücke Spektralklasse und Spektralgruppe unterschieden.

2. Das Datenmaterial. Wir geben zunächst eine kurze Übersicht über das für statistische Untersuchungen heute zur Verfügung stehende Beobachtungsmaterial. In Tabelle 1 sind die hauptsächlichsten allgemeinen Kataloge wie auch die auf spezielle Typen oder Himmelsgegenden beschränkten Verzeichnisse von photographischen Sternspektren zusammengestellt.

Tabelle 1. Die wichtigeren Kataloge und Verzeichnisse von Sternspektren.

Autor	Veröffentlicht	Jahr	Zahl der Sterne	Bemerkungen
PICKERING	HA 27	1890	10498	The DRAPER Catalogue of Stellar Spectra.
MAURY	HA 28, I	1897	680	Spectra of Bright Stars.
CANNON	HA 28, II	1901	1122	Spectra of Bright Southern Stars.
PICKERING u. a.	HA 50	1908	9110	Revised Harvard Photometry.
CANNON	HA 56, IV	1912	1477	Nordhimmel.
CANNON	HA 56, V	1912	1688	Südhimmel.
FLEMING	HA 56, VI	1912	1761	Peculiar Spectra (P, O, N, R, var. usw.).

Tabelle 1. (Fortsetzung.)

Autor	Veröffentlicht	Jahr	Zahl der Sterne	Bemerkungen
CANNON	HA 56, VII	1912	745	Spectra of Double Stars.
ADAMS, KOHLSCHÜTTER	MWC 105	1915	500	Spektren und Radialgeschwindigkeiten.
CANNON	HA 76, III	1916	287	Spectra having Bright Lines.
CANNON, PICKERING	HA 91–99	1918–24	225300	HENRY DRAPER Catalogue of Stellar Spectra.
ADAMS, JOY, BURWELL	MWC 199	1921	1646	Spektren und sp. Parallaxen.
ADAMS, JOY	MWC 244	1922	544	A-Sterne (sp. Parallaxen).
MERRILL	MWC 252	1922	22	S-Sterne.
ADAMS, JOY	MWC 258	1923	1013	Spektren und sp. Parallaxen.
J. S. u. H. H. PLASKETT, HARPER, YOUNG	Vict. II, 1	1924	594	Spektren und Radialgeschwindigkeiten.
MERRILL, HUMASON, BURWELL	MWC 294	1925	95	Be-Sterne
LUYTEN	MN 86, 48	1925	84	M-Zwerge (Spektren und andere Daten).
CANNON	HA 100, I–IV	1925–28	16200	HENRY DRAPER Extension.
ADAMS, JOY, HUMASON	MWC 319	1926	410	M-Sterne (Spektren und sp. Parallaxen).
YOUNG, HARPER	Vict. III, 1	1927	1105	Spektren und sp. Parallaxen.
MERRILL	MWC 334	1927		Sterne mit Fe-Emission.
MAXWELL	LOB 390	1927	557	Cygnusregion.
SCHALÉN	Upsala 37	1928	4300	B- und A-Sterne in Milchstraßengegenden.
BECKER	Potsd. Publ. 88	1929	3639	KAPTEYN-Felder, Zone −75° und Südpol.
BECKER	Potsd. Publ. 89	1930	11753	KAPTEYN-Felder, Zone −60°.
PAYNE	HB 878	1930	238	O-Sterne.
SCHWASSMANN	Hamb. Mitt. 6; 31	1930	1238	Plejaden.
ÖHMAN	Upsala 48	1930	882	B-, A-, F-Sterne; Polzone.
BECKER	Potsd. Publ. 90	1931	ca. 18000	KAPTEYN-Felder, Zone −45°.

Abkürzungen: HA = Harvard Annals; HB = Harvard Bulletin; LOB = Lick Obs. Bulletin; MWC = Mt. Wilson Contributions; Vict. = Victoria Publications.

b) Statistik auf Grund der scheinbaren Helligkeit und Verteilung der Sterne.

3. Ältere Untersuchungen. Die Vollendung des ersten DRAPER-Kataloges im Jahre 1890 bot zum ersten Male die Möglichkeit einer umfassenden Statistik der Spektraltypen. E. C. PICKERING[1] führte sie durch, indem er die relative Häufigkeit der einzelnen Spektralklassen sowie ihre Verteilung am Himmel und nach der scheinbaren Helligkeit der Sterne untersuchte. Da jedoch das Klassifizierungssystem damals noch erhebliche Mängel an sich trug, konnte auch die Verteilung der Spektren nur in rohen Umrissen zum Vorschein kommen.

[1] Harv Ann 26, S. 145 (1891).

Zuverlässigere Ergebnisse brachte eine im Jahre 1912 vom Harvard-Observatorium veröffentlichte Statistik[1], bei der das Hauptgewicht auf die Verteilung der Spektralklassen nach galaktischer Breite gelegt war. Es zeigte sich schon die später noch klarer erkannte Erscheinung, daß die B- und A-Sterne die Milchstraße bevorzugen, die F-, G- und K-Sterne dagegen ziemlich gleichmäßig über den ganzen Himmel verteilt sind. Das Gesamtergebnis der Untersuchung, die sich bis zu den Sternen der 7. Größe erstreckte, ist in 12 Diagrammen übersichtlich dargestellt.

4. Statistische Auswertung des Henry-Draper-Kataloges. Der Draper-Katalog ist am Nordhimmel bis zur Größe $8^m,25$, am Südhimmel bis $8^m,75$ vollständig; innerhalb dieser Grenzen gibt er ein klares und endgültiges Bild der Verteilung und relativen Häufigkeit der einzelnen Spektralklassen. Über 99% der Draper-Sterne gehören den Klassen B bis M an, diese sind daher in erster Linie Gegenstand statistischer Untersuchungen geworden.

Eine ausführliche Diskussion des Draper-Kataloges haben Shapley und seine Mitarbeiter am Harvard-Observatorium durchgeführt; die Ergebnisse sind in verschiedenen Harvard-Zirkularen von 1921 bis 1925 mitgeteilt. Etwas später (1927) veröffentlichte Charlier[2] eine erneute Abzählung der Draper-Sterne. Er teilte die Sphäre in 48 gegen die galaktische Ebene orientierte Felder und bestimmte die Verteilung der Spektralklassen auf die einzelnen Felder, einmal für sämtliche Sterne des Draper-Kataloges, dann für die Sterne heller als $8^m,0$ gesondert. Die Veröffentlichung beschränkt sich auf die Wiedergabe des Zahlenmaterials. Schließlich liegt noch eine besonders eingehende Statistik der Sterne bis zur 7. Größe vor, die O. Seydl[3] in Prag geliefert hat (1929); diese Arbeit umfaßt außer den Zahlentabellen eine Reihe von Hemisphärenkarten, die in Schraffendarstellung die Verteilung der Spektralklassen am Himmel zeigen.

In den drei folgenden Abschnitten sind die Ergebnisse dieser verschiedenen Untersuchungen im einzelnen besprochen.

5. Die scheinbare Häufigkeit der einzelnen Spektralklassen. Tabelle 2 gibt zunächst nach Charlier die Anzahl der Sterne jeder Spektralklasse, und zwar in der 2. Zeile unter Einschluß aller Sterne des Draper-Kataloges, in der 3. Zeile nur für die Sterne heller als $8^m,0$. Da der Katalog bis zur Größe $8^m,25$ vollständig ist, sind nur die Zahlen der letzten Zeile als endgültig anzusehen.

Tabelle 2.
Häufigkeit der Spektralklassen nach dem Draper-Katalog. Absolute Anzahl.

Klasse	P	O	B—B5	B8/9	A	F	G	K	M	N	R	S	Kont.	Pec.	Σ
Alle Sterne	112	170	3675	13111	57850	42988	41999	60620	4397	149	84	3	62	85	225305
Heller $8^m,0$	—	**70**	**1800**	**3383**	**9344**	**7309**	**5332**	**13012**	**1305**	**11**	**7**	**2**	—	**3**	**41578**

Stärkere Aufteilung des Materials nach Größenklassen bietet eine Arbeit von Shapley und Cannon[4]. Allerdings sind hier nicht wie bei Charlier alle Sterne einzeln abgezählt, sondern nur die Sternanzahl in den ersten fünf Minuten jedes Intervalles von 40 Minuten Rektaszension bestimmt und die Ergebnisse mit 8 multipliziert. Die einzelnen Spektralklassen werden der Übersichtlichkeit halber wie folgt in größere Gruppen zusammengefaßt:

Gruppe	Klassen
B	B0, B1, B2, B3, B5
A	B8, B9, A0, A2, A3
F	A5, F0, F2
G	F5, F8, G0
K	G5, K0, K2
M	K5, Ma, Mb, Mc

[1] Harv Ann 56, Nr. 1 (1912). [2] Lund Medd Série II, Nr. 36 (1927).
[3] Publ Obs National de Prague Nr. 6 (1929). [4] Harv Circ 226 (1921).

Diese Gruppen sind willkürlich gewählt und bedeuten nicht etwa gleiche Abschnitte hinsichtlich irgendeines physikalischen Argumentes, wie des Farbenindex oder der Oberflächentemperatur der Sterne.

Nachstehende Tabelle gibt die relative Häufigkeit dieser Spektralgruppen in Prozenten für verschiedene Größenintervalle nach den SHAPLEY-CANNONschen Abzählungen.

Tabelle 3.
Prozentuale Häufigkeit der Spektralgruppen für verschiedene Größenintervalle.

Spektralgruppe	Helligkeitsintervalle						
	< 6,25	6,26–6,75	6,76–7,25	7,26–7,75	7,76–8.25	8,26–8,75	< 8,75
B	10,9	5,7	3,7	2,6	1,7	1,2	2,5
A	30,5	31,6	29,6	26,9	24,9	26,0	26,7
F	10,4	11,3	11,6	11,5	11,0	10,7	11,0
G	9,9	11,4	15,5	15,0	16,8	19,2	16,7
K	30,1	32,7	32,4	35,6	38,2	35,5	35,4
M	8,1	7,3	7,0	8,5	7,4	7,3	7,6

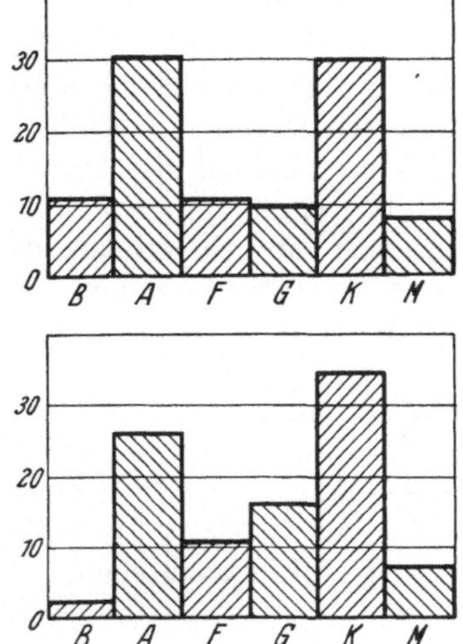

Abb. 1. Prozentuale Häufigkeit der Spektralgruppen B bis M für die Sterne heller als $6^m,25$ (oben) und heller als $8^m,75$ (unten). (Nach SHAPLEY und Miss CANNON, Harv Circ 226.)

Weitaus am häufigsten sind also in allen Helligkeitsintervallen bis $8^m,75$ die A- und K-Sterne. Mit abnehmender Helligkeit nimmt der Prozentsatz an G-Sternen beträchtlich zu, die Zahl der B-Sterne ab, während die Häufigkeit der Gruppen A, F, K und M nur geringe Änderungen zeigt. Die Zahlen für die Sterne heller als $6^m,25$ und heller als $8^m,75$ sind in Abb. 1 bildlich wiedergegeben; die beiden Diagramme können als beste zur Zeit erreichbare Darstellung der spektralen Mischung der helleren Sterne gelten.

6. Galaktische Verteilung der Spektralklassen. Es ist zu erwarten, daß die Milchstraße als die große Symmetrieebene unseres Sternsystems ihren Einfluß auch in der Verteilung der Spektralklassen am Himmel geltend macht.

Ein summarisches Bild der Spektralverteilung als Funktion der galaktischen Breite gibt Tabelle 4, die mit einigen Kürzungen der Arbeit SEYDLS[1] entnommen ist. Sie gilt für die Sterne heller als $7^m,0$ und zeigt die Anzahl der Sterne pro 100 Quadratgrad für Zonen von 10° in galaktischer Breite.

Tabelle 4. Verteilung der helleren Sterne ($< 7^m,0$) nach galaktischer Breite. Die am Kopf jeder Spalte angegebene Gradzahl bezeichnet den Mittelparallel einer 10° breiten Zone.

Spektralklasse	Galaktische Breite											
	+85°	+65°	+45°	+25°	+15°	+5°	−5°	−15°	−25°	−45°	−65°	−85°
B	0,0	0,2	0,1	1,5	2,9	7,0	11,4	5,6	2,0	0,6	0,5	0,3
A	8,0	6,6	6,6	10,5	14,0	17,3	21,4	16,3	12,4	6,6	4,2	3,9
F	3,2	3,7	3,5	3,9	3,6	4,3	4,8	4,1	4,2	3,3	2,4	2,2
G	4,5	3,2	2,9	4,3	3,9	4,5	4,7	3,9	4,5	3,8	4,1	1,6
K	9,3	8,0	9,8	12,1	13,8	14,0	14,3	13,3	12,3	10,7	12,7	8,7
M	1,3	2,3	2,8	3,0	2,6	3,4	3,5	2,8	2,3	1,3	1,7	1,6

[1] Publ Obs National de Prague Nr. 6 (1929).

Nach dieser Übersicht wächst mit Annäherung an die galaktische Ebene die Zahl der B- und A-Sterne stark an; viel geringer, aber noch merklich, ist die galaktische Konzentration der F-, K- und M-Sterne, während sich in der Verteilung der G-Sterne keinerlei Beziehung zur Milchstraße ausprägt. Ferner ergibt sich, daß das Häufigkeitsmaximum der B- und A-Sterne nicht in der galaktischen Ebene selbst, sondern einige Grade südlich davon erreicht wird; diese Sterne bilden den Kern des sog. lokalen Systems, dessen Zentralebene gegen die Ebene der Milchstraße geneigt ist. Vgl. auch unten Ziff. 7.

Bis zur Vollständigkeitsgrenze des DRAPER-Kataloges reichen SHAPLEYS Untersuchungen[1], die sich auf Abzählungen der Sterne in ausgewählten, über den ganzen Himmel verteilten Feldern stützen. Tabelle 5, die alle Sterne heller als $8^m,25$ einschließt, zeigt für verschiedene galaktische Breiten die Anzahl der Sterne pro 100 Quadratgrad für die in Ziff. 5 definierten Spektralgruppen.

Eine bildliche Darstellung dieser Zahlen geben die sechs oberen Diagramme der Abb. 2. Sie bestätigen mit kleinen Abweichungen die schon aus Tabelle 4 für die helleren Sterne gezogenen Schlüsse, nämlich:

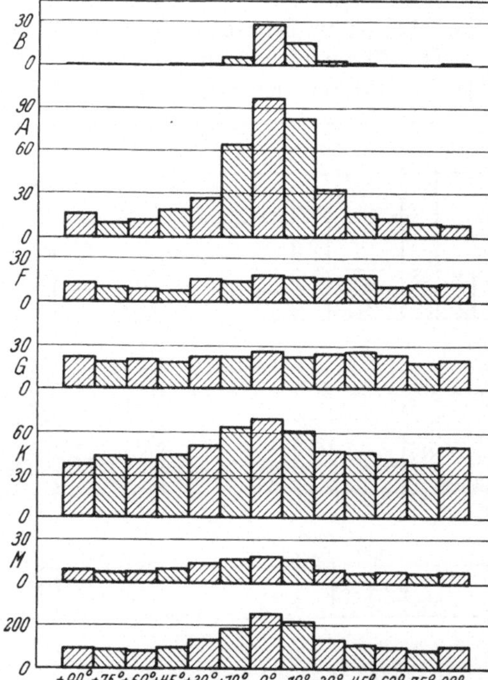

Abb. 2. Durchschnittliche Anzahl der Sterne heller als $8^m,25$ pro 100 Quadratgrad in verschiedenen galaktischen Breiten und Spektralgruppen. Das unterste Diagramm gilt für die Sterne der 6 Spektralgruppen zusammengenommen. (Nach SHAPLEY, Harv Circ 248.)

starke galaktische Konzentration der B- und A-Sterne,
mäßige galaktische Konzentration der K- und M-Sterne,
sehr geringe oder keine Konzentration der F- und G-Sterne.

Tabelle 5. Durchschnittliche Zahl der Sterne pro 100 Quadratgrad für verschiedene galaktische Breiten.

Galakt. Breite Grad	Sterne heller als $8^m,25$					
	B	A	F	G	K	M
+ 90	0	17	14	22	38	8
− 90	1	9	13	19	51	8
+ 75	0	10,3	11,7	17,7	43,3	7,3
− 75	0	10,0	13,0	17,0	38,3	6,7
+ 60	0	11,8	10,2	18,8	40,0	7,5
− 60	0	12,8	9,5	23,0	41,8	8,0
+ 45	0,5	18,5	9,2	17,8	44,8	8,2
− 45	1,0	15,8	19,2	23,8	45,8	7,2
+ 30	0,5	27,0	16,2	22,3	49,8	13,7
− 30	2,2	32,7	15,8	23,5	47,2	8,5
+ 10	5,7	65,4	15,7	22,2	63,7	15,7
− 10	16,1	82,7	18,1	22,0	61,1	15,9
0	29,7	96,9	18,7	26,0	69,0	17,5

[1] Harv Circ 248 (1923).

Auffallend ist nach SHAPLEY der scharfe Kontrast in galaktischer Konzentration zwischen den B8—A3- und den A5—F2-Sternen, worauf wir unter Ziff. 7c noch kurz zurückkommen werden.

Abb. 3 gibt das Zahlenmaterial noch einmal in anderer Gruppierung, die Häufigkeiten in Prozenten ausgedrückt.

Die Unterschiede in galaktischer Konzentration zwischen den einzelnen Klassen erklären sich wenigstens teilweise daraus, daß die K- und M-Sterne des DRAPER-Kataloges fast ausschließlich Giganten, also ebenso wie die B- und A-Typen absolut helle Sterne sind, während die Gruppe der F- und G-Sterne vorwiegend aus lichtschwachen Zwergen besteht. Bei gleicher scheinbarer Helligkeit füllen also die B-, A-, K- und M-Sterne einen weit größeren Raum aus als die F- und G-Sterne; sie bringen daher die flach geschichtete Form des Sternsystems klarer zum Ausdruck als die Sterne in dem kleinen Bereich rings um die Sonne, wo die Flächen gleicher Sterndichte im Durchschnitt nahezu sphärisch sind. Zum Teil sind aber die Unterschiede sicher reell; die A- und K-Sterne des DRAPER-Kataloges z. B. haben im Durchschnitt fast dieselbe absolute Helligkeit und Entfernung, aber verschiedene galaktische Konzentration, und ebenso ist bekannt, daß die B-Sterne nicht nur scheinbar, sondern wirklich ein überwiegend galaktisches Phänomen sind.

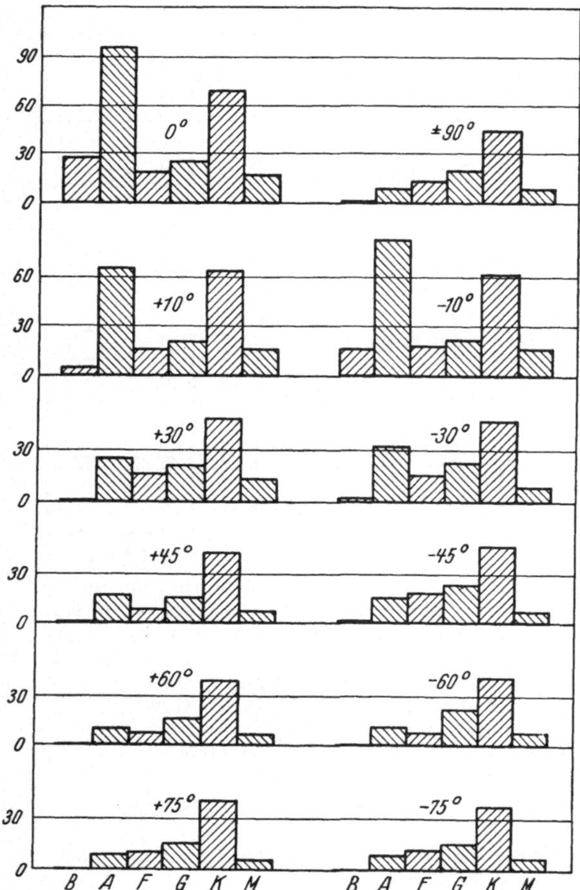

Abb. 3. Prozentuale Häufigkeit der Sterne heller als $8^m,25$ in Abhängigkeit von der galaktischen Breite und vom Spektraltypus. (Nach SHAPLEY, Harv Circ 248.)

Besondere Abzählungen längs des Milchstraßengürtels innerhalb $\pm 10°$ galaktischer Breite zeigen[1], daß von den Milchstraßensternen bis zur 8. Größe 38% der Klasse A, 29% der Klasse K angehören. Die Verteilung der Spektren längs dieser Zone ist ziemlich gleichförmig. Als Gebiet geringerer Sterndichte erscheint für alle Spektralgruppen die Spaltung der Milchstraße zwischen den Sternbildern Aquila und Sagittarius; Zentren größerer Häufigkeit sind die Cygnusregion für die A-Sterne, die Carinagegend für die B-Sterne. Die schwächeren A- und K-Sterne zeigen außerdem ein Häufigkeitsminimum in der ganzen Taurus-Sternleere, während hier die helleren Sterne dieser Klasse sowie die F- und G-Sterne unbeeinflußt bleiben.

7. Untersuchungen über einzelne Spektralklassen. a) Die B-Sterne. Die scheinbare Verteilung der B-Sterne des DRAPER-Kataloges (B0—B5) bis zur

[1] SHAPLEY, Harv Circ 240 (1922).

Vollständigkeitsgrenze $8^m,25$ ist von SHAPLEY und Miss CANNON[1] untersucht worden. Dabei zeigte sich eine starke Abhängigkeit der Verteilung von der scheinbaren Helligkeit. Teilt man die Sterne in vier Gruppen, nämlich

Gruppe 1	St. heller als $5^m,26$	346 Sterne
,, 2	5,26–6,25	367 ,,
,, 3	6,26–7,25	564 ,,
,, 4	7,26–8,25	719 ,,

so ordnen sich die Sterne der ersten Gruppe längs eines größten Kreises, der um etwa 15° gegen die galaktische Ebene geneigt ist. Die Sterne der folgenden Gruppen konzentrieren sich mit abnehmender Helligkeit immer stärker gegen die Milchstraßenebene selber, so daß in der letzten Gruppe 90% der Sterne innerhalb ± 10° galaktischer Breite liegen. Die allmähliche Drehung der Symmetrieebene ist in der zitierten Arbeit in vier Diagrammen veranschaulicht, sie kommt auch in nachstehender Tabelle 6 deutlich zum Ausdruck.

ıbelle 6. Galaktische Längen und Breiten der B-Sterne für vier Helligkeitsgruppen.

. Länge.	20°	60°	100°	140°	180°	220°	260°	300°	340°
tl. Breite, Gruppe 1	−2°,0	−1°,5	−5°,0	−15°,5	−14°,0	−7°,0	−1°,5	+10°,5	+14°,5

. Länge	15	45°	75°	105°	135°	165°	195°	225°	255°	285°	315°	345°
tl. Breite, Gruppe 2	+10°,0	+6°,5	−3°,5	−5°,0	−13°,5	−15°,0	−7°,5	−5°,5	−0°,5	−5°,0	0°,0	−3°,5
,, ,, 3	−0,5	0,0	−0,5	−3,5	−4,0	−11,5	−5,0	−5,0	−0,5	0,0	−3,5	−3,0
,, ,, 4	−3,5	−0,5	0,0	−2,5	+1,5	−2,5	−3,5	−1,5	−0,5	−1,5	−1,0	−1,5

Die Streuung der Einzelwerte ist ziemlich stark, auch zeigen die Sterne eine unverkennbare Tendenz zur Bildung getrennter Häufigkeitszentren, unter denen besonders die Oriongruppe hervortritt.

In höheren Breiten sind B-Sterne selten. Innerhalb 40° galaktischer Poldistanz verzeichnet der DRAPER-Katalog auf beiden Hemisphären insgesamt 28 Objekte, die SHAPLEY und CANNON in einer besonderen Liste zusammengestellt haben.

Die Verteilung der heißeren B-Sterne (B0—B3) und der O-Sterne mit Rücksicht auf die verschiedenen Strukturelemente der Milchstraße (Sternwolken, Sternleeren, diffuse Nebel) hat O. STRUVE[2] diskutiert. Danach findet man im Durchschnitt pro 100 Quadratgrad in den Sternwolken 17,5, in normalen Milchstraßengegenden 12,5, in hellen diffusen Nebeln 30,3 und in den Verdunklungsgebieten 7,3 Sterne der genannten Typen.

b) Die A-Sterne verhalten sich bis zu einem gewissen Grade analog den B-Sternen. Während die Symmetrieebene der schwächeren A-Sterne die galaktische Ebene ist, konzentrieren sich die helleren (bis $6^m,5$) gegen einen größten Kreis, der in demselben Sinne, aber nicht so stark (etwa 5°) wie die Zentralebene der hellen B-Sterne gegen den Milchstraßengürtel geneigt ist[3]. Im Gegensatz zu den B-Sternen ist die Verteilung in galaktischer Länge sehr gleichförmig, die Häufigkeitsabnahme in Richtung der galaktischen Pole weniger ausgeprägt als dort.

c) Die F-Sterne. Die Verteilung der F-Sterne (A5—F2) nach scheinbarer Helligkeit und galaktischer Breite geht aus dem Diagramm Abb. 4 hervor, das einer Untersuchung von SHAPLEY und Miss HOWARTH[4] entnommen ist. Merkliche galaktische Konzentration zeigen nur die Sterne schwächer als 8. Größe; die helleren sind, wie schon aus Abb. 2 zu ersehen war, mit nur geringer Bevor-

[1] Harv Circ 239 (1922).
[2] A N 231, S. 17 (1927).
[3] SHAPLEY u. CANNON, Harv Circ 229 (1922).
[4] Harv Circ 285 (1925).

zugung der Milchstraße fast gleichmäßig über den Himmel verteilt. Die absolut sehr hellen und galaktisch stark konzentrierten F-Giganten sind offenbar zu selten, um einen nennenswerten Einfluß auf die Verteilung der F-Sterne im allgemeinen auszuüben.

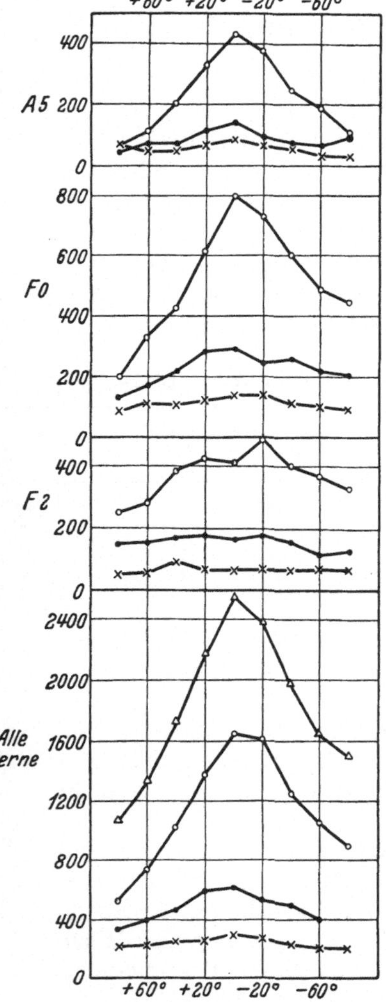

Abb. 4. Verteilung der F-Sterne (A5, Fo, F2) nach galaktischer Breite. Abszissen sind die galaktischen Breiten, Ordinaten die Anzahlen der Sterne für Intervalle von je 20° galaktischer Breite. Kreuze: Sterne heller als 7. Gr., Punkte: Sterne zwischen 7. und 8. Gr., Kreise: Sterne schwächer als 8. Gr., Dreiecke: Sterne aller Größen und der drei Spektraltypen zusammen. (Nach SHAPLEY und Miss HOWARTH, Harv Circ 285.)

Der früher erwähnte Kontrast in galaktischer Konzentration zwischen den Spektralgruppen A und F ist wahrscheinlich auf einen ziemlich steilen Abfall der absoluten Helligkeit von A0 nach F und die dadurch bedingte rasche Abnahme der mittleren Entfernung der Katalogsterne mit fortschreitendem Typus zurückzuführen[1].

d) Die K5- und M-Sterne. Auch die M-Sterne des DRAPER-Kataloges zeigen in ihrer Verteilung kein einheitliches Bild[2]. Von den beiden Gruppen Ma und Mb sind die Mb-Sterne heller als 8. Größe gleichmäßig verteilt, während die Ma-Sterne bis zur 8. Größe zwischen den galaktischen Breiten −10° und +30° um 30% pro Flächeneinheit zahlreicher sind als anderswo. Unterhalb der 8. Größe zeigen sowohl Ma- wie Mb-Sterne eine ausgeprägte galaktische Konzentration. Die Zahl der Mc-Sterne ist zu gering, um sichere Schlüsse zu gestatten. Die langperiodischen Veränderlichen der Klasse Md scheinen sich gegen den Breitengürtel −20° zu häufen, aber auffallender ist bei ihnen die starke Asymmetrie in galaktischer Länge. Die Taurusgegend zählt nur die Hälfte, die entgegengesetzte (Sagittarius-) Region das Doppelte der durchschnittlichen Häufigkeit. Auch die schwächeren Ma-, Mb- und Mc-Sterne zeigen eine ausgeprägte Bevorzugung der Sagittariusgegend.

Diese Ungleichmäßigkeit der Verteilung in galaktischer Länge erhellt auch aus einer Studie von INNA LEHMANN-BALANOWSKAJA[3] über die K5- und M-Sterne bis zur 8. Größe. Ihre Abzählungen ergeben Häufungsstellen in Richtung der galaktischen Längen 50° und 230°, ein tiefes Minimum zwischen 90° und 170° (Taurusgegend).

Tabelle 7. Anzahl der M- und K5-Sterne bis 8. Größe pro 100 Quadratgrad in verschiedenen Gürteln galaktischer Breite.

β	M	K5
0° ± 20°	4,75	9,25
± 20 ± 40	3,92	6,43
± 40 ± 60	3,65	4,48
± 60 ± 80	3,78	3,74
± 75 ± 90	4,21	4,12

Die Aufteilung nach galaktischer Breite (Tab. 7) zeigt für die M-Sterne zusammengenommen in Übereinstimmung mit SHAPLEY eine sehr schwache, für

[1] SHAPLEY, Harv Bull 796 (1923). [2] SHAPLEY u. CANNON, Harv Circ 245 (1923).
[3] A N 232, S. 73 (1928).

die K5-Sterne eine stärkere galaktische Konzentration und für beide Klassen ein leichtes Anwachsen der Sternzahlen gegen die Pole hin.

e) Die selteneren Spektraltypen. Von den planetarischen Nebeln (Klasse P) sind die Objekte mit kleinem scheinbaren Durchmesser vorzugsweise in der Milchstraße zu finden, die größeren (wahrscheinlich näheren) dagegen über den ganzen Himmel verteilt.

Die O-Sterne zeigen von allen Spektraltypen die stärkste galaktische Konzentration; läßt man die Objekte in den beiden Magellanschen Wolken beiseite, so liegen von 85, in Harv Ann 76 aufgeführten O-Sternen 82 innerhalb $\pm 10°$ und 69 innerhalb $\pm 5°$ galaktischer Breite.

Etwas geringer, aber immer noch stärker als die galaktische Konzentration der B-Sterne, ist die der N-Sterne, und an fünfter Stelle stehen in dieser Hinsicht hinter den B- und A-Sternen die der Klasse R. Innerhalb des galaktischen Gürtels, d. h. im Durchschnitt etwa in der Zone zwischen $\pm 30°$ Breite, finden sich vom Typus O 100%, vom Typus N 87% und vom Typus R 63% der uns bekannten Sterne dieser drei Klassen[1].

Die S-Sterne, meistens langperiodische Veränderliche, sind zu gering an Zahl für eine statistische Betrachtung.

8. Spektralstatistik der schwächeren Sterne. Obwohl der DRAPER-Katalog zahlreiche Sterne schwächer als 8. Größe enthält, bietet er wegen der mangelnden und noch dazu für die verschiedenen Spektralklassen ungleichmäßigen Vollständigkeit doch keine geeignete Grundlage für die Statistik dieser Sterne. Jenseits der 8. Größe wächst die Zahl der Sterne so rasch an, daß an eine wie der DRAPER-Katalog den ganzen Himmel umspannende Klassifizierung der Spektren überhaupt nicht mehr zu denken ist, vielmehr wird man sich hier mit Stichproben begnügen müssen. Solche Stichproben sind z. B. die einzelnen Teile der „DRAPER Extension"[2], die Auswahlfelder des McCORMICK-Observatoriums[3] und vor allem die 206 KAPTEYNschen Eichfelder, in denen zur Zeit die Spektren schwacher Sterne am Nordhimmel von A. SCHWASSMANN[4] und seinen Mitarbeitern in Bergedorf, am Südhimmel von FR. BECKER[5] und H. BRÜCK in Potsdam klassifiziert werden. Die Vollständigkeitsgrenze liegt hier in den nördlichen Feldern etwa bei der 12,5., in den südlichen, wo die Spektren mit größerer Dispersion aufgenommen wurden, bei der 12. photographischen Größe.

Als auffälligste Veränderung beim Übergang zu schwächeren Sternen zeigt die Statistik der Spektralklassen das starke Zurücktreten der A- und die Zunahme der G-Sterne, und zwar vor allem außerhalb, aber auch innerhalb des galaktischen Gürtels. Ein ungefähres Bild dieser Erscheinung gibt zunächst Tabelle 8, die nach VYSSOTSKY[6] die prozentuale Häufigkeit der Gruppen A (B8—A3) und G (F5—G0) in den Helligkeitsbereichen des HENRY-DRAPER-Kataloges, der DRAPER-Extension und der noch weiter ($11^m,2$ photovisuell) reichenden McCORMICK-Felder für drei verschiedene galaktische Zonen darstellt.

Tabelle 8. Prozentuale Häufigkeit der Klassen A und G für hellere und schwächere Sterne. (DRAPER-Katalog, DRAPER-Extension, McCORMICK-Felder.)

Galakt. Zone	A-Sterne			G-Sterne		
	HDC	DE	McC	HDC	DE	McC
10°	36,6%	34,2%	24,0%	10,9%	14,6%	21,5%
30	23,0	12,0	6,7	17,6	24,2	34,4
60	13,4	6,3	2,9	22,8	30,6	36,0

[1] Vgl. RUFUS, Publ Astron Obs Univ Michigan 2, S. 104 (1916).
[2] Harv Ann 100, Nr. 1—4 (1925/28). [3] Pop Astr 37, S. 394 (1929). [4] V J S 63, S. 269 (1928). [5] Potsd Publ Nr. 88 (1929); Nr. 89 (1930); Nr. 90 (1931). [6] Pop Astr 37, S. 394 (1929).

Von der Spektraldurchmusterung der KAPTEYN-Eichfelder, die nach ihrer Vollendung die wichtigste Quelle für die Spektralstatistik der schwachen Sterne sein wird, sind bisher erst einige der südlichen Zonen veröffentlicht[1]. Obwohl die in diesen Zonen enthaltenen Felder nur einen beschränkten Bereich in galaktischer Länge und Breite umfassen und den Spektren nur genäherte photographische Größen beigegeben sind, läßt doch das vorliegende Material die Verteilung der Spektren in den Hauptzügen schon erkennen. So zeigt Tabelle 9 die Änderung der prozentualen Häufigkeit verschiedener Spektralgruppen mit abnehmender scheinbarer Helligkeit der Sterne. Die Zahlen sind Mittelwerte aus sechs Eichfeldern zwischen den galaktischen Längen 240° und 300° und den galaktischen Breiten −30° und −60°.

Tabelle 9. Prozentuale Häufigkeit der Spektralklassen der Sterne 9. bis 12. phg. Größe (galaktische Zone − 30° bis −60°).

Größen	B8—A4	A5—F1	F2—F8	F9—G4	G5—G8	G9—K4
$9^m,1 - 10^m,0$	7,0	17,5	21,1	14,0	19,3	21,1
10 ,1 − 11 ,0	2,7	8,1	22,8	22,8	16,8	26,9
11 ,1 − 12 ,0	0,0	2,9	20,2	41,1	18,1	17,7

Der Anteil der Klasse A, die unter den helleren Sternen (vgl. Tab. 3, S. 112) mit etwa 30% neben K den am häufigsten vertretenen Spektraltypus darstellt, ist also schon bei der 10. phg. Größe auf 7% gesunken, und bei der 12. Größe sind in dieser Breitenzone so gut wie keine A-Sterne mehr zu finden. Ebenfalls sehr ausgeprägt, wenn auch nicht so kraß, ist die Abnahme der A5—F1-Sterne. Demgegenüber steht der starke Zuwachs an G-Sternen (F9—G4) von 14 auf 41%, während der Anteil der übrigen Klassen in Übereinstimmung mit den Ergebnissen VYSSOTSKYS sich nur wenig oder gar nicht ändert.

Zum weiteren Verständnis des Sachverhaltes ist zu bemerken, daß die A-Sterne 9. bis 12. Größe sich in Entfernungen befinden, in denen in mittleren galaktischen Breiten die Sterndichte schon stark abnimmt, und in geringerem Grade gilt dies auch für die Sterne der Gruppe A5 bis F1. Die G-Sterne dagegen, in erdrückender Mehrzahl Zwerge, stehen in einem Umkreis von nur etwa 150 Parsek um die Sonne, wo der Dichteabfall wenig oder gar nicht in Erscheinung tritt. Die K-Sterne endlich (G5—K4) stellen anscheinend eine Mischung von weit entfernten Riesen und nahen Zwergen dar; vermindert sich mit abnehmender scheinbarer Helligkeit die Zahl der Riesen, so nimmt dafür die der Zwerge zu, so daß der Prozentsatz sich im ganzen wenig ändert.

Erst mit Annäherung an die galaktische Ebene verschieben sich die Zahlenverhältnisse, wie schon aus Tabelle 8 hervorgeht, stärker zugunsten der frühen Klassen (B8—F1). Tabelle 10, in der die Sterne 10. bis 12. phg. Größe zusammengefaßt sind, gibt für dieselben Spektralgruppen wie in Tabelle 9 die prozentuale Häufigkeit in drei verschiedenen galaktischen Breiten.

Tabelle 10. Änderung der prozentualen Häufigkeit mit galaktischer Breite. Sterne 10. bis 12. phg. Größe.

Mittlere galakt. Breite	B8—A4	A5—F1	F2—F8	F9—G4	G5—G8	G9—K4
− 50°, 4 Eichfelder. . .	0,7	4,2	20,5	36,5	18,4	19,5
− 30°, 2 Eichfelder. . .	2,6	7,9	17,3	32,2	14,4	25,6
− 10°, 3 Eichfelder. . .	17,1	13,0	19,0	20,9	11,0	19,0

Ein endgültiges Bild der Sternverteilung läßt sich aus diesen wenigen Feldern natürlich noch nicht gewinnen. Erst wenn der ganze Plan ausgeführt ist, wird

[1] Potsd Publ Nr. 88 (1929); Nr. 89 (1930); Nr. 90 (1931).

man das Durchschnittsbild von den lokalen Unregelmäßigkeiten der Sternverteilung trennen können.

Als Grad der galaktischen Konzentration, letztere definiert durch das Verhältnis der Anzahl der Sterne in der Zone 10° zu der in der Zone 60° galaktischer Breite, findet VYSSOTSKY für die sechs Spektralgruppen (B, A, F, G, K, M wie in Ziff. 5) und den Bereich der McCORMICK-Felder:

B	A	F	G	K	M
60	25	4,1	1,8	1,9	3,9

Es ist aber wohl zu beachten, daß man aus diesen Zahlen noch nicht die wirklichen Unterschiede in der galaktischen Konzentration der einzelnen Klassen entnehmen kann, vielmehr überlagert sich noch ein Distanzeffekt, die Zunahme der Konzentration mit wachsender mittlerer Entfernung der Sterne, die in Klasse B am größten, in Klasse G am kleinsten ist.

Der Prozentsatz der selteneren Spektraltypen, besonders P, O, R, ist unter den schwachen Sternen noch geringer als unter den hellen; B-Sterne scheinen häufiger zu sein, als man erwartet hatte, bleiben aber auf bestimmte Milchstraßengebiete, z. B. die Carinagegend, beschränkt.

Längs des galaktischen Gürtels ist die Spektralverteilung für die schwächeren Sterne ungleichmäßiger als für die hellen. So fand SHAPLEY[1] bei der Diskussion der DRAPER-Extension in der Cygnusregion, daß dort die Zahl der A-Sterne mit abnehmender scheinbarer Helligkeit nicht ab-, sondern beständig zunimmt; die Gruppe B8—A3 überragt in diesem Gebiet mit 42% der klassifizierten Sterne an Häufigkeit alle anderen Spektralklassen bei weitem. Anscheinend lagert hier in verhältnismäßig geringer Entfernung von der Sonne eine lokale Kondensation von A-Sternen. Eine andere Anhäufung von A-Sternen findet man z. B. in der Carinagegend.

c) Spektralstatistik unter Berücksichtigung der absoluten Helligkeit und der räumlichen Verteilung der Sterne.

9. Trennung von Riesen- und Zwergsternen. Der unmittelbare Weg, die räumliche Verteilung der Sterne verschiedener Spektraltypen zu studieren, ist der, ihre Entfernungen zu messen. Aber dieser Weg führt uns nicht weit. Die trigonometrische Parallaxenbestimmung ist noch immer eine mühsame und zeitraubende Arbeit, so daß selbst in den engen Grenzen, bis zu denen zuverlässige Parallaxenmessungen möglich sind, nur ein kleiner Bruchteil der Sterne erfaßt werden kann. Die spektroskopischen Methoden der Parallaxenbestimmung gestatten wohl rascheres Arbeiten und sind auch nicht an diese Grenze gebunden, dafür aber aus technischen Gründen vorläufig nur auf hellere Sterne anwendbar.

Indessen kommt uns der Umstand zu Hilfe, daß wenigstens in einigen Spektralklassen die Streuung der absoluten Helligkeiten gering ist, so daß Abzählungen der Sterne nach scheinbarer Helligkeit unmittelbar ihre Anzahl und Verteilung in verschiedenen Entfernungen von der Sonne liefern. Dies gilt z. B. für die O-Sterne, die einzelnen Abteilungen der Klasse B und für die „frühen" A-Sterne (B8—A3). Etwa von A5 an wird allerdings infolge der Abzweigung des Giganten- und Übergigantenastes von der Hauptserie mit fortschreitendem Spektraltypus die Dispersion der absoluten Helligkeiten immer größer, und Spektralgruppen mit geringer Streuung in absoluter Helligkeit lassen sich dann nur noch bilden, wenn man Riesen und Zwerge gesondert behandelt.

[1] Harv Circ 278 (1925); vgl. auch OKUNEV, A Statistical Study of the Spectra ... in the Neighbourhood of the America-Nebula. Bull Obs Centr Poulkovo 10, S. 594 (1927).

Sind die absoluten Helligkeiten der einzelnen Sterne nicht bekannt, so läßt sich manchmal doch abschätzen, mit welchem Prozentsatz sich die Sterne eines Spektralkataloges auf den Giganten- und den Zwergast verteilen. So sind fast alle M-Sterne und zum weitaus größten Teil auch die K-Sterne des DRAPER-Kataloges Giganten, weil die Zwerge dieser beiden Klassen für den Katalog schon zu schwach sind. Daß geringe galaktische Konzentration der Sterne einer Spektralgruppe auf geringe Entfernung schließen läßt, und daß die F- und G-Sterne des DRAPER-Kataloges aus diesem Grunde in überwiegender Mehrzahl den Zwergen zugerechnet werden, wurde oben schon erwähnt. Doch können sich dieser Erscheinung wirkliche Unterschiede der galaktischen Konzentration zwischen den verschiedenen Spektralklassen überlagern, weshalb das Kriterium nicht unbedingt gilt.

Zuverlässigere Methoden, die Verteilung der absoluten Helligkeiten abzuschätzen, liefert die Statistik der Eigenbewegungen. So vergleicht GYLLENBERG[1] — um nur eine Untersuchung dieser Art zu erwähnen — die Verteilungskurve der Radialgeschwindigkeiten mit der der Eigenbewegungen der Sterne des „Greenwich Catalogue of Stars for 1900,0" und findet für die Sterne bis 8,5. Größe der drei Klassen F, G und K folgende Verteilung:

Klasse	Zahl der Sterne	Giganten	Zwerge
F	637	32,5%	67,5%
G	480	48,9	51,1
K	983	83,8	16,2

Das wirkliche Häufigkeitsverhältnis von Giganten und Zwergen kommt aber in diesen Zahlen wie auch im DRAPER- und den meisten anderen Sternkatalogen nicht annähernd zum Ausdruck, weil die Giganten wegen ihrer größeren Helligkeit in einem viel weiteren Bereich erfaßt werden als die Zwerge. In Wahrheit stellen die Riesen nur einen sehr geringen Bruchteil der Sterne dar. Innerhalb einer Entfernung von 10 Parsek von der Sonne gehört weniger als 1% der Sterne dem Gigantenzweig der Klassen F bis M an, obwohl wir in diesem Bereich aus den Parallaxenmessungen wahrscheinlich alle Giganten, aber nur etwa die Hälfte der Zwerge kennen[2]. Dabei besteht kein Grund zu der Annahme, daß jene Typen in unserer Gegend des Sternsystems besonders selten sind.

Die F-, G- und K-Typen unter den schwachen Sternen sind daher höchstwahrscheinlich in überwältigender Mehrzahl Zwerge, und das starke Anschwellen der Zahl dieser Spektren in der Durchmusterung der KAPTEYN-Felder ist keineswegs überraschend.

10. Die wahre Häufigkeitsverteilung der Spektralklassen. Geht aus den erwähnten Untersuchungen über die nächsten Sterne zunächst hervor, daß in der Umgebung der Sonne an 95% der Sterne der Hauptserie angehören, so zeigt dasselbe Datenmaterial weiterhin, daß längs der Hauptserie die Zahl der Sterne mit fortschreitendem Spektraltypus von B nach M hin zunimmt. Numerische Angaben über die Häufigkeit der einzelnen Klassen können allerdings nur mit Vorbehalt gemacht werden, solange wir nur eine beschränkte Kenntnis der Spektren schwacher Sterne besitzen. Außerdem herrscht sicher nicht in allen Teilen des Sternsystems dieselbe Verteilung der Spektraltypen. Danach sind die im folgenden mitgeteilten Ergebnisse zu bewerten.

a) VAN RHIJN hat für jede der Klassen A bis M gesondert die Häufigkeitsfunktion der absoluten Helligkeiten abgeleitet[3]. Die Tafel 68 seiner Veröffent-

---

[1] Lund Medd Série II, Nr. 41b (1927).

[2] Vgl. HAAS, Die nächsten Fixsterne. Veröff Berlin-Babelsberg 3, H. 3 (1923) und W. J. LUYTEN, Harv Ann 85, Nr. 5 (1923).

[3] Publ Astron Labor Groningen Nr. 38 (1925).

lichung gibt für die einzelnen Spektralklassen die Logarithmen der Sternanzahlen pro Kubikparsek in Intervallen von je einer Größenklasse in absoluter Helligkeit (visuell). Wir entnehmen dieser Tafel nachstehende Tabelle 11, in der für jede der fünf Klassen die Anzahl der Sterne pro 10000 Kubikparsek in demjenigen Intervall der absoluten Helligkeit angegeben ist, das die meisten Sterne der betreffenden Klasse enthält (M = 2,0 bedeutet das Intervall 1,45 bis 2,44 usw.).

Für die K- und M-Zwerge lassen sich nur Mindestzahlen angeben, weil wir keine genügende Kenntnis dieser schwachen Sterne besitzen, um die Häufigkeitskurve bis zum Maximum verfolgen zu können. Bei der 9. bzw. 12. Größe sind die betreffenden Kurven noch im Anstieg begriffen. Vergleichsweise sei bemerkt, daß nach derselben Arbeit die Zahl der M-Giganten im Helligkeitsintervall der maximalen Häufigkeit nur etwa 0,06 pro 10000 Kubikparsek beträgt.

Tabelle 11. Absolute Helligkeit und Zahl der Sterne pro 10000 Kubikparsek in der Gegend des Maximums der Häufigkeitsfunktion der Leuchtkräfte (Spektralreihe der Hauptserie).

A	$2^m,0$	1,4
F	4 ,0	11,5
G	6 ,0	22,4
K	> 9 ,0	> 25
M	> 12 ,0	> 120

Wenn die Tabelle auch nicht die Gesamtzahl der Sterne jeder Klasse gibt, sondern nur die Höhe der maximalen Ordinate der Häufigkeitskurve, so zeigt sie doch, daß die M-Zwerge den weitaus häufigsten Typus darstellen.

b) Unter der Voraussetzung, daß die F- und G-Sterne des Draper-Kataloges im wesentlichen Zwerge, die K- und M-Sterne Giganten sind, hat Shapley[1] auf einfache Weise die Häufigkeit der verschiedenen Spektralklassen abgeschätzt. Unter Annahme eines Durchschnittswertes der absoluten Helligkeit für jede der sechs Spektralgruppen (vgl. Ziff. 5) läßt sich zunächst berechnen, bis zu welcher Entfernung die Sterne heller als $8^m,25$, der Vollständigkeitsgrenze des Draper-Kataloges, in jeder Spektralklasse erfaßt werden. Mittels dieser Grenzentfernung können dann die Sterndichten pro Flächeneinheit an der Sphäre leicht in Sterndichten pro Volumeinheit im Raume umgewandelt werden. Tabelle 12 zeigt das Ergebnis der Rechnung, die sich übrigens nur auf die Sterne längs des galaktischen Gürtels bezieht.

Tabelle 12. Sterndichte für verschiedene Spektralgruppen in der Umgebung der Sonne. Nach Shapley.

Spektralgruppe	Sterne pro 100 □°	Grenzentfernung	Sterne pro 10000 Kubikparsek
B (B 0 – B 5)	29,7	880 Parsek	0,044
A (B 8 – A 3)	96,9	340 ,,	2,5
F (A 5 – F 2)	18,7	140 ,,	6,8
G (F 5 – G 0)	26,0	70 ,,	76
Giganten K (G 5 – K 2)	69,0	350 ,,	1,6
Giganten M (K 5 – Mc)	17,5	430 ,,	0,2

Der Schluß, den Shapley aus diesen Zahlen zieht, nämlich daß auf je einen B-Stern 4 M-Giganten und 1700 G-Zwerge kommen, ist allerdings nur dann berechtigt, wenn in jeder Spektralgruppe die mittlere Sterndichte mit zunehmender Entfernung konstant bleibt, was sicher nicht der Fall ist. Ein strenger Vergleich zwischen verschiedenen Spektralgruppen verlangt, daß es sich um Sterne in ungefähr demselben Entfernungsbereich handelt, wie es z. B. bei den A-Sternen und den K-Giganten der Fall ist.

[1] Harv Bull Nr. 792 (1923).

c) Derartige Vergleiche ermöglichte eine vorläufige Untersuchung des bisher vorliegenden Materials der Potsdamer Spektraldurchmusterung der KAPTEYN-Felder[1]. Die räumlichen Sterndichten wurden nach einem ähnlichen Verfahren berechnet, wie es SHAPLEY angewandt hat, und zwar für die Spektralgruppen B8—A4, A5—F1, F2—F8, F9—G4, unter der Voraussetzung, daß die Sterne der Hauptserie angehören. Danach kommen zwischen 300 und 500 Parsek auf je einen Stern der ersten Gruppe 2,2 Sterne der zweiten, zwischen 130 und 260 Parsek auf je einen Stern der zweiten Gruppe 3,7 Sterne der dritten und zwischen 60 und 135 Parsek auf einen Stern der dritten Gruppe 3,7 Sterne der vierten.

d) Für spezielle Gegenden im Cygnus (Entfernungen zwischen 125 und 200 Parsek) gelten die von MAXWELL[2] gefundenen Sterndichten. Sie sind in der nebenstehenden Tabelle zusammengestellt.

Tabelle 13. Sterndichte in verschiedenen Spektralklassen nach MAXWELL.

Klasse bzw. Gruppe	Sterne pro 10000 Kubikparsek
B	0,3
A	6,3
F	1,9
dG (F8—G5)	87
dK (G8—K5)	380
gG (F8—G5)	1,5
gK (G8—K5)	4,4
gM	0,3

Die Zwerge sind hier wahrscheinlich nicht einmal vollständig beobachtet; ihre Zahl dürfte in Wirklichkeit noch größer sein als angegeben.

e) Eine ähnliche Häufigkeitsverteilung der Spektralklassen wie in der Umgebung der Sonne haben KREIKEN[3] und KRIEGER[4] für die Sternwolke im Scutum gefunden. Nach KRIEGER ist dort der Prozentsatz der Zwerge noch etwas höher als in unserer Gegend des Sternsystems.

11. Die räumliche Verteilung der Sterne der verschiedenen Spektralklassen. Was wir auf diesem Gebiete an Erfahrungstatsachen besitzen, stützt sich fast ausschließlich auf die bereits erwähnte Methode, die Sterne in Spektralgruppen zusammenzufassen, die hinsichtlich der absoluten Helligkeit als hinreichend homogen gelten können, so daß die scheinbare Helligkeit unmittelbar die Entfernung gibt. Auf diese Weise sind zuerst und besonders ausführlich die B-Sterne untersucht worden; deshalb ist ihnen ein eigener Abschnitt gewidmet (vgl. Ziff. 12), während die übrigen Klassen summarischer behandelt werden können.

Den Verlauf der Dichtefunktion für die einzelnen Spektralklassen mit zunehmender Entfernung und mit Rücksicht auf die galaktische Ebene haben SHAJN[5], KREIKEN[6], FR. BECKER[1], PANNEKOEK[7] u. a. untersucht. Die räumliche Verteilung der Spektralklassen ist natürlich eng mit dem Aufbau des Sternsystems im ganzen verknüpft und ebenso wie dieser bisher erst teilweise erforscht.

Bei einer Vollständigkeitsgrenze von $12^m,0$ photographisch, wie sie in der Durchmusterung der südlichen KAPTEYN-Felder erreicht ist, werden die Sterne der bisher hauptsächlich diskutierten Spektralgruppen bis etwa zu folgenden Entfernungen erfaßt:

B8—A4	Hauptserie	1600	Parsek
A5—F1	„	660	„
F2—F8	„	330	„
F9—G4	„	180	„
G5—G8	„	115	„

[1] Potsd Publ Nr. 89 (1930).
[2] Lick Bull Nr. 390 (1927).
[3] B A N 5, S. 61 (1929).
[4] Lick Bull 14, S. 95 (1929).
[5] A N 232, S. 17 (1928).
[6] M N 85, S. 985 (1925).
[7] Publ Astron Inst Univers Amsterdam Nr. 2 (1929).

Die Statistik der einzelnen Klassen bezieht sich also auf ganz verschiedene Bereiche des Sternsystems, was bei einer Vergleichung der Einzelergebnisse untereinander zu beachten ist.

In der nächsten Umgebung der Sonne sind die Sterne aller Klassen ziemlich gleichmäßig und unabhängig von der Entfernung verteilt. Aber schon in Distanzen um 50 Parsek zeigt sich eine deutliche Abnahme der Sterndichte in Richtung senkrecht zur galaktischen Ebene. Diese Abnahme ist für die einzelnen Spektralklassen graduell verschieden, bei den A-Sternen z. B. viel stärker als bei den

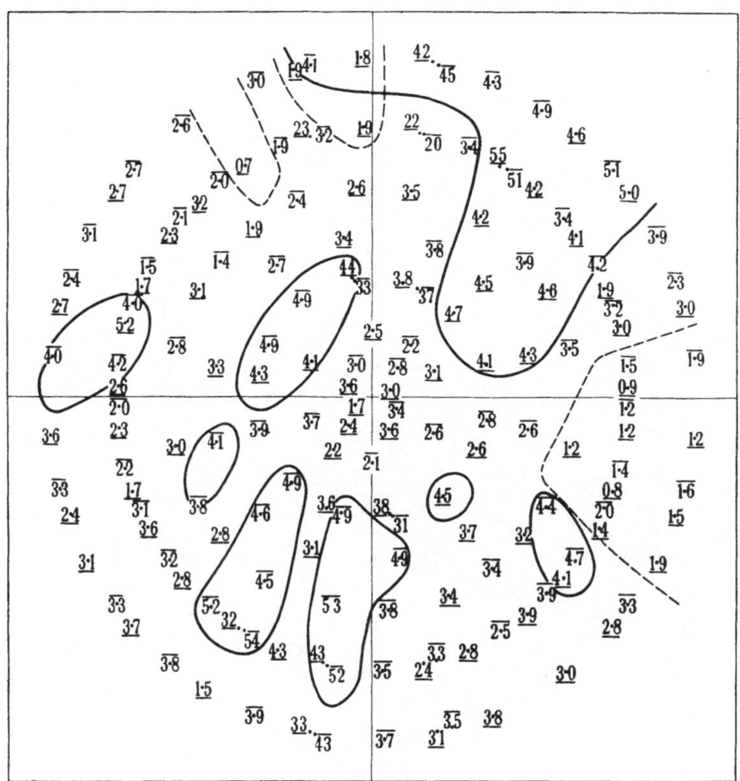

Abb. 5. Räumliche Verteilung der A-Sterne in konzentrisch um die Sonne gelegten Ringen einer 60 Parsek dicken galaktischen Mittelschicht. Die Zahlen geben die Sternhäufigkeit pro 100 000 Kubikparsek an den mit Punkten bezeichneten Stellen. Ein Strich oberhalb oder unterhalb der Zahlen deutet an, ob der betreffende Punkt auf der Ober- oder Unterseite der galaktischen Ebene liegt. Fette Linien umgrenzen Gebiete großer Sterndichte, gestrichelte solche geringer Dichte. Das dargestellte Gebiet mißt 600 Quadrat-Parsek. (Nach PANNEKOEK, Researches on the Structure of the Universe 2.)

K- und M-Riesen. So sind nach PANNEKOEK die mittleren Sterndichten (Sternanzahl pro 10000 Kubikparsek) in Distanzen von

	0	50	100	150	200	Parsek von der galakt. Ebene,
	3,5	2,4	1,4	0,7	0,3	für die A-Sterne,
und	1,8	1,6	1,3	1,0	0,8	für die K-Riesen.

Während also in der Milchstraßenebene die A-Sterne doppelt so zahlreich sind wie die K-Giganten, herrscht in 200 Parsek oberhalb oder unterhalb dieser Ebene bereits das umgekehrte Verhältnis.

Nicht so klar ist der Dichteverlauf in und parallel der galaktischen Ebene. Die A-Sterne z. B. zeigen nach PANNEKOEK innerhalb der Schicht zwischen

± 90 Parsek beiderseits der Milchstraßenebene bis zu Entfernungen von 300 Parsek im Durchschnitt eine fast gleichmäßige Verteilungsdichte. Erst in größeren Entfernungen nimmt hier, wie die Klassifizierung der schwächeren Sterne zeigt, die Häufigkeit ab; aber auch nicht überall, sondern nur in bestimmten Richtungen, während in anderen Gegenden, z. B. in Cygnus und Carina, die Zahl der A-Sterne mit wachsender Entfernung zunächst sogar zu- und erst später wieder abnimmt.

Ein Beispiel für den Dichteverlauf in verschiedenen Spektralgruppen gibt Tabelle 14, in der für das Eichfeld 192 ($\beta = -6°$) die Zahl der Sterne pro 10000 Kubikparsek in verschiedenen Entfernungsbereichen angegeben ist.

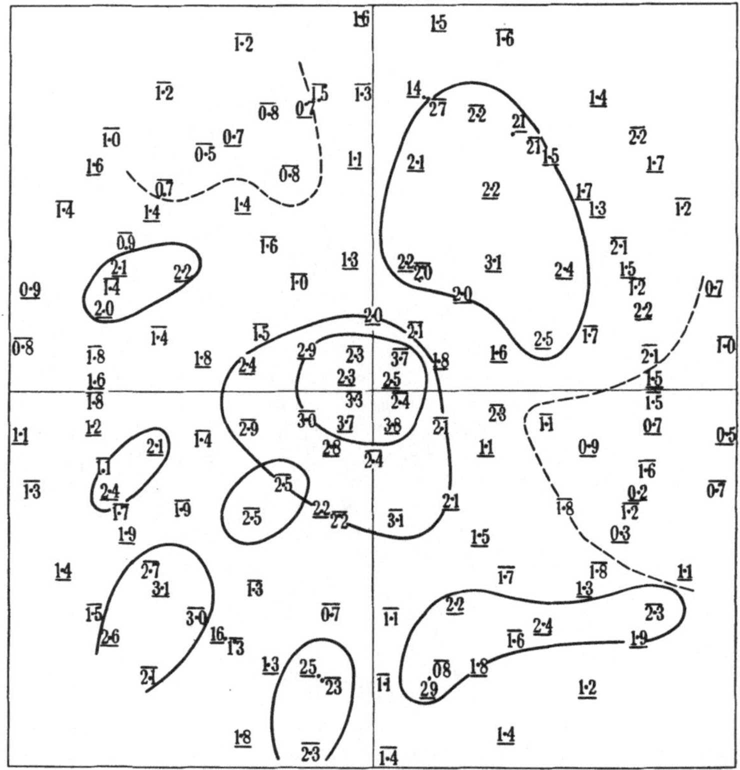

Abb. 6. Räumliche Verteilung der K-Giganten nach PANNEKOEK. Zur Erklärung vgl. Abb. 5.

Tabelle 14. Sterndichte in verschiedenen Entfernungsbereichen im Eichfeld 192. Sterne der Hauptserie (Nach FR. BECKER.)

B8—A4		A5—F1		F2—F8		F9—G4	
Parsek	Sterndichte	Parsek	Sterndichte	Parsek	Sterndichte	Parsek	Sterndichte
320— 400	1,4	170—210	7,2	83—105	21		
400— 500	1,1	210—260	2,7	105—130	22	57— 72	66
500— 630	1,7	260—330	3,1	130—170	18	72— 91	33
630— 800	1,3	330—420	4,2	170—210	31	91—115	173
800—1000	1,0	420—520	4,9	210—260	30	115—140	143
1000—1260	0,3						

Ein eindeutiger Gang der Sternhäufigkeit mit der Entfernung zeigt sich in keiner dieser Gruppen; vielmehr scheinen in der galaktischen Ebene Maxima und Minima der Sterndichte regellos miteinander abzuwechseln, wenn auch vielleicht im ganzen die Dichte nach außen hin abnimmt[1]. Während die Maxima wahrscheinlich reell sind, können die Minima auch durch Absorption des Sternlichtes in Dunkelnebeln vorgetäuscht sein. Eine Zusammenstellung verschiedener Kondensationszentren von A-Sternen und K-Riesen gibt PANNEKOEK in seiner oben zitierten Abhandlung. In der Umgebung der Sonne zeigen die A-Sterne eine mittlere Verteilungsdichte, die K-Giganten ein Häufigkeitsmaximum (vgl. die beiden Diagramme Abb. 5 und 6).

Die Verteilung der Spektren nach galaktischer Länge hat u. a. KREIKEN[2] untersucht. Er findet für die F- und G-Sterne gleichförmige Verteilung, für die A-Sterne und die K- und M-Giganten starke Schwankungen. Die Kurven gleicher Sterndichte zeigen Ausbuchtungen bei den Längen 60° und 270° und Einbiegungen bei 0° und 120°, die KREIKEN abweichend von PANNEKOEKS Auffassung als Anzeichen ungleichmäßiger Ausdehnung des Sternsystems im ganzen interpretiert.

Schließlich mögen hier als besondere Gruppe noch kurz die c-Sterne erwähnt werden, also die Sterne besonders hoher Leuchtkraft aus verschiedenen Spektralklassen, darunter vor allem viele δ Cephei-Veränderliche. Diese Sterne zeigen eine besonders starke galaktische Konzentration; als Verteilungsdichte in der Umgebung der Sonne findet SCHILT[3] $0{,}7 \cdot 10^{-6}$ Sterne pro Kubikparsek. Nimmt man als mittlere Gesamtdichte in der Umgebung der Sonne mit KAPTEYN 0,045 Sterne pro Kubikparsek, so kommt auf je 64000 Sterne ein c-Stern.

12. Das System der B-Sterne. Eine der ersten Arbeiten über die räumliche Verteilung der Spektraltypen waren CHARLIERS Untersuchungen über die B-Sterne. Unter der Voraussetzung, daß innerhalb der einzelnen Abteilungen B0, B1, B2, B3, B5 die Streuung der absoluten Helligkeiten zu vernachlässigen sei, bestimmte CHARLIER aus Eigenbewegungen und Radialgeschwindigkeiten die mittlere absolute Helligkeit der Sterne jeder dieser fünf Klassen, worauf durch Vergleichung mit der scheinbaren Helligkeit in bekannter Weise die Entfernungen der einzelnen Sterne berechnet werden konnten. CHARLIERS erste Abhandlung erschien im Jahre 1916[4], eine zweite, die sich auf das inzwischen zugänglich gewordene, weit umfangreichere Material des DRAPER-Kataloges stützt, folgte 10 Jahre später[5]. Wir beschränken uns hier auf die Ergebnisse dieser zweiten Untersuchung.

Danach bilden die B-Sterne ein gut definiertes abgeplattetes System, in dem die Sterndichte vom Zentrum nach außen hin allmählich abnimmt. Die Sonne steht ungefähr 16 Parsek nördlich der Hauptebene des Systems, dessen Zentrum nicht sehr weit entfernt in 243°,9 galaktischer Länge und — 13°,8 galaktischer Breite zu suchen ist. Im inneren Bereich des Systems zeigen die Sterne eine deutliche Tendenz zur Bildung einzelner voneinander getrennter Häufigkeitsmaxima, unter denen die Orion-, die Carina- und die Scutumgruppe besonders auffallen.

Nach CHARLIER haben KAPTEYN[6], SHAPLEY[7], GERASIMOVIČ[8] u. a. die Verteilung der B-Sterne untersucht mit teils ergänzenden, teils abweichenden Ergebnissen. So gehen nach SHAPLEY[9] in CHARLIERS Untersuchungen wichtige Details dadurch verloren, daß keine Trennung der hellen und schwachen Sterne vorgenommen wird, die nach ihrer scheinbaren Verteilung am Himmel als zwei verschiedene Gruppen zu betrachten sind; die schwachen B-Sterne konzentrieren

[1] Vgl. u. a. SCHALÉN, The Space Distribution of B and A Type Stars usw. Medd Astron Obs Upsala Nr. 37 (1928).
[2] M N 85, S. 985 (1925).
[3] B A N 2, S. 47 (1923/25).
[4] Lund Medd Série II, Nr. 14 (1916).
[5] Lund Medd Série II, Nr. 34 (1926).
[6] Mt Wilson Contr Nr. 82 (1914); Nr. 147 (1918).
[7] Mt Wilson Contr Nr. 157 (1918).
[8] V J S 61, S. 219 (1926).
[9] Harv Bull Nr. 846 (1927).

sich gegen die Milchstraße, die helleren gegen eine um 15° dagegen geneigte Ebene (vgl. oben Ziff. 7a), die als Zentralebene des sog. lokalen Sternsystems in der Stellarastronomie eine Rolle spielt.

Eine wesentlich andere Gestalt haben schließlich die Untersuchungen Pannekoeks[1] dem Verteilungsbild der B-Sterne gegeben. Pannekoek befolgt nicht

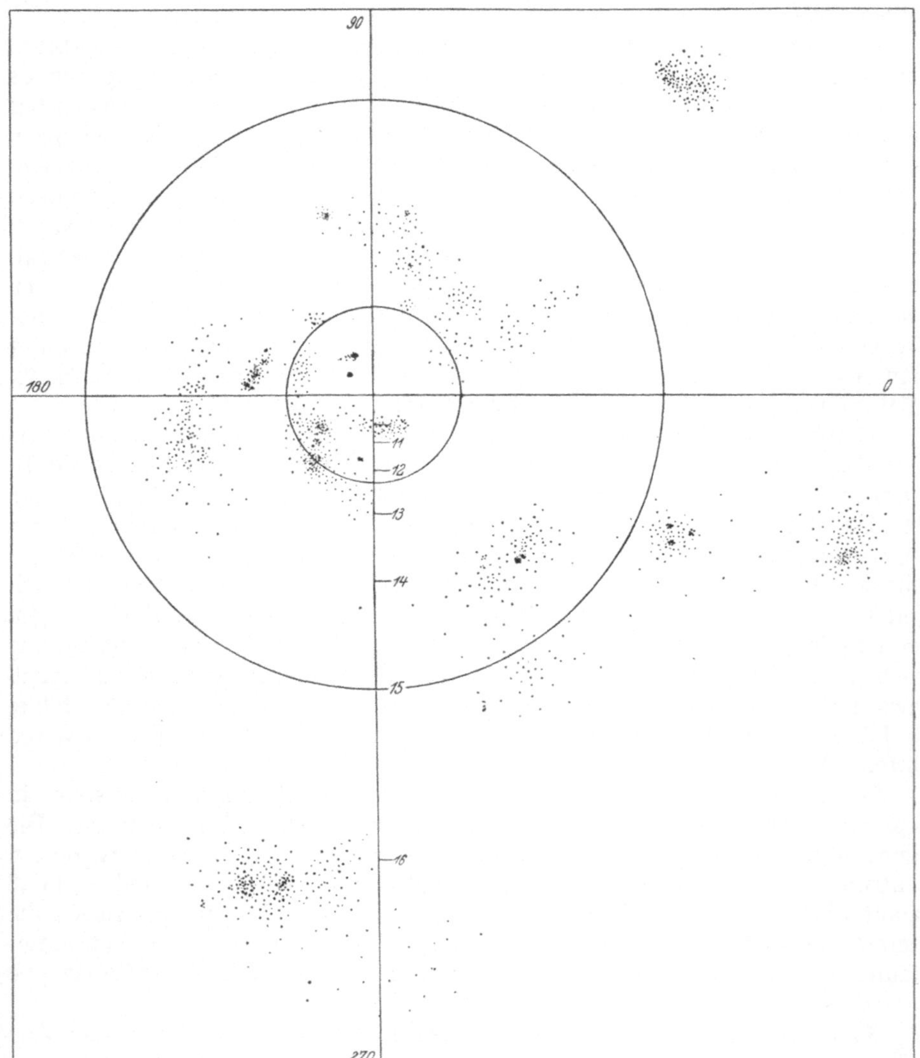

Abb. 7. Die verschiedenen Gruppen der B-Sterne auf die galaktische Ebene projiziert. Die beiden konzentrischen Kreise markieren Distanzen von 300 und 1000 Parsek von der Sonne. (Nach Pannekoek, Researches on the Structure of the Universe 2.)

die Charliersche Methode, für die verschiedenen Unterabteilungen der Klasse B Mittelwerte der absoluten Helligkeit anzunehmen, sondern geht, wie hier nicht im einzelnen ausgeführt werden kann, von der scheinbaren Verteilung der B-Sterne aus und kommt zu dem Ergebnis, daß ein zusammenhängendes System im Char-

[1] Publ Astron Inst Univers Amsterdam Nr. 2 (1929).

LIERschen Sinne überhaupt nicht existiert. Vielmehr verteilt sich die Hauptmasse der B-Sterne auf einzelne, unregelmäßig im Raume verstreute Gruppen, von denen die äußersten etwa 3000 Parsek von uns entfernt sind (vgl. Abb. 7). Eine bei CHARLIER bereits angedeutete, aber als sekundär bewertete Erscheinung wird hier also zum beherrschenden Faktor der Sternverteilung. Im ganzen zählt PANNEKOEK 37 Gruppen auf, die fast ausnahmslos in der Schicht zwischen 100 Parsek oberhalb und unterhalb der galaktischen Ebene liegen; die Anzahl der Sterne und die Verteilungsdichte variieren stark von Gruppe zu Gruppe.

Dieses Bild der räumlichen Verteilung der B-Sterne, das dem für die A- und K-Sterne gefundenen entspricht, verdient heute wohl den Vorzug, wenn auch die Möglichkeit, daß doch ein „lokales System" existiert, in dem die getrennten Verdichtungszentren zu einer höheren Einheit zusammengefaßt sind, nicht ganz von der Hand zu weisen ist.

Der Grund für die mangelnde Übereinstimmung der Ergebnisse CHARLIERS und PANNEKOEKS liegt wahrscheinlich darin, daß die Streuung der absoluten Helligkeiten der B-Sterne doch zu groß ist, als daß man auf Grund von Mittelwerten die Entfernungen der einzelnen Sterne berechnen könnte; die Unregelmäßigkeiten der wirklichen Verteilung werden dabei nivelliert.

Bemerkenswert ist noch ein von PANNEKOEK durchgeführter Vergleich der Kondensationen der B-Sterne mit denen der A- und K-Sterne. Jede bedeutende Gruppe von B-Sternen ist zugleich ein Häufigkeitszentrum von A-Sternen, aber nicht umgekehrt. Kondensationen von A-Sternen können sowohl mit solchen der B-Sterne wie auch denen der K-Sterne zusammenfallen, während sich anscheinend niemals an derselben Stelle zugleich ein Häufigkeitsmaximum von B-Sternen und von K-Giganten findet.

Kapitel 3.

Die Temperaturen der Fixsterne.

Von

A. BRILL-Neubabelsberg.

Mit 5 Abbildungen.

a) Die physikalischen Grundlagen.

1. Das KIRCHHOFFsche Gesetz und die PLANCKsche Strahlungsgleichung. Das Problem der Temperaturbestimmung der Fixsterne ist eng verknüpft mit der Lösung der physikalischen Aufgabe, den gesetzmäßigen Zusammenhang zwischen der Temperatur eines Körpers und seiner Strahlung zu finden. Erst die Kenntnis dieser Beziehung gestattet den Übergang von der durch die Beobachtung gegebenen Strahlungsintensität zur Temperatur des strahlenden Körpers.

Nach KIRCHHOFF ist für irgend eine Wellenlänge λ das Verhältnis des Emissionsvermögens $e_{\lambda T}$ eines Körpers von der absoluten Temperatur T zum Absorptionsvermögen $A_{\lambda T}$ unabhängig von dem Material und von der Beschaffenheit der strahlenden Oberfläche[1]; bei derselben Temperatur und für dieselbe Wellenlänge hat der Quotient für jede beliebige Substanz den gleichen Wert. Die mathematische Form des KIRCHHOFFschen Gesetzes lautet:

$$\frac{e_{\lambda T}}{A_{\lambda T}} = E_{\lambda T}\,. \tag{1}$$

Die Wellenlänge λ charakterisiert die Strahlungsart; $E_{\lambda T}$ ist das Emissionsvermögen eines Körpers, welcher die gesamte auf ihn fallende Strahlungsenergie absorbiert ($A_{\lambda T} = 1$). Körper dieser Art nennt man schwarze Körper[2]. Die Emission beliebiger Substanzen ($A_{\lambda T} < 1$) ist nach der Gleichung (1) nie größer als die Emission des schwarzen Körpers. Letztere ist demnach das Maximum der Strahlung, die ein Körper bei einer bestimmten Temperatur aussendet.

Das Emissionsvermögen $E_{\lambda T}$ des schwarzen Körpers hängt von der absoluten Temperatur T und von der Strahlungsart λ ab und wird die KIRCHHOFFsche Funktion genannt. KIRCHHOFF selbst kannte nur allgemeine Eigenschaften dieser Funktion; nach ihm waren die Bemühungen der Physiker darauf gerichtet, die mathematische Form der Funktion zu finden. Auf zwei verschiedenen Wegen konnte man zum Ziele gelangen: Entweder man schuf, anknüpfend an die modernen Prinzipien der Physik, eine Theorie der Strahlung, oder man maß im Laboratorium die Strahlungsintensität des schwarzen Körpers in ihrer Abhängigkeit von der Wellenlänge bei verschiedenen Temperaturen. Das enge

[1] G. KIRCHHOFF, Gesammelte Abhandlungen, S. 574. Leipzig 1882.
[2] G. KIRCHHOFF, Pogg Ann 109, S. 275 (1860).

Zusammenarbeiten der theoretischen und praktischen Physiker hat schließlich zum Erfolg geführt.

Die von WIEN[1] angegebene Form der KIRCHHOFFschen Funktion:

$$E_{\lambda T} = \frac{c_1}{\lambda^5} \cdot e^{-c_2/\lambda T} \tag{2}$$

stellte die Messungen von PASCHEN, LUMMER und PRINGSHEIM innerhalb eines weiten Temperatur- und Wellenlängenbereiches befriedigend dar; nur für die hohen Temperaturen und für die langwelligen Strahlen ergaben sich merkliche systematische Abweichungen. Nach WIEN hat PLANCK[2] auf Grund von quantentheoretischen Betrachtungen die KIRCHHOFFsche Funktion in die folgende Form gebracht:

$$E_{\lambda T} = \frac{c_1}{\lambda^5} \cdot \frac{1}{e^{c_2/\lambda T} - 1}. \tag{3}$$

Die Exponentialkonstante c_2 ist eine wichtige Naturkonstante, deren Wert nach neueren Untersuchungen der Zahl 1,43 nahekommt; die Konstante c_1 hängt von den jeweiligen Versuchsbedingungen ab. Die Gültigkeit der PLANCKschen Strahlungsgleichung wurde in der Folgezeit vielfach von den Physikern experimentell bestätigt.

2. Das KIRCHHOFFsche Gesetz und die Strahlung materieller Körper. Nach der Gleichung (1) ist nicht die Strahlung selbst, sondern der Quotient aus Emission und Absorption gleich einer bestimmten, von der Temperatur und von der Wellenlänge abhängigen, für alle strahlenden Körper identischen Funktion. Die Bestimmung der Temperatur eines materiellen Körpers aus der Strahlungsintensität erfordert mithin die Kenntnis des Absorptionsvermögens, das, abgesehen vom Material und von der Oberflächenbeschaffenheit, noch von der Temperatur und von der Wellenlänge abhängen kann.

Die PLANCKsche Strahlungsgleichung (3) gilt für Substanzen, deren Oberfläche die gesamte auffallende Strahlung absorbiert. Streng genommen ist dieser Fall in der Natur nie vollkommen realisiert; bei vielen Substanzen ist er mit weitgehender Annäherung erreicht. Die Energiekurven der meisten festen Körper (graphische Darstellung der Strahlungsintensität als Funktion der Wellenlänge) sind ähnlich der des schwarzen Körpers. Die Übereinstimmung ist gewöhnlich um so besser, je höher die Temperatur des Körpers ist.

Bei der Berechnung der Temperatur nicht schwarzer Körper ging man früher in der Weise vor, daß man zwar die Form der PLANCKschen Strahlungsgleichung für beliebige Körper beibehielt, der Exponentialkonstanten c_2 aber einen von dem normalen abweichenden Zahlenwert gab. Dieser wurde so gewählt, daß die Beobachtungen möglichst gut dargestellt werden. Von den Metallen weicht die Strahlung des blanken Platins am stärksten von der schwarzen Strahlung ab; die Konstante c_2 hat den Wert 1,28 statt 1,43.

Eine besondere Klasse von Körpern bilden die grauen Strahler. Bei ihnen ist das Absorptionsvermögen konstant; von jeder Strahlungsart wird der gleiche Bruchteil absorbiert. Die Gestalt der Energiekurve gleicht der des schwarzen Körpers von derselben Temperatur.

Leuchtende Körper in gasförmigem Zustand geben in der Regel ein Linienspektrum. Eine unendlich dicke emittierende und absorbierende Gasmasse, deren Schichten die gleiche Temperatur besitzen, strahlt wie ein schwarzer Körper.

[1] W. WIEN, Wied Ann 58, S. 662 (1896).
[2] M. PLANCK, Ann d Phys 4, S. 553 (1901).

3. Das KIRCHHOFFsche Gesetz und die Strahlung der Sterne. Vor Anwendung der physikalischen Gesetze auf die Strahlung der Sterne ist es nützlich, die Aufgabe schärfer zu umgrenzen und die Bedeutung der aus den Strahlungsmessungen zu berechnenden Temperaturen klarzulegen.

Die Bestimmung der Sterntemperaturen auf Grund des KIRCHHOFFschen Gesetzes setzt voraus, daß die Strahlung der Fixsterne als eine reine Wärmestrahlung betrachtet werden darf. Der Nachweis der Richtigkeit dieser Hypothese wird a posteriori durch die praktische Anwendbarkeit der physikalischen Gesetze erbracht.

Die PLANCKsche Strahlungsgleichung genügt den Messungen im Laboratorium bis $T = 1600°$; unterhalb dieser Grenze wird man die Gleichung ohne Bedenken anwenden dürfen. Die Extrapolation zu den viel höheren Sterntemperaturen erscheint gewagt, wenn man die PLANCKsche Gleichung nur als ein unterhalb der genannten Temperaturgrenze brauchbares Rechnungsschema ansieht. Die Ableitung der PLANCKschen Formel aus allgemeinen mechanischen Prinzipien spricht indessen dafür, daß sie mit der KIRCHHOFFschen Funktion identisch ist und für beliebige Werte λ und T gültig bleibt.

Auf Grund der Gleichungen (1) und (3) kann man in formaler Rechnung die absolute Temperatur T bestimmen, wenn die Strahlungsintensität e_λ für die Wellenlänge λ gemessen und das Absorptionsvermögen A_λ anderweitig bekannt ist. Die Bezeichnung dieses für den Stern charakteristischen Wertes T als seine Temperatur ist leicht mißzuverstehen, da man danach eine einheitliche Temperatur für die gesamte Sternmaterie erwartet. In Wirklichkeit bedeutet der Ausdruck „Fixsterntemperatur" nur einen andersgearteten Maßstab für die aus der Oberfläche des Sternes austretende und an der Erdoberfläche von uns gemessene Strahlung. Über die Frage, ob man die aus Gleichung (3) berechnete Temperatur einer bestimmten Schicht des Sternes zuordnen darf, kann man erst entscheiden, wenn ein den modernen Anschauungen entsprechendes Bild von dem Aufbau der Fixsterne vorliegt.

4. Der innere Aufbau der Sterne und der Ursprung der Strahlung. Die Fixsterne sind in erster Näherung Gaskugeln von enorm großen Dimensionen. In konzentrischen Schichten um den Kugelmittelpunkt haben die Zustandsgrößen: Gasdruck, Temperatur und Dichte den gleichen Wert. Die Temperaturen der äußeren Schichten sind niedriger als die der inneren. Gemäß der EDDINGTONschen Theorie des Strahlungsgleichgewichts liegen die Mittelpunktstemperaturen der Zwergsterne bei 30 Millionen, die der Riesensterne zwischen 4 und 10 Millionen Grad. Die Dichte nimmt von außen nach innen zu, und zwar von Null bis zu einem Maximalwert, welcher dem 54fachen der mittleren Dichte entspricht.

Jedes Teilchen der Sternmaterie emittiert und absorbiert Strahlung. Im Strahlungsgleichgewicht sind emittierte und absorbierte Strahlung einander gleich. Die Intensität für jede Strahlungsart hängt von der Temperatur des Teilchens ab.

Die konzentrisch um den Mittelpunkt liegenden Gasschichten geben einen verschieden großen Beitrag zu der aus der Oberfläche des Sternes austretenden Strahlung. Die des weiten Innern wird wegen des großen Absorptionskoeffizienten der Sternmaterie restlos absorbiert, ehe sie die oberflächennahen Schichten erreicht. Eine verhältnismäßig dünne Schicht, die „Photosphäre", ist der Sitz der im kontinuierlichen Spektrum des Sternes sichtbaren Strahlung. Da innerhalb dieser Schicht die Temperatur nach innen zu ansteigt, überlagern sich die Strahlungen von Schichten verschiedener Temperatur. Die über der photosphärischen liegenden „umkehrenden" und „chromosphärischen" Schichten geben wegen der niedrigen Dichte keinen merklichen Beitrag zur Strahlung; doch wirkt diese „atmosphärische" Hülle auf die von der Photosphäre kommende Strahlung in selektiver Weise lichtschwächend. Der Durchlässigkeitskoeffizient der photosphärischen Schicht ist für langwellige Strahlen größer als für kurzwellige. Daher wird langwellige Strahlung noch aus tieferen und heißeren Schichten sichtbar, deren kurzwellige bereits absorbiert ist, ehe sie an die Oberfläche des Sternes gelangt.

Eine weitere Komplikation tritt dadurch ein, daß man die von einer bestimmten Stelle der Sternscheibe kommende Strahlung nicht isolieren kann. Beobachtet wird die gesamte aus der sichtbaren Oberfläche des Sternes austretende Strahlung; diese ist ein Gemisch verschiedenartigster Strahlungen, herrührend von den ungleich zur Gesichtslinie orientierten Teilen der Sternphotosphäre. Je mehr die Richtung der aus der Sternoberfläche austretenden Strahlung von der radialen abweicht, um so länger ist der Weg, den sie in der absorbierenden Photosphäre und Atmosphäre zurücklegt. Am Rande der Sternscheibe genügt bereits eine Schicht von geringer Tiefe zur Absorption der Strahlung. In den Randzonen stammt diese daher aus den oberen und kühleren Schichten der Photosphäre, im zentralen Teil der Sternscheibe dagegen aus den tieferen und heißeren Schichten. Das beobachtete Spektrum des Gesamtlichtes der Sterne ist also durch Übereinanderlagerung zahlreicher Einzelspektren entstanden, die verschieden heißen Schichten zuzuordnen sind.

5. Die effektive Temperatur der Sterne. Will man aus der Natur der zu uns dringenden Strahlung Schlüsse über die Temperatur der Fixsterne ziehen, so handelt es sich um die Temperatur derjenigen Schicht, von der die Strahlung hauptsächlich ausgeht, d. h. der Sternphotosphäre. Da verschieden tiefe und heiße Schichten zu der Strahlung beitragen, ist die Temperatur nicht die einer bestimmten Schicht, sondern ein gewisser mittlerer Wert für eine Schichtenfolge.

Die Berechnung der wahren Temperatur eines Sternes aus der Strahlungsintensität nach den Gleichungen (1) und (3) setzt die Kenntnis des Absorptionsvermögens der die Photosphäre bildenden Gasmassen voraus. Dieses hängt im allgemeinen von der Art des Gasgemenges, von der Temperatur und von der Strahlungsart ab. Solange man nicht weiß, welches Gas oder Gasgemisch in der Photosphäre strahlt, läßt sich die wahre Temperatur eines Sternes nicht in exakter Weise bestimmen.

Genau bekannt ist das Strahlungsgesetz des absolut schwarzen Körpers ($A_{\lambda T} = 1$); es bleibt zur Zeit nur übrig, dieses Gesetz anzuwenden. Das Problem der Temperaturbestimmung der Fixsterne wird damit auf die Ermittlung der „effektiven" Temperatur zurückgeführt; hier bezeichnet man als effektiv diejenige Temperatur, bei welcher ein schwarzer Körper von den gleichen Dimensionen wie der Stern denselben Strahlungseffekt liefert.

Die effektive Temperatur ist grundsätzlich nur ein passendes physikalisches Arbeitselement, dessen Einführung notwendig ist, weil die physikalische und chemische Konstitution der Fixsterne in den photosphärischen Schichten nicht bekannt ist. Da die Emission des schwarzen Körpers ein Maximum der Strahlung bildet, welche ein Körper bei einer bestimmten Temperatur aussendet, so wird mit der effektiven Temperatur ein unterer Grenzwert der wahren Temperatur des Sternes erhalten.

6. Die Gesetze der Temperaturstrahlung des schwarzen Körpers. Die funktionelle Beziehung zwischen der Emission, der Temperatur und der Wellenlänge gibt für den schwarzen Körper das Planck sche Strahlungsgesetz. Nach ihm ist die Strahlungsenergie eines geradlinig polarisierten Strahlenbüschels in dem

Wellenlängengebiet zwischen λ und $\lambda + d\lambda$, die vom Oberflächenelement df des auf der absoluten Temperatur T befindlichen schwarzen Körpers senkrecht zur Oberfläche in das Vakuum in einen Raumwinkel $d\omega$ in der Zeit dt emittiert wird, gleich

$$E_{\lambda T} \cdot d\lambda \cdot d\omega \cdot df \cdot dt.$$

Hier hat $E_{\lambda T}$ für geradlinig polarisierte Strahlung den Wert:

$$(E_{\lambda T})_p = \frac{c_1}{\lambda^5} \cdot \frac{1}{e^{c_2/\lambda T} - 1}. \tag{3}$$

Das Emissionsvermögen für die unpolarisierte Strahlung ist:

$$E_{\lambda T} = \frac{2c_1}{\lambda^5} \cdot \frac{1}{e^{c_2/\lambda T} - 1}. \tag{4}$$

Die Konstanten c_1 und c_2 stehen zu den universellen Konstanten der Lichtgeschwindigkeit c im Vakuum, des PLANCKschen Wirkungsquantums h und der BOLTZMANNschen Gaskonstanten k in der folgenden Beziehung:

$$c_1 = c^2 \cdot h; \qquad c_2 = \frac{h}{k} \cdot c. \tag{5}$$

Bequemer für die logarithmische Rechnung ist die WIENsche Formel:

$$E_{\lambda T} = \frac{2c_1}{\lambda^5} \cdot \frac{1}{e^{c_2/\lambda T}}. \tag{4a}$$

Beide Ansätze der KIRCHHOFFschen Funktion, die WIENsche und die PLANCKsche Gleichung, sind von ähnlicher Bauart. Sie unterscheiden sich voneinander durch den Faktor $(1 - e^{-c_2/\lambda T})^{-1}$, der von der Einheit nur wenig abweicht, solange das Produkt $\lambda \cdot T$ gegenüber der Exponentialkonstanten c_2 klein ist. Für die mittlere Wellenlänge $0{,}5\,\mu$ sind die Zahlenwerte des Faktors, welche zu den darüberstehenden Temperaturen gehören, folgende:

$\lambda = 0{,}5\,\mu$

T	4000°	6000°	8000°	10000°	15000°	20000°
$(1 - e^{-c_2/\lambda T})^{-1}$	1,001	1,009	1,029	1,061	1,174	1,315

Die Integration der PLANCKschen Strahlungsgleichung über den gesamten Wellenlängenbereich $\lambda = 0$ bis $\lambda = \infty$ führt auf das experimentell von STEFAN[1] gefundene, thermodynamisch später von BOLTZMANN[2] abgeleitete Gesetz: Die Größe der Gesamtstrahlung des absolut schwarzen Körpers ist proportional der vierten Potenz der absoluten Temperatur. Wenn die gesamte Strahlungsenergie, welche ein Oberflächenelement df des schwarzen Körpers senkrecht zur Fläche im Raumwinkel $d\omega$ in der Zeit dt ausstrahlt, gleich $E \cdot d\omega \cdot df \cdot dt$ ist, so wird:

$$E = \frac{\sigma}{\pi} \cdot T^4. \tag{6}$$

Die Gesamtenergie, welche eine ebene 1 cm² große Fläche einseitig in der Zeiteinheit ausstrahlt, ist:

$$S = \sigma \cdot T^4. \tag{7}$$

---

[1] J. STEFAN, Wiener Sitzgsber 79, S. 391 (1879).

[2] L. BOLTZMANN, Wied Ann 22, S. 291 (1884).

Die Konstante σ drückt sich durch die universellen Konstanten in folgender Weise aus:

$$\sigma = \frac{2\pi^5 k^4}{15 c^2 h^3}. \tag{8}$$

Nach dem PLANCKschen Gesetz wird die Intensität der Strahlung für alle endlichen Werte der Temperatur T bei $\lambda = 0$ und $\lambda = \infty$ Null. Daraus folgt, daß $E_{\lambda T}$ für jeden endlichen Wert T einen Maximalwert besitzen muß. Bezeichnet man mit λ_{Max} die Wellenlänge des Intensitätsmaximums in der PLANCKschen Strahlungsgleichung, so folgt aus der Maximalbedingung:

$$\left(\frac{\partial E_{\lambda T}}{\partial \lambda}\right)_{\lambda = \lambda_{\text{Max}}} = 0 \tag{9}$$

das WIENsche[1] Verschiebungsgesetz:

$$\lambda_{\text{Max}} \cdot T = \frac{c_2}{\beta}, \tag{10}$$

wo $\beta = 4{,}9651$ (BRIGGSscher Logarithmus 0,69593) eine einfache Konstante ist.

Der Wert der maximalen Intensität für die unpolarisierte Strahlung ist dann nach dem PLANCKschen Gesetz:

$$E_{\text{Max}} = C \cdot T^5, \tag{11}$$

wo

$$C = \frac{2c_1}{e^{4{,}9651} - 1} \cdot \left(\frac{\beta}{c_2}\right)^5 \tag{12}$$

ist.

7. Die numerischen Werte der Strahlungskonstanten. Die Gültigkeit der Strahlungsgesetze steht heute außer allem Zweifel; über die in den Gleichungen auftretenden Konstanten bestehen noch Unsicherheiten. In der einschlägigen Literatur basieren die Temperaturbestimmungen auf mehr oder weniger verschiedenen numerischen Werten der Konstanten. Der Wirrwarr wird dadurch noch größer, daß bei Anwendung verschiedener Gesetze mit Konstanten gerechnet wird, die nicht in der von der Theorie und Rechnung geforderten Beziehung zueinander stehen.

Durch Festlegung der numerischen Werte der universellen Konstanten c, h und k sind die Konstanten in den Strahlungsgesetzen eindeutig bestimmt. An Stelle von k kann man auch über c_2 verfügen; k ist dann durch c, c_2 und h bestimmt.

Der Verfasser[2] hat vorgeschlagen, für die Lichtgeschwindigkeit im Vakuum den Wert:

$$c = 2{,}9985 \cdot 10^{10}\,\text{cm sec}^{-1} \quad [10{,}47690]$$

anzunehmen und die Konstanten c_2 und h in Übereinstimmung mit den Angaben von W. W. COBLENTZ (Bureau of Standards) festzulegen:

$$c_2 = 1{,}432\,\text{cm grad} \quad [0{,}15594],$$
$$h = 6{,}55 \cdot 10^{-27}\,\text{erg sec} \quad [0{,}81624-27];$$

in Klammern sind die BRIGGSschen Logarithmen der Konstanten beigefügt.

[1] W. WIEN, Sitzb Berlin 1893, S. 55.

[2] A. BRILL, Ein Vorschlag zur vorläufigen Festlegung der Strahlungskonstanten. A'N 235, S. 411 (1929).

Die BOLTZMANNsche Gaskonstante k und die übrigen Konstanten in den Strahlungsgesetzen werden dann:

$$k = 1{,}3715 \cdot 10^{-16}\,\text{erg grad}^{-1}\ [0{,}13720-16],$$
$$2c_1 = 1{,}1778 \cdot 10^{-5}\,\text{erg sec}^{-1}\,\text{cm}^2\ [0{,}07107-5],$$
$$\sigma = 5{,}7146 \cdot 10^{-5}\,\text{erg sec}^{-1}\,\text{cm}^{-2}\,\text{grad}^{-4}\ [0{,}75697-5],$$
$$c_2/\beta = 0{,}28841\,\text{cm grad}\ [0{,}46001-1],$$
$$C = 4{,}1474 \cdot 10^{-5}\,\text{erg sec}^{-1}\,\text{cm}^{-3}\,\text{grad}^{-5}\ [0{,}61777-5].$$

8. Die Tabellen zu den Gesetzen der Temperaturstrahlung des schwarzen Körpers. Mit den in Ziffer 7 angegebenen Konstanten hat der Verfasser Tabellen zu den einzelnen Strahlungsgesetzen berechnet, welche in der Praxis von Nutzen sind und deshalb im folgenden mitgeteilt werden. Die Anlage der Tafeln mit dem ungleichmäßig fortschreitenden Argument c_2/T basiert auf einer früher vom Verfasser[1] gemachten Zuordnung der c_2/T-Werte zu den Spektralklassen B0, B5 usf., welche nach dem heutigen Stande der Forschung nicht genau zutrifft. Mit den logarithmischen Tafelwerten kann man die Interpolation für jeden c_2/T-Wert graphisch oder numerisch leicht ausführen.

Die Tabelle 1 enthält zu den voranstehenden c_2/T-Werten die absolute Temperatur, die Wellenlänge des Intensitätsmaximums, den Wert der maximalen Intensität nebst dem BRIGGSschen Logarithmus und die Größe der gesamten Strahlungsintensität nebst ihrem Logarithmus. In der Tabelle 2 sind die Werte der Strahlungsintensität gemäß dem PLANCKschen Gesetz nebst ihren Logarithmen für die Wellenlängen $0{,}3 \cdot 10^{-4}$ bis $1{,}2 \cdot 10^{-4}$ cm in gleichmäßigen Intervallen von $0{,}05 \cdot 10^{-4}$ cm gegeben. Die Tabelle 3 enthält für dieselben Wellenlängen zu den voranstehenden c_2/T-Werten den BRIGGSschen Logarithmus des PLANCKschen Korrektionsfaktors $(1 - e^{-c_2/\lambda T})^{-1}$, die Tabelle 4 den Gradienten der logarithmischen Energiekurve $\frac{d}{d^1/\lambda}\log E_{\lambda T}$, wo der einfachen Schreibweise halber die Längeneinheit gleich 10^{-4} cm gesetzt ist.

Tabelle 1. Werte der absoluten Temperatur in Celsiusgraden, der Wellenlänge des Intensitätsmaximums: $\lambda_{\max} = \frac{1}{4{,}9651} \cdot \frac{c_2}{T}$ cm, der maximalen Intensität: $E_{\lambda\max} = C \cdot T^5$ erg sec^{-1} cm^{-3}, der gesamten Strahlungsintensität: $E = \frac{\sigma}{\pi} T^4$ erg sec^{-1} cm^{-2}; $c_2 = 1{,}432$ cm grad; $C = 4{,}1474 \cdot 10^{-5}$ erg sec^{-1} cm^{-3} grad^{-5}; $\sigma = 5{,}7146 \cdot 10^{-5}$ erg sec^{-1} cm^{-2} grad^{-4}.

$\frac{c_2}{T} \cdot 10^4$	T absolut	$\lambda_{\max} \cdot 10^7$	$E_{\lambda\max}$	$\log E_{\lambda\max}$	E	$\log E$
0,603	23750°	121,4	$3{,}132 \cdot 10^{17}$	17,49587	$5{,}785 \cdot 10^{12}$	12,76230
0,880	16270	177,2	$4{,}732 \cdot 10^{16}$	16,67507	$1{,}275 \cdot 10^{12}$	12,10566
1,216	11780	244,9	$9{,}393 \cdot 10^{15}$	15,97282	$3{,}498 \cdot 10^{11}$	11,54386
1,594	8985	321,0	$2{,}427 \cdot 10^{15}$	15,38502	$1{,}185 \cdot 10^{11}$	11,07362
1,816	7885	365,8	$1{,}264 \cdot 10^{15}$	15,10187	$7{,}032 \cdot 10^{10}$	10,84710
2,080	6885	418,9	$6{,}415 \cdot 10^{14}$	14,80717	$4{,}086 \cdot 10^{10}$	10,61134
2,392	5986	481,8	$3{,}189 \cdot 10^{14}$	14,50367	$2{,}336 \cdot 10^{10}$	10,36854
2,744	5219	552,7	$1{,}605 \cdot 10^{14}$	14,20557	$1{,}349 \cdot 10^{10}$	10,13006
3,140	4560	632,4	$8{,}181 \cdot 10^{13}$	13,91282	$7{,}868 \cdot 10^{9}$	9,89586
3,727	3842	750,6	$3{,}473 \cdot 10^{13}$	13,54067	$3{,}964 \cdot 10^{9}$	9,59814
4,020	3562	809,7	$2{,}379 \cdot 10^{13}$	13,37632	$2{,}929 \cdot 10^{9}$	9,46666
4,783	2994	963,3	$9{,}976 \cdot 10^{12}$	12,99897	$1{,}461 \cdot 10^{9}$	9,16478
5,740	2495	1156,1	$4{,}008 \cdot 10^{12}$	12,60292	$7{,}046 \cdot 10^{8}$	8,84794
7,175	1996	1445,1	$1{,}313 \cdot 10^{12}$	12,11837	$2{,}886 \cdot 10^{8}$	8,46030
9,567	1497	1926,9	$3{,}116 \cdot 10^{11}$	11,49357	$9{,}130 \cdot 10^{7}$	7,96046

[1] A. BRILL, Die Strahlung der Sterne. Veröff Univ Sternwarte Berlin-Babelsberg V, H. 1 (1924).

Tabelle 2.

Werte der spektralen Strahlungsintensität: $E_{\lambda T} = \frac{2c_1}{\lambda^5} \cdot \frac{1}{e^{c_2/\lambda T} - 1}$ erg sec^{-1} cm^{-3}; $c_2 = 1{,}432$ cm grad; $2c_1 = 1{,}1778 \cdot 10^{-5}$ erg sec^{-1} cm^2.

$1/\lambda$	$3{,}3333 \cdot 10^4$	$2{,}8571 \cdot 10^4$	$2{,}5000 \cdot 10^4$	$2{,}2223 \cdot 10^4$	$2{,}0000 \cdot 10^4$	$1{,}8182 \cdot 10^4$	$1{,}6667 \cdot 10^4$	$1{,}5385 \cdot 10^4$	$1{,}4286 \cdot 10^4$	$1{,}3333 \cdot 10^4$
$\frac{c_2}{T} \cdot 10^4$ \ λ cm	$0{,}30 \cdot 10^{-4}$	$0{,}35 \cdot 10^{-4}$	$0{,}40 \cdot 10^{-4}$	$0{,}45 \cdot 10^{-4}$	$0{,}50 \cdot 10^{-4}$	$0{,}55 \cdot 10^{-4}$	$0{,}60 \cdot 10^{-4}$	$0{,}65 \cdot 10^{-4}$	$0{,}70 \cdot 10^{-4}$	$0{,}75 \cdot 10^{-4}$
0,603	$7{,}4992 \cdot 10^{16}$ 16,87501	$4{,}8745 \cdot 10^{16}$ 16,68793	$3{,}2718 \cdot 10^{16}$ 16,51478	$2{,}2642 \cdot 10^{16}$ 16,35491	$1{,}61056 \cdot 10^{16}$ 16,20698	$1{,}1741 \cdot 10^{16}$ 16,06970	$8{,}7458 \cdot 10^{15}$ 15,94180	$6{,}6405 \cdot 10^{15}$ 15,82220	$5{,}1282 \cdot 10^{15}$ 15,70996	$4{,}0205 \cdot 10^{15}$ 15,60428
0,880	$2{,}7246 \cdot 10^{16}$ 16,43530	$1{,}9744 \cdot 10^{16}$ 16,29544	$1{,}4333 \cdot 10^{16}$ 16,15634	$1{,}05196 \cdot 10^{16}$ 16,02200	$7{,}8312 \cdot 10^{15}$ 15,89383	$5{,}9339 \cdot 10^{15}$ 15,77334	$4{,}5421 \cdot 10^{15}$ 15,65726	$3{,}5342 \cdot 10^{15}$ 15,54829	$2{,}7861 \cdot 10^{15}$ 15,44500	$2{,}2230 \cdot 10^{15}$ 15,34694
1,216	$8{,}5655 \cdot 10^{15}$ 15,93275	$7{,}1706 \cdot 10^{15}$ 15,85555	$5{,}7787 \cdot 10^{15}$ 15,76183	$4{,}5880 \cdot 10^{15}$ 15,66162	$3{,}6304 \cdot 10^{15}$ 15,55996	$2{,}8808 \cdot 10^{15}$ 15,45951	$2{,}2989 \cdot 10^{15}$ 15,36153	$1{,}8480 \cdot 10^{15}$ 15,26670	$1{,}4970 \cdot 10^{15}$ 15,17523	$1{,}2225 \cdot 10^{15}$ 15,08726
1,594	$2{,}3993 \cdot 10^{15}$ 15,38008	$2{,}3847 \cdot 10^{15}$ 15,37743	$2{,}1790 \cdot 10^{15}$ 15,33827	$1{,}9029 \cdot 10^{15}$ 15,27941	$1{,}62181 \cdot 10^{15}$ 15,21000	$1{,}36538 \cdot 10^{15}$ 15,13525	$1{,}14329 \cdot 10^{15}$ 15,05816	$9{,}5630 \cdot 10^{14}$ 14,98059	$8{,}0102 \cdot 10^{14}$ 14,90364	$6{,}7293 \cdot 10^{14}$ 14,82797
1,816	$1{,}1417 \cdot 10^{15}$ 15,05755	$1{,}2583 \cdot 10^{15}$ 15,09978	$1{,}2409 \cdot 10^{15}$ 15,09373	$1{,}1485 \cdot 10^{15}$ 15,06012	$1{,}0245 \cdot 10^{15}$ 15,01051	$8{,}9450 \cdot 10^{14}$ 14,95158	$7{,}7169 \cdot 10^{14}$ 14,88744	$6{,}6158 \cdot 10^{14}$ 14,82058	$5{,}6570 \cdot 10^{14}$ 14,75259	$4{,}8370 \cdot 10^{14}$ 14,68458
2,080	$4{,}7297 \cdot 10^{14}$ 14,67483	$5{,}9015 \cdot 10^{14}$ 14,77096	$6{,}3807 \cdot 10^{14}$ 14,80487	$6{,}3374 \cdot 10^{14}$ 14,80191	$5{,}9760 \cdot 10^{14}$ 14,77641	$5{,}4558 \cdot 10^{14}$ 14,73685	$4{,}8815 \cdot 10^{14}$ 14,68855	$4{,}3137 \cdot 10^{14}$ 14,63485	$3{,}7841 \cdot 10^{14}$ 14,57796	$3{,}3065 \cdot 10^{14}$ 14,51936
2,392	$1{,}6706 \cdot 10^{14}$ 14,22287	$2{,}4161 \cdot 10^{14}$ 14,38312	$2{,}9161 \cdot 10^{14}$ 14,46481	$3{,}1525 \cdot 10^{14}$ 14,49865	$3{,}1786 \cdot 10^{14}$ 14,50224	$3{,}0629 \cdot 10^{14}$ 14,48614	$2{,}8648 \cdot 10^{14}$ 14,45709	$2{,}6267 \cdot 10^{14}$ 14,41941	$2{,}3770 \cdot 10^{14}$ 14,37603	$2{,}13275 \cdot 10^{14}$ 14,32894
2,744	$5{,}1668 \cdot 10^{13}$ 13,71322	$8{,}8329 \cdot 10^{13}$ 13,94610	$1{,}20787 \cdot 10^{14}$ 14,08202	$1{,}4381 \cdot 10^{14}$ 14,15778	$1{,}5655 \cdot 10^{14}$ 14,19465	$1{,}5997 \cdot 10^{14}$ 14,20403	$1{,}58004 \cdot 10^{14}$ 14,19867	$1{,}51204 \cdot 10^{14}$ 14,17956	$1{,}4186 \cdot 10^{14}$ 14,15186	$1{,}3127 \cdot 10^{14}$ 14,11817
3,140	$1{,}3800 \cdot 10^{13}$ 13,13988	$2{,}8483 \cdot 10^{13}$ 13,45458	$4{,}4849 \cdot 10^{13}$ 13,65175	$5{,}9570 \cdot 10^{13}$ 13,77503	$7{,}0744 \cdot 10^{13}$ 13,84969	$7{,}7854 \cdot 10^{13}$ 13,89128	$8{,}1253 \cdot 10^{13}$ 13,90984	$8{,}1664 \cdot 10^{13}$ 13,91203	$7{,}9872 \cdot 10^{13}$ 13,90239	$7{,}6590 \cdot 10^{13}$ 13,88417
3,727	$1{,}9503 \cdot 10^{12}$ 12,29009	$5{,}3229 \cdot 10^{12}$ 12,72615	$1{,}03343 \cdot 10^{13}$ 13,01428	$1{,}6152 \cdot 10^{13}$ 13,20822	$2{,}1840 \cdot 10^{13}$ 13,33925	$2{,}6720 \cdot 10^{13}$ 13,42684	$3{,}0445 \cdot 10^{13}$ 13,48351	$3{,}2941 \cdot 10^{13}$ 13,51774	$3{,}4309 \cdot 10^{13}$ 13,53541	$3{,}4726 \cdot 10^{13}$ 13,54065
4,020	$7{,}3433 \cdot 10^{11}$ 11,86589	$2{,}3044 \cdot 10^{12}$ 12,36256	$4{,}9670 \cdot 10^{12}$ 12,69609	$8{,}4208 \cdot 10^{12}$ 12,92535	$1{,}2151 \cdot 10^{13}$ 13,08461	$1{,}5676 \cdot 10^{13}$ 13,19523	$1{,}8657 \cdot 10^{13}$ 13,27085	$2{,}0963 \cdot 10^{13}$ 13,32146	$2{,}2537 \cdot 10^{13}$ 13,35289	$2{,}3441 \cdot 10^{13}$ 13,36997
4,783	$5{,}7739 \cdot 10^{10}$ 10,76147	$2{,}6055 \cdot 10^{11}$ 11,41588	$7{,}3745 \cdot 10^{11}$ 11,86773	$1{,}5452 \cdot 10^{12}$ 12,18900	$2{,}6414 \cdot 10^{12}$ 12,42183	$3{,}9184 \cdot 10^{12}$ 12,59311	$5{,}2294 \cdot 10^{12}$ 12,71845	$6{,}4709 \cdot 10^{12}$ 12,81096	$7{,}5612 \cdot 10^{12}$ 12,87859	$8{,}4502 \cdot 10^{12}$ 12,92687
5,740	$2{,}3769 \cdot 10^{9}$ 9,37601	$1{,}6920 \cdot 10^{10}$ 10,22839	$6{,}7411 \cdot 10^{10}$ 10,82873	$1{,}8423 \cdot 10^{11}$ 11,26537	$3{,}8957 \cdot 10^{11}$ 11,59058	$6{,}8683 \cdot 10^{11}$ 11,83685	$1{,}06081 \cdot 10^{12}$ 12,02564	$1{,}4841 \cdot 10^{12}$ 12,17146	$1{,}9254 \cdot 10^{12}$ 12,28452	$2{,}3562 \cdot 10^{12}$ 12,37221
7,175	$1{,}9895 \cdot 10^{7}$ 7,29873	$2{,}7407 \cdot 10^{8}$ 8,43786	$1{,}8651 \cdot 10^{9}$ 9,27021	$7{,}5940 \cdot 10^{9}$ 9,88047	$2{,}2089 \cdot 10^{10}$ 10,34418	$5{,}0551 \cdot 10^{10}$ 10,70373	$9{,}7030 \cdot 10^{10}$ 10,98691	$1{,}6316 \cdot 10^{11}$ 11,21261	$2{,}4783 \cdot 10^{11}$ 11,39416	$3{,}4760 \cdot 10^{11}$ 11,54108
9,567	$6{,}8502 \cdot 10^{3}$ 3,83570	$3{,}0174 \cdot 10^{5}$ 5,47964	$4{,}7159 \cdot 10^{6}$ 6,67356	$3{,}7309 \cdot 10^{7}$ 7,57182	$1{,}8468 \cdot 10^{8}$ 8,26642	$6{,}5286 \cdot 10^{8}$ 8,81482	$1{,}8007 \cdot 10^{9}$ 9,25544	$4{,}1144 \cdot 10^{9}$ 9,61430	$8{,}1284 \cdot 10^{9}$ 9,91000	$1{,}4317 \cdot 10^{10}$ 10,15586

Tabelle 2 (Fortsetzung).

$1/\lambda$	$1,2500 \cdot 10^4$	$1,1765 \cdot 10^4$	$1,1111 \cdot 10^4$	$1,0526 \cdot 10^4$	$1,0000 \cdot 10^4$	$0,9524 \cdot 10^4$	$0,9091 \cdot 10^4$	$0,8696 \cdot 10^4$	$0,8333 \cdot 10^4$
λ cm / $\frac{c_2}{T} \cdot 10^4$	$0,80 \cdot 10^{-4}$	$0,85 \cdot 10^{-4}$	$0,90 \cdot 10^{-4}$	$0,95 \cdot 10^{-4}$	$1,00 \cdot 10^{-4}$	$1,05 \cdot 10^{-4}$	$1,10 \cdot 10^{-4}$	$1,15 \cdot 10^{-4}$	$1,20 \cdot 10^{-4}$
0,603	$3,1950 \cdot 10^{15}$ 15,50447	$2,5702 \cdot 10^{15}$ 15,40997	$2,0903 \cdot 10^{15}$ 15,32021	$1,71704 \cdot 10^{15}$ 15,23478	$1,42316 \cdot 10^{15}$ 15,15325	$1,18938 \cdot 10^{15}$ 15,07532	$1,00173 \cdot 10^{15}$ 15,00075	$8,4944 \cdot 10^{14}$ 14,92913	$7,2500 \cdot 10^{14}$ 14,86034
0,880	$1,7934 \cdot 10^{15}$ 15,25368	$1,4617 \cdot 10^{15}$ 15,16487	$1,20268 \cdot 10^{15}$ 15,08015	$9,9805 \cdot 10^{14}$ 14,99915	$8,3482 \cdot 10^{14}$ 14,92159	$7,0340 \cdot 10^{14}$ 14,84720	$5,9675 \cdot 10^{14}$ 14,77579	$5,0944 \cdot 10^{14}$ 14,70709	$4,3746 \cdot 10^{14}$ 14,64094
1,216	$1,0063 \cdot 10^{15}$ 15,00270	$8,3445 \cdot 10^{14}$ 14,92140	$6,9703 \cdot 10^{14}$ 14,84325	$5,8623 \cdot 10^{14}$ 14,76807	$4,9620 \cdot 10^{14}$ 14,69566	$4,2258 \cdot 10^{14}$ 14,62591	$3,6194 \cdot 10^{14}$ 14,55864	$3,1167 \cdot 10^{14}$ 14,49369	$2,6974 \cdot 10^{14}$ 14,43095
1,594	$5,6749 \cdot 10^{14}$ 14,75396	$4,8064 \cdot 10^{14}$ 14,68182	$4,0895 \cdot 10^{14}$ 14,61167	$3,4958 \cdot 10^{14}$ 14,54355	$3,0019 \cdot 10^{14}$ 14,47740	$2,5897 \cdot 10^{14}$ 14,41324	$2,2434 \cdot 10^{14}$ 14,35091	$1,9525 \cdot 10^{14}$ 14,29058	$1,7059 \cdot 10^{14}$ 14,23194
1,816	$4,1412 \cdot 10^{14}$ 14,61713	$3,5536 \cdot 10^{14}$ 14,55067	$3,0584 \cdot 10^{14}$ 14,48550	$2,6409 \cdot 10^{14}$ 14,42175	$2,2884 \cdot 10^{14}$ 14,35949	$1,9898 \cdot 10^{14}$ 14,29880	$1,7364 \cdot 10^{14}$ 14,23964	$1,5207 \cdot 10^{14}$ 14,18203	$1,3364 \cdot 10^{14}$ 14,12593
2,080	$2,8840 \cdot 10^{14}$ 14,46000	$2,5151 \cdot 10^{14}$ 14,40056	$2,1954 \cdot 10^{14}$ 14,34152	$1,9194 \cdot 10^{14}$ 14,28317	$1,68155 \cdot 10^{14}$ 14,22571	$1,47666 \cdot 10^{14}$ 14,16928	$1,30012 \cdot 10^{14}$ 14,11398	$1,14763 \cdot 10^{14}$ 14,05980	$1,01588 \cdot 10^{14}$ 14,00684
2,392	$1,9033 \cdot 10^{14}$ 14,27950	$1,6931 \cdot 10^{14}$ 14,22869	$1,5038 \cdot 10^{14}$ 14,17719	$1,3350 \cdot 10^{14}$ 14,12548	$1,1855 \cdot 10^{14}$ 14,07390	$1,0537 \cdot 10^{14}$ 14,02272	$9,3784 \cdot 10^{13}$ 13,97213	$8,3599 \cdot 10^{13}$ 13,92220	$7,4660 \cdot 10^{13}$ 13,87309
2,744	$1,20314 \cdot 10^{14}$ 14,08032	$1,09538 \cdot 10^{14}$ 14,03956	$9,9282 \cdot 10^{13}$ 13,99687	$8,9729 \cdot 10^{13}$ 13,95293	$8,0958 \cdot 10^{13}$ 13,90826	$7,2988 \cdot 10^{13}$ 13,86325	$6,5792 \cdot 10^{13}$ 13,81817	$5,9326 \cdot 10^{13}$ 13,77324	$5,3533 \cdot 10^{13}$ 13,72862
3,140	$7,2390 \cdot 10^{13}$ 13,85968	$6,7699 \cdot 10^{13}$ 13,83058	$6,2825 \cdot 10^{13}$ 13,79813	$5,7975 \cdot 10^{13}$ 13,76324	$5,3286 \cdot 10^{13}$ 13,72661	$4,8839 \cdot 10^{13}$ 13,68877	$4,4686 \cdot 10^{13}$ 13,65017	$4,0807 \cdot 10^{13}$ 13,61074	$3,7300 \cdot 10^{13}$ 13,57171
3,727	$3,4395 \cdot 10^{13}$ 13,53650	$3,3511 \cdot 10^{13}$ 13,52518	$3,2238 \cdot 10^{13}$ 13,50837	$3,0714 \cdot 10^{13}$ 13,48733	$2,9044 \cdot 10^{13}$ 13,46305	$2,7305 \cdot 10^{13}$ 13,43624	$2,5561 \cdot 10^{13}$ 13,40757	$2,3847 \cdot 10^{13}$ 13,37743	$2,2194 \cdot 10^{13}$ 13,34623
4,020	$2,3776 \cdot 10^{13}$ 13,37614	$2,3653 \cdot 10^{13}$ 13,37388	$2,3176 \cdot 10^{13}$ 13,36503	$2,24425 \cdot 10^{13}$ 13,35107	$2,15315 \cdot 10^{13}$ 13,33307	$2,0509 \cdot 10^{13}$ 13,31194	$1,9425 \cdot 10^{13}$ 13,28836	$1,8315 \cdot 10^{13}$ 13,26281	$1,7210 \cdot 10^{13}$ 13,23578
4,783	$9,1240 \cdot 10^{12}$ 12,96019	$9,5885 \cdot 10^{12}$ 12,98175	$9,8630 \cdot 10^{12}$ 12,99401	$9,9715 \cdot 10^{12}$ 12,99876	$9,9430 \cdot 10^{12}$ 12,99752	$9,8044 \cdot 10^{12}$ 12,99142	$9,5803 \cdot 10^{12}$ 12,98138	$9,2929 \cdot 10^{12}$ 12,96815	$8,9599 \cdot 10^{12}$ 12,95230
5,740	$2,7538 \cdot 10^{12}$ 12,43993	$3,1028 \cdot 10^{12}$ 12,49175	$3,3946 \cdot 10^{12}$ 12,53079	$3,6262 \cdot 10^{12}$ 12,55945	$3,7987 \cdot 10^{12}$ 12,57963	$3,9160 \cdot 10^{12}$ 12,59284	$3,9835 \cdot 10^{12}$ 12,60026	$4,0075 \cdot 10^{12}$ 12,60287	$3,9946 \cdot 10^{12}$ 12,60147
7,175	$4,5776 \cdot 10^{11}$ 11,66064	$5,7300 \cdot 10^{11}$ 11,75815	$6,8824 \cdot 10^{11}$ 11,83774	$7,9915 \cdot 10^{11}$ 11,90263	$9,0236 \cdot 10^{11}$ 11,95538	$9,9525 \cdot 10^{11}$ 11,99793	$1,07646 \cdot 10^{12}$ 12,03200	$1,14509 \cdot 10^{12}$ 12,05884	$1,20108 \cdot 10^{12}$ 12,07957
9,567	$2,3012 \cdot 10^{10}$ 10,36196	$3,4345 \cdot 10^{10}$ 10,53587	$4,8227 \cdot 10^{10}$ 10,68329	$6,4396 \cdot 10^{10}$ 10,80886	$8,2452 \cdot 10^{10}$ 10,91620	$1,01893 \cdot 10^{11}$ 11,00814	$1,22175 \cdot 10^{11}$ 11,08698	$1,42804 \cdot 10^{11}$ 11,15474	$1,6326 \cdot 10^{11}$ 11,21288

Tabelle 3. Werte des Planckschen Korrektionsfaktors log (1 − e)

$1/\lambda$	$3,3333 \cdot 10^4$	$2,8571 \cdot 10^4$	$2,5000 \cdot 10^4$	$2,2223 \cdot 10^4$	$2,0000 \cdot 10^4$	$1,8182 \cdot 10^4$	$1,6667 \cdot 10^4$	$1,5385 \cdot 10^4$	$1,4286 \cdot 10^4$	$1,3333 \cdot 10^4$
$\frac{c_2}{T} \cdot 10^4$ \ λ cm	$0,30 \cdot 10^{-4}$	$0,35 \cdot 10^{-4}$	$0,40 \cdot 10^{-4}$	$0,45 \cdot 10^{-4}$	$0,50 \cdot 10^{-4}$	$0,55 \cdot 10^{-4}$	$0,60 \cdot 10^{-4}$	$0,65 \cdot 10^{-4}$	$0,70 \cdot 10^{-4}$	$0,75 \cdot 10^{-4}$
0,603	0,06248	0,08543	0,10872	0,13185	0,15452	0,17658	0,19794	0,21858	0,23849	0,25769
0,880	0,02375	0,03665	0,05101	0,06626	0,08200	0,09794	0,11390	0,12974	0,14537	0,16074
1,216	0,00761	0,01367	0,02129	0,03014	0,03994	0,05041	0,06137	0,07264	0,08409	0,09563
1,594	0,00214	0,00459	0,00815	0,01276	0,01830	0,02463	0,03160	0,03910	0,04700	0,05388
1,816	0,00102	0,00243	0,00466	0,00775	0,01165	0,01629	0,02158	0,02742	0,03372	0,04039
2,080	0,00042	0,00114	0,00241	0,00429	0,00683	0,01001	0,01378	0,01807	0,02284	0,02801
2,392	0,00015	0,00047	0,00110	0,00214	0,00365	0,00565	0,00814	0,01109	0,01449	0,01827
2,744	0,00005	0,00017	0,00046	0,00098	0,00180	0,00296	0,00451	0,00642	0,00871	0,01134
3,140	0,00001	0,00006	0,00017	0,00040	0,00082	0,00145	0,00233	0,00348	0,00493	0,00665
3,727	0,00000	0,00001	0,00004	0,00011	0,00025	0,00050	0,00087	0,00141	0,00212	0,00303
4,020	0,00000	0,00000	0,00002	0,00006	0,00014	0,00029	0,00054	0,00090	0,00140	0,00205
4,783	0,00000	0,00000	0,00000	0,00001	0,00003	0,00008	0,00015	0,00028	0,00047	0,00074
5,740	0,00000	0,00000	0,00000	0,00000	0,00000	0,00001	0,00003	0,00007	0,00012	0,00021
7,175	0,00000	0,00000	0,00000	0,00000	0,00000	0,00000	0,00000	0,00001	0,00002	0,00003
9,567	0,00000	0,00000	0,00000	0,00000	0,00000	0,00000	0,00000	0,00000	0,00000	0,00000

$1/\lambda$	$1,2500 \cdot 10^4$	$1,1765 \cdot 10^4$	$1,1111 \cdot 10^4$	$1,0526 \cdot 10^4$	$1,0000 \cdot 10^4$	$0,9524 \cdot 10^4$	$0,9091 \cdot 10^4$	$0,8696 \cdot 10^4$	$0,8333 \cdot 10^4$
$\frac{c_2}{T} \cdot 10^4$ \ λ cm	$0,80 \cdot 10^{-4}$	$0,85 \cdot 10^{-4}$	$0,90 \cdot 10^{-4}$	$0,95 \cdot 10^{-4}$	$1,00 \cdot 10^{-4}$	$1,05 \cdot 10^{-4}$	$1,10 \cdot 10^{-4}$	$1,15 \cdot 10^{-4}$	$1,20 \cdot 10^{-4}$
0,603	0,27622	0,29409	0,31131	0,32797	0,34406	0,35962	0,37469	0,38929	0,40340
0,880	0,17580	0,19053	0,20492	0,21897	0,23269	0,24606	0,25910	0,27185	0,28426
1,216	0,10719	0,11871	0,13016	0,14149	0,15269	0,16374	0,17461	0,18533	0,19587
1,594	0,06367	0,07228	0,08100	0,08979	0,09861	0,10742	0,11622	0,12498	0,13366
1,816	0,04736	0,05457	0,06196	0,06948	0,07710	0,08480	0,09252	0,10027	0,10800
2,080	0,03352	0,03932	0,04535	0,05158	0,05796	0,06446	0,07106	0,07773	0,08444
2,392	0,02241	0,02686	0,03157	0,03651	0,04166	0,04696	0,05240	0,05796	0,06361
2,744	0,01430	0,01756	0,02110	0,02488	0,02887	0,03306	0,03741	0,04191	0,04654
3,140	0,00866	0,01094	0,01347	0,01624	0,01921	0,02240	0,02575	0,02928	0,03295
3,727	0,00414	0,00545	0,00697	0,00868	0,01058	0,01267	0,01492	0,01734	0,01991
4,020	0,00287	0,00385	0,00502	0,00636	0,00787	0,00955	0,01139	0,01338	0,01551
4,783	0,00110	0,00157	0,00214	0,00284	0,00365	0,00459	0,00565	0,00684	0,00814
5,740	0,00034	0,00051	0,00074	0,00103	0,00140	0,00184	0,00236	0,00296	0,00365
7,175	0,00006	0,00010	0,00015	0,00023	0,00034	0,00047	0,00064	0,00085	0,00110
9,567	0,00000	0,00000	0,00001	0,00002	0,00003	0,00005	0,00008	0,00011	0,00015

Tabelle 4. Werte des Gradienten der Energiekurve: $\frac{d}{d\,1/\lambda} \log E_{\lambda T}$ · Einheit der Länge 10^{-4} cm.

$1/\lambda$	3,3333	2,8571	2,5000	2,2223	2.0000	1,8182	1,6667	1,5385	1,4286	1,3333
c_2/T \ λ	0,30	0,35	0,40	0,45	0,50	0,55	0,60	0,65	0,70	0,75
0,000	+0,5212	+0,6080	+0,6949	+0,7817	+0,8686	+0,9554	+1,0423	+1,1292	+1,2160	+1,3029
0,603	+0,3490	+0,4412	+0,5322	+0,6224	+0,7119	+0,8010	+0,8898	+0,9783	+1,0665	+1,1546
0,880	+0,2478	+0,3442	+0,4388	+0,5320	+0,6241	+0,7155	+0,8061	+0,8962	+0,9859	+1,0752
1,216	+0,1140	+0,2150	+0,3140	+0,4111	+0,5068	+0,6012	+0,6946	+0,7872	+0,8791	+0,9704
1,594	−0,0442	+0,0604	+0,1632	+0,2643	+0,3637	+0,4616	+0,5584	+0,6540	+0,7487	+0,8425
1,816	−0,1391	−0,0331	+0,0714	+0,1743	+0,2756	+0,3755	+0,4740	+0,5714	+0,6677	+0,7630
2,080	−0,2528	−0,1457	−0,0398	+0,0649	+0,1681	+0,2699	+0,3704	+0,4697	+0,5680	+0,6651
2,392	−0,3877	−0,2799	−0,1729	−0,0668	+0,0381	+0,1419	+0,2444	+0,3458	+0,4460	+0,5451
2,744	−0,5404	−0,4321	−0,3244	−0,2172	−0,1109	−0,0055	+0,0987	+0,2020	+0,3042	+0,4054
3,140	−0,7123	−0,6038	−0,4956	−0,3878	−0,2805	−0,1739	−0,0681	+0,0368	+0,1408	+0,2439
3,727	−0,9672	−0,8586	−0,7502	−0,6419	−0,5338	−0,4262	−0,3190	−0,2124	−0,1065	−0,0013
4,020	−1,0945	−0,9859	−0,8774	−0,7690	−0,6607	−0,5527	−0,4452	−0,3380	−0,2315	−0,1256
4,783	−1,4258	−1,3172	−1,2086	−1,1001	−0,9916	−0,8833	−0,7751	−0,6671	−0,5594	−0,4522
5,740	−1,8414	−1,7328	−1,6242	−1,5157	−1,4071	−1,2986	−1,1901	−1,0818	−0,9735	−0,8654
7,175	−2,4646	−2,3560	−2,2474	−2,1389	−2,0303	−1,9217	−1,8131	−1,7047	−1,5961	−1,4877
9,567	−3,5035	−3,3949	−3,2863	−3,1778	−3,0692	−2,9606	−2,8520	−2,7435	−2,6349	−2,5263

$1/\lambda$	1,2500	1,1765	1,1111	1,0526	1,0000	0,9524	0,9091	0,8696	0,8333
c_2/T \ λ	0,80	0,85	0,90	0,95	1,00	1,05	1,10	1,15	1,20
0,000	+1,3897	+1,4766	+1,5634	+1,6503	+1,7372	+1,8240	+1,9109	+1,9977	+2,0846
0,603	+1,2425	+1,3302	+1,4180	+1,5056	+1,5932	+1,6806	+1,7681	+1,8554	+1,9428
0,880	+1,1643	+1,2531	+1,3417	+1,4301	+1,5184	+1,6065	+1,6946	+1,7825	+1,8704
1,216	+1,0612	+1,1516	+1,2416	+1,3314	+1,4209	+1,5101	+1,5992	+1,6880	+1,7767
1,594	+0,9356	+1,0281	+1,1201	+1,2116	+1,3028	+1,3935	+1,4839	+1,5741	+1,6640
1,816	+0,8576	+0,9514	+1,0447	+1,1373	+1,2296	+1,3213	+1,4127	+1,5037	+1,5944
2,080	+0,7613	+0,8568	+0,9516	+1,0456	+1,1392	+1,2322	+1,3247	+1,4168	+1,5085
2,392	+0,6433	+0,7406	+0,8371	+0,9329	+1,0280	+1,1226	+1,2166	+1,3101	+1,4031
2,744	+0,5056	+0,6049	+0,7033	+0,8009	+0,8978	+0,9941	+1,0897	+1,1848	+1,2794
3,140	+0,3460	+0,4472	+0,5477	+0,6472	+0,7461	+0,8442	+0,9416	+1,0384	+1,1345
3,727	+0,1030	+0,2067	+0,3096	+0,4116	+0,5129	+0,6135	+0,7134	+0,8127	+0,9112
4,020	−0,0203	+0,0843	+0,1881	+0,2913	+0,3936	+0,4953	+0,5963	+0,6967	+0,7964
4,783	−0,3453	−0,2390	−0,1332	−0,0280	+0,0767	+0,1808	+0,2842	+0,3870	+0,4892
5,740	−0,7576	−0,6500	−0,5428	−0,4359	−0,3294	−0,2234	−0,1178	−0,0127	+0,0919
7,175	−1,3793	−1,2710	−1,1628	−1,0548	−0,9471	−0,8394	−0,7320	−0,6250	−0,5182
9,567	−2,4177	−2,3092	−2,2007	−2,0922	−1,9838	−1,8754	−1,7670	−1,6588	−1,5506

b) Das Temperaturproblem in der Astrophysik.

9. Die Fixsterne als schwarze Strahler. Wenn die Energieverteilung im Spektrum eines Sternes bekannt ist, so scheint das einfachste Verfahren zu sein, die Temperatur nach dem WIENschen Verschiebungsgesetz zu bestimmen. Aus folgenden Gründen ist diese Methode im allgemeinen für die Praxis wenig geeignet: Das Maximum der Strahlungsintensität liegt bei vielen Sternen außerhalb des mit den üblichen Meßinstrumenten erreichbaren Spektralbereiches; nur die Sterne der mittleren Spektralklassen haben ihr Strahlungsmaximum innerhalb des visuell oder des photographisch wirksamen Wellenlängenbezirkes. Die Form der Energiekurve und damit die wahre Lage des Maximums wird durch die Empfindlichkeit der optischen Apparatur und durch die Absorption in der Erdatmosphäre stark beeinflußt; die unter verschiedenen Gesichtspunkten durchgeführte Elimination dieser die wahre Gestalt der Energiekurve verzerrenden Einflüsse hat eine Verschiebung des Energiemaximums zur Folge. Beobachtungsfehler rein zufälliger Art, herrührend von der Ungenauigkeit des Meßverfahrens, und systematische Auffassungsunterschiede, welche durch die nicht immer durchführbare Trennung der Absorptionslinien im Sternspektrum vom kontinuierlichen Untergrund bedingt sind, erschweren gleichfalls die exakte Festlegung des Maximums in der Energiekurve.

Die in der Praxis übliche Methode der Temperaturbestimmung fußt auf der Messung der Helligkeitsverteilung im kontinuierlichen Spektrum. Die Darstellbarkeit der Form der Energiekurve durch die PLANCKsche Strahlungsgleichung führt zu einem rechnerisch bestimmbaren Wert der effektiven Temperatur.

Die Messung der Gesamtstrahlung gibt bei Anwendung des STEFAN-BOLTZMANNschen Gesetzes ebenfalls einen Wert für die effektive Temperatur. Messungen der Strahlungsintensität innerhalb eines mehr oder minder weit begrenzten Spektralbereiches dienen dem gleichen Zweck. Die an die Gestalt der Energiekurve anknüpfende Methode besitzt gegenüber den letztgenannten den Vorzug, daß sie von der Kenntnis des scheinbaren Sterndurchmessers unabhängig ist.

Verhalten sich die Sterne wie schwarze Strahler, so liefern sämtliche Methoden den gleichen Wert der effektiven Temperatur; dieser ist dann gleich der wahren Temperatur der strahlenden Schicht.

10. Die Fixsterne als nicht schwarze Strahler. Wenn die Sternstrahlung von der des schwarzen Körpers abweicht, ist das WIENsche Verschiebungsgesetz im allgemeinen nicht anwendbar. Die formale Rechnung gibt mehr oder minder unsichere Werte für die Temperatur.

Wird die Temperatur aus der Gestalt der Energiekurve oder aus der Strahlungsintensität bestimmt, so führen diese Methoden bei nicht schwarzer Strahlung der Sterne gleichfalls meist nicht zu dem gleichen Wert. Über die Beziehung der wahren zur effektiven Temperatur kann man a priori nichts Bestimmtes aussagen. Die Darstellbarkeit der Intensitätsverteilung im Sternspektrum durch die PLANCKsche Strahlungsgleichung beweist noch nicht, daß die aus der Form der Energiekurve berechnete Temperatur eine für die wahren Temperaturverhältnisse des strahlenden Körpers charakteristische Konstante ist.

Die von WILSING und SCHEINER[1] ausgeführten Messungen in den Spektren des Mondes, des Jupiter und des Mars ergaben, daß sich auch für diese das Sonnenlicht reflektierenden Himmelskörper effektive Temperaturen aus der Gestalt der Energiekurve berechnen lassen. Da die Reflexionskoeffizienten in dem optischen Spektralgebiet mit einer den Beobachtungsfehlern entsprechenden Ge-

---

[1] J. WILSING u. J. SCHEINER, Spektralphotometrische Messungen an Gesteinen, am Monde, Mars und Jupiter. Publ Astrophys Obs Potsdam Nr. 77 (1921).

nauigkeit durch einen Ausdruck von der Form $R_\lambda = a \cdot e^{-c_2/\lambda b^2}$ dargestellt werden, wo a und b konstante Größen sind, so wird die Energie der reflektierten Sonnenstrahlung gemäß der WIENschen Strahlungsgleichung:

$$J_\lambda = \frac{c_1}{\lambda^5} \cdot e^{-\frac{c_2}{\lambda}\left(\frac{1}{T} + \frac{1}{b^2}\right)} = \frac{c_1}{\lambda^5} \cdot e^{-\frac{c_2}{\lambda T'}}, \tag{13}$$

wo T die Sonnentemperatur und T' die effektive Temperatur des das Sonnenlicht reflektierenden Himmelskörpers bedeuten. Es würde unsinnig sein, die Temperatur T' mit der wahren Temperatur in irgendeine Beziehung zu bringen.

Zu den selbstleuchtenden Körpern, welche die Voraussetzung der Darstellbarkeit durch die PLANCKsche Strahlungsgleichung erfüllen, gehören neben den schwarzen die grauen Strahler. Bei diesen ist das Absorptionsvermögen A_λ unabhängig von der Wellenlänge λ und kleiner als die Einheit. Die Gestalt der Energiekurve ist bei dem schwarzen und bei dem grauen Strahler gleich; aus der Intensitätsverteilung im Spektrum folgt daher derselbe Wert der effektiven Temperatur. Auch die wahren Temperaturen sind einander gleich.

Besondere Beachtung verdienen die selbstleuchtenden Körper, bei denen das Absorptionsvermögen für einen gewissen der Messung zugänglichen Teil des Spektrums in einer besonders einfachen Beziehung zur Wellenlänge steht. Gehört die Sternphotosphäre zu der Klasse von Strahlern, bei welchen nach der Gleichung $\log A_\lambda = -\alpha^2 + \frac{\beta^2}{\lambda}$ eine Abnahme des Absorptionskoeffizienten mit wachsender Wellenlänge stattfindet, so ist die aus der Form der Energiekurve berechnete effektive Temperatur größer als die wahre[1]. Umgekehrt ist, wenn der Absorptionskoeffizient nach der Gleichung $\log A_\lambda = -\alpha^2 - \frac{\beta^2}{\lambda}$ mit wachsender Wellenlänge zunimmt, die effektive Temperatur kleiner als die wahre.

Im allgemeinsten Fall läßt sich die Energiekurve eines Sternes aus verschiedenen Teilen zusammensetzen, für welche der Absorptionskoeffizient mit der Wellenlänge entweder zu- oder abnimmt. Demgemäß wird man aus der Gestalt der Energiekurve in den einzelnen Spektralbereichen verschieden hohe Temperaturen erhalten.

Die Messung der Strahlungsintensität innerhalb eines mehr oder weniger weit begrenzten Spektralbezirkes führt zu einer effektiven Temperatur, die bei nicht schwarzer Strahlung im allgemeinen von der aus der Intensitätsverteilung im Spektrum abgeleiteten abweicht und stets kleiner als die wahre Temperatur ist. Für selbstleuchtende Körper mit konstantem Absorptionsvermögen ist die Temperatur aus der Strahlungsintensität um $(A_\lambda)^{1/4}$ niedriger als aus der Gestalt der Energiekurve. Wenn der Absorptionskoeffizient mit wachsender Wellenlänge abnimmt, ist die aus der Strahlungsintensität berechnete effektive Temperatur durchweg kleiner; wenn jener dagegen zunimmt, kann sie kleiner oder auch größer als die aus der Form der Energiekurve berechnete Temperatur sein.

11. Die Definition der Sterntemperaturen. Erfahrungsgemäß strahlen die Sterne nur annähernd wie schwarze Körper; die Intensitätsverteilung im kontinuierlichen Spektrum der Sterne läßt sich nicht genau durch die PLANCKsche Strahlungsgleichung darstellen. Je nach dem Spektralbereich liefern die spektralphotometrischen Messungen desselben Sterns verschiedene Temperaturen. Die differentielle Verknüpfung der Integralhelligkeiten (photographischer, lichtelek-

[1] Die Folgerung in Ziffer 5, wonach die effektive Temperatur ein unterer Grenzwert der wahren Temperatur ist, fußt auf der Strahlungsintensität. Das Beispiel des mit wachsender Wellenlänge abnehmenden Absorptionskoeffizienten illustriert in markanter Weise, zu welch absurden Resultaten man kommen kann, wenn man die effektive Temperatur aus der Gestalt der Energiekurve bestimmt.

trischer Farbenindex, Wärmeindex, Wasserzellenabsorption) führt zu Temperaturen, die in der Regel nicht miteinander übereinstimmen. Die aus der Strahlungsintensität der Sterne berechneten Temperaturen weichen gleichfalls mehr oder weniger voneinander ab.

Da die Sterne sich nicht genau wie schwarze Körper verhalten, stellt die Bestimmung von Sterntemperaturen eine schwierige Aufgabe dar, die noch dadurch komplizierter wird, daß den Bedürfnissen der Theorie und der Beobachtung nicht mit der gleichen Temperatur genügt wird.

Die Temperaturdefinition, welche die nicht schwarze Strahlung der Sterne charakterisieren soll, hat naturgemäß an die Strahlung des schwarzen Körpers anzuknüpfen, deren Gesetze wohlbekannt sind. Die in der Praxis meist benutzte Methode der Temperaturbestimmung fußt auf der Messung der Helligkeitsverteilung im kontinuierlichen Spektrum der Sterne. Die Darstellbarkeit der Form der Energiekurve durch die PLANCKsche Strahlungsgleichung in einem gewissen Spektralbereich führt zu einem rechnerisch bestimmbaren Wert der Temperatur, die als „spektralphotometrische Farbtemperatur“ bezeichnet wird. Nur die ideal grauen Strahler, bei denen die Intensität der schwarzen Strahlung für alle Wellenlängen um den gleichen Bruchteil vermindert ist, besitzen eine für den gesamten Wellenlängenbereich gültige Farbtemperatur. Da die Sterne sich nicht wie graue Strahler verhalten, unterscheiden sich die spektralphotometrischen Farbtemperaturen je nach dem Spektralbereich, aus dem sie bestimmt sind (spektralphotometrische Farbtemperaturskalen von WILSING, ROSENBERG und SAMPSON). Wenn die Abweichungen der Sternstrahlung von der des schwarzen Körpers groß sind, wird man nur innerhalb eines kleinen Spektralbereichs mit einer den Beobachtungsfehlern entsprechenden Genauigkeit eine Angleichung der beobachteten Energiekurve an den schwarzen Strahler erzielen. Im Grenzfall des unendlich kleinen Spektralbereichs wird die Energiekurve des Sterns für jede Wellenlänge demjenigen schwarzen Strahler zugeordnet, für welchen die Richtung der Tangente an die Energiekurve des Sterns und des schwarzen Strahlers übereinstimmt. Die so definierte Farbtemperatur bezeichnet SIEDENTOPF[1] als „Gradationstemperatur“. Unebenheiten im Verlauf der Energiekurve des Sterns, wie sie, abgesehen von den eigentlichen Beobachtungsfehlern, durch die Absorptions- und Emissionslinien im Spektrum bedingt sind, erschweren die genaue Bestimmung der Gradationstemperatur.

Sind die systematischen Abweichungen der Energiekurve des Sterns von der des schwarzen Strahlers von wechselnder Größe und von verschiedenem Vorzeichen, so unterscheiden sich die spektralphotometrischen Farbtemperaturen je nach dem Spektralbereich voneinander. Will man einen für den Stern typischen Wert der Farbtemperatur aus der Gestalt der Energiekurve ableiten, so wird man einen möglichst großen Spektralbereich auswählen; die zugehörige Temperatur ist dann so zu bestimmen, daß ein Ausgleich zwischen den positiven und negativen Abweichungen der Energiekurven des Sterns und des schwarzen Strahlers stattfindet.

Die spektralphotometrischen Messungen beziehen sich auf Spektralbereiche, die mehr oder weniger frei von Absorptions- und Emissionslinien sind. Die Sternfarbe und ihre Äquivalente (effektive Wellenlänge, Farbenindex, Wärmeindex und Wasserzellenabsorption) bilden ein Kollektivmaß für die Intensitätsverteilung im Sternspektrum und begreifen die Wirkung der Absorptions- und Emissionslinien ein. Wenn auch bei Einführung des Begriffs der isophoten Wellenlänge (Ziffer 21) der Farbenindex, der Wärmeindex und die Wasserzellenabsorption

[1] H. SIEDENTOPF, Grundlagen der Kosmogonie. Veröff Universitätssternw Göttingen, Heft 3 (1928).

als Energiestufen in der Intensitätsverteilung des kontinuierlichen Spektrums sich deuten lassen und damit die Bezeichnung der aus jenen bestimmten Temperaturen als Gradationstemperaturen im Sinne SIEDENTOPFS angebracht erscheint, werden im folgenden allgemein die aus der Farbe oder ihren Äquivalenten berechneten Temperaturen kurz „Farbtemperaturen" genannt.

Die spektralphotometrische Farbtemperatur, die Gradationstemperatur und die Farbtemperatur charakterisieren die Form der Energiekurve. Fast alle bisher veröffentlichten Sterntemperaturen sind aus der Gestalt der Energiekurve abgeleitet. Die Abweichungen der Sternstrahlung von der des schwarzen Körpers prägen sich am deutlichsten in der Gradationstemperatur aus; diese ist daher ein geeignetes Maß, um die Abweichungen vom PLANCKschen Strahlungsgesetz zahlenmäßig festzulegen. Die Energiekurve des Sterns kann bei irgend einer Wellenlänge oder in einem gewissen Spektralbereich selbst durch einen schwarzen Strahler unendlich hoher Temperatur nicht darstellbar sein. Der auf der Gestalt der Energiekurve fußenden Temperatur ist daher a priori eine reale, die Intensität der Strahlung charakterisierende Eigenschaft abzusprechen.

Für die stellarastronomische Forschung hat meist nur die Strahlungstemperatur, welche durch die Intensität der Sternstrahlung bestimmt wird, Interesse. Man bezeichnet als Strahlungstemperatur für einen gewissen Spektralbereich diejenige Temperatur, welche ein schwarzer Körper haben muß, um die Strahlung dieses Bereichs in der gleichen Intensität zu emittieren wie der Stern. Wenn bolometrische, visuelle, photographische oder lichtelektrische Helligkeiten vorliegen, unterscheidet man die Strahlungstemperatur der bolometrischen, visuellen, photographischen und lichtelektrischen Helligkeit. Wenn der Spektralbereich unendlich klein ist, nennt man die zu der Strahlungsintensität der Wellenlänge λ gehörige Temperatur schwarze Temperatur.

Die Strahlungstemperaturen und die schwarzen Temperaturen lassen sich im allgemeinen nicht direkt bestimmen, weil man hierzu den scheinbaren Sternradius kennen muß. Die an die Gestalt der Energiekurve anknüpfende Methode der Temperaturbestimmung hat vor der die Strahlungsintensität benutzenden den Vorzug, daß sie von der Kenntnis des scheinbaren Sternradius unabhängig ist.

Bei schwarzer Strahlung entspricht der Emission jeder Schwingungszahl die gleiche Temperatur; Farb-, Gradations-, Strahlungs- und schwarze Temperatur sind im ganzen Wellenlängenbereich einander gleich. Bei ideal grauer Strahlung sind Farb- und Gradationstemperatur für alle Wellenlängen gleich. Die schwarze Temperatur des ideal grauen Strahlers ist in den einzelnen Wellenlängen verschieden groß; ihr reziproker Wert ist eine lineare Funktion der Wellenlänge.

Bei den Fixsternen ist die Voraussetzung der schwarzen oder der ideal grauen Strahlung im allgemeinen nicht erfüllt. Schwarze, Gradations-, Strahlungs- und Farbtemperatur sind mehr oder weniger voneinander verschieden. Wenn die Strahlungstemperaturen zweier Integralhelligkeiten einander gleich sind, gibt der aus letzteren gebildete Farbenindex den gleichen Wert der Farbtemperatur. Die Strahlungsintensitäten in der Energiekurve des Sterns und des schwarzen Strahlers sind für die isophoten Wellenlängen der Integralhelligkeiten einander gleich. Trotzdem kann der Verlauf der Energiekurve des Sterns in den Spektralbereichen, deren Grenzen durch die spektrale Empfindlichkeit der die Integralhelligkeiten messenden Apparaturen bestimmt sind, mehr oder weniger von der Energiekurve des schwarzen Strahlers abweichen. Nur muß innerhalb der Grenzen der Empfindlichkeitsbereiche ein Ausgleich zwischen den positiven und den negativen Abweichungen der Energiekurven des Sterns und des schwarzen Strahlers stattfinden. Wenn bei einer gewissen Wellenlänge Gradations- und schwarze Temperatur einander gleich sind, so haben hier die Energiekurven des Sterns und des schwarzen Strahlers die gleiche Intensität und den gleichen Gradienten. Der Stern verhält sich bei dieser Wellenlänge wie ein vollkommen schwarzer Körper.

Zwischen der Farb- und der Strahlungstemperatur wie auch zwischen der Gradations- und der schwarzen Temperatur bestehen analytische Beziehungen, welche den Übergang von der durch die Messung bestimmten Farb- oder Gradationstemperatur zu der für theoretische Untersuchungen wichtigen Strahlungs- oder schwarzen Temperatur vermitteln.

12. Die Beziehung zwischen der Farbtemperatur und der Strahlungstemperatur. Die aus dem Farbenindex, dem Wärmeindex oder der Wasserzellenabsorption abgeleitete Farbtemperatur ist dadurch bestimmt, daß die Energiekurven des Sterns und des schwarzen Strahlers zwischen den isophoten Wellenlängen der Integralhelligkeiten, aus denen der Farbenindex, der Wärmeindex oder die Wasserzellenabsorption gebildet ist, gleiche Energiestufen besitzen. Da der Nullpunkt der Integralhelligkeiten der empirischen Größensysteme in willkürlicher Weise festgelegt ist, sind die beobachteten Helligkeiten um eine Konstante zu korrigieren, die von photometrischem zu photometrischem System verschieden ist (vgl. Ziffer 21).

Es bedeuten λ_1 und λ_2 die isophoten Wellenlängen der Integralhelligkeiten, aus denen der Farbenindex, der Wärmeindex oder die Wasserzellenabsorption gebildet ist; $i(\lambda_1)$ und $i(\lambda_2)$ sind die zu den isophoten Wellenlängen gehörigen Intensitäten (Flächenhelligkeiten) der Sternstrahlung[1].

Für die weiteren Entwicklungen ist es vorteilhaft, statt der Intensitäten ihre natürlichen Logarithmen einzuführen. Die im System der Größenklassen aus den Beobachtungen bestimmten Werte des Farbenindex, des Wärmeindex und der Wasserzellenabsorption sind nach der Reduktion auf die durch die isophoten Wellenlängen gegebene Energiestufe gleich $-2{,}5 \log e [\ln i(\lambda_1) - \ln i(\lambda_2)]$, wo log den Briggsschen, ln den natürlichen Logarithmus und e die Basis der natürlichen Logarithmen bezeichnet.

Der analytische Ausdruck der Planckschen Funktion ist nach Gleichung (4):

$$E(\lambda, T) = \frac{2c_1}{\lambda^5} \cdot \frac{e^{-c_2/\lambda T}}{1 - e^{-c_2/\lambda T}}.$$

Die Konstanten $2c_1$ und c_2 bzw. ihre Briggsschen Logarithmen sind nach Ziffer 7:

$2c_1 = 1{,}1778 \cdot 10^{-5}\,[0{,}07107-5]$ erg cm² sec⁻¹, $\quad c_2 = 1{,}432\,[0{,}15594]$ cm grad.

Die Farbtemperatur T_F, welche zu der durch die Messung erfaßbaren Helligkeitsdifferenz $\ln i(\lambda_1) - \ln i(\lambda_2)$ gehört, wird durch die Gleichung bestimmt:

$$\ln i(\lambda_1) - \ln i(\lambda_2) = \ln E(\lambda_1, T_F) - \ln E(\lambda_2, T_F).$$

Führt man in diese Gleichung den analytischen Ausdruck von $E(\lambda, T)$ ein, so wird:

$$\left.\begin{aligned}\ln i(\lambda_1) - \ln i(\lambda_2) = -5 \ln \lambda_1 + 5 \ln \lambda_2 - \frac{c_2}{\lambda_1 T_F} + \frac{c_2}{\lambda_2 T_F} - \ln\left(1 - e^{-c_2/\lambda_1 T_F}\right)\\ + \ln\left(1 - e^{-c_2/\lambda_2 T_F}\right)\end{aligned}\right\}. \tag{14}$$

Bezeichnen T_{St_1} und T_{St_2} die Strahlungstemperaturen, welche zu den Integralhelligkeiten gehören, so ist gemäß Definition:

$$\ln i(\lambda_1) - \ln i(\lambda_2) = \ln E(\lambda_1, T_{\mathrm{St}_1}) - \ln E(\lambda_2, T_{\mathrm{St}_2})$$

oder:

$$\left.\begin{aligned}\ln i(\lambda_1) - \ln i(\lambda_2) = -5 \ln \lambda_1 + 5 \ln \lambda_2 - \frac{c_2}{\lambda_1 T_{\mathrm{St}_1}} + \frac{c_2}{\lambda_2 T_{\mathrm{St}_2}} - \ln\left(1 - e^{-c_2/\lambda_1 T_{\mathrm{St}_1}}\right)\\ + \ln\left(1 - e^{-c_2/\lambda_2 T_{\mathrm{St}_2}}\right)\end{aligned}\right\}. \tag{15}$$

[1] Die Wellenlängen λ_1 und λ_2 können natürlich auch zwei Stellen bezeichnen, an denen die Intensität des kontinuierlichen Spektrums gemessen worden ist.

Wenn man die rechten Seiten der Gleichungen (14) und (15) einander gleich setzt, erhält man die gesuchte Beziehung zwischen der Farbtemperatur T_F und den Strahlungstemperaturen T_{St_1} und T_{St_2}:

$$\frac{c_2}{\lambda_1 T_{St_1}} - \frac{c_2}{\lambda_2 T_{St_2}} + \ln\left(\frac{1 - e^{-c_2/\lambda_1 T_{St_1}}}{1 - e^{-c_2/\lambda_2 T_{St_2}}}\right) = \frac{c_2}{T_F}\left(\frac{1}{\lambda_1} - \frac{1}{\lambda_2}\right) + \ln\left(\frac{1 - e^{-c_2/\lambda_1 T_F}}{1 - e^{-c_2/\lambda_2 T_F}}\right). \tag{16}$$

Läßt man in erster Näherung die logarithmischen Glieder der Gleichung (16) fort, d.h. berücksichtigt man nicht die Abweichung des PLANCKschen vom WIENschen Strahlungsgesetz, so wird:

$$T_{St_1} = T_F \frac{1 - \frac{\lambda_1}{\lambda_2} \cdot \frac{T_{St_1}}{T_{St_2}}}{1 - \frac{\lambda_1}{\lambda_2}}. \tag{17}$$

Die Integralhelligkeiten der Sterne lassen sich deuten:

1. als Strahlung des schwarzen Körpers von der Temperatur T_{St},
2. als graue Strahlung der Farbtemperatur T_F und des Emissionsvermögens F.

Das Emissionsvermögen F hat hier nur rechnerische Bedeutung; es kann sowohl größer als auch kleiner als die Einheit sein und ist im allgemeinen verschieden je nach der Art der Verbindung der Integralhelligkeiten. Nach Definition ist:

$$i(\lambda_1) = E(\lambda_1, T_{St_1}); \; i(\lambda_2) = E(\lambda_2, T_{St_2}); \; i(\lambda_1) = F \cdot E(\lambda_1, T_F); \; i(\lambda_2) = F \cdot E(\lambda_2, T_F).$$

Die logarithmische Rechnung liefert dann:

$$\left.\begin{aligned} \frac{c_2}{T_{St_1}} + \lambda_1 \ln(1 - e^{-c_2/\lambda_1 T_{St_1}}) &= \frac{c_2}{T_F} + \lambda_1 \ln(1 - e^{-c_2/\lambda_1 T_F}) - \lambda_1 \ln F, \\ \frac{c_2}{T_{St_2}} + \lambda_2 \ln(1 - e^{-c_2/\lambda_2 T_{St_2}}) &= \frac{c_2}{T_F} + \lambda_2 \ln(1 - e^{-c_2/\lambda_2 T_F}) - \lambda_2 \ln F. \end{aligned}\right\} \tag{18}$$

13. Die Beziehung zwischen der Gradationstemperatur und der schwarzen Temperatur. Die Beziehung zwischen der Gradationstemperatur und der schwarzen Temperatur führt zu interessanten Schlußfolgerungen, welche für die Theorie wie auch für die Praxis der Beobachtung gleich wertvoll sind. Es sei $i(\lambda)$ die Flächenhelligkeit der Sternstrahlung für die Wellenlänge λ. Die schwarze Temperatur T_S des Sterns für die Wellenlänge λ ist dadurch bestimmt, daß die Flächenhelligkeiten des Sterns und des schwarzen Strahlers einander gleich sind. In logarithmischer Form schreibt sich die Definitionsgleichung der schwarzen Temperatur T_S:

$$\ln i(\lambda) = \ln E(\lambda, T_S)$$

oder:

$$\ln i(\lambda) = \ln 2c_1 - 5\ln\lambda - \frac{c_2}{\lambda T_S} - \ln(1 - e^{-c_2/\lambda T_S}). \tag{19}$$

Die schwarze Temperatur ist im allgemeinen eine Funktion der Wellenlänge. Die Gradationstemperatur T_G wird dadurch bestimmt, daß die Energiekurven des Sterns und des schwarzen Strahlers für jedes λ den Gradienten gemeinsam haben. In logarithmischer Form schreibt sich die Definitionsgleichung der Gradationstemperatur:

$$\frac{d}{d\,1/\lambda} \ln i(\lambda) = \frac{\partial}{\partial\,1/\lambda} \ln E(\lambda, T_G),$$

wo $\frac{\partial}{\partial\,1/\lambda} \ln E(\lambda, T_G)$ die partielle Ableitung von $\ln E(\lambda, T_G)$ nach der reziproken Wellenlänge $1/\lambda$ bedeutet. Setzt man für $E(\lambda, T_G)$ seinen analytischen Ausdruck, so wird:

$$\frac{d}{d\,1/\lambda} \ln i(\lambda) = 5\lambda - \frac{c_2}{T_G} \cdot \frac{1}{1 - e^{-c_2/\lambda T_G}}. \tag{20}$$

Die Gradationstemperatur T_G wird im allgemeinen ebenfalls eine Funktion der Wellenlänge sein[1].

Erfahrungsgemäß sind die Abweichungen der Sternstrahlung von der schwarzen Strahlung meist klein; deshalb sind auch die Änderungen der schwarzen Temperatur mit der Wellenlänge von geringem Betrage. Kleine Abweichungen der Sternstrahlung von der schwarzen Strahlung können aber große Änderungen der Gradationstemperatur zur Folge haben, die nach Definition von dem Gradienten der Energiekurve des Sterns abhängt.

Um die Beziehung zwischen der Gradationstemperatur und der schwarzen Temperatur abzuleiten, deutet man das Differential $d \ln i(\lambda)$ in verschiedener Weise:

1. als Variation von $\ln E(\lambda, T_S)$, wo sich die Wellenlänge λ um $d\lambda$ und die schwarze Temperatur T_S um dT_S ändern:

$$d \ln i(\lambda) = d \ln E(\lambda, T_S) = \frac{\partial}{\partial \lambda} \ln E(\lambda, T_S)\, d\lambda + \frac{\partial}{\partial T_S} \ln E(\lambda, T_S)\, dT_S$$

oder:

$$d \ln i(\lambda) = -\frac{5}{\lambda} d\lambda + \frac{c_2}{T_S} \cdot \frac{1}{\lambda^2} \cdot \frac{1}{1 - e^{-c_2/\lambda T_S}} d\lambda + \frac{c_2}{T_S^2} \cdot \frac{1}{\lambda} \cdot \frac{1}{1 - e^{-c_2/\lambda T_S}} \cdot dT_S; \qquad (21)$$

2. als Variation von $\ln E(\lambda, T_G)$, wo sich die Wellenlänge λ um $d\lambda$ ändert:

$$d \ln i(\lambda) = \frac{\partial}{\partial \lambda} \ln E(\lambda, T_G)\, d\lambda$$

oder

$$d \ln i(\lambda) = -\frac{5}{\lambda} d\lambda + \frac{c_2}{T_G} \cdot \frac{1}{\lambda^2} \cdot \frac{1}{1 - e^{-c_2/\lambda T_G}} d\lambda,$$

welches die Definitionsgleichung (20) der Gradationstemperatur ist.

Die Verbindung der Gleichungen (20) und (21) gibt die folgende Differentialgleichung erster Ordnung in c_2/T_S:

$$\frac{d\, c_2/T_S}{d\lambda} = \frac{c_2}{\lambda T_S} - \frac{c_2}{\lambda T_G} \cdot \frac{1 - e^{-c_2/\lambda T_S}}{1 - e^{-c_2/\lambda T_G}}. \qquad (22)$$

c_2/T_G ist durch die Beobachtungen als Funktion der Wellenlänge gegeben; die Differentialgleichung stellt die gesuchte Beziehung zwischen der Gradationstemperatur und der schwarzen Temperatur her. Die Diskussion der Differentialgleichung führt zu folgendem Resultat: Ist $c_2/T_G = c_2/T_S$, so wird $\frac{d\, c_2/T_S}{d\lambda} = 0$. Wenn die Differentialgleichung (22) gelöst und c_2/T_S als Funktion von λ gefunden ist, so schneiden sich die Kurven, welche die Abhängigkeit der Funktionen c_2/T_S und c_2/T_G von der Wellenlänge zeichnerisch zur Darstellung bringen, in denjenigen Wellenlängen, in welchen die Temperatur T_S ein Maximum oder ein Minimum hat. Solange $c_2/T_S > c_2/T_G$ ist, nimmt c_2/T_S mit wachsender Wellenlänge zu; umgekehrt, solange $c_2/T_S < c_2/T_G$ ist, nimmt c_2/T_S mit wachsender Wellenlänge ab.

Durch Einführung der neuen Veränderlichen $z = c_2/\lambda T_S$ lassen sich die Variablen z und λ der Differentialgleichung (22) separieren:

$$\frac{dz}{1 - e^{-z}} = \frac{c_2/T_G}{1 - e^{-c_2/\lambda T_G}} \cdot d\,\frac{1}{\lambda}. \qquad (23)$$

[1] Bei der Reduktion der Beobachtungen erweist sich eine Tabelle der $\frac{d}{d\, 1/\lambda} \ln E(\lambda, T)$ mit den Argumenten λ und c_2/T als vorteilhaft. In diesem Falle ist der relative Gradient der Energiekurven zweier Sterne nach Gleichung (20) in erster Näherung gleich der Differenz der zugehörigen c_2/T_G-Werte. Da natürliche Logarithmen in der Praxis wenig benutzt werden, enthält die Tabelle 4 die Werte der $\frac{d}{d\, 1/\lambda} \log E(\lambda, T)$.

Die Integration der Differentialgleichung (23) gibt:

$$z + \ln(1 - e^{-z}) = \int \frac{c_2/T_G}{1 - e^{-c_2/\lambda T_G}} \cdot d\,\frac{1}{\lambda}\,. \tag{24}$$

Der Wellenlänge $\lambda = 0$ entspricht eine Singularität; man wird die Nachbarschaft dieser Stelle ausschließen dürfen, da sie durch die Messung nicht erreichbar ist.

Es sei λ_0 die untere Grenze des Integrals in der Gleichung (24), wo $\lambda_0 > 0$ ist. In der Praxis ist λ_0 in der Regel nicht kleiner als 300 $\mu\mu$, d. i. die durch die Messung erreichbare ultraviolette Grenze des Sternspektrums. Die Beobachtung gibt c_2/T_G als Funktion von λ; der Wellenlänge $\lambda = \lambda_0$ entspricht der Wert $(c_2/T_G)_{\lambda=\lambda_0}$. Zu $\lambda = \lambda_0$ soll weiterhin der Wert $z = z_0$ gehören. Die obere Grenze des Integrals wird unbestimmt gelassen. Nach Einführung der unteren Grenze λ_0 und der oberen Grenze λ schreibt sich die Gleichung (24):

$$z + \ln(1 - e^{-z}) = z_0 + \ln(1 - e^{-z_0}) + \int_{\lambda_0}^{\lambda} \frac{c_2/T_G}{1 - e^{-c_2/\lambda T_G}}\, d\,\frac{1}{\lambda}\,. \tag{25}$$

14. Die Sterne als ideal graue oder als selektive Strahler. Man kann die Gleichung (25) in eine andere, für die Diskussion bequemere Form bringen, wenn man das Emissionsvermögen einführt. Die Emission $e(\lambda, T)$ eines Körpers ist mit dem Absorptionsvermögen $A(\lambda, T)$ nach Ziffer 1 durch das KIRCHHOFFsche Gesetz verknüpft:

$$\frac{e(\lambda, T)}{A(\lambda, T)} = E(\lambda, T)\,,$$

wo $E(\lambda, T)$ die von dem schwarzen Körper der absoluten Temperatur T emittierte Strahlung ist. Die Emission $e(\lambda, T)$ wird auf die Emission des schwarzen Körpers $E(\lambda, T)$ als Einheit bezogen. Man bezeichnet das Verhältnis der Emission des beliebigen Körpers zu der des schwarzen Körpers als sein Emissionsvermögen. Je nach der Abhängigkeit des Emissionsvermögens von der Wellenlänge unterscheidet man grau strahlende und selektiv strahlende Körper. Ist das Emissionsvermögen für alle Wellenlängen konstant, dann strahlt der Körper ideal grau; ändert sich das Emissionsvermögen mit der Wellenlänge, dann strahlt der Körper selektiv.

Beim ideal grauen Strahler ist c_2/T_G für alle Wellenlängen konstant. Führt man im Integral der Gleichung (25) statt $\frac{1}{\lambda}$ die Veränderliche $x = \frac{c_2}{\lambda T_G}$ ein und ist $x = x_0 = \frac{c_2}{\lambda_0 T_G}$ für $\lambda = \lambda_0$, so wird:

$$z + \ln(1 - e^{-z}) = z_0 + \ln(1 - e^{-z_0}) + \int_{x_0}^{x} \frac{dx}{1 - e^{-x}}$$

oder:

$$z + \ln(1 - e^{-z}) - x - \ln(1 - e^{-x}) = z_0 + \ln(1 - e^{-z_0}) - x_0 - \ln(1 - e^{-x_0}) = C, \tag{26}$$

wo C eine Konstante ist, deren Bedeutung im folgenden erklärt wird. Führt man in der Gleichung (26) statt z und x wieder die ursprünglichen Größen $c_2/\lambda T_S$ und $c_2/\lambda T_G$ ein, so wird:

$$\frac{c_2}{T_S} + \lambda \ln(1 - e^{-c_2/\lambda T_S}) = \frac{c_2}{T_G} + \lambda \ln(1 - e^{-c_2/\lambda T_G}) + C \cdot \lambda\,. \tag{27}$$

Wenn man in erster Näherung die logarithmischen Glieder, welche von der Abweichung des PLANCKschen vom WIENschen Strahlungsgesetz herrühren, unberücksichtigt läßt, wird die reziproke schwarze Temperatur eine lineare Funktion der Wellenlänge λ.

Bezeichnet man beim ideal grauen Strahler das für alle Wellenlängen konstante Emissionsvermögen mit F, so ist:

$$F = \frac{E(\lambda, T_S)}{E(\lambda, T_G)}. \tag{28}$$

Führt man für die PLANCKsche Funktion $E(\lambda, T)$ ihren analytischen Ausdruck ein und logarithmiert die Gleichung (28), so erhält man:

$$\frac{c_2}{T_S} + \lambda \ln(1 - e^{-c_2/\lambda T_S}) = \frac{c_2}{T_G} + \lambda \ln(1 - e^{-c_2/\lambda T_G}) - \lambda \ln F. \tag{29}$$

In der Lösung der Differentialgleichung (26) für den ideal grauen Strahler hat die Konstante C die Bedeutung $-\ln F$, wo F das Emissionsvermögen des grauen Strahlers von der Farbtemperatur T_G ist.

In dem allgemeinen Fall der nicht schwarzen Strahlung wird die Sternstrahlung für jede Wellenlänge λ gedeutet:

1. als Strahlung des schwarzen Körpers von der Temperatur T_S,
2. als selektive Strahlung der Gradationstemperatur T_G und des Emissionsvermögens F.

T_G, T_S und F sind Funktionen der Wellenlänge λ. Das Emissionsvermögen $F(\lambda)$ hat nur rechnerische Bedeutung; es charakterisiert die Abweichungen der Sternstrahlung von der schwarzen Strahlung. $F(\lambda)$ kann sowohl größer als auch kleiner als 1 sein. In denjenigen Wellenlängen, für welche $F(\lambda)$ gleich 1 ist, strahlt der Stern wie ein vollkommen schwarzer Körper; Gradationstemperatur und schwarze Temperatur sind einander gleich.

Nach Definition ist:

$$F(\lambda) = \frac{E(\lambda, T_S)}{E(\lambda, T_G)}. \tag{30}$$

Führt man wieder für die PLANCKsche Funktion $E(\lambda, T)$ ihren analytischen Ausdruck ein und logarithmiert die Gleichung (30), so wird:

$$\frac{c_2}{\lambda T_S} + \ln(1 - e^{-c_2/\lambda T_S}) = \frac{c_2}{\lambda T_G} + \ln(1 - e^{-c_2/\lambda T_G}) - \ln F(\lambda). \tag{31}$$

Subtrahiert man die Gleichungen (24) und (31) voneinander und differenziert total nach λ, so wird, da $z = \frac{c_2}{\lambda T_S}$ ist:

$$\frac{d}{d\lambda} \ln F(\lambda) = \frac{1}{\lambda} \cdot \frac{1}{1 - e^{-c_2/\lambda T_G}} \cdot \frac{d\, c_2/T_G}{d\lambda}. \tag{32}$$

Durch Integration erhält man:

$$\ln F(\lambda) - \ln F(\lambda_0) = \int_{\lambda_0}^{\lambda} \frac{1}{\lambda} \cdot \frac{1}{1 - e^{-c_2/\lambda T_G}} \cdot \frac{d\, c_2/T_G}{d\lambda}\, d\lambda, \tag{33}$$

wo $F(\lambda_0)$ das Emissionsvermögen für die Wellenlänge $\lambda = \lambda_0$ bedeutet.

Man kann das Integral auf der rechten Seite der Gleichung (33) nach numerischen oder graphischen Integrationsmethoden berechnen, wenn c_2/T_G durch spektralphotometrische Messungen als Funktion von λ bekannt ist. Die Bestimmung der Konstanten $F(\lambda_0)$ bereitet Schwierigkeiten und verlangt gewisse Voraussetzungen, die Erfahrungstatsachen verwerten, mit dem Temperaturproblem aber an und für sich nichts zu tun haben. Die Sterne strahlen erfahrungsgemäß annähernd wie schwarze Körper; die Sternstrahlung weicht im allgemeinen nur wenig von der schwarzen Strahlung ab. Man kann die Konstante $F(\lambda_0)$ durch die Forderung bestimmen, daß:

$$\int_{\lambda_1}^{\lambda_2} \ln F(\lambda)\, d\lambda = 0 \tag{34}$$

ist; dann wird:

$$\ln F(\lambda_0) = \frac{1}{\lambda_1 - \lambda_2} \int_{\lambda_1}^{\lambda_2} \int_{\lambda_0}^{\lambda} \frac{1}{\lambda} \cdot \frac{1}{1 - e^{-c_2/\lambda T_G}} \cdot \frac{d\, c_2/T_G}{d\lambda}\, d\lambda\, d\lambda . \tag{35}$$

Notwendige Bedingung müßte natürlich sein, daß in den Grenzen λ_1 und λ_2 ein möglichst großer Spektralbereich eingeschlossen ist. Gleichwohl entbehrt diese Art der Bestimmung von $\ln F(\lambda_0)$, allgemein angewandt, nicht einer gewissen Willkür[1].

Einfacher liegt der Fall, wenn bei einer gewissen Wellenlänge λ_a Gradationstemperatur und schwarze Temperatur einander gleich sind; dann ist:

$$\ln F(\lambda_a) = 0 \tag{36}$$

und:

$$\ln F(\lambda) = \int_{\lambda_a}^{\lambda} \frac{1}{\lambda} \cdot \frac{1}{1 - e^{-c_2/\lambda T_G}} \cdot \frac{d\, c_2/T_G}{d\lambda}\, d\lambda . \tag{37}$$

c) Die Farbtemperatur der Fixsterne aus der Gestalt der Energiekurve.

15. Die Ergebnisse der spektralphotometrischen Messungen, welche auf die Strahlung des schwarzen Körpers bezogen sind. Das Ideal jeder exakten spektralphotometrischen Methode besteht in der Verbindung von photometrischen Messungen der Energieverteilung im Sternspektrum mit entsprechenden an schwarzen Strahlern bekannter Temperatur. Die hieraus resultierende absolute Energiekurve des Sterns gibt die spektralen Intensitäten bis auf einen für alle Wellenlängen konstanten Faktor im absoluten Maß. Die Schwierigkeit der praktischen Anwendung dieser Methode zur Temperaturbestimmung der Fixsterne ist vor allem darin begründet, daß die spektralen Intensitäten der Sternstrahlung von dem vollen Betrag der Absorption in der Erdatmosphäre befreit werden müssen und daß der Übergang vom Sternspektrum zum irdischen Strahler bekannter Temperatur ein besonders exaktes Beobachtungsverfahren verlangt.

Die im Jahre 1905 von WILSING und SCHEINER[2] begonnenen, später von WILSING und MÜNCH[3] fortgeführten spektralphotometrischen Messungen an 199 helleren Sternen sind bisher die einzige umfangreiche Programmarbeit dieser Art[4]. Als Vergleichslichtquelle bei den Messungen am Himmel diente eine Kohlenfadenlampe, die im Laboratorium an den schwarzen Körper angeschlossen wurde. Die Messungen erfolgten anfangs an 5 Stellen im Spektrum; später wurde ihre Zahl auf 10 erhöht. Sie verteilen sich ziemlich gleichmäßig über das visuell wirksame Spektralgebiet λ 451 bis 642 $\mu\mu$. Im Bereich der zu messenden Stellen sollten sich keine stärkeren Absorptionslinien des Sternspektrums befinden. Die

---

[1] Eine andere mathematisch einwandfreiere Bedingung ist:

$$\int_{\lambda_1}^{\lambda_2} [\ln F(\lambda)]^2\, d\lambda = \text{Minimum}.$$

[2] J. WILSING u. J. SCHEINER, Temperaturbestimmung von 109 helleren Sternen aus spektralphotometrischen Beobachtungen. Publ Astrophys Obs Potsdam Nr. 56 (1909).

[3] J. WILSING, Effektive Temperaturen von 199 helleren Sternen nach spektralphotometrischen Messungen von J. WILSING, J. SCHEINER und W. MÜNCH. Publ Astrophys Obs Potsdam Nr. 74 (1919).

[4] Wegen Einzelheiten in der Meßmethode vgl. Handb. der Astrophysik II, Teil 1, Kap. 2, Ziffer 23 bis 36.

Reduktion der beobachteten scheinbaren Energiekurve auf die wahre Energiekurve erfolgte nach mittleren Extinktionswerten für Potsdam. Wenn auch bei der Berücksichtigung der Extinktion nur der Teil des Lichtverlustes in Betracht kommt, welcher sich stetig mit der Wellenlänge ändert, so können doch Abweichungen von der normalen Durchsichtigkeit sowohl die Zenitreduktion als auch die Reduktion auf den leeren Raum merklich beeinflussen.

Die Bestimmung der spektralphotometrischen Farbtemperatur aus der Energieverteilung im Sternspektrum fußt auf der Anwendung der PLANCKschen Gleichung:

$$J_\lambda = \frac{K}{\lambda^5} \cdot \frac{1}{e^{c_2/\lambda T} - 1}.$$

J_λ ist der spektrale Intensitätswert der Sternstrahlung für die Wellenlänge λ bis auf einen unbestimmten Faktor, der in die Konstante K einbezogen ist. Um die PLANCKsche Gleichung nach der Unbekannten T aufzulösen, schreibt WILSING sie in logarithmischer Form und eliminiert die Konstante durch Subtraktion des Mittels aller Gleichungen, welche den m beobachteten Stellen des Spektrums entsprechen:

$$\psi(\log J_\lambda) + 5\,\psi(\log\lambda) + \frac{c_2}{T}\log e\,\psi\left(\frac{1}{\lambda}\right) + \psi(\log[1 - e^{-c_2/\lambda T}]) = 0; \qquad (38)$$

unter der Funktion $\psi(x_\lambda)$ ist die Operation $x_\lambda - \frac{1}{m}\sum_1^m x_\lambda$ zu verstehen. WILSING berücksichtigt das letzte Glied der Gleichung (38) nur näherungsweise. Sind c_2/T_1 und c_2/T_2 diejenigen Werte, welche sich aus der WIENschen Gleichung:

$$\psi(\log J_\lambda) + 5\,\psi(\log\lambda) + \frac{c_2}{T}\log e\,\psi\left(\frac{1}{\lambda}\right) = 0 \qquad (39)$$

für die beiden Stellen an den Enden des vermessenen Spektralgebietes ergeben, so wird als Näherungswert für T:

$$\frac{c_2}{T_0} = \frac{1}{2}\left(\frac{c_2}{T_1} + \frac{c_2}{T_2}\right) \qquad (40)$$

angenommen. Die Bedingungsgleichung für die Unbekannte T geht dann über in:

$$\psi(\log J_\lambda) + 5\,\psi(\log\lambda) + \frac{c_2}{T}\log e\,\psi\left(\frac{1}{\lambda}\right) + \psi(\log[1 - e^{-c_2/\lambda T_0}]) = 0 \qquad (41)$$

oder

$$a x + n = 0, \qquad (41a)$$

wenn:

$$n = \psi(\log J_\lambda) + 5\,\psi(\log\lambda) + \psi(\log[1 - e^{-c_2/\lambda T_0}]), \qquad (42)$$

$$x = \frac{c_2}{T} \quad \text{und} \quad a = \log e\,\psi\left(\frac{1}{\lambda}\right)$$

gesetzt werden. Die Auflösung der 5 bzw. 10 Bedingungsgleichungen für jeden Stern erfolgte nach der Methode der kleinsten Quadrate.

Bei der Neureduktion der WILSINGschen spektralphotometrischen Messungen an 10 Stellen des Spektrums hat der Verfasser[1] ein Rechnungsverfahren angewandt, das von der Einschränkung der WILSINGschen Methode, betreffend den Näherungswert für T, frei ist und durch die graphische Interpretation anschaulich wirkt. Für die Wellenlängen der WILSINGschen Messungen und für das Temperaturintervall 2000° bis 40000° wurden in einer Tabelle die Zahlenwerte von:

$$12{,}5\,\psi(\log\lambda) + 2{,}5\,\psi(\log[e^{c_2/\lambda T} - 1]) \qquad (43$$

[1] A. BRILL, Spektralphotometrische Untersuchungen. Teil III, Tabelle I. A N 219, S. 353 (1923).

Tabelle 5.

Stern	Spektrum	Coblentz	Abbot	Wilsing, Scheiner, Münch	Plaskett	Rosenberg	Sampson
ι Orionis . . .	Oe 5						
λ Orionis	Oe 5			14900°			
γ Cassiopeiae . .	B 0p			6800	15000°	50000°	15500°
ε Orionis	B 0	13000°–14000°				46000	18300
δ Orionis	B 0					46000	18300
σ Orionis	B 0						
ε Persei	B 1			7400	15000	23000	18300
ζ Persei	B 1			7600		46000	9700
η Orionis	B 1						
γ Pegasi	B 2			10500		>400000	20000
γ Orionis	B 2			15200		42000	18500
α Virginis . . .	B 2					23000	
η Ursae maioris .	B 3			11000		33000	17000
δ Persei	B 5			6300		15500	
η Tauri	B 5p			9100			12200
β Orionis	B 8p	10000–12000	16000°			20500	13400
β Tauri	B 8			9900		25000	18000
β Canis minoris .	B 8			9800			
α Leonis	B 8			10100		20000	14000
α Lyrae	A 0	8000–10000	14000	9400		22000	13000
α Canis maioris .	A 0	8000–11000	11000			27500	13800
γ Geminorum . .	A 0			8600		16000	12300
β Ursae maioris .	A 0			8600		17500	12700
γ Ursae maioris .	A 0			10600		27500	13500
α Coronae bor. .	A 0			11900		∞	12800
δ Cygni	A 0			8300		23000	14500
ε Ursae maioris .	A 0p			12300		19500	14700
12 Canum venat.	A 0p			9000		19000	13800
α Andromedae .	A 0p			9400		33000	14500
α Geminorum . .	A 0					20500	13000
γ Lyrae	A 0			10100			
ζ Aquilae . . .	A 0			11800			12800
α Pegasi	A 0			10500		27500	13000
β Leonis	A 2			9400		14500	10500
ζ_1 Ursae maioris .	A 2p			10000		∞	13400
α Cygni	cA 2	8000–10000		9400	9000	20500	11000
δ Leonis	A 3			7800		10500	10000
β Arietis	A 5			7900		16500	10700
δ Cassiopeiae . .	A 5			5800	9000	10500	11300
α Aquilae . . .	A 5	7000–9000		8100		10500	8800
α Cephei	A 5			6900		10500	9100
α Ophiuchi . .	A 5			8100		14000	10700
α Canis minoris .	F 4	5500–7500	8000	7200		7000	7700
α Ursae minoris .	cF 9			5600		5200	6500
η Bootis	F 9			5500		5500	6300
α Aurigae . . .	gG 0 + F 5	5300–6500	5800	7100	5500–6000	4500	5500
η Pegasi	G 1			4700			
η Draconis . . .	G 6			4800		3750	5200
β Herculis . . .	G 7			5000		5100	4900
ε Cygni	G 9			4400		3600	4700
β Geminorum . .	K 0	4500–7000		4900	5000–5500	3500	4800
α Bootis	K 0	3500–4500		3700		3100	4300
ζ Cephei . . .	K 0						3800
γ Tauri	K 0			4800			
κ Ophiuchi . . .	K 0			4500			
δ Andromedae .	K 0			3700			
α Arietis	K 2			3900		2750	4400
ε Pegasi	cK 2			4500		3100	3700
γ Draconis . . .	K 5			3500		2350	3700
α Tauri	K 8	2800–4500	3000	3500		2150	3700
β Andromedae .	M 0	3500–4500		3200		2650	3500
α Orionis	M 1	2800–3300	2600	3000		2200	3200
α Scorpii	M 2	2500–3200					
β Pegasi	M 3	2500–3200	2850	2800			3400
μ Geminorum . .	M 3	2500–3300		3100			
α Herculis . . .	M 3		2500	3000			
ϱ Persei	M 4						

Tabelle 5 (Fortsetzung).

Stern	CH'ING-SUNG YÜ	GREAVES, DAVIDSON, MARTIN	WILSING	FESSENKOFF	NORDMANN	HERTZSPRUNG	BOTTLINGER	BRILL	PETTIT NICHOLSON
ι Orionis . . .			21000°						25000°
λ Orionis . . .	20000°		11400			11400°	20500°	23100°	
γ Cassiopeiae . .		10000°		8100°		9800	16700	17300	
ε Orionis . . .	22000		18400			11100	23900	20800	
δ Orionis . . .			24000			12100	22400	34200	
σ Orionis . . .			12100			10000			
ε Persei		13500		8500	15200°	10200	17100	17500	
ζ Persei		8300		6900		7900	12500	10600	
η Orionis. . . .			16500			11500	20200	21700	
γ Pegasi		15200	17700	9400		12000	21100	20500	
γ Orionis . . .			16300			11300	17700	21100	
α Virginis . . .	20000								17000
η Ursae maioris.		13500		10600		11100	15300	16700	
δ Persei		11800		9100	18500	8900	13800	13300	
η Tauri	13000	10900		8900		9700	13800	13500	
β Orionis . . .			23600				15100		13000
β Tauri	17000	12300		11600		10200	13200	15300	
β Canis minoris.	16000		10200			10300	12700	15800	
α Leonis	15000	11500				9700	14100	13400	12500
α Lyrae	16000	10500		8900	12200	9300	10700	11900	10600
α Canis maioris.	13200		16600						10100
γ Geminorum .	13000					8800	10400	10700	
β Ursae maioris.	16800	10800		9400		9400	10800	11900	
γ Ursae maioris.	13000	11000		8800		9600	10500	12300	
α Coronae bor. .	17000	10500	20000	8200		10000	10900	12500	
δ Cygni	13000	11200		10400		8500	11900	11900	
ε Ursae maioris.	17400	11000		9400		8300	10500	10200	
12 Canum venat.	15800	11100		10000		8400	11400	11800	
α Andromedae .		11600	9700	8900		9800	13900	13900	
α Geminorum .		9700				8800	10500	10600	10100
γ Lyrae	10600		9400	9300	14500	8900	12200	12500	
ζ Aquilae . . .	13700	10400	10200			8700	10100	11000	
α Pegasi		11200	9800	8900		10000	12300	13000	
β Leonis	12500	9400				8400	9300	10200	
ζ_1 Ursae maioris		9700		8400		8800	10000	10900	
α Cygni	14000	8600		11100		8500	11500	11000	10100
δ Leonis	9600	8600				7980	8860	9570	
β Arietis		8500	6800	6800		7500	9570	9500	
δ Cassiopeiae . .		8800		7600		7640	9570	9570	
α Aquilae . . .	10100	8000	7700	7300		7470	8260	8640	10100
α Cephei	9700	7700		8200		7790	9140	9380	
α Ophiuchi . . .	12000	8400	8300	7600		7760	9000	9320	
α Canis minoris .	8000		6200			6500	7560	7550	5620
α Ursae minoris.				5400	8200	5680	6210	6350	
η Bootis	6500	5800		5700		5570	6300	6380	
α Aurigae . . .				4370		5300	5460	5630	
η Pegasi			4780	4660		4610	5140	5260	
η Draconis . . .	5000			4260		4490	4910	5000	
β Herculis . . .			5100	4810		4520	4780	4860	
ε Cygni			4940	4530		4280	4430	4720	
β Geminorum .						4260	4480	4610	4260
α Bootis	4200		3800	3680		3990	3840	4340	3520
ζ Cephei . . .				3860	4260	3660	3260		
γ Tauri					7250	4540	4590		
$\varkappa$ Ophiuchi . . .			4090			3950	4060		
δ Andromedae .			4330	3920		3820	3770		
α Arietis			3920	3940		3960	4040	4400	3780
ε Pegasi			3360	3580		3540	3250	3680	
γ Draconis . . .	4000			3590		3280	3270	3600	3160
α Tauri			3300			3250	3210	3560	3120
β Andromedae .				3390		3390	3210	3630	3000
α Orionis. . . .			3220			3150	3010	3400	2640
α Scorpii				3540					2660
β Pegasi				3430		3210	3220	3460	2710
μ Geminorum .						3240	3280	3560	2800
α Herculis . . .				3560		3220			2340
ϱ Persei					2870	3220	3570		2550

zusammengestellt. Die durch die Beobachtungen im System der Größenklassen gegebene Intensitätsverteilung im Spektrum eines Sterns ist dann in die nach steigenden Temperaturen geordnete, rechnerisch festgelegte Energieverteilungsskala der Tabelle einzufügen. Die praktische Ausführung geschieht zweckmäßig in der folgenden Weise: Zu der spektralen Intensitätsverteilung des Sterns wählt man aus der Tabelle drei in der Temperaturskala aufeinanderfolgende Reihen spektraler Energiewerte aus, zwischen welche sich die spektralen Intensitäten des Sterns einordnen lassen. Für die drei durch die zugehörigen Farbtemperaturen charakterisierten Reihen erhält man durch Subtraktion die Unterschiede zwischen Rechnung und Beobachtung. Um die spektralphotometrische Farbtemperatur des Sterns zu erhalten, bildet man in jeder der drei genannten Reihen für die erste und für die zweite Hälfte der WILSINGschen Wellenlängen das Mittel der Unterschiede. In einer zeichnerischen Darstellung geben die mittleren Unterschiede, aufgetragen als Funktion der reziproken Farbtemperatur, 6 Punkte, von denen je drei zusammengehörige annähernd auf einer geraden Linie liegen. Der Schnittpunkt beider Geraden gibt die zu der spektralen Intensitätsverteilung des Sterns gehörige spektralphotometrische Farbtemperatur.

WILSING findet einen systematischen Unterschied der c_2/T-Werte, je nachdem sie aus Beobachtungsreihen mit 5 oder 10 Messungen im Spektrum erhalten sind. Um das gesamte Material der c_2/T-Werte homogen zu machen, verteilt er den Unterschied von 0,50 auf beide Reihen gleichmäßig. In der Tabelle 5 stehen unter dem Kopf „WILSING, SCHEINER und MÜNCH" für eine Reihe von Sternen die von WILSING mit der Konstanten $c_2 = 1{,}46$ berechneten spektralphotometrischen Farbtemperaturen. In der Tabelle 6 ist eine allgemeine Übersicht über die von verschiedenen Autoren benutzten Werte der Konstanten c_2 — soweit solche Angaben nicht fehlen — nebst Literaturnachweisen gegeben.

Tabelle 6.

Autorität	c_2	Literaturnachweis
WILSING, SCHEINER und MÜNCH . .	1,46	Publ Astrophys Obs Potsdam Nr. 56 und 74.
Neureduktion von BRILL	1,435	A N 219, S. 353.
H. H. PLASKETT	1,435	Publ Dominion Astrophys Obs Victoria 2, Nr. 12.
GREAVES und DAVIDSON	1,432	M N 86, S. 33.
ROSENBERG	1,435	Kais Leop-Carol Akademie Nova Acta 101, Nr. 2.
Neureduktion von BRILL	1,435	A N 219, S. 353.
SAMPSON	1,43	M N 83, S. 174; 90, S. 636.
SAMPSON	1,46	M N 85, S. 212.
GREAVES, DAVIDSON und MARTIN .	1,432	M N 87, S. 352; 90, S. 104.
WILSING	1,46	Publ Astrophys Obs Potsdam Nr. 76.
NORDMANN	1,46	C R 149, S. 1039.
HERTZSPRUNG	1,46	Ann v d Sterrewacht te Leiden XIV 1. Stuk.
BRILL	1,435	A N 223, S. 105.

Die Energieverteilung in den Sternspektren wird in dem visuell wirksamen Spektralbereich λ 451 bis 642 $\mu\mu$ durch die Plancksche Strahlungsformel im allgemeinen gut dargestellt. Bei den Wellenlängen λ 642, 494 und 472 $\mu\mu$ treten merkliche Abweichungen von der schwarzen Strahlung auf[1]. Die genannten Stellen liegen nicht weit von den Wasserstofflinien $H\alpha$ und $H\beta$. Die Messung

[1] A. BRILL, Spektralphotometrische Untersuchungen. Teil II. A N 219, S. 27; Teil III, ebenda, S. 356 nebst Abb. 3 auf Tafel 4 und 5 (1923).

gibt durchweg größere Intensitätswerte als die Rechnung. Da die Abweichungen für Sterne aller Spektraltypen von gleichem Betrag und von gleichem Vorzeichen sind, ist die Ursache der Fehlerquelle wahrscheinlich in dem Spektrum der Vergleichslampe zu suchen. Die Korrektionen der in Größenklassen ausgedrückten Intensitätswerte bei den Wellenlängen λ 642, 494 und 472 $\mu\mu$ sind $+ 0^m,14$, $+ 0^m,12$ und $+ 0^m,05$.

Die Arbeiten von H. H. Plaskett, J. Baillaud, Greaves und Davidson sind als vorläufige Untersuchungen über die Zweckmäßigkeit spezieller spektralphotometrischer Methoden zu werten. Nur wenige Sterne sind von ihnen gemessen worden.

Plaskett[1] erhält die Intensitätsverteilung im Sternspektrum aus Aufnahmen mit einem Spektrographen, vor dessen Spalt ein neutraler Keil variabler Dichte gesetzt ist. Als Vergleichslichtquelle dient der positive Krater des Kohlebogens, dessen spektralphotometrische Farbtemperatur durch Messungen im Laboratorium bestimmt ist. Die Arbeit betrifft die Sonne und 6 Sterne. Bei den Sternen von frühem Spektraltypus (γ Cassiopeiae, ε Persei, α Cygni und δ Cassiopeiae) reicht der vermessene Spektralbereich von λ 390 bis 500 $\mu\mu$, bei α Aurigae und β Geminorum von λ 400 bis 680 $\mu\mu$. Nur solche Stellen im Spektrum wurden gemessen, an denen nach Rowlands Atlas des Sonnenspektrums ein Minimum der Absorptionslinien zu erwarten war. Die selektive Absorption in der Erdatmosphäre ist aus Tagesbeobachtungen der Sonne bestimmt.

Die Energiekurven der Sterne lassen sich innerhalb der genannten Spektralgrenzen durch schwarze Strahler bestimmter Temperatur darstellen (vgl. Tab. 5 unter „Plaskett"). Die Sterne α Aurigae und β Geminorum geben im Ultraviolett jenseits λ 460 $\mu\mu$ merkliche Abweichungen von der schwarzen Strahlung im Sinne einer Depression der Energiekurve. Nach Plaskett ist die Senkung der Energiekurve darauf zurückzuführen, daß bei der kleinen Dispersion der Sternspektren die Wirkung der Absorptionslinien nicht vollständig eliminiert ist.

J. Baillaud[2] vergleicht die Energieverteilung im Sternspektrum mit der Strahlung einer Glühlampe nach der photometrischen Methode der „échelle de teintes". Die Prismenkamera ist so konstruiert, daß die Aufnahmen des Sterns und der Vergleichslampe gleichzeitig erfolgen. Das Spektrum der Glühlampe wird mit dem gleichen optischen System an den positiven Krater des Kohlebogens angeschlossen. Die atmosphärische Extinktion auf der Beobachtungsstation des Pic du Midi war Gegenstand einer besonderen Untersuchung.

Die beobachteten Sterne sind vom B-, A- und F-Spektraltypus. Die Spektren von α Lyrae und β Orionis erstrecken sich von λ 330 bis 650 $\mu\mu$, die der anderen Sterne von λ 330 bis 475 $\mu\mu$. Die Energiekurven der Sterne vom B- und A-Typus lassen sich nicht mit der Planckschen Gleichung in Einklang bringen; ihr Gradient ist in einzelnen Spektralgebieten größer als der des schwarzen Strahlers von unendlich hoher Temperatur. Die Farbtemperatur von α Canis minoris (7000°) stellt die beobachtete Intensitätsverteilung in wenig befriedigender Weise dar.

Greaves und Davidson[3] bestimmen die Energieverteilung im Sternspektrum photographisch nach der Methode Objektivprisma + Objektivgitter. Als Ver-

---

[1] H. H. Plaskett, The Intensity Distribution in the Continuous Spectrum and the Intensities of the Hydrogen Lines in γ Cassiopeiae. M N 80, S. 771 (1920); The Wedge Method and its Application to Astronomical Spectrophotometry. Publ Dom Astrophys Obs Victoria 2, Nr. 12 (1923); vgl. auch Handb. der Astrophysik II, Teil 1, Kap. 2, Ziffer 48 bis 53.

[2] J. Baillaud, Étude de spectrophotométrie stellaire. B A 4, S. 275 (1924); vgl. auch Handb. der Astrophysik II, Teil 1, Kap. 2, Ziffer 45 bis 47.

[3] W. M. H. Greaves and C. R. Davidson, Preliminary Note on the Determination of Effective Stellar Temperatures by the "Prism-crossed-by-grating" Method. M N 86, S. 33 (1925); vgl. auch Handb. der Astrophysik II, Teil 1, Kap. 2, Ziffer 55 und 56.

gleichslichtquelle dient eine Halb-Watt-Lampe, deren Spektrum an den Kohlebogen angeschlossen ist. Die Sternspektren erstrecken sich über den Spektralbereich λ 400 bis 570 $\mu\mu$.

Die Energiekurve von α Lyrae gleicht der eines schwarzen Strahlers, die von α Bootis weicht im Violett jenseits λ 460 $\mu\mu$ merklich von der schwarzen Strahlung ab, weniger die von α Ursae minoris. Die spektralphotometrische Farbtemperatur wurde aus dem mittleren Gradienten der Energiekurve bestimmt. Gemäß der PLANCKschen Strahlungsgleichung ist:

$$c_2/T_1 = X - r, \tag{44}$$

wo

$$X = \frac{c_2}{T_3} - \frac{1}{\log e}(\varDelta G_1 - \varDelta G_3) \quad \text{und} \quad r = \frac{c_2}{T_1} \cdot \frac{1}{e^{c_2/\lambda_0 T_1} - 1} \tag{45}$$

sind[1]. T_1 und T_3 sind die spektralphotometrischen Farbtemperaturen des Sternes und des Kohlebogens. Die Größe r entspricht dem PLANCKschen Korrektionsfaktor und ist für die mittlere Wellenlänge $\lambda_0 = 480\ \mu\mu$ mit dem Argument X in eine Tabelle gebracht. $\varDelta G_1 = \frac{d}{d\, 1/\lambda}\left(\log \frac{J_\lambda^1}{J_\lambda^2}\right)$ ist der Gradient der relativen Energiekurve von Stern und Vergleichslampe, $\varDelta G_3 = \frac{d}{d\, 1/\lambda}\left(\log \frac{J_\lambda^3}{J_\lambda^2}\right)$ der von Kohlebogen und Vergleichslampe. Da das Beobachtungsmaterial nicht hinreicht, um den Einfluß der atmosphärischen Absorption zu bestimmen, ist die Reduktion von dem scheinbaren auf den wahren Gradienten der Energiekurve nach mittleren von ABBOT angegebenen Extinktionswerten durchgeführt. GREAVES und DAVIDSON erhalten aus zwei Platten folgende spektralphotometrischen Farbtemperaturen:

	α Lyrae	α Ursae minoris	α Bootis
Platte A . .	15200°	5750°	4130°
Platte B . .	17500	5550	4620

16. Differentielle spektralphotometrische Messungen. Bei dem spektralphotometrischen Vergleich eines Sterns mit dem schwarzen Körper sind die gemessenen spektralen Intensitäten des Sterns wegen der selektiven Absorption des Lichtes in der Erdatmosphäre zu korrigieren. Die Reduktion der scheinbaren auf die wahre Energiekurve ist von den einzelnen Autoren nicht immer mit der wünschenswerten Schärfe ausgeführt. Die Berücksichtigung der Extinktion nach mittleren Werten ist nur ein Notbehelf, auch wenn sie am gleichen Ort aus Tagesbeobachtungen der Sonne abgeleitet sind. Um ein gesichertes System von spektralphotometrischen Farbtemperaturen zu erhalten, ist es unbedingt erforderlich, daß mit der eigentlichen Aufgabe, der Bestimmung der spektralphotometrischen Farbtemperatur, eine genaue Extinktionsuntersuchung verbunden wird. Zeitliche Änderungen in der selektiven Durchlässigkeit der Erdatmosphäre können besonders bei den Sternen hoher Temperatur zu irrigen Resultaten führen.

Die Bestimmung der Farbtemperatur aus spektralphotometrischen Messungen im Anschluß an die Strahlung des schwarzen Körpers braucht sich nur auf eine verhältnismäßig kleine Zahl von Sternen zu beschränken. Die Beobachtung dieser fundamentalen Temperatursterne nach verschiedenartigen Methoden ist für die Sicherung der Temperaturskala von grundlegender Bedeutung. Die Energiekurven aller anderen Sterne sind dann auf die der Fundamentalsterne zu beziehen.

[1] Die Gleichungen (44) und (45) folgen aus der fundamentalen Gleichung (20); die PLANCKschen Korrektionsfaktoren für die niedrigen Temperaturen des Kohlebogens und der Vergleichslampe sind zu vernachlässigen.

Bei relativen Messungen ist das Beobachtungs- und das Reduktionsverfahren wesentlich einfacher. Die Extinktion geht nur differentiell in die Messungen ein. Der Nullpunkt der Temperaturskala bleibt zunächst unbestimmt; er wird entweder durch passende Wahl der Temperatur eines Einzelsterns oder noch besser der mittleren Temperatur einer Gruppe von Sternen festgelegt.

ROSENBERGS[1] „Photographische Untersuchung der Intensitätsverteilung in Sternspektren" betrifft extrafokale Aufnahmen mit einer für den ultravioletten Teil des Spektrums durchlässigen Prismenkamera. Das Programm umfaßt außer der Sonne 70 hellere Sterne von verschiedenem Spektraltypus. Die spektralen Helligkeiten der Sterne gelten für den Spektralbereich λ 340 bis 575 $\mu\mu$ und sind auf die von α Aquilae als Nullpunkt bezogen. Der Einfluß der atmosphärischen Extinktion wird aus besonders angestellten Beobachtungen abgeleitet.

Innerhalb des Spektralgebietes λ 400 bis 500 $\mu\mu$ lassen sich die Logarithmen der Intensitätsverhältnisse jedes einzelnen Sterns gegen α Aquilae, wenn man sie nach reziproken Wellenlängen ordnet, angenähert durch eine Gerade darstellen, wie es die WIENsche Strahlungsgleichung fordert. Für Wellenlängen, die größer als 500 $\mu\mu$ oder kleiner als 400 $\mu\mu$ sind, ergeben sich merkliche Abweichungen von der schwarzen Strahlung. ROSENBERG benutzt zur Bestimmung der spektralphotometrischen Farbtemperatur nur Intensitätsmessungen zwischen den Wellenlängen λ 400 und 500 $\mu\mu$. Der Gradient der die Beobachtungen in diesem Spektralbezirk darstellenden Geraden gibt ein Maß für den Temperaturunterschied jedes Sterns gegen α Aquilae.

Um den Nullpunkt der Temperaturskala festzulegen, versucht ROSENBERG, den Anschluß an das Sonnenspektrum durch spektralphotometrischen Vergleich der Sonne mit α Aquilae zu erreichen. Er entnimmt den Angaben von ABBOT die Intensitätswerte im Sonnenspektrum für die Wellenlängen λ 400 und λ 500 $\mu\mu$ und bestimmt aus dem Gradienten der Energiekurve in diesem Spektralgebiet die spektralphotometrische Farbtemperatur der Sonne. Mit der PLANCKschen Gleichung berechnet ROSENBERG dann den Gradienten der Energiekurve zwischen den Wellenlängen λ 400 und 500 $\mu\mu$ für die Temperaturfolge 2000° bis 100000°. Die für die Sterne aus den Beobachtungen abgeleiteten sowie die für die Temperaturreihe berechneten Gradienten werden auf die Sonne als Nullpunkt bezogen. Durch Einordnen der den Sternen entsprechenden Gradienten in die nach steigenden Temperaturen fortschreitende Gradientenskala wird die spektralphotometrische Farbtemperatur jedes einzelnen Sterns erhalten.

Tabelle 7.

Spektrum	WILSING I	ROSENBERG I	WILSING II	ROSENBERG II	SAMPSON I	SAMPSON II
			Neureduktion von BRILL			
B0	10500°	70000°	12300°	30000°	25000°	20000°
B5	10000	45000	11450	18000	16400	15900
A0	9300	28000	10250	12000	13100	12600
A5	8100	17000	9000	9000	10700	10300
F0	7000	11000	7950	7850	8900	8600
F5	6100	8000	6880	6930	7500	7300
G0	5300	6100	5980	6000	6200	6200
G5	4650	4800	5250	5200	5100	5200
K0	3900	3600	4570	4570	4200	4400
K5	3300	2850	3860	3840	3500	3600
M0	3100	2600	3550	3580	3400	3500

[1] H. ROSENBERG, Photographische Untersuchung der Intensitätsverteilung in Sternspektren. Abh d Kais Leop-Carol Deutsch Akad d Naturforscher Nova Acta 101, Nr. 2 (1914); vgl. Handb. der Astrophysik II, Teil 1, Kap. 2, Ziffer 57 bis 60.

Die Temperaturen, welche ROSENBERG auf diese Weise ableitet (Tab. 5 unter „ROSENBERG“), erreichen bei einigen Sternen des frühen A-Typus unendlich hohe Werte (ζ_1 Ursae maioris, α Coronae borealis); die WILSINGschen Temperaturen überschreiten bei den heißesten Sternen 15000° nur in ganz seltenen Fällen. Die Unterschiede zwischen beiden Temperaturreihen, der WILSINGschen und der ROSENBERGschen, verschwinden bei etwa 5000° und treten bei den späten Spektraltypen mit entgegengesetztem Vorzeichen auf. Die Tabelle 7 enthält unter „WILSING I“ und „ROSENBERG I“ die zu den einzelnen Spektralklassen gehörigen mittleren spektralphotometrischen Farbtemperaturen WILSINGS und ROSENBERGS in der ursprünglichen Skala.

Wie ROSENBERG gezeigt hat, lassen sich beide Temperaturskalen durch eine lineare Transformation ineinander überführen. Stellt man den reziproken Wert der Temperatur nach WILSING c_2/T_W als Funktion des entsprechenden nach ROSENBERG c_2/T_R graphisch dar, so gruppieren sich die zugehörigen Punkte sehr nahe um eine Gerade, deren Gleichung lautet:

$$\frac{c_2}{T_R} = \frac{10}{6,2} \cdot \frac{c_2}{T_W} - 2,00 . \tag{46}$$

Die WILSINGschen Werte c_2/T_W und die ROSENBERGschen c_2/T_R lassen sich in hinreichend gute Übereinstimmung bringen, wenn man entweder die Potsdamer Skala im Verhältnis 10 : 6,2 erweitert oder die ROSENBERGsche im umgekehrten Verhältnis zusammendrückt.

Um den Widerspruch zwischen den Temperaturskalen von WILSING und ROSENBERG aufzuklären, hat der Verfasser[1] die ROSENBERGschen Messungen in ähnlicher Weise wie die WILSINGschen einer erneuten Diskussion unterzogen. Die Tafel der Funktion (43) in Ziffer 15: $12,5\,\psi(\log\lambda) + 2,5\,\psi(\log[e^{c_2/\lambda T} - 1])$ wurde für die ROSENBERGschen Wellenlängen ergänzt (Tab. V der in AN 219, S. 356 enthaltenen Abhandlung des Verfassers); der Nullpunkt der Energiewerte wird durch das Mittel derjenigen für die WILSINGschen Wellenlängen festgelegt. Mit dem Näherungswert 8100° für die spektralphotometrische Farbtemperatur von α Aquilae (entnommen aus Publ. Astrophys. Obs. Potsdam Nr. 74) wurde die Tafel so umgeformt, daß die Eingänge der ROSENBERGschen Wellenlängen und der Farbtemperaturen bestehen blieben, die Tafelwerte hingegen auf die Energieverteilung eines schwarzen Strahlers von der Temperatur 8100° bezogen sind. In diese Tafel wurden nach dem oben erläuterten Verfahren (Ziffer 15) die beobachteten logarithmischen Intensitätsverhältnisse jedes Sterns gegen α Aquilae eingeordnet. Bei den Sternen der Spektralklassen B0, B5, A0 und A5 wurde eine Darstellung der beobachteten Werte für alle Wellenlängen größer als 380 $\mu\mu$, bei F0 und F5 für $\lambda > 400\,\mu\mu$ und bei den späten Spektraltypen für $\lambda > 450\,\mu\mu$ angestrebt. Auf diese Weise wollte der Verfasser bei den Sternen von frühem Spektraltypus dem Einfluß der Wasserstoffabsorptionslinien und der sich anschließenden kontinuierlichen Wasserstoffabsorption Rechnung tragen, bei den Sternen von spätem Spektraltypus dem mit fortschreitendem Spektralcharakter sich verstärkenden Einfluß der Absorptionslinien im Ultraviolett, die wegen der extrafokalen Aufnahme eine allgemeine Depression der Energiekurve im kurzwelligen Teil des Spektrums zur Folge haben.

Die spektralphotometrischen Farbtemperaturen der Sterne vom Typus B0, B5 und A0 bleiben auch nach der Neureduktion der ROSENBERGschen Messungen größer als die entsprechenden WILSINGschen. Die c_2/T-Werte der Sterne vom Spektraltypus F0 ab sind nahe um den gleichen Betrag größer als die WILSINGschen in der Neureduktion. Im Mittel ist der Unterschied gleich der bereits

[1] A. BRILL, Spektralphotometrische Untersuchungen. Teil III. AN 219, S. 356 (1923).

früher (Ziffer 15) erwähnten Korrektion 0,25, welche WILSING an die aus der Beobachtungsreihe mit 10 Messungen abgeleiteten c_2/T-Werte anbrachte, um sie mit den aus 5 Messungen erhaltenen homogen zu machen. Wird die spektralphotometrische Farbtemperatur von α Aquilae gleich 9200° gesetzt, so stimmen, abgesehen von den Sternen der frühen Spektraltypen B0, B5 und A0 die aus den WILSINGschen und ROSENBERGschen Messungen abgeleiteten Temperaturskalen miteinander überein (Tab. 7 unter „WILSING II" und „ROSENBERG II").

Nach den Untersuchungen des Verfassers[1] bestehen keine systematischen Unterschiede zwischen den visuellen Messungen WILSINGS und den photographischen ROSENBERGS. In die Abb. 1 sind die relativen spektralen Farbenindizes eines B0-Sternes gegen einen M0-Stern, d. s. ihre spektralen Helligkeitsunterschiede, bezogen auf ihre visuelle Helligkeitsdifferenz als Nullpunkt, in der Skala von WILSING • und in der Skala von ROSENBERG × als Funktion der reziproken Wellenlänge eingetragen. Man erkennt zunächst, daß in dem beiden Messungsreihen gemeinsamen Spektralbereich λ 451 bis 571 $\mu\mu$ keine merklichen systematischen Unterschiede zwischen den visuellen und den photographischen Beobachtungen vorhanden sind. Wenn man also die rechnerische Darstellung der ROSENBERGschen und der WILSINGschen Messungen durch das PLANCKsche Gesetz auf den beiden gemeinsamen Spektralbereich beschränkt, so ist die Weite der Temperaturskalen von WILSING und von ROSENBERG einander gleich.

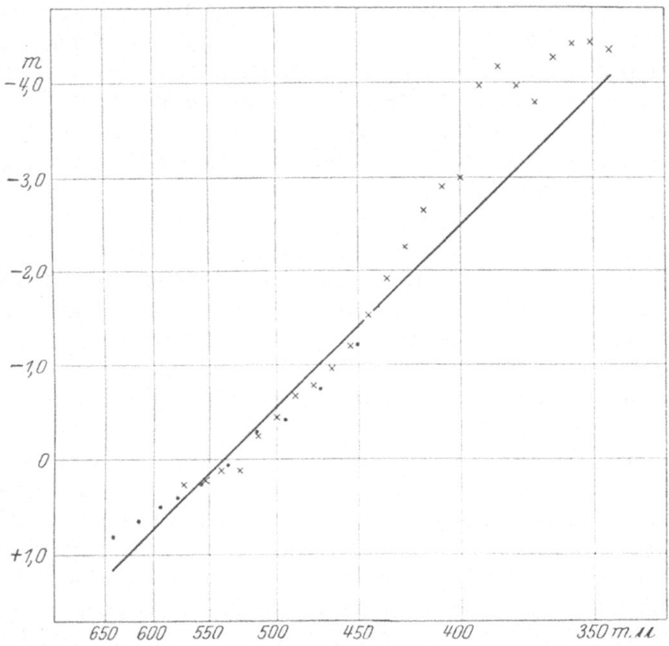

Abb. 1. Relativer spektraler Farbenindex eines B0-Sternes gegen einen M0-Stern; • WILSINGS Messungen, × ROSENBERGS Messungen. (Nach BRILL, Z. f. Physik, Bd. 52, S. 777.)

Die Unstimmigkeit zwischen den ursprünglichen Temperaturskalen von WILSING und von ROSENBERG ist daraus zu erklären, daß das PLANCKsche Gesetz in dem relativ weiten Spektralbereich der visuell und der photographisch wirk-

[1] A. BRILL, Spektralphotometrische Untersuchungen. Teil I. A N 218, S. 209 (1923); Die Temperaturstrahlung der Fixsterne. Z f Phys 52, S. 775 (1929).

samen Strahlen nicht mehr erfüllt ist. Würden sich die Sterne wie schwarze Strahler verhalten, so müßte die Farbenindexkurve annähernd eine gerade Linie von der in der Abb. 1 gezeigten Gestalt sein. In Wirklichkeit ist dies nicht der Fall. Für die kurzen Wellenlängen zwischen λ 400 und 500 $\mu\mu$ ist der Gradient der beobachteten Farbenindexkurve größer als für die langen Wellenlängen zwischen λ 500 und 600 $\mu\mu$. Man kann aus der Abbildung direkt ablesen, wie sich die Gradienten der Farbenindexkurve in dem WILSINGschen und in dem ROSENBERGschen Spektralbereich zueinander verhalten. Die spektralphotometrischen Farbtemperaturen WILSINGS beziehen sich auf den Spektralbereich λ 451 bis 642 $\mu\mu$; die in diesem Spektralbereich zu den relativen spektralen Farbenindizes gehörigen • liegen nahezu auf einer Geraden, welche unter dem Winkel 32° gegen die Horizontale geneigt ist. Die spektralphotometrischen Farbtemperaturen ROSENBERGS beziehen sich auf den Spektralbereich λ 400 bis 500 $\mu\mu$; die relativen spektralen Farbenindizes von ROSENBERG × werden in diesem Spektralbereich angenähert durch eine Gerade dargestellt, welche den Winkel 46° mit der Horizontalen bildet. Die Tangenten der Winkel verhalten sich wie 6 : 10, in guter Übereinstimmung mit dem von ROSENBERG angegebenen Überführungsverhältnis beider Temperaturskalen 6,2 : 10.

Bei den Sternen der mittleren und späten Spektralklassen ist es nicht leicht, die Temperaturen aus der Form der Energiekurve abzuleiten, da die zahlreichen Absorptionslinien und -banden stören, und zwar je nach der Art der Messung in verschieden starker Weise. WILSING war bemüht, die Helligkeitsverteilung in den Sternspektren an möglichst absorptionsfreien Stellen zu messen. Diese Bedingung läßt sich bei den Sternen von mittlerem und spätem Spektraltypus nicht immer erfüllen; die gemessene Strahlung gehört einem mehr oder minder eng begrenzten Spektralbereich an, in dem die Absorptionslinien verschieden dicht beieinander stehen. In den extrafokalen Aufnahmen ROSENBERGS vermischen sich die Absorptionslinien mit dem kontinuierlichen Untergrund. Beim Übergang von den frühen zu den späten Spektralklassen treten neue Absorptionslinien auf, deren Zahl in der Regel mit abnehmender Wellenlänge wächst. Die dadurch in dem extrafokalen Sternspektrum hervorgerufene Absorptionswirkung, welche eine mit fortschreitendem Spektraltypus stärker ausgeprägte Depression der Energiekurve im Ultraviolett jenseits λ 450 $\mu\mu$ zur Folge hat, besitzt einen Gang von der Form const/λ und hat somit eine Dehnung der Temperaturskala zur Folge.

In welchem Ausmaß die Divergenz in den ursprünglichen Temperaturskalen von WILSING und ROSENBERG auf die selektive Absorption der besonders im Ultraviolett zahlreichen Absorptionslinien, die bei dem von ROSENBERG benutzten Beobachtungsverfahren der extrafokalen Aufnahmen eine Schwächung des kontinuierlichen Untergrundes bewirken, zurückzuführen ist, läßt sich erst angeben, wenn man die Breite und die Intensität der in Betracht kommenden Linien kennt. Neben dieser rein selektiven Wirkung der Sternabsorption ist wahrscheinlich noch eine allgemeine mit der Wellenlänge veränderliche, in der Sternatmosphäre oder auch im interstellaren Raum wirksame Absorption oder Beugung vorhanden, welche die Anwendbarkeit des PLANCKschen Strahlungsgesetzes in strenger Weise auf die Strahlung der Sterne in Frage stellt.

Bei der Neureduktion der WILSINGschen und der ROSENBERGschen Messungen war der Verfasser bestrebt, diejenigen Messungen bei der rechnerischen Darstellung durch das PLANCKsche Gesetz unberücksichtigt zu lassen, welche durch die Wirkung der Absorptionslinien merklich verfälscht erscheinen. Die Divergenz zwischen der WILSINGschen und der ROSENBERGschen Temperaturskala in der Neureduktion bei den frühen Spektralklassen läßt sich daraus erklären, daß der

Verfasser die Beobachtungen nicht einheitlich, sondern für die einzelnen Spektralklassen in verschiedenen Spektralbereichen durch die PLANCKsche Strahlungsgleichung darzustellen versuchte: Die WILSINGsche Skala bezieht sich auf den Spektralbereich λ 451 bis 642 $\mu\mu$, die ROSENBERGsche für die Sterne der Spektralklassen B0, B5, A0 und A5 auf λ 380 bis 571 $\mu\mu$, für F0 und F5 auf λ 400 bis 571 $\mu\mu$ und für die späten Spektralklassen auf λ 450 bis 571 $\mu\mu$. Die Verschiedenheit der Spektralbezirke wirkt sich, wie das Resultat der Neureduktion zeigt, nur in den Temperaturen der frühen Spektraltypen B0, B5 und A0 aus, und zwar in dem Sinne, daß die spektralphotometrischen Farbtemperaturen ROSENBERGS für das weiter ins Ultraviolett sich erstreckende Spektralgebiet λ 380 bis 571 $\mu\mu$ größer sind als bei WILSING für den Spektralbereich λ 451 bis 642 $\mu\mu$. Dies deutet darauf hin, wie auch neuere Beobachtungen bestätigt haben, daß die Sterne von frühem Spektraltypus einen abnormen Verlauf der Energiekurve besitzen: verhältnismäßig flacher Gradient im visuell wirksamen Spektralgebiet und steiler Anstieg der Energiekurve nach dem Ultraviolett. Daß sogar in dem visuell wirksamen Wellenlängenbereich λ 451 bis 642 $\mu\mu$ die Temperaturskala sich nicht genau gleich bleibt, zeigt der Verlauf der relativen Farbenindexkurve eines B0-Sternes gegen einen M0-Stern (Abb. 1) in dem WILSINGschen Spektralbereich. Der Gradient der Kurve, welcher die Weite der Temperaturskala bestimmt, ist zwischen λ 550 und λ 642 $\mu\mu$ kleiner als zwischen λ 451 und λ 550 $\mu\mu$.

SAMPSON[1] bestimmt die Energieverteilung in den Sternspektren aus Aufnahmen mit einem photovisuellen Refraktor, vor dessen Objektiv sich ein Prisma von 12° brechendem Winkel befindet. Bezugsstern ist α Ursae minoris. Die kleinen Spektren, welche den Bereich λ 385 bis 700 $\mu\mu$ umfassen, wurden mit einer Art von KOCHschem Mikrophotometer ausgemessen. Die auf automatischem Wege erhaltene Registrierkurve spiegelt den Schwärzungsverlauf im Spektrum des Sterns wieder; die Absorptions- und Emissionslinien markieren sich durch Einsenkungen und Spitzen. SAMPSON verbindet die höchsten Punkte der Registrierkurve durch einen möglichst glatten Kurvenzug und gründet auf den geglätteten Verlauf die Bestimmung der spektralphotometrischen Farbtemperatur. Den differentiellen Einfluß der selektiven Absorption in der Erdatmosphäre eliminiert SAMPSON nachträglich in einer wenig exakten Weise: das Beobachtungsmaterial liefert eine Beziehung zwischen dem c_2/T-Wert und dem Spektraltypus. SAMPSON erklärt die Abweichungen des einzelnen c_2/T-Wertes von dem zu dem Spektraltypus gehörigen Normalwert durch den Einfluß der selektiven Absorption in der Erdatmosphäre und deutet sie als Zenitdistanzeffekt der Extinktion. Die zeitliche Konstanz der atmosphärischen Durchlässigkeit für Licht verschiedener Wellenlänge wird auf jeder Platte durch je eine Aufnahme von Polaris am Anfang und am Ende kontrolliert.

Die Energiekurven der Sterne zeigen eine mit fortschreitendem Spektraltypus zunehmende Depression im Ultraviolett. Hierdurch entstehen Schwierigkeiten bei der Bestimmung der spektralphotometrischen Farbtemperatur, wenn man die Messungen unter λ 450 $\mu\mu$ benutzen will. SAMPSON nimmt an, daß die Abweichungen der Sternstrahlung von der schwarzen Strahlung allein durch die Summation der von den verschiedenen Zonen der Sternoberfläche zu uns dringenden Teilstrahlungen bedingt sind (vgl. Ziffer 4). Die zentrale Strahlung eines Sterns ist nach der Voraussetzung SAMPSONS schwarz; die Abweichungen vom PLANCKschen Gesetz sind durch die zunehmende Rötung der Randzonen verursacht.

[1] R. A. SAMPSON, On the Estimation of the Continuous Spectrum of Stars. M N 83, S. 174 (1923); Effective Temperatures of Sixty-four Stars. M N 85, S. 212 (1925); vgl. Handb. der Astrophysik II, Teil 1, Kap. 2, Ziffer 61 bis 66.

Bezeichnet J_λ die spektrale Intensität der von der gesamten Sternscheibe ausgehenden Strahlung, I_λ die des zentralen Teiles, so ist:

$$J_\lambda = \psi(\lambda, T) \cdot I_\lambda = \psi(\lambda, T) \cdot \frac{\lambda^{-5}}{e^{c_2/\lambda T} - 1}. \tag{47}$$

Der Korrektionsfaktor ψ hängt von der Wellenlänge und von der Temperatur der Strahlung ab. Durch logarithmische Differentiation nach $1/\lambda$ erhält man:

$$\frac{d}{d\,1/\lambda}\log J_\lambda = \frac{d}{d\,1/\lambda}\log\psi(\lambda, T) - 5\,\lambda\log e - \frac{c_2}{T}\cdot\frac{1}{1 - e^{-c_2/\lambda T}}\cdot\log e. \tag{48}$$

Sind J'_λ und $\psi(\lambda, T')$ spektrale Intensität und Korrektionsfaktor des Bezugssterns, so ergibt die logarithmische Differentiation nach $1/\lambda$ in gleicher Weise:

$$\frac{d}{d\,1/\lambda}\log J'_\lambda = \frac{d}{d\,1/\lambda}\log\psi(\lambda, T') - 5\,\lambda\log e - \frac{c_2}{T'}\cdot\frac{1}{1 - e^{-c_2/\lambda T'}}\cdot\log e. \tag{49}$$

Aus der Verbindung der Gleichungen (48) und (49) folgt:

$$\left(\frac{c_2}{T}N + \chi_T\right) - \left(\frac{c_2}{T'}N' + \chi_{T'}\right) = -\frac{d}{d\,1/\lambda}\log\frac{J_\lambda}{J'_\lambda}.$$

$$N = \frac{\log e}{1 - e^{-c_2/\lambda T}}, \qquad N' = \frac{\log e}{1 - e^{-c_2/\lambda T'}}; \tag{50}$$

$$\chi_T = -\frac{d}{d\,1/\lambda}\log\psi(\lambda, T), \qquad \chi_{T'} = -\frac{d}{d\,1/\lambda}\log\psi(\lambda, T').$$

Wenn der aus dem spektralphotometrischen Vergleich zweier Sterne erhaltene Wert von $\log\frac{J_\lambda}{J'_\lambda}$ als Ordinate mit der Abszisse $\frac{1}{\lambda}$ aufgetragen wird, so ist bei schwarzer Strahlung der Gradient der relativen Energiekurve $\Delta G = \frac{d}{d\,1/\lambda}\log\frac{J_\lambda}{J'_\lambda}$ in erster Näherung konstant.

SAMPSON bestimmt die Funktion $\psi(\lambda, T)$ aus der spektralen Helligkeitsverteilung auf der Sonnenscheibe, wie sie uns für die Wellenlängen λ 323 bis 2097 $\mu\mu$ aus den Messungen von ABBOT und SCHWARZSCHILD bekannt ist. Der Korrektionsfaktor $\psi(\lambda, T)$ gibt den Einfluß der Randverdunklung für jede Wellenlänge λ und ist für die Sonne durch die Gleichung bestimmt:

$$\psi(\lambda, T_\odot) = \int_0^1 2r\,dr\,\frac{\varphi(r, \lambda)}{\varphi(0, \lambda)}. \tag{51}$$

Die Zahlenwerte $\frac{\varphi(r, \lambda)}{\varphi(0, \lambda)}$ geben die spektrale Helligkeit einer jeden Stelle der Sonnenscheibe in Prozenten der Helligkeit des Zentrums; der Abstand r von der Mitte ist in Bruchteilen des Sonnenradius gerechnet. Um den Korrektionsfaktor $\psi(\lambda, T)$ der Sonne auf die Fixsterne zu übertragen, macht SAMPSON die hypothetische Annahme, daß in der Funktion $\psi(\lambda, T)$ die Temperatur und die Wellenlänge nur in der Verbindung $\lambda\cdot T$ vorkommen, daß also $\psi(\lambda, T)$ für gleiche Werte $\lambda\cdot T$ konstant ist. Zwischen den Riesen- und Zwergsternen der gleichen Temperaturklasse soll kein Unterschied in der Größe des Korrektionsfaktors vorhanden sein.

SAMPSON beschränkt die praktische Anwendung der Gleichung (50) auf den mittleren Gradienten der relativen Energiekurve für die Wellenlänge λ 500 $\mu\mu$. Für diese hat SAMPSON eine nach steigenden Temperaturen geordnete Tafel der Werte $\Phi = \frac{c_2}{T}N + \chi_T$ berechnet (Tab. 8). Die Gleichung (50) liefert für jeden Stern mit dem aus den Beobachtungen abgeleiteten Gradienten der relativen

Tabelle 8. $c_2 = 1{,}46$; Einheit der Länge 10^{-4} cm.

T	c_2/T	Φ	T	c_2/T	Φ
3000°	4,87	4,88	12000°	1,22	1,32
3500	4,17	4,21	12500°	1,17	1,28
4000	3,65	3,71	13000	1,12	1,24
4500	3,24	3,29	13500	1,08	1,20
5000	2,92	2,99	14000	1,04	1,17
5500	2,65	2,74	14500	1,01	1,14
6000	2,43	2,52	15000	0,98	1,12
6500	2,25	2,33	15500	0,94	1,10
7000	2,09	2,16	16000	0,91	1,08
7500	1,95	2,02	16500	0,88	1,06
8000	1,83	1,89	17000	0,85	1,04
8500	1,72	1,78	17500	0,83	1,02
9000	1,62	1,69	18000	0,81	1,01
9500	1,54	1,61	18500	0,79	0,99
10000	1,46	1,54	19000	0,77	0,98
10500	1,39	1,47	19500	0,75	0,96
11000	1,33	1,41	20000	0,73	0,95
11500	1,27	1,36	30000	0,48	0,77
			∞	0,00	0,50

Energiekurve gegen Polaris die spektralphotometrische Farbtemperatur der zentralen Sternstrahlung. Die Farbtemperatur von Polaris findet SAMPSON durch Anschluß an Capella, für die eine Temperatur von 5500° angenommen wird. Die Tabelle 7 enthält unter dem Kopf „SAMPSON I" mittlere spektralphotometrische Farbtemperaturen der einzelnen Spektralklassen.

Nach SAMPSON ist die Strahlung der Sterne in ihrem zentralen Teil, d. h. wenn man die Randverdunklung in Rechnung zieht, nahezu schwarz. In geringem Maße ist noch eine Senkung der Energiekurve bei den Sternen vom späten Spektraltypus für Wellenlängen kleiner als 400 $\mu\mu$ durch die Messungen angedeutet. Anormale Fälle zeigen die Energiekurven von γ Cassiopeiae (B0p) und α Cygni (A2p); für Wellenlängen größer als 480 bzw. 450 $\mu\mu$ ist der Gradient der Energiekurve klein und nimmt nach den kurzen Wellenlängen stark zu.

Neuerdings hat SAMPSON seine sämtlichen spektralphotometrischen Messungen von Januar 1923 bis Mai 1929 in einheitlicher Weise bearbeitet[1], wobei er den Einfluß der atmosphärischen Extinktion in exakterer Weise zu berücksichtigen sucht. Das Programm umfaßt 80 Sterne, auf die ausschließlich der Extinktionssterne Polaris, α und β Ursae maioris durchschnittlich je 9,5 Beobachtungen kommen.

Nach den Aufnahmen von Polaris hat die Ultraviolettdurchlässigkeit der Luft im Laufe des Abends fast immer zugenommen. Die Reduktion von dem scheinbaren auf den wahren Gradienten der relativen Energiekurve in bezug auf Polaris erfolgte nach der Formel:

$$\Delta G = \Delta G' + A\,(\sec z - \sec Z) + B\,(\sec^3 z - \sec^3 Z)\,, \tag{52}$$

wo Z die Zenitdistanz von Polaris ist. Die Werte der Konstanten A und B wurden abgeleitet aus Spektrogrammen von Sirius, β, α Ursae maioris und der Gesamtheit der A0-Sterne. Die Voraussetzung, daß alle A0-Sterne den gleichen Gradienten der Energiekurve besitzen, trifft allerdings nicht zu.

Das Ergebnis der Untersuchung SAMPSONs ist folgendes: Die spektralphotometrische Farbtemperaturskala der Sterne nach dem umfangreicheren Material von M N 90 (in der Tabelle 7 mit „SAMPSON II" bezeichnet) ist ein wenig enger als die in M N 85 angegebene (mit „SAMPSON I" bezeichnet). Die spektral-

[1] R. A. SAMPSON, Effective Temperatures of Stars. Second Paper. M N 90, S. 636 (1930).

photometrischen Farbtemperaturen der zentralen Sternstrahlung liegen bei den K-Sternen 200°, bei den A-Sternen 300° über den für die durchschnittliche Strahlung geltenden. Die B- und A-Sterne der Mount Wilson-Unterklasse[1] s sind im allgemeinen heißer als die der Unterklasse n. Wenn auch eigentliche Zwergsterne in der Beobachtungsliste fehlen, so zeigt doch die Gruppierung der Sterne nach der absoluten Helligkeit bei den Sternen der Spektraltypen G, K und M eine Abnahme des Gradienten um 0,08 pro Größenklasse. Die Sterne von früherem Spektraltypus als Polaris haben systematisch im Violett eine stärkere Emission, als der schwarzen Strahlung entspricht, die Sterne von späterem Spektraltypus eine schwächere Emission. Häufig liegt die Energiekurve der frühen Sterne unter der Gradientenlinie bei λ 470 $\mu\mu$ und oberhalb bei λ 520 $\mu\mu$; für die Sterne von spätem Spektraltypus kehrt sich die Oszillation um. Die Tabelle 5 enthält für eine Reihe von Sternen unter dem Kopf „SAMPSON" die spektralphotometrischen Farbtemperaturen nach M N 90.

CH'ING-SUNG YÜ bestimmt die Intensitätsverteilung im Spektrum von 91 Sternen aus Aufnahmen mit einem Zweiprismen-Quarzspektrographen, der an den CROSSLEY-Reflektor des Lick-Obersvatoriums montiert ist[2]. Der Hauptzweck der Untersuchung bestand darin, den Einfluß der kontinuierlichen Wasserstoffabsorption in den Spektren der A-Sterne zu studieren; deshalb wurden vorzugsweise Sterne der Spektralklassen B5 bis A9 beobachtet. Der meßbare Bereich der Spektren erstreckt sich von λ 330 bis 510 $\mu\mu$. Die Absorption in der Erdatmosphäre wurde mit den meteorologischen Daten nach der Methode von FOWLE rechnerisch bestimmt.

Die Energiekurven der Sterne sind auf ζ Ophiuchi (Spektraltypus B0) als Nullstern bezogen. YÜ nimmt an, daß dieser Stern wie ein schwarzer Körper von der Temperatur 22000° strahlt. Die absoluten Energiekurven der Sterne von frühem Spektraltypus werden dann, abgesehen von der kontinuierlichen Wasserstoffabsorption, die ungefähr bei λ 380 $\mu\mu$ einsetzt, durch das PLANCKsche Gesetz gut dargestellt. Die zugehörigen spektralphotometrischen Farbtemperaturen sind für eine Reihe von Sternen in der Tabelle 5 unter dem Kopf „CH'ING-SUNG YÜ" zusammengestellt. Die Abb. 2 zeigt einige der YÜschen Energiekurven nach Spektrogrammen, die mit dem spaltlosen Spektrographen erhalten wurden.

Die Methode, welche GREAVES, DAVIDSON und MARTIN bei der Bestimmung der Temperaturen von 22 Sternen des frühen Spektraltypus anwenden[3], unterscheidet sich nicht wesentlich von der in Ziffer 15 besprochenen. Der Anschluß der Sternspektren an den Kohlebogen fehlt; es werden immer paarweise ein B-Stern und ein A-Stern miteinander verglichen. Die Reduktion von dem scheinbaren auf den wahren Gradienten der relativen Energiekurve zweier Sterne erfolgt nach der Formel:

$$\Delta G = \Delta G' + A\,(\sec z_1 - \sec z_2)\,. \tag{53}$$

ΔG ist der wahre, $\Delta G'$ der scheinbare Gradient der relativen Energiekurve; z_1 und z_2 sind die Zenitdistanzen, in denen die beiden Sterne photographiert sind. Die Konstante A der selektiven atmosphärischen Extinktion wurde aus besonderen Beobachtungen abgeleitet. In Übereinstimmung mit den Gleichungen (44) und (45) ist:

$$\frac{c_2}{T_1} = \frac{c_2}{T_2} - \frac{1}{\log e}\,\Delta G - \frac{c_2}{T_1}\cdot\frac{1}{e^{c_2/\lambda_0 T_1} - 1} + \frac{c_2}{T_2}\cdot\frac{1}{e^{c_2/\lambda_0 T_2} - 1}\,. \tag{54}$$

[1] Ap J 56, S. 242 (1922) = Mt Wilson Contr 244.

[2] CH'ING-SUNG YÜ, On the Continuous Hydrogen Absorption in Spectra of Class A Stars. Lick Bull 12, S. 104 (1926); vgl. Handb. der Astrophysik II, Teil 1, Kap. 2, Ziffer 43—44.

[3] W. M. H. GREAVES, C. DAVIDSON and E. MARTIN, The Relative Effective Temperatures of Twenty-two Stars of Early Type. M N 87, S. 352 (1927).

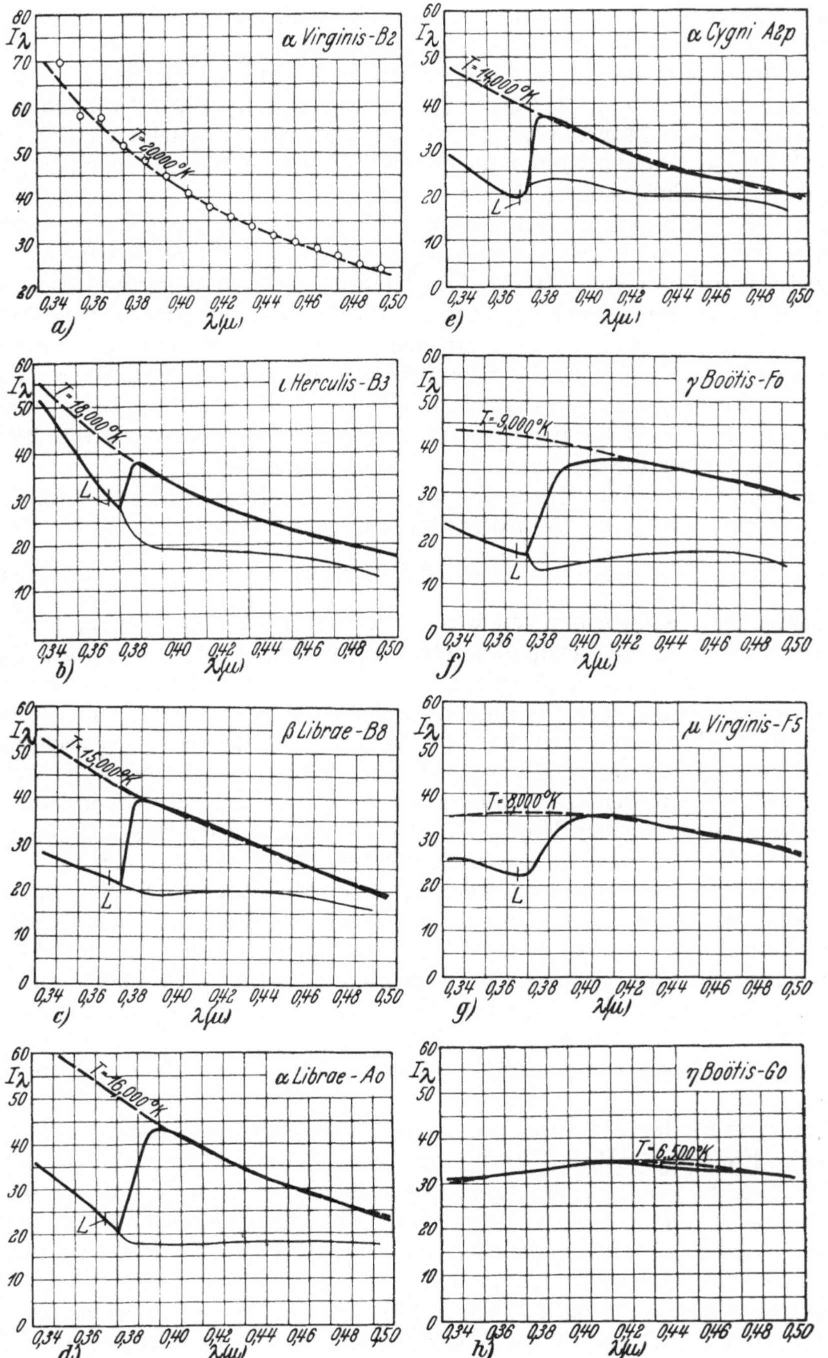

Abb. 2. Energiekurven von Sternen.
(Nach Ch'ing-Sung Yü, Lick Observatory Bull. Vol. XII, No. 375, Tafel VIII.)

T_1 und T_2 sind die spektralphotometrischen Farbtemperaturen der miteinander verglichenen Sterne; λ_0 ist die mittlere Wellenlänge des vermessenen Spektralbereiches. Der Nullpunkt der Temperaturskala bleibt unbestimmt; es wird angenommen, daß die mittlere Farbtemperatur der Sterne vom Spektraltypus A0 entweder gleich 10000° oder gleich 13000° ist.

In Fortsetzung obiger Messungsreihen bestimmten neuerdings GREAVES, DAVIDSON und MARTIN die spektralphotometrischen Farbtemperaturen einer ausgewählten Zahl von 24 möglichst gleichmäßig über den Himmel verteilten hellen Sternen[1]. Diese Standardsterne sollen die Grundlage für den spektralphotometrischen Anschluß anderer Sterne bilden.

Um von vornherein die Schwierigkeit bei der Bestimmung des ungestörten Verlaufes der Energiekurve auszuschalten, wurden die 24 Standardsterne von frühem Spektraltypus gewählt. Das Beobachtungsverfahren bestand darin, daß die 24 Standardsterne immer paarweise miteinander verglichen wurden; im ganzen erhielten die Autoren 325 Vergleiche zwischen 85 Paaren. Um möglichst unabhängig von dem Einfluß der selektiven atmosphärischen Extinktion zu sein, wurden die Sterne jedes Paares in nahezu gleichen Zenitdistanzen photographiert. Die photometrische Skala zur Umwandlung der Schwärzungen in Intensitäten lieferten wie früher die Aufnahmen mit Prisma und Gitter.

Die photometrischen Messungen eines Sternpaares, welche sich über den Spektralbereich λ 400 bis 650 $\mu\mu$ erstrecken, geben den relativen Gradienten $\Delta G = \frac{d}{d\,1/\lambda} \log \frac{J_\lambda^1}{J_\lambda^2}$ der Energiekurven beider Sterne, welcher auf die mittlere Wellenlänge $\lambda_0 = 500\ \mu\mu$ bezogen wird. Die Gleichung (54) gibt dann die Beziehung zwischen den spektralphotometrischen Farbtemperaturen T_1 und T_2 der beiden Sterne. Durch die willkürliche Festsetzung, daß im Mittel die Sterne des Spektraltypus A0 die spektralphotometrische Farbtemperatur 11000° besitzen, werden die Farbtemperaturen der 24 Standardsterne bestimmt. Anschließend teilen die Autoren die spektralphotometrischen Farbtemperaturen von 19 weiteren Sternen mit, die mit wenigstens 2 Standardsternen verglichen worden sind. Die Tabelle 5 enthält unter dem Kopf „GREAVES, DAVIDSON und MARTIN" die spektralphotometrischen Farbtemperaturen für eine Reihe von Sternen.

17. Thermoelektrische und radiometrische Messungen. Die Messungen der Strahlungsintensität der Sterne nach absoluten Methoden befinden sich noch im ersten Stadium der Entwicklung: Die Genauigkeit ist sehr beschränkt, das Reduktionsverfahren noch unsicher; der Einfluß der atmosphärischen Extinktion ist bisher nur oberflächlich berücksichtigt. Immerhin sind in der Zukunft von der absoluten Methode wesentliche Fortschritte in unserer Kenntnis von der Strahlung der Sterne, insbesondere bei den Sternen von spätem Spektraltypus, zu erwarten.

COBLENTZ bestimmt die Energieverteilung im Sternspektrum mit dem Thermoelement[2]. Ein oder mehrere Farbfilter werden vor dasselbe gesetzt, um die Strahlungskomponenten in begrenzten Spektralgebieten zu erhalten, vom äußersten Ultraviolett, wo die Undurchlässigkeit der Erdatmosphäre und das geringe Reflexionsvermögen der Silberspiegel eine Grenze ziehen, bis zum Ultrarot. Die Spektralbereiche, deren Strahlung gemessen wurde, reichen von 0,3 bis 0,43, von 0,43 bis 0,6, von 0,6 bis 1,4, von 1,4 bis 4,1 und von 4,1 bis 10 μ. Die Messungen begreifen den Einfluß der Absorptionslinien und -banden ein. Das Maximum der Strahlung liegt bei den A-Sternen im Ultraviolett zwischen 0,3 und 0,4 μ, bei den K- und M-Sternen im Infrarot zwischen 0,7 und 0,9 μ. Die Energieverteilung in den Spektren der 16 beobachteten Sterne läßt sich nach der PLANCK-

[1] W. M. H. GREAVES, C. DAVIDSON and E. MARTIN, The Colour Temperatures of Twenty-four Stars, Suitable for Use as Spectrophotometric Standards. M N 90, S. 104 (1929).

[2] W. W. COBLENTZ, Tests of Stellar Radiometers and Measurements of the Energy Distribution in the Spectra of 16 Stars. Scient Pap of Bureau of Stand Nr. 438 (1922).

schen Formel durch die in der Tabelle 5 unter „COBLENTZ" aufgeführten Temperaturen darstellen.

ABBOT mißt die Energieverteilung im Sternspektrum mit einem Radiometer[1]. Die Beobachtungen überdecken den Wellenlängenbereich λ 437 bis 2224 $\mu\mu$. ABBOT verbindet die radiometrischen Messungen mit den von WILSING, SCHEINER und MÜNCH im visuellen und von ROSENBERG im photographischen Teil des Spektrums angestellten. In den übereinandergreifenden Gebieten des Spektrums ergibt sich eine befriedigende Übereinstimmung. Die Darstellung der aus den kombinierten Beobachtungen erhaltenen Energiekurven durch den schwarzen Strahler führt zu spektralphotometrischen Farbtemperaturen, die bei den Sternen von frühem Spektraltypus größer sind als die von COBLENTZ berechneten (vgl. Tabelle 5 unter „ABBOT").

18. Die spektralphotometrische Farbtemperatur und die Gradationstemperatur der Sonne. Von allen Sternen ist die Energiekurve der Sonne am sichersten bekannt. Für die spektralphotometrische Farbtemperatur der Sonne geben die einzelnen Autoren mehr oder weniger verschiedene Werte. Der Verfasser hat für das visuell wirksame Spektralgebiet λ 451 bis 642 $\mu\mu$ die Messungsresultate von ABBOT und FOWLE auf dem Mount Wilson und Mount Whitney, von MÜLLER und KRON in Teneriffa und von WILSING (photographisch und bolometrisch) in Potsdam zu Mittelwerten vereinigt[2]. Mit dem PLANCKschne Strahlungsgesetz wird die spektralphotometrische Farbtemperatur der Sonne 6650° [3]. Die Reste sind verschwindend klein, im Maximum $0^{m},02$ (abgesehen von der Wellenlänge λ 451 $\mu\mu$ mit der Abweichung $0^{m},06$). MÜLLER und KRON[4] finden aus der Form der Energiekurve im Spektralbereich λ 430 bis 679 $\mu\mu$ die Farbtemperatur 6300°. WILSING[5] erhält aus photographischen und bolometrischen Messungen, die sich über das Spektralgebiet λ 400 bis 2128 $\mu\mu$ erstrecken, die Farbtemperatur 5900°. Diese Zahl ist ein Durchschnittswert von zwei der rechnerischen Ausgleichung zugrunde gelegten Temperaturen. Bei der einen (5810°) ist die Verteilung der Vorzeichen in den Resten gleichmäßig, bei der anderen (6070°) überwiegen die positiven Abweichungen im Sinne Beobachtung minus Rechnung.

Die spektralphotometrische Farbtemperatur der mittleren Sonnenstrahlung ist nach den obigen Beobachtungsresultaten um so kleiner, je größer der Spektralbereich ist. Die Unterschiede zwischen Messung und Rechnung nehmen mit wachsendem Spektralbereich zu. Die Energiekurve der mittleren Sonnenstrahlung läßt sich also nicht in ihrem ganzen Verlauf durch die PLANCKsche Gleichung darstellen. Die Farbtemperatur 6650° ist berechnet aus der Gestalt der Energiekurve im Spektralbereich λ 451 bis 642 $\mu\mu$. Die Energiekurve der Sonne besitzt in diesem Spektralbereich, in den das Maximum der Strahlung fällt, eine markante Spitze, deren Form der Temperatur eines schwarzen Strahlers von der Temperatur 7000° entspricht. H. H. PLASKETT[6] hat bei der Bestimmung der spektralphotometrischen Farbtemperatur der Sonne nach der Keilmethode besonderen Wert

[1] C. G. ABBOT, Radiometer Measurements of Stellar Energy Spectra. Ap J 60, S. 87 (1924).

[2] A. BRILL, Spektralphotometrische Untersuchungen. Teil I. A N 218, S. 225 (1923).

[3] A. BRILL, Spektralphotometrische Untersuchungen. Teil III. A N 219, S. 358 (1923).

[4] G. MÜLLER u. E. KRON, Die Extinktion des Lichtes in der Erdatmosphäre und die Energieverteilung im Sonnenspektrum nach spektralphotometrischen Beobachtungen auf der Insel Teneriffa. Publ Astrophys Obs Potsdam Nr. 64 (1912).

[5] J. WILSING, Über die Helligkeitsverteilung im Sonnenspektrum nach bolometrischen Messungen und über die Temperatur der Sonnenphotosphäre. Publ Astrophys Obs Potsdam Nr. 72 (1917).

[6] H. H. PLASKETT, The Wedge Method and its Application to Astronomical Spectrophotometry. Publ Dom Astrophys Obs Victoria 2, Nr. 12 (1923).

darauf gelegt, möglichst von Absorptionsstreifen freie Teile des Spektrums zu messen. Die Beobachtungen beziehen sich auf die Strahlung des zentralen Teiles der Sonne und werden in dem Spektralbereich λ 400 bis 670 $\mu\mu$ durch den schwarzen Strahler von der Temperatur 6700° bis 7000° dargestellt.

Durch die Arbeiten des Smithsonian Institutes sind uns von der Sonne bekannt: die Gesamtstrahlung, die Strahlung in den einzelnen Wellenlängen sowie der Helligkeitsabfall von der Mitte bis zum Rande der Sonnenscheibe (bolometrisch und spektral gemessen). MINNAERT hat diese Daten dazu benutzt, absolute Energiekurven der mittleren und der zentralen Sonnenstrahlung zu zeichnen[1] (Abb. 3). Die Größe der Ordinaten gibt für jede Wellenlänge die

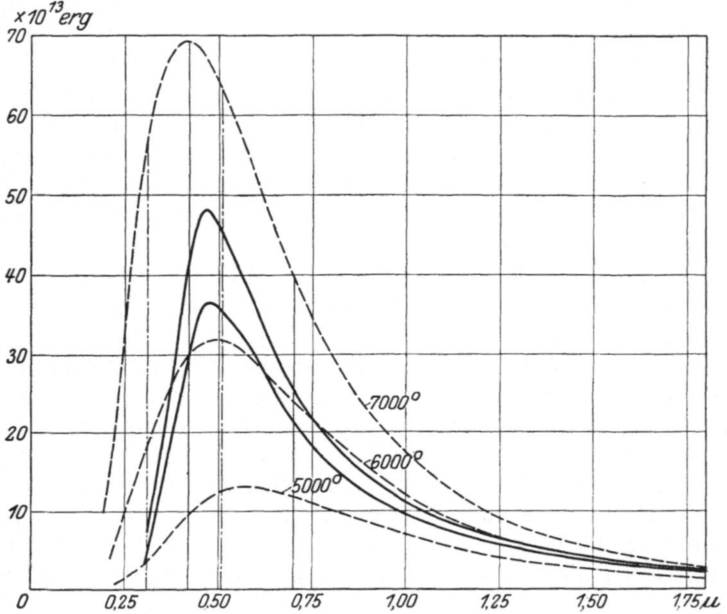

Abb. 3. Energiekurve der mittleren und zentralen Sonnenstrahlung. (Nach BRILL, Z. f. Physik Bd. 52, S. 773.)

Strahlung in erg pro sec pro cm² der Sonnenoberfläche. Die gestrichelten Energiekurven sind die Energiekurven der schwarzen Strahler von der Temperatur 5000°, 6000° und 7000°.

Schon die oberflächliche Betrachtung der Figur lehrt, daß es nicht leicht ist, aus der Form der Energiekurve der Sonne die Temperatur zu bestimmen. Solange der Spektralbereich klein ist, kann man immer den Verlauf der Energiekurve durch die PLANCKsche Strahlungsgleichung darstellen. Je nach der Lage im Spektrum gleicht dann die Form der Sonnenenergiekurve schwarzen Strahlern verschieden hoher Temperatur. Wenn man von der Depression der Energiekurve im Ultraviolett absieht, die zum Teil durch den Einfluß der Absorptionslinien bedingt ist, gleicht die Energiekurve der mittleren Sonnenstrahlung im Spektralbereich 0,4 bis 1,0 μ einem schwarzen Strahler von der Temperatur 6000° bis 7000°, im Spektralbezirk 1,0 bis 2,0 μ einem solchen von der Temperatur 5000° bis 6000°. Je weiter die Grenzen des Spektralbereiches gewählt sind, in dem man die Sonnenenergiekurve durch die PLANCKsche Strahlungsgleichung annähern will, um so größer sind die Abweichungen im Sinne Beobachtung minus Rechnung. Schwierig-

[1] M. MINNAERT, Recent Data on Solar Radiation Converted into Absolute Measure. B A N 2, S. 75 (1924).

keiten ergeben sich bei der Frage, wie man innerhalb der mehr oder weniger weit gesteckten Spektralgrenzen die schwarze Strahlungskurve definiert, welche die Sonnenenergiekurve am besten darstellt. Im wesentlichen bleibt es in das Belieben des Rechners gestellt, welche Annäherung ihm am günstigsten erscheint (vgl. die beiden WILSINGschen Ansätze für die spektralphotometrische Farbtemperatur der Sonne nach Publ Astrophys Obs Potsdam Nr. 72, S. 27).

Kennt man die Energiekurve der Sonne in ihrem ganzen Verlauf vom Ultraviolett bis zum Ultrarot, so könnte man als spektralphotometrische Farbtemperatur der Sonne diejenige bezeichnen, bei der die von der Energiekurve der Sonne bzw. von der ihr angeglichenen schwarzen Strahlungskurve umschlossenen Flächenstücke, je nachdem sie einen Überschuß oder Defekt ergeben, einander ergänzen. Gemäß dieser Definition wäre die Farbtemperatur der Sonne in dem Grenzfall unendlich weit gesteckter Spektralgrenzen gleich der Strahlungstemperatur der Gesamtstrahlung (vgl. WILSINGS ersten Ansatz 5810° für die spektralphotometrische Farbtemperatur der Sonne, die nur wenig größer ist als die von ABBOT aus der Solarkonstanten bestimmte Strahlungstemperatur 5740° der bolometrischen Helligkeit).

Bei dieser Umschreibung der Farbtemperatur wird ihr Charakteristikum, die Gestalt der Energiekurve, ganz außer acht gelassen; lehrt doch schon der bloße Augenschein, daß die Form der Sonnenenergiekurve mehr einem schwarzen Strahler von der Temperatur 7000° als von 6000° entspricht. Wenn auch eine passende Definition für die Ähnlichkeit zweier Kurven nicht leicht anzugeben ist, so scheint mir doch der folgende Vorschlag zweckmäßig zu sein, da er einen Ausweg aus dem Dilemma möglich macht: Man betrachtet die Sonne als einen grauen Strahler und bildet für die Energiekurve der mittleren und der zentralen Sonnenstrahlung in Abständen von 0,25 μ den Quotienten der Energiewerte gegen die schwarzen Strahler von 6000° und 7000°:

Mittlere Sonnenstrahlung:

$\lambda(\mu)$	0,50	0,75	1,00	1,25	1,50	1,75	2,00
6000°	1,12	0,85	0,79	0,87	0,93	1,00	1,18
7000	0,55	0,54	0,56	0,63	0,68	0,77	0,93

Zentrale Sonnenstrahlung:

$\lambda(\mu)$	0,50	0,75	1,00	1,25	1,50	1,75	2,00
6000°	1,45	0,96	0,90	0,98	1,02	1,02	1,18
7000	0,71	0,63	0,63	0,71	0,74	0,79	0,93

Es ist dann derjenige schwarze Strahler der Energiekurve der Sonne in der Form am ähnlichsten, bei der die Variation des Quotienten mit der Wellenlänge am kleinsten ist. Dabei wird man dem Quotienten ein um so höheres Gewicht geben, je größer der absolute Energiewert ist. Nach der obigen Tabelle entspricht die Energiekurve der mittleren Sonnenstrahlung im Hauptteil einem grauen Strahler von der Farbtemperatur 7000° mit dem Absorptionskoeffizienten 0,55, die der zentralen einem grauen Strahler von annähernd der gleichen Farbtemperatur mit dem Absorptionskoeffizienten 0,64.

Wenn man innerhalb eines weiten Spektralbereiches die schwarze Strahlungskurve bestimmt, welche die Energiekurve der Sonne am besten darstellt, so erhält man nur einen rohen Durchschnittswert der spektralphotometrischen Farbtemperatur. Will man eine bessere Annäherung erreichen, so wird man sich auf schmale Spektralbereiche beschränken; es ist dann zu jeder spektralphotometrischen Farbtemperatur der Spektralbereich anzugeben, auf den sich jene bezieht. Im Grenzfall des unendlich kleinen Spektralbereichs wird die Energiekurve der Sonne für jede Wellenlänge demjenigen schwarzen Strahler zugeordnet, für welchen die Richtung der Tangente an die Energiekurve der Sonne und des schwarzen Strahlers übereinstimmt. Die in dieser Weise definierten Gradationstemperaturen der mittleren Sonnenstrahlung wurden vom Verfasser für eine Reihe von Wellenlängen nach der Gleichung (20) berechnet[1]:

Gradationstemperaturen der mittleren Sonnenstrahlung nach Beobachtungen auf dem Mount Wilson-Observatorium in den Jahren 1903 bis 1910.

$\lambda\,(\mu\mu)$	300	325	350	375	400	425	450	475
$c_2/T_G \cdot 10^4$	4,77	5,14	4,20	2,69	3,43	3,60	2,85	2,30
$\lambda\,(\mu\mu)$	500	525	550	600	650	700	750	800
$c_2/T_G \cdot 10^4$	2,05	2,18	2,13	1,90	1,71	1,84	1,78	1,94
$\lambda\,(\mu\mu)$	850	900	950	1000	1100	1200	1450	1800
$c_2/T_G \cdot 10^4$	2,13	2,24	2,81	2,74	2,65	2,86	3,17	1,67

Die Abweichungen der Sonnenenergiekurve von der des schwarzen Strahlers werden bei der Gradationstemperatur deutlich sichtbar. Kleine Buckel der Sonnenenergiekurve (z. B. bei $\lambda\,375\ \mu\mu$) haben große Schwankungen in der Gradationstemperatur zur Folge; diese bildet demnach ein vorzügliches Maß für die Abweichungen der Sonnenstrahlung von der des schwarzen Körpers. Für Wellenlängen größer als $2{,}0\ \mu$ läßt sich kein schwarzer Strahler noch so hoher Temperatur angeben, der eine dem Gradienten der Energiekurve der Sonne entsprechende Gradationstemperatur besitzt. Die aus der Intensitätsverteilung im kontinuierlichen Spektrum der Sonne vom Verfasser für den Spektralbereich λ 451 bis 642 $\mu\mu$ abgeleitete spektralphotometrische Farbtemperatur $c_2/T_F = 2{,}16 \cdot 10^{-4}$ stimmt nahezu mit der für die mittlere Wellenlänge λ 545 $\mu\mu$ geltenden Gradationstemperatur $c_2/T_G = 2{,}14 \cdot 10^{-4}$ überein.

d) Die Farbtemperatur der Sterne aus der Farbe oder aus einem Farbenäquivalent.

19. Die Sternfarbe, die effektive und die Minimalwellenlänge. Die Farbenschätzung ist als eine primitive Spektralphotometrie zu betrachten, bei der die Farbe als Summe der Eindrücke aufzufassen ist, welche die einzelnen Teile des Spektrums auf das Auge ausüben. Die Sternfarbe ist demnach ein Kollektivmaß für die Intensitätsverteilung im Spektrum und folgt aus ihr im Zusammenhang mit der Farbenempfindlichkeit des menschlichen Auges. Farbe und Temperatur sind aufs engste miteinander verknüpft; den Farbenskalen liegt eine Abkühlungsskala zugrunde. Die Einordnung der Sternfarben in eine nach der Strahlung schwarzer Körper geeichte Farbenskala scheitert an der Unzulänglichkeit der subjektiven Farbenangaben, die vom Beobachter, vom Instrument und von der Helligkeit des Objektes abhängen[2]. In der Praxis eicht man die Farbenskala nach Sternen anderweitig bestimmter Farbtemperatur.

Wenn die Sternstrahlung im optischen Spektralgebiet mit der Strahlung eines schwarzen Körpers bestimmter Temperatur übereinstimmt, kann man die Farbtemperatur unmittelbar aus der Sternfarbe erhalten, wenn man einem terrestrischen Strahler, dessen Energiekurve ebenfalls der PLANCKschen Strahlungsgleichung genügt, die gleiche Farbtemperatur erteilt. WILSING verwandelt die Sternstrahlung durch Einschalten eines Rotfilters in den Strahlengang in meßbarer Weise in eine Strahlung niedrigerer Farbtemperatur, ehe er sie mit einer terrestrischen Lichtquelle in bezug auf Farbe und Intensität vergleicht[3].

[1] A. BRILL, Das Temperaturproblem in der Astrophysik. Veröff Universitätssternw Berlin-Babelsberg VII, Heft 6 (1930).

[2] Vgl. die Ausführungen von K. F. BOTTLINGER über „Physiologische Farben" im Handb. der Astrophysik II, Teil 1, Kap. 3, Ziffer 6.

[3] J. WILSING, Messungen der Farben, der Helligkeiten und der Durchmesser der Sterne mit Anwendung der PLANCKschen Strahlungsgleichung. Publ Astrophys Obs Potsdam Nr. 76 (1920).

WILSING bringt die PLANCKsche Strahlungsgleichung in die folgende für die analytische Diskussion bequeme Form:

$$c_1 \cdot \lambda^{-5} \cdot e^{-\gamma_0 - \frac{1}{\lambda}(c_2/T + \gamma_1)}.$$

Der Teilfaktor $e^{-\gamma_0 - \gamma_1/\lambda}$ entspricht dem PLANCKschen Korrektionsglied. γ_1 ist mit der absoluten Temperatur T durch die Gleichung

$$\gamma_1 \log e = 0{,}0075 \cdot 10^{-3}(T - 3000^\circ)$$

verknüpft. $\gamma_0 \log e$ hat für die mittlere Wellenlänge λ 575 $\mu\mu$ folgende Werte:

5000°	10000°	15000°	20000°
−0,029	−0,127	−0,245	−0,366

Die Variation mit der Wellenlänge ist verschwindend klein.

Nach dem Durchgang durch die Erdatmosphäre wird die Energie der Sternstrahlung von der Wellenlänge λ:

$$c_1 \cdot \lambda^{-5} \cdot e^{-\gamma_0 - \alpha_0 l_z} \cdot e^{-\frac{1}{\lambda}(c_2/T + \gamma_1 + \alpha_1 l_z)}.$$

l_z ist die Weglänge des Lichtstrahles in der Erdatmosphäre bei der Zenitdistanz z des Sterns. Der Transmissionskoeffizient der Erdatmosphäre läßt sich in dem optischen Spektralgebiet λ 450 bis 680 $\mu\mu$ durch die Gleichung $p_\lambda = e^{-\alpha_0 - \alpha_1/\lambda}$ darstellen, wo die Konstanten $\alpha_0 \log e = -0{,}138$ und $\alpha_1 \log e = 0{,}1463$ aus den Beobachtungen bestimmt sind. Nach dem Durchgang durch den Rotkeil ist die gesamte visuell wirksame Sternstrahlung:

$$c_1 \cdot e^{-\gamma_0 - \alpha_0 l_z - \beta_0 d} \int\limits_{\lambda=0{,}45\,\mu}^{\lambda=0{,}68\,\mu} \lambda^{-5} \cdot e^{-\frac{1}{\lambda}(c_2/T + \gamma_1 + \alpha_1 l_z + \beta_1 d)}\, d\lambda.$$

Der Transmissionskoeffizient des Rotkeiles genügt im optischen Teil des Spektrums dem Ausdruck $e^{-(\beta_0 + \beta_1/\lambda)d}$, wo d die Keildicke und β_0, β_1 die empirisch bestimmten Keilkonstanten sind.

Wenn T' die Farbtemperatur der Photometerlampe, φ die Ablesung am Intensitätskreis des Photometerapparates bei Gleichheit der Farbe und Helligkeit von Stern und Vergleichslampe und c_1' die von den Versuchsbedingungen abhängige Lampenkonstante im PLANCKschen Gesetz bezeichnen, so wird:

$$c_1 \cdot e^{-\gamma_0 - \alpha_0 l_z - \beta_0 d} \int\limits_{\lambda=0{,}45\,\mu}^{\lambda=0{,}68\,\mu} \lambda^{-5} \cdot e^{-\frac{1}{\lambda}(c_2/T + \gamma_1 + \alpha_1 l_z + \beta_1 d)}\, d\lambda = c_1' \cdot \sin^2\varphi \int\limits_{\lambda=0{,}45\,\mu}^{\lambda=0{,}68\,\mu} \lambda^{-5} \cdot e^{-c_2/\lambda T'}\, d\lambda. \quad (55)$$

Diese Gleichung zerfällt in die beiden folgenden:

$$\frac{c_2}{T} + \gamma_1 + \alpha_1 l_z + \beta_1 d = \frac{c_2}{T'}; \qquad c_1 \cdot e^{-\gamma_0 - \alpha_0 l_z - \beta_0 d} = c_1' \cdot \sin^2\varphi. \quad (56)$$

Die erste Gleichung gibt die Farbtemperatur des Sterns, die zweite die vom Durchmesser und von der Entfernung des Sterns abhängige Konstante c_1.

Bei der Berechnung der Farbtemperatur der Sterne nach der kolorimetrischen Methode eliminiert WILSING das von der Lampentemperatur T' abhängige Glied. Der Nullpunkt der Temperaturskala wird so gewählt, daß das Mittel der für die Sterne β Herculis, α Ophiuchi, ζ Aquilae und α Aquilae kolorimetrisch bestimmten c_2/T-Werte mit dem Mittel der spektralphotometrisch von ihm abgeleiteten zusammenfällt. Die Farbtemperaturen der Sterne sind nach den kolorimetrischen Beobachtungen bei den Sternen von frühem Spektraltypus höher, als die spektralphotometrischen Messungen WILSINGS aus der Form der Energiekurve ergeben haben (vgl. Tabelle 5 unter „WILSING“).

FESSENKOFF schlägt die Anwendung eines Kolorimeters vor, bei dem die Angleichung der Farben des natürlichen und des künstlichen Sterns mit einem blauen Glaskeil erzielt wird[1]. Dieser wird vor der Glühlampe aufgestellt, welche den künstlichen Stern bildet. Wenn die Messungen zur Bestimmung der Sterntemperaturen unabhängig von anderen Methoden sein sollen, muß die Farbtemperatur der Lampe für eine bestimmte Stromspannung durch Vergleich mit der Strahlung des absolut schwarzen Körpers ermittelt werden.

Im Sommer 1928 hat FESSENKOFF bei Odessa kolorimetrische Messungen mit einem blauen Glaskeil an 122 Sternen angestellt. Die Beobachtungen wurden an einem ZEISSschen Refraktor von 80 mm Öffnung ausgeführt, der zu dem speziellen Zweck mit einem ROSENBERGschen Flächenphotometer ausgerüstet war. Die Farbe des extrafokalen Sternbildes wurde mit der Farbe des Photometerfleckes verglichen, die sich beim Verschieben des blauen Keiles vor der Photometerlampe ändert. Die optischen Eigenschaften des blauen Glaskeiles, der im Laboratorium sorgfältig untersucht wurde, genügten nicht der theoretischen Forderung, daß die Lampenstrahlung im optischen Spektralbereich durch Vorschalten des blauen Keiles sich in eine Strahlung höherer Farbtemperatur verwandeln läßt. Deshalb war es notwendig, die Abhängigkeit der Farbtemperaturen der Sterne von den Keileinstellungen auf empirischem Wege zu bestimmen, was FESSENKOFF auf Grund der Dreifarbentheorie des Sehens durchgeführt hat. Die gesuchte Abhängigkeit wird mit hinreichender Genauigkeit durch eine lineare Gleichung dargestellt:

$$\frac{1}{T} = \alpha + \beta (s - 5{,}7)\,, \tag{57}$$

wo die Keileinstellung s in Millimetern ausgedrückt ist. Die Konstanten α und β ändern sich mit der Farbtemperatur T_0 der Vergleichslampe, wie folgende kleine Tabelle zeigt:

T_0	2000°	2300°	2500°	2700°	3000°
α	0,4923	0,4277	0,3951	0,3667	0,3313
β	−0,00862	−0,00839	−0,00841	−0,00838	−0,00836

Die selektive Absorption der Erdatmosphäre wurde auf Grund spezieller Beobachtungen der Mondoberfläche, des Jupiter und einiger heller Sterne bestimmt und durch Korrektion der Keilablesung berücksichtigt. Da es nicht möglich war, die Farbtemperatur der Lampe während der Beobachtungen sicher zu bestimmen, wurde der Nullpunkt des Temperatursystems mit den dem WILSINGschen kolorimetrischen Katalog gemeinsamen Sterntemperaturen festgelegt. In der Tabelle 5 sind unter dem Kopf „FESSENKOFF" für eine Reihe von Sternen die im System FESSENKOFF bestimmten Farbtemperaturen aufgeführt.

Setzt man vor das Objektiv eines Fernrohres ein Parallelgitter, so entstehen zu beiden Seiten der Zentralbilder der Sterne kurze Beugungsspektren. Ihr optischer Schwerpunkt, das ist diejenige Wellenlänge im Spektrum des Sterns, welche den größten Eindruck im Auge oder auf der photographischen Platte hervorruft, wird als effektive Wellenlänge bezeichnet. Die Grenze der Beugungsspektren nach der Seite der kurzen Wellenlängen nennt LINDBLAD Minimalwellenlänge. Beide Farbenäquivalente hängen von der Beschaffenheit des lichtaufnehmenden Apparates und von der Energieverteilung im Sternspektrum ab.

[1] B. FESSENKOFF, Über die Bestimmung der Temperatur der Sterne. R A J 6, Nr. 2 (1929); Determination of Star Temperatures. A N 236, S. 297 (1929). Bei seinen ersten Farbenschätzungen nach der obenerwähnten Methode stellte FESSENKOFF eine empirische Beziehung seiner Farbenskala zur WILSINGschen spektralphotometrischen Farbtemperaturskala an Hand von 112 gemeinsamen Sternen auf. [B. FESSENKOFF, Bestimmung der effektiven Temperaturen von 193 Sternen. R A J 4, Nr. 3 (1927).]

Die direkte Bestimmung der Farbtemperatur aus beiden Farbenäquivalenten stößt auf Schwierigkeiten, da die Bildstärke, die Verschiedenheit in der Achromasie des Fernrohrobjektives, die Fokussierung, die Empfindlichkeit des Strahlungsempfängers und die Luftunruhe in einer teilweise noch ungeklärten Weise die effektive wie die Minimalwellenlänge beeinflussen. Will man das entsprechende Beobachtungsmaterial zur Temperaturbestimmung verwerten, so muß man die Skala der beiden Farbenäquivalente nach Sternen bekannter Farbtemperatur eichen.

Die spektralphotometrischen Messungen im engeren Sinne beziehen sich auf schmale Spektralbereiche, die von Absorptions- und Emissionslinien möglichst frei sind. Die Sternfarbe und die Farbenäquivalente der effektiven und der Minimalwellenlänge sind mehr oder minder durch die selektive Absorption und Emission im Sternspektrum beeinflußt. Man wird daher a priori nicht erwarten dürfen, daß die aus der Farbe oder aus ihren Äquivalenten abgeleitete Temperaturskala genau mit der aus den spektralphotometrischen Messungen erhaltenen übereinstimmt.

20. Der Farbenindex. Die Sternfarbe und die Äquivalente der effektiven und der Minimalwellenlänge nehmen Bezug auf die Qualität des Sternenlichtes. Das Helligkeitsverhältnis der Sternstrahlung in zwei weiten, mehr oder minder scharf begrenzten Spektralbereichen hat einen quantitativen Charakter und bildet ebenso wie die anderen Farbenäquivalente ein Maß für die Farbtemperatur des Sterns. Die Wirkung der selektiven Absorption und Emission im Sternspektrum ist in dem neuen Farbenäquivalent einbegriffen.

NORDMANN bestimmt mittels des „photomètre hétérochrome" das Helligkeitsverhältnis R/B der durch ein Rot- und ein Blaufilter gehenden Sternstrahlung[1]. Für eine Reihe von künstlichen Lichtquellen, deren Farbtemperaturen aus Laboratoriumsmessungen bekannt sind, verläuft der Logarithmus von R/B nahezu linear mit der reziproken Temperatur. Durch Extrapolation der Eichkurve für die höheren Temperaturen gemäß dem PLANCKschen Gesetz leitet NORDMANN aus dem Helligkeitsverhältnis R/B die Farbtemperaturen von 15 Sternen ab (vgl. Tab. 5 unter „NORDMANN").

Seit SCHWARZSCHILD ist man dazu übergegangen, die Farbe eines Sterns durch eine quantitativ auch für die schwächeren Sterne bestimmbare Eigenschaft festzulegen, nämlich durch die F a r b t ö n u n g oder den F a r b e n i n d e x, d. i. durch das Intensitätsverhältnis zweier verschiedenen Spektralbezirke. Beim photographischen Farbenindex ist beispielsweise die photographische Helligkeit des Sterns, beim Wärmeindex die bolometrische mit der visuellen Helligkeit differentiell verknüpft.

Um aus dem photographischen Farbenindex die Farbtemperatur zu bestimmen, nimmt SCHWARZSCHILD an, daß die PLANCKsche Strahlungsgleichung für die Sternstrahlung gilt, und daß das menschliche Auge wie auch die photographische Platte monochromatisch empfinden[2]. Sind λ_v und λ_{ph} die wirksamen Wellenlängen der visuellen und der photographischen Helligkeit, so werden die zugehörigen Strahlungsintensitäten:

$$J_v = c_1 \lambda_v^{-5} (e^{c_2/\lambda_v T} - 1)^{-1}; \qquad J_{ph} = c_1 \lambda_{ph}^{-5} (e^{c_2/\lambda_{ph} T} - 1)^{-1}. \tag{58}$$

[1] CH. NORDMANN, Méthode permettant la mesure des températures effectives des étoiles. C R 149, S. 557 (1909); Nouvelle approximation dans l'étude des températures effectives des étoiles. Ebenda 149, S. 1038 (1909).

[2] K. SCHWARZSCHILD, Aktinometrie der Sterne der BD bis zur Größe $7^m,5$ in der Zone 0° bis +20° Deklination. Teil B. Abhandl Königl Ges d Wiss Göttingen Math phys Kl Neue Folge 8, Nr. 4, S. 29 (1912).

Die entsprechenden Helligkeiten im System der Größenklassen sind:

$$m_v = 2{,}5 \log [\lambda_v (e^{c_2/\lambda_v T} - 1)] + k_1; \qquad m_{ph} = 2{,}5 \log [\lambda_{ph} (e^{c_2/\lambda_{ph} T} - 1)] + k_2. \quad (59)$$

Die Konstanten k_1 und k_2, zu denen der Bequemlichkeit halber die Beträge $10 \log \lambda_v$ bzw. $10 \log \lambda_{ph}$ geschlagen sind, bleiben so lange unbestimmt, als über den Nullpunkt der visuellen und der photographischen Größenskala keine Festsetzung getroffen ist. Der photographische Farbenindex wird damit:

$$F.I. = m_{ph} - m_v = 2{,}5 \log [\lambda_{ph} (e^{c_2/\lambda_{ph} T} - 1)] - 2{,}5 \log [\lambda_v (e^{c_2/\lambda_v T} - 1)] + k. \quad (60)$$

Mit wachsender Temperatur nähert sich der Farbenindex asymptotisch dem Grenzwert k. Wenn der Nullpunkt der Größenskalen so festgelegt wird, daß k gleich Null ist, so zeichnet sich der in dieser Weise gezählte Farbenindex dadurch aus, daß er für den unendlich heißen Strahler verschwindet. Dieser Farbenindex mißt aber nicht die Energiestufe, welche der visuellen und der photographischen Helligkeit entspricht, weshalb die von SCHWARZSCHILD vorgeschlagene Bezeichnung „absoluter" Farbenindex nicht glücklich gewählt ist. Wenn die wirksamen Wellenlängen λ_v und λ_{ph} bekannt sind, kann man den Farbenindex $F.I.$ für $k = 0$ nach Gleichung (60) mit dem Argument der Temperatur T tabulieren.

In der Beobachtungspraxis ist man übereingekommen, den Sternen $5^m{,}5$ bis $6^m{,}5$ vom Spektraltypus A0 in allen photometrischen Systemen die gleiche visuelle und photographische Helligkeit zu geben. Den Übergang von dem rechnerisch festgelegten Farbenindex zu dem aus den Beobachtungen abgeleiteten vermittelt bei SCHWARZSCHILD die Sonne. Der sog. absolute Farbenindex der Sonne folgt aus ABBOTS Messungen der Energieverteilung im Sonnenspektrum bzw. aus dem Intensitätsverhältnis der spektralen Helligkeiten für die Wellenlängen λ_v und λ_{ph}. Den Farbenindex der Sonne im Beobachtungssystem der visuellen und der photographischen Helligkeiten setzt SCHWARZSCHILD gleich dem mittleren Farbenindex der Sterne vom Spektraltypus G0. Nachdem der Zusammenhang zwischen dem beobachteten Farbenindex und der Temperatur hergestellt ist, läßt sich mit Zuhilfenahme einer kleinen Tabelle die zu dem Farbenindex eines jeden Sternes gehörige Farbtemperatur bestimmen.

Je nachdem SCHWARZSCHILD die Potsdamer oder die Harvardskala der visuellen Helligkeiten benutzt, findet er mit den von ihm gemessenen photographischen Helligkeiten systematische Unterschiede in den Temperaturen. Übereinstimmung läßt sich nur erzielen, wenn man annimmt, daß die wirksamen Wellenlängen der Potsdamer und der Harvardhelligkeiten voneinander verschieden sind.

Der Ansatz von SCHWARZSCHILD bedarf aus folgenden Gründen noch einer Korrektur: Die Sterne verhalten sich nur näherungsweise wie schwarze Strahler. Das menschliche Auge und die photographische Platte empfinden nicht monochromatisch; es besteht eine Abhängigkeit der wirksamen Wellenlänge von der Farbe des Sterns. Der Anschluß an die Sonne ist ungenau, da ihr photographischer Farbenindex durch die Beobachtungen nicht sichergestellt ist.

Die visuelle und die photographische Helligkeit kommen bekanntlich in der Weise zustande, daß sich die Intensitäten schmaler Wellenlängenbereiche additiv zu einem Gesamteindruck vereinigen. Bezeichnet man mit $f(\lambda)$ die Empfindlichkeitsfunktion des menschlichen Auges, mit $g(\lambda)$ die der photographischen Platte (einschließlich der Farbendurchlässigkeit der optischen Apparatur), so sind unter der Voraussetzung, daß die Strahlung der Sterne die Eigenschaften des schwarzen

Körpers besitzt, die visuelle und die photographische Flächenhelligkeit durch die folgenden Ausdrücke gegeben:

$$\left.\begin{aligned} M_v &= -2{,}5 \log \left[\int_0^\infty 2\, c_1\, \lambda^{-5} \left(e^{c_2/\lambda T} - 1\right)^{-1} \cdot f(\lambda) \cdot p(\lambda)\, d\lambda \right], \\ M_{ph} &= -2{,}5 \log \left[\int_0^\infty 2\, c_1\, \lambda^{-5} \left(e^{c_2/\lambda T} - 1\right)^{-1} g(\lambda) \cdot p(\lambda)\, d\lambda \right]. \end{aligned} \right\} \quad (61)$$

$p(\lambda)$ ist der Transmissionskoeffizient der Erdatmosphäre; die Konstante $2\,c_1$ hat nach Ziffer 7 den Wert $1{,}1778 \cdot 10^{-5}$. Die Berechnung der Integrale erfolgte durch planimetrische Ausmessung der entsprechenden Flächenstücke[1].

Der Nullpunkt der rechnerisch bestimmten visuellen und photographischen Flächenhelligkeiten ist durch die Annahme festgelegt, daß in der Wellenlänge der maximalen Empfindlichkeit die volle Strahlung sowohl im menschlichen Auge als auch in der photographischen Platte wirksam ist. Der Nullpunkt der empirischen Größenskala ist willkürlich festgelegt; man ist übereingekommen, den Sternen des Spektraltypus A0 die gleiche visuelle und photographische Helligkeit zu geben. Verhalten sich die Sterne wirklich wie schwarze Strahler, so dürfen sich die rechnerisch ermittelten und die beobachteten Farbenindizes nur um eine Konstante unterscheiden. Der Versuch des Verfassers, mit den durch seine spektralphotometrischen Untersuchungen bestimmten Farbtemperaturen die beobachteten KINGschen Farbenindizes darzustellen, blieb ohne Erfolg; die Unterschiede ergeben einen merklichen Gang mit dem Spektraltypus (Tab. 9a).

Tabelle 9.

Spektrum	a				b		c		
	$\frac{c_2}{T} \cdot 10^4$	*F. I.* KING	$M_{ph}-M_v$	$B-R$	$M_{ph}^a-M_v^a$	$B-R$	*F. I.* KING verbessert	$B-R$	$\frac{c_2}{T} \cdot 10^4$
B0	0,603	$-0^m{,}330$	$-0^m{,}736$	$+0^m{,}406$	$-0^m{,}736$	$+0^m{,}406$	$-0^m{,}328$	$+0^m{,}408$	0,631
B5	0,880	−0 ,170	−0 ,658	+0 ,488	−0 ,597	+0 ,427	−0 ,173	+0 ,424	0,943
A0	1,216	0 ,000	−0 ,504	+0 ,504	−0 ,419	+0 ,419	−0 ,015	+0 ,404	1,236
A5	1,594	+0 ,210	−0 ,329	+0 ,539	−0 ,234	+0 ,444	+0 ,183	+0 ,417	1,634
F0	1,816	+0 ,330	−0 ,209	+0 ,539	−0 ,092	+0 ,422	+0 ,299	+0 ,391	1,809
F5	2,080	+0 ,470	−0 ,083	+0 ,553	+0 ,053	+0 ,417	+0 ,434	+0 ,381	2,054
G0	2,392	+0 ,670	+0 ,081	+0 ,589	+0 ,245	+0 ,425	+0 ,631	+0 ,386	2,376
G5	2,744	+0 ,930	+0 ,254	+0 ,676	+0 ,445	+0 ,485	+0 ,884	+0 ,439	2,819
K0	3,140	+1 ,120	+0 ,450	+0 ,670	+0 ,673	+0 ,447	+1 ,070	+0 ,397	3,143
K5	3,727	+1 ,580	+0 ,758	+0 ,822	+1 ,005	+0 ,575	+1 ,525	+0 ,520	3,946
M0	4,020	+1 ,730	+0 ,908	+0 ,822	+1 ,181	+0 ,549	+1 ,667	+0 ,486	4,180

Die Unstimmigkeit zwischen Rechnung und Beobachtung ist darauf zurückzuführen, daß die Energiekurve der Sterne von der PLANCKschen Kurve abweicht, und daß die visuelle und die photographische Helligkeit als Integralhelligkeiten die abschwächende Wirkung der Absorptionslinien einbegreifen. Die hauptsächlichste Abweichung der Sternstrahlung von der schwarzen Strahlung besteht in einer Depression der Sternenergiekurve im Ultraviolett; sie wurde vom Verfasser empirisch nach den ins Ultraviolett sich erstreckenden spektralphotometrischen Messungen ROSENBERGS bestimmt. Die Berücksichtigung der Senkung der Energiekurve bei der planimetrischen Ausmessung führt zu

[1] A. BRILL, Die Strahlung der Sterne. Veröff Univ Sternw Berlin-Babelsberg V, Heft 1 (1924). Die Zahlenwerte der universellen Konstanten $2c_1$ und c_2 in Veröff Berlin-Babelsberg V, 1 sind verschieden von den in Ziffer 7 mitgeteilten. Der Unterschied bleibt ohne Einfluß auf den Farbenindex und die Skala der c_2/T-Werte.

einer fast vollkommenen Übereinstimmung zwischen Rechnung und Beobachtung (Tab. 9b).

Die beobachteten Farbenindizes entsprechen den auf Grund der PLANCKschen Strahlungsgleichung berechneten nur dann, wenn das System der visuellen und der photographischen Helligkeiten, wie es durch die Rechnung festgelegt ist, dem der Beobachtungen vollkommen gleichartig ist. Wie bereits in Ziffer 11 und 12 erwähnt wurde, kann man die Integralstrahlung der visuellen bzw. der photographischen Helligkeit durch die monochromatische Helligkeit einer bestimmten Wellenlänge ersetzen. In der Bezeichnungsweise des Verfassers ist für die „isophote" Wellenlänge die spektrale Intensität in der PLANCKschen Energiekurve (gerechnet im System der Größenklassen) bis auf eine Konstante gleich der visuellen oder der photographischen Helligkeit. Damit die Skala der beobachteten Farbenindizes mit den auf Grund des PLANCKschen Gesetzes berechneten übereinstimmt, müssen die Energiestufen, welche beiden Farbenindizes in der PLANCKschen Energiekurve entsprechen, einander gleich sein. Die isophoten Wellenlängen der visuellen und der photographischen Helligkeit können in Rechnung und Beobachtung mehr oder minder voneinander abweichen.

Von den beiden Systemen isophoter Wellenlängen stützt sich das eine, welches der Verfasser als das fundamentale bezeichnet, auf die Annahme, daß die Sterne nahezu wie schwarze Körper einer bestimmten Farbtemperatur strahlen. Die Empfindlichkeitsfunktion des menschlichen Auges und der photographischen Platte sind in bestimmter Weise vorgeschrieben. Das andere System der isophoten Wellenlängen ist aus den Beobachtungen erhalten und bezieht sich auf die PLANCKsche Energiekurve, welche die spektralphotometrischen Messungen am besten darstellt. Die Energiestufe in der PLANCKschen Energiekurve zwischen den zusammengehörigen isophoten Wellenlängen liefert den absoluten Farbenindex in dem rechnerisch festgelegten und in dem durch die Beobachtungen bestimmten System. Mit den Unterschieden in der Größe der Energiestufen lassen sich die beobachteten Farbenindizes auf die mit dem PLANCKschen Gesetz berechneten reduzieren. Damit wird dann entschieden, in welche Farbtemperaturskala sich die beobachteten Farbenindizes einordnen lassen.

Nach der Untersuchung des Verfassers in Veröff Berlin-Babelsberg V, 1 sind die isophoten Wellenlängen der visuellen Helligkeiten der Revised Harvard Photometry und der photographischen Helligkeiten von KING merklich von denjenigen der rechnerisch festgelegten photometrischen Systeme verschieden. Die Energiestufe, welche den Farbenindizes in Rechnung und Beobachtung entspricht, ist dagegen nahezu von der gleichen Größe. Die Korrektionen, welche an die KINGschen Farbenindizes anzubringen sind, um sie auf das rechnerisch festgelegte System zu beziehen, erreichen nicht den Betrag von $0^{m},1$. In der letzten Spalte der Tabelle 9c sind die Farbtemperaturen zusammengestellt, wie sie für die einzelnen Spektraltypen aus der Diskussion der spektralphotometrischen Messungen von WILSING und ROSENBERG und der Farbenindizes von KING folgen.

Die von BOTTLINGER auf der Babelsberger Sternwarte bestimmten lichtelektrischen Farbenindizes von 459 Sternen wurden mit einer Kaliumzelle erhalten, vor die wechselweise ein Blau- bzw. ein Gelbfilter gesetzt wurde[1]. Da die Empfindlichkeitskurve der Zelle und die Absorption der verwendeten Filter nicht bestimmt worden sind, kommt der aus den lichtelektrischen Farbenindizes abgeleiteten Temperaturskala keine selbständige Bedeutung zu. BOTTLINGER leitete zwischen seinen Farbenindizes und den zugehörigen c_2/T-Werten eine funktionelle Beziehung ab, welche für die Sterne der frühen Spektraltypen mit

[1] K. F. BOTTLINGER, Lichtelektrische Farbenindizes von 459 Sternen. Veröff Univ-Sternw Berlin-Babelsberg III, Heft 4 (1923).

den Temperaturen SAHAS, für die Sterne der späten Spektraltypen mit den von WILSING aus seinen spektralphotometrischen Messungen berechneten Temperaturen übereinstimmt. Unabhängig davon, daß mit den lichtelektrischen Farbenindizes eine absolute Temperaturskala nicht erhalten wurde, sind die individuellen c_2/T-Werte wegen der hohen Genauigkeit der Meßmethode in dem von BOTTLINGER gewählten Temperatursystem als die zur Zeit besten aus Farbenindizes abgeleiteten anzusprechen (Tab. 5 unter „BOTTLINGER").

Auf das WILSINGsche System der spektralphotometrisch bestimmten Farbtemperaturen reduziert HERTZSPRUNG in dem 734 Sterne heller als 5^m umfassenden Katalog alle ihm zugänglichen Farbenäquivalente[1], indem er für die Beziehung zwischen den verschiedenen Arten von Farbenäquivalenten und dem WILSINGschen c_2/T-Wert eine lineare Gleichung ansetzt. HERTZSPRUNG erteilt den einzelnen Farbkatalogen Gewichte, die von der Zahl der benutzten Beobachtungen und von der Größe der bei der linearen Reduktion übrigbleibenden Reste abhängen, und leitet dann aus den zur Verfügung stehenden Bestimmungen Mittelwerte der c_2/T für die einzelnen Sterne ab (Tab. 5 unter „HERTZSPRUNG").

Auf das gleiche System der WILSINGschen spektralphotometrischen Farbtemperaturen bringt HERTZSPRUNG mittels einer linearen Beziehung die Farbenindizes von 658 Sternen, deren photographische Helligkeiten mit dem Leidener 33 cm-Refraktor bestimmt und mit den visuellen Helligkeiten des Potsdamer Generalkataloges kombiniert wurden[2].

21. Die isophoten Wellenlängen der Integralhelligkeiten der Fixsterne in Rechnung und Beobachtung. Unter der effektiven Wellenlänge versteht man diejenige Wellenlänge im Spektrum, welche den stärksten Eindruck im Auge bzw. auf der photographischen Platte hervorruft. Die effektive Wellenlänge hängt von der Empfindlichkeit des menschlichen Auges bzw. der photographischen Platte und von der Energieverteilung im Spektrum ab; die Fernrohroptik, die Fokussierung, die Schichtdicke der Plattenemulsion, die Art der Entwicklung und die Bildstärke beeinflussen in schwer feststellbarer Weise den Wert der effektiven Wellenlänge. Wenn man die Strahlung der effektiven Wellenlänge, sei es für die visuelle oder für die photographische Helligkeit, äquivalent der zugehörigen Integralstrahlung setzt, so entspricht diese Substitution nicht dem wahren Sachverhalt.

Will man die Integralstrahlung der visuellen oder der photographischen Helligkeit durch die monochromatische Strahlung einer bestimmten Wellenlänge ersetzen, so bietet sich von selbst der Begriff der isophoten Wellenlänge dar, welche Bezeichnungsweise ihren besonderen Charakter hervortreten läßt[3]. Die isophote Wellenlänge der visuellen oder der photographischen Helligkeit ist durch diejenige Wellenlänge bestimmt, für welche bis auf eine allen Sternen gemeinsame Konstante die spektrale Helligkeit in der Energiekurve (gerechnet im System der Größenklassen) gleich der visuellen oder gleich der photographischen Helligkeit ist. Der in dieser Weise eingeführte Begriff der isophoten Wellenlänge wurde später vom Verfasser auch auf die bolometrische und auf die von der Wasserzelle durchgelassene bolometrische Strahlung angewandt.

---

[1] E. HERTZSPRUNG, Mean Colour Equivalents and Hypothetical Angular Semi-Diameters of 734 Stars Brighter than the Fifth Magnitude and within 95° of the North Pole. Ann v d Sterrewacht te Leiden Deel 14, Eerste Stuk (1922).

[2] E. HERTZSPRUNG, Photographic Magnitudes of 658 Stars from Plates taken mainly by W. H. VAN DEN BOS with the 33 cm Leiden Refractor. B A N 1, S. 201 (1923).

[3] A. BRILL, Spektralphotometrische Untersuchungen. Teil I—III. A N 218, S. 209; 219, S. 21, 353 (1923); Die Strahlung der Sterne. Veröff Univ-Sternw Berlin-Babelsberg V, Heft 1 (1924); Die isophoten Wellenlängen der Integralhelligkeiten der Fixsterne in Rechnung und Beobachtung. Ebenda VII, Heft 5 (1929); Das photometrische System der photovisuellen Helligkeiten von E. S. KING. A N 237, S. 225 (1929).

Nach internationalem Übereinkommen ist der Nullpunkt der Zählung der Integralhelligkeiten in der Weise festgelegt, daß die photometrische Größe der Sterne vom Spektraltypus A0 und von der scheinbaren Helligkeit $5^m,5$ bis $6^m,5$ in allen photometrischen Systemen gleich ist. Die Willkür, welche in dieser empirischen Festlegung des Nullpunktes der Integralhelligkeiten liegt, läßt sich durch die Einführung von absoluten Farbenindizes vermeiden, d. i. von solchen Farbenindizes, welche direkt die Energiestufe zwischen den isophoten Wellenlängen der zugehörigen photometrischen Helligkeiten messen.

Die isophoten Wellenlängen bilden ein wichtiges Arbeitselement für alle Untersuchungen, welche sich auf die gesamte innerhalb eines gegebenen Spektralbereiches wirksame Strahlung beziehen. Die isophoten Wellenlängen charakterisieren die Farbenempfindlichkeit des photometrischen Systems. Die Farbtemperatur T_F, welche zu der durch die Messung bestimmten Helligkeitsdifferenz eines Sterns in zwei verschiedenen photometrischen Systemen gehört, wird nach Reduktion dieser Helligkeitsdifferenz auf die durch die isophoten Wellenlängen gegebene Energiestufe aus der Gleichung (14) in Ziffer 12 berechnet.

Im Anschluß an die Neureduktion der WILSINGschen und der ROSENBERGschen spektralphotometrischen Messungen hat der Verfasser in seinen „Spektralphotometrischen Untersuchungen" die isophoten Wellenlängen der visuellen Helligkeiten der Revised Harvard Photometry und der photographischen Helligkeiten von KING bestimmt; die isophoten Wellenlängen beziehen sich hier auf den beobachteten Verlauf der Energiekurve. Die folgende Tabelle gibt die Abhängigkeit der isophoten Wellenlänge ($\mu\mu$) vom Spektraltypus:

Spektraltypus	B 0	A 0	F 0	G 0	K 0	M 0
Revised Harvard Photometry	525	530	532	534	535	535
Photographische Helligkeiten von KING	415	424	430	433	435	436

Die praktische Verwendung der beobachteten Integralhelligkeiten verlangt ihre Reduktion auf ein gemeinsames fehlerfreies, fundamentales System. Um von dem Einfluß des Skalenfehlers unabhängig zu sein, wählte der Verfasser in Veröff Berlin-Babelsberg V, 1 ein durch Rechnung festgelegtes System als grundlegendes System der Photometrie. Es wird angenommen, daß die Sterne annähernd wie schwarze Körper strahlen; die hauptsächlichste Abweichung der Sternstrahlung von der schwarzen Strahlung besteht in einer Depression der Energiekurve im Ultraviolett, deren Einfluß empirisch aus den weit ins Ultraviolett sich erstreckenden spektralphotometrischen Messungen ROSENBERGS bestimmt wurde. Das Transmissionsvermögen der Erdatmosphäre, die Farbenempfindlichkeit des menschlichen Auges und der photographischen Platte (einschließlich der Farbendurchlässigkeit der optischen Apparatur) sind in bestimmter Weise vorgeschrieben. Die isophoten Wellenlängen der durch Rechnung festgelegten visuellen und photographischen Flächenhelligkeit ${}^R\lambda_v$ bzw. ${}^R\lambda_{ph}$ sowie die der visuellen Helligkeit der Revised Harvard Photometry ${}^H\lambda_v$ und der photographischen Helligkeit von KING ${}^K\lambda_{ph}$ sind auf die PLANCKsche Energiekurve der zugehörigen Farbtemperatur bezogen, d. h. die visuelle und die photographische Helligkeit, welche die Senkung der Energiekurve im Ultraviolett einbegreifen, werden der PLANCKschen Energiekurve zugeordnet, welche in dem Spektralbereich λ 451 bis 642 $\mu\mu$ die spektralphotometrischen Messungen am besten darstellt. Die folgende Tabelle enthält zu den in der Zeile 1 stehenden c_2/T_F-Werten die isophoten Wellenlängen ($\mu\mu$) der visuellen und der photographischen Helligkeit in Rechnung und Beobachtung:

$c_2/T_F \cdot 10^4$	0,603	0,880	1,216	1,594	1,816	2,080
${}^R\lambda_v$	539	542	545	547	548	550
${}^H\lambda_v$	525	528	530	531	532	533
${}^R\lambda_{ph}$	414	417	419	421	422	423
${}^K\lambda_{ph}$	402	405	407	409	410	411

$c_2/T_F \cdot 10^4$	2,392	2,744	3,140	3,727	4,020	4,783	5,740
${}^R\lambda_v$	551	552	553	553	554	558	562
${}^H\lambda_v$	534	534	535	535	535		
${}^R\lambda_{ph}$	423	424	425	425	426	430	436
${}^K\lambda_{ph}$	412	412	413	413	413		

Die Farbtemperatur, auf welche die isophoten Wellenlängen der obigen Tabelle bezogen sind, ist wohl ein Charakteristikum der Form der Energiekurve, aber nicht der Intensität der Strahlung. Für die stellarastronomische Forschung hat meist nur die Strahlungstemperatur, sei es der visuellen, der photographischen oder der bolometrischen Helligkeit, Bedeutung. In Veröff Berlin-Babelsberg VII, 5 sind die isophoten Wellenlängen der durch Rechnung festgelegten visuellen, photographischen, bolometrischen und der von der Wasserzelle durchgelassenen bolometrischen Helligkeit auf die Energiekurve der zugehörigen Strahlungstemperatur bezogen:

$c_2/T_{St} \cdot 10^4$	0,603	0,880	1,216	1,594	1,816	2,080	2,392
${}^R\lambda_v$	537	539	542	544	546	547	549
${}^R\lambda_{ph}$	422	424	426	428	429	431	432
${}^R\lambda_w$	600	605	612	631	644	659	674
${}^R\lambda_b$	786	806	834	881	914	955	1006

$c_2/T_{St} \cdot 10^4$	2,744	3,140	3,727	4,020	4,783	5,740	7,175	9,567
${}^R\lambda_v$	550	552	554	555	558	561	566	573
${}^R\lambda_{ph}$	434	435	437	438	441	444	449	456
${}^R\lambda_w$	692	710	735	745	778	814	862	921
${}^R\lambda_b$	1064	1128	1226	1277	1403	1565	1803	2200

Wenn die isophote Wellenlänge einer Integralhelligkeit mit λ_Φ, die Empfindlichkeitsfunktion der optischen Apparatur mit $R(\lambda)$, der Transmissionskoeffizient der Erdatmosphäre mit $p(\lambda)$ und die Strahlungstemperatur der betreffenden Integralhelligkeit mit T_{St} bezeichnet werden, so ist die isophote Wellenlänge ganz allgemein durch die Gleichung definiert:

$$2\,c_1\,\lambda_\Phi^{-5}\,(e^{c_2/\lambda_\Phi T_{St}}-1)^{-1}\int\limits_0^\infty R(\lambda)\cdot p(\lambda)\,d\lambda = \int\limits_0^\infty 2\,c_1\,\lambda^{-5}\,(e^{c_2/\lambda T_{St}}-1)^{-1}\,R(\lambda)\cdot p(\lambda)\,d\lambda. \quad (62)$$

Mit der Form der Empfindlichkeitsfunktion $R(\lambda)$ ändert sich die isophote Wellenlänge: In dem extremen Fall, daß der lichtaufnehmende Apparat monochromatisch empfindet, ist die isophote Wellenlänge konstant für alle Strahlungstemperaturen und gleich der wirksamen Wellenlänge der optischen Apparatur. Andererseits, wenn das Produkt $R(\lambda)\cdot p(\lambda)$ in einem weiten Spektralbereich variiert, ist der Spielraum der isophoten Wellenlängen ziemlich groß. Die Änderung der isophoten Wellenlänge hat eine Farbengleichung zur Folge. In der Zukunft wird eine wichtige Aufgabe der photometrischen Untersuchungsmethoden darin bestehen, für die Hauptkataloge der Sternhelligkeiten aus der Farbengleichung die Empfindlichkeitsfunktion $R(\lambda)$ bzw. das Produkt $R(\lambda)\cdot p(\lambda)$ als Lösung der Integralgleichung (62) zu finden.

In der Veröff Berlin-Babelsberg VII, 5 wurden die isophoten Wellenlängen der empirischen Helligkeitssysteme unter der Annahme bestimmt, daß die zugehörigen Empfindlichkeitskurven genau die gleiche Gestalt haben wie die der rechnerisch festgelegten photometrischen Systeme. Trotz dieser Einschränkung

Tabelle 10.

$\frac{c_2}{T}\cdot 10^4$	${}^{R}M_{\lambda_{ph}}$	${}^{R}M_{\lambda_v}$	${}^{R}M_{\lambda_{ph}}-{}^{R}M_{\lambda_v}$	${}^{R}M_{\lambda_w}$	${}^{R}M_{\lambda_b}$	${}^{R}M_{\lambda_v}-{}^{R}M_{\lambda_b}$	${}^{R}M_{\lambda_w}-{}^{R}M_{\lambda_b}$
0,40	−41^{m},79	−40^{m},86	−0^{m}.93	−40^{m},49	−39^{m},39	−1^{m},47	−1^{m},10
0,50	−41 ,45	−40 ,56	−0 ,89	−40 ,16	−39 ,11	−1 ,45	−1 ,05
0,60	−41 ,13	−40 ,28	−0 ,85	−39 ,86	−38 ,84	−1 ,44	−1 ,02
0,70	−40 ,81	−40 ,00	−0 ,81	−39 ,58	−38 ,58	−1 ,42	−1 ,00
0,80	−40 ,50	−39 ,73	−0 ,77	−39 ,32	−38 ,33	−1 ,40	−0 ,99
0,90	−40 ,20	−39 ,47	−0 ,73	−39 ,08	−38 ,09	−1 ,38	−0 ,99
1,00	−39 ,91	−39 ,22	−0 ,69	−38 ,85	−37 ,86	−1 ,36	−0 ,99
1,10	−39 ,62	−38 ,98	−0 ,64	−38 ,62	−37 ,63	−1 ,35	−0 ,99
1,20	−39 ,34	−38 ,74	−0 ,60	−38 ,39	−37 ,41	−1 ,33	−0 ,98
1,30	−39 ,06	−38 ,51	−0 ,55	−38 ,16	−37 ,19	−1 ,32	−0 ,97
1,40	−38 ,79	−38 ,29	−0 ,50	−37 ,94	−36 ,98	−1 ,31	−0 ,96
1,50	−38 ,52	−38 ,07	−0 ,45	−37 ,72	−36 ,78	−1 ,29	−0 ,94
1,60	−38 ,26	−37 ,86	−0 ,40	−37 ,51	−36 ,58	−1 ,28	−0 ,93
1,70	−38 ,00	−37 ,65	−0 ,35	−37 ,31	−36 ,39	−1 ,26	−0 ,92
1,80	−37 ,74	−37 ,44	−0 ,30	−37 ,11	−36 ,20	−1 ,24	−0 ,91
1,90	−37 ,48	−37 ,23	−0 ,25	−36 ,92	−36 ,01	−1 ,22	−0 ,91
2,00	−37 ,22	−37 ,02	−0 ,20	−36 ,73	−35 ,83	−1 ,19	−0 ,90
2,10	−36 ,96	−36 ,81	−0 ,15	−36 ,54	−35 ,66	−1 ,15	−0 ,88
2,20	−36 ,71	−36 ,61	−0 ,10	−36 ,35	−35 ,49	−1 ,12	−0 ,86
2,30	−36 ,46	−36 ,41	−0 ,05	−36 ,17	−35 ,33	−1 ,08	−0 ,84
2,40	−36 ,21	−36 ,21	−0 ,00	−35 ,99	−35 ,17	−1 ,04	−0 ,82
2,50	−35 ,96	−36 ,01	+0 ,05	− 35 ,82	−35 ,01	−1 ,00	−0 ,81
2,60	−35 ,71	−35 ,81	+0 ,10	−35 ,65	−34 ,86	−0 ,95	−0 ,79
2,70	−35 ,46	−35 ,61	+0 ,15	−35 ,48	−34 ,71	−0 ,90	−0 ,77
2,80	−35 ,21	−35 ,41	+0 ,20	−35 ,31	−34 ,56	−0 ,85	−0 ,75
2,90	−34 ,97	−35 ,22	+0 ,25	−35 ,14	−34 ,42	−0 ,80	−0 ,72
3,00	−34 ,72	−35 ,02	+0 ,30	−34 ,97	−34 ,28	−0 ,74	−0 ,69
3,10	−34 ,47	−34 ,82	+0 ,35	−34 ,81	−34 ,14	−0 ,68	−0 ,67
3,20	−34 ,22	−34 ,62	+0 ,40	−34 ,65	−34 ,00	−0 ,62	−0 ,65
3,30	−33 ,97	−34 ,42	+0 ,45	−34 ,50	−33 ,87	−0 ,55	−0 ,63
3,40	−33 ,72	−34 ,23	+0 ,51	−34 ,34	−33 ,74	−0 ,49	−0 ,60
3,50	−33 ,48	−34 ,04	+0 ,56	−34 ,19	−33 ,61	−0 ,43	−0 ,58
3,60	−33 ,24	−33 ,85	+0 ,61	−34 ,04	−33 ,48	−0 ,37	−0 ,56
3,70	−32 ,99	−33 ,65	+0 ,66	−33 ,90	−33 ,36	−0 ,29	−0 ,54
3,80	−32 ,74	−33 ,45	+0 ,71	−33 ,75	−33 ,24	−0 ,21	−0 ,51
3,90	−32 ,50	−33 ,25	+0 ,76	−33 ,60	−33 ,12	−0 ,14	−0 ,48
4,00	−32 ,25	−33 ,06	+0 ,81	−33 ,46	−33 ,00	−0 ,06	−0 ,46
4,10	−32 ,00	−32 ,86	+0 ,86	−33 ,32	−32 ,89	+0 ,03	−0 ,43
4,20	−31 ,76	−32 ,67	+0 ,91	−33 ,17	−32 ,78	+0 ,11	−0 ,39
4,30	−31 ,52	−32 ,48	+0 ,96	−33 ,03	−32 ,67	+0 ,19	−0 ,36
4,40	−31 ,28	−32 ,29	+1 ,01	−32 ,89	−32 ,56	+0 ,27	−0 ,33
4,50	−31 ,04	−32 ,09	+1 ,05	−32 ,75	−32 ,46	+0 ,37	−0 ,29
4,60	−30 ,80	−31 ,89	+1 ,09	−32 ,61	−32 ,36	+0 ,47	−0 ,25
4,70	−30 ,56	−31 ,70	+1 ,14	−32 ,48	−32 ,26	+0 ,56	−0 ,22
4,80	−30 ,31	−31 ,50	+1 ,19	−32 ,34	−32 ,16	+0 ,66	−0 ,18
4,90	−30 ,07	−31 ,31	+1 ,24	−32 ,21	−32 ,06	+0 ,75	−0 ,15
5,00	−29 ,83	−31 ,12	+1 ,29	−32 ,07	−31 ,96	+0 ,84	−0 ,11
5,10	−29 ,59	−30 ,93	+1 ,34	−31 ,94	−31 ,86	+0 ,93	−0 ,08
5,20	−29 ,34	−30 ,73	+1 ,39	−31 ,81	−31 ,77	+1 ,04	−0 ,04
5,30	−29 ,10	−30 ,54	+1 ,44	−31 ,68	−31 ,68	+1 ,14	0 ,00
5,40	−28 ,86	−30 ,35	+1 ,49	−31 ,55	−31 ,59	+1 ,24	+0 ,04
5,50	−28 ,62	−30 ,16	+1 ,54	−31 ,42	−31 ,50	+1 ,34	+0 ,08
5,60	−28 ,38	−29 ,97	+1 ,59	−31 ,30	−31 ,41	+1 ,44	+0 ,11
5,70	−28 ,14	−29 ,78	+1 ,64	−31 ,18	−31 ,32	+1 ,54	+0 ,14
5,80	−27 ,90	−29 ,59	+1 ,69	−31 ,06	−31 ,24	+1 ,65	+0 ,18
5,90	−27 ,67	−29 ,41	+1 ,74	−30, 94	−31 ,16	+1 ,75	+0 ,22
6,00	−27 ,43	−29 ,22	+1 ,79	−30 ,82	−31 ,08	+1 ,86	+0 ,26
6,10	−27 ,19	−29 ,03	+1 ,84	−30 ,70	−31 ,00	+1 ,97	+0 ,30
6,20	−26 ,95	−28 ,84	+1 ,89	−30 ,59	−30 ,91	+2 ,07	+0 ,32
6,30	−26 ,71	−28 ,65	+1 ,94	−30 ,47	−30 ,83	+2 ,18	+0 ,36
6,40	−26 ,48	−28 ,46	+1 ,98	−30 ,35	−30 ,75	+2 ,29	+0 ,40

Tabelle 10. (Fortsetzung.)

$\frac{c_2}{T}\cdot 10^4$	${}^{R}M_{\lambda_{ph}}$	${}^{R}M_{\lambda_v}$	${}^{R}M_{\lambda_{ph}}-{}^{R}M_{\lambda_v}$	${}^{R}M_{\lambda_w}$	${}^{R}M_{\lambda_b}$	${}^{R}M_{\lambda_v}-{}^{R}M_{\lambda_b}$	${}^{R}M_{\lambda_w}-{}^{R}M_{\lambda_b}$
6,50	$-26^m,25$	$-28^m,27$	$+2^m,02$	$-30^m,23$	$-30^m,67$	$+2^m,40$	$+0^m,44$
6,60	−26 ,01	−28 ,08	+2 ,07	−30 ,12	−30 ,59	+2 ,51	+0 ,47
6,70	−25 ,78	−27 ,90	+2 ,12	−30 ,00	−30 ,52	+2 ,62	+0 ,52
6,80	−25 ,55	−27 ,72	+2 ,17	−29 ,89	−30 ,44	+2 ,72	+0 ,55
6,90	−25 ,31	−27 ,53	+2 ,22	−29 ,78	−30 ,37	+2 ,84	+0 ,59
7,00	−25 ,08	−27 ,34	+2 ,26	−29 ,67	−30 ,29	+2 ,95	+0 ,62
7,10	−24 ,85	−27 ,15	+2 ,30	−29 ,55	−30 ,22	+3 ,07	+0 ,67
7,20	−24 ,62	−26 ,96	+2 ,34	−29 ,43	−30 ,15	+3 ,19	+0 ,72
7,30	−24 ,39	−26 ,77	+2 ,38	−29 ,31	−30 ,09	+3 ,32	+0 ,78
7,40	−24 ,15	−26 ,58	+2 ,43	−29 ,20	−30 ,02	+3 ,44	+0 ,82
7,50	−23 ,92	−26 ,39	+2 ,47	−29 ,08	−29 ,95	+3 ,56	+0 ,87
7,60	−23 ,69	−26 ,21	+2 ,52	−28 ,97	−29 ,88	+3 67	+0 ,91
7,70	−23 ,45	−26 ,02	+2 ,57	−28 ,86	−29 ,82	+3 ,80	+0 ,96
7,80	−23 ,22	−25 ,84	+2 ,62	−28 ,75	−29 ,75	+3 ,91	+1 ,00
7,90	−22 ,99	−25 ,66	+2 ,67	−28 ,64	−29 ,69	+4 ,03	+1 ,05
8,00	−22 ,76	−25 ,47	+2 ,71	−28 ,53	−29 ,62	+4 ,15	+1 ,09
8,10	−22 ,53	−25 ,29	+2 ,76	−28 ,42	−29 ,56	+4 ,27	+1 ,14
8,20	−22 ,30	−25 ,10	+2 ,80	−28 ,31	−29 ,50	+4 ,40	+1 ,19
8,30	−22 ,07	−24 ,92	+2 ,85	−28 ,20	−29 ,43	+4 ,51	+1 ,23
8,40	−21 ,84	−24 ,73	+2 ,89	−28 ,09	−29 ,37	+4 ,64	+1 ,28
8,50	−21 ,61	−24 ,55	+2 ,94	−27 ,98	−29 ,31	+4 ,76	+1 ,33
8,60	−21 ,38	−24 ,37	+2 ,99	−27 ,87	−29 ,25	+4 ,88	+1 ,38
8,70	−21 ,15	−24 ,18	+3 ,03	−27 ,76	−29 ,19	+5 ,01	+1 ,43
8,80	−20 ,92	−23 ,99	+3 ,07	−27 ,65	−29 ,14	+5 ,15	+1 ,49
8,90	−20 ,69	−23 ,80	+3 ,11	−27 ,54	−29 ,08	+5 ,28	+1 ,54
9,00	−20 ,46	−23 ,62	+3 ,16	−27 ,44	−29 ,02	+5 ,40	+1 ,58
9,10	−20 ,23	−23 ,43	+3 ,20	−27 ,34	−28 ,96	+5 ,53	+1 ,62
9,20	−20 ,00	−23 ,24	+3 ,24	−27 ,23	−28 ,90	+5 ,66	+1 ,67
9,30	−19 ,77	−23 ,06	+3 ,29	−27 ,13	−28 ,84	+5 ,78	+1 ,71
9,40	−19 ,54	−22 ,88	+3 ,34	−27 ,02	−28 ,79	+5 ,91	+1 ,77
9,50	−19 ,31	−22 ,70	+3 ,39	−26 ,92	−28 ,73	+6 ,03	+1 ,81
9,60	−19 ,08	−22 ,52	+3 ,44	−26 ,82	−28 ,67	+6 ,15	+1 ,85

wird den Beobachtungen in ausreichendem Maße Genüge getan. Die isophoten Wellenlängen des ZINNERschen Kataloges der visuellen Helligkeiten sind um 10 μ, die der photovisuellen Helligkeiten von KING um 7 μ nach dem Rot und die der Revised Harvard Photometry um 15 μ nach dem Violett verschoben gegenüber denen des durch Rechnung festgelegten fundamentalen visuell-photometrischen Systems. Die isophoten Wellenlängen der photographischen Helligkeiten von HERTZSPRUNG sind gegenüber denen des durch Rechnung festgelegten fundamentalen photographisch-photometrischen Systems um 5 μ, die der KINGschen photographischen Helligkeiten nur um 2 μ nach dem Violett verschoben. Die photometrischen Systeme der bolometrischen und der von der Wasserzelle durchgelassenen bolometrischen Strahlung nach den Messungen von PETTIT und NICHOLSON auf dem Mount Wilson-Observatorium besitzen keine merkliche Farbengleichung gegenüber den durch Rechnung festgelegten fundamentalen photometrischen Systemen.

Beim Vergleich der beobachteten Integralhelligkeiten der Sterne mit den durch Rechnung aus der PLANCKschen Energiekurve ermittelten Flächenhelligkeiten wird man stets auf die durch Rechnung festgelegten photometrischen Systeme der visuellen, der photographischen, der bolometrischen und der von der Wasserzelle durchgelassenen bolometrischen Helligkeit zurückgehen. Bezeichnet m die scheinbare beobachtete Helligkeit eines Sterns (bolometrisch, mit der Wasserzelle, visuell oder photographisch gemessen), M die zugehörige Flächenhelligkeit, so sind m und M mit dem scheinbaren Sternhalbmesser ϱ durch die Gleichung verbunden:

$$M = M_0 + m + \Delta m + 5 \log \varrho. \tag{63}$$

Die Konstante M_0 drückt aus, daß der Nullpunkt der empirischen Größenskala der scheinbaren Helligkeiten nicht mit dem der Flächenhelligkeiten identisch ist. Die Größe Δm, welche sich mit der Farbe des Sterns ändert, deutet an, daß die photometrischen Systeme der scheinbaren Helligkeiten und der Flächenhelligkeiten in bezug auf die Farbenempfindlichkeit nicht vollkommen gleichartig sind[1].

Die Tabelle 10 gibt mit dem Argument der Strahlungstemperatur bzw. mit dem zugehörigen c_2/T-Wert die zu den entsprechenden isophoten Wellenlängen gehörigen spektralen Energiewerte M der bolometrischen, der von der Wasserzelle durchgelassenen bolometrischen, der visuellen und der photographischen Helligkeit in den durch Rechnung festgelegten fundamentalen photometrischen Systemen. T ist die Strahlungstemperatur des schwarzen Körpers, welcher die gleiche Flächenhelligkeit hat wie der Stern.

Da die isophoten Wellenlängen der beobachteten Integralhelligkeiten und der durch Rechnung ermittelten Flächenhelligkeiten im allgemeinen nicht übereinstimmen, sind die beobachteten Helligkeiten noch mit der Größe Δm auf die rechnerisch festgelegten photometrischen Systeme zu beziehen. Die Tabelle 11 gibt mit dem Argument der reziproken Strahlungstemperatur die Korrektionen, welche an die visuellen Helligkeiten von ZINNER und der Revised Harvard Photometry, sowie an die photovisuellen Helligkeiten von KING anzubringen sind, um sie auf das durch Rechnung festgelegte visuell-photometrische System zu beziehen. Weiterhin enthält die Tabelle 11 die Korrektionen, mit denen die photographischen Helligkeiten von KING und von HERTZSPRUNG auf das durch Rechnung festgelegte System der photographischen Helligkeiten reduziert werden.

Die durch Rechnung festgelegten photometrischen Systeme der bolometrischen, der von der Wasserzelle durchgelassenen, der visuellen und der photographischen Flächenhelligkeit bilden ein natürliches homogenes System, bei dem die Flächenhelligkeiten in der PLANCKschen Energiekurve zu den isophoten Wellenlängen der entsprechenden Strahlungstemperatur gehören. Der Nullpunkt der scheinbaren visuellen Helligkeit der Beobachtung ist nach empirischen Daten in willkürlicher Weise festgelegt. Der Nullpunkt der bolometrischen, der von der Wasserzelle durchgelassenen bolometrischen und der photographischen Helligkeit der Beobachtung ist an den Nullpunkt der schein-

Tabelle 11.

$\frac{c_2}{T}\cdot 10^4$	$\Delta\,{}^{Z}m_v$	$\Delta\,{}^{K}m_{pv}$	$\Delta\,{}^{Ha}m_v$	$\Delta\,{}^{K}m_{ph}$	$\Delta\,{}^{He}m_{ph}$
0,603	$-0^{\mathrm{m}},069$	−0 ,047	+0 ,097	+0 ,019	+0 ,042
0,880	$-0^{\mathrm{m}},061$	−0 ,041	+0 ,086	+0 ,016	+0 ,035
1,216	$-0^{\mathrm{m}},051$	−0 ,034	+0 ,071	+0 ,011	+0 ,025
1,594	$-0^{\mathrm{m}},039$	−0 ,027	+0 0,54	+0 ,007	+0 ,014
1,816	$-0^{\mathrm{m}},032$	−0 ,021	+0 ,043	+0 ,003	+0 ,008
2,080	$-0^{\mathrm{m}},023$	−0 ,015	+0 ,030	+0 ,001	+0 ,002
2.392	$-0^{\mathrm{m}},012$	−0 ,008	+0 ,014	−0 ,003	−0 ,007
2,744	$-0^{\mathrm{m}},002$	−0 ,001	−0 ,002	−0 ,006	−0 ,017
3,140	$+0^{\mathrm{m}},013$	+0 ,007	−0 ,022	−0 ,014	−0 ,027
3,727	$+0^{\mathrm{m}},032$	+0 ,019	−0 ,055	−0 ,022	−0 ,043
4,020	$+0^{\mathrm{m}},040$	+0 ,025	−0 ,072	−0 ,027	−0 ,052
4,783	$+0^{\mathrm{m}},064$	+0 ,042	−0 ,112	−0 ,037	−0 ,072
5,740	$+0^{\mathrm{m}},094$	+0 ,062	−0 ,160	−0 ,050	−0 ,099
7.175	$+0^{\mathrm{m}},139$	+0 ,092	−0 ,229	−0 ,070	−0 ,137
9,567	$+0^{\mathrm{m}},215$	+0 ,143	−0 ,336	−0 ,101	−0 ,202

[1] Es wird vorausgesetzt, daß die beobachteten Integralhelligkeiten frei von Skalenfehlern sind und auch keinen PURKINJE-Effekt zeigen.

baren visuellen Helligkeit dadurch angeschlossen, daß für die Sterne des Spektraltypus A0 alle Arten von Integralhelligkeiten einander gleich sind. Da der Nullpunkt der rechnerisch bestimmten Flächenhelligkeiten nicht an diese Übereinkunft gebunden ist, weicht das System der Flächenhelligkeiten von dem der scheinbaren Helligkeiten um verschiedene Werte der Konstanten M_0 voneinander ab.

In der Veröff Berlin-Babelsberg VII, 5 wählte der Verfasser als grundlegendes System der Beobachtung das des ZINNERschen Helligkeitsverzeichnisses. Die Revised Harvard Photometry, die photovisuellen Helligkeiten von KING, die photographischen Helligkeitskataloge von KING und von HERTZSPRUNG und das bolometrische Helligkeitsverzeichnis von PETTIT und NICHOLSON wurden an das ZINNERsche System angeschlossen. Will man von den durch die Beobachtung gegebenen Systemen der bolometrischen, der von der Wasserzelle durchgelassenen bolometrischen, der visuellen und der photographischen Helligkeit zu den durch die Rechnung festgelegten photometrischen Systemen der Flächenhelligkeit übergehen, so wird man die Reduktionskonstante M_0 gleich setzen der Summe aus einem für alle Arten von Integralhelligkeiten konstanten und einem für die betreffende Integralhelligkeit charakteristischen Term:

$$M_0 = M_1 + M_2.$$

Mit der Konstanten M_2 werden die beobachteten Integralhelligkeiten, wenn sie auf das ZINNERsche System als Nullpunkt bezogen sind, an die durch Rechnung festgelegten fundamentalen Systeme angeschlossen; mit der Reduktionskonstanten M_1 wird der Nullpunkt des ZINNERschen Systems bestimmt. Die Werte der Konstanten M_2 sind nach der Veröff Berlin-Babelsberg VII, 6 und nach A N 237, S. 225 für mehrere photometrische Systeme der Beobachtung:

Bolometrisch	Wasserzelle	Visuell	Photovisuell
PETTIT und NICHOLSON		Revised Harvard Photometry	KING
$+1^m,58$	$+0^m,38$	$+0^m,07$	$+0^m,24$

Photographisch	
KING	HERTZSPRUNG
$-0^m,59$	$-0^m,58$

Die Konstante M_1 wurde aus den beobachteten Integralhelligkeiten der Sonne gleich $+ 2^m,21$ bestimmt (vgl. Ziffer 24).

22. Der Wärmeindex und die Wasserzellenabsorption. WILSING prüft seine Temperaturskala nach den thermoelektrischen Messungen von COBLENTZ an den Reflektoren des Lick- und des Flagstaff-Observatoriums[1]. Die eine Messungsreihe gibt die Galvanometerausschläge, wenn die Gesamtstrahlung des Sterns auf die Lötstelle des Thermoelementes fällt. Bei der anderen wird der langwellige Teil des Spektrums von etwa 0,9 μ ab durch eine Wasserzelle absorbiert. Die erste Serie von Beobachtungen erlaubt das Energieverhältnis der bolometrischen Helligkeit von zwei Sternen zu berechnen. Die von den scheinbaren Sternhalbmessern abhängige Konstante eliminiert WILSING mit dem visuellen Helligkeitsverhältnis der beiden Sterne. Die zweite Serie der Beobachtungen gibt das Verhältnis der im kurzwelligen Teil des Spektrums vorhandenen Energie zur Gesamtstrahlung. Diese Relativzahlen stellen eine Art Farbenindex dar und sind genauer als die Beobachtungen der ersten Reihe. Da die Relativzahlen bei einem Temperaturintervall von 2000° bis 20000° in den verhältnismäßig engen Grenzen 0,24 bis 0,88 bleiben, müssen sie bis auf wenige Hundertstel

---

[1] J. WILSING, Über die Extinktion der Strahlung in der Erdatmosphäre. A N 220, S. 1 (1923).

bekannt sein, um einigermaßen sichere Werte für die Farbtemperatur zu erhalten. In der Reduktion von WILSING weichen die Relativzahlen nach den Messungen auf dem Lick-Observatorium beträchtlich von denen des Flagstaff-Observatoriums ab.

PETTIT und NICHOLSON bezeichnen die Differenz der visuellen und der bolometrischen Helligkeit, wenn beide auf die zenitale Extinktion bezogen sind, als Wärmeindex, die Differenz der von der Wasserzelle durchgelassenen bolometrischen und der bolometrischen Helligkeit selbst als Wasserzellenabsorption. Die bolometrische und die von der Wasserzelle durchgelassene bolometrische Helligkeit werden im absoluten Maßsystem gegeben; der Nullpunkt der visuellen Größenskala ist willkürlich. Nach Übereinkunft wird in der Beobachtungspraxis der Wärmeindex ebenso wie der photographische Farbenindex für die Sterne des Spektraltypus A0 gleich Null gesetzt.

Der Verfasser hat nach dem beim photographischen Farbenindex auseinandergesetzten Verfahren (Ziffer 20) zu den aus den WILSINGschen und ROSENBERGschen spektralphotometrischen Messungen abgeleiteten Farbtemperaturen die Differenz der visuellen und der bolometrischen Flächenhelligkeit berechnet, unter der Voraussetzung, daß die Senkung der Energiekurve im Ultraviolett die einzige Abweichung der Sternstrahlung von der schwarzen Strahlung ist[1]. Die Korrektion wegen der selektiven Reflexion der Sternstrahlung an den Silberspiegeln des Beobachtungsinstrumentes wurde bei der Rechnung berücksichtigt, dagegen nicht die selektive Wasserdampfabsorption der Erdatmosphäre im Ultrarot.

Wenn die bolometrischen Helligkeiten im absoluten Maßsystem gemessen werden, sind der Nullpunkt und die Skala in dem durch die Beobachtung und in dem durch die Rechnung festgelegten System einander gleich. Der Nullpunkt der durch Rechnung bestimmten visuellen Helligkeit ist durch die Annahme festgelegt, daß in der Wellenlänge der maximalen Empfindlichkeit die volle Strahlung im menschlichen Auge wirksam ist; der Nullpunkt der empirischen Größenskala ist willkürlich. Da die visuellen photometrischen Systeme der Beobachtung und der Rechnung nicht vollkommen gleichartig sind, hat man die beobachteten Helligkeiten auf das durch Rechnung festgelegte fundamentale visuell-photometrische System zu reduzieren.

Wenn sich die Sterne annähernd wie schwarze Strahler der in der Spalte 2 der Tabelle 12 aufgeführten Farbtemperaturen verhalten, müssen die Differenzen zwischen den beobachteten und den berechneten Wärmeindizes konstant sein. Die beobachteten Wärmeindizes für die einzelnen Spektralklassen (normale

Tabelle 12.

Spektrum	$\frac{c_2}{T} \cdot 10^4$	*W. I.* PETTIT	*W. I.* verbessert	$M_v - M_b$	$B-R$
B0	0,603	$0^m,05$	$0^m,15$	$1^m,36$	$-1^m,21$
B5	0,880	0 ,01	0 ,10	1 ,32	−1 ,22
A0	1,216	0 ,00	0 ,07	1 ,40	−1 ,33
A5	1,594	0 ,02	0 ,07	1 ,46	−1 ,39
F0	1,816	0 ,15	0 ,19	1 ,53	−1 ,34
F5	2,080	0 ,30	0 ,33	1 ,65	−1 ,32
G0	2,392	0 ,47	0 ,48	1 ,79	−1 ,31
G5	2,744	0 ,65	0 ,65	1 ,99	−1 ,34
K0	3,140	0 ,90	0 ,88	2 ,25	−1 ,37
K5	3,727	1 ,57	1 ,51	2 ,67	−1 ,16
M0	4,020	1 ,86	1 ,79	2 ,92	−1 ,13

[1] A. BRILL, Die Strahlung der Sterne. Veröff Univ-Sternw Berlin-Babelsberg V, Heft 1, S. 17 (1924).

Riesensterne) sind der Arbeit von PETTIT und NICHOLSON entnommen[1]. An die visuellen Helligkeiten der Revised Harvard Photometry wurden nach der Tabelle 11 in Ziffer 21 Korrektionen angebracht, um sie auf das durch Rechnung festgelegte visuell-photometrische System zu reduzieren („*W. I.* verbessert" der Tabelle 12). Die Differenzen Beobachtung minus Rechnung sind nahezu konstant; die Wärmeindizes lassen sich also annähernd durch die Folge der spektralphotometrisch bestimmten Farbtemperaturen darstellen. Doch erkennt man, daß der Wärmeindex ebenso wie der photographische Farbenindex niedrigere Farbtemperaturen für die Sterne der Spektralklassen K5 und M0 verlangt. Die Diskrepanz bei den Sternen der Spektraltypen B0 und B5 entspricht der in Ziffer 16 an den WILSINGschen und ROSENBERGschen spektralphotometrischen Messungen festgestellten Unstimmigkeit. Diese Sterne besitzen im visuellen Spektralbereich einen schwachen Gradienten der Energiekurve und damit auch eine niedrige Farbtemperatur.

PETTIT und NICHOLSON lassen bei der Bestimmung der Farbtemperatur aus dem Wärmeindex die Abhängigkeit der isophoten Wellenlänge der visuellen Harvardhelligkeit von der Spektralklasse außer acht; überdies wird erst nachträglich in ziemlich willkürlicher Weise über die Größe der mittleren isophoten Wellenlänge verfügt[2]. Der Verfasser hat nach der Gleichung (63) aus den von PETTIT und NICHOLSON angegebenen mittleren Wärmeindizes mit Zuhilfenahme der Tabelle 10 Farbtemperaturen berechnet, einmal unter der Voraussetzung, daß sich die Sterne verhalten wie absolut schwarze Körper, zum andern bei Berücksichtigung der Depression der Energiekurve im Ultraviolett nach dem in der Tabelle 16, Ziffer 25 mitgeteilten Betrag für die bolometrische Helligkeit. Im ersten Fall täuscht die kontinuierliche Wasserstoffabsorption bei den A-Sternen

Tabelle 13.

Spektrum	*W. I.*	$\frac{c_2}{T_1} \cdot 10^4$	$\frac{c_2}{T_2} \cdot 10^4$	*W. C.*	$\frac{c_2}{T} \cdot 10^4$
B0	$0^m,05$	1,00	0,98	$0^m,20$	(0,76)
B5	0 ,01	0,75	1,14	0 ,23	1,36
A0	0 ,00	0,60	1,24	0 ,26	1,56
A5	0 ,02	0,60	1,29	0 ,30	2,03
F0	0 ,15	1,30	1,87	0 ,36	2,33
F5	0 ,30	2,02	2,21	0 ,41	2,63
gG0	0 ,47	2,42	2,55	0 ,50	2,99
gG5	0 ,65	2,78	2,85	0 ,60	3,43
gK0	0 ,90	3,18	3,22	0 ,70	3,85
gK5	1 ,57	4,07	4,08	0 ,93	4,56
gM0	1 ,86	4,41	4,42	1 ,01	4,79
gM2	2 ,2	4,72		1 ,14	5,16
gM4	3 ,1	5,60		1 ,30	5,59
gM6	4 ,2	6,58		1 ,46	6,02
gM8	5 ,2	7,42		1 ,62	6,46
Me Max.	4 ,4	6,77		1 ,5	6,13
Me Min.	8 ,9			2 ,2	7,81
dG0	0 ,32	2,05		0 ,42	2,68
dG5	0 ,39	2,22		0 ,47	2,89
dK0	0 ,55	2,56		0 ,54	3,18
dK5	1 ,10	3,46		0 ,76	4,09
dM0	1 ,40	3,83		0 ,87	4,42
dM2	2 ,1	4,63		1 ,14	5,16
☉	0 ,28	1,92	2,12	0 ,41	2,63

[1] E. PETTIT and S. B. NICHOLSON, Stellar Radiation Measurements. Ap J 68, S. 297 (1928).

[2] A. BRILL, Die Temperaturstrahlung der Fixsterne; Naturwiss 1929, S. 418.

außergewöhnlich hohe Farbtemperaturen vor (c_2/T_1 in Tab. 13). Wird die Abweichung der Sternstrahlung von der schwarzen Strahlung in Rechnung gestellt, so folgt aus den Wärmeindizes die vom Verfasser aus WILSINGS spektralphotometrischen Messungen abgeleitete Farbtemperaturskala in befriedigender Annäherung (c_2/T_2 in der Tab. 13); nur die Farbtemperaturen der Spektralklassen K5 und M0 sind auf Grund der Wärmeindizes wesentlich niedriger, ein Resultat, welches durch die neueren spektralphotometrischen Beobachtungen und durch die Diskussion der aus den photographischen Farbenindizes berechneten Farbtemperaturen bestätigt wird. Die nach der Gleichung (63) mit Hilfe der Tabelle 10 aus der Wasserzellenabsorption abgeleiteten Farbtemperaturen sind im allgemeinen niedriger als die aus dem Wärmeindex bestimmten. In der Natur des Problems liegt es begründet, daß die Farbtemperatur für die Sterne der Spektralklasse B0 sich nur ganz unsicher aus der Wasserzellenabsorption bestimmen läßt. Die Divergenz in beiden Temperaturskalen, der aus dem Wärmeindex und der aus der Wasserzellenabsorption, ist wahrscheinlich darauf zurückzuführen, daß die Energiekurve der Sterne im Ultrarot flacher verläuft als im Bereich der visuell wirksamen Strahlen. Die Umkehr in dem Temperaturverhältnis bei den Riesensternen vom Spektraltypus gM4 ab läßt sich daraus erklären, daß die Titanoxydbanden bei diesen späten Sternen eine merkliche Minderung der visuellen Helligkeit bedingen.

23. Die Farbtemperatur der Sonne aus dem photographischen Farbenindex, aus dem Wärmeindex und aus der Wasserzellenabsorption. Die visuelle und die photographische Gesamthelligkeit der Sonne sind noch relativ unsicher bestimmt. Nach HERTZSPRUNG ist die visuelle Helligkeit der Sonne gleich $-26^m,90$, nach BIRCK die photographische Helligkeit in der Reduktion von RUSSELL $-26^m,12$. Die bolometrische Helligkeit der Sonne ist nach PETTIT und NICHOLSON $-27^m,18$, die von der Wasserzelle durchgelassene bolometrische Strahlung $-26^m,77$.

Unter der Annahme, daß die visuelle Helligkeit der Sonne $-26^m,90$ sich auf das System der Revised Harvard Photometry, die photographische Helligkeit $-26^m,12$ sich auf das System von KING bezieht, gibt die Tabelle 11 in Ziffer 21 mit den Strahlungstemperaturen der visuellen und der photographischen Helligkeit als Argument (vgl. Ziffer 24) die Korrektionen, welche wegen der Farbengleichung an die beobachteten Helligkeiten anzubringen sind. Gemäß der Gleichung (63) sind die aus den Beobachtungen abgeleiteten $m + \Delta m + M_2$:

	Bolometrisch	Wasserzelle	Visuell	Photographisch
m	$-27^m,18$	$-26^m,77$	$-26^m,90$	$-26^m,12$
Δm	0 ,00	0 ,00	+ 0, 01	0 ,00
M_2	+ 1 ,58	+ 0 ,38	+ 0 ,07	− 0 ,59
$m + \Delta m + M_2$	−25 ,60	−26 ,39	−26 ,82	−26 ,71

bis auf die allen Integralhelligkeiten der Sonne gemeinsame Konstante $M_1 + 5 \log \varrho$ gleich den Flächenhelligkeiten M. Die mit den Größen $m + \Delta m + M_2$ bestimmten Werte des photographischen Farbenindex, des Wärmeindex und der Wasserzellenabsorption geben dann direkt auf Grund der Tabelle 10 in Ziffer 21 die zugehörigen Farbtemperaturen:

	Photographischer Farbenindex	Wärmeindex	Wasserzellenabsorption
Energiestufe . . .	$+0^m,11$	$-1^m,22$	$-0^m,79$
c_2/T_F	$2,62 \cdot 10^{-4}$	$1,90 \cdot 10^{-4}$	$2,60 \cdot 10^{-4}$
T_F	5460°	7550°	5520°

An Hand der Energiekurve der mittleren Sonnenstrahlung lassen sich die Strahlungstemperaturen und damit gemäß der Tabelle 10 die Flächenhelligkeiten der bolometrischen, der von der Wasserzelle durchgelassenen, der visuellen und der photographischen Helligkeit berechnen (vgl. Ziffer 24). Mit den Zahlenwerten für die Flächenhelligkeit der Integralhelligkeiten werden der photographische Farbenindex, der Wärmeindex und die Wasserzellenabsorption im absoluten System der isophoten Wellenlängen bestimmt. Die zweite und dritte Zeile der folgenden Tabelle enthält gemäß Tabelle 10 die zu der Energiestufe des photographischen Farbenindex, des Wärmeindex und der Wasserzellenabsorption gehörige Farbtemperatur T_F bzw. den c_2/T_F-Wert:

	Photographischer Farbenindex	Wärmeindex	Wasserzellenabsorption
Energiestufe . . .	$+0^m,143$	$-1^m,258$	$-0^m,848$
c_2/T_F	$2{,}686 \cdot 10^{-4}$	$1{,}710 \cdot 10^{-4}$	$2{,}260 \cdot 10^{-4}$
T_F	5330°	8375°	6335°

Die aus den beobachteten Integralhelligkeiten abgeleiteten Farbtemperaturen stimmen in befriedigender Weise mit den aus der Energiekurve der mittleren Sonnenstrahlung gefundenen überein.

e) Die Strahlungstemperatur aus der Strahlungsintensität in mehr oder minder weit begrenzten Spektralbereichen.

24. Die schwarze und die Strahlungstemperatur der Sonne. Die bisher genannten Methoden geben die Farbtemperatur aus der Gestalt der Energiekurve, aus der Farbe oder aus einem Farbenäquivalent. Über die schwarze Temperatur und über die Strahlungstemperatur sei es der bolometrischen, visuellen oder photographischen Helligkeit, weiß man vorläufig nur wenig. Aus dem noch sehr dürftigen Beobachtungsmaterial scheint hervorzugehen, daß sich die Sterne annähernd wie schwarze Strahler verhalten.

Daß die in verschiedener Weise definierten Temperaturen merklich voneinander abweichen, lehrte bereits das Beispiel der Sonne (Ziffer 18 und 23). Die Energiekurve der mittleren Sonnenstrahlung ist in ihrem gesamten Verlauf ähnlich einem schwarzen Strahler von der Temperatur 7000°, wenn man eine derartige Feststellung überhaupt als berechtigt gelten lassen darf. Die Gradationstemperatur ändert sich mit der Wellenlänge und schwankt innerhalb weiter Grenzen. Die Temperaturen aus den Farbenäquivalenten unterscheiden sich gleichfalls beträchtlich voneinander.

Das Material für die Berechnung der absoluten Energiewerte der mittleren Sonnenstrahlung lieferten die Messungen des SMITHSONIAN-Institutes. Mit den Daten für die Gesamtstrahlung und für die Strahlung in den einzelnen Wellenlängen bestimmte MINNAERT die absolute Energiekurve der mittleren Sonnenstrahlung (Abb. 3 in Ziffer 18). Die Spalten 2 und 5 der Tabelle 14 enthalten die Flächenhelligkeit der mittleren Sonnenstrahlung, ausgedrückt in erg pro sec pro cm² der Sonnenoberfläche, für eine Reihe von Wellenlängen. Die Gleichung (19) gibt zu jeder Wellenlänge die schwarze Temperatur bzw. den c_2/T_S-Wert

Tabelle 14. Schwarze Temperaturen der mittleren Sonnenstrahlung nach Beobachtungen auf dem Mount Wilson-Observatorium in den Jahren 1903—1910.

$\lambda\mu\mu$	$i(\lambda)$ erg	c_2/T_S	$\lambda\mu\mu$	$i(\lambda)$ erg	c_2/T_S
300	$3{,}17 \cdot 10^{13}$	$2{,}890 \cdot 10^{-4}$	550	$33{,}11 \cdot 10^{13}$	$2{,}350 \cdot 10^{-4}$
325	$7{,}48 \cdot 10^{13}$	$2{,}722 \cdot 10^{-4}$	600	$29{,}69 \cdot 10^{13}$	$2{,}371 \cdot 10^{-4}$
350	$15{,}80 \cdot 10^{13}$	$2{,}540 \cdot 10^{-4}$	700	$21{,}46 \cdot 10^{13}$	$2{,}462 \cdot 10^{-4}$
375	$20{,}37 \cdot 10^{13}$	$2{,}497 \cdot 10^{-4}$	800	$15{,}69 \cdot 10^{13}$	$2{,}538 \cdot 10^{-4}$
390	$21{,}28 \cdot 10^{13}$	$2{,}504 \cdot 10^{-4}$	1000	$9{,}76 \cdot 10^{13}$	$2{,}573 \cdot 10^{-4}$
420	$30{,}92 \cdot 10^{13}$	$2{,}384 \cdot 10^{-4}$	1300	$5{,}29 \cdot 10^{13}$	$2{,}531 \cdot 10^{-4}$
430	$31{,}33 \cdot 10^{13}$	$2{,}386 \cdot 10^{-4}$	1600	$3{,}13 \cdot 10^{13}$	$2{,}438 \cdot 10^{-4}$
450	$35{,}49 \cdot 10^{13}$	$2{,}339 \cdot 10^{-4}$	2000	$1{,}45 \cdot 10^{13}$	$2{,}53 \cdot 10^{-4}$
470	$36{,}74 \cdot 10^{13}$	$2{,}325 \cdot 10^{-4}$	2500	$0{,}25 \cdot 10^{13}$	$4{,}41 \cdot 10^{-4}$
500	$35{,}70 \cdot 10^{13}$	$2{,}335 \cdot 10^{-4}$	3000	$0{,}08 \cdot 10^{13}$	$5{,}9 \cdot 10^{-4}$

(Spalte 3 und 6 der Tab. 14). Die c_2/T_S-Werte der mittleren Sonnenstrahlung schwanken im Wellenlängenbereich 0,4 bis 2,0 μ innerhalb verhältnismäßig enger Grenzen. Nur im Ultraviolett und im Ultrarot nehmen die schwarzen Temperaturen schnell ab; doch bleibt vorläufig zweifelhaft, ob diese niedrigen Temperaturen in Anbetracht der kleinen Energiewerte der Sonnenstrahlung oder auch wegen der Unsicherheit in der Berücksichtigung der atmosphärischen Extinktion reell sind.

Die graphische Darstellung der funktionellen Beziehung zwischen den c_2/T_S- bzw. den c_2/T_G-Werten (Ziffer 18) einerseits und der Wellenlänge λ andererseits zeigt, daß minimale Abweichungen der Sonnenenergiekurve von der des schwarzen Strahlers bei den Gradationstemperaturen sehr stark, bei den schwarzen Temperaturen kaum merklich in Erscheinung treten (Abb. 4).

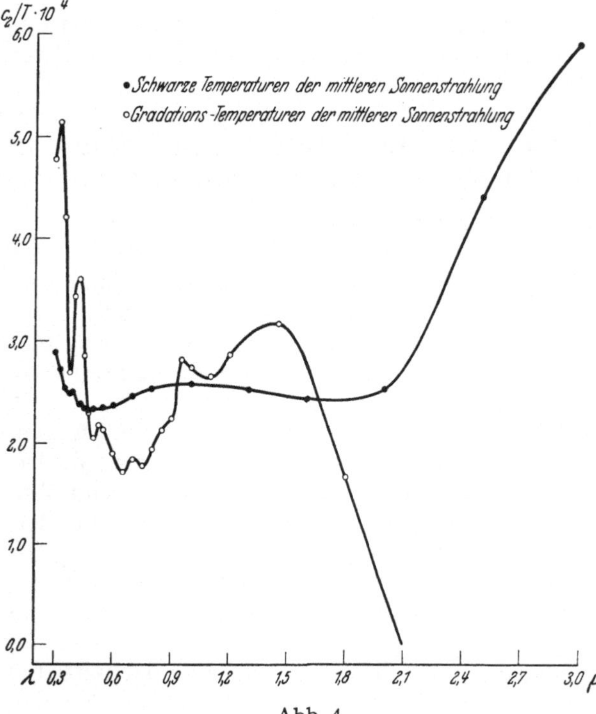

Abb. 4.
(Aus Veröff Univ-Sternw Berlin-Babelsberg VII, Heft 6.)

Die Kurven, welche bei der Sonne die Abhängigkeit der Funktionen c_2/T_S und c_2/T_G von der Wellenlänge zur Darstellung bringen, schneiden sich bei den Wellenlängen λ 470, 930 und 1710 $\mu\mu$; diese Stellen entsprechen zwei Maxima und einem Minimum der schwarzen Temperatur. Die Energiekurve des schwarzen Strahlers hat in den genannten Wellenlängen Strahlungsintensität und Gradient mit der Energiekurve der Sonne gemeinsam. Die Sonne strahlt hier nach Definition (Ziffer 11) wie ein vollkommen schwarzer Körper. Die Stellen λ 930 und 1710 $\mu\mu$ interessieren nicht besonders; die Energiekurve der Sonne besitzt auf dem nach Rot abfallenden Ast zunächst einen steilen, dann einen flachen und schließlich wieder einen steilen Gradienten.

Sehr wichtig für die Untersuchungen über die Sonnenstrahlung ist die Feststellung der Tatsache, daß die erstgenannte Stelle λ 470 $\mu\mu$ mit dem Maximum der Energiekurve der Sonne zusammenfällt. Im Maximum strahlt also die Sonne wie ein vollkommen schwarzer Körper. Es scheint deshalb die Ausdrucksweise, von einer Überhöhung der Sonnenenergiekurve im Maximum zu sprechen, nicht angebracht zu sein. Die anomale Intensitätsverteilung im Sonnenspektrum läßt sich besser folgendermaßen charakterisieren: Die Energiekurve der Sonne fällt vom Maximum, wo Normalstrahlung herrscht, nach Rot und Violett steiler ab, als es der Energiekurve des schwarzen Normalstrahlers entspricht, und bleibt sowohl im Ultraviolett wie im Ultrarot unterhalb der Energiekurve des Normalstrahlers.

Da sich die Sonne im Maximum der Energiekurve wie ein schwarzer Körper verhält, lassen sich auch alle anderen Gesetze der schwarzen Strahlung anwenden.

Setzt man in der Gleichung (11) von Ziffer 6 für E_{Max} die Flächenhelligkeit der Sonne bei λ 470 $\mu\mu$, so erhält man $c_2/T = 2{,}325 \cdot 10^{-4}$ in Übereinstimmung mit dem aus dem PLANCKschen Gesetz für λ 470 $\mu\mu$ berechneten Wert. Das WIENsche Verschiebungsgesetz [Gleichung (10) in Ziffer 6] gibt mit $c_2/T = 2{,}325 \cdot 10^{-4}$ als Wellenlänge des Energiemaximums: $\lambda_{\mathrm{Max}} = 468{,}3\ \mu\mu$. Dies Resultat steht in Widerspruch zu der von E. A. MILNE begründeten Schlußfolgerung, daß die Intensitätsverteilung im kontinuierlichen Spektrum der Sonne, wenn diese als eine Gaskugel im Strahlungsgleichgewicht aufgefaßt wird, zwar der eines schwarzen Körpers entspricht, im allgemeinen aber eine Verschiebung gegen die kleinen Wellenlängen zeigt[1]. Wenn letzteres zutreffen würde, könnten im Maximum der Sonnenenergiekurve Gradationstemperatur und schwarze Temperatur nicht einander gleich sein.

An Hand der von MINNAERT gezeichneten Energiekurve der mittleren Sonnenstrahlung (Abb. 3) kann man die Strahlungstemperaturen der bolometrischen, der von der Wasserzelle durchgelassenen bolometrischen, der visuellen und der photographischen Helligkeit berechnen. Dabei gilt als Strahlungstemperatur der Sonne diejenige, welche ein schwarzer Körper haben muß, um die Strahlung des für die Integralhelligkeit in Betracht kommenden Spektralbereichs in gleicher Intensität zu emittieren wie die Sonne. Die bolometrische Strahlungstemperatur bezieht sich auf den gesamten der Messung zugänglichen Bereich des Spektrums; die ultrarote Grenze der von der Wasserzelle durchgelassenen bolometrischen Strahlung liegt bei $\lambda = 1{,}3\ \mu$. Für das menschliche Auge wurden die Empfindlichkeitsgrenzen λ 420 und λ 700 $\mu\mu$ angenommen, für die photographische Platte λ 310 und λ 507 $\mu\mu$. Die planimetrische Ausmessung entsprechender Flächenstücke und ihre Einordnung in die Skala der zu den schwarzen Strahlern von 5000°, 6000° und 7000° gehörigen Flächenstücke liefert die Strahlungstemperaturen und die zugehörigen c_2/T_{St}-Werte:

	Bolometrisch	Wasserzelle	Visuell	Photographisch
T_{St}	5740°	5795°	6050°	5895°
c_2/T_{St} . . .	$2{,}494 \cdot 10^{-4}$	$2{,}472 \cdot 10^{-4}$	$2{,}366 \cdot 10^{-4}$	$2{,}430 \cdot 10^{-4}$
λ_{Φ}	1024,0 $\mu\mu$	678,1 $\mu\mu$	548,4 $\mu\mu$	432,4 $\mu\mu$
M	$-35^{\mathrm{m}},020$	$-35^{\mathrm{m}},868$	$-36^{\mathrm{m}},278$	$-36^{\mathrm{m}},135$

In der vorletzten Reihe der obigen Tabelle stehen die isophoten Wellenlängen der durch die planimetrische Ausmessung der Sonnenenergiekurve ermittelten Flächenhelligkeiten; sie sind mit dem Argument c_2/T_{St} den Tabellen 3, 4, 6 und 7 der Veröff Berlin-Babelsberg VII, 5 entnommen (vgl. Ziffer 21). In der letzten Zeile der obigen Tabelle sind die zu den isophoten Wellenlängen und zu den Strahlungstemperaturen gehörigen spektralen Energiewerte (Flächenhelligkeiten) nach der Tabelle 10 in Ziffer 21 aufgeführt. Mit den Angaben für die Flächenhelligkeit der Integralhelligkeiten wurden der photographische Farbenindex, der Wärmeindex und die Wasserzellenabsorption im absoluten System der isophoten Wellenlängen berechnet und in Ziffer 23 zur Bestimmung der Farbtemperaturen der Sonne benutzt.

Der Nullpunkt der empirischen Größensysteme gegenüber den durch Rechnung festgelegten fundamentalen photometrischen Systemen, d. h. die Konstante M_0 der Gleichung (63) in Ziffer 21, wird durch Vergleich der beobachteten Integralhelligkeiten der Sonne mit den durch Rechnung aus der Energiekurve der Sonne ermittelten Flächenhelligkeiten bestimmt. Die Reduktionskonstante M_0 ist nach Ziffer 21 gleichgesetzt der Summe aus einem für alle Arten von

[1] E. A. MILNE, Radiative Equilibrium and Spectral Distribution. M N 81, S. 375 (1921).

Integralhelligkeiten konstanten und einem für die betreffende Integralhelligkeit charakteristischen Term:

$$M_0 = M_1 + M_2.$$

Mit der Konstanten M_2 werden die beobachteten Integralhelligkeiten, wenn sie auf das ZINNERsche System als Nullpunkt bezogen sind, an die durch Rechnung festgelegten fundamentalen Systeme angeschlossen; mit der Konstanten M_1 wird der Nullpunkt des ZINNERschen Systems bestimmt. Die Konstanten M_2 wurden in Veröff Berlin-Babelsberg VII, 5 und in A N 237, S. 225 für mehrere photometrische Systeme der Beobachtung abgeleitet (vgl. Ziffer 21). Die Konstante M_1 wird mit den beobachteten Integralhelligkeiten der Sonne und mit den durch Rechnung aus der Sonnenenergiekurve ermittelten Flächenhelligkeiten durch Umkehrung der Gleichung (63):

$$M_1 = M - (m + \Delta m + M_2) - 5 \log \varrho$$

erhalten (vgl. Ziffer 23). Wird der scheinbare Halbmesser der Sonne gleich 961″,2 gesetzt, so gibt obige Gleichung folgende Werte für die Reduktionskonstante M_1:

Bolometrisch	Wasserzelle	Visuell	Photographisch
$+2^m,24$	$+2^m,18$	$+2^m,20$	$+2^m,24$

Mit dem Mittelwert $M_1 = +2^m,21$ findet man die folgenden, durch Rechnung aus Gleichung (63) bestimmten Integralhelligkeiten der Sonne, welche sich nur um wenige Hundertstel Größenklassen von den beobachteten Integralhelligkeiten unterscheiden:

	Bolometrisch	Wasserzelle	Visuell	Photographisch
m_R	$-27^m,15$	$-26^m,80$	$-26^m,92$	$-26^m,10$
$B-R$. . . .	$-0\ ,03$	$+0\ ,03$	$+0\ ,02$	$-0\ ,02$

25. Die Strahlungstemperatur der Fixsterne. Die Untersuchungen über die Temperatur der Fixsterne betreffen fast ausnahmslos die Farbtemperatur. Die Strahlungstemperatur der visuellen, der photographischen, der bolometrischen und der von der Wasserzelle durchgelassenen bolometrischen Helligkeit läßt sich nur dann berechnen, wenn man den scheinbaren Durchmesser des Sterns kennt, d. i. den Winkel, unter dem von der Sonne aus gesehen der wahre Durchmesser des Sterns erscheint. Bisher sind die scheinbaren Halbmesser für 7 Sterne von PEASE auf dem Mount Wilson-Observatorium interferometrisch bestimmt. Die beobachtete Integralhelligkeit m führt auf Grund der Gleichung (63) zur Flächenhelligkeit und damit nach der Tabelle 10 zur Strahlungstemperatur.

Die Spalte 2 der Tabelle 15 enthält neben dem Spektraltypus den scheinbaren Halbmesser des voranstehenden Sterns nach den interferometrischen Messungen von PEASE auf dem Mount Wilson-Observatorium. In der 3. bis 6. Spalte stehen die scheinbare photographische, die visuelle, die von der Wasserzelle durchgelassene und die bolometrische Helligkeit nebst den zugehörigen Strahlungstemperaturen. Hier beziehen sich die photographischen Helligkeiten auf das KINGsche System, die visuellen auf das System von ZINNER, ausgenommen die Sonne und der veränderliche Stern o Ceti, deren visuelle Helligkeiten im System der Revised Harvard Photometry gegeben sind. Die bolometrische und die von der Wasserzelle durchgelassene bolometrische Helligkeit sind der Arbeit von PETTIT und NICHOLSON entnommen[1]. Die Strahlungstemperaturen der visuellen und der photographischen Helligkeit sind nur wenig voneinander verschieden. Bei der Sonne ist die Strahlungstemperatur der visuellen Helligkeit, bei den Sternen vom späten Spektraltypus die der bolometrischen Helligkeit am größten. Bei o Ceti ist an-

[1] E. PETTIT and S. B. NICHOLSON, Stellar Radiation Measurements. Ap J 68, S. 289 (1928).

Tabelle 15.

Stern	Spektraltypus	Photographisch	Visuell	Wasserzelle	Bolometrisch
Sonne	G0 961″,2	$-26^m,12$ 5910°	$-26^m,90$ 6030°	$-26^m,77$ 5740°	$-27^m,18$ 5780°
α Bootis	K0 0″,0100	$+1^m,38$ 4140°	$+0^m,14$ 4180°	$-0^m,18$ 4070°	$-0^m,98$ 4230°
α Tauri	K5 0″,0100	$+2^m,66$ 3600°	$+1^m,08$ 3670°	$+0^m,36$ 3680°	$-0^m,60$ 3880°
α Scorpii	M1 0″,0200	$+3^m,03$ 3020°	$+1^m,35$ 2970	$-0^m,11$ 3100°	$-1^m,32$ 3280°
α Orionis	M2 0″,0235	$+2^m,65$ 3030°	$+1^m,00$ 2970°	$-0^m,48$ 3110°	$-1^m,67$ 3280°
β Pegasi	M2 0″,0105	$+4^m,40$ 3030°	$+2^m,66$ 2990°	$+1^m,45$ 3020°	$+0^m,27$ 3140°
α Herculis A	M5 0″,0150	$+5^m,02$ 2700°	$+3^m,52$ 2530°	$+0^m,75$ 2980°	$-0^m,76$ 3320°
o Ceti A Max.	M6e 0″,0235		$+3^m,6$ 2340°	$+1^m,10$ 2460°	$-0^m,20$ 2410°
o Ceti A Min.			$+9^m,3$ 1580°	$+2^m,60$ 2010°	$+0^m,70$ 2020°

genommen, daß sich der scheinbare Halbmesser vom Maximum zum Minimum der Helligkeit nicht ändert.

Wenn die Sterne annähernd wie schwarze Körper strahlen, geben die Abweichungen der Energiekurve des Sterns von der des zugehörigen schwarzen Normalstrahlers im visuellen oder im photographisch wirksamen Wellenlängenbereich oder auch für das gesamte Spektrum die Strahlungstemperatur der visuellen, der photographischen oder der bolometrischen Helligkeit. Rechnungen dieser Art wurden vom Verfasser auf Grund der Temperaturskala durchgeführt, welche sich aus Wilsings und Rosenbergs spektralphotometrischen Messungen im Spektralbereich λ 451 bis 642 $\mu\mu$ ergeben hatte[1]. Wenn die Abweichungen der Sternstrahlung von der schwarzen Normalstrahlung im visuellen Wellenlängenbereich unbedeutend sind, wird bei Annahme einer annähernd schwarzen Strahlung der Sterne die Strahlungstemperatur der visuellen Helligkeit gleich der spektralphotometrisch bestimmten Farbtemperatur. Für Wellenlängen kleiner als 450 $\mu\mu$ geben die spektralphotometrischen Messungen eine mit fortschreitendem Spektraltypus zunehmende Depression der Energiekurve im Ultraviolett; die Strahlungstemperaturen der photographischen und der bolometrischen Helligkeit sind demnach niedriger als die Strahlungstemperatur der visuellen Helligkeit. Die Tabelle 16 gibt den Einfluß der Depression der Energiekurve im Sternspektrum, wenn angenommen wird, daß sich die Sterne der einzelnen Spektralklassen annähernd wie schwarze Körper der in der Spalte 2 stehenden spektralphotometrisch bestimmten Farbtemperaturen verhalten. Der in Größenklassen ausgedrückte Betrag der Senkung der Sternenergiekurve gilt in der 3. Spalte für die photographische, in der 5. Spalte für die bolometrische Helligkeit. Um die Strahlungstemperatur der photographischen und der bolometrischen Helligkeit zu erhalten, ist die spektralphotometrische Farbtemperatur der Spalte 2, welche nach Voraussetzung gleich der Strahlungstemperatur der visuellen Helligkeit ist, um die in der 4. und 6. Spalte der Tabelle 16 stehenden Beträge zu ändern. Die Strahlungstemperatur der photographischen Helligkeit

[1] A. Brill, Die Strahlung der Sterne. Veröff Univ-Sternw Berlin-Babelsberg V, Heft 1 (1924).

Tabelle 16.

Spektrum	$\frac{c_2}{T} \cdot 10^4$	Δm_{ph}	$\Delta \frac{c_2}{T_{ph}} \cdot 10$	Δm_b	$\Delta \frac{c_2}{T_b} \cdot 10^4$
B0	0,60	$0^m,00$	0,00	$0^m,00$	0,00
B5	0,88	0 ,06	0,02	0 ,07	0,03
A0	1,22	0 ,10	0,04	0 ,11	0,05
A5	1,59	0 ,11	0,04	0 ,12	0,06
F0	1,82	0 ,14	0,05	0 ,09	0,05
F5	2,08	0 ,15	0,06	0 ,06	0,04
G0	2,39	0 ,17	0,07	0 ,05	0,03
G5	2,74	0 ,20	0,08	0 ,04	0,03
K0	3,14	0 ,22	0,09	0 ,02	0,02
K5	3,73	0 ,25	0,10	0 ,01	0,01
M0	4,02	0 ,28	0,11	0 ,01	0,01

ist für die späten Spektralklassen merklich tiefer als die Strahlungstemperatur der visuellen Helligkeit. Der Einfluß der Senkung der Energiekurve im Ultraviolett auf die bolometrische Helligkeit nimmt bei den frühen Spektraltypen infolge der kontinuierlichen Wasserstoffabsorption schnell zu, um dann nach den späten Spektralklassen langsam wieder abzunehmen.

Die Farbtemperaturen in Spalte 2 der Tabelle 16 waren vom Verfasser aus den spektralphotometrischen Messungen von WILSING und ROSENBERG bestimmt. Da die spektralphotometrischen Farbtemperaturen der Sterne von spätem Spektraltypus niedriger sind als die in der Veröff Berlin-Babelsberg V, 1 benutzten (vgl. Ziffer 20 und 22), ist der Einfluß der Sternabsorption im Ultraviolett auf Grund der ROSENBERGschen Messungen kleiner, als in der Tabelle 16 angegeben ist. Die Strahlungstemperaturen der photographischen und der visuellen Helligkeit unterscheiden sich dann für die Sterne der späten Spektralklassen ebenfalls um kleinere Beträge.

Die Rechnung in der Veröff Berlin-Babelsberg V, 1 war auf Grund der Annahme durchgeführt, daß die Depression im ultravioletten Teil des Spektrums die einzige Abweichung der Sternstrahlung von der normalen schwarzen Strahlung ist und daß insbesondere die Sterne der Spektralklasse B0 sich vollkommen wie schwarze Körper verhalten. Wie das Beispiel der Sonne zeigt (Ziffer 18, 23 und 24), liegen die Verhältnisse hinsichtlich der Abweichungen von der schwarzen Strahlung anscheinend nicht so einfach, daß man sie der Depression des ultravioletten Teiles des Spektrums gleichsetzen kann. Andererseits ist durch eine Reihe von Untersuchungen der Beweis erbracht, daß man gerade die B-Sterne nicht als vollkommen schwarze Strahler betrachten darf (Ziffer 27).

Unter der Voraussetzung, daß die Strahlungstemperatur der visuellen Helligkeit gleich der spektralphotometrisch bestimmten Farbtemperatur ist, lassen sich auch mit Hilfe der Gleichung (63) aus dem KINGschen Farbenindex, aus dem Wärmeindex und aus der Wasserzellenabsorption die Abweichungen der Sternenergiekurve von der schwarzen Strahlungskurve in den entsprechenden Spektralbereichen berechnen. Die Diskussion des gesamten bisher vorliegenden Beobachtungsmaterials an spektralphotometrischen Messungen und an jeglicher Art von Farbenäquivalenten führt nach Ziffer 26 zu der in der 2. Spalte der Tabelle 17 aufgeführten Farbtemperaturskala. Mit diesen Farbtemperaturen, als Strahlungstemperaturen der visuellen Helligkeit, wurde nach der Gleichung (63) die Größe der Abweichungen von der schwarzen Strahlungskurve für den photographisch wirksamen, für den bolometrischen und für den von der Wasserzelle erfaßten Spektralbereich bestimmt (Spalten 4, 6, 8 der Tab. 17). Man erkennt, daß selbst ohne Berücksichtigung der Abweichungen der Sternstrahlung von der schwarzen Strahlung die Übereinstimmung zwischen Beobachtung und Rech-

Tabelle 17.

Spektrum	T	$\frac{c_2}{T}\cdot 10^4$	Δm_{ph}	$\Delta\frac{c_2}{T_{ph}}\cdot 10^4$	Δm_b	$\Delta\frac{c_2}{T_b}\cdot 10^4$	Δm_w	$\Delta\frac{c_2}{T_w}\cdot 10^4$
B0	22000°	0,65	$-0^m,09$	$-0,03$	$-0^m,07$	$-0,03$	$-0^m,05$	$-0,02$
B5	17700°	0,81	$+0$,01	0,00	$+0$,01	0,00	$+0$,04	$+0,02$
A0	13500	1,06	$+0$,08	$+0,03$	$+0$,08	$+0,03$	$+0$,13	$+0,06$
A5	10500	1,36	$+0$,16	$+0,06$	$+0$,11	$+0,05$	$+0$,18	$+0,08$
F0	8550	1,68	$+0$,13	$+0,05$	$+0$,04	$+0,02$	$+0$,13	$+0,06$
F5	7000	2,05	$+0$,10	$+0,04$	$+0$,01	0,00	$+0$,12	$+0,06$
G0	5800	2,47	$+0$,11	$+0,04$	$+0$,02	$+0,01$	$+0$,14	$+0,08$
G5	4860	2,95	$+0$,14	$+0,06$	$+0$,10	$+0,07$	$+0$,21	$+0,12$
K0	4370	3,28	$+0$,18	$+0,07$	$+0$,07	$+0,05$	$+0$,22	$+0,14$
K5	3460	4,14	$+0$,23	$+0,10$	$+0$,08	$+0,07$	$+0$,23	$+0,15$
M0	3240	4,42	$+0$,26	$+0,11$	$+0$,03	$+0,03$	$+0$,17	$+0,12$

nung noch befriedigend ist. Im Gang der Reste prägt sich die kontinuierliche Wasserstoffabsorption aus, welche bei den Sternen des Spektraltypus A5 ihr Maximum erreicht. Die negativen Korrektionen für den Spektraltypus B0 weisen auf eine Depression der Sternenergiekurve im visuell wirksamen Spektralbereich hin, die bereits durch WILSINGs spektralphotometrische Beobachtungen angezeigt war. In den Spalten 5, 7, 9 der Tabelle 17 stehen die Änderungen der reziproken Strahlungstemperatur der visuellen Helligkeit, um die Strahlungstemperatur der photographischen, der bolometrischen und der von der Wasserzelle durchgelassenen Strahlung zu erhalten. Wenn die Reste in den Spalten 4, 6, 8 der Tabelle 17 typisch für die einzelnen Spektralklassen sind, wird man bei einem Einzelstern die für die betreffende Spektralklasse geltende Korrektion mit umgekehrtem Vorzeichen an den photographischen Farbenindex, an den Wärmeindex und an die Wasserzellenabsorption anbringen, um dann nach der Gleichung (63) mit Hilfe der Tabellen 10 und 11 die Farbtemperatur, d. i. die Strahlungstemperatur der visuellen Helligkeit, zu bestimmen.

f) Zusammenfassende Darstellung der Beobachtungsresultate.

26. Die Temperaturskala der Fixsterne. Fast alle bisher publizierten Sterntemperaturen sind Farbtemperaturen. In der Form der Energiekurve spiegelt sich die Farbe des Sterns wider. Zahlreiche Absorptionslinien und -banden durchziehen, besonders bei den Sternen von spätem Spektraltypus, das Spektrum und bedingen Unebenheiten im Verlauf der Energiekurve. Wenn der zur Bestimmung der Farbtemperatur benutzte Spektralbereich schmal ist, geben derartige Störungen — dazu kommen die Fehler der Beobachtung — einen unrichtigen Gradienten der Energiekurve und täuschen damit eine falsche Farbtemperatur vor. Die Farbenindizes beziehen sich auf mehr oder minder weite Spektralbereiche und begreifen die Wirkung der Absorptionslinien mit ein. Die durch die eigentlichen Beobachtungsfehler bedingten Abweichungen der einzelnen Messungen lassen sich bei den Farbenindizes in engen Grenzen halten. A priori wird man nicht erwarten dürfen, daß das inhomogene Material der spektralphotometrischen Messungen und der Farbenäquivalente zu der gleichen Temperaturskala führt.

WILSING und PLASKETT waren bemüht, die Helligkeitsverteilung in den Sternspektren an möglichst absorptionsfreien Stellen zu messen. Bei den Sternen vom späten Spektraltypus läßt sich diese Bedingung im allgemeinen nicht erfüllen: die beobachtete Strahlung gehört einem mehr oder minder eng begrenzten Spektralbereich an, in dem die Absorptionslinien beliebig dicht beieinander stehen. SAMPSON und CH'ING-SUNG YÜ verbinden die Spitzen der Registrierkurve durch einen möglichst glatten Kurvenzug und bestimmen so den ungestörten

Verlauf der Energiekurve. Bei den nur wenig extrafokalen Aufnahmen ROSENBERGS gehen die schmalen Absorptionslinien im kontinuierlichen Spektrum unter und sind daher als solche nicht mehr zu erkennen. Wenn die Absorptionslinien dicht beieinander stehen, überlagern sie sich gegenseitig und haben eine allgemeine Schwächung des kontinuierlichen Untergrundes zur Folge. Die radiometrischen, thermoelektrischen und bolometrischen Messungen im Sternspektrum, mögen sie direkt oder durch Zwischenschalten von Farbfiltern erhalten sein, beziehen sich auf mehr oder minder weite Spektralbereiche und begreifen die Wirkung der selektiven Sternabsorption mit ein. Das gleiche gilt von den Sternfarben, von den photographischen Farbenindizes, von den Wärmeindizes und von den übrigen Farbenäquivalenten.

Beim Vergleich der Temperaturskalen, wie sie von den einzelnen Autoren aus dem Beobachtungsmaterial abgeleitet sind, muß man darauf achten, ob jene als ein in sich abgeschlossenes Ganzes eine selbständige Bedeutung haben oder ob sie sich an andere Temperaturskalen anlehnen. WILSING und seine Mitarbeiter haben die spektralphotometrischen Messungen der Sterne an den schwarzen Körper angeschlossen und so eine absolute Temperaturskala geschaffen. PLASKETT bezieht die Energieverteilung in den Sternspektren auf die des Kohlebogens, dessen Farbtemperatur im Laboratorium bestimmt worden ist. Die radiometrischen Messungen von ABBOT und die thermoelektrischen von COBLENTZ geben direkt die Energieverteilung in den Sternspektren und damit die Farbtemperatur in der absoluten Skala.

Die meisten photographisch-spektralphotometrischen Beobachtungsverfahren sind differentieller Natur. Der Nullpunkt der Temperaturskala bleibt unbestimmt und wird nachträglich durch passende Wahl der Temperatur des Vergleichssterns festgelegt. SAMPSON setzt die Temperatur von Capella gleich 5500°, CH'ING-SUNG YÜ die von ζ Ophiuchi gleich 22000°; GREAVES, DAVIDSON und MARTIN machen drei verschiedene Annahmen und geben den Sternen vom Spektraltypus A0 die Farbtemperatur 10000° bzw. 13000° bzw. 11000°. WILSING wählt in den kolorimetrischen Untersuchungen den Nullpunkt der Temperaturskala so, daß das Mittel der für 4 Sterne kolorimetrisch bestimmten c_2/T-Werte mit dem Mittel der von ihm spektralphotometrisch abgeleiteten zusammenfällt. ROSENBERG erstrebt den Anschluß der Sternspektren an das Sonnenspektrum; doch ist diese Reduktion, wie WILSING gezeigt hat, nicht einwandfrei durchgeführt[1].

Bei der Neureduktion der WILSINGschen und der ROSENBERGschen spektralphotometrischen Messungen bestimmte der Verfasser die Intensitätsverteilung im Spektrum des ROSENBERGschen Vergleichssterns α Aquilae nach zwei Methoden[2]:

1. aus den spektralphotometrischen Beobachtungen von 38 Fixsternen, die den Listen von WILSING und ROSENBERG gemeinsam sind,

2. aus ROSENBERGS spektralphotometrischen Beobachtungen der Sonne (ohne Berücksichtigung der Beugungskorrektion) in Verbindung mit der von anderen Forschern bestimmten Energieverteilung im Sonnenspektrum.

Die nach beiden Reduktionsverfahren erhaltenen spektralen Energiewerte von α Aquilae stimmen gut miteinander überein. Bei der Sicherung der Temperaturskala durch die KINGschen Farbenindizes in Veröff Berlin-Babelsberg V, 1 hat der Verfasser die Konstante, um welche sich die beobachteten und die berechneten Farbenindizes voneinander unterscheiden, durch Anschluß an die bei der Neureduktion der WILSINGschen und ROSENBERGschen spektralphotometrischen Messungen erhaltenen Temperaturen bestimmt. Das Verfahren, eine

[1] J. WILSING, Über effektive Sterntemperaturen. A N 204, S. 153 (1917).

[2] A. BRILL, Spektralphotometrische Untersuchungen. Teil II. A N 219, S. 21 (1923).

Gruppe von Sternen bzw. die in der Spektralfolge zusammengefaßte Gesamtheit der Sterne zur Festlegung des Nullpunktes der Temperaturskala zu verwenden, ist der Einordnung nach einem einzelnen Stern vorzuziehen.

Das umfangreichste Material an Farbtemperaturen enthalten zur Zeit die Kataloge von HERTZSPRUNG, BOTTLINGER und BRILL. Wenn auch im allgemeinen die einzelnen c_2/T-Werte im jeweiligen Temperatursystem genauer sind als die entsprechenden spektralphotometrisch bestimmten, so haben doch die Temperaturskalen keine selbständige Bedeutung, da sie auf andere Skalen bezogen sind. HERTZSPRUNG reduziert in dem 734 Sterne heller als 5^m umfassenden Katalog alle ihm zugänglichen Farbenäquivalente auf das WILSINGsche System der spektralphotometrisch bestimmten Temperaturen[1]. Auf das gleiche System bringt er die Farbenindizes von 658 am Leidener 33 cm-Refraktor photographierten Sternen[2]. BOTTLINGER legt die aus den lichtelektrischen Farbenindizes von 459 Sternen abgeleitete Temperaturskala in der Weise fest, daß die Temperaturen der heißesten Sterne bei 20000° liegen, während die der kühlsten den von WILSING spektralphotometrisch bestimmten Temperaturen angepaßt sind[3]. Der Katalog von BRILL (Tab. 5 unter „BRILL") umfaßt 134 Sterne, welche den Farbenindexverzeichnissen von KING und BOTTLINGER sowie dem Katalog der mittleren Farbenäquivalente von HERTZSPRUNG gemeinsam sind[4]. Die Temperaturen sind bezogen auf das an die KINGschen Farbenindizes sich anlehnende System der Farbtemperaturen, wie es von dem Verfasser aus dem Vergleich von beobachteten und berechneten Farbenindizes in der Veröff Berlin-Babelsberg V, 1 erhalten ist. Mit den von SAMPSON aus der Intensitätsverteilung in 64 Sternspektren bestimmten spektralphotometrischen Farbtemperaturen prüfte der Verfasser die Temperaturskala der 134 Sterne[5]. Für die Sterne von frühem und mittlerem Spektraltypus ergab sich eine befriedigende Übereinstimmung. Die Temperaturen der Sterne von spätem Spektraltypus sind bei SAMPSON kleiner. Die radiometrischen und die thermoelektrischen Messungen verlangen, wie bereits in Ziffer 22 und 25 erwähnt ist, gleichfalls niedrigere Farbtemperaturen für die Sterne von spätem Spektraltypus.

Die mittleren spektralphotometrischen Farbtemperaturen für die Sterne der einzelnen Spektralklassen (normale Riesensterne) sind nach einer noch nicht abgeschlossenen, bisher unveröffentlichten Untersuchung des Verfassers in der folgenden Tabelle zusammengestellt:

B0	B5	A0	A5	F0	F5	G0
22000°	17700°	13500°	10500°	8550°	7000°	5800°

G5	K0	K5	M0
4860°	4370°	3460°	3240°

Es ist wenig wahrscheinlich, daß hiermit eine endgültige Festlegung der Farbtemperaturskala der Fixsterne erzielt ist. Spektralphotometrische, vor allem auch spektralbolometrische Messungen von einer kleinen Zahl von ausgewählten Sternen nach verschiedenartigen Methoden sind weiterhin sehr erwünscht. Für den Astrophysiker wird es eine wenn auch mühevolle, so doch lohnende Aufgabe sein,

[1] E. HERTZSPRUNG, Mean Colour Equivalents and Hypothetical Angular Semi-Diameters of 734 Stars Brighter than the Fifth Magnitude and within 95° of the North Pole. Ann v d Sterrewacht te Leiden Deel 14, Eerste Stuk (1922).

[2] E. HERTZSPRUNG, Photographic Magnitudes of 658 Stars from Plates taken mainly by W. H. VAN DEN BOS with the 33 cm Leiden Refractor. B A N 1, S. 201 (1923).

[3] K. F. BOTTLINGER, Lichtelektrische Farbenindizes von 459 Sternen. Veröff Univ-Sternw Berlin-Babelsberg III, Heft 4 (1923).

[4] A. BRILL, Temperaturen und scheinbare Halbmesser von 134 Sternen. A N 223, S. 105 (1924).

[5] A. BRILL, Die Temperaturskala der Sterne. A N 225, S. 161 (1925).

das gesamte Material an einschlägigen Beobachtungen nach einheitlichen Gesichtspunkten zu verarbeiten und mit Berücksichtigung der absoluten Helligkeit der Sterne eine Temperaturskala aufzubauen, welche den spektralphotometrischen, den spektralbolometrischen, den radiometrischen und den thermoelektrischen Messungen wie auch den Farben, den Farbenindizes und den sonstigen Farbenäquivalenten in gleicher Weise gerecht wird.

27. Systematische Fehlerquellen bei der Bestimmung der Temperatur aus der Form der Energiekurve. Die spektralphotometrischen Beobachtungen liefern je nach der Wahl des Spektralbereichs für den gleichen Stern verschiedene Temperaturen. Die mittleren Wellenlängen der von WILSING, SAMPSON und ROSENBERG vermessenen Spektralgebiete sind λ 530, 500 und 430 $\mu\mu$; die zugehörigen Temperaturskalen zeigen eine mit der Verschiebung der mittleren Wellenlänge nach dem Ultraviolett zunehmende Verbreiterung. Die aus den WILSINGschen und ROSENBERGschen spektralphotometrischen Messungen abgeleiteten Temperaturskalen brachte der Verfasser in Übereinstimmung dadurch, daß er auf die rechnerische Darstellung der spektralphotometrischen Beobachtungen von ROSENBERG im Violett jenseits λ 450 $\mu\mu$ verzichtete. Die mittlere Wellenlänge der vom Verfasser bei der Neureduktion benutzten ROSENBERGschen Messungen fällt nahezu mit der der SAMPSONschen (λ 500 $\mu\mu$) zusammen, was die Gleichheit beider Temperaturskalen erklärt.

Die Unstimmigkeit in den verschiedenen Temperaturskalen ist bedingt durch Abweichungen der Sternstrahlung von der Strahlung des schwarzen Körpers. Berücksichtigt man rechnerisch die Ungleichheiten, so lassen sich unter gewissen Voraussetzungen außer der Farbtemperatur, welche die ideelle, sozusagen störungsfreie Energiekurve des Sterns charakterisiert, auch die Strahlungstemperaturen der visuellen, der photographischen und der bolometrischen Helligkeit bestimmen. Man kann jede Art von Farbenäquivalent dazu benutzen, den für den Stern charakteristischen Wert der Farbtemperatur zu berechnen und auf diese Weise zur Sicherung der Temperaturskala beitragen. Die Form der Energiekurve mit ihren Abweichungen von der schwarzen Strahlungskurve bildet weiterhin ein Charakteristikum für den Aufbau des Sterns in den photosphärischen Schichten; man wird Zusammenhänge mit den für den einzelnen Stern typischen Elementen: der absoluten Helligkeit, der Masse und der Dichte erwarten dürfen.

Die Intensitätsverteilung im mittleren Sonnenspektrum für den Spektralbereich λ 451 bis 642 $\mu\mu$, in welchem die Energiekurve der Sonne eine markante Spitze besitzt, gibt die spektralphotometrische Farbtemperatur von nahezu 7000°. Aus der Solarkonstanten folgt die Strahlungstemperatur der bolometrischen Helligkeit 5740°. Im Bereich der visuell wirksamen Strahlen besitzt die Sonnenenergiekurve gegenüber der Energiekurve des schwarzen Strahlers von der Temperatur 5740° eine starke Überhöhung. Im kurzwelligen Teil des Spektrums ist eine Depression der Sonnenenergiekurve vorhanden, im langwelligen sind die Unterschiede klein.

Bei den Fixsternen läßt sich die durch die ROSENBERGschen spektralphotometrischen Messungen angezeigte Depression der Energiekurve im Ultraviolett jenseits λ 450 $\mu\mu$ wenigstens zum Teil durch die selektive Absorption der in dem ultravioletten Teil des Spektrums besonders zahlreichen Absorptionslinien und -banden erklären, die bei dem ROSENBERGschen Verfahren der extrafokalen Aufnahmen eine Schwächung des kontinuierlichen Untergrundes hervorrufen. Bis zu einem gewissen rechnerisch nur angenähert feststellbaren Betrag ist die Depression eine Folge der Randverdunklung von der Art, wie sie SAMPSON bei der Diskussion seiner Messungen berücksichtigt hat. Um den störenden Einfluß der Absorptionslinien möglichst zu vermeiden, verbindet SAMPSON die Spitzen der

Registrierkurve, welche ein getreues Abbild der Schwärzungsverteilung im Sternspektrum ist, durch einen glatten Kurvenzug. Die Energiekurven von α Bootis (K0) und β Andromedae (M0) zeigen nach SAMPSON im Ultraviolett bis λ 390 $\mu\mu$ keine merkliche Einsenkung der Energiekurve. SAMPSON schließt daraus, daß bei Berücksichtigung der Randverdunklung die Strahlung eines „normalen" Sterns schwarz ist.

Die Neureduktion der WILSINGschen und ROSENBERGschen spektralphotometrischen Messungen durch den Verfasser ergab für die Sterne von späterem Spektraltypus als A0 gleiche Farbtemperaturen. Die Unstimmigkeit bei den Sternen von frühem Spektraltypus läßt sich daraus erklären, daß der Verfasser entgegen seiner ursprünglichen Absicht bei diesen Sternen die Darstellung der ROSENBERGschen Messungen durch das PLANCKsche Gesetz selbst für Wellenlängen kleiner als 450 $\mu\mu$ anstrebte. Nach der Abb. 1 in Ziffer 16 ist selbst für das visuell wirksame Spektralgebiet λ 451 bis 642 $\mu\mu$ die Weite der Temperaturskala ungleich; sie ist enger zwischen λ 550 bis 642 $\mu\mu$ als zwischen λ 451 bis 550 $\mu\mu$.

SAMPSON und ROSENBERG finden mit ihren spektralphotometrischen Messungen außergewöhnlich hohe Farbtemperaturen für einige Sterne der frühen Spektraltypen B5 bis A5. In der Temperaturskala der 134 Sterne, welche sich an die KINGschen Farbenindizes anlehnt, sind die Temperaturen wesentlich niedriger. Der Grund hierfür ist wahrscheinlich darin zu suchen, daß die Senkung der Energiekurve im Ultraviolett bei den Sternen dieser Spektraltypen verschieden groß ist. Bekanntlich besitzen die Spektren mit vorherrschendem Wasserstoff (später B-Typus und Spektraltypus A bis F) im Ultraviolett eine ziemlich plötzlich einsetzende allgemeine Absorption, die ungefähr bei λ 380 $\mu\mu$ beginnt und sich bis etwa λ 340 $\mu\mu$ erstreckt. Die Erscheinung ist bei dem Spektraltypus B nur schwach angedeutet und fehlt bei den späten Spektralklassen vollständig. Die Energiekurven CH'ING-SUNG YÜs (Abb. 2 in Ziffer 16) geben in qualitativer und quantitativer Beziehung ein anschauliches Bild von der für die einzelnen Sterne verschieden starken kontinuierlichen Wasserstoffabsorption. Der Spektralbereich, in dem die Wasserstoffabsorption liegt, wird bei den spektralphotometrisch bestimmten Farbtemperaturen nicht berücksichtigt. Bei den photographischen Farbenindizes täuscht die Senkung der Energiekurve im Ultraviolett, wenn sie größer oder kleiner als die durchschnittliche ist, eine niedrigere oder eine höhere Farbtemperatur des Sterns vor.

In einer vor einigen Jahren erschienenen Arbeit wies der Verfasser darauf hin, daß die B-Sterne und auch zum Teil die A-Sterne anscheinend im Ultraviolett eine Überhöhung der Energiekurve besitzen[1]. ROSENBERG bestimmte aus der Intensitätsverteilung im Spektrum zwischen λ 400 und 500 $\mu\mu$ außergewöhnlich hohe Temperaturen für einzelne Sterne von frühem Spektraltypus. J. BAILLAUD fand, daß sich zu einigen Sternen kein schwarzer Strahler noch so hoher Temperatur angeben ließ, welcher die beobachtete Intensitätsverteilung darstellte. Die Berechnung der Farbtemperatur aus dem photographischen Farbenindex lieferte bei einigen sehr weißen Sternen ebenfalls außergewöhnlich hohe Farbtemperaturen. Die Sterne α Lupi und ζ Centauri besitzen einen Farbenindex, der absolut größer ist als der des unendlich heißen Strahlers. Ein Analogon zur Sonnenenergiekurve ist hier unverkennbar. Auf dem vom Maximum nach Rot hin abfallenden Ast der Sonnenenergiekurve nimmt die Gradationstemperatur zu, um dann von der Wellenlänge λ 650 wieder abzunehmen. Der steile Abfall der Sonnenenergiekurve nach dem Rot hin bedingt verhältnismäßig hohe Gradations-

[1] A. BRILL, Die Temperaturskala der Sterne. A N 225, S. 161 (1925).

temperaturen, wie man sie auch auf dem nach Rot abfallenden Ast der Energiekurve der B- und A-Sterne beobachtet.

Wie der Verfasser bereits an letztgenannter Stelle ausgeführt hat, spricht die starke Streuung der SAMPSONschen spektralphotometrisch bestimmten Farbtemperaturen bei den frühen Spektraltypen dafür, daß die Überhöhung der Energiekurve bei den einzelnen Sternen verschieden groß ist. In den normalen Fällen von γ Orionis und β Trianguli ist nach SAMPSON eine geringe Überhöhung der Energiekurve jenseits λ 400 $\mu\mu$ angedeutet. Die anomalen Energiekurven von γ Cassiopeiae und α Cygni geben für Wellenlängen größer als 480 bzw. 450 $\mu\mu$ einen schwachen Gradienten, der nach den kurzen Wellenlängen stark zunimmt. Die Überhöhung der Energiekurve nimmt anscheinend von den A- nach den B- und O-Sternen zu; bei den Neuen Sternen ist sie besonders groß und übertrifft beispielsweise im Falle der Nova Geminorum 2 die der B-Sterne um mehr als eine Größenklasse[1].

Bei einigen Sternen des frühen Spektraltypus wird eine auffallende Gelbfärbung beobachtet; der photographische Farbenindex ist für den Spektraltypus der Sterne zu groß. Der Stern ζ Persei, der nach dem Liniencharakter zur Spektralklasse B1 gehört, ist spektralphotometrisch dem Spektraltypus A2 zuzuordnen[2]. Neuere spektralphotometrische Untersuchungen haben ergeben, daß die anomalen gelben Sterne im äußersten Ultraviolett jenseits λ 370 $\mu\mu$ eine starke Überhöhung der Energiekurve besitzen, anscheinend zum Ausgleich des schwachen Gradienten im visuell und photographisch wirksamen Spektralgebiet[3]. GERASIMOVIČ stellte fest, daß der Gradient der Energiekurve bei den frühen B- und bei den O-Sternen ein Charakteristikum der absoluten Helligkeit ist. Die gelben B-Sterne, die Be-Sterne und die O-Sterne mit Absorptionslinien sind absolut hell und besitzen im visuell und photographisch wirksamen Spektralgebiet einen schwachen Gradienten der Energiekurve, die weißen B-Sterne und die Wolf-Rayet-Sterne sind absolut lichtschwächer und haben einen steilen Gradienten der Energiekurve[4].

Die von GERASIMOVIČ festgestellte Tatsache, daß die Farbtemperaturen von den O- zu den B5-Sternen zunehmen, wird vielleicht verständlich, wenn man sich den Verlauf der c_2/T_G-Kurve der Sonne in der Abb. 4 der Ziffer 24 vor Augen hält. Mit fortschreitendem Spektraltypus (O nach B5) fällt der den Messungen von GERASIMOVIČ zugrunde liegende Spektralbereich auf Teile der Energiekurve, welche höheren Gradationstemperaturen entsprechen. Der maximale Wert der Gradationstemperatur, welcher für die Sonne ungefähr bei λ 650 $\mu\mu$ liegt, rückt bei den Sternen, welche im Spektraltypus der Sonne vorausgehen, nach den kurzen Wellenlängen und kommt dann mehr und mehr in die Mitte des visuell wirksamen Spektralbereiches zu liegen. Dies hat zur Folge, daß die spektralphotometrisch bestimmten Farbtemperaturen bei diesen Sternen verhältnismäßig große Werte annehmen. Das Maximum der Abweichung wird ungefähr beim Spektraltypus F0 erreicht sein. Bei den Sternen von noch früherem Spektraltypus rückt die Stelle maximaler Gradationstemperatur weiter nach kürzeren Wellenlängen. Die Abweichungen werden wieder kleiner; ungefähr für die Sterne des Spektraltypus A0 herrscht wieder im Spektralbereich λ 451 bis 642 $\mu\mu$ annähernd normale schwarze Strahlung. Für die Sterne der Spektraltypen B

[1] A. BRILL, Spektralphotometrische Untersuchungen. Teil II. A N 219, S. 353 (1923).

[2] A. BRILL, Temperaturen und scheinbare Halbmesser von 134 Sternen. A N 223, S. 105 (1924).

[3] H. KIENLE, On the Low Temperature of the B1-Type Star ζ Persei. M N 88, S. 700 (1929).

[4] B. P. GERASIMOVIČ, Spectrophotometric Temperatures of Early Stars. Harv Circ 339 (1929).

und O streben die spektralphotometrisch bestimmten Farbtemperaturen niedrigeren Werten zu.

Die Diskussion der Strahlungstemperaturen der bolometrischen und der von der Wasserzelle durchgelassenen Strahlung, der Wärmeindizes und der Wasserzellenabsorption nach den Messungen von PETTIT und NICHOLSON macht auch bei den Sternen vom späten Spektraltypus eine Überhöhung der Energiekurve im Maximum wahrscheinlich[1]. Zusammenfassend kann man also sagen, daß anscheinend die Energiekurven aller Sterne vom Energiemaximum nach beiden Seiten steiler abfallen, als der schwarzen Strahlung entspricht. Der Abfall ist bei den O- und B-Sternen um so stärker, je größer die absolute Helligkeit des Sterns ist. Der steilere Abfall der Energiekurve bei den absolut sehr hellen Sternen wird durch einen schwachen Gradienten im visuell und photographisch wirksamen Spektralgebiet ausgeglichen.

28. Die Temperaturen der Riesen- und Zwergsterne. Nach den Untersuchungen zahlreicher Autoren besteht eine Abhängigkeit des photographischen Farbenindex von der absoluten Helligkeit. Bei gleichem Charakter der Absorptionslinien, d. i. bei gleichem Spektraltypus, ist der photographische Farbenindex der absolut lichtschwachen Sterne (Zwergsterne) kleiner als derjenige der absolut hellen (Riesensterne). In den Spektralklassen G, K und M sind die Zwergsterne weißer als die Riesensterne.

Bei Gelegenheit von quantitativen Bestimmungen des Spektraltypus hat A. KOHLSCHÜTTER einige Absorptionslinien gefunden, deren Intensität bei gleichem allgemeinen Spektralcharakter von der absoluten Helligkeit abhängt. Auf diesen Zusammenhang haben ADAMS und KOHLSCHÜTTER eine Methode zur spektroskopischen Bestimmung der absoluten Helligkeit gegründet, die trotz des zunächst noch rohen Schätzungsverfahrens der Linienintensitäten bereits große Erfolge gezeitigt hat. Da der Einfluß der Absorptionslinien im Spektrum für die Sterne der gleichen Spektralklasse verschieden groß ist, je nachdem es sich um einen Riesen- oder einen Zwergstern handelt, so werden die beiden Gattungen von Sternen selbst bei gleicher spektralphotometrischer Farbtemperatur voneinander verschiedene photographische Farbenindizes besitzen. In der Veröff. Berlin-Babelsberg V, 1 hatte der Verfasser die Frage zur Diskussion gestellt, ob die Unterschiede in den photographischen oder sonstwie definierten Farbenindizes der Riesen- und Zwergsterne auf verschiedene spektralphotometrische Farbtemperaturen zurückzuführen sind oder ob sie sich, wenigstens zu einem gewissen Bruchteil, durch die verschieden große Abweichung der Energiekurve des Sterns von der schwarzen Strahlungskurve und durch einen verschieden starken Einfluß der Absorptionslinien erklären lassen. Eine Trennung der Wirkung ungleicher spektralphotometrischer Farbtemperatur und ungleich starken Absorptionsgebietes allein auf Grund der photographischen Farbenindizes ist nicht möglich. Dazu braucht man entweder spektralphotometrische Messungen oder Farbenindizes, welche sich auf absorptionsfreie Spektralbereiche beziehen.

SAMPSON fand in M N 85 relativ hohe spektralphotometrische Farbtemperaturen für die Sterne großer absoluter Helligkeit, niedrige für die Sterne kleiner absoluter Helligkeit. Das Material, auf das SAMPSON seine Folgerung gründete, war recht dürftig; unschwer konnte man auf Grund desselben auch zu dem entgegengesetzten Schluß kommen[2]. Nach der neuesten Veröffentlichung SAMPSONS in M N 90 nimmt der c_2/T-Wert im Durchschnitt der Spektralklassen G, K und M um 0,08 ab, wenn die absolute Helligkeit sich um $+1^{\mathrm{m}}$ ändert.

[1] A. BRILL, Die Temperaturstrahlung der Sterne. Naturwiss S. 418 (1929).
[2] A. BRILL, Die Temperaturskala der Sterne. A N 225, S. 161 (1925).

Spezialuntersuchungen von STORER und HUFNAGEL über die Intensitätsverteilung im kontinuierlichen Spektrum der Riesen- und Zwergsterne haben wesentlich zur Klärung des Problems beigetragen[1]. STORER photographierte mit dem Spaltspektrographen des Crossley-Reflektors auf rotempfindlichen Platten — um einen möglichst weiten Spektralbereich zu erfassen — zweimal das Spektrum von 55 ausgewählten Sternen, die vorzugsweise den Spektralklassen F, G, K und M angehören[2]. Die Belichtungszeit der zweiten Aufnahme war im allgemeinen viermal so lang wie die der ersten, um meßbare Schwärzungen auch für die lichtschwachen Teile des Sternspektrums zu erhalten, und schwankte zwischen 2 Min. und 60 Min. für die verschiedenen Sterne. Außer den beiden Sternaufnahmen trägt jede Platte vier Vergleichsspektra einer irdischen Lichtquelle. Die Aufnahmen der Vergleichsspektren waren von gleicher Dauer (4 Min.); durch Änderung der Entfernung der Vergleichslichtquelle wurden die Intensitäten der vier Vergleichsspektren im Verhältnis 1 : 4 : 16 : 64 abgestuft. Sämtliche sechs Spektra jeder Platte wurden mit dem MOLLschen selbstregistrierenden Mikrophotometer ausgemessen. Die sechs Diagramme wurden an 14 Stellen des Spektrums zwischen λ 378 und λ 742,5 $\mu\mu$ ausgemessen; der Abstand der Marken für absolute Dunkelheit und für den allgemeinen Plattenuntergrund bildet den Einheitsmaßstab für die Höhe der Registrierkurve über der Marke des Plattenschleiers. Mit der durch die vier Vergleichsspektren gegebenen Schwärzungsskala wurde für jede Wellenlänge die Intensität der beiden Sternspektren bestimmt und in Einheiten der Intensität des schwächsten Vergleichsspektrums ausgedrückt. STORER erstrebte in ähnlicher Weise wie ROSENBERG den Anschluß des Spektrums der Vergleichslampe an das Sonnenspektrum. Die Aufnahmen des zentralen Teiles der Sonnenscheibe erfolgten in gleicher Weise wie die der Sterne, nur wurde der Spiegel auf 1 mm abgeblendet, und die Belichtungszeit betrug $^1/_4$ bis 2 Sek. Die Intensitätsverteilung im Sonnenspektrum lieferten die neuesten ABBOTschen Werte der mittleren Sonnenstrahlung. Die atmosphärische Absorption wurde nach der Formel $a_\lambda = a_{0\lambda}^{\sec z}$ berücksichtigt, wo z die Zenitdistanz des Sterns oder der Sonne zur Zeit der Beobachtung ist; die Transmissionskoeffizienten $a_{0\lambda}$ sind die auf die Seehöhe des Mount Hamilton umgerechneten mittleren Werte des Mount Wilson-Observatoriums.

Die 55 Sterne, deren Intensitätsverteilung im kontinuierlichen Spektrum STORER bestimmte, waren so ausgewählt, daß jede Spektralklasse annähernd gleichviel Riesen- und Zwergsterne enthielt; dazu kamen noch einige heißere Sterne, um die Skala der spektralphotometrischen Farbtemperaturen womöglich auch für die Sterne der frühen Spektraltypen festzulegen. Die roten Zwergsterne der Spektralklasse M waren nicht hell genug, um beobachtet werden zu können. Die Energiekurven jedes Sterns wurden mit den 14 spektralen Intensitätswerten gezeichnet. Im Spektralbereich λ 440 bis 690 $\mu\mu$ lassen sich erstere durch die PLANCKsche Gleichung gut darstellen; im violetten Teil des Spektrums zeigten sich erhebliche Abweichungen von der schwarzen Strahlung, von der Art und Größe, wie sie der Verfasser früher aus den spektralphotometrischen Messungen ROSENBERGS gefunden hatte. Die Depression der Energiekurve im Ultraviolett kann auch hier zum Teil durch das Verfahren bei der Ausmessung der Registrierkurve bedingt sein: Im allgemeinen wurde die Höhe der Registrierkurve selbst gemessen. Wenn sich gelegentlich eine Absorptionslinie an der zu messenden

[1] Da bei Drucklegung des Bandes II, 1 des Handb. der Astrophysik die Veröffentlichungen von STORER und HUFNAGEL noch nicht vorlagen, soll an dieser Stelle auch auf ihre Methode ausführlicher eingegangen werden.

[2] N. W. STORER, A Photometric Study of the Continuous Spectra of Giant and Dwarf Stars. Lick Bull Nr. 410 (1929).

Stelle befand, wurde diese überbrückt, wenn es genügend genau geschehen konnte. Bei den Sternen der mittleren und späten Spektraltypen sind die Absorptionslinien und -banden zu zahlreich, um den Verlauf des ungestörten kontinuierlichen Spektrums feststellen zu können; die Höhe der Registrierkurve wurde dann einschließlich der selektiven Sternabsorption gemessen.

Diejenige PLANCKsche Energiekurve, welche die empirische am besten darstellt, liefert für den betreffenden Stern die spektralphotometrische Farbtemperatur. Die übrigbleibenden Reste müßten graphisch aufgetragen eine Horizontale ergeben, was für den Spektralbereich λ 440 bis λ 690 $\mu\mu$ auch nahezu der Fall ist. Im violetten Teil des Spektrums bedingen die sich häufenden Absorptionslinien und -banden eine mit fortschreitendem Spektraltypus verstärkte Verzerrung des kontinuierlichen Spektrums. Nach der Untersuchung STORERs sind die spektralphotometrischen Farbtemperaturen der Zwergsterne höher als die der Riesensterne von gleichem Spektraltypus (Tab. 18); sonstige systematische Unterschiede der Intensitätsverteilung im kontinuierlichen Spektrum zwischen Riesen- und Zwergsternen sind nicht erkennbar.

Tabelle 18.

Spektrum	STORER	HUFNAGEL	SEARES	HERTZSPRUNG	BOTTLINGER	Mittel $\Delta\frac{c_2}{T}\cdot 10^4$	$\frac{c_2}{T}\cdot 10^4$	T
F5	0,00		0,00	0,00	0,00	0,00	2,05	7000°
G0	0,44		0,22	0,24	0,08	0,24	2,23	6400
G5	0,22	0,60	0,50	0,49	0,45	0,45	2,50	5700
K0	0,83	0,62	0,77	0,42	0,53	0,63	2,65	5400
K5	0,31	1,20	0,90		0,76	0,79	3,35	4280
M0			0,35		0,40	0,38	4,04	3530

Die STORERschen Sterntemperaturen sind durchweg höher als die von anderen Autoren bestimmten; STORER glaubte hierfür nicht irgendwelche Ungenauigkeit in den Beobachtungen verantwortlich machen zu dürfen. Bei Festlegung einer absoluten Temperaturskala ist die Reduktion von Tages- und Nachtbeobachtungen nach den gleichen mittleren Extinktionswerten nicht zulässig. Die zu hohen Sterntemperaturen STORERs lassen darauf schließen, daß die Erdatmosphäre am Tage weniger durchlässig war als bei Nacht, was auch der allgemeinen Erfahrung entspricht. Die spektralphotometrischen Farbtemperaturskalen hängen nach Ziffer 27 von dem Wellenlängenbereich ab, auf den sich die Temperatur bezieht. Die mittlere Wellenlänge der STORERschen Messungen liegt nahe bei derjenigen der WILSINGschen spektralphotometrischen Messungen; die STORERschen Farbtemperaturen für die frühen Spektralklassen sollten danach zwischen 10000° und 12000° liegen. Beobachtungstechnisch ist die Konstanz des SCHWARZSCHILDschen Exponenten p_λ für das weite Wellenlängengebiet bei den enormen Unterschieden in der Belichtungszeit ($^1/_4$ Sek. bis 1 Stunde) durch die STORERsche Untersuchung nicht genügend geklärt; die Variation von p_λ mit der Belichtungszeit wird von STORER fälschlicherweise mit dem PURKINJE-Effekt in Zusammenhang gebracht[1].

HUFNAGEL bestimmte die spektralphotometrischen Farbtemperaturen von Riesen- und Zwergsternen nach Harvard-Spektrogrammen, die mit dem 16zölligen METCALF-Teleskop für den HENRY DRAPER-Katalog aufgenommen waren[2]. Die Reduktion der mit einem SCHILTschen Photometer ausgemessenen Platten wurde

---

[1] Die Kritik KIENLEs an STORERs Messungen [H. KIENLE, Bemerkungen zur Temperaturskala der Fixsterne. Z f Astrophys 1, S. 332 (1930)] betrifft im wesentlichen die gleichen Punkte.

[2] L. HUFNAGEL, Temperatures of Giants and Dwarfs. Harv Circ 343 (1929).

unter der Annahme ausgeführt, daß die Energieverteilung im kontinuierlichen Spektrum aller auf den einzelnen Platten befindlichen A0-Sterne die gleiche ist; die spektralen Helligkeitsdifferenzen der A0-Sterne werden gleich ihren photographischen Größendifferenzen gesetzt. Um Riesen- und Zwergsterne voneinander zu trennen, benutzt HUFNAGEL das Cyanogen-Kriterium, das in der spektralphotometrischen Messung der Intensitätsdifferenzen zwischen besonders charakteristischen Teilen des Spektrums und zwischen gewissen Standardregionen besteht: Die Farbe des Sterns wird bestimmt durch das Helligkeitsverhältnis G_m der Spektralbereiche λ 423 bis 426,7 $\mu\mu$ und λ 433 bis 437 $\mu\mu$, die absolute Helligkeit des Sterns durch das Helligkeitsverhältnis C_m der Spektralregionen λ 414 bis 417,4 und λ 423 bis 426,7 $\mu\mu$. Die von LINDBLAD empirisch gewonnenen Diagramme geben mit den Argumenten G_m und C_m die absolute Helligkeit des Sterns.

Die spektralphotometrischen Farbtemperaturen leitet HUFNAGEL aus dem relativen Gradienten der Energiekurve des Riesen- bzw. des Zwergsterns gegen das Mittel der A0-Sterne ab. Der Gradient der relativen Energiekurve wurde aus Messungen an 5 Stellen im Spektrum [$H\beta$, $\frac{1}{2}(H\beta + H\gamma)$, $H\gamma$, $H\delta$, $H\varepsilon$] bestimmt; die Einstellungen erfolgten mit dem SCHILTschen Photometer zu beiden Seiten der Wasserstofflinien. Die Standardtemperatur der A0-Sterne wurde gleich 10000° gesetzt. Der Katalog umfaßt die spektralphotometrischen Farbtemperaturen von 82 Sternen, von denen 46 Riesen- und 36 Zwergsterne sind. Die Tabelle 18 enthält unter dem Kopf „HUFNAGEL“ die mittleren c_2/T-Differenzen der Riesen- und Zwergsterne für die Spektralklassen G5, K0 und K5.

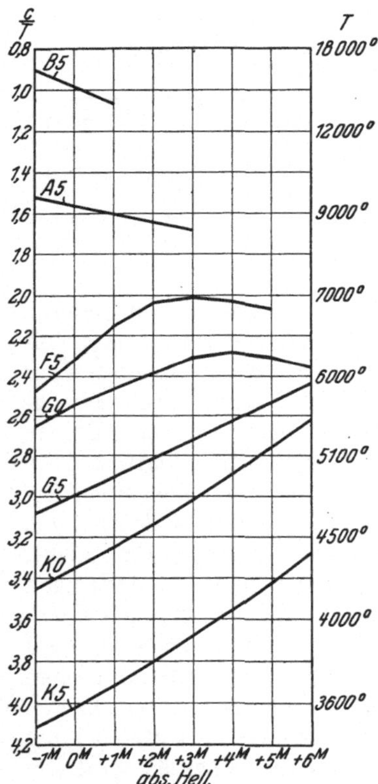

Abb. 5. Beziehung zwischen c_2/T und absoluter Helligkeit für die Spektralklassen B bis K. (Nach STICKER, Z. f. Physik, Bd. 61, S. 561.)

SEARES bestimmte aus den photographischen Farbenindizes des Mount Wilson-Observatoriums die Farbtemperaturen der Riesen- und Zwergsterne; HERTZSPRUNG benutzte hierfür das gesamte ihm verfügbare Material an Farben und Farbenäquivalenten. Die Temperaturskalen von SEARES und HERTZSPRUNG lehnen sich an die WILSINGsche Skala der spektralphotometrischen Farbtemperaturen an. Die 4. und 5. Spalte der Tabelle 18 enthalten für die einzelnen Spektralklassen die Unterschiede in den c_2/T-Werten der Riesen- und Zwergsterne nach SEARES und HERTZSPRUNG. Die BOTTLINGERschen c_2/T-Differenzen, abgeleitet aus den lichtelektrischen Farbenindizes von 459 Sternen, gelten für das von ihm angenommene Temperatursystem. Die 7. Spalte der Tabelle 18 enthält die mittleren c_2/T-Differenzen der Riesen- und Zwergsterne gemäß den Angaben der vorausgehenden Spalten. In den beiden letzten Spalten sind die mittleren c_2/T-Werte und die Farbtemperaturen der Zwergsterne aufgeführt, wenn für die normalen Riesensterne die Durchschnittswerte der Farbtemperatur nach der 2. Spalte der Tabelle 17 in Ziffer 26 gesetzt sind.

Während sich die bisher genannten Autoren mit einer qualitativen Feststellung des absoluten Helligkeitseffektes begnügten, hat STICKER die Untersuchung auf eine streng numerische Basis gestellt[1]. Das Ausgangsmaterial lieferten die photographischen Helligkeiten der Göttinger Aktinometrie und die visuellen Helligkeiten der Potsdamer Durchmusterung. Die Herleitung der absoluten Helligkeiten erfolgte nach trigonometrischen, spektroskopischen, Gruppen- und statistischen mittleren Parallaxen. Der Einfluß der absoluten Helligkeit auf den photographischen Farbenindex wurde für die einzelnen Spektralklassen getrennt bestimmt. Die Umrechnung der photographischen Farbenindizes auf die c_2/T-Skala des Verfassers in AN 223, S. 105 führt zu dem in Abb. 5 gezeigten Zusammenhang zwischen dem c_2/T-Wert und der absoluten Helligkeit für die verschiedenen Spektralklassen[2]. Bei den B- und A-Sternen haben die absolut helleren Sterne die größere Farbtemperatur. Bei den F-Sternen kehrt sich das Verhältnis um; mit fortschreitendem Spektraltypus prägt sich die Abhängigkeit des c_2/T-Wertes von der absoluten Helligkeit immer mehr in dem Sinne aus, daß die absolut hellen Sterne kühler sind als die schwachen.

29. Die Temperaturen der veränderlichen Sterne. Mit dem Lichtwechsel der veränderlichen Sterne ist meist ein entsprechender Gang in der Farbe und im Spektraltypus verbunden. Die Farbenänderung kann verursacht sein entweder durch die Variation der Temperatur der strahlenden Schicht oder auch durch das mehr oder weniger starke Hervortreten der Absorptions- bzw. Emissionslinien und -banden während der Lichtwechselperiode. Wegen der Abweichungen der Sternstrahlung von der des schwarzen Körpers ist eine kritiklose Verwertung von spektralphotometrisch oder aus irgendwelchen Farbenäquivalenten berechneten Sterntemperaturen zur Prüfung gewisser Theorien, wie z. B. der Pulsationstheorie der δ Cephei-Veränderlichen, vom streng wissenschaftlichen Standpunkt aus abzulehnen. Das Studium des Lichtwechsels der Bedeckungsveränderlichen nach spektralphotometrischen Methoden führt für jede Wellenlänge zu einer Beziehung zwischen der Gradationstemperatur und der schwarzen Temperatur; die Messung zweier zu verschiedenen Spektralgebieten gehörigen Integralhelligkeiten gibt das Verhältnis von Farb- und Strahlungstemperatur. Die Diskussion der einschlägigen Beobachtungsresultate gehört in die Kapitel über veränderliche und neue Sterne (Handb. der Astrophysik Bd. 6, Kap. 2 u. 3). Hier sollen nur kurz die Verfahren dargelegt werden, nach denen bisher die Temperaturen der veränderlichen Sterne bestimmt wurden; ferner ist eine tabellarische Übersicht der Ergebnisse für zwei Hauptgruppen von Veränderlichen beigefügt.

Spektralphotometrische Beobachtungen von veränderlichen Sternen sind nur in ganz vereinzelten Fällen angestellt worden. EBERHARD erhielt mit einem Spiegel, vor dem ein Prisma von $2^1/_2{}^\circ$ brechendem Winkel gesetzt war, Aufnahmen der Nova Geminorum 2 und des δ Cephei-Veränderlichen S Sagittae. Das Plattenmaterial der Nova Geminorum 2 diskutierte der Verfasser, das des Sterns S Sagittae GYLLENBERG[3]. Der Anschlußstern ist in beiden Fällen ein Stern vom A-Typus. Die relativen Energiekurven der beiden Veränderlichen lassen sich nicht in ihrem ganzen Verlauf durch das PLANCKsche Strahlungsgesetz darstellen. Die abnorme Gestalt der Energiekurven ist zum Teil auf den Vergleichsstern zurückzuführen, welcher als A-Stern im Ultraviolett die kontinuierliche Wasserstoffabsorption besitzt. Der Veränderliche S Sagittae zeigt den gleichen Effekt in verstärktem

[1] B. STICKER, Untersuchungen über Sternfarben. Veröff Univ-Sternw Bonn Nr. 23 (1930).

[2] B. STICKER, Temperaturen von Riesen- und Zwergsternen. Z f Phys 61, S. 557 (1930).

[3] A. BRILL, Die Helligkeitsschwankungen im Spektrum der Nova Geminorum 2 nach Aufnahmen von G. EBERHARD. Publ Astrophys Obs Potsdam Nr. 70 (1914); A N 211, S. 1 (1920); 212, S. 457 (1921); W. GYLLENBERG, Über die Intensitätsverteilung im Spektrum des δ Cephei-Veränderlichen S Sagittae. Lund Medd. Série II, Nr. 24 (1920).

Ausmaß. Die Depression der Energiekurve im Ultraviolett ist mit der Phase des Lichtwechsels veränderlich; in der Maximalhelligkeit ist sie am größten, in der Minimalhelligkeit am kleinsten. Der Änderung des Spektraltypus vom Maximum zum Minimum (F8 bis G7) entspricht in der Temperaturskala der Riesensterne nach Ziffer 26 die Differenz 0,78 der c_2/T-Werte, die mit der von GYLLENBERG angegebenen Amplitude 0,81 übereinstimmt. Die Überhöhung der Energiekurve der Nova Geminorum 2 im Ultraviolett ist wahrscheinlich durch die kontinuierliche Wasserstoffemission bedingt. Die Überhöhung ist gleichfalls mit der Phase des Lichtwechsels veränderlich; im Maximum ist sie kleiner als im Minimum. Um die Temperatur der Nova Aquilae 3 in den einzelnen Phasen ihrer Entwicklung zu bestimmen, hat WILSING die Energieverteilung im visuell wirksamen Teil des Spektrums gemessen[1]. Die Temperaturen am Anfang und am Ende der Beobachtungsreihe entsprechen derjenigen eines Sterns vom Spektraltypus B. Die Temperaturänderungen laufen den Helligkeitsänderungen parallel. HERTZSPRUNG verglich nach der Objektivprisma-Gitter-Methode die Intensitätsverteilung im Spektrum der Nova Aquilae 3 zur Zeit ihrer Maximalhelligkeit mit dem benachbarten A-Stern α Aquilae[2]. Die Temperatur der Nova ist nur wenig höher als die des Vergleichssterns; im Ultraviolett ergeben sich starke Abweichungen von der Strahlung des schwarzen Körpers.

HOPMANN bestimmt die Schwankungen in der Temperatur von veränderlichen Sternen nach der WILSINGschen kolorimetrischen Methode. Vom beobachtungstechnischen Standpunkt aus scheint diese Methode für das Studium des Temperaturwechsels der veränderlichen Sterne besonders geeignet zu sein. Die Apparatur ist äußerst einfach und unterscheidet sich von den üblichen Photometern nur durch den Rotkeil. Die Reduktion der Messungen erleichtern besondere von SCHNAUDER berechnete Tafeln. Mit den visuellen Helligkeitsschwankungen der roten Veränderlichen sind starke Änderungen der Temperatur und des scheinbaren Durchmessers verknüpft. Die bolometrische Helligkeit weist im Vergleich zur visuellen Helligkeit nur geringe Schwankungen auf, was durch die radiometrischen Messungen von PETTIT und NICHOLSON bestätigt wird.

Die von TERKÁN angewandte Methode zur Untersuchung der periodischen Helligkeits- und Temperaturänderungen von β Lyrae besteht in der Beobachtung des Verschwindens des extra- und intrafokalen Sternbildes bei verschiedener Feldbeleuchtung[3]. Diese wird entweder in der natürlichen Farbe benutzt, oder es werden passende Farbfilter vor das Okular gesetzt. Bei konstanter Feldbeleuchtung wird der Skalenwert des Okulartriebes abgelesen, welcher dem Verschwinden des Sternscheibchens entspricht. Wenn das Verschwinden bei zwei konstanten Feldbeleuchtungen beobachtet wird, geben die Skalenablesungen entsprechend einem von SCHWARZSCHILD vorgeschlagenen Reduktionsverfahren die Helligkeiten in den einzelnen Farben. Die Unterschiede der in den Farben gemessenen Helligkeiten sind charakteristisch für die Farbe des Sterns. Bedeuten m, m_r, m_g und m_b die Helligkeiten des Sterns in der natürlichen, der roten, gelben und blauen Farbe, so ist nach TERKÁN:

$$m_r - m_b = f\left(T, \frac{1}{m}\right); \qquad m_g - m_b = \varphi\left(T, \frac{1}{m}\right), \tag{64}$$

wo T die Farbtemperatur des Sterns ist. Die Form der Funktionen f und φ bestimmt TERKÁN empirisch; er beschränkt sich dabei in der Entwicklung der

[1] J. WILSING, Bestimmung der effektiven Temperatur der Nova Aquilae 3. A N 208, S. 191 (1919).

[2] E. HERTZSPRUNG, Photographisch-spektralphotometrischer Vergleich zwischen Altair und Nova Aquilae 3 in der Nähe ihrer maximalen Helligkeit. A N 207, S. 75 (1918).

[3] L. TERKÁN, Über die periodischen Temperaturveränderungen von β Lyrae, deren Verlauf nahe das Spiegelbild der Lichtkurve ist. A N 226, S. 345 (1926).

Funktionen auf die Glieder erster Ordnung in T und $1/m$, deren Koeffizienten er nach den Temperaturen aus WILSINGs spektralphotometrischen Messungen berechnet.

Die Temperaturschwankungen der veränderlichen Sterne lassen sich auch aus dem photographischen Farbenindex, aus dem Wärmeindex oder aus der Wasserzellenabsorption bestimmen. Zu dem Zweck sind die entsprechenden Integralhelligkeiten auf bestimmte rechnerisch festgelegte Systeme zu reduzieren, wie es der Verfasser beispielsweise für die visuellen Helligkeiten von ZINNER und der Revised Harvard Photometry, für die photovisuellen Helligkeiten von KING, für die photographischen Helligkeiten von KING und von HERTZSPRUNG und für die radiometrischen Messungen von PETTIT und NICHOLSON durchgeführt hat. Die Kombination von zeitlich und örtlich auseinanderliegenden Beobachtungen gibt leicht zu Fehlerquellen Veranlassung. Die idealste Beobachtungsweise gründet die Bestimmung jeder Art von Farbenindex auf die gleichzeitige Messung der Intensität in zwei mehr oder minder weiten Spektralgebieten mit dem gleichen Instrument an demselben Ort, wodurch insbesondere Änderungen des Luftzustandes ausgeschaltet werden.

Die Farbe und ihre Äquivalente begreifen die Wirkung der selektiven Sternabsorption mit ein; erst wenn der mit dem Lichtwechsel veränderliche Einfluß der selektiven Sternabsorption und -emission bekannt ist, wird die aus der Farbe oder ihren Äquivalenten abgeleitete Farbtemperatur die Form der störungsfreien Energiekurve des Veränderlichen charakterisieren. Die in formaler Rechnung aus kolorimetrischen und radiometrischen Messungen bestimmten Farbtemperaturen der δ Cephei-Veränderlichen (Tab. 19) und der Mira-Sterne (Tab. 20) sind

Tabelle 19.

Stern	Spektrum Max. bis Min.	Temperatur Max. bis Min.	Beobachtungsverfahren	Literaturnachweis
δ Cephei . . .	F4 bis G6	6980° bis 5780°	spektralphotometrisch	SAMPSON, M N 85, S. 240 (1925).
		6400 bis 5400	„	PANNEKOEK u. REESINCK B A N 3, S. 47 (1925).
		6700 bis 4780	kolorimetrisch	HOPMANN, A N 226, S. 1 (1925).
		6200 bis 4580	radiometrisch	PETTIT u. NICHOLSON, Ap J 68, S. 288 (1928).
ζ Geminorum .		5310 bis 5030	spektralphotometrisch	PANNEKOEK u. REESINCK, B A N 3, S. 47 (1925).
		7200 bis 5000	„	CH'ING-SUNG YÜ, Publ A S P 38, S. 357 (1926).
		4570 bis 3580	kolorimetrisch	HOPMANN, A N 227, S. 257 (1926).
η Aquilae . .	F2 bis G9	5240 bis 3960	kolorimetrisch	HOPMANN, A N 222, S. 1 (1924).
		4950 bis 3900	radiometrisch	PETTIT u. NICHOLSON, Ap J 68, S. 288 (1928).
S Sagittae . .	F8 bis G7	5700 bis 4300	spektralphotometrisch	GYLLENBERG Lund Medd (2), 24 (1920)
RT Aurigae .	F1 bis G5	5880 bis 4160	kolorimetrisch	HOPMANN, A N 227, S. 257 (1926).
T Vulpeculae.	F5 bis G1	4670 bis 3290	„	HOPMANN, A N 221, S. 337 (1924).
SU Cassiopeiae	F2 bis F9	6440 bis 5110	„	HOPMANN, A N 226, S. 1 (1925).

Tabelle 20.

Stern	Spektrum	Temperaturintervall Max. bis Min.	Beobachtungsverfahren	Literaturnachweis
o Ceti . . .	M6e	3780° bis < 2000°	kolorimetrisch	HOPMANN, A N 226, S. 1 (1925).
		2540 bis 2020	radiometrisch	PETTIT u. NICHOLSON, Ap J 68, S. 291 (1928).
χ Cygni . . .	M6e	2230 bis 1200	kolorimetrisch	HOPMANN, AN 222, S. 233 (1924).
		2260 bis 1580	radiometrisch	PETTIT u. NICHOLSON, Ap J 68, S. 291 (1928).
X Ophiuchi .	M6e	2260 bis 1890	,,	,,
R Cancri . .	M7e	2450 bis 1890	,,	,,
R Aquilae .	M7e	2360 bis 1890	,,	,,
R Aquarii .	M7e	2180 bis 1760	,,	,,
R Leonis minoris . . .	M8e	2260 bis 1830	,,	,,
R Leonis . .	M8e	2260 bis 1760	,,	,,
R Hydrae . .	M8e	2360 bis 1950	,,	,,

deshalb verfälscht durch die Wirkung der selektiven Absorption und Emission im Sternspektrum. Die für die Mira-Sterne in ihrer Minimalhelligkeit aus dem photographischen Farbenindex abgeleitete höhere Farbtemperatur ist darauf zurückzuführen, daß die Titanoxydbanden, welche im visuell wirksamen Spektralbereich liegen, vom Maximum zum Minimum der Sternhelligkeit stark zunehmen.

g) Die Temperaturen der Sterne nach der Theorie der thermischen Ionisation.

30. Die Spektralklassifikation der Fixsterne. Die Energieverteilung im Spektrum eines Sterns führt zum Begriff der spektralphotometrischen Farbtemperatur, d.i. die Temperatur des schwarzen Strahlers, dessen Spektrum die gleiche Intensitätsverteilung zeigt wie der Stern. Die visuelle, die photographische und die bolometrische Helligkeit geben die Strahlungstemperatur der visuellen, der photographischen und der bolometrischen Helligkeit. Die Strahlungstemperatur eines Sterns bezüglich eines mehr oder weniger weiten Spektralbereiches ist hierbei definiert als diejenige Temperatur, welche ein schwarzer Körper haben muß, um die Strahlung dieses Bereiches in gleicher Intensität zu emittieren wie der Stern. Wenn die Energieverteilung im Sternspektrum genau der eines schwarzen Strahlers entspricht, sind Farb- und Strahlungstemperatur einander gleich. Die aus Farbe und Helligkeit abgeleitete Temperatur bezeichnet man sinngemäß auch als „photosphärische“ Temperatur.

Die Theorie der thermischen Ionisation gibt die Temperatur in der „umkehrenden“ Schicht, welche das Absorptionsspektrum hervorbringt. Die Spektren der Fixsterne lassen sich erfahrungsgemäß in eine kontinuierliche Reihe einordnen, wobei für die Zuteilung eines Sternspektrums zu einer bestimmten Spektralklasse das Vorhandensein und die Stärke des Auftretens von einzelnen Liniengruppen maßgebend ist. Diese Spektraleinteilung war ursprünglich rein formaler Natur und vertrat gewissermaßen die ausführliche Beschreibung der Sternspektren. Die auf der Atomtheorie von NIELS BOHR aufgebaute, von MEG NAD SAHA begründete und von DARWIN, FOWLER und MILNE weiterentwickelte Theorie der Sternspektren hat die tiefere Bedeutung dieser empirisch gewonnenen Spektralanordnung erkennen lassen. Die allgemein bekannte Tatsache, daß zwischen der Farbe und dem Spektraltypus ein enger Zusammenhang besteht, ist darauf zurückzuführen, daß beide Sterncharakteristika durch den thermischen Aufbau der äußeren Schichten, der Sternphotosphäre und der umkehrenden Schicht, bedingt sind.

Die Theorie der thermischen Ionisation ist von A. PANNEKOEK im Handb. der Astrophysik Bd. III, Teil 1, Kap. 3, behandelt. An dieser Stelle sollen nur kurz ohne analytisches Beiwerk die Grundlagen der Theorie entwickelt und die neuesten Forschungsergebnisse den Resultaten aus den Strahlungsmessungen gegenübergestellt werden.

Das Auftreten und die Stärke der Absorptionslinien im Sternspektrum hängt von der Konzentration an ionisierten, angeregten und neutralen Atomen ab, welche jene erzeugen. Der Prozentsatz neutraler, angeregter, einfach und mehrfach ionisierter Atome ist durch den Ionisationszustand bedingt, in welchem sich das Atom in der umkehrenden Schicht befindet. Die Lage der wirksamen Schicht wird für alle Atome nicht die gleiche sein; die Beobachtungsresultate scheinen jedoch dafür zu sprechen, daß die Unterschiede unerheblich sind. Temperatur und Gasdruck (Partialdruck der freien Elektronen) in der umkehrenden Schicht sowie das Ionisationspotential des Atoms in dem besonderen Zustand sind die bestimmenden Faktoren. Mit steigender Temperatur nimmt die Ionisierung zu, mit steigendem Druck und mit wachsendem Ionisationspotential nimmt sie ab. Je größer das Ionisationspotential ist, um so höher muß die Temperatur oder um so geringer muß der Druck sein, der zu der gleichen prozentualen Ionisierung führt. Die Dissoziationsformel der physikalischen Chemie, in folgerichtiger Weise auf das vorliegende Problem angewandt, gibt die funktionelle Abhängigkeit der Ionisierung des Atoms von Temperatur, Druck und Ionisationspotential. Die Annahme von SAHA für den Wert der minimalen Konzentration, bei welcher die Atome des betreffenden Zustandes gerade noch beobachtbare Absorptionslinien hervorbringen, ist ziemlich willkürlich; daher sind die Resultate, welche er aus dem Erscheinen und Verschwinden der Absorptionslinien in der Spektralreihe gezogen hat, in quantitativer Beziehung nicht einwandfrei. DARWIN und FOWLER haben auf Grund statistischer Überlegungen eine Formel für die maximale Konzentration eines Atoms in neutralem, angeregtem oder ionisiertem Zustand abgeleitet. Das Maximum der Konzentration wird mit der maximalen Intensität der entsprechenden Absorptionslinien in der Spektralreihe der Sterne identifiziert. Bei gegebenem Ionisationspotential ergibt sich eine Beziehung zwischen Druck und Temperatur in der die Absorptionslinien erzeugenden Schicht. FOWLER und MILNE setzen die Temperatur in der umkehrenden Schicht der A-Sterne gleich 10000° und erhalten damit für den Druck in der umkehrenden Schicht (Partialdruck der freien Elektronen) $1{,}3 \cdot 10^{-4}$ Atm. Benutzt man den für die A0-Sterne erhaltenen mittleren Druck in der umkehrenden Schicht durchweg für alle Absorptionslinien und für alle Spektralklassen — wofür ein zwingender Grund allerdings nicht vorliegt —, so findet man, wie die Arbeiten von FOWLER, MILNE und Miss PAYNE gezeigt haben, für die Spektralreihe der Sterne nahezu dieselbe Temperaturskala, welche aus der Strahlung der Sterne folgt (Ziffer 26) — ein Resultat, das a priori nicht zu erwarten war:

Spektrum . .	O	B0	B5	A0	A5
PAYNE[1]. . .	25000°	20000°	15000°	10000°	8400°
BRILL . . .		22000	17700	13500	10500

F0	F5	G0	G5	K0	K5	M0
7500°	7000°	5600°	5000°	4000°	3000°	3000°
8550	7000	5800	4860	4370	3460	3240

31. Die Theorie des Nebelleuchtens. Die systematischen Untersuchungen von HUBBLE auf dem Mount Wilson-Observatorium über die Beziehungen der diffusen Nebel zu den in ihnen oder in ihrer Nähe gelegenen Sternen haben

[1] C. H. PAYNE, Stellar Atmospheres. Harv Monogr Nr. 1, S. 33 (1925).

gelehrt, daß zwischen den Spektren der Nebel und der mit ihnen verbundenen Sterne eine so ausgesprochene Korrelation besteht, daß man das Nebelspektrum auf Grund des Spektrums der mit ihm verbundenen Sterne voraussagen kann und umgekehrt. Hat der Nebel ein kontinuierliches Spektrum, so gehören die Sterne den verschiedenen Unterklassen des B-Typus an, jedoch in keinem Fall einem früheren Spektraltypus als B1. Die Sterne, welche mit Nebeln verbunden sind, die Emissionsspektra besitzen, sind durchweg von früherem Spektraltypus als B1. Nimmt man noch die planetarischen Nebel hinzu, so ergibt sich folgende bedeutungsvolle Sequenz, welche nach den Spektren der mit den Nebeln verbundenen Sterne geordnet ist:

Stern	Nebel
Wolf-Rayet-Sterne und Übergangstypen zu Oe5.	Planetarische Nebel (Emissionsspektrum mit kontinuierlichem Untergrund).
Spektraltypen Oe5 bis B0.	Diffuse Nebel mit reinem Emissionsspektrum.
Spektraltypen B1 und später.	Diffuse Nebel mit kontinuierlichem Spektrum.

Dieser enge Zusammenhang zwischen dem Licht der galaktischen Nebel und der mit ihnen verknüpften Sterne legte den Gedanken nahe, daß die Ursache des Leuchtens der Nebel in der Strahlung der Sterne zu suchen ist. Allerdings kann es sich hier nicht um eine einfache Reflexion oder Streuung des Lichtes handeln, da das Nebelspektrum von dem Sternspektrum verschieden ist. Der physikalische Prozeß muß vielmehr in einer Absorption des Sternenlichtes und in einer Reemission diskreter Strahlungen bestehen. Die Aufstellung des allgemeinen Prinzips, daß die Ursache des Leuchtens aller galaktischen Nebel in der Strahlung der mit ihnen verbundenen Sterne liegt, setzt für jeden bekannten hellen Nebel einen oder mehrere passend gelegene Sterne voraus, die das Leuchten des Nebels bewirken. Die planetarischen Nebel, bei denen ein stellarer Kern die allgemeine Regel ist, bieten in dieser Beziehung keine Schwierigkeiten. HUBBLE ist es gelungen, auch für die diffusen Nebel Sterne zu finden, deren Lage ihre Zusammengehörigkeit mit dem Nebel wahrscheinlich erscheinen läßt.

Der Mechanismus der Absorption der stellaren Energiestrahlung durch den Nebel und der nachfolgenden Reemission wurde von MENZEL und ZANSTRA[1] diskutiert. Die Beobachtungsdaten lassen sich durch die Annahme einer photoelektrischen Ionisation der Nebelatome befriedigend darstellen. ZANSTRA betrachtet den idealen Fall, daß der Nebel nur aus Wasserstoff besteht; der anregende Stern verhalte sich wie ein schwarzer Strahler einer bestimmten Temperatur. Durch die Absorption der gesamten kurzwelligen Sternstrahlung jenseits der Grenze der LYMAN-Serie werden die neutralen Wasserstoffatome des Nebels photoelektrisch ionisiert. Bei der Wiedervereinigung der freien Elektronen mit den Atomen werden die kontinuierlichen Spektra an der Grenze der LYMAN- und der BALMER-Serie emittiert, beim Herabfallen des Elektrons von den höheren zu den tieferen Energiestufen die Linien der verschiedenen Wasserstoffserien, darunter auch die im beobachtbaren Bereich gelegene BALMER-Serie. Die Intensität der von den H-Atomen nach der Wiedervereinigung emittierten Energie wird durch photometrische Messungen der monochromatischen Nebelbilder bestimmt. Da die Zahl der absorbierten gleich der Zahl der emittierten Quanten sein muß, gibt die Gesamthelligkeit der BALMER-Serie ein Maß für die Helligkeit im Spektrum des das Nebelleuchten anregenden Sterns unterhalb λ 912 A und damit auch für seine Temperatur.

[1] D. H. MENZEL, Publ Astr Soc Pacific 38, S. 295 (1926); H. ZANSTRA, Ap J 65, S. 50 (1927).

Außer der BALMER-Serie des Wasserstoffs und des anschließenden Grenzkontinuums besitzen die Spektra der diffusen und der planetarischen Nebel eine Reihe von Emissionslinien, die man früher einem hypothetischen Nebulium zuschrieb. Es war eine wissenschaftliche Glanzleistung von BOWEN[1], für die Hauptnebellinien, deren Erzeugung in Laboratoriumslichtquellen bisher nicht gelungen ist, eine Deutung gegeben zu haben, die nicht nur auf die Vorgänge in den Nebeln ein neues Licht wirft, sondern auch vom atomphysikalischen Standpunkt ein hervorragendes Interesse beansprucht. Bei den Hauptnebellinien handelt es sich um sog. verbotene Übergänge aus einem metastabilen Niveau im Spektrum zweifach ionisierten Sauerstoffs. Die Lebensdauer eines metastabilen Zustandes ist sehr viel größer als die eines normalen Zustandes, so daß unter gewöhnlichen Laboratoriumsbedingungen das Atom durch Zusammenstöße schon wieder aus dem metastabilen Zustand herausgebracht wird, ehe es Gelegenheit hat, die verbotene Linie auszusenden. Nach der BOWENschen Hypothese lebt ein völlig ungestörtes metastabiles Atom nicht unendlich lange. Seine Lebensdauer ist dadurch begrenzt, daß es nach einer bestimmten Zeit spontan unter Emission einer verbotenen Linie in einen Zustand kleinerer Energie übergeht. Die Erklärung für das Auftreten der verbotenen Linien in den Nebeln ist in den besonderen physikalischen Bedingungen, vor allem in der geringen Dichte der Nebelmaterie, zu suchen, so daß die metastabilen Ionen bis zum Einsetzen der spontanen Emission ungestört bleiben.

Nach BOWEN können die Hauptnebellinien N_1 und N_2 nicht durch den bisher diskutierten Mechanismus der Wiedervereinigung hervorgerufen sein, da dieser für die planetarischen Nebel eine viel zu hohe Temperatur der Zentralsterne verlangen würde. Bei den diffusen Nebeln kam ZANSTRA durch Anwendung einer angenäherten Methode auf die Hauptnebellinien zu einer Sterntemperatur von etwa 90000°, die mit dem aus der Wasserstoffwiedervereinigung abgeleiteten Wert von etwa 33000° nicht übereinstimmt.

BOWEN und ZANSTRA haben als Hauptmechanismus für die sog. Nebuliumlinien den Elektronenstoß vorgeschlagen, der damit zu einer unabhängigen Temperaturbestimmung des anregenden Sterns verwandt werden kann. Die durch den Mechanismus der photoelektrischen Ionisation freigemachten Elektronen des Wasserstoffs, welche eine gewisse kinetische Energie besitzen, regen die Nebuliumlinien durch Elektronenstoß an, bevor sie rekombinieren. Die Intensität der Nebuliumlinien entspricht der nur von der Temperatur abhängigen Differenz zwischen der Ultraviolettintensität des Sterns unterhalb λ 912 A und der zur Ionisation des Wasserstoffs benötigten Energie; die gesamte Energie der Nebuliumlinien wird durch Messung im Spektrum des Nebels bestimmt.

Die Formel zur Temperaturbestimmung nach dem Mechanismus der photoelektrischen Ionisation und der Wiedervereinigung kann außer auf Wasserstoff auch auf beliebige andere Rekombinationsspektra des Nebels, z. B. He I und He II, angewandt werden; man bekommt dann ebenso viele Temperaturbestimmungen, als es Rekombinationsspektra im Nebel gibt. BERMAN hat für einige planetarische Nebel die Temperatur aus der Intensitätsverteilung im kontinuierlichen Spektrum des Zentralsterns im Anschluß an die Sonne zu ermitteln versucht, so daß auch damit eine Prüfung der Theorie möglich wird.

Die Tabelle 21 enthält für die Zentralsterne von mehreren planetarischen Nebeln die Resultate der nach den drei verschiedenen Methoden ausgeführten Temperaturbestimmungen auf Grund der spektralphotometrischen Messungen

[1] J. S. BOWEN, Publ Astr Soc Pacific 39, S. 295 (1927).

Tabelle 21.

Nebel	Rekombination H	Rekombination He I	Rekombination He II	Elektronenstoß Nebulium	Kontinuierliches Spektrum	Autor
NGC 6572	41000°	34000° bis 41000°		38000°		ZANSTRA
	43500	43500		41500		BERMAN
NGC 6543	37500			35000		ZANSTRA
	36100	36400		32000	30000°	BERMAN
I C 4593	25000			24000	25000	,,
NGC 6826	26500	29500		27000	25000	,,
NGC 7009	40000	39000	70000°	40000	40000	,,
NGC 7662	43000		81000	51000		,,
NGC 7027	>52000	>40000	>86000	>53000		,,

von ZANSTRA und BERMAN[1]. Die Temperaturen aus H, He I und He II stimmen gut untereinander überein, was für den Mechanismus der Wiedervereinigung spricht; noch wichtiger ist, daß für die einzelnen Nebel die Temperaturen aus der Wiedervereinigung von H, He I und He II und des Elektronenstoßes nahezu einander gleich sind. Die beiden Temperaturbestimmungen sind experimentell ganz unabhängig: die eine beruht auf einer Abzählung von Quanten, gedeutet als herrührend vom ultravioletten Sternlicht, die andere auf einer Bestimmung der Energie der Nebuliumlinien, die von den Photoelektronen angeregt werden, welche dieses Sternlicht erzeugt.

Tabelle 22.

N. G. C.	m_*	m_n	T
6445	19	10,4	140000°
1952	15,9	8,4	100000
2438	16,6	9,8	85000
650,1	16,6	9,9	85000
6853	13,6	7,3	75000
6818	14,9	8,8	70000
6720	14,7	8,8	70000
3587	14,3	9,4	55000
3242	11,7	7,1	55000
7009	11,7	7,2	50000
7662	12,7	8,4	50000
6905	14,5	10,7	45000
6309	14,1	10,8	40000
6210	11,7	8,5	40000
6543	11,3	8,1	40000
1535	11,6	8,8	38000
4361	12,8	10,1	37000
6826	10,8	8,4	35000
6572	10,5	8,4	34000

ZANSTRA benutzt auch die photographische Helligkeitsdifferenz zwischen Nebel und anregendem Stern zur Temperaturbestimmung nach dem Mechanismus der Wiedervereinigung und fand für die Temperaturen der mit den diffusen Nebeln verbundenen Sterne, geordnet nach dem Spektraltypus[2]:

O 34000° B0 28000° B1 21000°

[1] H. ZANSTRA, Untersuchungen über planetarische Nebel. I. Teil. Der Leuchtprozeß planetarischer Nebel und die Temperatur der Zentralsterne. Z f Astrophys 2, S. 1 (1931); L. BERMAN, A Spectrophotometric Study of Certain Planetary Nebulae. Lick Bull 15, S. 86 (1930).

[2] H. ZANSTRA, An Application of the Quantum Theory to the Luminosity of Diffuse Nebulae. Ap J 65, S. 50 (1927).

Diese Zahlen sind in guter Übereinstimmung mit den Temperaturen, wie sie Miss PAYNE auf Grund der Theorie der thermischen Ionisation in den Sternatmosphären berechnet hat (Ziffer 30), und schließen sich auch eng an die spektralphotometrisch bestimmte Farbtemperaturskala in Ziffer 26 an.

Für die Zentralsterne der planetarischen Nebel, wo die Helligkeitsdifferenzen zwischen Stern und Nebel bis 9^m betragen, liefert der Mechanismus der Wasserstoffwiedervereinigung unwahrscheinlich hohe Temperaturen. Dies Resultat kann nach der vorausgehenden Diskussion nicht überraschen, da in den planetarischen Nebeln nicht wie in den diffusen Nebeln die BALMER-Serie die Hauptrolle spielt, sondern die Nebuliumlinien. Die Temperaturbestimmung nach dem Mechanismus des Elektronenstoßes führt zu genäherten Temperaturen der Zentralsterne, die mit den auf Grund der monochromatischen Nebelbilder erhaltenen befriedigend übereinstimmen. Die Tabelle 22 enthält in der ersten Spalte die NGC-Nummer des planetarischen Nebels, in der zweiten die photographische Größe m_* des Zentralsterns und in der dritten die visuelle Helligkeit m_n des Nebels. In der letzten Spalte der Tabelle 22 steht die von ZANSTRA aus der beobachteten Differenz $d = m_* - m_n$ berechnete Sterntemperatur; bei der Beurteilung der Temperaturwerte ist zu berücksichtigen, daß der Fehler in d sogar 1^m betragen kann, was insbesondere für hohe Temperaturen einer großen Abweichung in T entspricht.

Chapter 4.

Luminosities, Colours, Diameters, Densities, Masses of the Stars.

By

Knut Lundmark-Lund.

With 147 illustrations.

Abbreviations used in this Chapter.

m = apparent magnitude.
M = absolute magnitude at a distance corresponding to 10 parsecs.
$\mathfrak{M}$ = the mass of a star, expressed in that of the Sun as unit.
m_v = apparent visual magnitude.
m_p = apparent photographic magnitude.
m_{pv} = apparent photovisual magnitude.
C = colour index.
S = spectral index.
$\varphi(M)$ = frequency of absolute magnitudes.
$\overline{M}$ = arithmetic mean of absolute magnitudes.
$\underline{M}$ = geometric mean of absolute magnitudes.
ε = mean error of one observation $= \frac{\sigma_x}{\sqrt{n-1}}$.
σ_x = dispersion around the mean of x: $\pm\sqrt{\frac{\Sigma(\bar{x}_i - x_i)^2}{n}} = \pm\sqrt{\frac{\Sigma x_i^2}{n} - \bar{x}_i^2}$.
d = apparent diameter of a star.
T = effective temperature of a star.
n = number of observations or number of objects.
π_c = ratio circumference of circle/diameter.
π = parallax of a star.
ϱ = mean density of a star.
H S = Harvard Scale.
P D = Potsdam Durchmusterung.
B D = Bonner Durchmusterung.
C P D = Cape (photographic) Durchmusterung.
R H P = Revised Harvard Photometry.
C d C = Carte du Ciel.
C D M = Cordoba Durchmusterung.
N H D = New Henry Draper Catalogue.

Introduction. When treating the results of the determinations of stellar magnitudes it is impossible to present them in any other way than the historical. A division of the existing material into ancient and modern results is quite impossible, because there is no such division. Besides, just as the positions of the stars when proper motions are derived, on account of the division of the differences in position with time, with increasing time reach a higher value, thus also the magnitudes when deriving the secular change in the star light reach a higher value according to the time elapsed since they were determined. The extremely important question of the secular increase or decrease of starlight is entirely dependent on direct observations. No astrophysical theory can contribute anything to this problem, unless it has a solid foundation resting on empirical data.

The main problems to be considered in this chapter are the relations of the magnitudes to established scales, the general accuracy in different sets, the relation to colour, the relation between number of stars and magnitudes, etc. An excellent history of visual photometry for stars brighter than $5^m,50$ is given in the valuable memoir by E. ZINNER[1], a work which will be quoted very often in the following. As this work is restricted to bright stars, and does not deal with photographic or other magnitudes such as radiative, photoelectric, and others, it has been considered to be of importance to sum up the general results hitherto reached in stellar photometry. The different determinations are scattered throughout considerable parts of the astronomical literature, and accounts of numerous investigations have been written concerning the general results. The student of photometry will undoubtedly find it troublesome work if he wants to get acquainted with the development within this branch, as well as with the present situation.

In treating the colours of the stars and our knowledge as to the masses, densities, and dimensions of the stars, we have also preferred to use the historical order.

a) Apparent Stellar Magnitudes.

a_1) Estimates of Magnitudes. Uranometries.

1. Definitions of Light-Units. At the 1921 Meeting of the International Commission on Illumination the following definitions were adopted.

1. Luminous Flux is the rate of passage of radiant energy evaluated by reference to the luminous sensation produced by it.

Although luminous flux should be regarded, strictly, as the rate of passage of radiant energy as just defined, it can, nevertheless, be accepted as an entity for the purposes of practical photometry, since the velocity may be regarded as being constant under those conditions.

2. The unit of luminous flux is the Lumen. It is equal to the flux emitted in unit solid angle by a uniform point source of one international candle.

3. Illumination. The illumination at a point of a surface is the density of the luminous flux at that point, or the quotient of the flux by the area of the surface, when the latter is uniformly illuminated.

4. The practical unit of illumination is the Lux. It is the illumination of a surface 1 square metre in area, receiving a uniformly distributed flux of 1 lumen, or the illumination produced at the surface of a sphere having a radius of 1 metre by a uniform point source of 1 international candle situated at its centre.

In view of certain recognised usages illumination may also be expressed in terms of the following units:

Taking the centimetre as the unit of length, the unit of illumination is the lumen per square centimetre; it is known as the "Phot". Taking the foot as the unit of length, the unit of illumination is the lumen per square foot; it is known as the Foot-Candle.

1 Foot-Candle = 10,764 Lux = 1,0764 Milli-Phot.

[1] Helligkeitsverzeichnis von 2373 Sternen bis zur Größe 5,50. Veröffentl d Sternw Bamberg. II (1926).

5. Luminous Intensity (Candle-Power). The luminous intensity in any direction (candle-power) of a point source is the luminous flux per unit solid angle, emitted by that source in that direction (the flux emanating from a source whose dimensions are negligible in comparison with the distance from which it is observed may be considered as coming from a point).

6. The unit of Luminous Intensity (Candle-power) is the International Candle, such as resulted from agreements effected between the three National Standardising Laboratories of France, Great Britain, and the United States in 1909. This unit has been maintained since then by means of incandescent electric lamps in these laboratories, which continue to be entrusted with its maintenance.

Of the definitions adopted at the 1924 meeting of the same commission we quote the following.

7. Brightness. The brightness in a given direction of a surface emitting light is the quotient of the luminous intensity measured in that direction by the area of this surface projected on a plane perpendicular to the direction considered.

8. The unit of brightness is the candle per unit area of surface.

The Lambert is the average brightness of any surface emitting or reflecting 1 lumen per square centimetre or the uniform brightness of a perfectly diffusing surface emitting or reflecting 1 lumen per square centimetre.

For most purposes the Millilambert (0,001 Lambert) is the preferable practical unit.

Brightness expressed in candles per square centimetre may be reduced to Lamberts by multiplying by π_c.

Brightness expressed in candles per square inch may be reduced to Lamberts by multiplying by $\pi_c/6,45$.

C. FABRY has pointed out in his work, "Leçons de Photométrie" (Paris 1924) (not accessible to me), that the most rational method of expressing the light-giving power of a star is in terms of the illumination which it produces at the earth's surface under given conditions of atmospheric absorption[1]. FABRY finds the value of this illumination of a first-magnitude star at the zenith on a clear night to be:

$$8,3 \cdot 10^{-7}\,\text{lux}.$$

Illumination Units.

	Lux	Milliphots	Foot-candles
Lux	1	0,1	0,092903
Milliphots	10	1	0,92903
Foot-candles	10,764	1,0764	1

2. Definition of Magnitude and Colour. The problem of determining the light of the stars involves the problem of measuring the intensities of luminous points. Very few of the stars have an angular diameter equal to or exceeding 0″,010; in fact the main bulk of stars have diameters of an order of magnitude of 0″,001 or less. Thus the starlight will be more strictly parallel than any light from sources in our laboratories. The photometric laws concerning point-sources can be strictly applied in the case of the stars.

The chief photometric quantity to be considered in the following is the intensity or the amount of luminous energy emitted per second. The most common unit of intensity is defined as the intensity of a HEFNER lamp in a horizontal direction, the dimensions of the HEFNER lamp being prescribed by the "Reichsanstalt" at Berlin.

[1] J. W. T. WALSH, Photometry. London 1926.

The light sources to be compared do not in general exhibit the same colour. It is necessary to make the colours as equal as possible when comparing intensities, or rather to reduce all photometric determinations to spectral-photometric measurements. In a number of cases this will not be possible, and then a certain by no means negligible influence due to the colour will prevail in photometric measurements and estimates.

The determinations of the intensities of the stars cannot in general be made directly. What can be measured or estimated is the amount of energy received by our eye. It has been customary to determine the magnitude of the stars, that is a quantity, m, which is related to the intensity i in the following way:

$$i = 10^{a\,m}.$$

According to POGSON's suggestion a is taken as exactly equal to $-0{,}4$ and thus:

$$m = -2{,}5 \log i.$$

The main difficulty in using that formula is that in order to obtain accurate results it is necessary to know which wave lengths are included in the (integrated) intensity. I will here attempt to define the magnitude and colour in a stricter way.

The energy curve of a radiating black body can be written in the form:

$$e(\lambda) = \varphi(\lambda) + A_3\,\varphi^{III}(\lambda) + A_4\,\varphi^{IV}(\lambda) + \cdots,$$

where:

$$\varphi(\lambda) = \frac{C}{\sigma\sqrt{2\pi}}\, e^{-\frac{(\lambda-\lambda_0)^2}{2\sigma^2}},$$

where σ is the dispersion around the mean and λ the wave length. The constants λ_0, σ, C, A_3 (depending on the Skewness), and A_4 (depending on the Excess) can be expressed in terms of the absolute effective temperature T.

According to the law of STEFAN:

$$\int_0^\infty e(\lambda)\, d\lambda = a\,T^4 \quad (a \text{ is a constant}).$$

According to the first law of WIEN, $e(\lambda)$ reaches a maximum value when $\lambda = b/T$, b being a constant, and according to the second law of WIEN $e(\lambda)^{\max} = c\,T^5$, where c is a constant.

If we apply the radiation formula of PLANCK we can write

$$e(\lambda) = \frac{\alpha}{\lambda^5} \cdot \frac{1}{e^{\frac{\beta}{\lambda T}} - 1}.$$

In this way CHARLIER[1] has found that:

$$e(\lambda) = c\,T^5\, e^{-\frac{1}{2\sigma^2}\left(\log\frac{\lambda T}{b}\right)^2}$$

where σ, c, and b are constants independent of T.

In a subsequent paper CHARLIER[2] has pointed out that it will be convenient to use as independent variable $y = \log\lambda$ instead of λ. The energy curve then takes the form:

$$\varepsilon(\lambda) = \varphi(y) + A_3\,\varphi^{III}(y) + A_4\,\varphi^{IV}(y) + \cdots.$$

---

[1] Lund, Meddelanden Ser I No. 41 (1912).

[2] Lund, Meddelanden Ser I No. 55 (1913).

Using the numerical values of LUMMER and PRINGSHEIM, CHARLIER has found the following expression for the energy curve:

$$e(\lambda) = 1{,}184 \cdot 10^{-14}\, T^5\, e^{-\frac{1}{2(0{,}236)^2}\left[\log\frac{\lambda T}{2940}\right]^2}.$$

Now, this curve will be modified when light passes the eye. The $e(\lambda)$ curve has to be multiplied by another curve $e'(\lambda)$ expressing the sensibility of the eye for the different wave lengths. The product of two frequency curves will, as can easily be shown, always be another frequency curve, $\Phi(\lambda)$. Hence:

$$\Phi(\lambda) = e(\lambda) \cdot e'(\lambda).$$

The same applies to any telescope having a certain sensibility curve $e''(\lambda)$. There are several determinations of $e'(\lambda)$, of which I mention the following:

$$e'(\lambda) = 0{,}999\left(\frac{0{,}556}{\lambda}\, e^{1-\frac{0{,}556}{\lambda}}\right)^{200} + 0{,}04\left(\frac{0{,}465}{\lambda}\, e^{1-\frac{0{,}465}{\lambda}}\right)^{400} + 0{,}095\left(\frac{0{,}610}{\lambda}\, e^{1-\frac{0{,}610}{\lambda}}\right)^{1000}.$$

For $e(\lambda)$ the following formula, as adopted by CHARLIER, can be used:

$$e(\log\lambda) = \frac{K}{\alpha\sqrt{2\pi_e}}\, e^{-\frac{(\log\lambda + 0{,}340)^2}{2(\sigma^2+\alpha^2)}}$$

where α is the dispersion of the $e'(\lambda)$ curve, σ the dispersion of the real energy curve, and K is a coefficent expressing the general degree of sensibility of the photometric instrument. CHARLIER[1] has derived the following relation between photometric intensity I and absolute temperature (B_1 being a constant):

$$I = \frac{K a e^{4\log T}}{\sqrt{2\pi_e(\sigma^2+\alpha^2)}}\, e^{-\frac{(\log\lambda - B_1 + \log T)^2}{2(\sigma^2+\alpha^2)}}$$

or, introducing the absolute magnitude $M = -2{,}5\log I$ and $b = \frac{0{,}2}{\text{Mod.}} = 0{,}4605$:

$$bM = \frac{(-0{,}340 - B_1 + \log T)^2}{4(\sigma^2+\alpha^2)} - 2\log T + \text{const}.$$

On account of the form of $\Phi(\lambda)$, it is evident that the value of I or M as integrated from the formula:

$$M = -2{,}5\log\left[\int_{\lambda_1}^{\lambda_2}\Phi(\lambda)\, d\lambda\right]$$

will be dependent on the spectral interval $\lambda_1 - \lambda_2$ used as limits for the integration. This means that the spectral sensitiveness of the eye or of the photometric instruments used has influence on the value of M. This fact explains the existence of the colour equation, that is the fact that objects not having the same colour do not define the same magnitude or intensity scale. In such a way also the PURKINJE effect can be explained.

It will be very difficult to derive accurate numerical values of the constants in the above equations, e.g., in that connecting M and T. At least for the present we have not data extensive and accurate enough for such a calculation. Neither have we enough data of the numerical value of $\Phi(\lambda)$ for different observers or different types of optical instruments. The deductions given here are only intended to help towards a clearer understanding of the intimate relationship between light intensity and colour — a connection very much commented upon, but of which the nature very often does not seem to have been quite understood.

[1] Lund, Meddelanden Ser I No. 67, p. 9 (1915).

3. The Colours of the Stars. If the magnitudes have been determined with regard to two photometric systems, which have a different distribution of spectral energy, systematic deviations between the results will appear. If we restrict the two systems to the wave lengths around λ 4000 and λ 5500, the difference between the intensities of these two spectral regions will be very closely related to the colour expressed in some numerical scale, or to the effective temperature of the stars. The difference between the magnitude as derived from the intensities of photographic rays, m_p, and the magnitude derived from visual rays, m_v, is called the colour index:

$$C = m_p - m_v .$$

A red star appears comparatively faint on photographic plates and thus m_p will be numerically larger than m_v. The value of C varies from $-0^m,4$ for the B stars to $+1^m,5$ for some very red stars (M stars) and reaches, in extremely exceptional cases, $5^m,0$.

Let $e(\lambda)$ be the spectral distribution curve and $e'(\lambda)$ and $e''(\lambda)$ the visibility curves of two optical instruments (eye and camera). The colour index can then be written:

$$C = a + bC'$$

$$C' = -2,5 \left[\log \frac{\int_{\lambda_1}^{\lambda_2} e(\lambda)\, e'(\lambda)\, d\lambda}{\int_{-\infty}^{+\infty} e(\lambda)\, d\lambda} - \log \frac{\int_{\lambda_3}^{\lambda_4} e(\lambda)\, e''(\lambda)\, d\lambda}{\int_{-\infty}^{+\infty} e(\lambda)\, d\lambda} \right].$$

It should thus be possible to compute the colour indices. At present we have not determinations of $e'(\lambda)$ or $e''(\lambda)$ of such accuracy that a theoretical derivation of C is feasible.

When C has a small value, it is termed a colour equation. Compare, e. g., the A and K stars in a certain photometric catalogue. The two groups do not define quite the same magnitude scale and the difference $m_K - m_A = C$ is then said to be the colour equation of the catalogue, and gives a measurement of the deviation of the two scales.

When investigating the colours of the stars it is, of course, desirable to have as large values of C as is possible. This condition is reached when $\frac{\lambda_1 + \lambda_2 - \lambda_3 - \lambda_4}{2}$ is as large as possible, e. g. if measurements of spectral intensities in ultra-violet are compared with radiometric measurements.

4. The Eye as a Photometric Instrument. It has been said by HELMHOLTZ that the eye is a rather imperfect optical instrument. This sentence seems to be somewhat paradoxical, and was certainly not meant as disqualifying the human eye as a physiological instrument. The eye certainly has its limitations and imperfections. It has to be studied as every other instrument, and its peculiarities carefully investigated. Physiology has gained much, especially from the photometric observations of the stars, because these light-points are much more perfect than any that it is possible to realize in the laboratory. Thus the light of the stars, in a far higher degree than any of the terrestrial sources of light, fulfil the condition of arriving to the instrument or human eye in strictly parallel rays.

A point (star) seen by the ordinary eye will not in general be seen as a minute spot. Around a nucleus are seen a number of rays, varying for different individuals from 4 to 9[1]. The star image is not the same for both the eyes and its form varies with the time.

From operations for cataract it has been found that the star image has its origin in the crystal lens. After the operation the rays are not seen any more, but the stars look like round spots. The star image is a product of the complicated refraction in the crystal lens, forming a caustica, which is projected on the cones of the retina.

The conception of star-light is regulated mainly by two factors. One is the size of the effective part of the image on the retina, i. e. the parts of the image reaching above the threshold value. The other is the average value of the incitement within the effective part of the image. This does not increase the intensity of the image of the object, but it increases the irradiation, which according to PLATEAU and G. GÖTHLIN has its origin in the influence of the effective image on the elements bordering it. The explanation is probably that the individual elements on the retina are collected into groups, and from a strong incitement an impression of the same kind follows as when the individual elements are incited.

G. FECHNER[2] has found that the size of the star image is proportional to the intensity of the light. If the impression on the retina is ε and i is the intensity of the light, the equation:

$$\Delta\varepsilon = \frac{\Delta i}{i}$$

will give an approximately correct relation. This formula can be integrated and thus we obtain:

$$\varepsilon - \varepsilon_0 = \log (i - i_0) .$$

On account of a constant excitation of the optic nerves this equation takes the form:

$$\varepsilon' - \varepsilon = \log \frac{i' + i_0}{i + i_0}$$

where i_0 is the "self-luminous" intensity of the eye and i and i' two intensities.

The light of the stars, as we observe it, is also dependent on the intensity of the background. Thus the stars cannot be seen in the daylight. Very much has been written about this question in early stages of astronomical development, because of its importance for the conceptions of the Universe. From observations during moonshine and during solar eclipses it was early recognized that the stars shine in the daytime as well as at night. Then the question arose why the stars could not be seen in the daytime. Already ARISTOTELES and PLINIUS stated that the stars could be seen from the bottom of deep wells or cisterns. Later on, we find a number of accounts as to supposed observations of stars from wells and mines.

In recent times A. KÜHL[3] has made experiments with a tube directed parallel to the large refractor of the Observatory at Munich. Although the head of the observer was kept in a dark cloth no stars could be seen in full daylight.

It may have happened on very rare occasions that stars have been seen in the daytime. Several of the supposed observations very probably refer to planets, but most are entirely illusory.

That part of the retina which is the place for the sharpest vision consists only of cones, then the rods begin to appear and at greater distances these

---

[1] Ögats intryck af fixstjärnorna. Vetenskapen och livet. 1919.

[2] Leipzig, Ges d Wiss Abh 4 (1859).

[3] Über die Leistungsgrenze und Empfindungsstärke unseres Sehorgans bei Abbildung von Fixsternen und Planeten mit und ohne Fernrohr. Dissertation. München 1909.

dominate. In the dark the threshold value of the rods decreases slowly at first, but later on rapidly. The minimum of the threshold value is reached after one hour in darkness.

The apparatus of the cones does not change its sensitiveness with the amount of excitation. With very high intensities blindness occurs, but after a comparatively short time the normal sensitiveness returns. There is a considerable difference in the threshold value for the rods and the cones, in the sense that the latter have the higher value.

The rods are colour-blind and give the light of different wave lengths in different grades of white light. The maximum of intensity for the rods is between 5000—5300 A, whereas for the cones it is between 5800—6000 A. The origin of the PURKINJE phenomenon certainly lies in this fact. When the intensity of light of different wave lengths decreases, the colour becomes more milky, on account of the white-sensitivity of the rods, and at last the colour disappears entirely and the object appears white. When the intensity decreases the impression of colour corresponding to the mean wave length 5900 A is preserved longest.

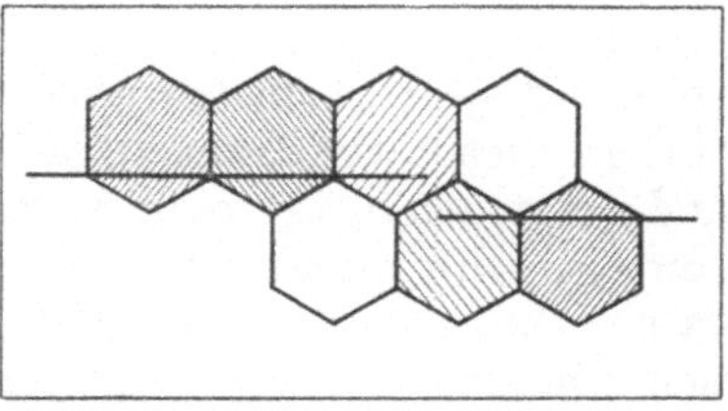

Fig. 1. The rods of the human eye.

A certain number of rods have the same connection with the centre of vision, whereas each cone is independently connected with this centre. The law of the individual action of the rods has a certain limitation, on account of the fact that neighbouring rods also react to a certain extent to a sensation, not directly applied to them.

The rods are of hexagonal shape and in the mean their diameter is 3 μ. In order to see two luminous points separately it is necessary that at least one unexcited rod is situated between the excited ones. This gives as a lower limit of separation, 40″, which is in good agreement with experience. Individual peculiarities in the structure of a certain object disappear when it is viewed under an angle smaller than 40″. Thus it might happen in a few exceptional cases that it is possible to observe the crescent form of Venus since the diameter of the planet sometimes reaches 60″.

According to theory the radius of an aberration-circle of a luminous point on the retina will be:

$$\varrho = f\,\frac{1{,}22\,\lambda}{2p}\cdot 206265.$$

The value of f is 15 mm; λ is 0,589 and p, the radius of the diaphragm, is 2,5 mm thus $\varrho = 2{,}15\ \mu$. We find that the diameter of the aberration-circle is slightly larger than the mean diameter of the rods. Let the centre of the rod coincide with the centre of the circle. The amount of energy within the first diffraction disc is then:

$$4\,A^2\,\lambda^2\,p_0^2\int_0^{2\pi} d\psi\int_0^{i}\frac{y_1^2(i)}{i}\,d\,i\,,$$

where the variable i is equal to $\frac{2\,\pi_c\,p}{\lambda f}$ if ψ is a linear element on the retina measured in mm and $y_1^2(i)$ a Besselian function.

The amount of energy falling on the rod is equal to the same expression integrated between 0 and 2π or 0 and $\frac{30}{43}\,i'$, where $i' = 1{,}220\,\pi_c$ is the radius of the diffraction disc in "optical units". Evaluating the integral, A. KÜHL has concluded that a rod receives at most 92 per cent of the energy emitted by a luminous point.

The size of the rods might be the result of natural selection; anyhow, our eye is wonderfully adopted for the observations of very faint luminous points or extremely small surfaces.

5. The General Aim of Stellar Photometry. Knowledge of the intensities in certain spectral regions is a problem of fundamental importance for astrophysics. As soon as the distances are known, the light power or its equivalent, the absolute magnitude, can be derived. This quantity depends on the surface temperature and the dimensions of the star. It seems that the absolute magnitude is related to the mass in some comparatively simple way and thus we find an indirect method of approaching this very important stellar characteristic which, in general, cannot be determined directly.

Stellar photometry also gives important contributions to the problem of determining the general structure of the Universe. Two of the coordinates defining the position of the stars in space can be found with great accuracy viz. the right ascension and the declination. The third coordinate, the radius vector, or the distance from our Sun, can now be approximated for several thousands of the stars. In spite of all the important advances within this branch we cannot in this way reach the distances of the multitudes of stars composing the Milky Way and situated at distances of ten thousand parsecs or more. Here stellar photometry comes in with valuable assistance. The problem involves the solution of a special case of the fundamental equation in stellar statistics. Suppose we have n independent characters, observed to have the apparent intensities $i_1, i_2, \ldots, i_n$, corresponding to the absolute intensities, $J_1, J_2, \ldots, J_n$, and let the two sets of variables depend on the distance r, in some known and simple way, $i_1 = f_1(r, J)$, etc. Then we have:

$$a(i_1, i_2, \ldots i_n) = \omega \int_0^\infty D(r)\, r^2 \frac{\partial J}{\partial i_1} \cdot \frac{\partial J_2}{\partial i_2} \cdots \frac{\partial J_n}{\partial i_n} \varphi_1(J_1)\, \varphi_2(J_2) \cdots \varphi_n(J_n)\, dr\,.$$

In this equation, $a(i_1, i_2, \ldots, i_n)$ is the frequency of stars within the intervals $di_1, di_2, \ldots, di_n$, and $D(r)$ is the number of stars in unit space (density-function); the functions $\varphi(J)$ express the frequency-distribution of the n independent characters, and ω is a certain solid angle equal to the part of the sky considered. If the equation is restricted to the case of apparent and absolute light intensities of the stars, we have the relation:

$$J = k i^{-2}$$

which leads to the well-known equation giving a relation between the spatial density of the stars, the distance, the luminosity function, and the numbers of stars of a certain intensity. A very important task of stellar photometry is to determine the form or properties of $a(i)$ or $a(m)$, and in order to get accurate results the underlying photometric scale must be very accurately established.

A third problem of much importance is to determine the changes in stellar brightness. Much work has already been performed within the field of variable stars with a comparatively short period. In cases of periods amounting to centuries or more we have not observations extensive enough to investigate the nature of the variations.

In this connection also the problem of a secular change in the light of the stars presents itself. There are two possible causes for such a change. If the stars at first become brighter and cooler with increasing age, we can expect that the mean brightness will undergo changes. The motion of the Sun through space will change its situation with respect to the nearest stars. In the course of time some of these become apparently fainter and others grow brighter.

Although the effects are small, it seems quite possible that in the future a photometric determination of the elements of the solar apex will be possible.

Important problems are met with also in the branch of eclipsing variables, which stars present no real light variation as far as is known, but which have been included by tradition among the variable stars. The nearness of the double-star components to each other causes changes in their figure, and the light is influenced both on account of reflexion and on account of excitation. When the photometric data of the eclipsing stars have reached an accuracy of the order of $0^{m},001$, the problem of deriving their orbits will have a foundation nearly as solid as the classical problem of the determination of orbital elements.

6. Photometric Parallaxes. Accurate photometry of the stars will also give a means of deriving parallaxes. Suppose that we consider so short an element of time that the absolute magnitude of a star is invariable and suppose that its motion in space can be taken as equal to its motion in the line of sight. If m_1 and m_2 are the apparent magnitudes at two epochs t_1 and t_2 and if further V is the radial velocity, and π_1 and π_2 the parallaxes, we have the relations:

$$\pi_1 = \pi_2\, 10^{0,2\,(m_2 - m_1)},$$

$$\frac{1}{\pi_1} - \frac{1}{\pi_2} = 0,212 \sin 1''\, V\,(t_1 - t_2)$$

which can easily be improved to correspond to the actual orbital motion of the star. Using numerical data it is found that in the case of BARNARD's proper motion star its motion during a time of 450 years will cause a change of $0^{m},20$ in the apparent magnitude. Of course, this is a favourable case, but still it shows that it will be possible to determine photometric parallaxes some centuries from now, even if our photometric methods of today should not be essentially improved.

7. The Oldest Conceptions of Stellar Magnitude. The charting of the stars can be traced back to primitive ages. On the rocks of some parts of Scandinavia

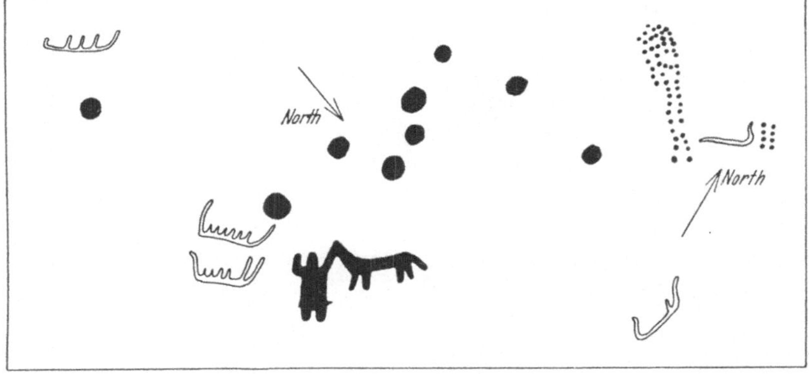

Fig. 2. Rock engraving from Bohuslän, Sweden, showing the constellation Cassiopeia.

and especially in the county of Bohuslän in Sweden, a number of rock-sculptures from the bronze age have been found[1]. These show people, animals, boats, and primitive arms. Among the figures are scattered a number of round cavities, which certainly have had some ritual importance. As a rule these cavities (in Swedish "offerskålar", that is sacrificial cups) are distributed without any certain plan. But in several cases the outlines of some of the well-known constellations

[1] A description is found in the work, Hällristningar från Bohuslän (Glyphes des Rochers [Suède]) samlade och utgivna av L. Baltzer med ett förord av V. Rydberg. Göteborg (1881–1908).

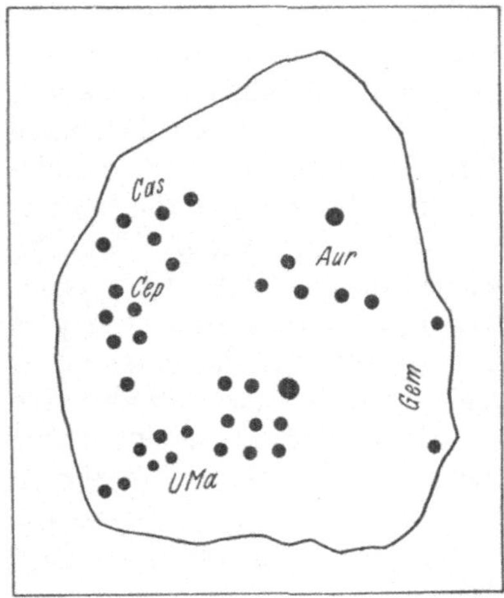

Fig. 3. Stone-engraving from Venslev, Danemark, showing a number of northern constellations.

can be traced, suggesting that the cavities have originally been designed to form rough "charts" of stellar groups. This theory is confirmed by several facts. To mention only one, the figures of animals and men can be identified with the signs of the Zodiac[1].

The magnitudes of the stars of the rock-sculptures are as a rule not discernible. This is to be expected, when we remember that the sculptures certainly had to be made in the day-time. But in a few cases we find that the stars are evidently of unequal size in such a way that one must admit the possibility that primitive man had reached the conception of stellar magnitude. On the sculpture given in fig. 5 there is a decided difference between MIZAR and ALCOR, and it is hardly possible to assume this to be due to chance. Other documents are known to exist from

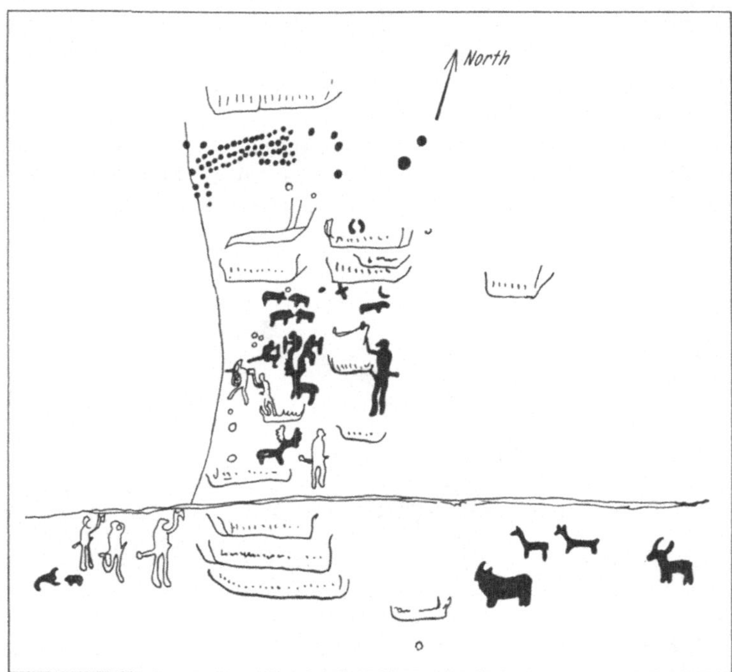

Fig. 4. Rock-engraving from Bohuslän showing at right the Big Dipper. This engraving is of a peculiar interest because a number of constellation figures can be indentified from the animals and men represented on the drawing. Very interesting is the occurrence of the Scorpion in the left lower corner.

the stone age, where the difference in magnitude between these two stars is evidently noted.

[1] See SCHOENFELD, La Nature 1920.

The oldest observational documents we possess from the civilized ages are the observations by the Chinese. These have proved to be very valuable, as they extend our knowledge regarding Comets and Novae to a long time ago. As for the ordinary stars there seem to be no observations preserved that give magnitudes, nor even a rough classification of the stars with respect to their brightness. In the work Pou-Tienho from the 7th century A.D., Tan Yuen-Tse has described 283 constellations containing 1464 stars. The charts from that time, which show a somewhat smaller number of stars, do not distinguish between different magnitudes.

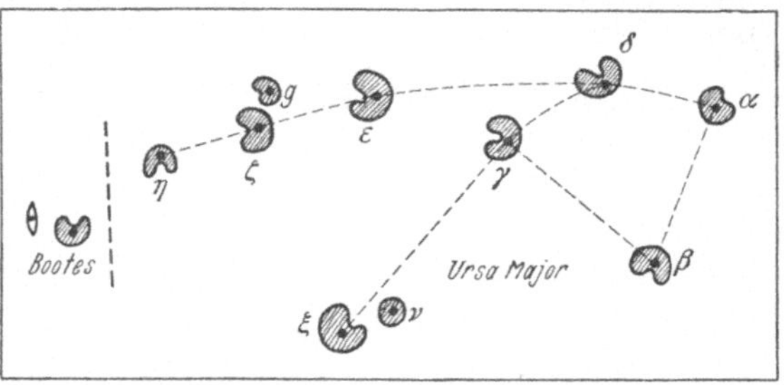

Fig. 5. Representation of stars in Ursa Major and Bootes on an amulet from the stone-age. The different size of Mizar and Alcor is noteworthy. The form of the Big Dipper suggests a rather high age for the amulet.

8. The Babylonian Observations of Stars. There exists a comparatively rich material for a reconstruction of the star sky of the Babylonians. A number of star lists have been published mainly by King in the work, "Cuneiform Texts from Babylonian Tablets in the British Museum", London 1896 and following years. The most important of the Babylonian star lists have been discussed by E. F. Weidner in his Handbuch der Babylonischen Astronomie and, besides, a summary is given of the knowledge of the Babylonian Star Sky.

We do not need to treat here the very much discussed problem of the age of Babylonian astronomy. It seems that Weidner has presented good reasons for assuming that the star catalogue giving 25 differences in right ascension between pairs of stars originated about 2500 B. C.

In the Babylonian star lists "faint stars" seem to be spoken about now and then, but there is so far no actual division into classes, according to the brightness (magnitudes). As far as our knowledge now goes it is evident that the conception "stellar magnitude" has its origin in Alexandria.

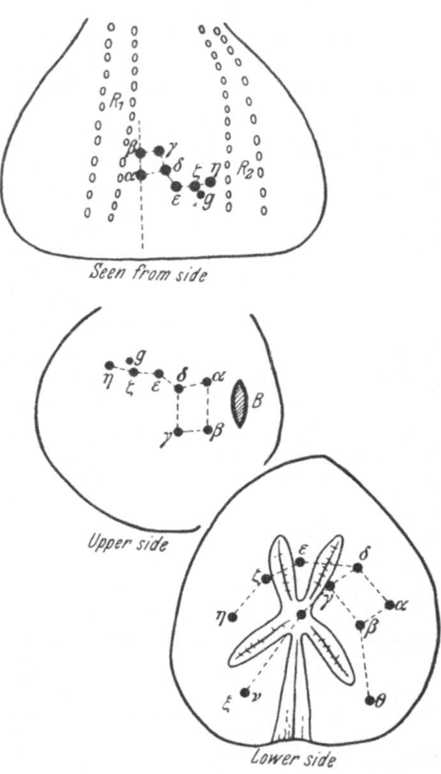

Fig. 6. Representation of the Big Dipper on a fossil and silicified seeurchin (Echinus) from the stone-age (Baudouin).

9. The Egyptian Observations. A number of representations of the constellations have been discovered (Brugsch, Thesaurus Inscriptionum Aegypticarum). Rebekka Biegel has reconstructed in her thesis[1] the Egyptian star

[1] Zürich (1924).

sky and made a successful identification of the constellations. In several cases images of stars are given in connection with representations of the constellations. We reproduce here a couple of well-known representations of stars. These are given without differentiation as regards their magnitudes. In

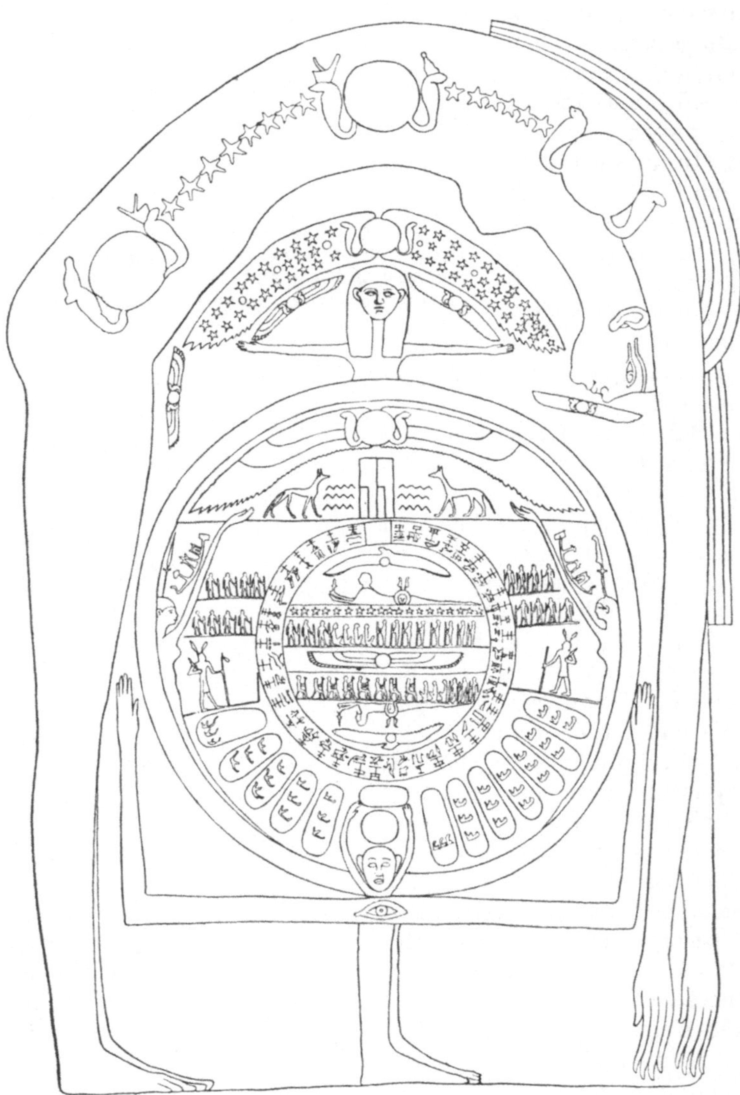

Fig. 7. Egyptian conception of the starry Heaven. The winged Sun might be a representation of the Sun with its corona.

some other cases several stars are made larger than the others, but evidently without taking into consideration their actual magnitudes.

10. HIPPARCHOS and PTOLEMAIOS. The oldest observational document we possess that gives more of critical information on the relative light of the stars is the catalogue of stars contained in the 7th and 8th book of KLAUDIOS PTOLEMAIOS's *Μεγίστη* or *Μεγάλη Σύνταξις*, often called "The Almagest" from the Arabian translation of the title, Al Magesti.

This catalogue is the first ancient document known that gives a description of the Heavens of sufficient exactness to admit comparison with modern observations. The positions of the catalogue are no longer of any use for the derivation of the proper motions of the stars on account of the low accuracy in the position-determinations. The situation with regard to the magnitudes is somewhat different. Although the estimates of PTOLEMAIOS are not very accurate in individual cases, they are, as a whole, of extreme value on account of the early epoch at which they were observed. The very important question whether there is a secular change in the light of the stars can be tested, to a certain degree, by using the magnitudes in the Almagest.

The conception "stellar magnitude" has, in fact, its origin in the catalogue of PTOLEMAIOS. As regards the mode of observation we have some information. It is not incredible that in antique times the observers thought that the magnitudes expressed the real size of the stars. This was very natural, as it seems that it was sometimes assumed that the stars did not shine by their own light. Such an opinion is expressed, e. g., by IBN AL HAITHAM in the 11th century.

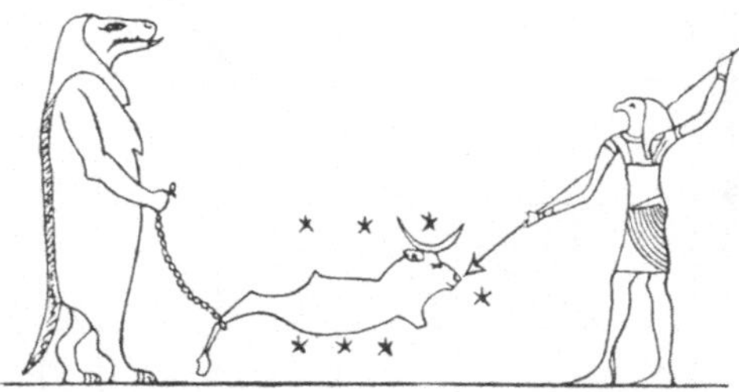

Fig. 8. Egyptian representation from the Ptolemean time in the Temple at Dendera of a stellar constellation.

The history of the Almagest is of much interest. The oldest of the Greek manuscripts known are from the 9th century. The work was translated from the Greek into Arabic by AL MAMUN about A. D. 827. The Arabic manuscripts are very valuable, as the errors of the scribe are of a different kind in the two languages. A comparison of the manuscripts will give a means for a reconstruction of the lost original catalogue. Now, the problem is complicated on account of the fact that the two independent translations known are written in the character called Neskhi. This form of Arabic writing was invented at the beginning of the 10th century. Thus the original translation of AL MAMUN was probably in Cufic Arabic, and rewriting this into Neskhi would certainly provide possibilities for many errors.

The immense and difficult work of collating and translating the different manuscripts of the Almagest has been performed by C. H. F. PETERS and E. B. KNOBEL[1]. In their work historical notes have been given and the "sources of errors" in the Greek and Arabic have also been discussed. The Greek text has been edited by J. L. HEIBERG, and a translation of this edition into German has been published by K. MANITIUS.

In PETERS's and KNOBEL's edition the magnitudes of PTOLEMAIOS are given according to three Greek (Paris 2389, Vatican 1594, Venice 313) and three Arabic manuscripts (BM. Reg. 16, BM. 7475, and Bod 369). Besides these the magnitudes adopted by MANITIUS from several Greek manuscripts examined by him are given. The agreement is very good, and it is likely that the original catalogue is satisfactorily reconstructed with regard to its magnitudes.

Much has been written about the relation between the star catalogue of PTOLEMAIOS and the lost one by HIPPARCHOS. The astronomical opinion generally held has been inclined to believe that PTOLEMAIOS has borrowed all his material from HIPPARCHOS. According to DELAMBRE and others the proof should be that

[1] PTOLEMAIOS's Catalogue of Stars. A Revision of the Almagest. Carnegie Institution of Washington, Publication No. 86. Washington 1915.

PTOLEMAIOS's longitudes are on an average a degree too small, because he assumed the constant of the precession to be 36″ instead of 50″. The epoch of the catalogue of PTOLEMAIOS is A. D. 137, that of HIPPARCHOS 129 B. C. The difference in the precession constant therefore explains the error in longitude. This simple explanation does not seem to be the right one.

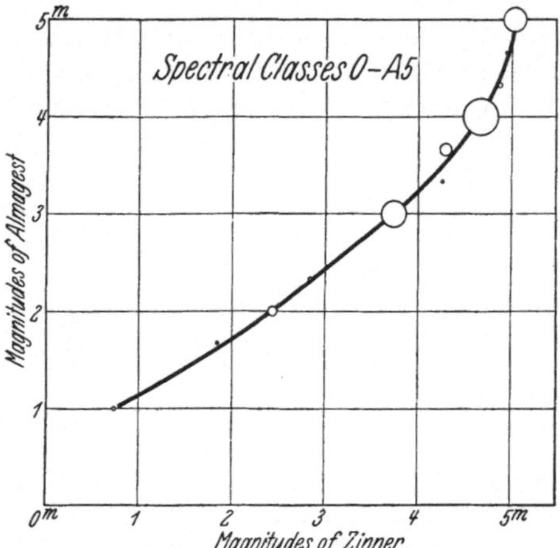

Fig. 9. Curve for reducing the magnitudes of white stars in Almagest to magnitudes in Zinner's system. Zinner's magnitudes are practically on the scale of PD. The size of the circles in this and a number of following curves are proportionate to the number of the stars in each group and thus indicates the weights to be assigned to the groups at the solution for the curve.

F. BOLL has investigated a Greek list of the constellations attributed to HIPPARCHOS. It seems that this list gives the number of stars observed by HIPPARCHOS. From calculations and comparisons with PTOLEMAIOS it seems very probable that the star catalogue of HIPPARCHOS did not contain more than about 850 stars. If PTOLEMAIOS has made use of these 850 star places, then at least the remaining 175 stars must have been observed by himself or taken from some other source.

DREYER [MN 77, p. 528 (1917); 78, p. 343 (1918)] has pointed out that it may be possible to distinguish between the two (main) sources. In the Almagest the positions are given to fractions of a degree, the fractions being generally multiples of $^1/_6$, suggesting that the instrument has been divided into 10′. But the fractions $^1/_4$ and $^3/_4$ also occur, as if a

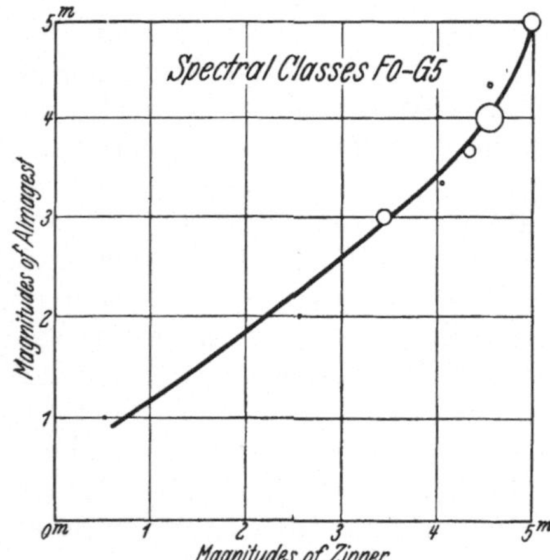

Fig. 10. Curve for reducing the magnitudes of yellow stars in Almagest to the magnitudes in Zinner's system.

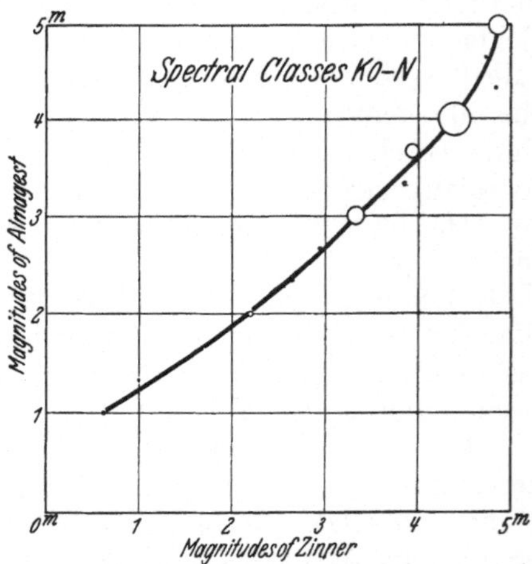

Fig. 11. Curve for reducing the magnitudes of red stars in Almagest to the magnitudes in Zinner's system.

different instrument had been used. DREYER finds that 142 stars distributed fairly well throughout the constellations have latitudes with fractions $^1/_4$ or $^3/_4$, while only three stars, all in Virgo, have their longitude distinguished in the additional catalogue, because of the relations: $^1/_2 = {}^2/_4 = {}^3/_6$, and $1 = {}^4/_4 = {}^6/_6$. Thus the number added to the earlier catalogue might very well have been 175.

It is also interesting to note that of the 145 stars two-thirds are of magnitudes $4^m—6^m$, and there is only one star of the second magnitude, POLLUX, which may have the honour of being a connecting link between the earlier catalogue and the additional stars.

DREYER concludes that we have no reason to believe that PTOLEMAIOS borrowed the additional star places from MENELAOS. Neither does it seem likely that the Arabs have possessed any knowledge about observations known to him, other than those quoted in the Syntaxis. As MENELAOS is the only name after HIPPARCHOS of whom it has been asserted that he supplied PTOLEMAIOS with star places, it seems quite possible that PTOLEMAIOS may at least have observed the additional stars not observed by HIPPARCHOS.

Another very interesting investigation as to the origin of the oldest star catalogues has recently been published by H. VOGT[1]. Only two of the contributions of HIPPARCHOS have been preserved to our time. One is his comments on notes to the celestial phenomena by ARATOS and EUDOXOS[2] and the other a list of 24 time-stars, which follow each other at intervals of exactly one hour[3]. SCHJELLERUP has shown that these stars were so well chosen, that they gave the time every hour, within a minute, which is a fine testimony to the accuracy reached by HIPPARCHOS. The comments to ARATOS's and EUDOXOS's Phaenomena are accompanied by descriptions of the constellations containing numerous data of the coordinates of the stars. The α and δ are not given, but instead the point where the declination circle reaches the Ecliptic. Data are also given of the rising and setting of some of the stars in the constellations together with the point of the ecliptic rising or culminating at the same time as a certain star. By the aid of spherical trigonometry VOGT[1] has been able to reconstruct the values of the latitudes and longitudes for 122 of the stars observed by HIPPARCHOS. More coordinates could not be derived, if certain assumptions are to be avoided. It will not be possible to follow the interesting paper by VOGT in detail. He shows that PTOLEMAIOS from his starting point could not obtain another value for the precession constant than 36″. The detailed comparison between the coordinates of the two observers, made by VOGT in his paper, shows that it is not likely that PTOLEMAIOS borrowed even the coordinates for the 850 stars observed by HIPPARCHOS. The Almagest gives no reason to doubt the statement of PTOLEMAIOS that his catalogue is a product of his own observations. Evidence is now very strongly in favour of placing PTOLEMAIOS, who has for more than a century been considered at best a mere compilator and at worst a plagiarist, among the first-rank observers of all ages!

As far as we know, HIPPARCHOS did not introduce the conception of magnitudes. He classified the stars with regard to their lustre in the following categories:

		Magn.
Brilliant light	λαμπρότατος	$1^m — 3^m$
Second degree	ἐκφάνης	4 — 5
Faint	ἐκφανέςτατος	6

PTOLEMAIOS introduced the magnitudes 1—6 and besides the letter μ or ἐ is added in several cases, as an abbreviation for the word μείζων or ἐλάσσων, meaning more or less. Generally these grades have been assumed as corresponding

[1] Versuch einer Wiederherstellung von HIPPARCH's Fixsternverzeichnis. A N 224, p. 17 (1925).

[2] Hipparchi in Arati et Eudoxi Phaenomena Commentariorum libri tres. Griechisch und Deutsch von KARL MANITIUS. Leipzig 1894.

[3] COPERNICUS: 1, p. 25 (1881).

to half or third intervals of magnitudes, which is not correct. According to PTOLEMAIOS's own work there are 15 stars of 1^m, 45 of 2^m, 208 of 3^m, 474 of 4^m, 217 of 5^m, and 49 of 6^m.

The question of the relation between the magnitudes in the Almagest and modern determinations has been discussed by PEIRCE and PICKERING, and more recently by ZINNER and by the present writer.

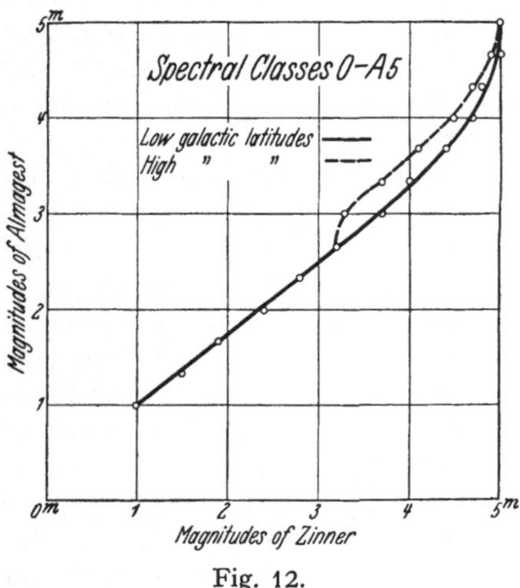

Fig. 12.

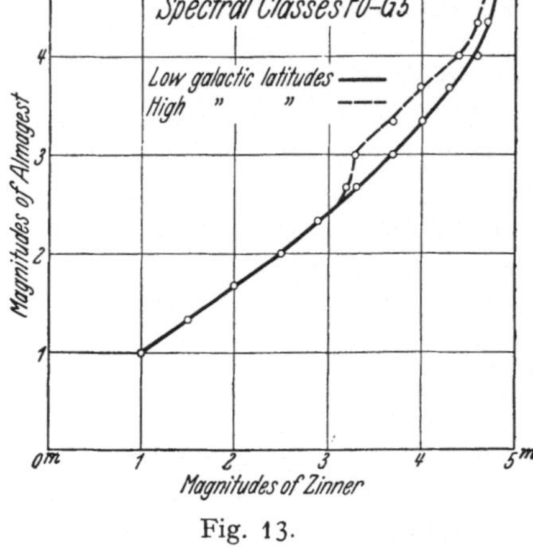

Fig. 13.

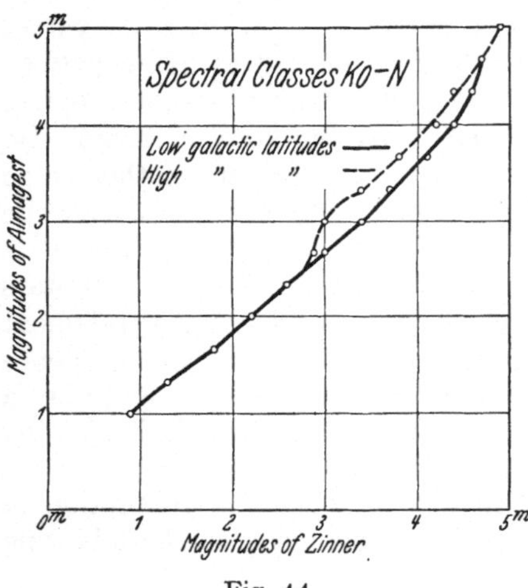

Fig. 14.

Fig. 12—14. Curves showing the influence of the Milky Way upon the estimates of magnitudes in Almagest.

11. AL SÛFI. The Middle Ages. A very important investigation concerning the magnitudes of the stars was made in the 10th century by the Persian astronomer AL SÛFI. The astronomical literature of the Middle Ages does not contain many contributions regarding the positions of stars, and thus also such by-products as star magnitudes are missing even in the work of the Arabs. The catalogue of AL SÛFI possesses unique interest on account of its character as a revision of the Almagest. The main purpose was to obtain a description of the sky, and much care was given to the estimates of the magnitudes. Nearly all the 1022 stars of PTOLEMAIOS were identified in the sky and carefully examined. Several stars are added, so that the total number of stars is 1151.

This valuable document on stellar magnitudes has been translated by the Danish scholar, H. C. F. C. SCHJELLERUP. For the translation two manuscripts were used, one is in the R. Library at Copenhagen, signed LXXXIII, and written in A. D. 1601. The other manuscript is preserved in Leningrad. SCHJELLERUP has given a literal translation, and a study of this leaves a strong impression of the critical mind and the extreme carefulness of AL SÛFI. In fact, one might mention AL SÛFI as being one of the predecessors of those, who later on used the method of estimating magnitudes by means of a comparison of different objects. It seems though that AL SÛFI has restricted the method to comparing stars having the same brightness. When the magnitudes differed by a considerable amount

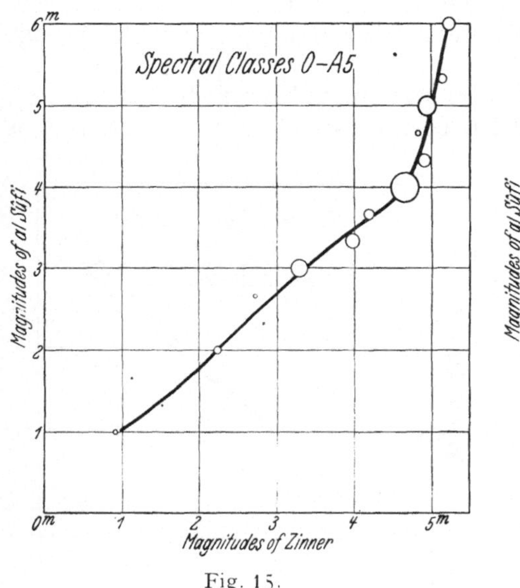

Fig. 15.

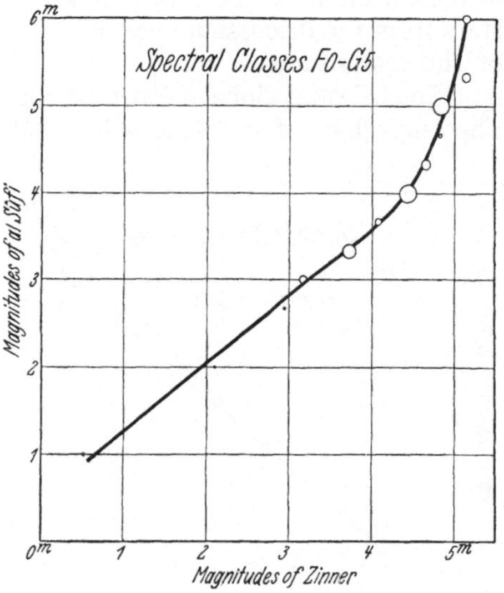

Fig. 16.

he probably worked with some magnitude scale in his mind[1].

The famous astronomer, ULUGH BEGH at Samarkand, did not observe the magnitude of the stars. The values he uses in his catalogue[2] are simply copied from the magnitudes in SÛFI's description. According to E. B. KNOBEL the magnitudes of ULUGH BEGH show "many errors of the copyist, and these variants have not seemed to be of importance, seeing that SÛFI in his text describes the magnitudes in words". The differences between ULUGH BEGH's magnitudes in KNOBEL's edition and those of AL SÛFI are statistically unimportant.

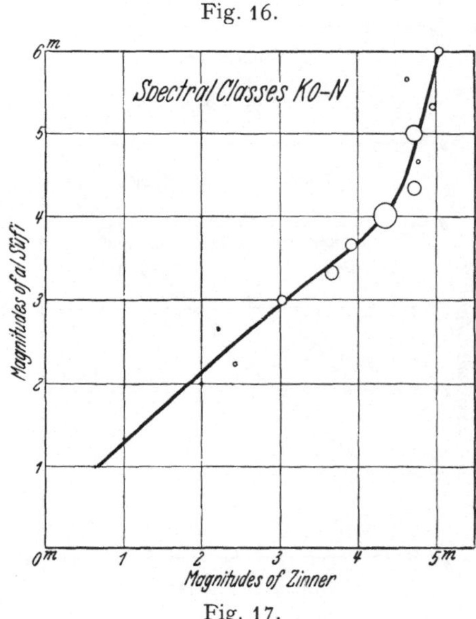

Fig. 17.

Fig. 15—17. Curves for reducing the white, yellow, and red stars in AL SÛFI's catalogue into magnitudes in ZINNER's system.

Of the star catalogues from the Middle Ages we will now briefly consider the one contained in the Alfonsine Tables. Not very much was known about this work until the publication of the great work "Libros del Saber de astronomia del REY DON ALFONSO X de Castilla" ed. by MANUEL RICO Y SINOBAS (Madrid 1863—67). The catalogue of stars seems to contain positions and magni-

[1] Description des étoiles fixes par . . . ABD-AL RAHMAN-AL-SÛFI avec des notes par H. C. F. C. SCHJELLERUP. St. Petersburg 1874.

[2] Wash Carn Inst Publ No. 250 (1917).

tudes taken from AL SÛFI's description, though dropping the sub-classes of the latter. Also some stars not observed by PTOLEMAIOS (or AL SÛFI) are mentioned. It does not seem possible to decide if these give original values or not. On account of their limited number, and the low accuracy of the data, the exclusion of these stars from the discussion does not materially detract from our general knowledge of the magnitudes.

The celestial globe of PTOLEMAIOS which was still in Cairo in 1043 is now lost. The only globe of scientific value that has been preserved from classical time is

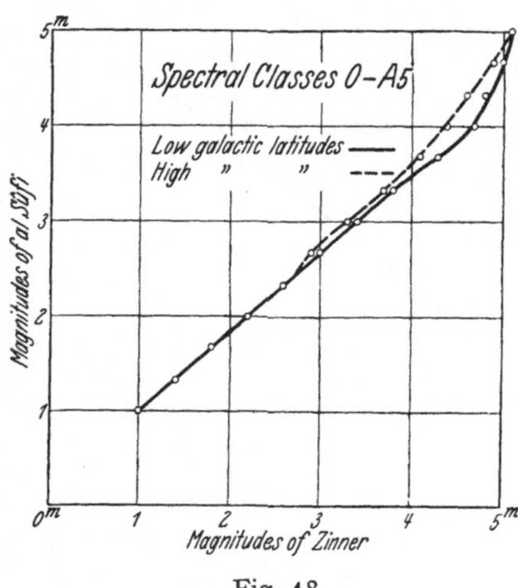

Fig. 18.

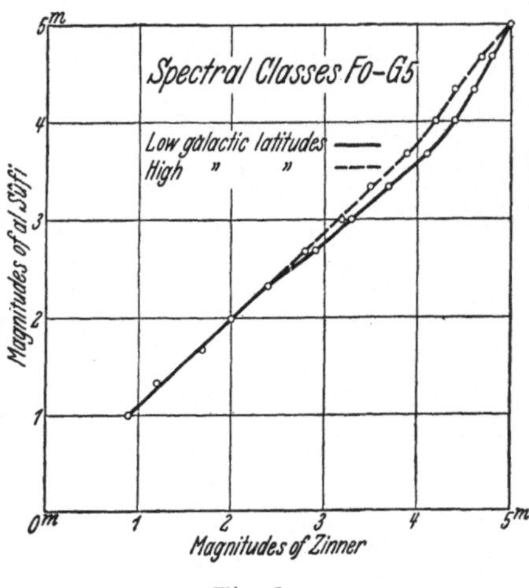

Fig. 19.

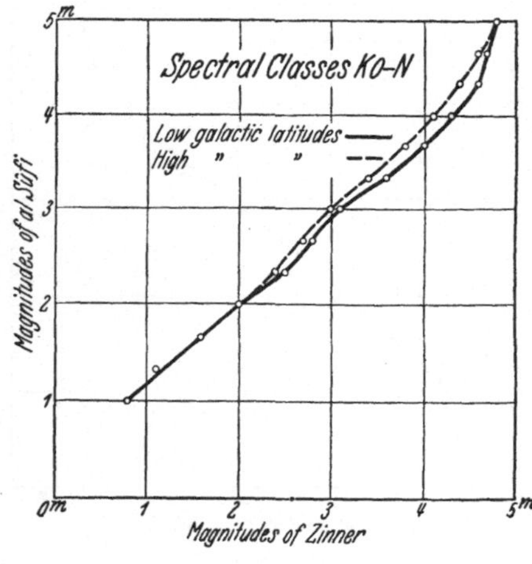

Fig. 20.

Fig. 18—20. Curves showing the influence of the Milky Way upon the estimates of magnitudes in AL SÛFI's catalogue of stars.

the one shown in the Farnesian Atlas (Atlas Farnese), now in the Museum of Naples. According to G. THIELE[1] there is no doubt that the globe is a copy of one of the globes of HIPPARCHOS. It is to be regretted that it does not give any stars, but only the figures of the constellations, and thus is of no use for our purpose.

From the Middle Ages there are a number of representations of the constellations in accordance with the descriptions of these as given by ARATOS. The most important of these celestial atlases is the manuscript from the 9th century known as Codex Vossianus Lat. 0 to 79, now in the possession of the University Library at Leyden. The artistic value of these pictures of the constellations is certainly very high. Stars are also shown, but evidently no care was taken to represent the magnitudes accurately. Neither do other medieval representations of the constellations seem to give any contribution to a knowledge of the star magnitudes.

[1] Antike Himmelsbilder mit Forschungen zu Hipparchos, Aratos und seinen Fortsetzern und Beiträgen zur Kunstgeschichte des Sternhimmels. Berlin 1898.

AL BATTANI and other astronomers of the Middle Ages reproduce the magnitudes of PTOLEMAIOS without adding any actual determinations of their own. In only three cases in AL BATTANI's catalogue do deviations occur in the magnitudes which may be new estimates.

12. TYCHO BRAHE. It was certainly the appearance of the famous Nova B Cassiopeiae that inspired TYCHO BRAHE to observe the positions and magnitudes of the stars. Most of his observations for determining accurate positions of stars were made on Hven before the end of 1592, and the results are contained in a catalogue of 777 stars for the equinox of 1600, printed in his work Progymnasmata[1]. In 1595, his observations of fixed stars were resumed in order to bring the number in the catalogue up to 1000[2]. Several of the observations were carried out immediately before TYCHO's leaving Hven. It is clear that the hurried way in which the data had to be collected did not place the star catalogue on a level with the best that could be produced from Uraniborg.

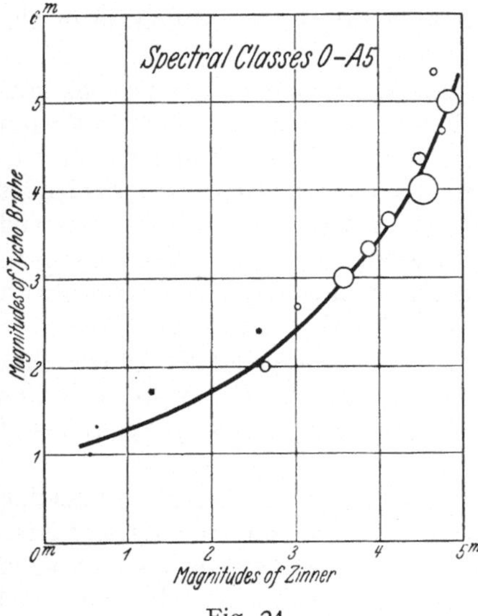

Fig. 21.

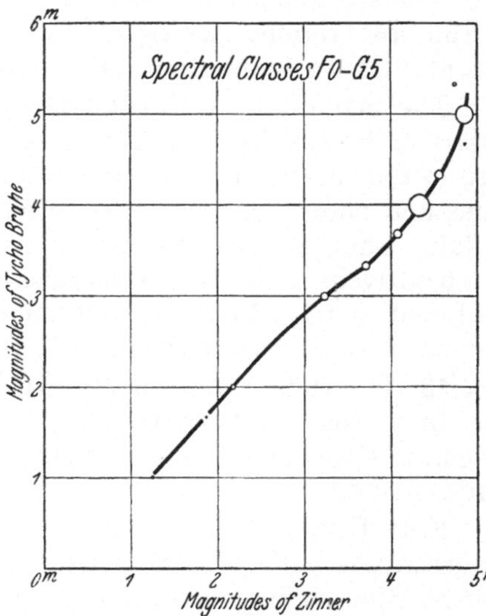

Fig. 22.

The second catalogue was distributed in a number of manuscript copies, and after TYCHO's death was published by KEPLER in his Tabulae Rudolphinae In this edition several errors were corrected by KEPLER, and an improved description given of the stars in Ophiuchus, observed in connection with the apparition of the Nova of 1604.

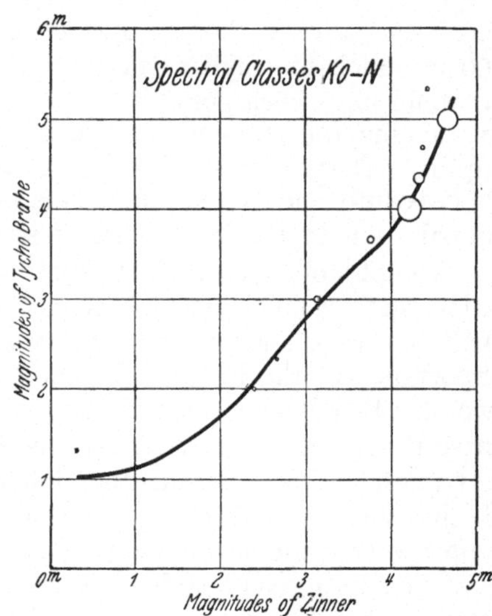

Fig. 23.

Fig. 21—23. Curves for reducing the magnitudes in TYCHO BRAHE's first catalogue (in Progymnasmata; 777 stars) into the magnitudes of ZINNER.

[1] Opera Omnia 2, p. 258, Havniae (1915).

[2] Opera Omnia 3, p. 344 (1916).

The two catalogues of TYCHO BRAHE have been edited by DREYER, and in the following discussion the magnitudes given in Opera Omnia have been adopted.

In the first catalogue sub-classes of magnitudes are given and denoted by · and : In the second catalogue these are not used any longer, and TYCHO admits that the magnitudes have not been determined with the uttermost care. This statement is also confirmed by the determination of the mean error of the observations. As regards the method followed at the estimation of the magnitudes we know practically nothing.

The magnitudes of TYCHO BRAHE are the foundation of the famous star charts of BAYER. Very apparent errors in TYCHO's catalogue were transferred also to the charts. This fact has inspired a number of observers to make corrections and notes. BAYER seems to have corrected some of BRAHE's magnitudes of faint stars, and also to have added several faint stars and to have observed comparatively faint stars, sometimes down to $8^m,0$. These observations are to be found in the edition by SCHILLER of the Uranometria, which contains a number of improved star places in comparison with the edition by BAYER.

13. The Oldest Star Charts. It does not seem to be generally known that the first star charts were published by PICCOLOMINI[1]. His work contains 48 charts showing the stars brighter than 4^m according to the magnitudes of PTOLEMAIOS.

E. B. KNOBEL has pointed out that PICCOLOMINI preceded BAYER in referring to descriptions of stars by means of letters, and in arranging the stars of each constellation mainly according to magnitude. E. C. PICKERING[2] has examined the degree of precision to be attributed to such an arrangement. He identified 262 stars from PICCOLOMINI's charts and compared them with known photometric magnitudes. Out of these there were 143 cases in which the brighter of two successive stars in PICCOLOMINI's list had been placed by him before the other, 100 in which the fainter star preceded, and 19 in which the stars were equal. It would have been possible to derive the error in the estimates of PICCOLOMINI by ranging the stars in sequences, but PICKERING did not make any further inquiry. There is without doubt considerable uncertainty in the case of both PICCOLOMINI and BAYER, and their sequences of stars do not give any essential contribution to the knowledge of stellar magnitudes.

The photometric work of JOHANN BAYER consisted mainly of a mere compilation, and not much attempt seems to have been made to improve the magnitudes of the sources. The reason why BAYER's Uranometria of 1603 has been of such importance in the history of photometry, as it undoubtedly has, seems to be twofold: BAYER conceived the idea of printing the stars on charts, which represented the inside of the celestial sphere, and he assigned Greek letters to the stars as names, instead of using the clumsy antique notation of describing the part of the constellation-figure where a certain star appeared. It has often been supposed that the letters were assigned in order of decreasing brightness, and on that account a good deal has been written about the subject. Many unsuccessful attempts have been made to use the BAYER numbers as sequences in order to derive improved values of the magnitudes used. This is not possible because the sequence of the letters depends upon the position of the stars in the figure, beginning usually at the head, and following its course until all the stars of that order of magnitudes have been exhausted. This fact seems to be rediscovered now and then, but after a short time it is evidently forgotten again.

[1] Delle stelle fisse. Venezia 1540.

[2] Harv Ann 14, Pt II (1885).

It does not seem to be generally recognized that the drawings of the forty-eight ancient constellation-figures, as we have accepted them from BAYER's work and up to date, have originated from the hand of ALBRECHT DÜRER (1515).

An amusing incident in the history of science is the publication of Caelum Stellatum Christianum in 1627 by JULIUS SCHILLER and BAYER. It seems that it was in reality the second edition of BAYER's work almost ready for the press at his death in 1625 but appropriated by SCHILLER to embody his new Christian scheme of constellations.

14. General Discussion of the Oldest Photometric Estimates. A card has been written out for each star in the four oldest catalogues of magnitude which we possess, containing name, position, R H P magnitude, spectral class, colour, and the estimates in the old catalogues. The identification of the stars is a troublesome part of the work. Advantage has been taken of the investigations of BAILY, PETERS, KNOBEL, PEIRCE, PICKERING, DREYER, and others, but there still remain a number of very difficult cases. In some cases it is impossible to identify the objects in question. Although a few uncertain objects have been included in the final catalogue, the conclusions derived will not be materially altered when excluding them.

The following material was used:

	Objects identified	Total number of obj.
PTOLEMAIOS	1020	1028
AL SÛFI	1014	1045
TYCHO BRAHE I	773	777
TYCHO BRAHE II	958	1000
In total	1239 different objects	

The evaluation of the old magnitudes gave the results:

Magn. acc. to Pt or AS	PTOLEMAIOS Harv. magn.		n	AL SÛFI Harv. magn.		n	Magn. acc. to TB	First cat. Harv. magn.		n	Second cat. Harv. magn.		n
1	$0^m,55$	$\pm 0^m,21$	13	$0^m,59$	$\pm 0^m,22$	13	1	$0^m,39$	$\pm 0^m,26$	7	$0^m,81$	$\pm 0^m,20$	12
1ε̉	1,55	0,44	2	1,05	0,09	2	1.	1,40	0,24	5			
2μ	1,63	0,21	3	0,90	0,00	2	1:						
2	2,21	0,09	35	1,89	0,05	26	2	2,31	0,06	29	2,17	0,07	41
2ε̉	2,93	0,18	7	2,44	0,07	7	2.	2,27	0,18	7			
3μ	2,67	0,09	11	2,44	0,07	11	2:	1,18	0,21	5			
3	3,28	0,03	175	2,93	0,04	95	3	3,37	0,04	99	3,44	0,03	159
3ε̉	3,81	0,11	21	3,54	0,03	93	3.	3,60	0,07	34			
4μ	3,96	0,04	89	3,77	0,03	75	3:	3,06	0,09	13			
4	4,35	0,02	362	4,26	0,02	265	4	4,21	0,03	187	4,25	0,02	329
4ε̉	4,76	0,09	18	4,53	0 03	86	4.	4,32	0,04	54			
5μ	4,78	0,12	10	4,71	0,06	22	4:	3,81	0,10	22			
5	4,91	0,03	205	4,72	0,02	167	5	4,68	0,03	122	4,75	0,02	232
5ε̉	—			5,11	0,04	67	5.	4 87	0,09	27			
6μ	—			4,80		1	5:	4,51	0,06	26			
6	5,29	0,03	50	5,25	0,03	70	6	5,22	0,03	100	5,24	0,03	180
6ε̉				5,65	0,05	8	6.	5,20	0,07	20			
neb	4,70	0,45	5	5,08	±0,47	4	6:	5,21	0,06	11			
αμ	4,76	±0,10	14				neb	5,80	±0,15	5	5,80	±0,15	5

It is clear that the three scales are not uniform ones. The intermediate magnitudes do not correspond to intervals of $^1/_3{}^m$. Another feature of the evaluation is that the magnitudes having the highest relative weight do not differ considerably for the four catalogues. This means that the best determined magnitudes in the old sources have a rather constant systematic error. Remembering that AL SÛFI's magnitudes are the result of a revision of PTOLEMAIOS's, this is to be expected. On the other hand, as TYCHO's material does not quite coincide with that of his predecessors, it is rather remarkable that the zero-point of his scale is so close to that of AL SÛFI and PTOLEMAIOS.

Groupings have been made according to rightascension and declination. We do not know in which way AL SÛFI observed the stars, but in the case of PTOLEMAIOS it seems to have been a special system of coordinates that was employed. Anyhow, the groupings according to α and δ do not show anything of peculiar interest. The curves of PTOLEMAIOS's and AL SÛFI's residuals agree pretty closely, whereas that of TYCHO BRAHE has a different course. The influence of extinction is appreciable only for the most southern stars.

The observations of the magnitudes give an interesting contribution to the knowledge of the persistency of the colour sensation of the human eye. Now and then it has been suggested that the sensitiveness of the eye to different wave lengths has changed since classical times. The residuals of the old magnitudes corrected for systematic errors have been grouped according to the colour index as follows:

Colour index	Colour equations							
	PTOLEMAIOS		AL SÛFI		TYCHO BRAHE			
		n		n	1st Cat.	n	2nd Cat.	n
$-0^m,4$ to $-0^m,2$	$+0^m,16$	163	$+0^m,10$	162	$+0^m,03$	114	$-0^m,01$	137
$-0,1$ „ $+0,2$	$-0,06$	368	$-0,09$	364	$0,00$	295	$+0,02$	354
$+0,3$ „ $+0,5$	$+0,04$	103	$+0,03$	101	$-0,05$	79	$0,00$	100
$+0,6$ „ $+0,8$	$+0,08$	80	$+0,10$	82	$+0,01$	58	$-0,02$	75
$+0,9$ „ $+1,4$	$-0,01$	269	$0,00$	264	$-0,06$	201	$-0,04$	252
$+1,5$ „ $+1,8$	$-0,01$	35	$+0,06$	35	$+0,23$	26	$+0,22$	34

The smallness of the colour equations makes it probable that no great change has taken place in our eye, or in the relative colours of the stars, during the time passed since the beginning of our era. It seems that TYCHO BRAHE had an appreciable colour equation for red stars, although not of any unusual amount. Many observers in recent times have had much larger colour equations.

It is an enigmatic question to decide according to which principle the faintest objects in the catalogues were selected. A determination of the frequencies of different spectral classes among all stars, and among stars of 5^m and 6^m in the old catalogues, does not show anything like a systematic difference in the frequencies of brighter and fainter stars. It was probably some other criterion than the brightness alone of the star that determined which objects should be measured or estimated.

This suggestion gains in strength from a very peculiar feature of the catalogue in the Almagest. When the stars in this catalogue are plotted in a chart there is, of course, an area around the South Pole within which no stars fall. This area ought to define the situation of the horizon at the time, and at the place the observations were made. The precession has changed the circle defined by the most southern stars so that its centre is rather far away from the present pole. Computing the time when the most southern stars in the Almagest were at the same mean altitude above the horizon, the surprising result of 1900 B. C. is found. Now, it is impossible that the observations could be as old as that, because, as is well known, PTOLEMAIOS adopted too low a value of the constant of precession, and thus a reduction to the epoch of the catalogue would give very erroneous values as regards the longitudes. In fact, the epoch of the observations has been determined by E. PACI[1] from an investigation of the systematic error in

[1] Alcuni scandagli sulla esattezza del catalogo di Tolemeo. Palermo 1913.

λ according to PTOLEMAIOS which is found to be 1°,18, which means that the epoch of the catalogue should be 52 A. D. On the other hand it is evident that PTOLEMAIOS did not observe the stars down to a certain altitude above the horizon. The selection of the stars must have followed some unknown principle, but what?

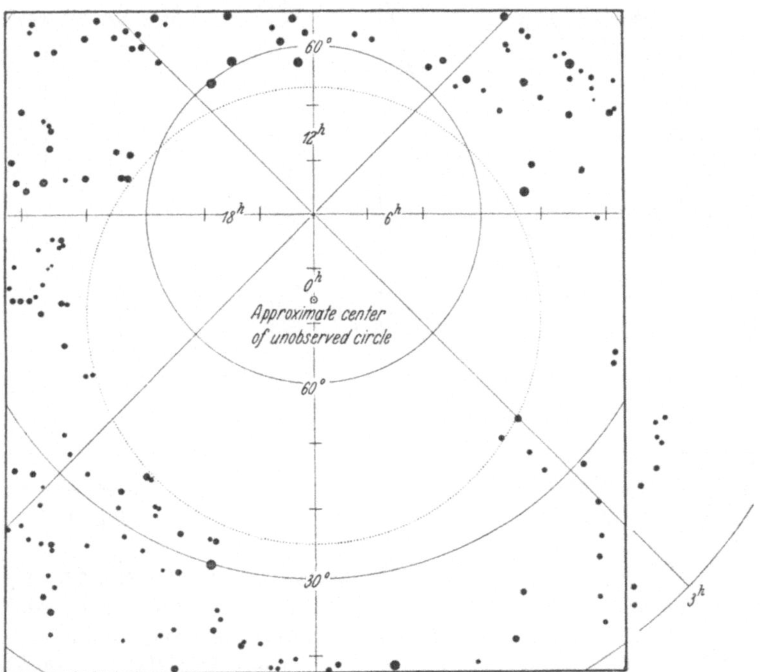

Fig. 24. The unobserved part of the sky in the oldest preserved star catalogue (Ptolemaios's Almagest). The chart reproduced in the figure is given in a stereographic (polar) projection which represents all circles on the sphere as circles. The approximate centre given is not the geometrical center on the figure but the real center on the sphere. The unobserved area represents such part below the horizon of the place where the catalogue has been observed that a much earlier date of the observations than the epoch when even Hipparchos flourished is suggested. It seems that the observers of the catalogue did not observe stars which had not been included in the constellations when these were named about 4800 B.C.

15. First Photometric Data from the Southern Sky. Nothing is known with any certainty about the origin and the first explorations of the constellations around the South Pole. They are generally attributed to navigators from Spain, Portugal, Italy, or Holland, but, the original data being lost, it is not possible to give any strict account of the first mapping and estimates of the southern magnitudes.

E. B. KNOBEL[1] has made an investigation in the British Museum in order to throw light on this matter, and has given the result in a valuable paper, from which we borrow the following account.

The earliest existing map of the Southern Heavens is found in the map of the world by P. PLANCIUS in 1594 given by LINSCHOTEN. This map shows the 48 Ptolemean constellations and of the southern ones Columba, Crux, Eridanus,

[1] M N 77, p. 414 (1917).

Triangulum Australe, and a large constellation named Polophilax in the shape of a man. Also the two Nubeculae appear on the chart, but nothing else.

The earliest publications containing the constellations around the South Pole, as we now know them, appeared in 1603 and are three in number: FREDERICK DE HOUTMAN's Catalogue of stars, BAYER's Uranometria, and JANSZOON BLAEU's Celestial Globe.

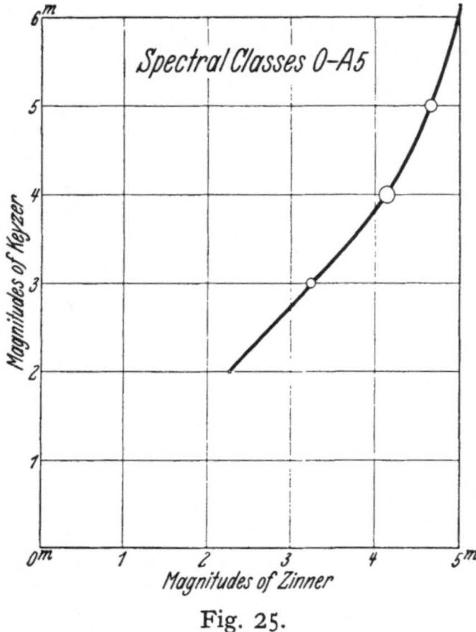

Fig. 25.

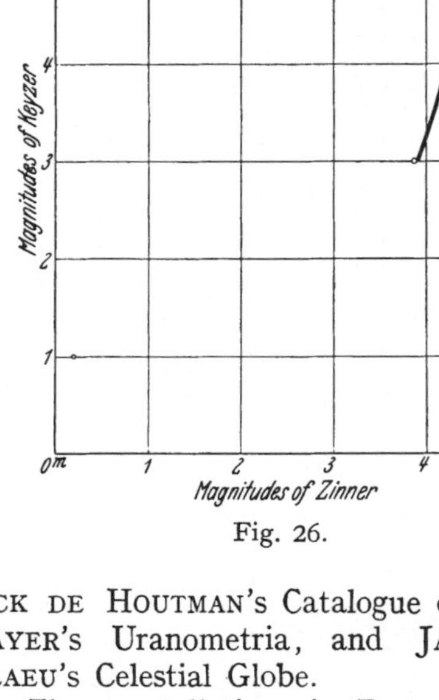

Fig. 26.

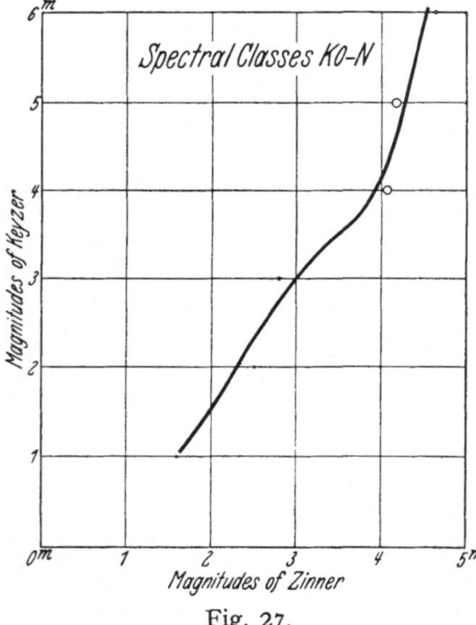

Fig. 27.

Fig. 25—27. Curves for reducing KEYZER's magnitudes into those in ZINNER's system.

The constellations in BAYER's Uranometria are ascribed to AMERIGO VESPUCCI, ANDREA CORSALI, and PEDRO DE MEDINA. The star places have been determined by that "most learned" seaman, PETRUS THEODORUS. In BLAEU's Celestial Globe the same constellations are ascribed to FREDERICK DE HOUTMAN.

KNOBEL has found that AMERIGO VESPUCCI[1] contributed little or nothing to the knowledge of the Southern Sky. He mentions no constellations and his observations are very unsatisfactory. A rough sketch of a few stars cannot be used because of the impossibility of identifying the stars.

ANDREA CORSALI[1] describes the two Nubeculae and the Southern Cross. Some of the stars mentioned in connection with this description cannot be identified. No further astronomical references are known in his letters in the quoted work.

PEDRO DE MEDINA[2] observed α Crucis and mentions other stars in Crux, but gives no other contribution to our knowledge of the southern constellations.

[1] Rasmusio, Navigazione e viaggi. 1563—74. [2] Arte de navegar. 1545.

As for PETRUS THEODORUS, KNOBEL finds that he must be identical with PIETER DIRCKSZ (VON) KEYZER, a pupil of PLANCIUS.

KEYZER took part as chief pilot in the first Dutch expedition to the East Indies in 1595. In the same expedition DE HOUTMAN also took part as sub-commissioner on the mercantile side. KEYZER "sought comfort in science and enriched his knowledge of astronomy by improving the position of old and new constellations". DE HOUTMAN prepared a portion of a vocabulary published in 1603, dealing with Malayan and Madagassic dialects[1]. At the end of the work he gives a catalogue of southern stars "observed with efficient instruments" by him in the island of Sumatra, their positions (corrected), and their numbers. He also says "Magnitude is the size of the stars: often a star is of first size or greatest light: thus there are seven degrees of size (with regard to) light. Further there will be found at the end the declinations of several fixed stars in the region of the South Pole which I had observed on my first voyage and which on my second voyage I revised and corrected with more care and brought up to the number of 300 as may be seen on the Celestial Globe constructed by W. JANSENN (BLAEU)".

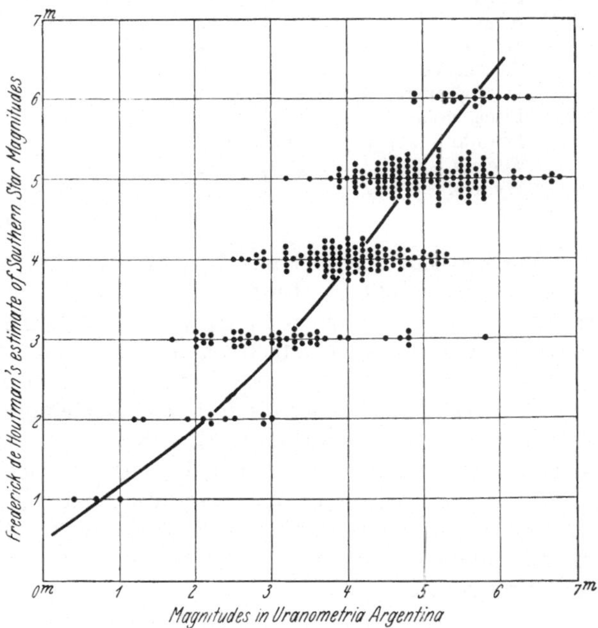

Fig. 28. Comparison between the estimates of HOUTMAN and those in Uranometria Argentina. The dots are individual stars.

The catalogue has been edited by KNOBEL[2] and earlier by A. MARRE[3]. It consists of 303 stars, of which 107 stars are contained in PTOLEMAIOS, leaving 196 newly discovered ones.

KNOBEL doubts if this catalogue was made by DE HOUTMAN, because he could not have seen all the stars from his observational place, Achin ($\varphi = +5°{,}6$). It seems possible that the whole catalogue and the formation of the twelve constellations must be attributed to KEYZER, and the catalogue was sent to PLANCIUS in Holland in 1597, who probably communicated the catalogue to BAYER, who made use of it in his Uranometria.

DE HOUTMAN obtained an imperfect copy of it, which, as KEYZER was dead, he published as his own work in the above-mentioned vocabulary.

ZINNER points out that already on the globe constructed in 1601 by HONDIUS there are 12 southern constellations designated with 124 stars according to KEYZER. This globe was used by BAYER for his star chart of the Southern Sky. J. BARTSCH, too, has used the observations of KEYZER as a foundation for his southern constellations. He may have obtained the data from the charts in

[1] Spraeckende woordboeck Inde Maleysche ende Madagaskarche Talen met vele Arabische ende Turksche woorden. Amsterdam.

[2] M N 77 p. 421 (1917).

[3] Bull des Sc Mathm tome 5 (1881).

BAYER's Uranometria. On the other hand, on all the globes constructed by W. JANSZOON BLAEU, the observations of the southern constellations are ascribed to DE HOUTMAN.

ZINNER gives the following table of the numbers in the new constellations according to the different authors.

Constellations	HONDIUS 1601	BAYER 1603	BARTSCH 1624	BARTSCH 1662	DE HOUTMAN 1603
Grus	13	13	13	13	12
Phoenix	14	14	13	13	13
Indus	12	12	12	12	11
Pavo	16	16	21	21	19
Apus	12	12	11	11	9
Musca	4	4	4	4	4
Chamaeleon	10	9	8	10	9
Triang. austr.	5	5	5	5	4
Piscis volans	7	7	7	7	5
Dorado	7	7	6	6	4
Tucana	8	8	8	8	6
Hydrus	22	22	22	23	15
All	130	129	130	133	111

We see that KEYZER has 19 more objects than DE HOUTMAN. BAYER is probably dependent on HONDIUS, but BARTSCH uses in his catalogue one more star than BAYER, which is in accordance with the fact that he also had access to manuscript notes. The fact that DE HOUTMAN sometimes gives more accurate places than BARTSCH does is important. As to the magnitudes the globes of HONDIUS and the catalogue of BARTSCH do not as a rule agree with those of DE HOUTMAN. It seems from the synopsis below that the magnitudes of DE HOUTMAN are slightly more accurate than those founded on the authority of KEYZER.

For a comparison of the magnitudes of the last-mentioned investigator, ZINNER has used the catalogue of BARTSCH of 1624 and has excluded the data on the globes of HONDIUS as not being of proper accuracy. He has found the following result:

Spectral Classes.

Magn. acc. to KEYZER	O—A0	n	A2—G5	n	K0—N	n	Mean error	
1			$-0^m,80$	1	$+0^m,60$	1		
2	$+0^m,28$	6	—	—	$+0\ ,50$	1	$\pm 0^m,64$ (1–3)	
3	$+0\ ,22$	12	$+0\ ,88$	6	$-0\ ,20$	4		
4	$+0\ ,14$	35	$+0\ ,27$	7	$+0\ ,08$	12		$\pm 0^m,54$
5	$-0\ ,35$	23	$-0\ ,60$	4	$-0\ ,81$	13	$\pm 0\ ,51$ (4–6)	
6	$-0\ ,93$	4	$-1\ ,33$	4	$-1\ ,38$	4		

Using the catalogue of DE HOUTMAN as edited by KNOBEL, ZINNER has derived:

Magn. acc. to DE HOUTMAN	O—A0	n	A2—G5	n	K0—N	n	Mean error	
1	$-0^m,10$	1	$-1^m,25$	2	—	—		
2	$+0\ ,28$	5	$+1\ ,20$	1	$+0^m,65$	4	$\pm 0^m,75$ (1–3)	
3	$-0\ ,01$	20	$+0\ ,45$	10	$-0\ ,03$	8		
4	$+0\ ,02$	43	$+0\ ,10$	11	$+0\ ,02$	22		$\pm 0^m,52$
5	$-0\ ,32$	32	$-0\ ,64$	9	$-0\ ,85$	15	$\pm 0\ ,43$ (4–5)	
6	$-1\ ,02$	5	—	—	—	—	—	

The remarkable agreement between stars of intermediate magnitudes in the two comparisons as well as the agreement of the general course of the scale makes it very likely that the two sources are not independent of each other, but that one (DE HOUTMAN's) is an improvement based on the other.

It is not very easy to decide as to the priority of the explorations of the southern constellations, but it seems indubitable that the Dutch seafarers should be given the palm.

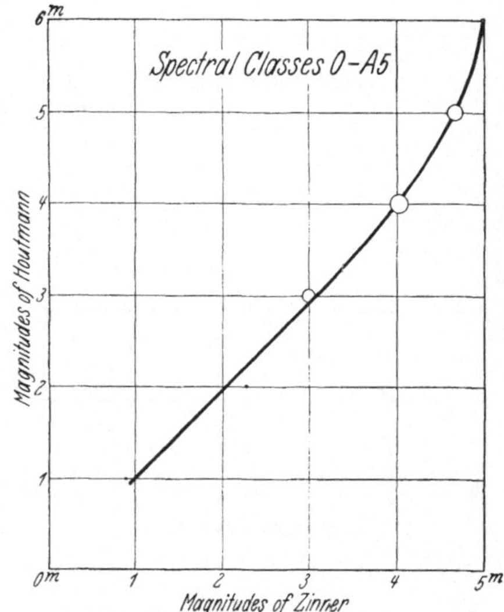

Fig. 29.

The Linschoten Society has planned to publish a full account of the life and work of PIETER DIRCKSZ KEYZER.

A photographic copy of the catalogue has been privately edited by the late H. H. TURNER.

During his journey to St. Helena, in the years 1676—78, E. HALLEY observed the positions of 341 southern stars. His results have been published under the title of Catalogus stellarum australium[1]. His observations were made without a telescope, and the magnitudes of the stars are given in whole magnitudes and now and then also in half intervals. HALLEY's observations are very important, as they are the second photometric contribution from the Southern Sky. From ZINNER's work we borrow the adjoining evaluation of the scale.

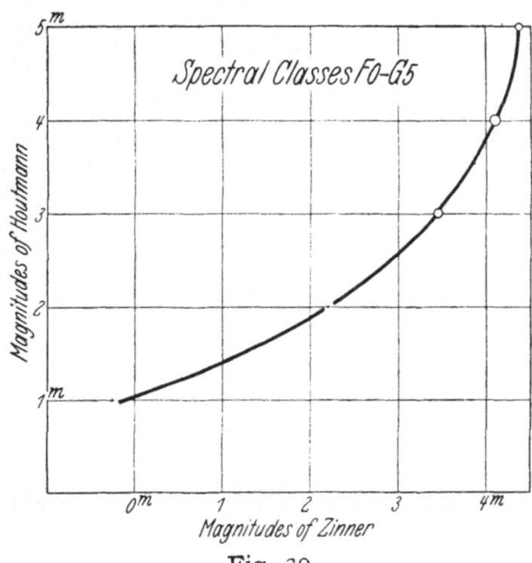

Fig. 30.

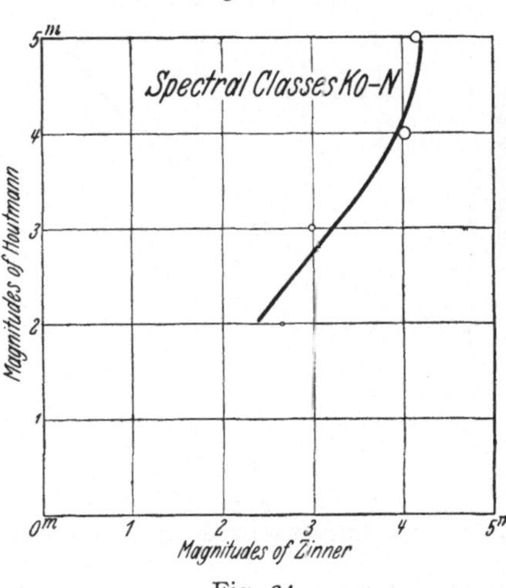

Fig. 31.

Fig. 29—31. Curves for reducing the magnitudes of HOUTMAN into the magnitudes of ZINNER.

Spectral class.

HALLEY	O—A 0	A 2—G 5	K 0—N	Mean error
1 m		0m,64	—	
2	1m,96	2 ,40	2m,27	
2 1/2	2 ,58	2 ,15	2 ,75	±0m,43
3	3 ,26	3 ,62	3 ,26	
3 1/2	3 ,82	3 ,33	—	
4	4 ,00	3 ,98	4 ,01	
4 1/2	4 ,20	—	—	±0 ,41
5	4 ,65	4 ,56	4 ,26	
6	5 ,06	4 ,80	4 ,87	

There is the same general behaviour in the scale of HALLEY as in that of BRAHE. The

[1] Catalogus stellarum australium. London 1769.

value of the colour equation cannot be derived with accuracy, it seems that it is considerable for stars fainter than $5^m,0$.

16. HEWELKE (HEVELIUS). This famous astronomer made observations of variable stars from 1638 onwards and noted many differences between the charts and the sky. In his extremely rare work, Machina Coelestis, there are numerous observations, and his work, Prodromus Astronomiae cum Catalogo Fixarum et Firmamentum Sobieskianum, Gedani 1690, gives the results of some of his comparisons. Most of the estimates were destroyed in the lamentable fire that laid his observatory in ruins.

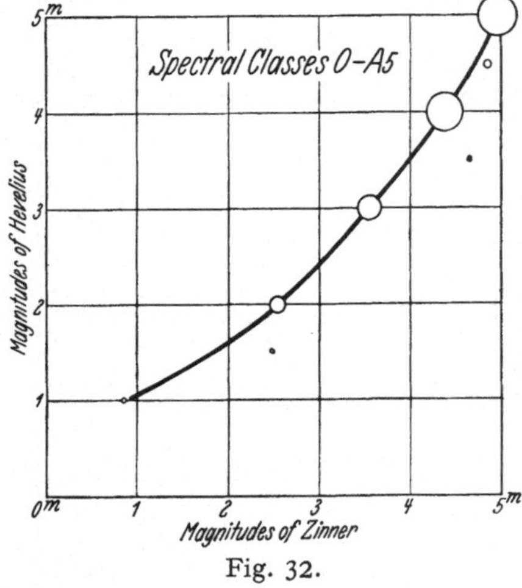

Fig. 32.

In his Atlas he gives, as a rule, only whole magnitudes. But sometimes also quarter, half, and three quarter magnitudes are given, probably as a result of taking means. HEWELKE extended TYCHO BRAHE's scale, so that he also used the seventh magnitude. ZINNER has evaluated his magnitudes in modern measurements and finds the results given in the adjoining table.

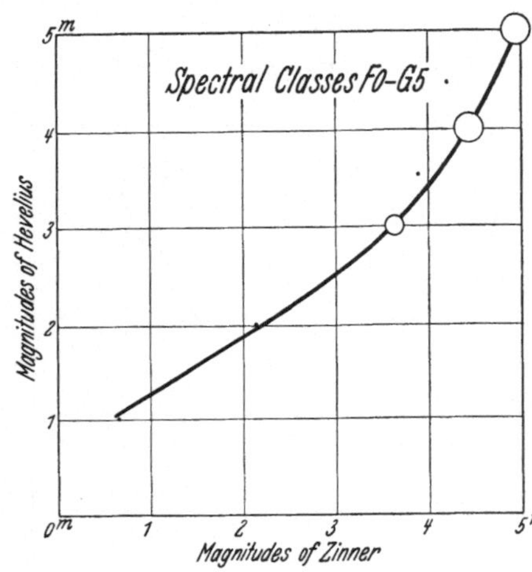

Fig. 33.

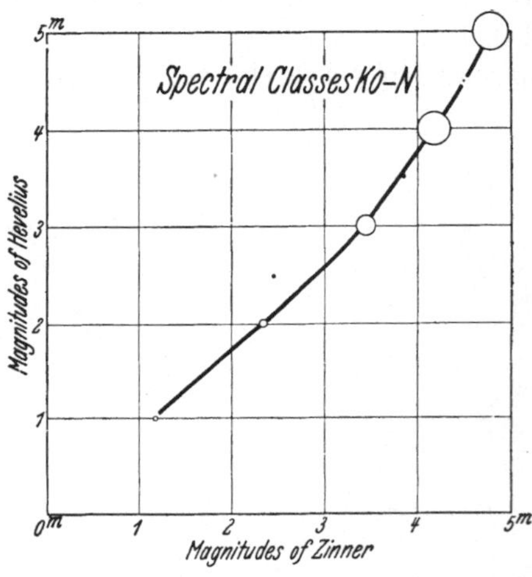

Fig. 34.

Fig. 32—34. Curves for reducing the magnitudes of HEVELIUS into the magnitudes of ZINNER.

Magn. of HEWELKE	O—A 5	F—G	K—N	Mean error
1^m	$0^m,86$	$0^m,65$	$1^m,18$	
$1^1/_2$	2 ,50	—	—	$\pm 0^m,52$
2	2 ,53	2 ,13	2 ,33	
$2^1/_2$	—	—	2 ,50	
3	3 ,55	3 ,63	3 ,43	±0 ,52
$3^1/_2$	4 ,60	3 ,90	3 ,85	
4	4 ,39	4 ,32	4 ,18	±0 ,48
$4^1/_2$	4 ,85	4 ,20	4 ,50	
5	4 ,96	4 ,94	4 ,79	±0 ,32
$5^1/_2$	5 ,06	5 ,00	4 ,84	±0 ,41
6	5 ,60	5 ,40	5 ,21	±0 ,41
$6^1/_2$	6 ,07	5 ,80	5 ,79	
7	6 ,34	—	5 ,77	

The scale of HEWELKE differs in its general course from the scale of the earlier observers. HEWELKE has revised the magnitudes of TYCHO BRAHE and in several cases improved the values of the latter. It is not quite clear if HEWELKE has made several comparisons, so that his magnitudes are the results of taking means. I have not had an opportunity to investigate his manuscript catalogue, preserved in the Archives of the Society of Natural History in Danzig. It seems that the magnitudes have been corrected and improved several times, but one cannot be sure that these alterations have not been made after the death of HEWELKE. The catalogue of HEWELKE is more complete than that of TYCHO BRAHE. The brightest star north of $-30°$ not given is of the magnitude $4^m,52$.

17. Other advances during the 17th Century. It cannot be said that the astronomers of the 17th Century were keen to adopt the telescope as an astronomical instrument. GALILEO GALILEI describes the appearance of the fixed stars in his Sidereus Nuncius and also divides the telescopic scale into 12^m corresponding to $8^m,5$ in the modern photometric scale[1]. But the remarkable discoveries of GALILEI do not seem to have inspired new studies and new observations. The photometric contributions of his century can be mainly characterized as reobservations of stars in BAYER's charts. The inaccuracy of the magnitudes in the Uranometria inspired the astronomers more than GALILEI's extension of the stellar world to hundreds of thousands of members.

G. D. CASSINI was one of the main revisers of BAYER's star charts. He made observations in Bologna as well as in Paris and made some use of the telescope in his observations. He also seems to have inspired MONTANARI to revise the charts of BAYER. The observations of MONTANARI in Bologna from 1665 to 1681 are very important in this respect. Most of the observations were made with the unaided eye. The charts of MONTANARI which were thought to be lost, have been recently discovered by F. PORRO in the library of the Chapter at Verona[2]. MONTANARI and BIANCHINI seem, indeed, to have been the first to use the JOHN HERSCHEL method of ranging into sequences. In this way many uncertainties in the work of BAYER were discovered. As for the estimates of magnitudes ZINNER has shown that the accuracy is not very high for stars between $4^m,0$ and $5^m,5$, and that the colour difference between the spectral classes O—A5 and K0—N is $0^m,5$ and thus of considerable importance. The individual values of MONTANARI's estimates for stars brighter than $5^m,5$ have been used by ZINNER and are given in the notes to his catalogue.

Most of FLAMSTEED's observations were made without a telescope. He estimated the stars in intervals of half a magnitude. F. BAILY has given attention to the magnitudes in his edition of FLAMSTEED's catalogue. The observations have not much value with regard to the magnitudes, and it was the poor agreement between FLAMSTEED's catalogue and the sky that made HERSCHEL start his observations, using at first the method of sequences, and later on the step-degree method.

MARALDI observed variable stars and also other stars in the years 1692—1719. A catalogue giving the zodiacal stars and some others has been published. In the Archives of the Observatory at Upsala there is a manuscript (A 518) of another star catalogue which was never published. The magnitudes are given as a rule in whole magnitudes and, comparatively seldom, in half magnitudes. The scale embraces magnitudes down to 7^m and there are besides a number of

[1] GALILEI said: "Verum, infra stellas magnitudinis sextae, adeo numerosum gregem aliarum naturalem intuitum fugientium, per Perspicillum intueberis, ut vix credibile sit, plures enim, quam sex aliae magnitudinum differentiae videas licet; quarum maiores, quas magnitudinis septimae, seu primae invisibilium appellare possumus, Perspicilli beneficio maiores et clariores apparent, quam magnitudinis secundae Sidera, acie naturali visa."

[2] Sugli schizzi di carte celesti eseguite da FRANCESCO BIANCHINI nel secolo XVII sopra osservazioni proprie e di Geminiano Montanari. V J S 33, p. 284 (1898).

"telescopic stars". The faintest of the stars, visual to the eye, or MARALDI's 7th magn., are in modern magnitudes $5^m,9$, and the telescopic stars are of the magnitude $6^m,4$. The number of stars in the catalogue is 506.

18. The School of A. CELSIUS. It is evident that A. CELSIUS, the energetic and skilful director of Upsala Observatory, had a plan of investigating the sky from a photometric point of view. It is to be regretted that the death of CELSIUS at an early age in 1741 interrupted this splendid plan. Had it been executed we should have had in our possession a Uranometry from the beginning of the 18th century, which would have filled in an excellent way the gap between the estimates of the 16th century and those by SIR WILLIAM HERSCHEL.

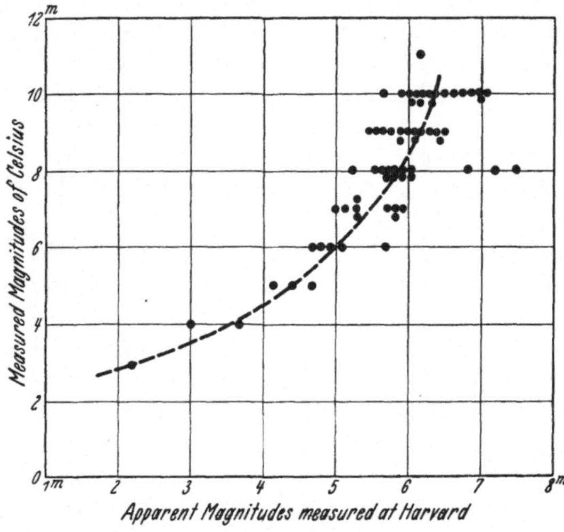

Fig. 35. Relation between the magnitudes as measured by CELSIUS at Upsala and the Harvard magnitudes (RHP). The dots represent individual stars.

CELSIUS's activity was not restricted to the estimates of stellar magnitudes. He also started actual measurements of magnitudes. The first attempts at measuring the light of the stars were performed at Upsala by CELSIUS's pupil, TULENIUS, who wrote a thesis concerning the constellation Aries[1]. For these observations a telescope of a length of one foot was used. The magnitudes were determined by extinguishing the light of the stars by means of glasses interposed in the way of the light rays. One magnitude corresponded roughly to two glasses, and Sirius was not extinguished before 24 glasses or more were put in. The faintest stars, visible in the instrument, were called 12^m, but are, in fact, not fainter than $6^m,9$ in P D scale. Thus the scale of CELSIUS was far too narrow in the same way as was the scale introduced by Sir WILLIAM HERSCHEL for the faint "telescopic" stars (12^m—20^m). The catalogue given in the thesis of TULENIUS contains 64 stars, and some fainter stars are added in the chart. ZINNER has investigated these stars and finds that they cannot be used for a derivation of the colour equation. He finds the mean error to be $\pm 0^m,38$ for the mean magnitude, and for the faintest stars $\pm 0^m,41$. The relation between the magnitudes and the Harvard scale is given in fig. 35.

This work was not continued, probably on account of the difficulties which presented themselves at the first experiments to measure photometrically the light of the stars. TULENIUS[2] published another survey of one of the constellations, Taurus, but used estimated magnitudes, which were mainly in agreement with FLAMSTEED. The same applies to the survey of the constellation Leo by HODELL[3]. The successor of CELSIUS, M. STRÖMER, tried to continue the work as regards the estimates of star magnitudes. His pupil, MELANDER[4], published a survey of the constellation Virgo and used in his paper the revised values of the FLAMSTEED

[1] Dissertatio astronomica de constellatione Arietis. Holmiae (1740).
[2] Dissertatio de constellatione Tauri. Holmiae (1743).
[3] Dissertatio de constellatione Leonis. Upsala (1741).
[4] Dissertatio astr. de Virgine sexta zodiaci constellatione. Upsala (1755).

magnitudes. He says that it would be of value to measure the light of the stars, but states that several difficulties are involved in the problem, so that estimates are to be preferred. ZINNER has derived for 6^m the colour equation $+0^m,36$. According to the same authority the mean error is $\pm 0^m,41$ and $\pm 0^m,43$ respectively.

The observatory at Upsala also possesses a chart of the constellations Sagittarius and Pisces, on which the stars have been compared and estimates put down in A. CELSIUS's hand.

19. The Peking Catalogue. This catalogue was edited in 1914 by S. CHEVALIER[1]. It is contained in a Chinese work, Kin-ting-i-Siang-has-tscheng, i. e. Astronomical instruments constructed by order of the Emperor.

This catalogue also contains estimates (or measures?) of the stellar magnitudes. Although these are rough it has been thought worth while on account of their epoch to reduce them to a photometric system. The 3083 stars have as far as possible been identified by me, and the estimated magnitudes compared with the photometrical ones in Harv Ann 50 and 54. The following are the results of the comparison:

Chinese magnitude	Harvard scale	Mean error of the mean	n
1^m	$0^m,91$	$\pm 0^m,08$	16
2	2 ,23	±0 ,03	68
3	3 ,52	±0 ,03	209
4	4 ,31	±0 ,02	457
5	4 ,97	±0 ,01	678
6	5 ,75	±0 ,01	1355

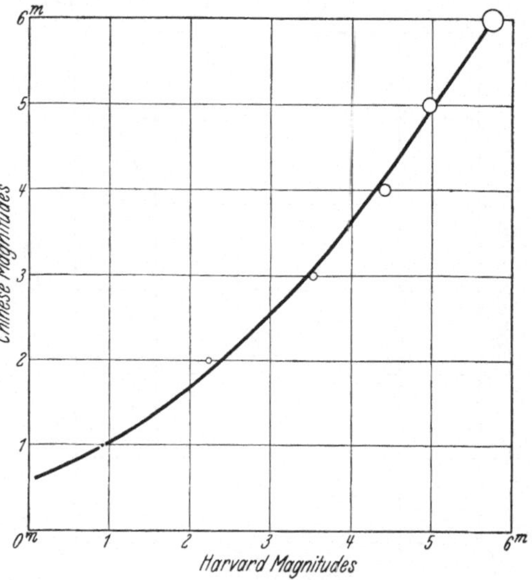

Fig. 36. Relation between the estimates of magnitudes performed by the Jesuits in China in the first part of the 18th century and the Harvard magnitudes (RHP).

It is evident that the observers have established their own scale. The observational work was probably performed by P. KÖGLER (1680—1746), A. DE HALLERSTEIN (1703—1774), A. GOGEIST (1701—1771), and A. DA ROCHA (1728—1781), assisted by Chinese scholars of high rank. It is not clear whether the magnitudes have been taken over from some older source or have been observed about 1750, but the second possibility seems to be the more probable one. A notable feature is the comparatively wide photometric interval of $0^m,78$ of the faintest magnitude. This suggests that the observations were not made with the unaided eye, and if this is the case the estimates were not transported from some very ancient source.

20. Further Explorers of the Southern Sky. In two catalogues Father NOËL has given contributions to the photometry of the Southern Sky. The observations were made in Rachal in 1684, in Macao from 1685 to 1700, and also in Goa from 1706 to 1707. The first catalogue[2], containing the results of the observations from 1685 to 1687, gives places and magnitudes for 220 stars. In the second catalogue[3] the results of all the observations or 352 stars are brought

[1] Annales de l'Observatoire Zô-Sé T VII. Shang-Hai (1914).

[2] Paris Mém. Ac. 1699, T VII 2, p. 221 (1729).

[3] Observationes mathematicæ et physicæ in India et China factæ a Patre FR. NOËL S. J. etc. Prag (1710).

together. The positions are not very accurate, which makes the identification of faint stars difficult. The magnitudes are estimated in whole and half magnitudes. ZINNER has derived the following comparison between the magnitudes in the second catalogue (excluding doubles and variables).

NOËL's magn.	O—A	n	F—G	n	K—N	n	Mean error
1	—	—	$0^m,43$	7	—	—	
2	$2^m,48$	13	2 ,30	2	$2^m,32$	6	
2—3	2 ,90	5	—	—	2 ,50	1	$\pm 0^m,47$
3	3 ,31	9	3 ,13	3	2 ,82	5	
3—4	3 ,43	3	3 ,50	2	3 ,17	3	
4	3 ,74	17	3 ,73	6	3 63	10	
4—5	4 ,46	10	4 ,60	4	3 ,85	6	
5	4 ,55	33	4 ,29	14	4 ,30	20	± 0 ,43
5—6	4 ,70	1	—	—	4 ,15	4	
6	4 ,80	25	4 ,85	17	4 ,84	20	

There is general agreement between the course of the magnitude curve and that of PTOLEMAIOS. The colour equation cannot very well be derived on account of the scanty material.

ZINNER remarks that in NOËL's second catalogue most of the stars of HALLEY have been included. In several cases, but not in all, the magnitudes agree. In order to test the question whether NOËL has made his magnitudes fit those of HALLEY, ZINNER has investigated 68 stars, where the magnitudes differ. In 36 cases NOËL has kept his original magnitudes, in

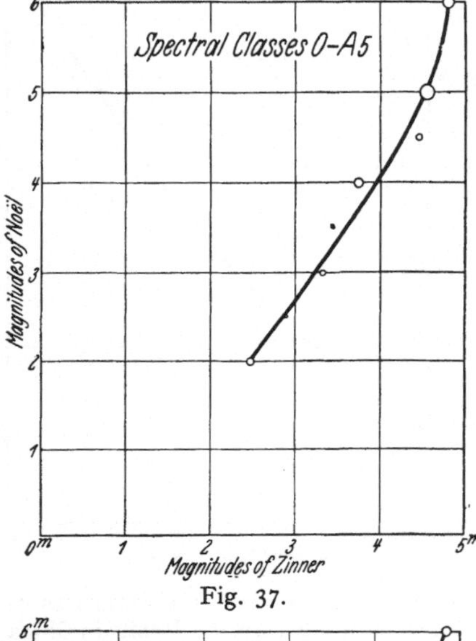

Fig. 37.

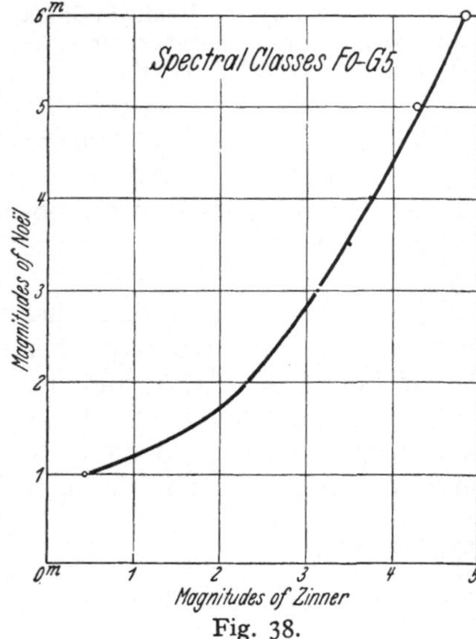

Fig. 38.

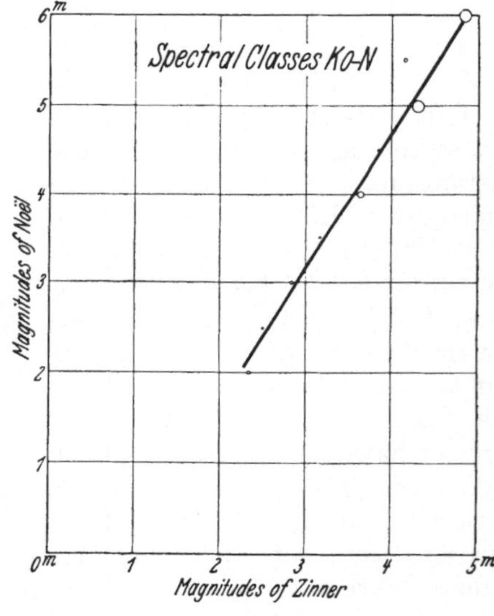

Fig. 39.

Fig. 37—39. Curves for reducing the magnitudes of NOEL into the magnitudes of ZINNER.

17 cases a change of direction causing better agreement with HALLEY has been made, but in 13 cases the changes pointed in the other direction, and two cases cannot be decided. It seems that in any case NOËL has not à tout prix made his magnitudes agree with those of HALLEY (ciph. 15).

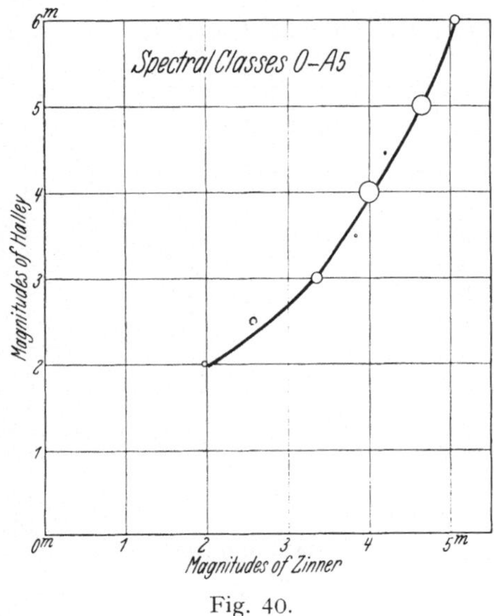

Fig. 40.

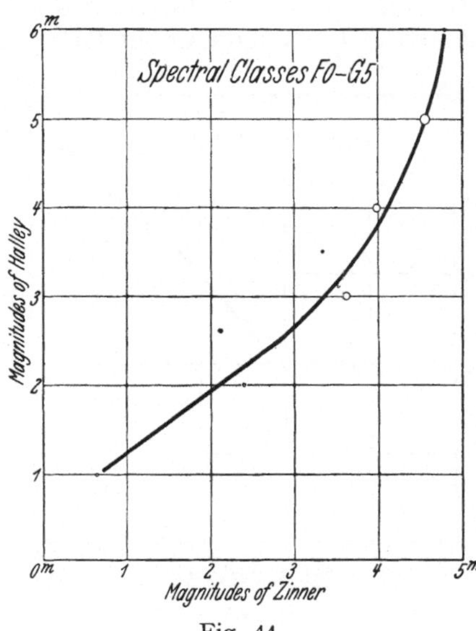

Fig. 41.

It is peculiar that HALLEY observed the southern stars slightly more accurately than NOËL, as is seen from the mean error of stars between 1^m—4^m viz. $\pm 0^m,42$, $\pm 0^m,45$.

LACAILLE. This astronomer has set down his exploration of the southern sky in three sources: catalogue of 398 stars, i.e., the fundamental stars, the zones (Caelum Australe Stelliferum) containing 9766 stars, and the catalogue of 515 zodiacal stars[1]. The three catalogues have been reduced by BAILY[2]. LACAILLE observed from 1746 to 1750 in Paris and from 1751 to 1752 at the Cape. He estimated the magnitudes of the stars in connection with his determination of their position. He used intervals of whole magnitudes and called the faintest stars visible to the unaided eye 7^m. He also used intervals between the magnitudes, but these do not correspond to half magnitudes, but to the intensity half way between two successive magnitudes. ZINNER has also investigated LACAILLE's estimates. He finds that BAILY's edition of the catalogue is not well adapted for a derivation

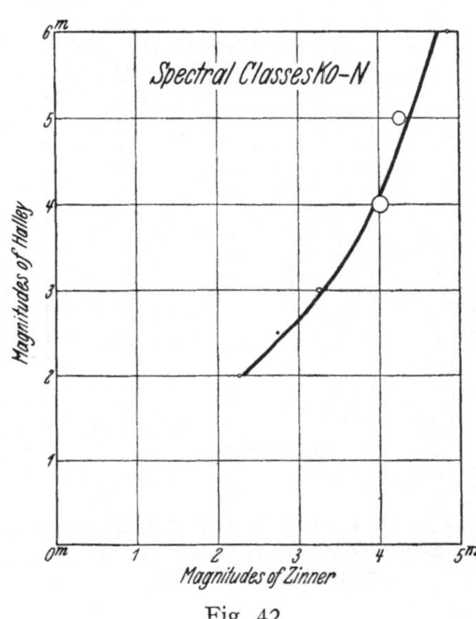

Fig. 42.

Fig. 40—42. Curves for reducing the magnitudes of HALLEY into the magnitudes of ZINNER.

[1] Coelum australe stelliferum. Paris (1763). — Stellarum ascensiones rectæ veræ etc. Astronomiæ fundamenta. Paris (1757). Observations sur 515 étoiles du zodiaque. Paris (1763).

[2] A Catalogue of 9766 Stars in the Southern Hemisphere. London (1847).

of LACAILLE's magnitude and colour curve, because, while most of the magnitudes are taken from the catalogue of 1763, some were also taken from other catalogues. A comparison with the magnitudes in HALLEY's catalogue shows that LACAILLE has made his magnitudes fit the earlier estimates. Because of the deviations of the magnitudes of HALLEY from the photometric scales, the forcing of the magnitudes to fit those of the latter has introduced increased uncertainty. This

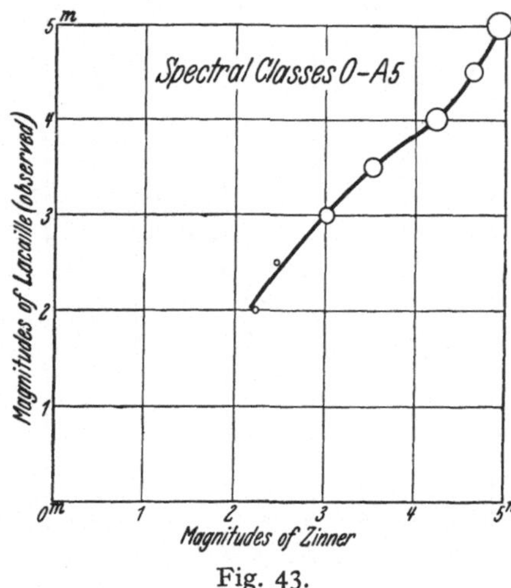

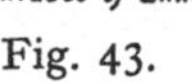
Fig. 43.

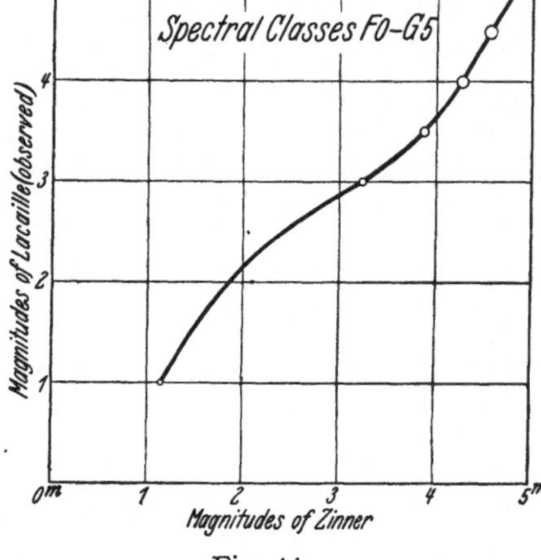

Fig. 44.

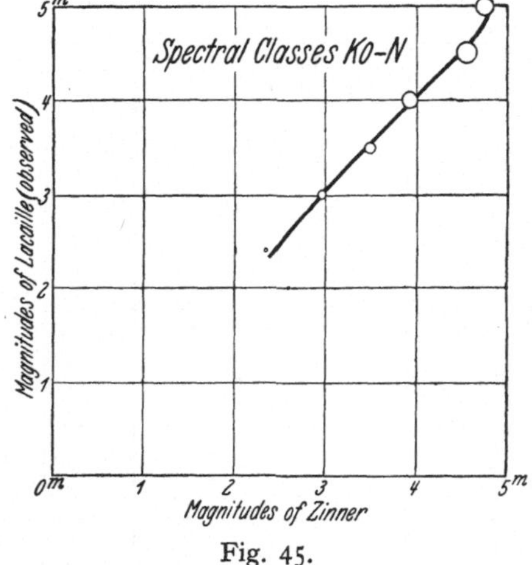

Fig. 45.

Fig. 43—45. Curves for reducing the magnitudes of LACAILLE (Lacaille I) into the magnitudes of ZINNER.

fact was recognized by GOULD when he compared LACAILLE's magnitudes with those in Uranometria Argentina. In the following evaluation of the magnitudes ZINNER uses the data from the observed magnitudes, together with those in the catalogue of 1757.

Lac. magn.	O—A	F—G	K—N	Mean error
1	—	$1^m,15$	—	
2	$2^m,23$	2 ,10	$2^m,60$	
$2^1/_2$	2 ,47	2 ,07	2 ,35	$\pm 0^m,43$
3	3 ,01	3 ,25	2 ,98	
$3^1/_2$	3 ,52	3 ,89	3 ,50	
4	4 ,23	4 ,29	3 ,93	
$4^1/_2$	4 ,65	4 ,57	4 ,57	± 0 ,39
5	4 ,93	4 ,87	4 ,77	

ZINNER has concluded that, if the whole magnitudes are used for a derivation of the mean error, the resulting value is $\pm 0^m,39$, but if the intermediate intervals are used, $\pm 0^m,45$, which is in harmony with the above-mentioned fact that these intervals do not correspond to half magnitudes but to half intervals of brightness. It ought to be remembered that LACAILLE estimated his magnitudes, when the stars passed through his field. No comparisons with standards could be made, and the degree of accuracy could not be expected to be much higher than it is found to be.

The catalogue of 1757 gave on the whole a result in agreement with those above. The mean error was slightly larger, viz., $\pm 0^m,46$.

21. Other Pre-Herschelian Determinations of Magnitudes. There are a number of contributions from the 18th century, but generally the observations are few or uncertain. As contributions to the general knowledge of stellar photometry these observations do not carry much weight.

In BRADLEY's famous observations at Greenwich the magnitudes were also observed. A. AUWERS has given a synopsis of the results in his reduction, separately for different instruments and observers. The number of estimated stars is nearly 400. The intervals used are 1^m, $^1/_2{}^m$, $^1/_4{}^m$, and sometimes $^1/_3{}^m$. A. AUWERS has given a comparison between the estimated magnitudes and values founded on GOULD's and HOUZEAU's determinations. A reduction to a modern scale is desirable.

TOBIAS MAYER. BAILY has edited the star catalogue of T. MAYER, corrected and enlarged[1]. The number of positions is 998 + 42, and in most cases the magnitude is given, estimated in 1^m and $^1/_2{}^m$. As far as I know, no discussion has been undertaken of the magnitudes and their relation to a photometric scale.

22. WILLIAM HERSCHEL. The first estimates of stellar magnitudes by Sir WILLIAM were published in a paper from 1796[2]. Comparing the charts of FLAMSTEED with the sky, he had found numerous differences which he thought were to be ascribed to errors in the estimates and not to real changes in the brightness of the stars. At first he "ranged" or ordered the stars within the constellations; for instance:

Order of the stars in Ophiuchus: $\alpha\ \beta\ \delta\ \zeta\ \eta\ \varkappa\ \gamma\ \varepsilon$.

He says that a defect of this arrangement is that in such cases we have not always a proper connection of the steps of the series that may be formed by rank, there may be too much difference in the lustre of some of the stars, and too little

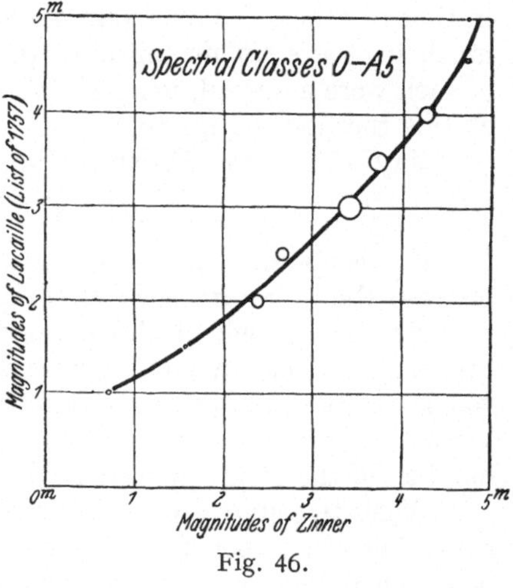

Fig. 46.

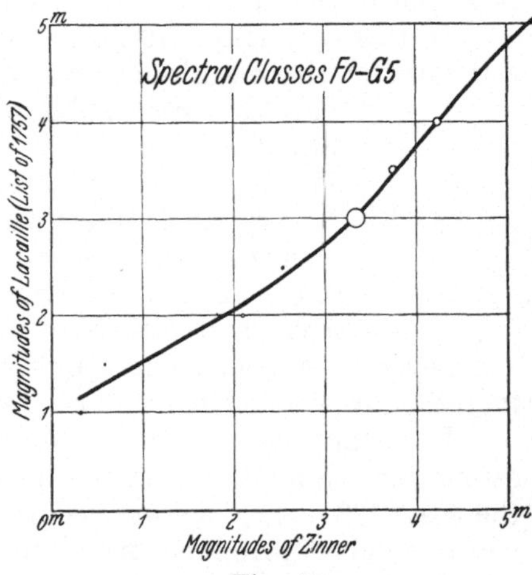

Fig. 47.

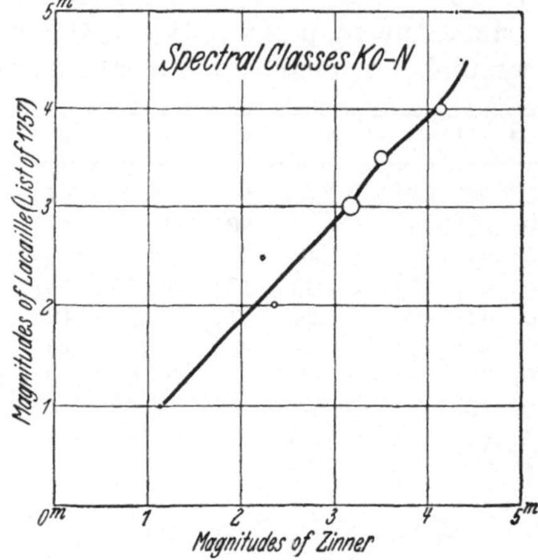

Fig. 48.

Fig. 46—48. Curves for reducing the magnitudes of LACAILLE (LACAILLE II) into the magnitudes of ZINNER.

[1] London Mem RAS 4, p. 391 (1830).

[2] On the method of observing the changes that happen to the fixed stars; with some remarks on the stability of the light of our Sun. To which is added, a catalogue of comparative brightness, for ascertaining the permanency of the lustre of stars. Phil Trans 1796, p. 166.

in that of others. To overcome the difficulties he marked such stars that differed much in lustre by magnitudes, or degrees of difference. Three different sorts of each were assumed, viz., 1′, 1″, 1‴, 2′, 2″, 2‴ etc. As an example of the use of this method we quote:

May 12th 1783. Order of the stars in Bootes:

$$\alpha\, 1'\, \varepsilon\, 2''\, \eta\, 2'''\, \gamma\, \beta\, \delta\, 3'\, \varrho\, 3'''\, \pi\, 4''.$$

He also points out that this is not recurring to the old method of magnitudes because the stars are strictly compared.

He soon extended the method. When two stars were perfectly alike in brightness so that by looking often and for a long while at them he could not tell which was the brightest or occasionally the one or the other appeared largest, and sometimes, not long after, he must give the preference to the other, he noted the number of the stars separated with a point.

When two stars are so nearly alike in lustre that they may almost be called equal, but when upon a longer inspection of them, one always returns to a decision in favour of the same one, he separated the numbers of the stars with a comma.

In this way HERSCHEL introduced the following characters to be used the estimates of the magnitudes of the stars:

‘ The least perceptible difference less bright.
. Equality.
, The least perceptible difference more bright.
— A very small difference more bright.
—, A small difference more bright.
— — A considerable difference more bright.
— — — Any great difference more bright in general.

Several compound characters were also used and can be found explained in the above-quoted paper by HERSCHEL.

HERSCHEL was very well aware of the difficulties to be overcome when estimating the brightness of the stars and took minute care in the work. During the years 1781—1798 he surveyed the visible sky in England. From his estimates, which number 3010, the magnitudes can be derived for 1251 stars.

The results of his estimates are given in the collected works I:530—584, II:22—30, 628—649. They were first discussed by PEIRCE and later by PICKERING [Harv Ann 14, pt. II, p. 350 (1885)], who found the following values for the different symbols, expressed in Harvard magnitudes.

mbol	I	n	II	n	III	n	IV	n	V	n	VI	n	All	Adop
.	+0m,07	99	+0m,10	67	−0m,02	54	+0m,11	63	+0m,08	52	+0m,00	50	+0m,06	0m,1
,	0 ,25	274	0 ,25	150	0 ,20	112	0 ,23	128	0 ,21	82	0 ,20	122	0 ,23	0 ,2
—	0 ,41	113	0 ,47	87	0 ,46	79	0 ,30	88	0 ,42	56	0 ,23	82	0 ,38	0 ,4
;	0 ,01	17	0 ,09	13	0 ,04	18	0 ,10	24	0 ,12	17	0 ,02	12	0 ,07	0 ,1
—̦	0 ,44	19	0 ,24	11	0 ,34	10	0 ,23	18	0 ,45	8	0 ,29	13	0 ,33	0 ,3
—,	0 ,54	21	0 ,54	22	0 ,60	26	0 ,53	29	0 ,66	26	0 ,58	41	0 ,57	0 ,6
—	0 ,77	15	0 ,94	8	0 ,91	9	0 ,85	16	1 ,37	6	0 ,89	8	0 ,90	0 ,8
⁏	0 ,32	4	0 ,00	1	0 ,08	4	0 ,40	2	—	0	—	0	0 ,16	0 ,0
=,	0 ,10	2	0 ,80	1	0 ,30	3	0 ,70	1	0 ,30	1	0 ,30	1	0 ,36	0 ,5
—̦	1 ,00	1	0 ,80	1	0 ,90	1	0 ,80	3	0 ,85	2	0 ,80	1	0 ,95	0 ,7

With these values a catalogue of magnitudes can be derived from HERSCHEL's estimates. When the symbols are grouped according to apparent magnitude remarkable constancy in the values of the symbols is found.

It was stated by G. MÜLLER in his Photometrie der Gestirne that a discussion of the catalogue of HERSCHEL was a very important task. The discussion by

PICKERING could not be taken as definite, and the very high accuracy of the estimates, all at an early epoch of photometric observations, are a subject for a discussion of a more extensive nature.

Such a discussion has recently been undertaken by E. ZINNER. He used for this purpose the Potsdam magnitudes and colours. The difference in the colours of the stars compared has been considered. The colour difference was said to be 1 if the colours differed: $(W + GW) - WG$, and 2 if the difference was $(W + GW) - (G + R)$. With grouping according to magnitude the following relation between the scale of PD and that of ZINNER and the colour was found.

	Colour diff. 1		Colour diff. 2	
Magnitude	PD	ZINNER	PD	ZINNER
$0^m,0 - 1^m,9$		$+0^m,06$		$+0^m,01$
2 ,0 − 2 ,9	$-0^m,01$		$+0^m,13$	
3 ,0 − 3 ,9		+0 ,08		+0 ,06
4 ,0 − 4 ,9	+0 ,10	+0 ,09	+0 ,17	+0 ,18
5 ,0 − 5 ,9	+0 ,18	+0 ,06	+0 ,31	+0 ,19
6 ,0 − 6 ,9	+0 ,21	—	+0 ,35	—

Since these differences had been smoothed out by empirical formulae the following values for the Herschelian symbols expressed in PD magnitudes were found:

$$
\begin{aligned}
. &= +0{,}020\,m - 0{,}016\\
; &= +0{,}030\,m - 0{,}055\\
, &= +0{,}075\,m - 0{,}132\\
\bar{,} &= +0{,}067\,m + 0{,}008\\
- &= +0{,}100\,m - 0{,}045\\
-, &= +0{,}230\,m - 0{,}394\\
-- &= +0{,}125\,m + 0{,}47
\end{aligned}
$$

From the use of the Potsdam and the ZINNER magnitudes together the following formulas resulted:

$$
\begin{aligned}
. &= +0{,}020\,m - 0{,}020\\
; &= +0{,}01\,(m - 2{,}0)^2\\
, &= +0{,}01\,(m - 1{,}0)^2 + 0^m{,}07\\
\bar{,} &= +0{,}085\,m - 0{,}085\\
- &= +0{,}104\,m - 0{,}074\\
-, &= +0{,}027 m^2 + 0{,}01\\
-- &= +0{,}244\,m - 0{,}187
\end{aligned}
$$

The characters of HERSCHEL are called in order 0, $^1/_2$, 1, $1^1/_2$, 2, 3, and 4. The different values of the characters from the second system of formulae were computed for the values 3,0, 4,0, 5,0 and 6,0, and reduced to the degree 1 in the scale just introduced. The following values were found: $0^m,10$, $0^m,17$, $0^m,24$, $0^m,30$, coresponding to the said magnitudes. These numbers satisfy the equation:

$$\text{one degree} = 0{,}10 + 0{,}07\,(m - 3{,}0).$$

The following table gives the definite values of the Herschelian characters when the colour difference in different combinations is taken into account. The stars are taken according to the magnitudes in the order that b denotes the brighter stars and f the fainter. Three cases have to be discerned: 1. the stars are of equal colour; 2. between b and f is a difference in colour of one degree. 3. between the said stars is a difference in colour of two degrees. If the colour difference is one and b is the redder, the designation is 1 b r and in case of f being the redder 1 f r etc. Equal colour means a difference below $0^m,3$ in colour index. A difference of one degree means a difference in colour index between $0^m,4$ and $0^m,8$ and two degrees a difference in colour index between $0^m,9$—$1^m,7$ approximately.

Herschelian character	Colour difference					
	0	1 f r	1 b r	2 f r	2 b r	Magnitude
.	$+0^m,03$	$0^m,00$	$+0^m,06$	$+0^m,00$	$+0^m,06$	
;	0 ,06	0 ,03	0 ,09	0 ,03	0 ,09	
,	0 ,09	0 ,06	0 ,12	0 ,06	0 ,12	
‾,	0 ,14	0 ,11	0 ,17	0 ,11	0 ,17	$2^m,0 - 2^m,9$
—	0 ,19	0 ,16	0 ,22	0 ,16	0 ,22	
—,	0 ,30	0 ,27	0 ,33	0 ,27	0 ,33	
— —	0 ,42	0 ,39	0 ,45	0 ,39	0 ,45	
.	0 ,05	−0 ,01	0 ,11	−0 ,03	0 ,13	
;	0 ,09	+0 ,03	0 ,15	+0 ,01	0 ,17	
,	0 ,13	0 ,07	0 ,19	0 ,05	0 ,21	
‾,	0 ,21	0 ,15	0 ,27	0 ,13	0 ,29	3 ,0 − 3 ,9
—	0 ,29	0 ,23	0 ,35	0 ,21	0 ,37	
—,	0 ,48	0 ,42	0 ,54	0 ,40	0 ,56	
— —	0 ,67	0 ,61	0 ,73	0 ,59	0 ,75	
.	+0 ,07	−0 ,04	0 ,18	−0 ,10	0 ,24	
;	0 ,13	+0 ,02	0 ,24	−0 ,04	0 ,30	
,	0 ,19	0 ,08	0 ,30	+0 ,02	0 ,36	
‾,	0 ,29	0 ,18	0 ,40	0 ,12	0 ,46	4 ,0 − 4 ,9
—	0 ,40	0 ,29	0 ,51	0 ,23	0 ,57	
—,	0 ,64	0 ,53	0 ,75	0 ,47	0 ,81	
— —	0 ,91	0 ,80	1 ,02	0 ,74	1 ,08	
.	0 ,08	−0 ,08	0 ,24	−0 ,16	0 ,32	
;	0 ,17	+0 ,01	0 ,33	−0 ,07	0 ,41	
,	0 ,25	0 ,09	0 ,41	+0 ,01	0 ,49	
‾,	0 ,36	0 ,20	0 ,52	0 ,12	0 ,60	5 ,0 − 5 ,9
—	0 ,47	0 ,31	0 ,63	0 ,23	0 ,71	
—,	0 ,78	0 ,62	0 ,94	0 ,54	1 ,02	
— —	1 ,09	0 ,93	1 ,25	0 ,85	1 ,33	

From the use of the above tables all the comparisons of HERSCHEL can be reduced to the corresponding magnitudes in the scale of ZINNER. As for the rules used in the formation of mean values we must refer to the paper by ZINNER, in which the values of HERSCHEL for stars brighter than $5^m,50$ are given. The values of HERSCHEL's magnitudes as derived from the Herschelian characters by PICKERING are given in the Harvard Revised Photometry.

Some of the stars, suspected by HERSCHEL to be variable and therefore observed many times, can be used for a derivation of the accuracy of the observations. ZINNER has found:

		n			n
α Herc.	$\pm 0^m,22$	65	β Herc.. . . .	$\pm 0^m,16$	68
δ Herc.	± 0 ,15	27	ζ Herc.. . . .	± 0 ,18	66
π Herc.	± 0 ,20	44	α Urs. Min. . .	± 0 ,17	101

Excluding the suspected objects α Herc. and α Ursae Min. the mean error of one observation is found to be $\pm\, 0^m,17$, and as the mean number of observations is 2,4, the mean error of the catalogue magnitudes is $\pm 0^m,11$.

HERSCHEL was, in fact, the inventor and founder of the step-degree method (Stufenschätzungsmethode). His method is practically identical with that of ARGELANDER, which has undoubtedly more convenient designations.

It is to be regretted that in ZINNER's elaborate evaluation of the estimates of HERSCHEL no account has been taken of the equation of position, or of the systematic error arising from the fact that the objects compared did not have the same position with regard to the line joining the eyes of the observer. It is quite possible that the scale is as accurately derived as is possible and that the influence of the equation of position is shoveled over in the colour equation and zero-point correction. From a methodical point of view it would have been of much interest to have a determination of the influence of the equation of position in the naked eye estimates.

23. JOHN HERSCHEL. During his stay at the Cape, Sir JOHN HERSCHEL attacked the subject of southern stellar photometry[1]. By the method of sequences he determined the relative brightness of a number of stars visible to the eye. On the homeward voyage the southern stars were connected with the northern ones. His "astrometer" giving the lustre of α Centauri as $\frac{1}{275000}$ of that of the moon gave the first connection to the light scale established from measurements of terrestrial sources.

HERSCHEL continued the estimates in England, but did not finish the work on account of the appearance of ARGELANDER's Uranometria Nova. In 1867 he presented to the R. Astronomical Society his MS. Charts, in number 113, containing the estimated magnitudes of stars visible to the naked eye in both hemispheres. An explanatory note to the charts is given in MN 27 p. 213.

At first, stars were taken together in groups and arranged with regard to their order within the groups. Later on the stars were arranged in order, but the faintest were taken together in groups, without any order of sequence within the same group. The stars were observed at night by moonlight, but not in the immediate vicinity of the moon. Stars at lower altitudes than 20° were not observed. The observations in England were not made by using the method of sequences alone, but were also directly estimated to $^1/_3$ of a magnitude.

HERSCHEL has computed the magnitudes of the stars from the sequences in the way explained at length in his Cape Observations, pp. 327, 377.

At first the sequences were reduced to normal sequences. The frequencies of different combinations were plotted in a kind of scattering diagram. If the order was different for two letters it could easily be decided, from an inspection of this diagram, which order had the majority of the stars. For the evaluation of the normal sequences into magnitudes those of the Catalogue of the British Astronomical Association were used, and the magnitudes of the other stars read off from curves. In this reduction no attention was paid to the colours of the stars.

W. DOBERCK, assisted by J. I. PLUMMER and F. G. FIGG[2], has discussed the magnitudes of 1919 stars in the sequences of J. HERSCHEL. The corrected normal sequences of JOHN HERSCHEL have been used, and, as standards for the magnitudes, those of Uranometria Argentina. No consideration was given to the effect of the colour.

Recently the sequences of J. HERSCHEL have been reduced by ZINNER[3]. He found considerable influence from the colour as regards the fainter stars.

Magn.	F0—G0	K0—N
$2^m,0 - 2^m,5$	$+0^m,02$	$+0^m,04$
2 ,5 — 3 ,0	0 ,07	0 ,09
3 ,0 — 3 ,5	0 ,13	0 ,15
3 ,5 — 4 ,0	0 ,18	0 ,20
4 ,0 — 4 ,5	0 ,23	0 ,26
4 ,5 — 5 ,0	0 ,28	0 ,31
5 ,0 — 5 ,5	+0 ,34	+0 ,37

The formation of the magnitudes is described at length in ZINNER's paper. The mean errors as found by ZINNER and DOBERCK are:

Magn.	DOBERCK	ZINNER
0^m	$\pm 0^m,16$	$\pm 0^m,15$
1	0 ,15	0 ,12
2	0 ,10	0 ,09
3	0 ,12	0 ,09
4	0 ,18	0 ,12
5	0 ,28	(0 ,18)
6	0 ,44	

[1] Results of Astronomical Observations made during the Years 1834, 5, 6, 7, 8 at the Cape of Good Hope. London (1847).

[2] Ap J 11, p. 192 (1900).

[3] Bamberg Veröff Band II, p. 46 (1926).

The influence of the extinction was not computed, as care was taken not to use objects at low altitudes. ZINNER states that the reduction shows that the inclusion of a colour equation gives more accurate magnitudes than those obtained by earlier reductions. An influence is also exerted by the Milky Way which amounts to $-0^m,06$ for stars of 3^m, and to $-0^m,10$ for fainter stars. Thus the Milky Way stars have been estimated too faint.

The brightest stars are most frequently observed. Stars between $0^m,0-1^m,9$ are on an average founded on 10,0 observations, $2^m,0-2^m,9$ on 6,7, $3^m,0-3^m,9$ on 3,5, $4^m,0-4^m,9$ on 1,6, and $5^m,0-5^m,5$ on 1,1. Magnitudes have been derived by ZINNER to the number of 1904 for 658 different stars. The mean error of a catalogue magnitude averages $\pm 0^m,068$.

TULENIUS was the first and Sir JOHN HERSCHEL was the second to measure the magnitudes of the stars and the first who received results of accuracy. He used the "astrometer", invented by himself, for the derivation of a catalogue of 69 stars by comparing them with the light of the Moon. Stars at low altitudes were not measured. HERSCHEL found that the measurements showed a dependence on the position of the Moon with regard to the star. When the extinction is computed there is no influence left, according to ZINNER.

HERSCHEL has given his observations in such a form that they can be reduced again. ZINNER adopts the apparent magnitude of the Moon, $-12^m,55$, according to RUSSELL's investigation[1], and computes the apparent magnitude, $M☾$, of the Moon for each evening of observation. He finds the following formula for computing the magnitudes, m_H, of HERSCHEL:

$$m_H = M☾ - 10 + 5 \log d + \Delta M☾ - \Delta m,$$

where d is the measured distance of the lens and $\Delta M☾$ and Δm are the corrections on account of the extinction, for which the Potsdam table can be used[2].

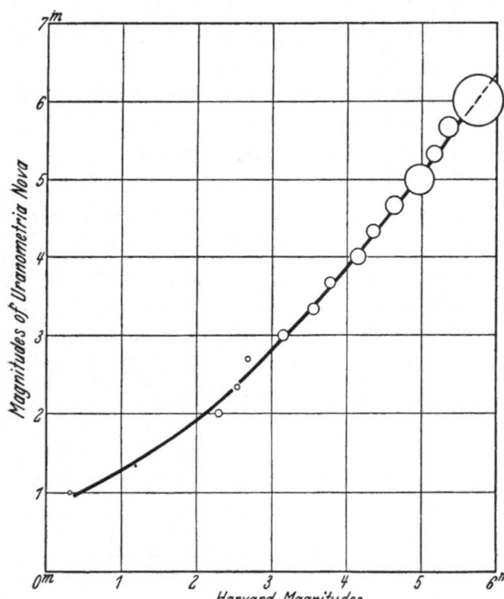

Fig. 49. Comparison between magnitudes of Uranometria Nova and those of HARVARD (RHP).

The magnitudes of HERSCHEL have been reduced to ZINNER's scale with a formula of the type:

$$m_Z = f + c(m_H - 3{,}0).$$

The constants f and c are derived from the material of each evening.

From the stars frequently observed by HERSCHEL ZINNER finds the mean error to be $\pm 0^m,090$, and, from deviations from the evening means for all the observed stars to be $\pm 0^m,10$, or of the same order of magnitude. As the stars are observed in the mean on 2,6 evenings, the average mean error of the magnitudes will be $\pm 0^m,056$.

RUSSELL has used the magnitudes of HERSCHEL, reduced to the Harvard system, for deriving the value of the magnitude of the Moon[3]. From HERSCHEL's observations alone he finds $M☾ = -10{,}79$.

[1] Ap J 43, p. 103 (1916). [2] MÜLLER, Photometrie der Gestirne, p. 515 (1897).
[3] Loc. cit.

During the first part of the 19th century the contributions to stellar photometry were few and generally only restricted to a few stars. Extensive observations were made by SCHWERD, but his observations do not seem to be very accurate.

24. ARGELANDER. Uranometria Nova. This work was published in 1843 and founded on observations between 1838 and 1843. ARGELANDER estimated all the stars visible to the naked eye at the horizon of Bonn and divided the light grades into magnitudes from 1—6. Besides this notations 1,2 and 2,1 were also used, and thus the total number of photometric degrees is 19. There are altogether 3256 objects delineated in the Uranometria Nova[1].

ARGELANDER does not say enough about his method, but it seems probable that objects were compared with each other. Although the accuracy is higher than in the earlier catalogues, it is evident that the scale of Uranometria Nova is also

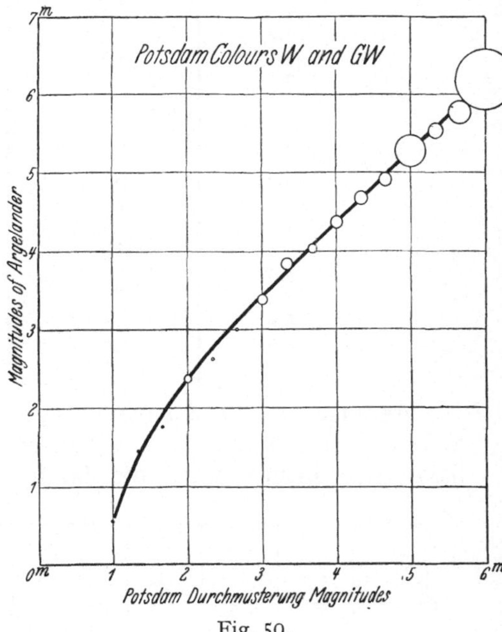

Fig. 50.

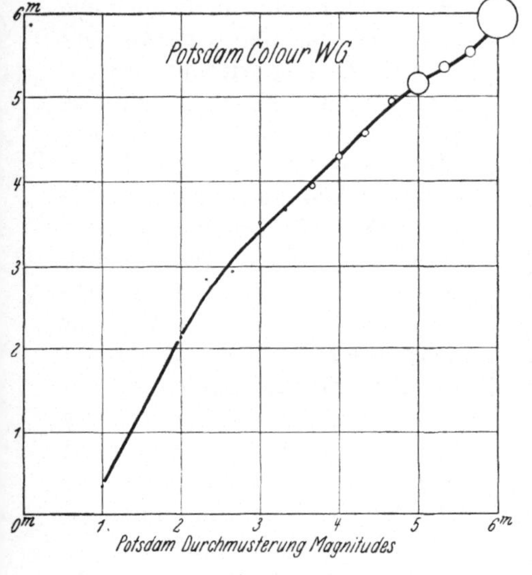

Fig. 51.

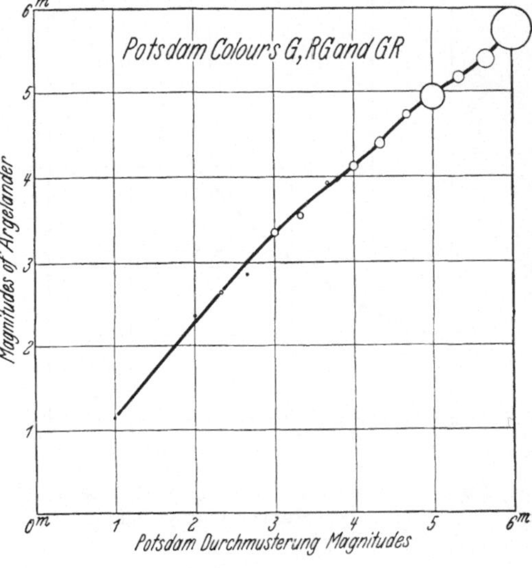

Fig. 52.

Fig. 50—52. Relation between the magnitudes of ARGELANDER and those of Potsdam Durchmusterung (PD).

not a uniform one. PICKERING has found the following general relation between the scale of Uranometria Nova and H P[2]:

---

[1] Uranometria Nova. Berlin 1843.

[2] Harv Ann 14, pt. II, p. 360 (1885).

Argelander magn.	0h—8h	8h—16h	16h—24h	All stars
1	0m,25	0m,60	0m,20	0m,32
1,2	—	1 ,40	1 ,13	1 ,20
2,1	1 ,40	—	1 ,50	1 ,42
2	2 ,08	2 ,29	2 ,40	2 ,30
2,3	2 ,28	2 ,73	2 ,52	2 ,53
3,2	2 ,57	2 ,93	2 ,68	2 ,69
3	3 ,20	3 ,16	3 ,12	3 ,15
3,4	3 ,47	3 ,63	3 ,55	3 ,55
4,3	3 ,70	3 ,92	3 ,78	3 ,78
4	4 ,15	4 ,29	4 ,01	4 ,14
4,5	4 ,28	4 ,30	4 ,40	4 ,34
5,4	4 ,49	4 ,72	4 ,66	4 ,62
5	4 ,88	5 ,06	4 ,94	4 ,96
5,6	5 ,11	5 ,28	5 ,21	5 ,17
6,5	5 ,26	5 ,36	5 ,43	5 ,35
6	5 ,70	5 ,84	5 ,69	5 ,74

ARGELANDER did not pay any attention to the influence of the extinction, and the effect of this is clearly shown in the estimates. Stars at low altitude have been estimated as too bright, and the faintest stars are comparatively low. ZINNER has made extensive investigations concerning the relation of the scale of ARGELANDER to that of PD[1]. The results are illustrated in the graphs.

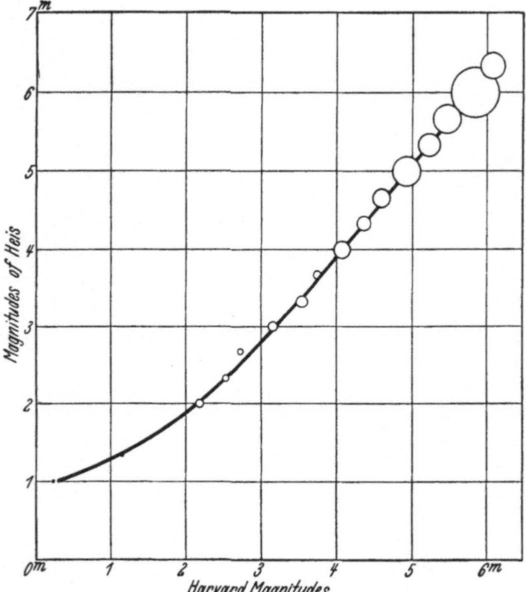

Fig. 53. Relation between estimates of HEIS and Harvard magnitudes.

25. HEIS. This observer is famous for his unusually keen eye sight. He was able to see 10 stars in the Pleiades without any instrument, and he saw the stars as points without any rays. He completed and extended the Uranometry of ARGELANDER during the years 1845—72. At first he identified the stars of Uranometria Nova and revised the magnitudes, but he also added the new designation — of $6\frac{1}{3}^m$. He did not use any artificial light in his observations, nor did he use any instrument when estimating the magnitudes. The number of stars observed by him from the horizon of Münster is 5421[2].

ZINNER has found the same general phenomena in the estimates of HEIS as in those of ARGELANDER. Thus there is an influence of the colour, of the apparent star density, and of the extinction. The stars at low altitudes are too faint. ARGELANDER and HEIS have tried to balance the influence of extinction in such a way that they estimate the stars near the horizon as somewhat brighter than the others. ZINNER's Table 40 quoted below shows in what degree this was possible.

[1] Bamberg Veröff Bd. II (1926).

[2] De magnitudine relativa numeroque accurato stellarum fixarum quae solis oculis conspiciuntur. Monasterii 1852. — Atlas coelestis novus. Köln 1872.

Mean δ	Extinction	ZINNER—HEIS	Excess in the estimates
$-3^{\circ},5$	$-0^{m},17$	$+0^{m},05$	$-0^{m},22$
$-8,5$	$-0,24$	$0,00$	$-0,24$
$-13,5$	$-0,33$	$-0,05$	$-0,28$
-18	$-0,45$	$-0,10$	$-0,35$
$-22,5$	$-0,62$	$-0,10$	$-0,52$

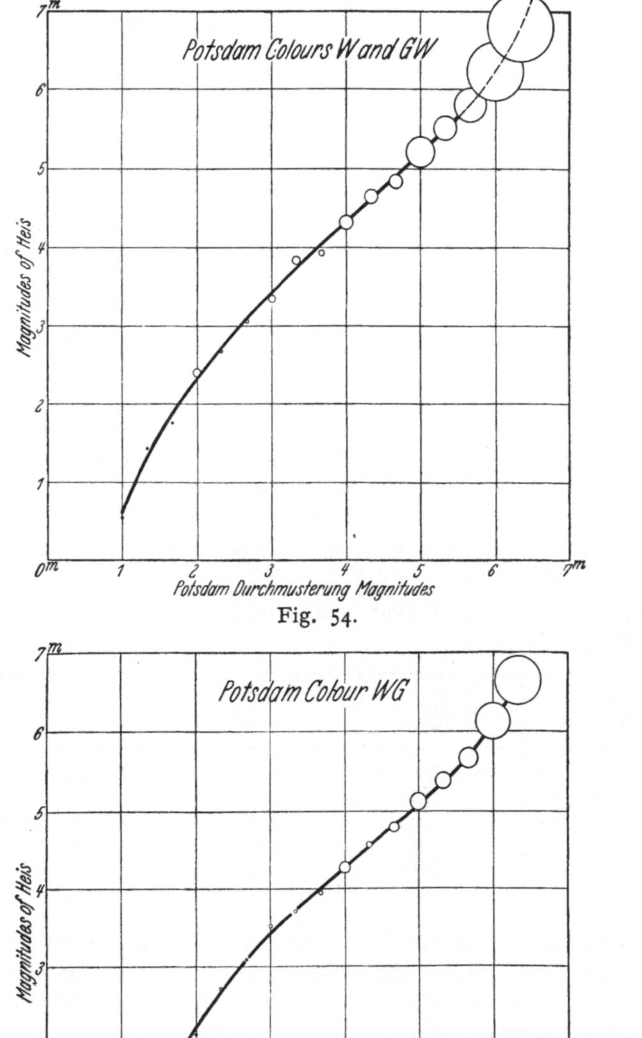

Fig. 54.

Fig. 55.

As is seen from an inspection of the figures 12—14, and 18—20, there is a decided influence of the Milky Way in the oldest estimates of magnitudes in the sense that a star, situated in the Milky Way, is assigned too high a magnitude, probably on account of the circumstance that the bright background acts as a veil and thus adds its intensity to that of the star.

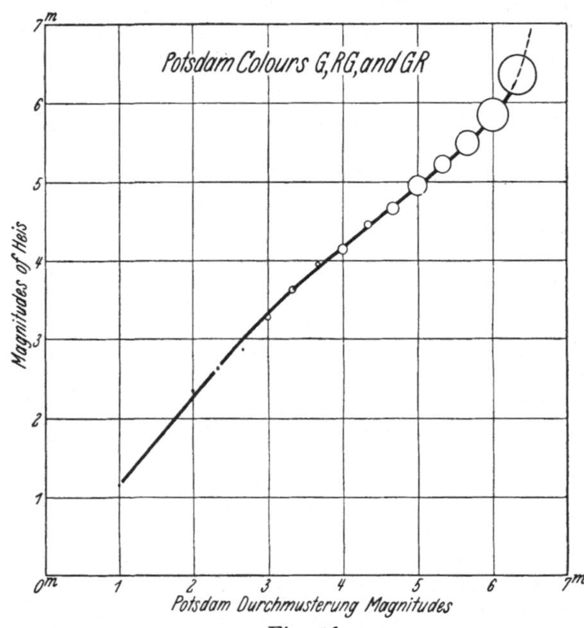

Fig. 56.

Fig. 54—56. Curves for reducing magnitudes of HEIS into magnitudes of PD.

The influence of the Milky Way in the estimates of HEIS is illustrated from Table 41 in ZINNER's work.

HEIS magnitude	Density equation Dist.—Near	n	HEIS mag.	Density equation Dist.—Near	n
3	$+0^m,24$	9	5	$+0^m,20$	76
$3^1/_2$	0 ,17	10	$5^1/_3$	0 ,25	67
$3^2/_3$	0 ,00	9	$5^2/_3$	0 ,22	114
4	0 ,16	32	6	0 ,15	326
$4^1/_3$	0 ,14	25	$6^1/_3$	0 ,16	408
$4^2/_3$	0 ,20	25			

26. Uranometria Argentina. The introduction of the Uranometria Argentina[1] gives a vivid and very interesting account of the work of B. A. GOULD. It is indeed a human document which will be read with much advantage not only by students of photometry but by all astronomers.

The work was started at Albany. During the year 1858 the collaborators of GOULD—TOOMER, SEARLE, WINSLOW, and MCLANE TILTON—estimated the magnitudes of stars between declinations $-5°$ and $+50°$. No instrument was used, and the estimations were put down in $0^m,1$. The limiting magnitude is $6^m,1-6^m,2$. Some 6000 estimates of 2070 stars were performed. The observations have partly been destroyed at Albany, but a catalogue of resulting magnitudes has been printed as a working list.

During the years 1870—1871 M. ROCK and W. M. DAVIES estimated at Cordoba the stars brighter than 4^m between $\delta + 10°$ and $-90°$. From these

[1] Cordoba Publ 1 (1879)

observations of stars a number of standards were obtained for testing suspected stars for variability. According to ZINNER the mean error is $\pm 0^m,12$ for stars between $1^m,4$ and $2^m,9$ and $\pm 0^m,13$ for stars between $2^m,9$ and $4^m,1$.

M. ROCK, W. M. DAVIES, C. L. HATHAWAY, and J. M. THOME, took part in the estimates for the Uranometria Argentina during the years 1871—73. The last-mentioned observer went over all the magnitudes again during the years 1874—79. Being extremely short-sighted, GOULD could not take part in the observations himself. The catalogue does not give the dates of the separate observations.

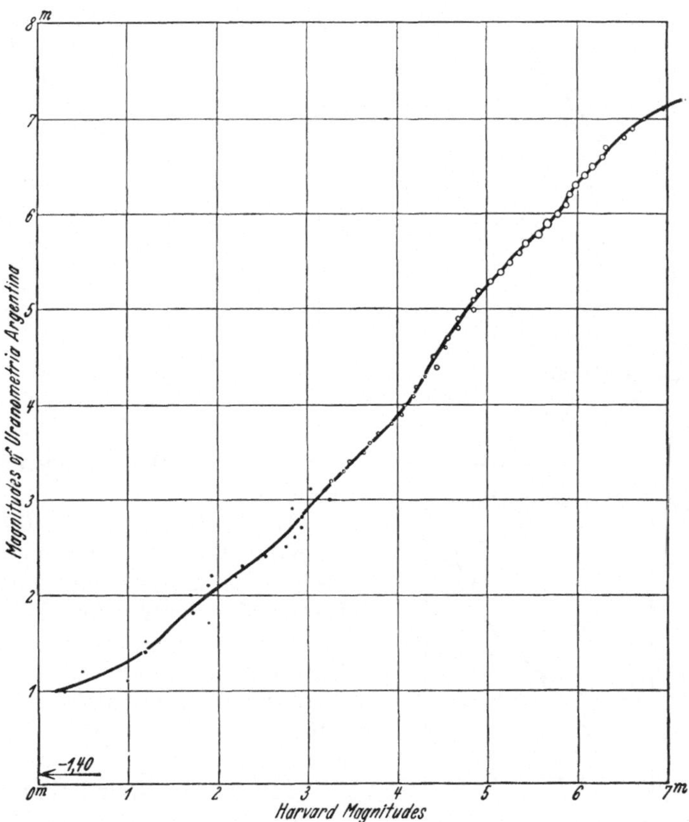

Fig. 57. Comparison between the magnitudes of Uranometria Argentina and those of Harvard (R H P).

The estimates were originally made in tenths of a magnitude, and the mean results, also given in tenths, are given in the catalogue. Vulgar fractions are also employed in the case of groups of stars, or doubles, or multiples. The catalogue is arranged by constellations, the limits of which are defined by meridians and parallels. The scale of magnitudes is intended to agree with ARGELANDER's. The connection with the scale of Uranometria Nova was found by observations in the zone $+5°$ to $+15°$ which is at the same altitude at Bonn and at Cordoba.

Altogether 1800 stars were observed, but only 722 of these were used for establishing the scale. Two regions close to the South Pole were also used.

The behaviour of the eyes of the different observers is illustrated in the Uranometria.

The influence of the colour was also taken into account, and GOULD is of the opinion that the resulting magnitudes are more accurate than those resulting from the arrangement in sequences.

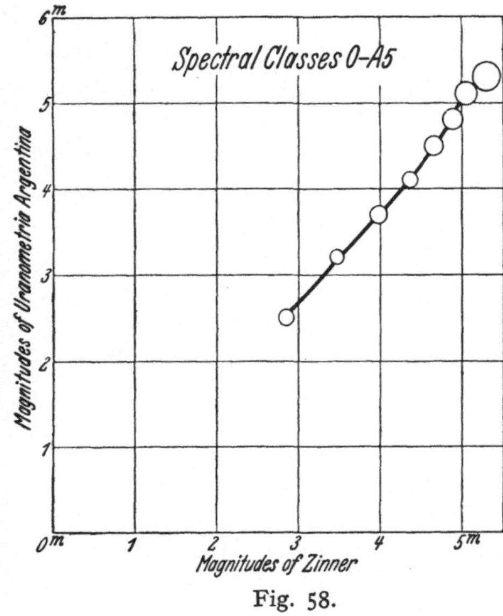

Fig. 58.

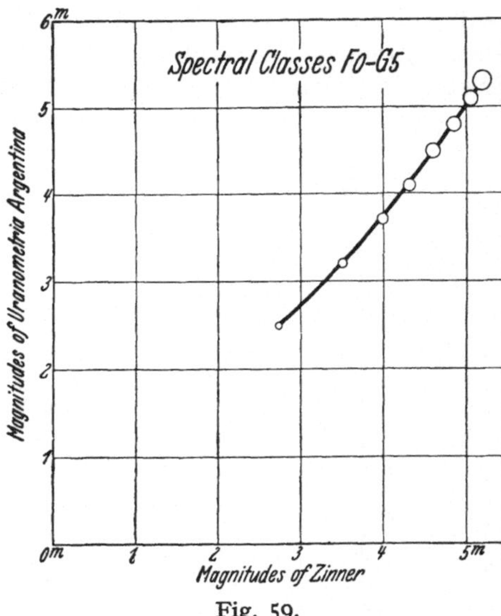

Fig. 59.

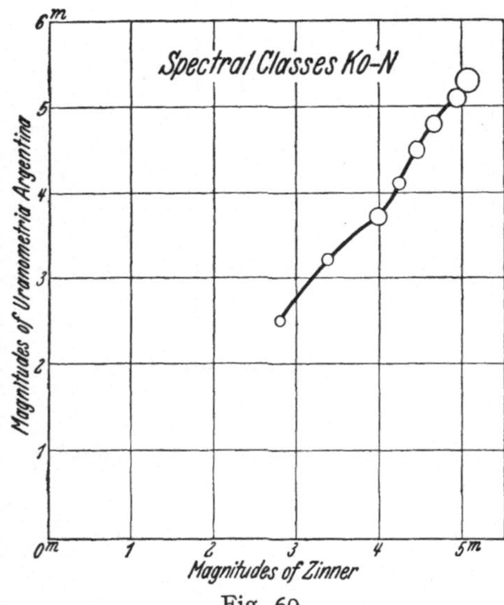

Fig. 60.

Fig. 58—60. Curves for reducing the magnitudes of Uranometria Argentina into the magnitudes of ZINNER.

GOULD had certain difficulties in extending the scale to lower grades than $6^m,0$. The faintest stars observed on favourable nights were close to $7^m,0$. In order to test the faintest grades of light GOULD introduced a scheme to diaphragm the aperture of a telescope until a star was just visible at a glance.

ZINNER has found that there is a certain influence from the Milky Way. For spectral classes O—A5 the correction is $+0^m,24$ and for F—G and K—M $+0^m,25$, and $+0^m,18$ respectively.

The magnitudes were corrected for the influence of the apparent star density (Milky Way), and then the dependence on the colour has been determined. The following empirical formulae were found:

$$C_1 = +\,0{,}016\,(m - 2{,}0) \quad \text{P D colour: } (W + GW) - WG$$
$$C_2 = +\,0{,}057\,(m - 2{,}0) \quad \text{,, ,, : } (W + GW) - G.$$

The following table gives the reduction of magnitudes in Uranometria Argentina to the Harvard scale.

U.A.	0m,0	0m,1	0m,2	0m,3	0m,4	0m,5	0m,6	0m,7	0m,8	0m,9
1	0 ,7	0 ,8	0 ,9		1 ,2	1 ,3		1 ,6	1 ,7	
2	2 ,0	2 ,1	2 ,2	2 ,3	2 ,4	2 ,6	2 7	2 ,8	2 ,9	3 ,0
3	3 ,1	3 ,2	3 ,3	3 ,4	3 ,5	3 ,6	3 ,7	3 ,8	3 ,9	4 ,0
4	4 ,1	4 ,2	4 2	4 ,3	4 ,4	4 ,4	4 ,5	4 ,6	4 ,7	4 ,7
5	4 ,8	4 ,9	4 ,9	5 ,0	5 ,1	5 ,2	5 ,3	5 ,4	5 ,6	5 ,7
6	5 ,8	5 ,9	6 ,0	6 ,1	6 ,2	6 ,3	6 ,4	6 ,5	6 ,6	6 ,7
7	6 ,8	6 ,9			7 ,2					

27. HOUZEAU and Others. During the years 1875—76 I. C. HOUZEAU, working at Jamaica, estimated the magnitudes of all the stars visible to the unaided eye. The faintest stars were called $6^1/_2$ in his Uranométrie Générale[1]. No instrument was used during his estimates. The stars were graded in half magnitudes. According to ZINNER the mean error of the magnitudes for the limits $1^m—4^1/_2{}^m$ is $\pm 0^m,34$. The magnitudes of the bright stars seem to deviate considerably from the law of FECHNER.

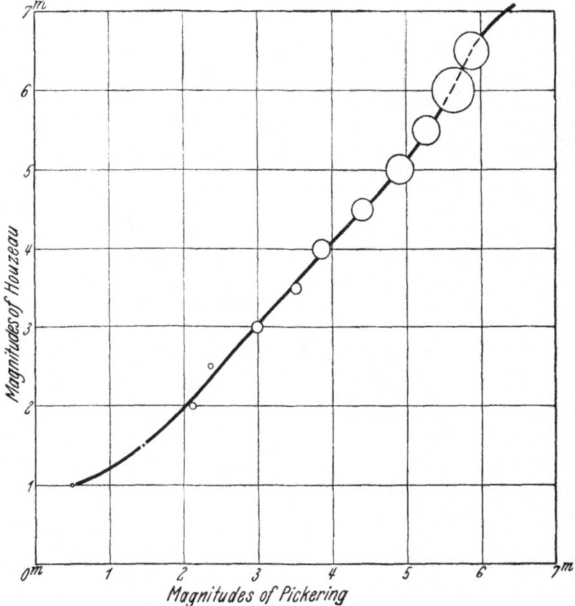

Fig. 61. Curve showing the relation between magnitudes of HOUZEAU and those of PICKERING (Harv Ann 14 = HP).

FLAMMARION. In his famous work, Les Étoiles[2], C. FLAMMARION gives magnitudes of the stars within the different constellations, which were estimated to a considerable degree by himself. The estimates seem to have been performed with the aid of an opera-glass, and the magnitudes are expressed in $0^m,1$. The observations cover the sky from the north pole to $-35°$. For the stars south of that declination the magnitudes in the Uranometria Argentina are used. ZINNER has derived the following mean error for the different magnitudes of FLAMMARION:

Magn.	Mean error	
$2^m,0-2^m,9$	$\pm 0^m,23$	$\pm 0^m,24$ (for $2^m,0$–$4^m,9$)
3 ,0—3 ,9	0 ,28	
4 ,0—4 ,4	0 ,25	
4 ,5—4 ,9	0 ,21	
5 ,0—5 ,4	0 ,24	

The decimals 0 and 5 have a high frequency in the estimates of FLAMMARION. The first decimal occurs but seldom.

ROGERS. During 1878 W. A. ROGERS[3] in connection with meridian work at the Harvard Observatory made careful estimates of magnitudes. The estimates were so performed that his scale should be in concordance with that of the B D. An interesting source of systematic errors affects these observations. The stars were selected from the point of view of determining A R and Decl of stars of 6^m (or brighter) not recently observed for position. Of these stars those which are really fainter would be more likely to

[1] Ann de l'Obs de Bruxelles, Nouv Ser 1 (1878).

[2] See also Atlas de toutes les étoiles visibles à l'œil nu, dans les deux hemisphères. First edition Paris 1872.

[3] Harv Ann 14, p 368 (1874).

be included among the stars observed by ROGERS than those brighter. Accordingly the scale of the B D as inferred from these stars would be too faint. The course of the systematic error is illustrated in fig. 63.

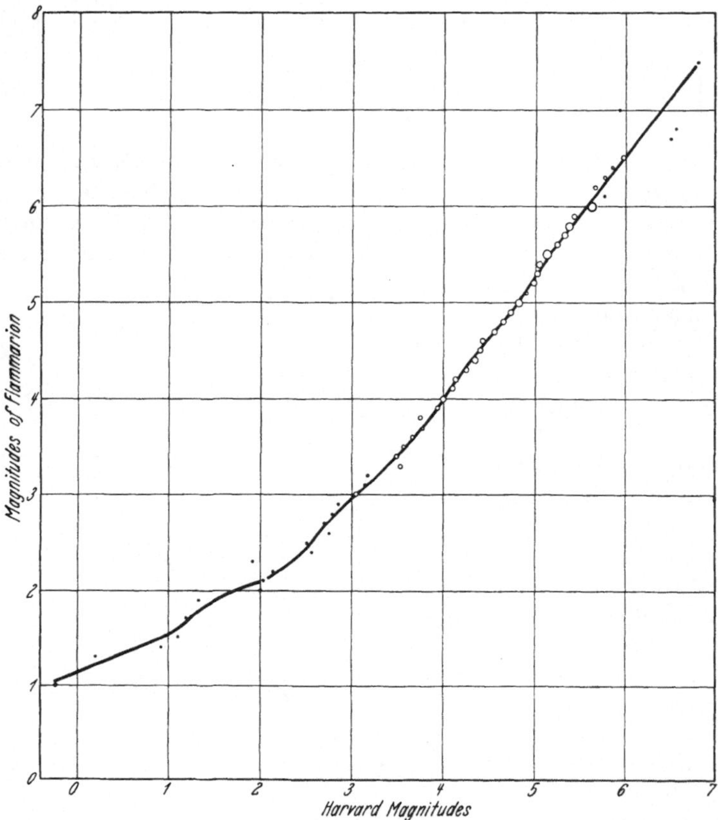

Fig. 62. Relation between the magnitudes of FLAMMARION and the Harvard Photometry (HP).

BEHRMANN. In 1866—67 during a journey to the southern hemisphere BEHRMANN[1] estimated magnitudes of southern stars which were used as foun-

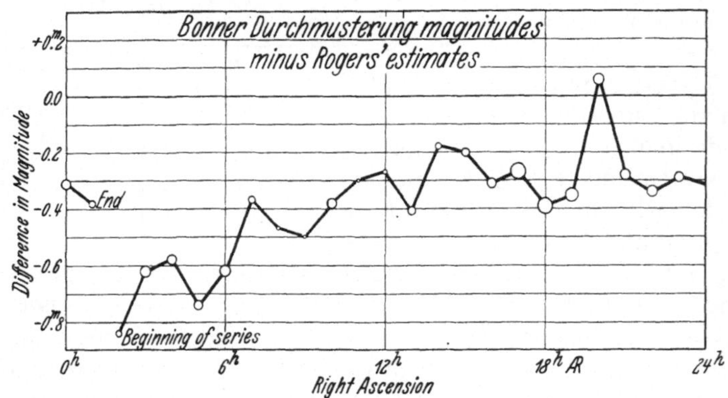

Fig. 63. Comparison between B D magnitudes and ROGERS's estimates. As the observations were started around 2^h and carried on during the course of a year it seems that the observer partly adjusted his scale during the course of observing.

[1] Atlas des südlichen gestirnten Himmels (1874).

dation for his Atlas. The region observed included stars visible to the unaided eye situated within 70° distance from the South Pole. The same designations as in Uranometria Nova are used and the faintest stars are named 6^m.

EDMANDS. When the extensive photometric work at Harvard inaugurated by E. C. PICKERING[1] was started, PICKERING found it desirable that the photometric determinations should be compared with similar results by direct estimates of relative brightness. As no such precise and systematically performed estimates of magnitudes of northern stars were available it was decided to procure careful estimates of the brightness of the stars visible to the naked eye in the northern hemisphere. The general conduct of the work was assigned to J. R. EDMANDS. The estimates were carried out by EDMANDS, S. C. CHANDLER, J. C. HOWARD and E. C. PICKERING of which the two first mentioned have made the main part of the work. This series of observations embraces altogether 2750 stars.

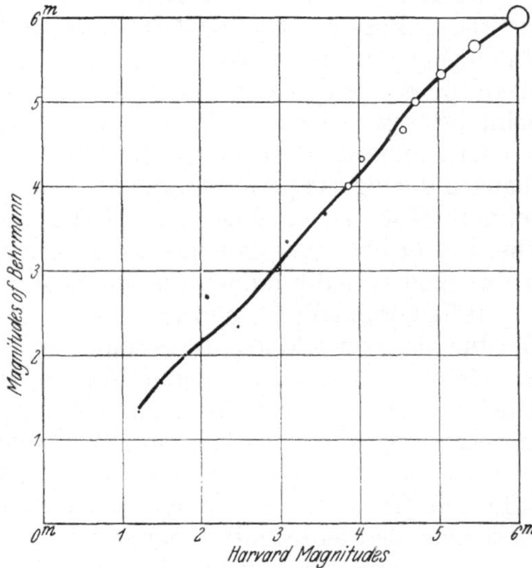

Fig. 64. Relation between the magnitudes of BEHRMANN and the Harvard Photometry (H P).

A. STANLEY WILLIAMS[2] has estimated the magnitudes of 1081 stars situated between —30° decl. and the South Pole. The observations were made in the years 1885—86 with the aid of an opera-glass and each star was directly compared with two others. The residuals between WILLIAMS's estimates and the Harvard scale are found in the R H P and could be used for a determination of the mean error of the magnitudes.

28. Bonner Durchmusterung (B D). The extensive work of F. W. ARGELANDER and his pupils at Bonn, generally known by the name of Bonner Durchmusterung, is of very great importance as regards stellar photometry. The work was performed during the years 1852—68 and consists of a series of charts and a catalogue giving approximate coordinates of stars between the declinations +90° and —2°, and their magnitudes. The catalogue and the charts are complete down to $9^m,2$. Fainter stars down to $10^m,5$ are also included, but only in a smaller proportion.

ARGELANDER's work was extended by his successor SCHÖNFELD, who observed and reduced the stars between the declinations —2° and —23° during the years 1875—1884. The catalogue was designed to be complete to $9^m,5$ and to include partially fainter grades down to 10^m.

The following table gives references to the catalogues of the B D.

Name				Limits of Decl.	Number of stars	Source
Bonner Sternverzeichnis	I.	Section	1859	—2° 0′ to +20°0′	110985	Bonn Astr Beob Vol III
,, ,,	II.	,,	1861	+20 0 ,, +41 0	105075	,, ,, ,, ,, IV
,, ,,	III.	,,	1862	+41 0 ,, +90 0	108129	,, ,, ,, ,, V
,, ,,	IV.	,,	1886	— 2 0 ,, —23 0	133659	,, ,, ,, ,, VIII

[1] Harv Ann 14, p. 70 (1884).

[2] Catalogue of 1081 Stars. London 1898.

Of course, it is very important to know the relation between the scale of the B D and other fundamental scales, and the accuracy of the work. As has been pointed out by G. MÜLLER in his "Photometrie der Gestirne", astronomers seem to have had, sometimes, a tendency to overestimate the accuracy of the magnitudes in B D. The main purpose of the work was to obtain an index catalogue of the stars. Thus it was unnecessary to estimate the magnitudes with the utmost accuracy. Besides, the inevitable rush in doing such a work, especially in rich star regions, did not make it possible to give much time to the estimates of the magnitudes. Neither were the nights of observation selected from a photometric point of view, which is also of some importance.

One must also remember the fact that the method of observation slightly changed during the performance of the work. The tenths of magnitudes were not actually observed in some parts of the work, but were results of computation. The way of observing is explained in a letter from SCHÖNFELD to PEIRCE, and the method actually followed is practically the same as that given in the letter[1].

It had originally been intended to estimate the brightness in half magnitudes, and the observer adopted the scale 1 mg, 1,5 mg, 2 mg, 2,5 mg, but still used the symbols 1,2 and 2,3 to denote the magnitudes between 1 and 2, 2 and 3, etc., which in this case were the exact half magnitudes 1,5 and 2,5. Two years after the beginning of the work, or in 1854, SCHÖNFELD and KRUEGER, who made by far the greater number of the observations, began to take account of a perceptible difference from half a magnitude. Thus, if a star belonged to the faintest of those classified as seventh magnitude, that is, was less than the division 7·8 or 7,5 mg, it was distinguished by the addition of the letter s, schwach (faint) or 7·8 s, which made it a little fainter than 7,5 mg or the equivalent of 7,7 mg. Similarly, a star that belonged to the brighter half of the class was designated by gt, gut, and 7·8 gt meant a star somewhat brighter than 7,5 or 7,3 mg. One magnitude was thus divided at the observations into six parts, which is clearly seen from the adjoining synopsis.

Designation	Corresponding to magnitudes
7^m	$6^m,9$; $7^m,0$; $7^m,1$
7s	7 ,2
7·8 gt	7 ,3
7·8	7 ,4; 7 ,5; 7 ,6
7·8 s	7 ,7
8 gt	7 ,8
8	7 ,9; 8 ,0; 8 ,1

SCHÖNFELD points out that it was not so easy to make such a distinction of $0^m,1$ as of $0^m,2$ in the rush of observing; this explains why three tenths are included in three groups. Besides, the stars were often coming so rapidly that the observers had no time to write the necessary notes. Hence it resulted that in the regions where the stars were fewer finer degrees of difference were noted than among the stars in the Milky Way.

In the year 1857 when the observers had reached the more northern declinations and the stars were on the average less numerous, they became accustomed to distinguish tenths of magnitudes directly. But it is noticeable that the fractions $0^m,1$, $0^m,6$, and $0^m,9$ occur less frequently than the others, especially the two first. The magnitudes which were published in the B D are the means of the separate determinations, some of them depending on two observations, and some on three. When they depended on two observations, the nearest even tenth was chosen, so that the star was estimated as fainter in general than the arithmetical mean. For example, if we have the observations $8{\cdot}9 = 8^m,5$ and

[1] The methods of observing and the results are given at length in the paper by ARGELANDER, „Anzeige von einer auf der Königlichen Universitätssternwarte zu Bonn unternommenen Durchmusterung des nördlichen Himmels als Grundlage neuer Himmelskarten". Bonn 1856.

$9 = 9^m,0$, the mean is $8^m,8$ and not $8^m,7$. As a result of this variation in the method of determining the magnitudes of the B D, it appears that they are not homogeneous throughout the entire catalogue.

A count of the number of stars for each tenth of a magnitude in the B D has been made by LITTROW[1] and also by PEIRCE[2]. From the counts of the latter the following frequency table has been condensed.

Decimal	Frequency	Decimal	Frequency
$0^m,0$	8681	$0^m,5$	14168
0 ,1	2290	0 ,6	5006
0 ,2	5061	0 ,7	8956
0 ,3	6373	0 ,8	14593
0 ,4	3771	0 ,9	8895

Although the accuracy of B D cannot compete in general with the accuracy of Sir WILLIAM HERSCHEL's estimates of the stars, still the B D magnitudes are of much importance. They not only give a very valuable contribution to the knowledge of the function $B(m)$, i. e., the number of stars brighter than a certain magnitude m, but they are in several cases the only decent visual estimate of the magnitude we have.

Thus it is clear that the most extreme care must be employed when discussing the scale of the B D. This question was discussed at an early date and numerous contributions have been made. PEIRCE remarked that a glance at the well-known map by R. A. PROCTOR[3], where all the stars are plotted on a single map, shows that the scale of magnitudes for all stars north of 81° declination is different from that used more southwards. The photometric measurements by P. G. ROSÉN[4] show the same thing. Thus he finds for the light ratio $\varrho = i_m/i_{m+1}$, where $m = -2,5 \log i$ and m an integral number:

$$\log \varrho = 0,433 \pm 0,012 \text{ for the polar zone, and}$$

$$\log \varrho = 0,380 \pm 0,008 \text{ for the others.}$$

PEIRCE was also aware of the fact that the difference of scale cannot be nearly as large as ROSÉN makes it on account of the few objects measured in his investigations.

ROSÉN measured only 110 stars and the subsequent measurements of E. LINDEMANN[5] embrace only 290 stars; this is also too low a number to give a foundation for a derivation of the scale.

The following table gives a collection of a number of photometric determinations of the value of the POGSON-ratio ($\log \varrho$) for different magnitude intervals. It has not been possible to distinguish between the estimates in the B D and those in ARGELANDER's Uranometria Nova. Details about the photometric catalogues used are found in the text.

Source \ Estimates at Bonn	2^m—3^m	3^m—4^m	4^m—5^m	5^m—6^m	6^m—7^m	7^m—8^m	8^m—9^m	2^m—6^m	6^m—9^m
ZÖLLNER	0,406	0,283	0,315	0,209	—	—	—	0,341	
SEIDEL	0,487	0,362	0,342	—	—	—	—	—	
PEIRCE	—	0,391	0,340	0,437	—	—	—	0,371	
WOLFF	0,368	0,328	0,230	0,178	—	—	—	0,305	
Harv Phot	0,396	0,368	0,328	0,382	—	—	—	0,356	
Uranometria	0,424	0,368	0,364	0,377	—	—	—	0,385	
Potsdam D M	0,329	0,329	0,329	0,329	0,400	(0,400)	—	0,329	(0,400)
ROSÉN		—	—	0,388	0,388	0,363	0,379	—	0,380
LINDEMANN	0,291			0,303	0,394	0,392	0,437	0,280	0,402

[1] Wien Sitzgsber d K Akad d Wiss 59 (1869); 61 (1870).
[2] Harv Ann 9, pt. 1 (1878).
[3] M N 31, p. 175 (1871).
[4] St Pétersbourg Ac Sc Bull 14 (1870).
[5] St Pétersbourg Ac Sc Bull 20 (1874).

It is clear that the B D scale is too narrow or that it gives somewhat too low a value for one magnitude. On the other hand, it will be difficult to base a determination of the properties and behaviour of the Bonn scale on the photometric measurements quoted above. There are perceptible differences of an accidental and systematic nature between the different series, on account of the fact that the photometric values are not free from systematic errors, or do not depend on a photometric scale fulfilling the condition $m = -2{,}5 \log i$.

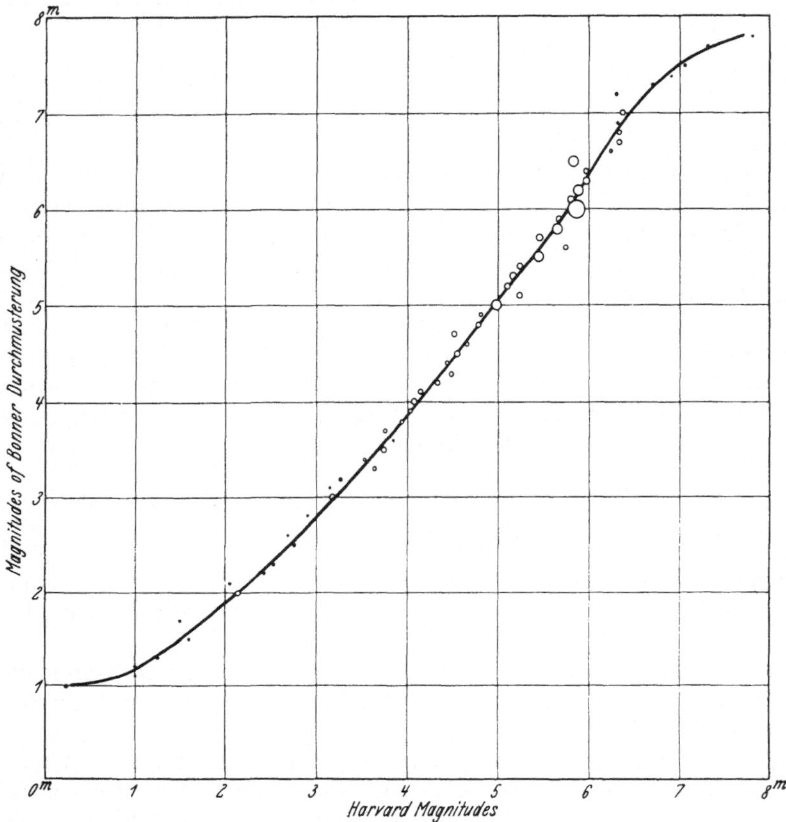

Fig. 65. Relation between the magnitudes of Bonner Durchmusterung and those of Harvard (H P).

The most rational way will undoubtedly be to reduce the scale of the B D to the Harvard scale (H S). In Harv Ann Vol. 24 E. C. PICKERING has published a number of 16865 magnitudes of B D stars measured with the meridian photometer for the purpose of a photometric revision of the B D.

PICKERING returned later on to a complete discussion of the scale of the B D. In Vol. 70 of the Harv Ann he has published magnitudes of 11139 stars of B D magnitudes $9^m{,}0$ to $9^m{,}5$. The earlier results in Harv Ann 24 giving the magnitudes of 14258 stars, of B D magnitudes $6^m{,}3$ to $9^m{,}0$, were also used. The detailed discussion is given in Harv Ann 72, No. 6. Table XVIII in PICKERING's paper gives the resulting corrections in such a form that the Harvard magnitude can be read off at once with the B D magnitude and the right ascension as argument. In the table on the next page is given a condensed table from the extensive table in the Harv Ann 72, No. 6.

Reduction of B D-magnitudes to the H S. (Condensed table from PICKERING's investigations.) Tabulated values are B D-magnitudes reduced into H S-magnitudes.

BD	I	II	III	IV	Mean.	BD	I	II	III	IV	Mean.
$7^m,1$	$7^m,1$	$7^m,0$	$7^m,2$	$7^m,2$	$7^m,1$	$8^m,6$	$8^m,8$	$8^m,6$	$8^m,8$	$8^m,8$	$8^m,8$
,2	7 ,2	7 ,1	7 ,3	7 ,3	7 ,2	,7	8 ,9	8 ,7	8 ,9	8 ,9	8 ,9
,3	7 ,3	7 ,2	7 ,4	7 ,4	7 ,3	8 ,8	9 ,1	8 ,9	9 ,1	9 ,0	9 ,0
,4	7 ,4	7 ,3	7 ,5	7 ,5	7 ,5	,9	9 ,2	9 ,0	9 ,2	9 ,1	9 ,1
7 ,5	7 ,5	7 ,4	7 ,6	7 ,6	7 ,5	9 ,0	9 ,4	9 ,2	9 ,4	9 ,3	9 ,3
7 ,6	7 ,6	7 ,5	7 ,7	7 ,7	7 ,6	9 ,1	9 ,6	9 ,4	9 ,5	9 ,5	9 ,5
,7	7 ,7	7 ,6	7 ,8	7 ,8	7 ,7	,2	9 ,8	9 ,6	9 ,8	9 ,6	9 ,7
,8	7 ,9	7 ,7	7 ,9	7 ,9	7 ,8	,3	10 ,0	9 ,9	10 ,0	9 ,7	9 ,9
,9	8 ,0	7 ,8	8 ,0	8 ,0	7 ,9	,4	10 ,3	10 ,2	10 ,2	10 ,0	10 ,2
8 ,0	8 ,1	7 ,9	8 ,1	8 ,1	8 ,0	9 ,5	10 ,5	10 ,6	10 ,7	10 ,3	10 ,5
8 ,1	8 ,2	8 ,0	8 ,2	8 ,2	8 ,2	9 ,6	10 ,6	—	—	—	10 ,6
,2	8 ,3	8 ,1	8 ,3	8 ,3	8 ,3	,7	10 ,7	—	—	—	10 ,7
,3	8 ,5	8 ,3	8 ,5	8 ,5	8 ,4	,8	10 ,9	—	—	—	10 ,9
,4	8 ,6	8 ,4	8 ,6	8 ,6	8 ,5	,9	11 ,0	—	—	—	11 ,0
8 ,5	8 ,7	8 ,5	8 ,7	8 ,7	8 ,6	10 ,0	11 ,1	—	—	—	11 ,1

H. v. SEELIGER[1] has used the earlier material for a derivation of the corrections to the B D scale. His general results can be summed up in the following statements.

Within the separate zones of declination the scale is not a uniform one. The magnitudes in B D are dependent on the apparent density of stars in the regions considered, that is to say, the scale is not the same in the Milky Way regions and outside them. This fact was first brought to light by SCHÖNFELD.

H. v. SEELIGER has found the following general relation between the Harvard magnitudes, m_H, and the B D magnitudes m_{BD}.

$$m_{BD} - m_H = C_m - 0{,}014\Delta - 0{,}043\,(m\cdot\Delta) + 0{,}0368\,\frac{\Delta}{i_m}$$

where C_m is a constant depending on the magnitude m; Δ is a measure of the star density, and i_m is the intensity of the magnitude m. As for Δ it is found from the formula: $\Delta = D - 0{,}7$, where D is the star density expressed in units of the star density D_0 in galactic zone V.

The following simplified formulas resulted:

$$\begin{aligned} m_{BD} - m_H &= -0^m{,}016 + 0^m{,}023\,\Delta; && 6^m{,}5\\ &= -0\ {,}058 + 0\ {,}024 && 7\ {,}0\\ &= -0\ {,}067 + 0\ {,}035 && 7\ {,}5\\ &= -0\ {,}067 + 0\ {,}068 && 8\ {,}0\\ &= -0\ {,}118 + 0\ {,}131 && 8\ {,}5\\ &= -0\ {,}199 + 0\ {,}246 && 9\ {,}0 \end{aligned}$$

The parameter Δ is considered constant within each galactic zone. It has the following values for the eight zones used in the investigations by VON SEELIGER:

Zone β	I	II	III	IV	V	VI	VII	VIII
Δ	−0,29	−0,23	+0,07	+0,30	−0,02	−0,25	−0,33	−0,35

Numerous comparisons between the magnitudes of B D and those in older catalogues have been made by a number of investigators. We refrain from giving a detailed account of these investigations here inasmuch as direct comparisons with P D or Harvard magnitudes are to prefer. In order to give the reader a general view of the relationship between the B D magnitudes and those

[1] Sitzber k bayer Akad d Wiss 28, p. 147 (1898).

in old catalogues we give here a table borrowed from MÜLLER's Photometrie der Gestirne:

Comparison between the BD-magnitudes and those in a number of old photometric catalogues.

Catalogue \ BD	$3^m,0$	$3^m,5$	$4^m,0$	$4^m,5$	$5^m,0$	$5^m,5$	$6^m,0$	$6^m,5$	$7^m,0$	$7^m,5$	$8^m,0$	$8^m,5$	$9^m,0$
PTOLEMAIOS . . .	3,1	3,6	4,4	4,7	5,0	5,3	5,5						
AL SÛFI	3,0	3,5	4,1	4,6	4,9	5,1	5,4	5,9					
Uranometria Nova (ARGELANDER) .	3,0	3,4	4,0	4,5	5,0	5,4	6,0						
HEIS	3,0	3,4	4,0	4,5	5,0	5,5	6,0	(6,5)					
HOUZEAU	2,8	3,3	3,9	4,4	4,9	5,4	5,9	6,3	6,6				
Uranometria Argentina	2,9	3,4	4,0	4,4	4,9	5,4	6,0	6,5	7,0				
LALANDE	2,9	3,3	3,9	4,5	4,9	5,5	6,15	6,75	7,15	7,55	7,9	8,3	8,5
BESSEL	3,1	3,4	3,7	4,2	4,7	5,2	5,7	6,2	6,8	7,4	7,9	8,4	8,8
STRUVE	3,2	3,7	4,4	4,8	5,2	5,7	6,2	6,6	7,2	7,7	8,3	8,8	9,3
SCHÖNFELD (S D M)					4,9	5,4	6,0	6,6	7,2	7,6	8,1	8,5	9,1
SCHJELLERUP . . .							5,7	6,6	7,0	7,5	8,2		8,9

29. HOPMANN's Reduction of the B D-Scale[1]. In order to reduce the scale of B D to a photometric system J. HOPMANN has used the catalogue of KÜSTNER of 10663 stars observed at Bonn. Although the magnitudes m_B in question are estimated, they have been reduced to the P D system through the use of gratings during the observations. As practically no mistake with regard to the identification is thus possible, and as the magnitudes depend on the most accurate photometric scale that has been established by means of visual measurements, they ought to be well adapted for a derivation of the B D scale. The comparison was limited to stars fainter than $7^m,5$, as the brighter stars in B D have been already superseded by the observations in the P D catalogue.

It is known that the magnitudes of the B D stars depend on the relative number of stars in the background, in such a way that the stars have been estimated as too bright in the regions poor in stars and as too faint in the Milky Way regions. The density of the background was determined from the numbers of stars in B D brighter than $9^m,5$.

The mean error of the difference $m_B - m_{BD}$ was found to be as follows:

Magnitude	Mean error	n	Declination	Mean error	n
$7^m,5 - 7^m,9$	$\pm 0^m,29$	441	$0° - 9°$	$\pm 0^m,26$	1868
8 ,0 − 8 ,4	0 ,30	949	$10° - 19°$	0 ,25	1947
8 ,5 − 8 ,9	0 ,26	1497	20 − 29	0 ,25	1878
9 ,0 − 9 ,2	0 ,24	4418	30 − 39	0 ,26	1547
9 ,3 − 9 ,5	0 ,24	1096	40 − 50	0 ,26	1161

According to KÜSTNER the mean error in the Bonn catalogue is $\pm 0^m,14$ and thus the mean error of one magnitude in the B D catalogue is $\pm 0^m,215$. This may be compared with PICKERING's determination by means of two other methods, viz. $\pm 0^m,214$ and $\pm 0^m,225$, respectively. The internal mean error of B D will thus be $\pm 0^m,22$ in the mean.

The scale correction for mean density and for reduction to zenith is represented by the following formula:

$$\Delta m_{BD} = +0^m,422 - 0^m,00488\,(50° - \delta) + 0^m,367\,(m_{BD} - 9,0) + 0^m,161\,(m_{BD} - 9,0)^2.$$

All the coefficients are accurately determined. The maximum value of the correction is $+0^m,63$.

[1] J. HOPMANN, Neue Untersuchungen über die Größenskala der schwachen Sterne der nördlichen Bonner Durchmusterung. Inaugural-Dissertation Bonn (1914).

The density coefficient is expressed by the formula:

$$\Delta m_D = -0,178 - 0,00107\,(\delta - 25°) + 0,000219\,(\delta - 25°)^2 - 0^m,151\,(m - 9,0) - 0,103\,(m - 9,0)^2 .$$

The largest numerical value of Δm_D according to this formula is $-0^m,28$.

An extensive table is given in HOPMANN's paper giving the value of the scale correction to the magnitudes of B D for each 5° zone in declination, each 40^m strip in A R and each tenth of a magnitude from $8^m,6$ to $9^m,5$. For the $9^m,5$ grade in B D the highest value of the total scale correction is $0^m,7$.

HOPMANN finds that the mean error of a corrected B D magnitude will be $0^m,23$. This corresponds to the external mean error.

At the observatory of Lund extensive comparisons have been made by CHARLIER and his pupils between the magnitudes of B D and those of Harvard. Since these investigations cannot be published in the immediate future we shall not give any review of them on account of their general agreement with the results of PICKERING, HOPMANN, and STICKER. Only one of the results obtained in the evaluation of the B D scale at Lund will be mentioned. A. CORLIN has used the density of the background according to the star counts of the FRANKLIN-ADAMS plates and charts made at the observatory of Lund, and computed the coefficient of correlation between the density and the scale correction.

He finds:

r	BD magn.	r	BD magn.
0,123	7,5	0,490	8,9
0,162	8,0	0,552	9,0
0,446	8,5	0,474	9,1
0,552	8,8	0,445	9,2

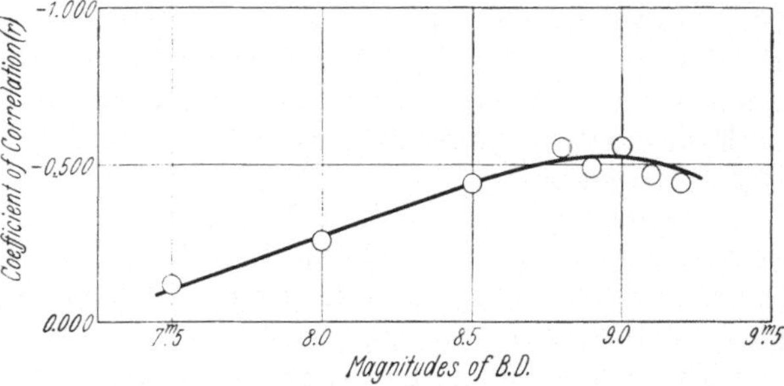

Fig. 66. Course of correlation between the apparent star density according to counts at Lund on basis of FRANKLIN-ADAMS charts, and the scale correction of the B D magnitudes according to CORLIN's unpublished determination.

This shows the important influence of the density of the background on the size of the scale correction for stars between 8,5 and 9,2. A considerable part of the correction for the B D stars is undoubtedly caused by the luminosity of the background acting in the same way as the veil on the photographic plates.

30. The Cordoba Durchmusterung (C D M). The aim of the C D M was to estimate the magnitudes of southern stars down to 10^m, and this has also been carried out. The work is an extension of the B D of ARGELANDER and SCHÖNFELD, but includes somewhat fainter stars. An equatorial was used that showed $10^m,5$ stars without loss of field. Among the improvements on B D it may be mentioned that the transits have been recorded. The mode of observation has not been changed during the work in such a way as in the case of B D, and thus the photometric data should be more uniform, but of course the first observations do not match the later ones with regard to their general accuracy. The size of the work is shown by the following summary:

Vol.	Zone	Cordoba Publ.	Epoch of observation	Number of stars	Published
1	−22° to −32°	vol. 16	1885—1891	179800	1892
2	−32 „ −42	vol. 17	1885—1891	160415	1894
3	−42 „ −52	vol. 18	1894—1897	149447	1900
4	−52 „ 62	vol. 21	1900—1908	89140	1914

The work hitherto published has been carried out under the direction of J. THOME, who has made the observations together with R. H. TUCKER. The observations of the stars in the remaining parts were made under the leadership of C. D. PERRINE.

The positions are given for 1875,0. The accuracy of the coordinates of the three visual Durchmusterungs is: for ARGELANDER $\pm 0^s,60$, $\pm 0',43$, for SCHÖNFELD $\pm 0^s,38$, $\pm 0',16$, and for Cordoba $\pm 0^s,42$, $\pm 0',23$.

As to the scale it has been closely connected with the B D scale by aid of extensive comparisons with SCHÖNFELD zones. E. C. PICKERING has reduced the C D M scale to the Harvard system, on the basis of the photometric measurements at Harvard. The reduction cannot be considered definite on account of the comparatively few photometric determinations available. We cannot at present expect to derive very accurate magnitudes on the photometric scale from the C D M estimates especially in cases of stars fainter than $9^m,0$, because of the possible existence of considerable systematic deviations in certain regions of the sky.

31. PANNEKOEK's Reduction of the three Visual Durchmusterungen. A thorough and extensive investigation of the B D and C D M scales has been undertaken by A. PANNEKOEK[1]. The author has adopted the method of SCHWARZSCHILD for the derivation of the $D(r)$, but discusses the deviations from his curves when a star cloud is superposed on the regular strata of stars, or when empty spaces or local clouds of absorbing matter cause other kinds of irregularities in the star distribution.

For the reduction of the B D magnitudes to a photometric scale the author decided to make use of the Harvard catalogues. In this connection his theoretical researches concerning the influence on the scales which are reviewed in another place in this book, proved necessary.

The behaviour of the decimal error was first investigated. The suspected variation with the declination proved to be insignificant.

The influence of density is very marked, as was also found by other investigators. The adjoining table which is formed from the data in Harv Ann 72 will show its general character:

Volume of BD	Hours of AR	Apparent density of stars	Scale correction to B D (= Harv—Bonn)					
			$9^m,0$	$9^m,1$	$9^m,2$	$9^m,3$	$9^m,4$	$9^m,5$
III	10—15	9,0	$+0^m,24$	$+0^m,33$	$+0^m,33$	$+0^m,79$	$+0^m,86$	$+1^m,11$
V	7—6	9,2	0 ,27	0 ,52	0 ,43	0 ,62	0 ,70	0 ,93
IV	4, 9—16	9,7	0 ,45	· ,57	0 ,67	0 ,78	1 ,03	1 ,42
III	0—3, 9, 23	10,3	0 ,41	0 ,48	0 ,58	0 ,69	0 ,82	1 ,22
III	4, 8, 16, 21, 22	13,3	0 ,25	0 ,28	0 ,45	0 ,62	0 ,74	1 ,07
IV	0—3, 8, 23	13,5	0 ,28	0 ,36	0 ,52	0 ,62	0 ,77	1 ,15
V	3—6, 17, 18	15,0	0 ,16	0 ,37	0 ,33	0 ,50	0 ,66	0 ,74
IV	7, 17, 18, 21,22	19,3	0 ,29	0 ,38	0 ,50	0 ,59	0 ,70	1 ,09
III	5, 7, 17, 20	19,6	0 ,03	0 ,16	0 ,14	0 ,37	0 ,60	0 ,92
V	0, 1, 2, 19	21,1	0 ,24	0 ,25	0 ,47	0 ,46	0 ,59	0 ,79
V	20—23	23,8	0 ,24	0 ,41	0 ,31	0 ,45	0 ,52	0 ,73
III	6, 18, 19	26,2	0 ,01	0 ,15	0 ,17	0 ,31	0 ,50	0 ,82
IV	5, 6, 19, 20	28,6	+0 ,19	+0 ,23	+0 ,55	+0 ,39	+0 ,62	+0 ,89

The coefficients of density defined as the change in m for unity of apparent density of stars have been smoothed out by a quadratic function of the photometric magnitudes. In this way the following values have been found:

[1] Publ of the Astr Inst of the Univ of Amsterdam No. 1 (1924).

BD m	Coeff. of density	BD m	Coeff. of density
6,5	−0 ,001	9,1	−0 ,014
7,0	−0 ,003	9,2	−0 ,016
7,5	−0 ,005	9,3	−0 ,018
8,0	−0 ,007	9,4	−0 ,022
8,5	−0 ,012	9,5	
9,0	−0 ,013		

The scale used in observing the DM magnitudes exists only in the mind of the observer; thus it may be subject to variations with the state of his mind as well as with the state of the sky. On account of the arrangement of the observation the work could only proceed comparatively slowly, during years, from lower to higher declinations. Thus variations of the scale corrections with time will appear as in reality being variations with place in the sky. The results of VON SEELIGER earlier mentioned have shown that the corrections for the zones 0°—10° deviate considerably from those for the higher zones; between 10° and 15° decl. there is a sudden and strong variation.

As a final result of his reduction of the BD-scale, taking account of the decimal error and the influence of the density and using the photometric measurements in Harv Ann 24, 45 and 70 and taking due account to the multitude and incompleteness of these catalogues PANNEKOEK finds the adjoining table which gives the photometric magnitudes of the B D scale with B D magnitudes as argument:

Declination zone	$6^m,5$	$7^m,0$	$7^m,5$	$8^m,0$	$8^m,5$	$9^m,0$
−2°						
	6,05	6,58	7,12	7,72	8,42	9,16
0						
	6,08	6,57	7,11	7,69	8,40	9,17
+5						
	6,06	6,59	7,14	7,74	8,42	9,16
10						
	6,27	6,76	7,30	7,87	8,47	9,11
15						
	6,32	6,89	7,43	7,97	8,56	9,18
20						
	6,36	6,92	7,48	8,03	8,62	9,27
25						
	6,36	6,91	7,47	8,03	8,63	9,30
30						
	6,35	6,93	7,50	8,06	8,65	9,27
35						
	6,42	7,03	7,60	8,16	8,74	9,36
40						
	6,42	7,04	7,63	8,18	8,76	9,40
45						
	6,44	7,07	7,63	8,14	8,72	9,38
50						
	6,41	7,02	7,59	8,13	8,67	9,30
55						
	6,37	6,98	7,57	8,09	8,64	9,25
60						
	6,41	7,03	7,58	8,09	8,64	9,24
65						
	6,46	7,08	7,59	8,06	8,56	9,11
70						
	6,50	7,10	7,58	8,01	8,49	9,02
75						
	6,46	7,04	7,50	7,98	8,46	9,05
80						
	6,39	6,99	7,47	7,94	8,45	9,14
90						

Next, PANNEKOEK reduces the „Südliche Durchmusterung" or the extension of the BD by SCHÖNFELD to a photometric scale.

In his extension of the BD with a fourth section from −2° to −23° SCHÖNFELD (1876—81) has used a somewhat larger instrument and made observations in a field faintly illuminated by red light in which stars were still visible whose magnitudes are a whole class below the limit of the catalogue. Completeness was only aimed at for stars as faint as $9^m,3$ or $9^m,4$, but 10^m were nominally included.

As the SBD has not been counted for all tenth magnitudes PANNEKOEK tried to deduce statistical magnitudes and decimal error by counting a part of the whole work.

The decimals show the same general behaviour as in the northern B D.

The density error was derived from the differences Harv Ann 24 — B D.

Hours \ SBD	D	$9^m,0$	$9^m,1$	$9^m,2$	$9^m,3$	$9^m,4$	$9^m,5$	$9^m,6$	$9^m,7$	$9^m,8$	$9^m,9$	$10^m,0$
0, 1, 11–13, 23	8,3	$+0^m,39$	$+0^m,56$	$+0^m,74$	$+0^m,78$	$+0^m,92$	$+1^m,15$	$+1^m,22$	$+1^m,12$	$+1^m,40$	$+1^m,06$	$+1^m,25$
2, 3, 10, 14, 16, 22	9,1	42	58	66	80	92	1 ,12	1 ,06	1 ,18	1 ,22	1 ,26	1 ,34
4, 9, 15, 17, 21	10,9	31	51	61	76	78	1 ,09	1 ,12	1 ,11	1 ,18	1 ,05	1 ,12
5, 8, 19, 20	15,5	23	34	55	54	60	0 ,88	0 ,83	0 ,81	0 ,90	0 ,96	0 ,91
6, 7, 18	20,2	+0 ,14	+0 ,14	+0 ,36	+0 ,42	+0 ,56	+0 ,72	+0 ,61	+0 ,67	+0 ,70	+0 ,79	+0 ,69
Total no. of stars		247	236	221	330	351	685	193	216	488	84	404
Coefficient of density (Unit to $0^m,01$)		−23	−37	−28	−34	−33	−38	−49	−45	−56	−31	−53
Adopted value		−26	−28	−30	−33	−36	−39	−42	−46	−49	−52	−56

Whereas one would expect that by the use of the illuminated field, the influence of the multitude or the scantiness of the stars on the estimates would vanish, or at least be strongly diminished, just the contrary takes place. The coefficients of apparent density are much larger than in the case of the northern B D.

The corrections depending on declination and right ascension were derived in the same way as for the northern B D.

The influence of density is shown in the adjoining table.

The following table gives the reduced photometric magnitudes of the SCHÖNFELD B D scale:

Zone \ SCHÖNFELD Magn.	6,5	7,0	7,5
− 2° to − 7°	$6^m,23$	$6^m,70$	$7^m,40$
− 7 „ −12	6 ,28	6 ,83	7 ,39
−12 „ −17	6 ,26	6 ,86	7 ,43
−17 „ −22	6 ,29	6 ,87	7 ,42

Zone \ SCHÖNFELD Magn.	8,0	8,5	9,0
− 2° to − 7°	$7^m,96$	$8^m,53$	$9^m,18$
− 7 „ −12	7 ,96	8 ,54	9 ,17
−12 „ −17	7 ,97	8 ,54	9 ,20
−17 „ −22	7 ,94	8 ,50	9 ,16

Finally PANNEKOEK determines the scale corrections to the Cordoba DM.

An extension of the D M over the Southern Sky at Cordoba was begun during 1885–91. The instrument was used with a dark field. A great number of faint stars were observed, but the faintest stars were named 10^m. On an average the mean number of stars observed at Cordoba is three times the corresponding number at Bonn. A variation with AR is found also in this case. The corrections are given in PANNEKOEK's paper pp. 44 and 45. The important question arises whether these corrections may be assumed to be the same for all magnitudes. After careful consideration the author has abstained from representing the corrections in the form $\varphi(m, \alpha, \delta)$, though doubtless this would be the most correct form. It is only the insufficiency of data that prevents the introduction of this form for corrections of the C D M scale.

The long duration of the work at Cordoba was not favourable to its homogeneity. This has been shown by the comparison with the photometric data given by PICKERING in Harv

Ann 72, No 7 and 80, No 7. The decimal error has been investigated by PANNEKOEK, who found the following results among others. The decimals 5 and 0 occupy too much room at the expense of the adjacent decimals 9,1, 4,6. The range for decimals 3 and 7 is greater than for 2 and 8, especially in three zones III and IV, where they extend three times as much. These irregularities increase steadily during the first four zones. At $-42°$ they disappear and a sudden return to nearly normal values occurs.

The direct estimates have been made in quarters of a magnitude. The sudden change at $-42°$ coincides with a change of the same character in the reduction to a photometric scale. The observer had become aware of the existence of systematic errors in his estimated magnitudes. Then he tried to adapt his scale to the photometric scale.

The density error has been investigated and the following coefficients were found:

$6^m,5$	$-0^m,020$	$8^m,5$	$-0^m,015$
7 ,0	0 ,019	9 ,0	0 ,016
7 ,5	0 ,017	9 ,5	0 ,024
8 ,0	−0 ,016	10 ,0	−0 ,037

Further the corrections depending on declination were investigated. PANNEKOEK shows that THOME and TUCKER have tried to adapt their scales as closely as possible to the photometric magnitudes.

The following table shows the general relation between the CDM scale and the Harvard photometric measurements which are the tabulated values:

Zone \ CDM-scale	6,35	6,65	6,85	7,15	7,35	7,65	7,85	8,15	8,35	8,65	8,85	9.15	9,35
$-22°$	$6^m,01$	$6^m,36$	$6^m,60$	$6^m,97$	$7^m,22$	$7^m,61$	$7^m,87$	$8^m,27$	$8^m,54$	$8^m,95$	$9^m,24$	$9^m,77$	$10^m,18$
27	6 ,03	6 ,41	6 ,69	7 ,11	7 ,39	7 ,79	8 ,05	8 ,45	8 ,73	9, 14	9 ,44	9 ,94	10 ,36
32	6 ,04	6 ,39	6 ,63	7 ,01	7 ,28	7 ,69	7 ,97	8 ,38	8 ,65	9 ,06	9 ,35	9 ,82	10 ,20
37	6 ,06	6 ,41	6 ,66	7 ,07	7 ,35	7 ,74	8 ,00	8 ,38	8 ,64	9 ,03	9 ,30	9 ,71	10 ,04
42	6 ,05	6 ,43	6 ,71	7 ,10	7 ,34	7 ,65	7 ,86	8 ,15	8 ,34	8 ,64	8 ,85	9 ,18	9, 45
47	6 ,04	6 ,43	6 ,70	7 ,10	7 ,34	7 ,66	7 ,87	8 ,17	8 ,36	8 ,67	8 ,90	9 ,25	9 ,52
52	6 ,04	6 ,41	6 ,72	7 ,12	7 ,34	7 ,65	7 ,85	8 ,16	8 ,37	8 ,68	8 ,00	9 ,25	9, 52
57	6 ,07	6 ,44	6 ,72	7 ,11	7 ,33	7 ,65	7 ,86	8 ,17	8 ,39	8 ,72	8 ,94	9 ,30	9, 58

Lastly, the corrections depending on right ascension were investigated and separately derived for $6^m,5$ and for $7^m,5-9^m$.

32. Photometric Data of Meridian Catalogues. The oldest estimations of magnitudes were undertaken in connection with determinations of position, and generally not much attention was paid to the estimates of the magnitudes, which in most cases were only intended to serve as an aid for the identification. In the first extensive star catalogues, which were derived at the end of the 18th and the beginning of the 19th century, such as LALANDE's, PIAZZI's, BESSEL's and GROOMBRIDGE's, the magnitudes have been estimated independently of previous sources. As a rule, they are given in whole or half intervals of a magnitude and thus do not possess very high individual accuracy. On the other hand, being derived during the course of extensive series of observations by one or two observers, they possess a homogeneity that gives them a certain value as sources for statistical researches. The magnitudes in the BESSEL-WEISSE catalogue have also been used by F. G. W. STRUVE[1], who worked out an ingenious method of deriving, from the star numbers in overlapping zones, the most probable number of stars of different magnitudes.

The magnitudes in a number of the older star catalogues have been reduced to the B D scale by J. SCHEINER[2].

It is natural that after the appearance of the B D catalogues the older meridian catalogues have lost most of their importance as photometric sources. In a number of special cases their data will of course be useful.

In recent times very accurate estimates have been made by a number of meridian observers, e. g., R. H. TUCKER. It is impossible to give even a short review of the independent photometric data in the different catalogues on account of the many sources and the very different methods applied. Some observers have estimated all the stars in their programme anew; others have used a certain source of magnitudes and only made estimates of their own when the magnitude is unknown, others again have only noted conspicuous deviations from earlier estimates, and so on.

Some observers, for example, KÜSTNER and R. PRAGER, have made very careful estimates of the magnitudes, and by the aid of a grating have also reduced the estimates to a photometric scale. Such catalogues are of very high value, and the accuracy of the photometric data is strengthened on account of the fact that the accurate measurements of coordinates practically exclude the possibility of wrong identification of some objects, which cannot be avoided in the case of the photometric zone catalogues.

Of modern catalogues, where photometric estimates can be found in a more extensive way, we only mention the Cape, Pulkova, Cordoba, and Washington catalogues and the catalogues of the Astronomische Gesellschaft (A G).

The catalogue of the Astronomische Gesellschaft is a kind of revision of the B D down to $9^m,2$, giving accurate positions. Care was also taken to estimate the magnitudes as accurately as was possible during the work. Each star was estimated on two or more nights. At several observatories, and probably at most of them, after the observer had made his estimate, the B D magnitude was read to him by the recorder. He was thus enabled to keep his scale nearly the same as that of the B D. PICKERING has measured a number of stars in the overlapping zones in the section $+50°$ to $+55°$. Besides, the stars used for the scale of the B D could also be used for establishing the scale of the A G. There are some 175000 stars (subtracting the overlapping zones) of the B D stock for which we possess no other independent estimate, having a weight at least equal to that to be assigned to the B D magnitude.

There are very few investigations of the scale in A G; v. ARETIN[3] has compared the scale of A G OTTAKRING with the southern part of B D (SCHÖNFELD) for stars of magnitudes $8^m,7$ to $9^m,3$. The differences are dependent on the declination and on the relative density of the stars. The following general relations have been found:

$$m_{AG} = m_{BD} + 0^m,03 \qquad \delta = -6^0,$$
$$m_{AG} = m_{BD} + 0^m,06 \qquad \delta = -8^0.$$

The dependence on the apparent star density is illustrated by the adjoining small table.

Number of stars in 15 □°	−6° AG—BD	−8° AG—BD
1—100	$+0^m,06$	$+0^m,16$
101—200	+0 ,04	+0 ,06
201—300	0 ,00	+0 ,01

[1] Études d'astronomie stellaire. St Pétersbourg (1847).
[2] A N 116, p. 81 (1887).
[3] A N 197, p. 45 (1914).

J. POCOCK and BHASKARAN[1] have reduced the visual scale of the Washington zone of AG to the photographic scale of CHAPMAN and MELOTTE (m_p) and found:

$$m_p = 1{,}62\, m_w - 5^{\mathrm{m}}{,}1\,.$$

S. HOLM has done extensive research[2] with regard to the scale of A G Lund using available photometric measurements. He has found that the mean error of the estimates of one catalogue magnitude is $\pm 0^{\mathrm{m}}{,}20$, which shows that the accuracy is higher than could be expected. H. v. ZEIPEL is carrying out accurate measurements of the photovisual and photographic magnitudes of the 11000 stars in the Lund zone. This work has now advanced considerably.

Catalogue of the Astronomische Gesellschaft.

Zone	Designation	Limits in Declination	Equinox	Number of stars	Time when results were published
Kasan . . .	AG Kas	+80°20′ to +74°40′	1875,0	4281	1898
Berlin . . .	Berl C	+75 20 ,, +69 40	,,	3461 +37	1910
Kristiania . .	AG Kri	+70 10 ,, +64 50′	,,	3949	1890
Helsingfors-Gotha . .	AG H—G	+65 10′ ,, +54 55	,,	14680	1890
Cambr., Mass.	AG Harv	+55 10′ ,, +49 50	,,	8627	1892
Bonn	AG Bonn	+50 10 ,, +39 50	,,	18457	1894
Lund	AG Lu	+40 10 ,, +34 50	,,	11450	1902
Leiden . . .	AG Lei	+35 10 ,, +29 50	,,	10239	1902
Cambridge .	AG Cambr	+30 57 ,, +24 15	,,	14464	1897
Berlin . . .	AG Berl B	+25 10 ,, +20 0	,,	9208	1895
Berlin . . .	AG Berl A	+20 10 ,, +14 50	,,	9789 +372	1896
Leipzig . . .	AG Lpz I	+15 15 ,, +10 0	,,	9547 +125	1900
Leipzig . . .	AG Lpz II	+10 0 ,, + 4 42	,,	11875 +910	1899
Albany . . .	AG Alb	+ 5 10 ,, + 0 50	,,	8241	1890
Nicolajev . .	AG Nic	+ 1 10 ,, − 2 10	,,	5954	1900
Strassburg .	AG Str	− 1 42 ,, − 6 10	1900,0	8204 +107	1906
Wien-Ottakring	AG Wi	− 5 50 ,, −10 10	,,	8468	1904
Cambridge, Mass . . .	AG Harv	− 9 50 ,, −14 10	,,	8337	1912
Washington DC	AG Wash	−13 50 ,, −18 10	,,	8824	1908
(Algier[3] . .	AG Alg	−17 50 ,, −23 0	,,	9997	Paris 1924)

33. CARRINGTON's Zone. The magnitudes in the catalogue of CARRINGTON[4] have been independently estimated and are of higher accuracy than those in B D. All the stars are situated north of +80°50′ declination and include stars of $10^{\mathrm{m}}{,}8$ and brighter. The brighter as well as the faintest stars depend on estimates, but the intermediate stars have been measured by the aid of a photometer constructed by CARRINGTON. The magnitudes have been compared by HASSENSTEIN[5] with the Potsdam polar magnitudes. It is evident that the CARRINGTON magnitudes are not quite homogeneous. There is a decimal equation, in as much as $9^{\mathrm{m}}{,}7$

[1] A J 29, p. 126 (1916).
[2] Lund Medd Ser II, No. 55 (1930).
[3] Intended to be part of AG but on account of political reasons published as a separate publication.
[4] Catalogue of 3735 Circumpolar Stars observed at Redhill in the Years 1854, 55 and reduced to Mean Positions for 1855—1856.
[5] Publ Astrophys Obs Potsdam Nr. 85 (1927).

and $10^m,3$ are preferred, and $9^m,8$, $9^m,9$ occur very seldom if at all. It was necessary to divide the material into two sub-groups, $\delta < 86°$ and $\delta > 86°$.

The comparison gives the following results:

CARRINGTON magn.	$\delta < 86°$		$\delta > 86°$	
	$m_p - m_c$	n	$m_p - m_c$	n
$5^m,2-5^m,6$	$-0^m,1$	10		
5 ,7– 6 ,2	+0 ,2	14		
6 ,3– 6 ,8	+0 ,1	19		
6 ,9– 7 ,4	−0 ,1	31		
7 ,5– 8 ,3	−0 ,1	150	$+0^m,2$	37
8 ,4– 9 ,6	−0 3	452	−0 ,02	107
9 ,7–10 ,0	−0 ,1	835	+0 ,1	129
10 ,1–10 ,2	+0 ,1	184	+0 ,2	88
10 ,3–10 ,4	+0 ,2	1165	+0 ,4	262
10 ,5–10 ,6	+0 ,3	146	+0 ,4	19
10 ,7–10 ,8	−0 ,1	5	+0 ,2	2

A distribution of the differences $m_p - m_c$ corrected for systematic errors is given on page 26 of the paper of HASSENSTEIN. Only about 5% of the magnitudes in CARRINGTON's catalogue possess an accidental error of more than $0^m,5$, assuming that the error of the Potsdam measurements is insignificant.

34. The Magnitudes in PGC and the San Luis Catalogue. In the Preliminary General Catalogue of Boss (PGC), all the magnitudes of the stars of $6^m,5$ or brighter have been taken from a manuscript catalogue constructed by S. C. CHANDLER. The magnitudes are based upon a collation of the results of all the principal uranometries and photometric results. The scale adopted is approximately that of the Uranometria Nova of ARGELANDER, so that the magnitude scale should, according to CHANDLER, fulfil the condition:

$$m = -2{,}778 \log i$$

corresponding to a value of 0,36 for the intensity ratio $\varrho = i_n/i_{n-1}$. TUCKER has recently[1] communicated a detailed count of the frequency of m in PGC and finds, using the expression for the total number of stars to any definite point of the magnitude scale:

$$\sum m = a b^m,$$

where a and b are constants, and the value of the constant b can be derived by comparing the summations at differences of one unit. The catalogue is complete to $6^m,0$ and contains 3991 stars brighter than that magnitude. The coefficient in the equation connecting m and i is $2{,}90 \pm 0{,}08$ corresponding to the value 0,31 of the intensity ratio or the POGSON constant. This is not so far away from the "historical scale" or the scale of the old Uranometries, where the POGSON constant is 0,36, and thus the above relation is satisfied.

The San Luis Catalogue[2] gives accurate magnitudes of 15333 stars, of which 6725 have been measured photometrically by M. L. ZIMMER and H. JENKINS and reduced to the Harvard scale. In all 20758 observations of stars fainter than $6^m,5$ were made, by means of a wedge photometer (differential method) for which no photometric measurements existed. The mean error of a single observation is $\pm 0^m,12$.

35. The Star Catalogue of BACKHOUSE[3]. This catalogue is a compilation of all stars supposed to be visible to the unaided eye. The estimates in the Uranometries of HEIS, BEHRMANN, HOUZEAU, and GOULD have been included in the catalogue. Besides, the estimates at Harvard as published in Harv Ann 14 have

---

[1] Publ ASP 41, p 177 (1919).

[2] San Luis Catalogue of 15333 Stars for the Epoch 1910 prepared at the Dudley Observatory Albany, New York under the Direction of LEWIS BOSS (1908—12) and BENJAMIN BOSS. Carnegie Inst Wash Publ No. 386 (1928).

[3] Catalogue of 9842 Stars or all Stars very Conspicuous to the Naked Eye for the Epoch 1900. Sunderland (1911).

been used. The estimates of magnitudes in the catalogues of LALANDE, LALANDE-FEDORENKO, LACAILLE, PIAZZI, BESSEL-WEISSE, the First Radcliffe Catalogue, the Catalogo General Argentino, and several other catalogues have also been used and, of course, the B D magnitudes.

The magnitudes adopted in the catalogue were derived by weighting the different estimates. No corrections have been applied for systematic errors in the different sources. This circumstance makes the catalogue very valuable as a source catalogue. Its adopted magnitudes are certainly also very useful for different photometric purposes.

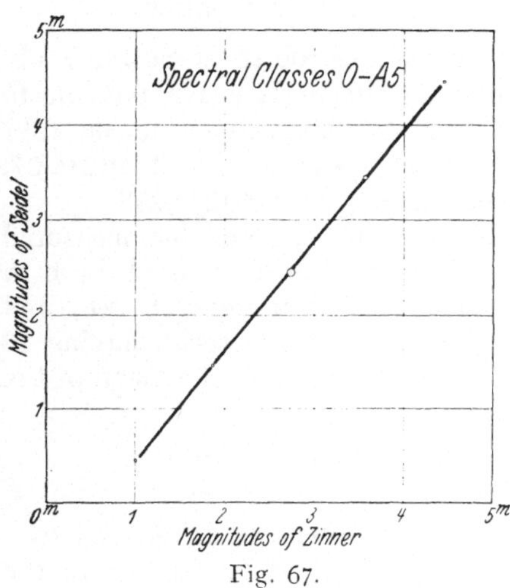

Fig. 67.

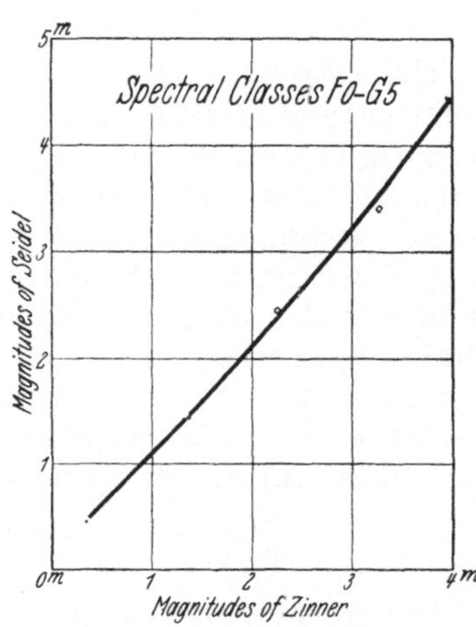

Fig. 68.

a_2) Visual Photometry.

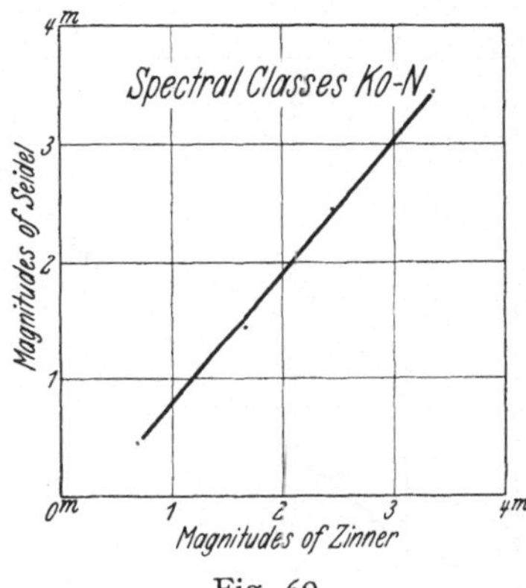

Fig. 69.

Fig. 67—69. Relation between the magnitudes measured by SEIDEL and those of ZINNER.

36. The First Photometric Measurements. SEIDEL, ZÖLLNER, WOLFF. The first modern catalogue of measured magnitudes is due to L. SEIDEL[1]. The observations were made in Munich from 1852 to 1860 by SEIDEL and E. LEONHARD. The instrument used was the Prism-Photometer invented by STEINHEIL and used earlier by him for a shorter series of observations. Altogether 208 stars were measured, based on 744 observations, each with 8 readings on an average. The extinction was carefully determined and corrected.

From his stock of observations SEIDEL selected those for 72 stars which had been determined very often, in order to have a standard. Vega was selected as a zero point, and the logarithms of the intensity of the other stars were given in the catalogue.

SEIDEL finds the mean error to be $\pm 0^m,11$ for all the observations, and, excluding the first 107, $\pm 0^m,09$. ZINNER, who has recomputed the mean error in a stricter manner, finds for all observations $\pm 0^m,096$, and for the first 107, $\pm 0^m,138$, and if these are excluded $\pm 0^m,071$.

[1] Resultate photometrischer Messungen. Abhandl Bayer Akad 9 (1862). — Helligkeitsmessungen an 208 Fixsternen. Ebenda 10 (1866).

If the POGSON scale is assumed, ZINNER finds the following differences between ZINNER's and SEIDEL's magnitudes:

Magnitude	O–A 5	F 0–G 5	K 0–N
$0^m,0-0^m,9$	$+0^m,55$	$+0^m,24$	$-0^m,11$
1 ,0–1 ,9	+0 ,44	+0 ,22	−0 ,09
2 ,0–2 ,9	+0 ,29	0 ,00	−0 ,20
3 ,0–3 ,9	+0 ,13	−0 ,08	−0 ,18
4 ,0–4 ,9	0 ,00		−0 ,54

As colour equations the values: B–K stars $= +0^m,48$ and B–F stars $= +0^m,24$ have been assumed by ZINNER. Taking into account the colour, the following equation has been used for computing the magnitudes m_s of SEIDEL

$$m_s = \log \text{SEIDEL} \times 2{,}19 + 0{,}585 - 0{,}\ 24 \times \text{colour equation.}$$

The Pole star was measured 45 times during the course of SEIDEL's observations, and in 1844—1845 it was measured 21 times. It is not possible to find a regular variation in the values. But from 19 observations during 1857 to 1860 ZINNER finds a maximum for that variable star at J D 2410235,07, whereas, according to the elements, it should be at J D 2410236,298.

J. F. C. ZÖLLNER, the inventor of the ZÖLLNER photometer, has measured the magnitudes of 226 bright stars. As a rule only objects situated in high altitudes were measured[1]. Every measurement consisted at least of two, and generally of 4, 6, 8, or 10 settings. From the differences of series having 20 observations with readings of the intensity, ZÖLLNER computed the mean of the error of observation. We quote the following results:

Bright background . . . . $\pm 0^m,146$; $\pm 0^m,135$; $\pm 0^m,187$
Dark ,, ... $\pm 0\ ,124$; $\pm 0\ ,166$; $\pm 0\ ,256$

The observations are given in logarithms of the intensity, as referred to a star of the series. ZINNER has compared the logarithms of ZÖLLNER with the magnitudes of his catalogue. The agreement between the magnitudes and the logarithms of the intensity was good and proves that the observations are not far from the ZINNER (Potsdam) magnitudes.

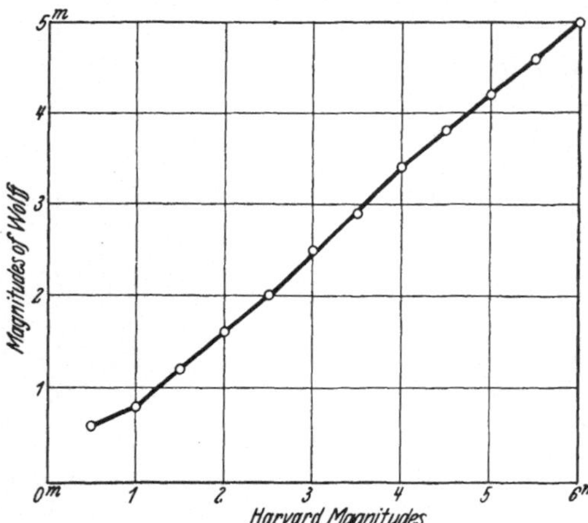

Fig. 70. Relation between the magnitudes measured by WOLFF and those measured at Harvard (R H P).

It would be of interest to apply SCHWARZSCHILD's differential method for a derivation of the connection between the intensity logarithms and the Potsdam magnitudes.

ZINNER finds the mean error of an intensity based on 6 observations to be $\pm 0^m,021$, and for one observation $\pm 0^m,051$. This seems to be altogether too small a value, as could be expected, as the measurements compared were made on the same evening. Anyhow, the accuracy of the determinations is certainly considerable.

WOLFF. His observations were also made with a ZÖLLNER photometer at Palermo and Bonn, during the years 1869—1883[2]. The catalogues of WOLFF contain 475 stars from the years 1869—1875 and 1130 stars from the years 1876

[1] Grundzüge einer allgemeinen Photometrie des Himmels. S. 102. Berlin (1861).
[2] Photometrische Beobachtungen an Fixsternen. Leipzig (1877). — Photometrische Messungen an Fixsternen. Berlin 1884.

to 1883, many of which refer to common objects. The stars are situated north of the equator and are brighter than $5^1/_3$. The light of an oil-lamp was used as a comparison star with the tacit assuming that after burning for a while the lamp gave a constant light for hours. It was intended to observe two polar stars each night, but this could not be done on account of the unsteadiness of the air in the polar region.

The first catalogue has no references to any comparison stars, and unless a reduction, according to SCHWARZSCHILD's differential method[1], is undertaken, it is necessary to search for the relation between the Harvard Photometry and WOLFF's measurements. The faint magnitudes are uncertain, but still a comparison will possess interest. ZINNER has found a very satisfactory relation between WOLFF's measurements and the modern magnitude scale.

In the second catalogue comparison stars from the equator and from the zone at $\delta = +40°$ have been used.

The photometer was diaphragmed from an aperture of 27,4 mm to 25,9 and 13,9 mm. It was found that the decrease observed in intensity was smaller than computed from the aperture.

ZINNER makes use of the diaphragming for the derivation of the mean error, and finds its value for one observation based on 4 settings to be $\pm 0^m,16$. WOLFF has derived, from the internal agreement, the corresponding values $\pm 0^m,11$ and $\pm 0^m,08$. The agreement between the Potsdam magnitudes and the logarithms of the intensity is not very good. The difference in intensity between stars within the limits $2^m - 6^m$ is smaller than the corresponding value from the Potsdam magnitudes, a fact also noticed by WOLFF when comparing his results with ZÖLLNER's. It is strange that this is not a question of a difference in scale, because the deviation from the photometric scale is fairly constant.

During the years 1872—1875 C. S. PEIRCE [Harv Ann 9 (1878)] used a ZÖLLNER photometer for measuring the magnitudes of 475 stars in the zone $+40°$ to $+50°$ decl. The measurements were performed without using comparison stars. The programme was so arranged that the artificial star could be eliminated when the magnitudes were derived.

There seems to be very little effect of colour in the measurements. The agreement with the P D scale is good, except that the stars brighter than 4^m are measured too bright. A correction of $0^m,12$ should be added to the magnitudes.

37. Uranometria Oxoniensis[2]. This determination, made during the years 1881—1885 under the supervision of C. PRITCHARD by W. PLUMMER and B. C. JENKINS, contains all the stars between $+90°$ to $-10°$ visible to the unaided eye. The number of stars is 2784. The observers used two wedge photometers.

The influence of the extinction was corrected by using the formula $0,253 \times (\sec z - \sec 40°)$. Stars observed south of $30°$ altitude were not used, but instead observations of the southern stars were performed during an expedition to Cairo.

ZINNER has derived the following relation between Uranometria Oxoniensis and the magnitudes of P D.

Potsdam magn.	Potsdam magnitudes — magnitudes of Uranometria Oxoniensis					
	O—A 0	n	A 2—G 5	n	K 0—N	n
$0^m,0-2^m,9$	$+0^m,26\pm0^m,03$	24	$+0^m,16\pm0^m,03$	5	$+0^m,10\pm0^m,03$	16
3 ,0—3 ,9	+0 ,15±0 ,02	44	+0 ,15±0 ,03	31	+0 ,09±0 ,03	50
4 ,0—4 ,9	+0 ,07±0 ,02	158	+0 ,04±0 ,02	88	−0 ,09±0 ,02	140
5 ,0—5 ,5	+0 ,12±0 ,01	191	+0 ,06±0 ,01	120	−0 ,07±0 ,02	178

[1] A N 172, p. 565 (1906). [2] Oxford 1885.

Thus there is a colour equation present, which is in accordance with the suspicion of PRITCHARD. According to him the accuracy of the measurements is very high and the mean error should be on an average $\pm 0^m,08$ for a star magnitude, depending on 10 settings. From a comparison with the Potsdam magnitudes ZINNER has found much higher values:

Magnitude	Mean error
$0^m,0-2^m,9$	$\pm 0^m,15$
3 ,0—3 ,9	±0 ,17
4 ,0—4 ,9	±0 ,21
5 ,0—5 ,5	±0 ,20

If consideration is paid to the improved values of the magnitudes, the mean error will still be $\pm 0^m,18$. The considerable divergence between the mean errors, computed in different ways, may be partly explained, as pointed out by MÜLLER and KEMPF, from the fact that the majority of the stars have been measured too bright, which will easily happen when the method of extinction is used. Much higher accuracy is obtainable with the wedge-photometer if the measurements are arranged differentially.

A consideration of the fact that the majority of the stars were observed during only one night will certainly explain, when considered together with the sources of errors peculiar to the extinction method, why the accuracy of the Uranometria Oxoniensis is so low.

38. The Revised Harvard Photometry (R H P). The principal results of the measurements of bright stars with Meridian Photometers during the years 1879 to 1906 are contained in the Harv Ann vol. 50. The magnitudes of 9110 stars, brighter than $6^m,50$, in all parts of the sky are given in that invaluable source. From the introduction we take the following statistical data, which are of interest:

Observers	Number of settings
S. I. BAILEY, Harv Ann 34, 46	235924
O. C. WENDELL, Harv Ann 14, 23, 24	123544
A. SEARLE, Harv Ann 14	28136
R. H. FROST (unpublished)	11352
E. C. PICKERING, H A 14, 23, 24, 44, 45, 46	683104
	1082060

Instrument	
2-inch photometer	94476
4-inch photometer	987584
12-inch photometer	646732
	1728729

All the observations with the 12-inch, and a large proportion of those with the other two instruments, are of stars fainter than $6^m,50$, and are not included in Harv Ann 50, but are given in Harv Ann 54.

The magnitudes given in Harv Ann 50 have been formed by taking the simple mean of the values given in the volumes Harv Ann 14, p. 272; 23, p. 98; 24, p. 102 and p. 204; 34, p. 108; 44, p. 5; and 46, p. 98. Equal weight is given to the results of each catalogue, independently of the number of observations from which they have been derived. The reason for this procedure is certainly sound. PICKERING says "that systematic errors are much more to be apprehended than accidental errors. The conditions vary so much in the different catalogues, as regards size of the instrument, observer's latitude and climate, that the systematic errors in one catalogue may well have an opposite sign in another".

The residuals of the individual catalogues are also given, which is very convenient, making it possible to obtain easily the different values on which the magnitude is based. The weight of a determination can also easily be estimated from the residuals.

Besides this a number of residuals from magnitudes derived in eight different earlier catalogues and the Harvard magnitudes are given in the column headed Catalogue Residuals. The earlier catalogues in question are: WILLIAM HERSCHEL'S and JOHN HERSCHEL'S catalogues, Uranometria Oxoniensis, Potsdam Durchmusterung, WILLIAMS'S Catalogue of 1083 stars, the catalogue given in Harv Ann 46, p. 58, and the not otherwise published Uranometry of BAILEY. The residuals showing the photographic magnitudes of the stars or the colour indices are also given. The spectral classes are, of course, from the old HENRY DRAPER (HD) Catalogue and are superseded by the spectral classes in the new H D. The catalogue is accompanied by useful remarks with regard to binaries, variables, parallax and proper motion stars, spectral peculiarities, etc. Among the other tables, No. VII, which is an index to the letters of BAYER and LACAILLE, is of much value.

Although it may be outside the scope of this chapter, it ought to be mentioned that PICKERING has pointed out in the introduction that the mistake has frequently been made of supposing that the measurements are dependent on the constancy of the light of the Polar star. Variations of low amplitude in its light, which are now known to be present, will have no effect on the derived magnitudes. The actual standard is the mean magnitude of one hundred circumpolar stars of about 5^m, each of which has been observed on two or three hundred nights. During the years 1879—1882 the stars were compared with α Ursae Min., but its variation is too small to be detected in the Harvard magnitudes. Since then the stars have been compared with λ Urs. Min. and σ Octantis, which are not known to be variable to such a degree that the accuracy of the measurements can be substantially influenced.

The following synopsis of the catalogues on which the R H P is founded may be of interest:

	Observer	Epoch	Source	Zones	Aperture of instr.
Pi I	PICKERING . .	1879—1882	Harv Ann 14	+90° to −30°	4 cm
II	PICKERING-WENDELL . . .	1882—1888	,, ,, 23, 24	+90 ,, −40	10,5 ,,
III	PICKERING . .	1891—1894	,, ,, 44	+90 ,, −47	,,
IV	,, . .	1895—1898	,, ,, 45	+90 ,, −47	,,
Ba I	BAILEY	1889—1891	,, ,, 34	−30 ,, −90	,,
II	,, ...	1899	,, ,, 46	0 ,, −90	,,
III	,, ...	1900—1902	,, ,, 46	+90 ,, −40	,,
IV	,, ...	1902—1906	,, ,, 50	+50 ,, −90	,,

For information about the individual measurements, on which the mean values in Harv Ann are founded, the reader is advised to consult the sources given above.

The extinction was computed from the formula:

$$\text{extinction} = d(\sec z - 1)$$

where d is a constant, and z the zenith-distance. The value of d was determined from the standard stars. MÜLLER and KEMPF have criticized this method[1]. ZINNER has also made a contribution to this discussion. He has used the stars between $\delta = -41°$ to $-47°$, measured at Cambridge as well as at Arequipa, and from the differences:

$$\overline{m}_{\text{Cambr.+Arequipa}} - \overline{m}_{\text{Arequipa}}$$

[1] Publ Astrophys Obs Potsdam 17, p. XXVI (1907).

the mean error of the mean has been computed with, and without, the Cambridge measurements. He has found that from $-44°$ onwards the inclusion of the Cambridge measurements increases the mean error. Thus the extinction has not been sufficiently corrected in the Harvard work.

The derivation of the magnitudes has been criticized too by KEMPF[1] and others. This criticism, although undoubtedly justified as regards its underlying principles, does not seem to be of such practical consequence that we need to review it in more detail.

The policy of excluding very large residuals has also been blamed. The way the work was organized made it impossible to avoid mistakes, and it seems better to radically exclude such measurements where the observer has a strong feeling that something is wrong, than to be encumbered with them in all future photometry.

ZINNER has computed the mean error, using the formula $\varepsilon = \pm 0{,}886 \frac{\Sigma v}{\sqrt{n(n-1)}}$ where v denotes the residuals: second measurement — first measurement, fourth measurement — third measurement etc., for one and the same star. The following table gives the results:

Harvard catalogue	$0^m,0$—$3^m,9$			$4^m,0$—$5^m,3$		
	Mean error $n=1$	Mean error of one catalogue magnitude	n	Mean error $n=1$	Mean error of one catalogue magnitude	n
Pi I	$\pm 0^m,23$	$\pm 0^m,08$	9,1	$\pm 0^m,22$	$\pm 0^m,10$	4,9
Pi II	± 0 ,29	± 0 ,14	4,0	± 0 ,19	± 0 ,11	2,8
Pi III	± 0 ,19	± 0 ,08	6,0	± 0 ,17	± 0 ,08	4,0
Pi IV	± 0 ,18	± 0 ,06	9,9	± 0 ,15	± 0 ,06	6,2
Ba I	± 0 ,19	± 0 ,08	6,0	± 0 ,18	± 0 ,10	3,4
Ba II	± 0 ,10	± 0 ,05	5,0	± 0 ,09	± 0 ,05	3,0
Ba III	± 0 ,13	± 0 ,03	17	± 0 ,15	± 0 ,03	18

ZINNER has also used the measurements in Ba III of stars of the magnitude 5. Each night the stars have been measured twice, with four settings. Although it is not quite justifiable, ZINNER makes use of the differences 2—1, 4—3, and finds for $0^m,0$—$3^m,9$, $\varepsilon = \pm 0^m,11$ and for $4^m,0$—$5^m,3$, $\varepsilon = \pm 0^m,10$. The mean of these can be considered as significant of the accuracy reached at Harvard.

The Harv Ann 54 gives magnitudes of 36682 stars fainter than $6^m,50$, observed with the 4-inch photometer, and is a continuation of R H P. The results are scattered through seven volumes of Harv Ann, viz., Vol. 23, 24, 34, 44, 45, 46, and 64. The catalogue is a supplement to the R H P. No residuals are given, but results from different catalogues are quoted on separate lines. The catalogue is followed by a series of remarks, which include additional facts regarding individual stars.

From a comparison of the magnitudes of stars observed in two or more catalogues, PICKERING finds that the average deviation of the separate magnitudes from the mean is about $\pm 0^m,07$. This corresponds to a mean error of approximately $\pm 0^m,09$. Thus the error in the Harvard magnitudes does not increase when fainter stars are measured.

SEARES has derived the following relation between the Harvard Visual Photometry and the International Scale (Mount Wilson standard magnitudes)

$$m_{MW} = m_H + 0{,}07\,(m_{MW} - 6) \qquad 6^m < m < 9^m{,}5$$
$$m_{MW} = m_H - 0{,}08\,(m_{MW} - 12{,}5) \qquad 9^m{,}5 < m.$$

[1] V J S 31, p. 191 (1896).

From a comparison between magnitudes in Harv Ann 44 and 45, MÜLLER and KEMPF[1] have derived the formula:

$$m_P - m_H = +0^m,0229 + 0,024\ (m_H - 2,25) - 0^m,027 \text{ Colour} - 0,008\ (m_H - 2,25) \text{ Colour}.$$

From this formula, the following table has been derived by the least-squares solution:

Harv. magnitudes	$m_{Pots.} - m_{Harv.}$				
	W	GW	WG	G	All
$2^m,00$ to $2^m,49$	$+0^m,23$	$+0^m,20$	$+0^m,12$	$+0^m,07$	$+0^m,16$
2 ,50 ,, 2 ,99	+0 ,24	+0 ,21	+0 ,12	+0 ,05	+0 ,16
3 ,00 ,, 3 ,49	+0 ,25	+0 ,22	+0 ,11	+0 ,04	+0 ,16
3 ,50 ,, 3 ,99	+0 ,27	+0 ,23	+0 ,11	+0 ,03	+0 ,16
4 ,00 ,, 4 ,49	+0 ,28	+0 ,23	+0 ,10	+0 ,02	+0 ,16
4 ,50 ,, 4 ,99	+0 ,29	+0 ,,24	+0 ,10	+0 ,01	+0 ,16
5 ,00 ,, 5, 49	+0 ,30	+0 ,25	+0 ,10	−0 ,01	+0 ,16
5 ,50 ,, 5 ,99	+0 ,31	+0 ,26	+0 ,09	−0 ,02	+0 ,16
6 ,00 ,, 6 ,49	+0 ,33	+0 ,27	+0 ,09	−0 ,03	+0 ,16
6 ,50 ,, 6 ,99	+0 ,34	+0 ,27	+0 ,09	−0 ,04	+0 ,16
7 ,00 ,, 7 ,49	+0 ,35	+0 ,28	+0 ,08	−0 ,05	+0 ,16
7 ,50 ,, 7 ,99	+0 ,36	+0 ,29	+0 ,08	−0 ,07	+0 ,16
8 ,00 ,, 8 ,49	+0 ,37	+0 ,30	+0 ,07	−0 ,08	+0 ,16

The question in which way the faintest stars in the Harvard Photometry are related to the international photovisual stars is not settled as yet. In order to correct the measurements of BAILEY with the RUMFORD photometer to the H S it was necessary to apply the correction: $+0,24\ (9,0 - m_H)$. BAILEY's scale agrees with that of PARKHURST, and if the two scales agree this would mean that the H S fulfils the relation: $m = -2,05 \log i$. VAN DER BILT found[2], using the Utrecht refractor, a result in substantial agreement with those of BAILEY and PARKHURST, but when he applied the wire-gauze screen method the scale was not confirmed. On the other hand considerable deviations appeared which suggest that the H S is not homogeneous for fainter magnitudes. A reduction of these magnitudes to the international scale is thus a very important task.

39. Potsdam Durchmusterung (P D). This catalogue, which is very important on account of its high accuracy, includes stars north of 0° and brighter than $7^m,5$ B D scale, and is based on observations of G. MÜLLER and P. KEMPF, performed during the years 1886—1905[3].

The ZÖLLNER photometer was adopted with certain modifications and several telescopes were used, it being arranged in such a way that observations with a certain instrument could be reduced to the scale of the others. Most of the stars were observed in four different positions with regard to the artificial star, i. e.: above, underneath, to the left, to the right. By the use of a number of diaphragms the measurement of differences as large as $2^m,5$ was avoided.

For the conversion of the measurements into magnitudes, 144 standard stars of magnitudes $4^m,5$—$7^m,3$ were used. Stars of extreme colour as well as suspected binaries or variables were not included among the standards. It seems that the influence of any appreciable colour equation on the standards is scarcely to be feared. The mean magnitude of the standards, $6^m,02$, was brought into agreement with the corresponding mean B D magnitude. The scale of the P D is thus anchored to the B D scale.

[1] Potsd Publ Bd 17, p. XXXIV (1907). [2] B A N 1, p. 167 (1922).
[3] Potsd Publ Bd. 9 (1894), 13 (1899), 14 (1903), 16 (1906) and 17 (1907).

Each star was measured twice, once by MÜLLER and once by KEMPF. In cases of differences between the observers exceeding $0^m,30$, two other measurements were taken. Among other precautions it may be mentioned that the stars were generally measured at an altitude of 40°—50° and that the extinction was carefully derived.

The colours were estimated in the Potsdam colour sequence: W, GW, WG, G, RG, GR, and R, subdivisions being introduced by adding the signs + or — to the literal symbols. According to MÜLLER and KEMPF the accuracy of the estimates is $\pm 0,9$ of a subdivision, or correspondig on an average to $\pm 0^m,17$.

MÜLLER and KEMPF derived the mean error of one observation to be $\pm 0^m,086$, when a slight systematic difference between the observers was eliminated. If no consideration is paid to that difference, the mean error is $\pm 0^m,092$. The catalogue magnitudes depend on at least two determinations, and then the mean error is $\pm 0^m,061$ in general and $\pm 0^m,052$ for stars brighter than $5^m,5$. There is no doubt that the Potsdam magnitudes are the most accurate of those hitherto measured on a large scale.

ZINNER has investigated the influence of the colour on the P D magnitudes. The following synopsis shows the general course of the relative colour equation:

Zone	Epoch	MÜLLER and KEMPF		
		O—A	F—G	K—M
0°—20°	1886—1890	$-0^m,01$		$+0^m,03$
20 —40	1890—1898	−0 ,03	$+0^m,03$	+0 ,06
40 —60	1897—1902	−0 ,02	+0 ,01	+0 ,06
60 —90	1902—1905	−0 ,03	+0 ,05	+0 ,04

The relative colour equation is thus of an order of magnitude $-0^m,07$ and depends primarily on the higher colour sensitivity of MÜLLER.

A number of comparisons between the P D magnitudes and those of BAILEY and PARKHURST were made. In order to come into agreement with the latter the Potsdam magnitudes should be corrected by $-0^m,01$ for F—G stars and by $-0^m,02$ for K—M stars.

It was found that the P D magnitudes need a very slight correction amounting to $-0^m,03$ or $-0^m,04$ for stars brighter than $2^m,9$ but insignificant for other magnitudes. The connection between the magnitudes in P D, m_P, and the magnitudes m_H, in Harv Ann 14, as well as the magnitudes in Uranometria Oxoniensis, is given through the formulas:

$$m_P = m_H + 0^m,01\,(m_P + 1,0)^2 - 0^m,07,$$

$$m_P = m_O + 0\ ,01\,(m_P - 2,0) + 0^m,01\,(m_P - 3,0)^2.$$

The P D is related to the international scale as follows[1]:

$$m_P - m_{MW} = +0^m,26 - 0,03\,(m_P - 5,0) - 0,07\,C$$

where m_{MW} stands for the international magnitude.

The relation to the Harvard visual scale, reduced to the international scale, is according to SEARES:

$$m_P = m_H + 0^m,25 + 0,03\,(m_H - 5,0).$$

If the spectral index S is used (A = 0; F = 1 etc.) the formula

$$S = 1,59\,C + 0,140\,m_P + 0,264$$

[1] Mount Wilson Contr 288 (1925) = Ap J 41, p. 284.

relates according to SEARES the HENRY DRAPER spectra to colour index C and PD magnitude.

Finally, the P D magnitudes are related to KING's photovisual magnitudes, m_{Kpv} in the following way:

$$m_P - m_{Kpv} = +0^m,14 + 0,04\,(m_P - 6).$$

40. ZINNER's Catalogue of Bright Stars[1]. ZINNER has constructed a fundamental catalogue of 2373 star magnitudes above $5^m,50$, using the determinations

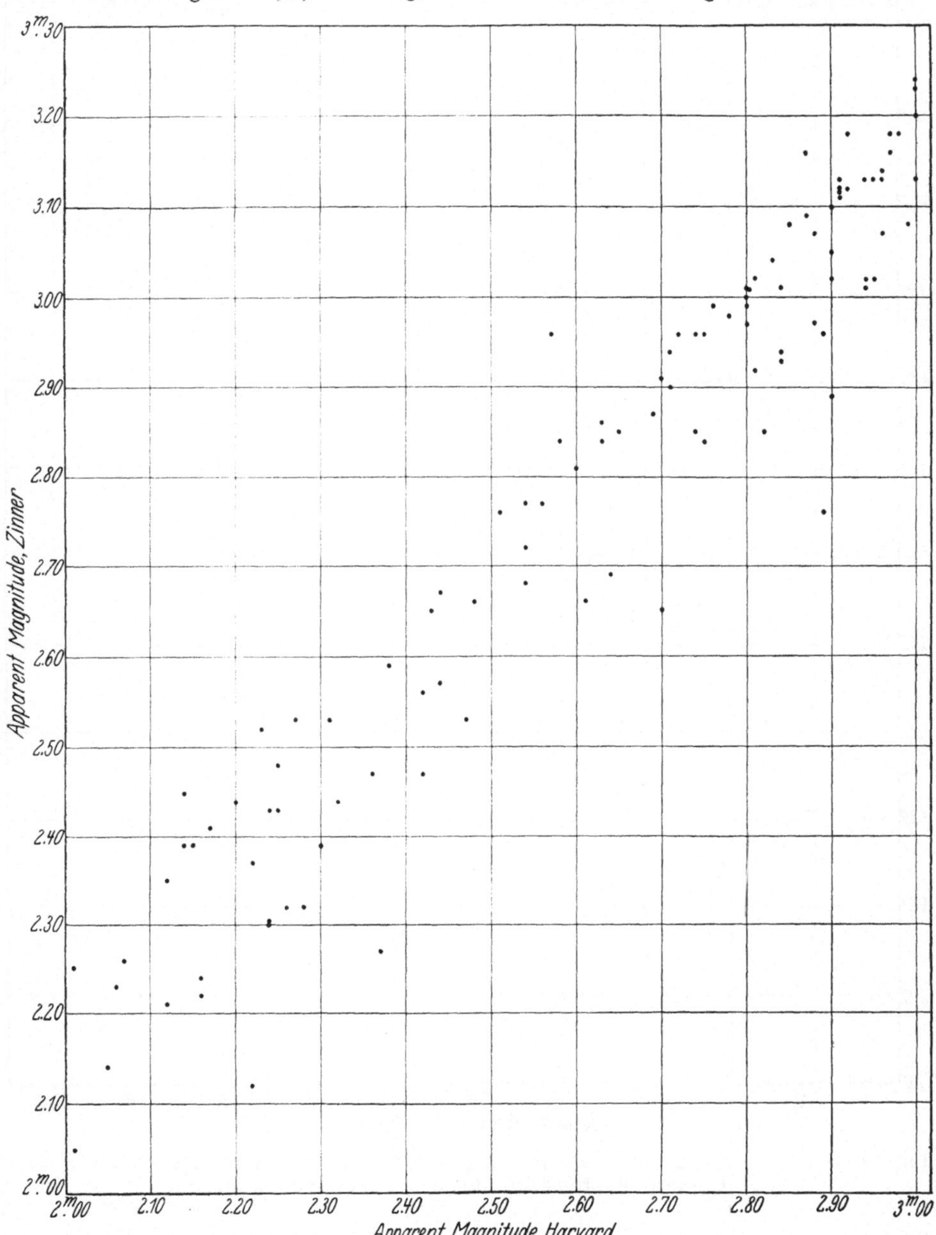

Fig. 71. Relation between the standard magnitudes of ZINNER and those in the Revised Harvard Photometry. Each dot represents a star.

[1] Veröffentl Sternwarte Bamberg, Bd. 2 (1926).

at Potsdam and at Harvard. The former are from the years 1879—1905. All the measurements have been performed using Nicol prisms, but in other respects they are not comparable. The mean error of one observation at Potsdam is $\pm 0^m,10$ and the corresponding value for Harvard varies between the limits

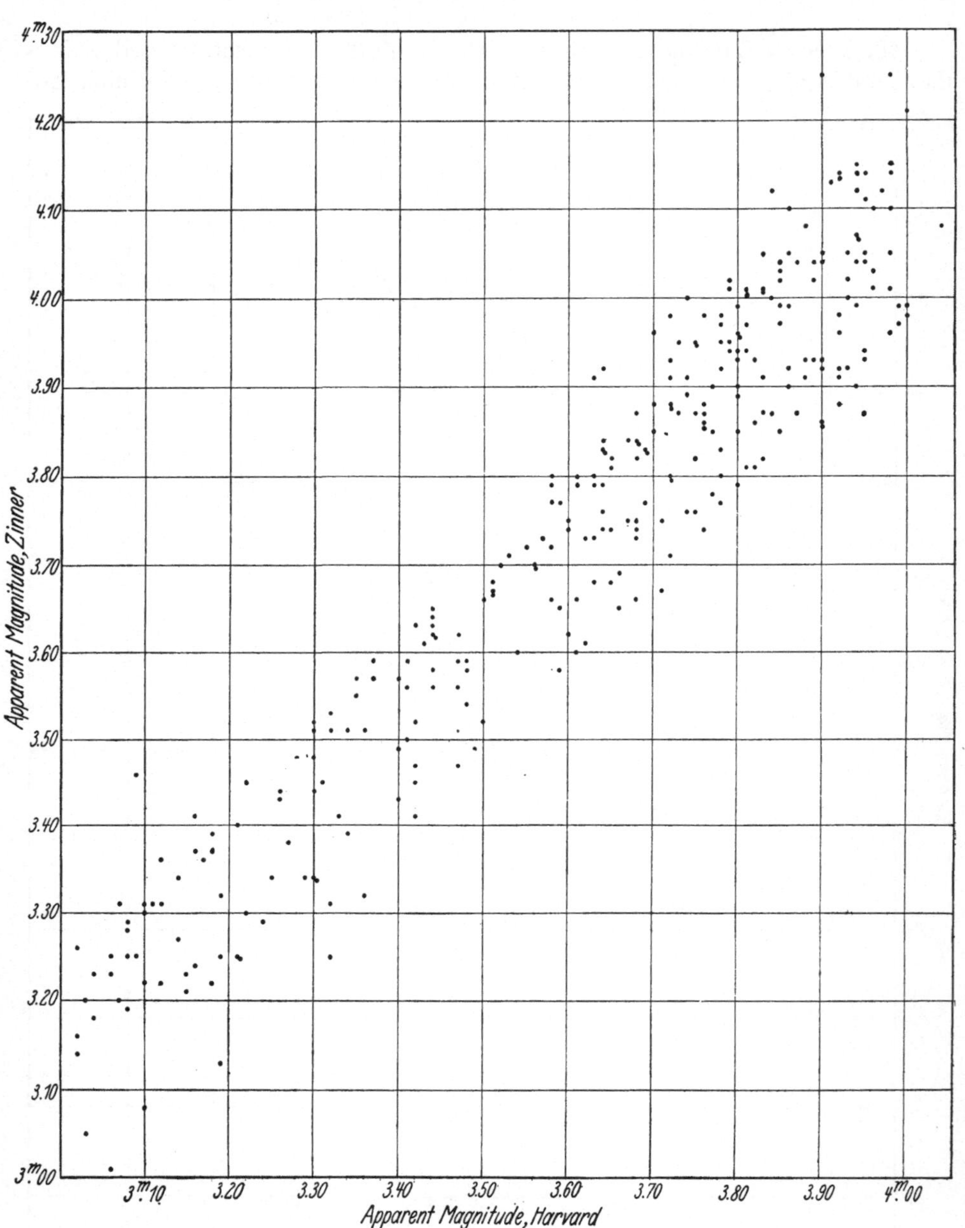

Fig. 72. Relation between the standard magnitudes of ZINNER and those in the Revised Harvard Photometry. Each dot represents a star.

$0^m,09$—$0^m,22$. The number of observations per magnitude at Potsdam is generally 2. The Harvard magnitudes are based on 1 to 40 separate observations. The observations have been combined and consideration has been given to the weights. In the comparison the objects have been grouped according to their colour and magnitudes. The differences between the two systems have been smoothed out by means of a graphical process.

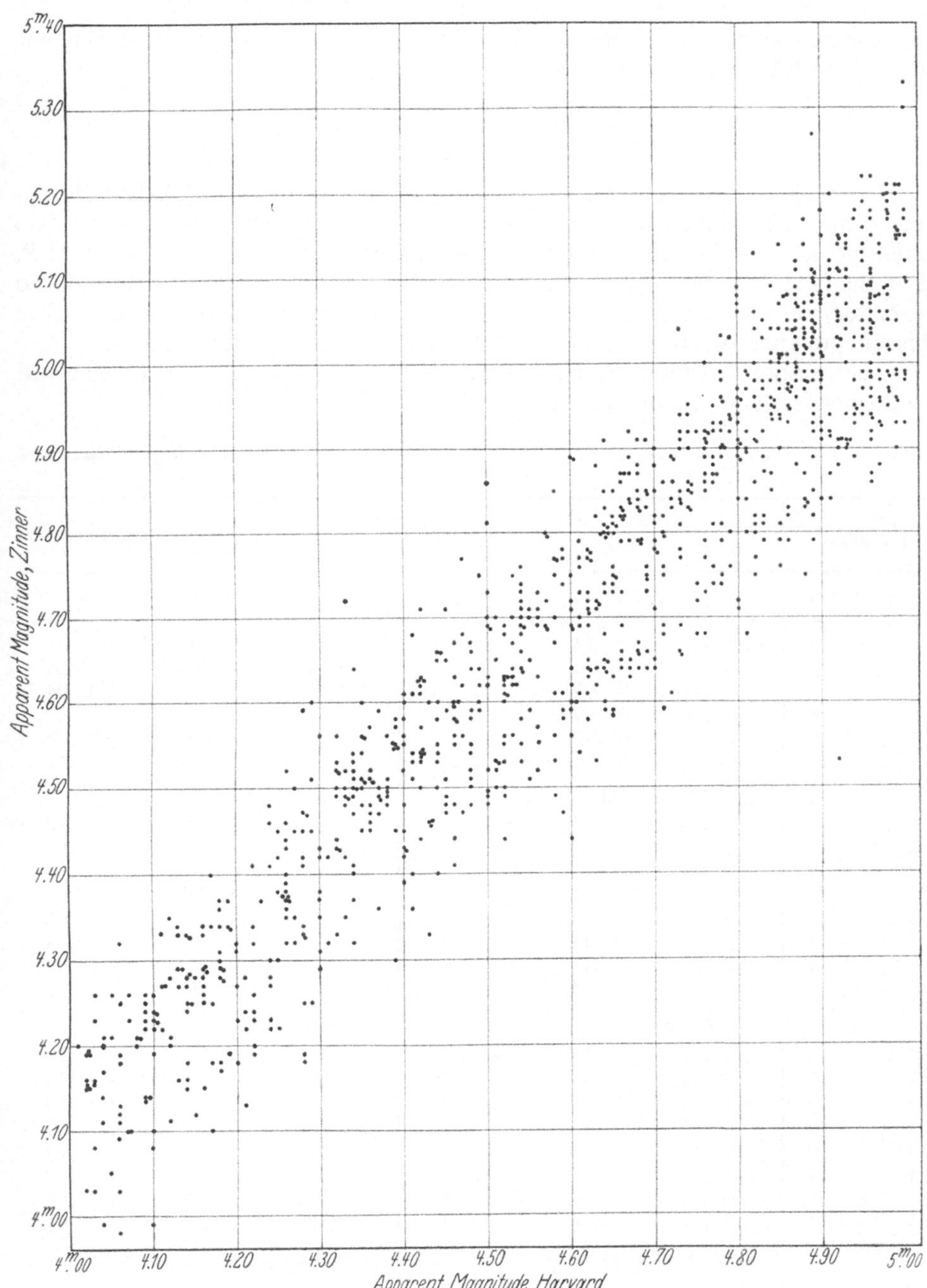

Fig. 73. Relation between the standard magnitudes of ZINNER and those in the Revised Harvard Photometry. Each dot represents a star.

The magnitudes of ZINNER m_Z and of the PD m_P are related in the following way:

Magnitude		Magnitude	
0	$m_Z = m_P + 0^m{,}050$	3	$m_Z = m_P - 0^m{,}007$
1	$m_Z = m_P - 0\ {,}005$	4	$m_Z = m_P - 0\ {,}005$
2	$m_Z = m_P - 0\ {,}008$	5	$m_Z = m_P - 0\ {,}003$

The mean error corresponding to weight 4—6 is on an average also dependent on the spectral class. ZINNER has found:

B	$\pm 0^m{,}045$	G	$\pm 0^m{,}040$
A	$\pm 0\ {,}041$	K	$\pm 0\ {,}042$
F	$\pm 0\ {,}040$	M—N	$\pm 0\ {,}050$

The scale of ZINNER is practically identical with that of the Potsdam Durchmusterung. This fact proves how small the deviations are between the two extensive main series of photometric measurements, the P D and the R H P, after the fairly constant systematic error has been applied. ZINNER has reduced his catalogue to the same scale as the P D. The different comparisons quoted in this chapter are based on either the Potsdam or the Harvard scale.

The relation between the magnitudes of Potsdam and those in the Harvard catalogues is shown in the adjoined table.

Systematic differences between Potsdam Durchmusterung and eight Harvard catalogues.

Potsdam magnitudes	PD—Pi I			PD—Pi II		
	O—A	F—G	K—N	O—A	F—G	K—N
$0^m{,}0-1^m{,}9$	$+0^m{,}21$	$0^m{,}19$	$+0^m{,}17$	$+0^m{,}39$	$0^m{,}19$	—
2 ,0—2 ,4	+0 ,18	0 ,15	+0 ,12	+0 ,33	0 ,16	—
2 ,5—2 ,9	+0 ,17	0 ,12	+0 ,08	+0 ,29	0 ,14	—
3 ,0—3 ,4	+0 ,16	0 ,12	+0 ,06	+0 ,25	0 ,12	$-0^m{,}01$
3 ,5—3 ,9	+0 ,15	0 ,09	+0 ,01	+0 ,21	0 ,10	−0 ,04
4 ,0—4 ,4	+0 ,16	0 ,09	−0 ,01	+0 ,18	0 ,08	−0 ,07
4 ,5—4 ,9	+0 ,18	0 ,09	−0 ,03	+0 ,17	0 ,06	−0 ,06
5 ,0—5 ,5	+0 ,20	0 ,10	−0 ,04	+0 ,19	0 ,07	−0 ,04
	P D—Pi III			P D—Pi IV		
0 ,0—1 ,9	+0 ,28	0 ,14	+0 ,07	+0 ,28	0 ,21	+0 ,12
2 ,0—2 ,4	+0 ,26	0 ,13	+0 ,04	+0 ,26	0 ,18	+0 ,07
2 ,5—2 ,9	+0 ,25	0 ,13	+0 ,42	+0 ,24	0 ,16	+0 ,04
3 ,0—3, 4	+0 ,24	0 ,12	+0 ,01	+0 ,23	0 ,14	+0 ,02
3 ,5—3 ,9	+0 ,23	0 ,12	+0 ,00	+0 ,21	0 ,11	−0 ,02
4 ,0—4 ,4	+0 ,22	0 ,11	−0 02	+0 ,19	0 ,09	−0 ,05
4 ,5—4 ,9	+0 ,22	0 ,11	−0 ,03	+0 ,19	0 ,09	−0 ,06
5 ,0—5 ,5	+0 ,22	0 ,11	−0 ,04	+0 ,21	0 ,10	−0 ,06
	P D—Ba I			P D—Ba II		
0 ,0—1 ,9	+0 ,31	0 ,14	+0 ,33	+0 ,24	0 ,08	+0 ,26
2 ,0—2 ,4	+0 ,27	0 ,15	+0 ,29	+0 ,20	0 ,08	+0 ,22
2 ,5—2 ,9	+0 ,24	0 ,15	+0 ,26	+0 ,17	0 ,08	+0 ,19
3 ,0—3 ,4	+0 ,20	0 ,16	+0 ,22	+0 ,14	0 ,08	+0 ,16
3 ,5—3 ,9	+0 ,17	0 ,16	+0 ,19	+0 ,11	0 ,09	+0 ,13
4 ,0—4 ,4	+0 ,14	0 ,16	+0 ,16	+0 ,08	0 ,09	+0 ,10
4 ,5—4 ,9	+0 ,12	0 ,15	+0 ,14	+0 ,06	0 ,09	+0 ,08
5 ,0—5 ,5	+0 ,09	0 ,14	+0 ,11	+0 ,05	0 ,09	+0 ,07
	P D—Ba III			P D—Ba IV		
0 ,0—1 ,9	+0 ,20	0 ,14	+0 ,22			
2 ,0—2 ,4	+0 ,16	0 ,11	+0 ,18			
2 ,5—2 ,9	+0 ,13	0 ,09	+0 ,15			
3 ,0—3 ,4	+0 ,09	0 ,06	+0 ,11	+0 ,27	0 ,29	+0 ,20
3 ,5—3 ,9	+0 ,05	0 ,04	+0 ,07	+0 ,23	0 ,29	+0 ,20
4 ,0—4 ,4	+0 ,03	0 ,04	+0 ,05	+0 ,21	0 ,18	+0 ,29
4 ,5—4 ,9	+0 ,04	0 ,06	+0 ,06	+0 ,19	0 ,18	+0 ,29
5 ,0—5 ,5	+0 ,07	0 ,10	+0 ,09	+0 ,17	0 ,20	+0 ,14

41. Extension of PD by TASS and TERKÁN[1]. These astronomers have extended the Potsdam Durchmusterung to the declinations 0° to −10°. It is intended to extend the work to −15°. Two photometers of the ZÖLLNER type were used for the measurements. As standards 139 stars, the magnitudes of which were determined from comparisons with 42 stars in the Potsdam zone, were employed.

The catalogue gives the photometric magnitudes of 2122 stars within the said zone, brighter than $7^m,5$. The same high accuracy as that of the Potsdam work was aimed at, but could certainly not be obtained. Still the magnitudes are, no doubt, very accurate and it is to be hoped that the work can be extended.

The following relations show that the general agreement between Ógyalla and Potsdam is good:

$$m_{\text{Pots}} - m_{\text{Pritchard}} = +0^m,13; \qquad m_{\text{Ogyalla}} - m_{\text{Pitchard}} = +0^m,12,$$
$$m_{\text{Pots}} - m_{\text{Harv 14}} = +0\ ,16; \qquad m_{\text{Ogyalla}} - m_{\text{Harv 14}} = +0\ ,14,$$
$$m_{\text{Pots}} - m_{\text{Harv 24}} = +0\ ,13; \qquad m_{\text{Ogyalla}} - m_{\text{Harv 24}} = +0\ ,07.$$

The average error of the measurements of the standard stars is $\pm 0^m,046$, which seems, perhaps, a little too low.

The bright stars have been compared by ZINNER with the magnitudes of his General Catalogue.

Magnitude	$m_{\text{Zinner}} - m_{\text{Ogyalla}}$
$0^m,00-2^m,99$	$+0^m,06$
3 ,00−3 ,99	+0 ,02
4 ,00−4 ,99	−0 ,15
5 ,00−5 ,49	−0 ,12

42. Potsdam Photometry of Stars in the Polar Zone. This catalogue contains B D stars in the polar zone, +80° to +90° decl., and also selected stars north of +73°. The measurements were made under the direction of G. MÜLLER during the years 1907—1920 and the reduction was begun by him but could not be finished before his death in 1925. W. HASSENSTEIN has performed the largest part of the reduction work[2].

In 1907 MÜLLER measured 253 stars between +73° and +90° in order to establish a photometric system for the polar zone. The results of this work are published in the paper: „Helligkeiten von 253 Fundamentalsternen zwischen den Größen $7^m,7$ and $9^m,6$ innerhalb der Zone von +75° Deklination bis zum Nordpol" [A N 182, p. 197 (1909)].

The main part of the programme is the measurements of the magnitudes in the Zone +80° to +90°. During this work MÜLLER was assisted by E. KRON and A. KOHLSCHÜTTER. The catalogue arranged according to declination zones gives the magnitudes of some 5000 stars.

Lastly the magnitudes of 26 stars of the North Polar sequence are given in the same publication.

The mean error of one observation is derived by HASSENSTEIN for both the principal observers from five series of measurements.

Epoch	n	MÜLLER	KRON	COMBINED
1909	19	$\pm 0^m,093$	$\pm 0^m,106$	$\pm 0^m,105$
1909	18	0 ,086	0 ,072	0 ,084
1911	16	0 ,111	0 ,082	0 ,113
1912	42	0 ,118	0 ,055	0 ,113
1914	19	0 ,135	0 ,093	0 ,132
1915	38	0 ,120		0 ,120
1913, 1918	42	0 ,080	0 ,092	0 ,102
1913, 1920	41	0 ,101	0 ,098	0 ,122

[1] Publ Ógyalla 1 (1916). [2] Publ Astrophys Obs Potsdam Nr. 85 (1927).

The average value of the mean error is found to be $\pm 0^m,111$.

Those magnitudes that are in common with the R H P (Harv Ann 50 and 54) have been compared. The results of this very important comparison is given in the table below.

$m_P - m_H$.

Harv. magn. \ Spectral class	B—A	n	F	n	G	n	K—M	n	All	n
4,4—6,0	$+0^m,26$	11	$+0^m,17$	3	$+0^m,07$	3	$-0^m,13$	13	$+0^m,06$	30
6,0—7,0	+0 ,32	22	+0 ,17	12	+0 ,16	10	−0 ,03	21	+0 ,16	65
7,0—8,0	+0 ,24	52	+0 ,19	39	+0 ,07	36	−0 ,02	53	+0 ,12	180
8,0—9,0	+0 ,37	30	+0 ,25	45	+0 ,16	41	+0 ,04	40	+0 ,20	156
9,0—10,6	+0 ,39	24	+0 ,34	24	+0 ,19	14	+0 ,11	16	+0 ,28	78
All	+0 ,31	139	+0 ,24	123	+0 ,13	104	0 ,00	143	+0 ,17	509

The mean correction to the Harvard magnitudes is thus $+0^m,17$, which is in very good accordance with the comparison, mentioned earlier, between the Harvard and Potsdam magnitudes.

182 stars have also been measured with the Harvard 12-inch meridian photometer (Harv Ann 70, 71, and 74). A comparison gave the following results:

Magn.	H A 70	n	H A 71	n	H A 74	n	All	n
4,5—7,0	$+0^m,08$	2	—	—	+0.06	14	$+0^m,07$	16
7,0—9,0	+0 ,11	19	—	—	+0,14	36	+0 ,13	55
9,0—10,0	+0 ,23	27	+0,20	2	+0,22	24	+0 ,22	53
10,0—11,4	+0 ,16	31	+0,18	7	−0,02	20	+0 ,10	58
4,5—11,4	+0 ,17	79	+0,18	9	+0,11	94	+0 ,14	182

If the stars are grouped according to the spectral classes it is found that the systematic difference for A—F stars is $+0^m,17$, whereas for G—K stars it is somewhat smaller, or $0^m,11$. The same behaviour is shown by the stars included in the above comparison.

The Potsdam magnitudes of the stars in the North Polar sequence confirm the existence of a systematic difference between Potsdam and Harvard. The 26 stars gave:

$$m_P - m_H = +0^m,17.$$

A comparison with the photovisual magnitudes of the North Polar sequence as established by SEARES gave:

$$m_P - m_{MW} = +0^m,09.$$

43. Atlas Stellarum Variabilium[1]. In 1890 J. G. HAGEN started the Atlas Stellarum Variabilium, intended to be used for identifying variables and also for supplying sequences of comparison stars. The work contains in its seven parts 311 charts and a catalogue sheet for each. Besides these 37 charts and catalogue data have been published in the Publications of the Specola Vaticana.

The charts of the fainter variables cover a region of one square degree with the variable in the centre. In this area all the B D stars are plotted and also verified with regard to their positions by HAGEN. All the stars visible in the 12-inch telescope of the Georgetown Observatory are also plotted, this telescope having a field of half a square degree.

The relative magnitudes of the stars in these fields are given in the atlas. The stars are observed, according to the method of JOHN HERSCHEL, in sequences.

[1] Freiburg i. Br. 1899—1908.

The degrees or grades, g, estimated by HAGEN, were changed into magnitudes, m_H, by the formula:

$$m_H = a + b\,(g - c)$$

in such a way that they should agree as nearly as possible with the scale of the B D between 7^m—10^m. In the case of stars fainter than 14^m the magnitudes given are merely relative and are not intended to be a continuation of the B D scale.

GINGRICH[1] has determined the photographic magnitudes for the comparison stars on the charts of HAGEN's Atlas Stellarum Variabilium. His paper contains a description of the Yerkes 6-inch camera, and a methodical part. The out-of-focus method was used. The zero-point was selected in accordance with the international system, and white stars as determined by MÜLLER and KEMPF at Potsdam, as well as stars in the Pleiades, were used as standards. The opacity of the inside focal images of the stars were measured by the use of the HARTMANN microphotometer. The correction to the centre was determined in terms of the distance from the centre. For a distance smaller than 1°,6 the correction is smaller than $0^m,09$; it increases more rapidly for greater distances, reaching $0^m,36$ for the angular distance 3°,6. For the reflector this correction is appreciably greater.

The catalogue contains magnitudes of the comparison stars for the following variables, U Cancri, R Camel, RU Herc., W Herc., X Cygni, S Cygni, V Delphini, R Lacertae, R Peg. and Y Cass.

In the Publications of the Yerkes Observatory, Vol. IV, part II, Miss HARRIET MCWILLIAMS PARSONS communicates results with regard to positions, proper motions, photographic and photovisual magnitudes of 187 stars between $7^m,66$ to $16^m,77$ situated in 8 fields in the RUMFORD regions as well as HAGEN's Atlas. Her results are from photographs with the 24-inch reflector at the McCormick Observatory.

The Leander McCormick Observatory has been engaged in a project similar to that of the Atlas Stellarum Variabilium[2]. By means of the 66 cm visual refractor the Harvard sequences have been extended to stars fainter than 15^m in 188 variable star regions. The standards have been derived jointly by H. L. ALDEN and S. A. MITCHELL[3]. The same observers made independent grade estimates, which were later on reduced to magnitudes with the aid of the standards. The MCCORMICK magnitudes seem to have a mean error of $\pm 0^m,06$, while the scale is one per cent less extensive than the photovisual scale of Mount Wilson, the latter being in substantial accord with the international scale.

Twenty-seven of the McCormick regions are in common with Series VII of Atlas Stellarum variabilium where the standards have been measured at Bergedorf. The mean difference between the 384 stars in common, which are mainly between 8^m—14^m, was found to be:

$$m_{Mc} - m_H = -0^m,016.$$

There is practically no run with decreasing magnitude. The average difference is $0^m,186$. Giving equal weights to both the series, either system of magnitudes is found to be reliable within $\pm 0^m,13$, a value which could be diminished by adjusting the zero-points. Considering the entirely different ways in which the magnitudes are observed and the different way the standards are determined at Bergedorf and McCormick the agreement is surprisingly good.

[1] Ap J 38, p. 209 (1913).

[2] Memoirs American Academy of Arts and Sciences 14, No. IV (1923).

[3] M N 89, p. 634 (1929).

44. J. A. PARKHURST, Researches in Stellar Photometry[1]. In order to obtain accurate and complete light-curves of 12 variable stars of long period, which have faint minima, photometric magnitudes have been derived for a number of comparison stars.

The photometer used was PICKERING's equalizing wedge photometer attached to the 6, 12, and 40-inch telescopes of the Yerkes Observatory. PARKHURST points out some essentials for obtaining good visual comparisons. Among these the error of position is very important. The two stars to be compared should be made parallel to the line joining the eyes. If PARKHURST placed two stars which were observed as equal in this position, in a vertical line, the lower appeared to be the brighter by more than half a magnitude. This is certainly an unusually large equation of position, but it emphasizes the importance of this source of error, which is not mentioned much in photometric literature.

Among other precautions necessary to ensure good visual observations the following may be mentioned.

The stars to be compared should be in the same field. Light in the eyes should be avoided by using a shielded lamp for recording.

A "patchy" sky is the signal to stop photometric work.

The real and artificial stars should resemble each other closely. This was possible in the work with the 6- and 12-inch telescopes, the two kinds of stars being indistinguishable, but it was not always possible to keep this rule with the 40-inch instrument.

From the agreement between the measurements the mean error seems to be of an order of magnitude $\pm 0^m,05$. The greater accuracy of the Potsdam measurements in comparison with Harvard is corroborated from the comparison made by PARKHURST. It seems reasonable to assume that the accuracy attained is at least equal to if not slightly higher than that of the Potsdam measurements. The accuracy of the measurements of the 40-inch telescope does not compete with the measurements of the smaller instruments.

Instrument	Limiting magnitude		
	Calculated from Pogson's formula	Observed	
		Harvard scale	Potsdam scale
6-inch	$12^m,89$	$12^m,90$	$13^m,02$
12-inch	14 ,40	14 ,27	14 ,57
40-inch	17 ,01	16 ,8	17 ,1

The limits of vision of the three instruments used could be determined with considerable accuracy.

The great thickness of the 40-inch objective ($4^1/_2$ inches) necessitates a correction for absorption. The CLARKES estimated that 0,85 of the light was transmitted by the lens. The corresponding correction is $0^m,18$, reducing the calculated limit to 16,83, which is in good agreement with the observed value.

It should be noted as a matter of the greatest importance that these limits of vision furnish a check on the adopted absorption curve of the photometer wedge used in the work. A change of 5 per cent in the curve would make a difference of half a magnitude in the range between the standard of 7^m and the limit of the 40-inch telescope. PARKHURST concludes that the absorption curve is not erroneous by as much as 5 per cent.

45. FESSENKOFF[2]. The observations of 1155 stars in the zone between $+79°,5$ to $+90°$ were made at the Charkow Observatory with a ZÖLLNER photo-

[1] Researches in Stellar Photometry during the Years 1894—1906 made chiefly at the Yerkes Observatory. Washington 1906. Carnegie Inst Washington Publ. No. 33.

[2] B. FESSENKOFF, Photometrical Catalogue of 1155 Stars. State Editorial Office of Ukraina. Charkov 1926.

meter attached to a 4-inch telescope. All the stars down to 9^m were observed. Stars from the Yerkes Actinometry (photo-visual), from MÜLLER's catalogue, and from those of CERASKI, all reduced to the scale of the first, were used as standards. After the observations had been reduced for atmospherical extinction, etc., it was assumed that the relation between the normal scale and the working scale could be expressed by the formula:

$$m = m_0 \alpha + K$$

where K is the correction to the zero-point and α the relation of the two scales. On determining the quantities for each day it was found that α was fairly constant, but K varied considerably. A second degree expression of α was also used. The 266 standard stars furnished the relation:

$$m_{\text{Yerkes}} = m_{\text{Fessenkoff}} + 0^m{,}047\,.$$

The mean residual was $0^m{,}066$, and thus the mean error of one observation $\pm 0^m{,}083$.

46. STICKER's Investigations[1]. The instrument used was the 6-inch refractor at Bonn, which has been used by HOPMANN for a number of photometrical and colorimetrical investigations. The photometer is of the wedge type. The measurements were made in order to test the scale of the catalogue of 10663 stars observed at Bonn by KÜSTNER, and the scale of the catalogue of 8803 stars observed by PRAGER at Berlin-Babelsberg. Altogether 620 stars of magnitudes 8^m—12^m were measured in the zone 0°—51° decl. The mean error of one observation was $\pm 0^m{,}192$ and of one magnitude in the catalogue $\pm 0^m{,}130$, which was in good agreement with the earlier results of HOPMANN.

The appended table gives the reduction of the KÜSTNER or PRAGER stars to the Potsdam system:

KÜSTNER or PRAGER	Correction to KÜSTNER			Correction to PRAGER	KÜSTNER or PRAGER	Correction to KÜSTNER			Correction to PRAGER
	δ 0°—27° 36 —51	δ 27°—36°	All stars			0°—27° 36 —51	27°—36°	All stars	
$4^m{,}6$			$-0^m{,}04$	$-0^m{,}07$	$7^m{,}6$	$+0^m{,}10$	$+0^m{,}18$	$+0^m{,}12$	$+0^m{,}12$
4 ,8			5	9	7 ,8	14	23	16	16
5 ,0			6	10	8 ,0	17	28	20	22
5 ,2			7	10	8 ,2	22	32	25	28
5 ,4			7	9	8 ,4	28	36	30	32
5 ,6			6	8	8 ,6	35	40	35	36
5 ,8			5	7	8 ,8	40	44	40	41
6 ,0			5	6	9 ,0	45	48	45	45
6 ,2			3	4	9 ,2	51	53	51	50
6 ,4			1	3	9 ,4	56	58	56	54
6 ,6			0	1	9 ,6	60	62	60	50
6 ,8	$+0^m{,}02$	$+0^m{,}03$	+ 2	2	9 ,8	63	66	64	63
7 ,0	3	6	4	4	10 ,0	67	70	69	68
7 ,2	5	10	6	6	10 ,2	70	74	74	72
7 ,4	+0 ,07	+0 ,14	+0 ,08	+0 ,08	10 ,4	+0 ,72	+0 ,78	+0 ,78	+0 ,76

The magnitudes of KÜSTNER have also been reduced to a number of other systems, as will be seen from the following comparison:

[1] B. STICKER, Photometrische Untersuchungen über die Größenskala des KÜSTNERschen Kataloges Bonn 1900. Dissertation Bonn (1928).

KÜSTNER's magn.	STICKER	YERKES[1]	GRAFF[2]	PANNEKOEK[3]	HOPMANN[4]
$7^m,5$	$7^m,60$		$7^m,60$		$7^m,57$
8 ,0	8 ,20	$8^m,13$	7 ,97	$8^m,03$	8 ,20
8 ,5	8 ,83	8 ,80	8 ,34	8 ,56	8 ,68
9 ,0	9 ,44	9 ,45	8 ,84	9 ,16	9 ,13
9 ,5	10 ,08	10 ,02	9 ,50	9 ,96	9 ,63
10 ,5	10 ,70	10 ,57	10 ,50	10 ,96	10 ,90

47. GALLISSOT's Investigations. In his thesis CH. GALLISSOT[5] has discussed very thoroughly the photometry of luminous points. The author first investigates the relation of Potsdam D M to Harvard Photometry and Uranometria Oxoniensis. Then he makes use of a photometer of the type constructed by NORDMANN (photomètre hétérochrome). The law of extinction is tested. The degree of sensibility is then defined as the inverse value of the quantity in the law of FECHNER or $\frac{J}{\Delta J}$. The following table gives the results obtained from measures using a ZÖLLNER photometer and red, green and blue screens:

Apparent intensity		Values of $\frac{J}{\Delta J}$			White light	
In magnitudes	In light units	Red	Green	Blue	$\frac{J}{\Delta J}$	$\frac{\Delta J}{J}$
0	2^3	—	—	—	5	0,211
3	2^{-1}	14	20	19	16	0,063
6	2^{-5}	22	17	18	34	0,029
9	2^{-9}	21	38	32	29	0,035
10,5	2^{-11}	23	26	28	22	0,045
11,25	2^{-12}	17	17	24	18	0,054
12	2^{-13}	11	10	10	12	0,081

The quantity $\frac{\Delta J}{J}$ is found from differentiating the law of MALUS:

$$J = K \sin^2 \alpha$$

and is equal to:

$$\frac{\Delta J}{J} = 2 \cot \alpha \, d\alpha .$$

The ratio $\left(\frac{J}{\Delta J}\right)_{\text{red}} : \left(\frac{J}{\Delta J}\right)_{\text{blue}}$ determining the PURKINJE-GALLISSOT effect has been investigated and the following important result has been found.

As is well known it is impossible to compare photometrically two light sources of different colour as long as their intensities are considerable. In case of faint sources a comparison is possible, because the colourisation is not perceptible. But the result of the measurement has not a strict physical significance, because of the fact that the photometric equilibrium is reached by means of weakening the two sources simultaneously without changing their spectral compositions. This phenomenon has been discovered by PURKINJE in case of luminous surfaces. GALLISSOT has studied the phenomenon in case of luminous points and has found the phenomenon to be present but showing a different behaviour from that of PURKINJE. When the simultaneous weakening of a red and blue surface makes the blue one to appear brighter than the red one the contrary takes place in case of stars. This extension of the

[1] Publ Yerkes Obs IV, part 6 (1927).
[2] A N 220, p. 135 (1924).
[3] Publ Astr Inst Amsterdam I (1924).
[4] A N 214, p 425 (1921).
[5] La photométrie du point lumineux appliquée aux déterminations des éclats stellaires. — Absorption atmosphérique — Scintillation — Coloration et températures. Thèse Lyon (1921).

PURKINJE phenomenon has been named the GALLISSOT phenomenon by DANJON. The results of some 1200 estimates are given in the adjoining table.

The effect is thus most appreciable between the magnitudes 7^m—11^m.

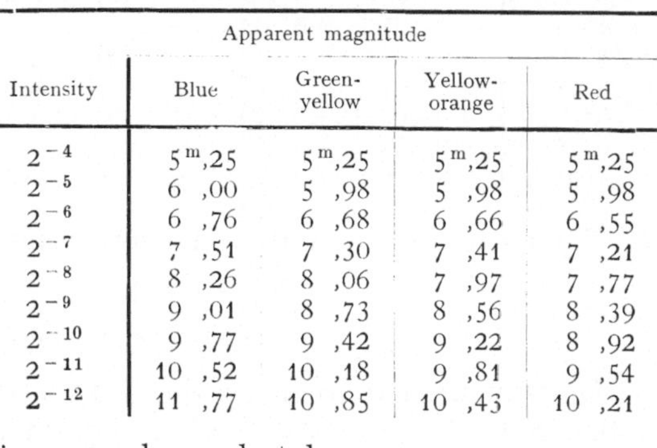

Intensity	Apparent magnitude Blue	Green-yellow	Yellow-orange	Red
2^{-4}	5^m,25	5^m,25	5^m,25	5^m,25
2^{-5}	6 ,00	5 ,98	5 ,98	5 ,98
2^{-6}	6 ,76	6 ,68	6 ,66	6 ,55
2^{-7}	7 ,51	7 ,30	7 ,41	7 ,21
2^{-8}	8 ,26	8 ,06	7 ,97	7 ,77
2^{-9}	9 ,01	8 ,73	8 ,56	8 ,39
2^{-10}	9 ,77	9 ,42	9 ,22	8 ,92
2^{-11}	10 ,52	10 ,18	9 ,81	9 ,54
2^{-12}	11 ,77	10 ,85	10 ,43	10 ,21

Among GALLISSOT's other studies of considerable interest from the methodical point of view we mention the influence of direct and oblique vision, and the phenomenon of simultaneous contrast.

A study of the value of the extinction for different wave-lengths (or colours) is also performed.

An experimental investigation of the nature of the scintillation was also undertaken.

From an inquiry of Harv Ann 44 GALLISSOT has found the mean error for the upper culmination to be $\pm 0^m,19$, whereas for the lower culmination it is $\pm 0^m,25$, thus clearly showing the tendency of the atmosphere to diminish the accuracy of the photometric measurements.

The author believes that it will be possible to "fuse" together the different photometric catalogues of precision. The discrepancies between two determinations, whether physiological or instrumental, can be divided into two main groups:

1. those which can be attributed to a variation of the photometric scale on account of intensity and colour.

2. those which can be attributed to displacements of the zero point, which also depend on the colour.

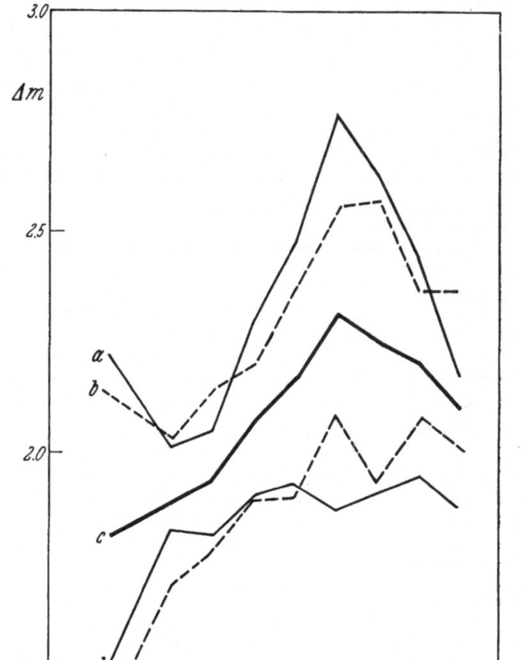

Fig. 74. Study of the PURKINJE-GALLISSOT effect according to ROUGIER. *a* Red star to the right and right eye. *b* Red star to the right and left eye. *c* Mean curve of *a*, *b*, *d*, and *e*. *d* Red star to the left and right eye, *e* Red star to the left and left eye.

48. DANJON's Investigations at Strasbourg[1]. DANJON et ROUGIER have used a number of photometers in order to test the photometric laws. Their results are given in an extensive paper, which covers many of the photometric problems and which ought to be studied by every student of this branch. Of special interest is the high accuracy reached with the cat-eye (œil-de-chat) photometer according to DANJON's construction. Also the R H P, the P D, and the Uranometria Oxoniensis (U O) have been compared with each other. The following result is quoted:

[1] Strasbourg Ann t II, 1er fasc (1929).

Oxford — Harvard.

Magnitude	Sp. class B	A	F	G	K	M	Mean
Bright stars	$0^m,00$	$0^m,00$	$+0^m,01$	$-0^m,03$	$-0^m,03$	$+0^m,09$	$-0^m,02$
$4^m,5$	+0 ,07	+0 ,11	+0 ,13	+0 ,08	+0 ,06	−0 ,01	+0 ,08
5 ,0	+0 ,12	+0 ,11	+0 ,07	+0 ,03	+0 ,04	+0 ,06	+0 ,07
5 ,5	+0 ,07	+0 ,08	+0 ,09	+0 ,09	+0 ,11	+0 ,07	+0 ,09
6 ,0	+0 ,07	+0 ,04	+0 ,08	+0 ,05	+0 ,06	−0 ,01	+0 ,05
6 ,5	−0 ,02	−0 ,09	−0 ,05	−0 ,04	−0 ,10	−0 ,02	−0 ,07
Mean	+0 ,06	+0 ,04	+0 ,06	+0 ,03	+0 ,03	+0 ,01	+0 ,04

The mean deviation after the correction for systematic errors is as follows:

$$\text{Potsd.} - \text{Harv.} = \pm 0^m,095$$
$$\text{Oxf.} - \text{Harv.} = \pm 0\ ,145$$

DANJON wanted to check the GALLISSOT phenomenon for stars being just on the limit of the unaided vision. For that purpose an artificial double star and a meridian photometer were used together with Wratten colour screens Nr. 76 and 86. In fact the artificial star resembles very much the binary system β Cygni.

It was started with a determination of Δm without using screens, having at first the red star to the right of the blue one and then the red star to the left of the blue one. Later a rotating sector was placed between the photometer and the double star and Δm was determined for various apertures of the sector. The adjoining table gives in the first 7 columns the results according to the measures of ROUGIER and in the 8th column the mean value of the corresponding results of DANJON.

Aperture of the sector	Magnitude for the unaided eye corresponding to the magnitude of the brighter artificial star	The values of Δm (ROUGIER)					Δm (DANJON) Mean values
		Red comp. to the right		Red comp. to the left		Mean values	
		left eye	right eye	left eye	right eye		
180°	$-1^m,1$	$2^m,13$	$2^m,22$	$1^m,40$	$1^m,41$	$1^m,81$	$2^m,21$
64	0 ,0	,03	,01	,69	,82	,89	,29
32	+0 ,7	,14	,05	,77	,81	,94	,43
16	1 ,5	,20	,30	,89	,90	2 ,07	,37
8	2 ,2	,37	,48	,90	,93	,17	,39
4	3 ,0	,56	,77	2 ,09	,87	,32	,28
2	3 ,7	,57	,64	1 ,94	,91	,26	,27
1	4 ,5	,37	,45	2 ,08	,95	,21	—
0,5	5 ,2	,37	,18	2 ,01	,88	,11	—

If there were no PURKINJE-GALLISSOT phenomenon present Δm should be constant. The table shows that for stars brighter than $3^m,0$ to the unaided eye there is a PURKINJE effect present and for stars between $3^m,1$—$5^m,2$ a GALLISSOT effect. The latter effect is not so pronounced in the measures of DANJON as in the ones of ROUGIER. The influence of the effect could be eliminated by using a suitable filter (Wratten 21) transmitting 0,5% at λ 5400, 52,5% at λ 5600, and 100% at λ 7200.

The material collected in the table given above contributes also to the question of the size of the equation of position. As is seen from fig. 76 DANJON has a rather constant equation of position, viz. $0^m,094$. ROUGIER always estimates the right star considerably brighter than the left. His observations give a mean value of 0^m25 when no screen is used and with screen No. 21, $0^m,28$. As is seen from fig. 76 the introduction of the screen does not reduce the equation of position but it gives it a more regular course. When one remembers that the equation of position of KEMPF is zero and that of MÜLLER $0^m,07$, it is clear that this quantity is liable to show considerable variations from observer to observer. Furthermore the equation is by no means constant in the case of ROUGIER. The conclusion must be that the utmost care should be given in photometric work to the elimination or determination of this quantity.

Numerous measures are given by DANJON using the photometer invented by him. As these measures relate to variable stars they will not be reviewed here. It should only be pointed out that the accurary reached with the said instrument seems to be very high. From the measures of δ Cephei it can be estimated that the mean error of one magnitude (dependent on some 40 readings) is:

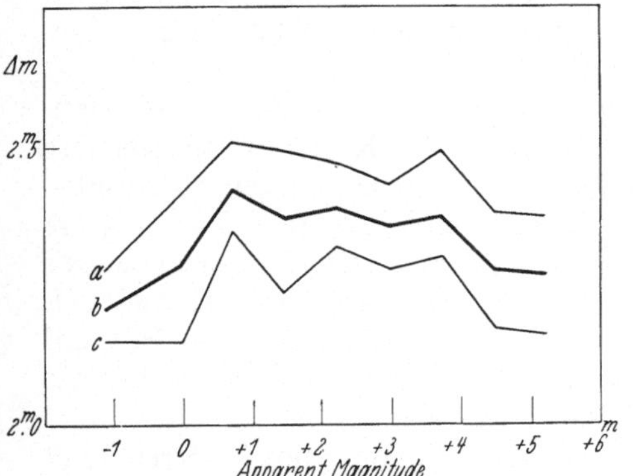

Fig. 75. Study of the PURKINJE-GALLISSOT effect according to DANJON. *a* Red star to the left. *b* Mean curve of *a* and *c*. *c* Red star to the right.

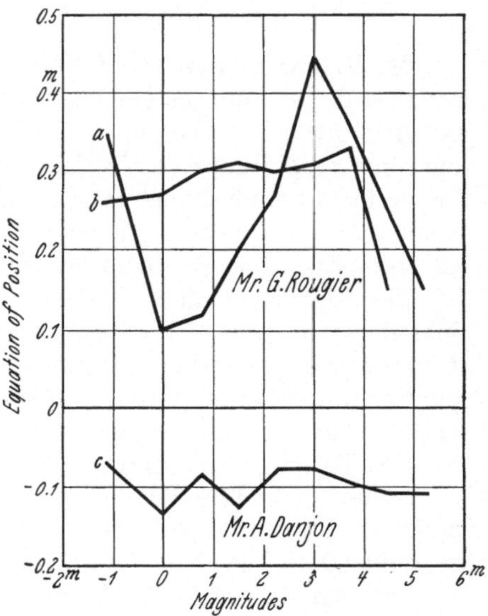

Fig. 76. Equation of position as derived from the measures of ROUGIER and DANJON. *a* ROUGIER's values of equation of position without screen. *b* With screen. *c* The values of DANJON without screen.

$\pm 0^m,027$ in yellow,

$\pm 0^m,034$,, red.

The observations of Algol suggest still smaller values. It seems that the accuracy of the cat-eye photometer it at least four times higher than that of the meridian photometer.

The catalogues in the different volumes of the Harv Ann have been investigated by DANJON, and the differences between the B and M stars (colour equation) formed:

Harv Ann	B stars—M stars	Harv Ann	B stars—M stars
14	$-0^m,02$	44	$-0^m,04$
23	$-0\ ,14$	45	$-0\ ,07$
24, 1	$+0\ ,02$	46, 1	$+0\ ,15$
24, 4	$-0\ ,01$	46, 2	$+0\ ,33$
34	$+0\ ,11$		

The same difference between Potsdam and Harvard amounts to $+0^m,34$.

Finally the following table will show the general relation between the three large catalogues:

	White stars		Coloured stars.	
	Potsd – Harv.	Oxf – Harv.	Potsd – Harv.	Oxf – Harv.
$4^m,25$	$+0^m,21$	$+0^m,01$	$+0^m,08$	$+0^m,04$
4 ,5	,21	,10	+ ,01	,03
5 ,0	,21	,10	+ ,01	,03
5 ,5	,24	,08	−0 ,01	,04
6 ,0	,28	,06	+0 ,02	,03
6 ,5	+0 ,31	+0 ,04	+0 ,02	+0 ,03

49. Visual Standards of Faint Stars. A very valuable extension of the visual scale to faint magnitudes has been made thanks to a plan of E. C. PICKERING[1]. In 1900 he inaugurated this plan with the aim of determining the photographic magnitudes of the faintest stars, visible in the largest refractors in the world. Five photometers of equalizing wedge type were constructed and the intention was to have these photometers identical in design, with their wedges cut from the same photographic plate.

The Harvard, Lick, Yerkes, and McCormick Observatories have participated in the work. The regions observed, called the RUMFORD regions, are 36 in number, and their locations are found in the memoir giving the results. They contain groups of stars surrounding variables of long period, which are very faint at their minimum. Twelve of the regions were selected from HAGEN's Atlas Stellarum Variabilium Ser. I, and twelve each from Series II and III.

A number of 12^m stars served as standards; they were measured at Harvard by E. C. PICKERING with the 12-inch horizontal telescope. The magnitudes are reduced to the scale underlying Vol. 24 of the Harv Ann.

The plan as regards conformity of the wedges used could not be carried out entirely[2]. Care was taken with the calibration of the wedges. An attempt had been made by KING to cause an increase in the absorption of the wedge, by a constant amount, with an increase in the scale readings, and this desire was also very nearly attained. In the early days of the use of the wedge photometer the very important role played by the human eye was overlooked. Even when a uniform wedge is used, the effect produced on the eye of the observer needs not be one of constant ratio. The result of the measurement is in fact a combination of the wedge absorption and the eye estimate. The observations in the present case consist of matching the real star with an artificial star placed close alongside it. As usual, it is also found here that observers are much better able to detect slight differences in magnitude between two stars when they are faint, than when they are bright.

A number of precautions have been taken at the determination of the calibration curves viz: 1. Resemblance in appearance and colour between the real and artificial star. 2. Necessity of keeping the stars to be compared in the same relative position. 3. Constancy of the light of the photometer lamp, with regard to illumination and with regard to the time. 4. Unchangeability of the photometric adjustments. 5. Uniform atmospheric conditions. 6. Measurements of a region immediately repeated in the reverse order. 7. Use of shade glasses.

Each of the RUMFORD regions gives visual magnitudes of 20 stars between 12^m and 16^m. It is unavoidable that the interval between two successive magnitudes sometimes reaches $0^m,50$, or even more, but generally it is $0^m,20$. The value of a number of visual sequences around the sky can scarcely be overestimated.

---

[1] Photometric Magnitudes of Faint Standard Stars measured visually at Harvard, Yerkes, Lick and McCormick Observatories by EDWARD C. PICKERING, J. A. PARKHURST, R. G. AITKEN, H. D. CURTIS, S. A. MITCHELL, H. L. ALDEN, T. McN. SIMPSON, F. W. REED, prepared for Publication by S. A. MITCHELL. Mem Amer Acad Arts Sciences 14, No. 4 (Cambr. 1923). [2] Loc. cit. p. 273.

The relation of the scales at the three observatories is illustrated by the relation:

0,933 Yerkes magn. = 1,020 Lick magn. = 1,047 McCormick magn.

The deviation in the scale of YERKES is considerable, but it would be premature to conclude that the scale is wrong, from the agreement between Lick and McCormick. Only further observations can settle the question of the course of the actual visual scale among the faint stars.

The values of the corrections to a mean scale for the different observers are;

	Observer — Mean scale
PARKHURST	$+0^m,072$
AITKEN	$-0\ ,019$
SIMPSON and REED	$+0\ ,044$
CURTIS	$-0\ ,029$
MITCHELL and ALDEN	$-0\ ,045$

The paper also contains photographs of the standard regions in order to make the identification of these regions easy.

The interesting and important question concerning the relation of the scale of the RUMFORD regions and that of the North Polar region has been discussed by MITCHELL[1]. For this purpose the photovisual magnitudes by SEARES for the region of Nova Persei could be used. The following equation was formulated:

$$m_{MC} = m_{MW} + 0^m,014 + 0,031\,(m_{MW} - 14,0) + 0^m,108\,(C - 1,00).$$

This applies to the photometer and the visual estimates. If only the photometer was used the following expression was obtained:

$$m_{MC} = m_{MW} + 0^m,020 + 0,044\,(m_{MW} - 14,0) + 0^m,144\,(C - 1,00)\,.$$

50. Magnitudes of Double Stars. In the various double star catalogues there are magnitudes estimated for a number of faint stars. These estimates represent an extension of the visual scales to faint objects and, thus, are of considerable interest. One of the first comparisons was made by Sir JOHN HERSCHEL[2], who found the following relation between the scales of STRUVE[3] and of the HERSCHEL's:

STRUVE	Average values	n	Smoothed values	STRUVE	Average values	n	Smoothed values
$4^m,0$	$5^m,42$	15	$5^m,42$	$8^m,0$	$9^m,01$	110	$9^m,05$
4 ,5	—		5 ,98	8 ,5	9 ,51	73	9 ,60
5	6 ,45	10	6 ,38	9	10 ,33	70	10 ,19
5 ,5	6 ,91	6	6 ,77	9 ,5	10 ,75	14	10 ,76
6	7 ,13	17	7 ,08	10	11 ,09	43	11 ,17
6 ,5	7 ,44	16	7 ,54	10 ,5	12 ,00	5	11 ,52
7 ,0	8 ,10	68	8 ,12	11 ,0	11 ,54	16	11 ,73
7 ,5	8 ,82	33	8 ,58	11 ,5	11 ,00	1	12 ,04
				12	12 ,70	5	12 ,70

For fainter stars the difference in the scales is considerable as is seen from the following comparison:

H	STRUVE (Σ)	H	Σ	H	Σ
$6^m,0$	$5^m,5$	$10^m,0$	$8^m,8$	$14^m,0$	$10^m,5$
6 ,5	5 ,9	10 ,5	9 ,1	14 ,5	10 ,7
7 ,0	6 ,4	11 ,0	9 ,3	15 ,0	10 ,9
7 ,5	6 ,8	11 ,5	9 ,6	16 ,0	11 ,1
8 ,0	7 ,3	12 ,0	9 ,8	17 ,0	11 ,4
8 ,5	7 ,7	12 ,5	10 ,0	18 ,0	11 ,6
9 ,0	8 ,1	13 ,0	10 ,2	19 ,0	11 ,8
9 ,5	8 ,5	13 ,5	10 ,4	20 ,0	12 ,0

[1] M N 86, p 356 (1926).

[2] Observations with a 20-Feet Reflecting Telescope. Third Series containing a Catalogue of 384 New Double and Multiple Stars. Mem R A S 3, p. 177 (1828).

[3] Mensurae Micrometricae. Petropoli 1837.

Practically the same scale has been employed by WILHELM STRUVE, OTTO STRUVE, DEMBOWSKI, BURNHAM, and all later observers. There is no doubt that this scale is closer to the International Scale than that of HERSCHEL. A comparison between the binary star scale and the photometric one has been undertaken by ÖPIK[1]. To mention the different contributions to double star magnitudes would be to give a list of all the observers of double stars. Particulars are to be found in ÖPIK's paper. Many of the estimates are affected by considerable systematic errors, but are nevertheless of rather high accuracy. If estimates of Δm are given, these are to be preferred, provided that this difference does not exceed 2^m.

a_3) Photographic Photometry.

51. Pioneer Work. The first attempts to use photography in the service of astronomical research were not at all encouraging. In their first experiments G. P. BOND and WHIPPLE[2] obtained a picture of the bright stars α Lyrae and β Geminorum. The time of exposure was considerable, about 2 minutes. Before the invention of dry-plates it was impossible to think of photography as a photometric method, because of the fact that only bright stars could be reached.

As soon as it was possible to picture numerous stars on photographic plates with moderate exposures it was thought that the photographic method could be used for photometry. During the first wave of enthusiasm the difficulties of the problem were not realized. Much literature has been written concerning what was supposed to be the "law of reciprocity", i. e. the proportionality between the photographic action expressed as the number of silver-grains precipitated, or the intensity of blackening, and the intensity of the light source.

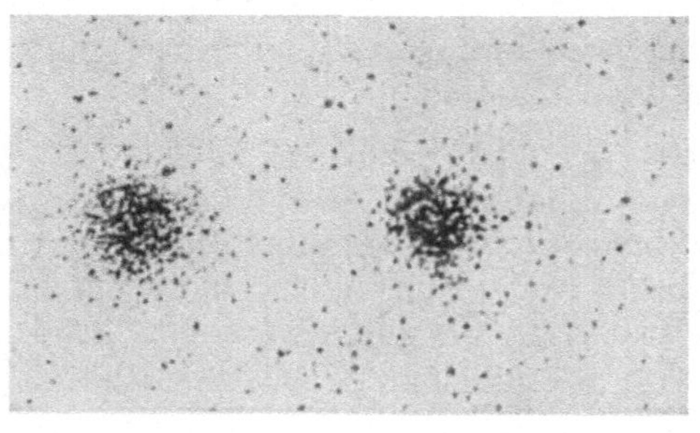

Fig. 77. Enlarged star images. Note the likeness between the structure of the star images and the structure of the star clusters. Using a small magnification the diameter can be measured or estimated with advantage and the magnitude derived with fair accuracy from the diameter laws.

One of the first experiments for establishing photographic photometry was that of CHARLIER[3]. As a measurement of the light intensity CHARLIER used the diameters as measured micrometrically. It was found that the formula $m = a - b \log D$ could be used for obtaining m; a and b are constants, which were determined by using the photometric values of LINDEMANN for the Pleiades. This formula has had a very wide application and many modifications, and a number of similar formulae have been derived. Because these formulae have not only the character of interpolation formulae, but also a physical meaning, we give here a summary of the principal ones hitherto deduced.

52. The Diameter Laws. The first to derive a formula connecting the diameter D and the magnitude m seems to have been G. P. BOND[2]. He found three important laws viz. 1. The first image of a star originates suddenly on the plate and that moment can be determined with considerable accuracy. 2. The

[1] Publ Obs Astr Tartu 25, No. 6 (1924).
[2] A N 47 p. 1 (1857); 48 p. 1 (1858); 49 p. 81 (1858).
[3] Publ d Astr Ges No. 19. Leipzig (1889).

surface of a star image increases in proportion to the time of exposure. 3. For each star and each plate we have the equation:

$$D^2 = a + bt,$$

where D is the diameter corresponding to the time t, and a and b are constants.

The magnitude could be derived from the reciprocity law:

$$It = \text{const.}$$

Nowadays it is difficult to understand how the important results of BOND given in 1857 could be so completely forgotten during three decades that they had to be rediscovered again.

In 1889, CHARLIER[1] and SCHEINER[2] simultaneously investigated the diameter relation. Using photographs of the Pleiades the former found:

$$D = e^{0,250 \log t - 0,148\, m + 2,55}$$

and:

$$m = a - b \log D$$

where:

$$D = D_0\, t^{\frac{1}{4}} \qquad \text{and} \qquad a = a_0 + \frac{b}{4} \log t.$$

SCHEINER[3] found the law:

$$m = a - bD.$$

At the same time SCHAEBERLE[4] suggested the formula:

$$D = a + b \log t.$$

Another formula was derived by E. WESSELL[5] in his dissertation:

$$\frac{1}{m} = a + bD + cD^2.$$

In the Greenwich astrographic work the following formula was introduced by CHRISTIE[6]:

$$m = a - bD^{\frac{1}{2}}.$$

Recently F. E. ROSS[7] has suggested the formula:

$$a + bm = (D + c)^{\frac{1}{2}}.$$

In his photometric work at Catania A. BEMPORAD[8] has adopted the formula:

$$D = a + b(m - 7) + c(m - 7)^2 + d(m - 7)^3.$$

All of these expressions are mainly interpolation formulae, but it can be shown that they also have a certain physical significance. It has been thought that the growth of images is a chemical phenomenon, grains of silver bromide being supposed to develop by infection from neighbouring grains. As has been pointed out by ROSS, it seems that there is a distribution of scattered light in the region immediately surrounding an image, which can account for its spread or increase in size. The simplest assumption then will be that the light distribution is exponential, following BOUGUER's law of intensity variation in a simple absorbing medium. Then:

$$I = I_0\, e^{-\varkappa r}$$

---

[1] Publ d Astr Ges No. 19. Leipzig (1889).
[2] Die Photographie der Gestirne. Leipzig (1897).
[3] Loc. cit. [4] Publ A S P 1, p. 51 (1889).
[5] Dissertation Helsingfors 1904.
[6] See various Greenwich publications and communications, especially M N 52, p. 125 (1892); 65, p. 755 (1905).
[7] Pop Astr 30, p. 5 (1922); Ap J 56, p. 345 (1922); The Physics of the Developed Photographic Image. New York 1925.
[8] Contrib Astr Capodimonte No. 19 (1924).

where I is the light intensity at a distance r measured perpendicularly from the edge of an image and I_0 the intensity just at the boundary of the image. If a series of images are impressed on the plate, the light intensity i, producing the boundary density, d_0, must be the same for all the images. If the central intensity is increased to I, the increase in size of the image is governed by the condition $i =$ constant. Applying the above equation we then have:

$$I e^{\varkappa r} = \text{constant}$$

which, introducing a constant a, can be written:

$$r = a + \frac{1}{\varkappa} \log I .$$

Applying this equation to star images whose total diameter is $d = g + 2r$, where g is thus the diameter of the disk of supposed uniform illumination, we get:

$$d = c + \frac{2}{\varkappa} \log I .$$

This expression is analogous to the equation of SCHEINER, which is thus a consequence of a law of light distribution following BOUGUER's law.

If the law of distribution is given by:

$$I = I_0 e^{-\varkappa\left(\left(r+\frac{g}{2}\right)^{\frac{1}{2}} - \left(\frac{g}{2}\right)^{\frac{1}{2}}\right)}$$

the corresponding diameter law is of the form used at Greenwich. In an analogous way other diameter formulae could also be derived.

C. HOFFMEISTER has shown[1] that even when an objective is used which is not achromatic, the magnitudes can be derived from the measured diameters with fair accuracy.

53. Further Pioneer Work. The weakness of the first studies of photographic photometry was that there was no scale established, fulfilling the condition:

$$m = -2{,}5 \log i .$$

Visual magnitudes were used in the work of CHARLIER, as well as in earlier photometric work in general, for computing the constants in the formulae connecting photographic magnitude and diameter. If all the stars used are normally distributed with regard to their colours, the effect of these reductions will be that the scale differs by a more or less constant value from the corresponding visual scale. If this condition is not fulfilled and the number of comparison stars is comparatively small, the effect will be more complicated and comparatively difficult to calculate, if the colours are not accurately known. Photographic photometry was employed for a long time, this method being applied, and the stock of photometric data contained in the different volumes of the international photographic chart (C d C) is founded on visual estimates. In recent times work has been done which will be discussed in connection with the standard photometric scales.

One of the first catalogues based on photographic magnitudes is that of E. C. PICKERING[2]. The work gives 3 different lists of magnitudes. The first is a catalogue of the Pleiades containing 420 stars. The second is a catalogue of 1009 stars within 1° of the North Pole, and there is a third catalogue with magnitudes for 1131 stars between Decl. $-2°$ and $+2°$. Part of the material is based on measurements of star trails. The catalogue of PICKERING, which was subjected to criticism at the time it appeared, is, indeed, very remarkable, as it is the first one derived on the basis of a real photographic scale. PICKERING

[1] Die Sterne, 1926, p. 35. [2] Harv Ann 18, p. 119 (1890).

declared that the general principle for determining the photographic brightness of the stars was to form photographic images of them either as points, lines, or surfaces. The stars are compared with a standard plate formed by photographing, in the same way, series of images of a bright star with varying apertures or exposures. It is also very interesting to note the suggestion that the stars can be thrown out of focus for photometric purposes, or that the spectra can be compared inter se. In this work we find in reality the origin of the fruitful idea of establishing standard photographic scales, which idea was also carried out later on by PICKERING. It is quite justifiable to say, as did MÜLLER in his "Photometrie der Gestirne", that the intensity scale obtained by giving standard stars different exposures on the same plates cannot represent a correct scale. He was right, too, when he criticized the scales established by using different apertures of the same instrument. But at the same time he overlooked the possibility of establishing a system of photometry, starting from the general principles indicated by PICKERING, who was aware of the weakness of the measurements and tried to avoid some of the errors by using a group of polar stars which were exposed on each plate. Thus the idea of a standard scale at the pole or a sequence was conceived for the first time in the remarkable paper by PICKERING, which did not yet receive proper recognition at first.

54. Cape Photographic Durchmusterung (CPD). This extensive work was performed at the Cape Observatory in the years 1895—1900 by SIR DAVID GILL together with J. C. KAPTEYN[1]. The three volumes, contain the magnitudes of altogether 454875 stars, mainly brighter than $9^m,2$. For each plate a special curve has been constructed connecting magnitudes and diameters. The basis for the magnitudes were the magnitudes of the stars as given in GOULD's zones. The fact that the curves were so different made the authors attribute the variations to a combination of three causes, viz., errors in GOULD's magnitudes, errors produced from the discordance between the visual and photographic magnitudes, and, finally, errors in the measurements of the diameters of the stars. In their introduction to the work the authors are also aware of the fact that there is no fixed relation between diameter and magnitude, even when the same telescope, the same plates, and the same exposures are used at the same altitudes. The scale of the CPD has been thoroughly discussed by E. C. PICKERING[2]. It was clear beforehand that the scale would deviate considerably from the photometric one (RHP), and this made the discussion difficult.

At first the differences for all the stars in the first 6 hours of AR were grouped, and the mean correction found for each Cape magnitude after correcting by subtraction the difference between the two magnitudes; it appeared that in some cases differences were mainly positive, in others negative. The explanation of this was the variable limiting magnitude, on account of differences of sensitiveness and the different transparency of the air. The mean of the positive residuals was taken and applied to the Cape magnitudes. The second measure consisted in revising the correction to the Harvard magnitude from groups extending over 3 hours of AR in each volume of the CDM.

The third measure was to plot points with abscissae equal to the Cape magnitudes and ordinates equal to the differences formed by subtracting the Cape from the photometric magnitude. The smoothed curves then give the correction to be applied to the Cape magnitudes in order to reduce them to the photometric scale.

The three volumes of the Cape catalogue include zones $-19°$ to $-37°$, $-38°$ to $-57°$, and $-58°$ to $-89°$, respectively. The last of these regions was

[1] Cape Ann III—V (1896—1900).

[2] Harv Ann 76, p. 243 (1913).

photographed first, and the photographic lens is said to have been repolished after its completion. Owing to the change in the scale of magnitudes of the stars it is probable that one or more of its surfaces was reground.

The following table gives the "mean" correction to CPD:

Correction	CPD	Correction	CPD
$+0^m,0$	$5^m,8$ to $6^m,2$	$+0^m,6$	$8^m,0$ „ $8^m,1$
0 ,1	6 ,3 „ 6 ,5	0 ,7	8 ,2 „ 8 ,3
0 ,2	6 ,6 „ 6 ,9	0 ,8	8 ,4 „ 8 ,5
0 ,3	7 ,0 „ 7 ,3	0 ,9	8 ,6 „ 8 ,7
0 ,4	7 ,4 „ 7 ,6	1 ,0	8 ,8 „ 8 ,9
0 ,5	7 ,7 „ 7 ,9	1 ,1	9 ,0 „ 9 ,2

To this correction should further be added the corrections given in the fifth column of table V in PICKERING's paper. The values of these corrections vary from $+0^m,4$ to $+0^m,5$. The average difference between the photometric value formed by adding the said corrections to the CPD and that derived from the photometric magnitude corrected for the spectral class is $\pm 0^m,27$. This includes the errors in the CPD measurements and in the photometric measurements, as well as in the assumed corrections for magnitudes, the spectral classification, and the correction for distance from the centre. The average values of the residuals according to the three different methods are $\pm 0^m,27$, $\pm 0^m,28$, and $\pm 0^m,24$. As the three methods of reduction give substantially the same results, it is probable that the deviations are real, and that they cannot be materially diminished by another method of reduction. While the deviations are rather large for individual stars they will probably not be serious in statistical studies. At the time the Cape magnitudes were derived only little material existed for forming a basis for an accurate scale, especially for the fainter stars.

55. The International Photographic Chart. Carte du Ciel. (CdC). When this immense undertaking was started, astronomical photography was in its embryonic state. Photometry was also in an early stage of development and this explains why little attention was given to the question of the scale. The connection between diameter and magnitude was thought to be a real law of Nature, which could be put into a mathematical form.

Now and then voices were heard, expressing more sceptical views. The late N. C. DUNÉR, who had found that the reciprocity law did not hold good, wrote in a paper in 1900[1] that he thought it would be wise to interrupt the work and to wait until the manufacturing of wide angle lenses had advanced considerably. It is quite certain that it would have been well to follow his advise. The CdC involves an enormous waste of time, work, and money. The small observatories that have taken part in this work have been tried up with the CdC for decades and have not had much opportunity to take up other work.

Of course, the energy put into it deserves very high praise and recognition. The work when once completed will be a monument of astronomy of position, which will certainly stand for ages as one of the greatest enterprises of a scientific nature.

There is a regrettable lack of homogeneity in the work. Each observatory has preferred to introduce variations in its method of reduction. In most of the zones the equatorial coordinates are not given. In the Potsdam zones approximate right ascensions and declinations are given of sufficient accuracy for purposes of identification The Helsingfors zone contains the accurately derived equatorial coordinates ($0^s,001$ and $0'',01$) and the volumes of the Catania zone have followed

[1] Stockholm, K Vet Akad Öfversigt Årg 57: Om den på fotografisk väg framställda stjärnkatalogen. 1900.

the same practice until 1925, but later on have given up the conversion of measurements into right ascensions and declinations.

The identification of a certain star is in many cases troublesome, and it has happened to the present writer that he has met with wrong formulae in the introduction to a certain catalogue[1]. I understand very well why some astronomers have said that they prefer to observe a star by means of their meridian circles rather than to plunge into the collection of the catalogues for the computation of a single position.

In some twenty papers H. H. TURNER has discussed the question of the magnitude scales of the different observatories taking part in the Astrographic Catalogue[2]. His method is simply that the number of images recorded under each unit of the magnitude scale should be counted and tabulated. Using in his first notes the value of $B(m)$ as tabulated by KAPTEYN in Groningen Publ No. 18 and later on the improved values of the same function according to CHAPMAN's and MELOTTE's investigation[3], TURNER derived a number of results which cannot be given here in full.

In the Bordeaux zone TURNER has found the following relation between the scales:

Bordeaux	KAPTEYN	Bordeaux	KAPTEYN	Bordeaux	KAPTEYN
$6^m,25$	$6^m,82$	$8^m,75$	$9^m,56$	$11^m,25$	$11^m,50$
6 ,75	7 ,24	9 ,25	10 ,15	11 ,75	11 ,64
7 ,25	7 ,83	9 ,75	10 ,59	12 ,25	11 ,68
7 ,75	8 ,38	10 ,25	10 ,99	—	—
8 ,25	9 ,00	10 ,75	11 ,22		

There is a systematic excess of bright stars near $AR = 12^h$, four times the galactic effect of KAPTEYN, but not localized to both sides of the Galaxy. Its origin is probably of a seasonal nature, not due to temperature, but perhaps to wind, which makes bright stars appear brighter and faint stars fainter.

The errors of the magnitude scale of the Algiers zone range from $-0^m,43$ to $+0^m,38$ according to KAPTEYN. Using harmonic analysis TURNER finds evidence against the relative increase of faint stars as the Galaxy is approached.

[1] The existence of the erroneous formulae in the Toulouse zones has been pointed out by LIVLÄNDER in A N 224, p. 95 (1925).

[2] M N 69, S. 392 (1908—1909). 1) (A proposal for comparison of the Stellar Magnitude Scales of the different Observatories taking part in the Astrographic Catalogue): 2) M N 72, p. 464 (1911—1912). (The Bordeaux Magnitudes); 3) M N 72, p. 700 (1911—1912). (The Algiers Magnitudes); 4) M N 75, p. 57 (1914—1915). (The Cape Magnitudes); 5) M N 75, p. 143 (1914—1915). (The Perth Magnitudes); 6) M N 75, p. 465 (1914—1915). (The Oxford Magnitudes, with a letter from F. G. BROWN); 7) M N 75, p. 601 (1914—1915.) (The Vatican Magnitudes); 8) M N 76, p. 2 (1915—1916). (The Cape Magnitudes for $-42°$); 9) M N 76, p. 149 (1915—1916). (The Toulouse Magnitudes); 10) M N 77, p. 35 (1916—1917). (The Melbourne Magnitudes); 11) M N 78, p. 54 (1917—1918). (The Cordoba Magnitudes); 12) M N 78, p. 578 (1917—1918). (The Hyderabad Magnitudes $-17°$ and $-18°$; The Perth Magnitudes $-32°$ and $-34°$; the Edinburgh Magnitudes $-38°$; and the Cape Magnitudes $-41°$ and $-42°$); 13) M N 79, p. 136 (1918—1919). (The Bordeaux Magnitudes); 14) M N 79, p. 565 (1918—1919). (The Tacubaya Magnitudes; The Cape Magnitudes); 15) M N 80, p. 620 (1919—1920). (The Perth Magnitudes for $-36°$ with some remarks on neighbouring zones); 16) M N 81, p. 525 (1920—1921). (The Cordoba Magnitudes for $-27°$, $-29°$ and $-31°$, with some remarks on the whole region $-15°$ to $-43°$. 17) M N 83, p. 385 (1922—1923). (The San Fernando Magnitudes for $-3°$ and $-5°$, and the Paris Magnitudes for $+22°$, $+23°$ and $+24°$); 18) M N 84, p. 735 (1923—1924). (The San Fernando Magnitudes for $-4°$ and $-6°$, and the Algiers Magnitudes for $-1°$); 19) M N 85, p. 471 (1924—1925). (The Oxford Magnitudes for $+31°$ to $+25°$); 20) M N 85, p. 610 (1924—1925). (Provisional general survey); 21) M N 87, p. 64 (1926—1927). (The Sydney Magnitudes for $-52°$ and The Melbourne Magnitudes for $-65°$ and $-67°$.)

[3] Mem R A S 60, part IV (1914).

This does not prevent KAPTEYN's figures from being essentially correct, but it can be put in this way:

1. The stars of the Galaxy, all spectral classes taken together, are bluer in colour than those near its poles (KAPTEYN). 2. The distant stars of the Galaxy are redder than those near its poles.

TURNER also finds that the evidence from the counts "is not unfavourable to the conclusion that there are more faint stars in the direction of the 94° vertex as demanded by the hypothesis of a finite stellar system".

The fifth note deals with the Perth magnitudes which are founded on a remarkably uniform scale. There is an excess of faint stars for low galactic latitudes, but other phenomena exist the origin of which it is rather difficult to explain.

When comparing the results of the Oxford zones +31° to +25° with the scale of CHAPMAN and MELOTTE[1], TURNER found that the catalogue scale derived from the formula $m = 16{,}00 - 1{,}18\sqrt{d}$ was satisfactorily uniform until just near the lower end. A very remarkable feature of the Oxford measurements is that the number of stars varied in a systematic way. There was a deficiency of certain stars which could be explained by assuming the presence of some obstructing cloud of matter in certain directions. The effect is not galactic and is certainly not due to seasonal effects, such as temperature or wind. TURNER has returned to the zones that were investigated earlier, and finds obscuration effects also here.

As for the scale of the Vatican plates it is also based on the square-root formula and is fairly regular, about one whole magnitude brighter than the scale of CHAPMAN and MELOTTE. It is clear from the grouping made by TURNER that the part in which faint stars are relatively scarce is in, and near, the Galaxy itself. The galactic condensation is at least swamped out by the obscuration. The value of $\frac{\Delta \log N}{\Delta \beta}$ for a degree is found to be $-0{,}00158$ for unobscured parts, but $-0{,}00078$ for obscured parts. The value for unobscured Cape regions is very high, viz. $-0{,}00390$.

The topic discussed in the next paper is the Cape magnitudes for −42°. The counts confirm the conclusions previously deduced. A large obscured region is present, extending from 22^h through 0^h to 10^h. The value of $\frac{\Delta \log N}{\Delta \beta}$ shows that near the Galaxy log N increases at a rate rising to 0,0015 per magnitude for one degree of galactic latitude.

In connection with further counts TURNER has found that the obscured patches form a spiral round the sky, the median line of which follows approximately the curve:

$$\alpha + 4\delta = 265°,$$

where α and δ are the right ascension and northern declination respectively.

The counts in the Oxford zones[2] +31° to +25°, analyzed according to the standard process of M N 78, p. 55, did not appear consistent with the formerly suggested spiral with regard to the rate of the first harmonic. The first, second, and fourth harmonics for all stars are large and consistent for the different zones, which indicates definite features of stellar distribution distinct from the galactic condensation in galactic latidude, which was eliminated before the analysis was made.

A collection of the results of 40 zones in different parts of the sky was given by TURNER in M N 85, p. 610 (1925). The mean value of log N is 2,26, corres-

[1] Mem R A S 60, part IV (1914). [2] M N 85, p. 471 (1925).

ponding to a magnitude 13,35 on the scale of CHAPMAN and MELOTTE. The means for each zones were represented by a series of spherical harmonics. As a standard value of log N, 1,50 has been derived, but has not been critically discussed. In the paper under review attention is concentrated on the distribution of the stars of mean magnitude 11^m.

Contrary to expectation it has been found that the third and fourth harmonics are by no means small and irregular. Thus the coefficient of sin 4 α is systematically positive in all the groups. Sometimes the computation of the sixth harmonic is required.

It appears that the galactic concentration is not the only systematic feature in the stellar distrubution. There are also others, which can be profitably studied by means of simple counts from the CdC plates.

The limiting magnitude of the charts of the Carte du Ciel has been determined at the Observatory at Lund. The heliogravures and the charts of HAGEN were compared with each other. Since the charts did not contain such faint stars as the photographs, a number of observers estimated the difference in photographic magnitudes. The results, giving $13^m,09 \pm 0^m,17$ as limiting magnitude, show rather fair agreement, but it is to be remembered that systematic errors are liable to enter into such a determination. If greater intervals of magnitudes have to be estimated, it is necessary that the same kind of objects are compared, that is, stars in the sky or in the field of the same instrument, or star-images with star-images on plates taken with the same instrument etc.

When dealing with many objects in the CdC the work of identification can be considerably facilitated. Besides, the Observatory of Bergedorf has edited convenient tables[1], which reduce to a minimum the work of converting rectilinear coordinates to equatorial ones in the catalogues.

In some zones the magnitudes are computed from formulae connecting diameter and magnitude; in other cases the diameters alone have been given.

The following summary of the present state of the work may be of interest:

The classification is according to TURNER (Transactions of the Internat Astr Union vol. III): F (finished); S (safe); D (doubtful).

Observatory	Zone	Classification	Charts	Catalogue
Greenwich	+90° to +65°	F	Completely published	Complete. The results published in four parts containing 78803 stars.
Specola Vaticana	+64° to + 55°	F	88 charts have been published	Completely printed in 10 volumes.
Catania	+54° to +47°	S	—	—
Helsingfors	+46° to +40°	S	—	All zones printed from 1^h30 to 18^h in six parts containing 156754 stars.
Potsdam	+39° to +32° The part +36° to +39° will be measured and reduced at Hyderabad	D	See Uccle and Hyderabad at the end of this table.	Seven parts containing 144342 stars have been published.
Oxford	+31° to +25°	F	—	Completely published in 8 parts containing 470783 stars

[1] Sammlung von Hilfstafeln der Hamburger Sternwarte in Bergedorf. G. (1924).

Continued.

Observatory	Zone	Classification	Charts	Catalogue
Paris	$+24°$ to $+18°$	S	Zone $+24°$ compl. Some 420 charts publ. of the zones $+18°$, $+20°$, $+22°$	Five zones out of seven ($+20°$ to $+24°$) completely printed. The last volume (VII) is expected in print pretty soon
Bordeaux	$+17°$ to $+11°$	S	126 charts of zone $+12°$, 153 charts of zone $+14°$, 160 charts of zone $+16°$, have been published	Four zones out of seven printed ($+11°$, $+9°$, $+7°$, $+5°$).
Toulouse	$+10°$ to $+5°$	S	147 Charts of zone $+5°$, 139 charts of zone $+7°$, 154 charts of zone $+9°$ have been published	Four zones out of seven printed ($+11°$, $+9°$, $+7°$, $+5°$)
Algiers	$+4°$ to $-2°$	F	Zones $-1°$, $+1°$, and $+3°$ are completely publ. Two charts in zone $+4°$ publ.	Completely printed.
San Fernando	$-3°$ to $-9°$	F	Zone $-9°$ completely publ. 173 charts in zone $-7°$, 153 charts in zone $-5°$, 81 charts in zone $-3°$ have further been publ.	The last volume $-9°$ in print (1929). Seven volumes printed.
Tacubaya	$-10°$ to $-16°$	D	—	$1^1/_4$ volumes out of 7 printed.
Hyderabad	$-17°$ to $-20°$	F	90 charts in zone $-16°$, 116 in zone $-15°$, 89 in zone $-13°$, 79 in zone $-11°$ have been publ., zone $-9°$ is compl.	Completely published
Hyderabad	$-21°$ to $-23°$	F	—	The six published volumes contain 433412 stars.
Cordoba	$-24°$ to $-31°$	S	Zone $-25°$ completely published	Five volumes out of eight printed.
Perth	$-32°$ to $-37°$	F	—	Completely printed.
Edinburgh	$-38°$ to $-40°$	D	—	Awaiting funds for printing.
Cape of Good Hope	$-41°$ to $-51°$	F	—	Completely printed in 11 volumes.
Sydney	$-52°$ to $-64°$	D	—	Six out of 52 (small) volumes have appeared. Material ready for four other volumes.
Melbourne	$-65°$ to $-90°$	D	—	Zones $-65°$ and $-66°$ published in two volumes.
Hyderabad	$+36°$ to $+39°$	S	—	
Uccle	$+33°$ to $+39°$	D	106 charts in zone $+33°$ 62 charts in zone $+35°$ 68 charts in zone $+37°$ 62 charts in zone $+39°$ have been published	No catalogue work has been undertaken.

56. Standards for the Astrographic Catalogue. Because of the need of photographic magnitudes on the International Scale MISS LEAVITT has undertaken the work of establishing secondary sequences for the stars in the catalogue series[1]. The regions compared are situated in the zones one degree wide that are common to adjacent zones of the Map. Each sequence includes from fifteen to twenty-two stars, brighter than 15^m. The same methods of observational measurement and reduction were used as in establishing the North Polar Sequence.

The adjoining table indicates the positions of the centres.

Zone	δ	0^h	1^h	2^h	3^h	4^h	5^h	6^h	7^h	8^h	9^h	10^h	11^h	12^h	13^h	14^h	15^h	16^h	17^h	18^h	19^h	20^h	21^h	22^h	2
eenwich																									
	+64°,5	×		×		×		×		×		×		×		×		×		×		×		×	
ma																									
	+54 ,5		×		×		×		×		×		×		×		×		×		×		×		×
tania																									
	+46 ,5	×		×		×		×		×		×		×		×		×		×		×		×	
elsingfors																									
	+39 ,5		×		×		×		×		×		×		×		×		×		×		×		×
tsdam																									
	+31 ,5	×		×		×		×		×		×		×		×		×		×		×		×	
ford																									
	+24 ,5		×		×		×		×		×		×		×		×		×		×		×		×
ris																									
	+17 ,5	×		×		×		×		×		×		×		×		×		×		×		×	
rdeaux																									
	+10 ,5		×		×		×		×		×		×		×		×		×		×		×		×
ulouse																									
	+ 4 ,5	×		×		×		×		×		×		×		×		×		×		×		×	
giers																									
	− 2 ,5		×		×		×		×		×		×		×		×		×		×		×		×
n Fernando																									
	− 9 ,5	×		×		×		×		×		×		×		×		×		×		×		×	
cubaya																									
	−16 ,5		×		×		×		×		×		×		×		×		×		×		×		×
yderabad																									
	−23 ,5	×		×		×		×		×		×		×		×		×		×		×		×	
rdoba																									
	−31 ,5		×		×		×		×		×		×		×		×		×		×		×		×
rth																									
	−40 ,5	×		×		×		×		×		×		×		×		×		×		×		×	
pe																									
	−51 ,5		×		×		×		×		×		×		×		×		×		×		×		×
dney																									
	−64 ,5	×		×		×		×		×		×		×		×		×		×		×		×	
elbourne																									

57. Reduction of thirty-nine Astrographic Zones to the International Photometric Scale. The Mount Wilson Contr. No. 305 contains a paper by SEARES and MARY C. JOYNER[2] giving the calibration of 39 astrographic zones. The N_m, as collected and published mainly by TURNER, were used on the basis of the counts in each zone. The limiting magnitude was derived from VAN RHIJN's distribution table (Groningen 27) and reduced to the International Scale by comparison with the Mount Wilson catalogue. The calibration is not the best possible,

[1] The sequences are published in Harv Ann 85, Nos 1, 7, and 8. The last catalogue was prepared for publication by Miss WALKER owing to the death of Miss LEAVITT in 1921. The average deviation of a single measurment is about $0^m,10$, which suggests a mean error of $\pm 0^m,12$. After the effect of systematic errors due to the condition of the sky, photographic film, etc. have been eliminated the mean error is about $\pm 0^m,09$.

[2] Also Ap J 62, p. 320 (1925).

because of the influence of residual errors in VAN RHIJN's table, but steps have been taken to minimize the influence of these errors.

Table IV in the said Mount Wilson Contr. gives the magnitudes corresponding to the different scale readings in the separate zones investigated. Although the calibrations refer to the zones as a whole, the residuals in the limiting magnitudes for different galactic latitudes indicate that, in many cases, the calibrations are directly applicable to small parts of a zone. These errors are not, however, a final test of consistency in the scales because errors of reduction and irregularities in stellar distribution both contribute to their progressive character.

58. Photometric Scale of the Cape Astrographic Zones and the Cape Photographic Durchmusterung. In the volume "Magnitudes of Stars contained in the Cape Zone Catalogue of 20843 Stars for Equinox 1900, Zones $-40°$ to $-52°$", London 1927, an account is given of the valuable photometric work at the Cape that aims at furnishing means to convert the measured diameter values in the Cape Zone Catalogue into photographic magnitudes. The whole of the work connected with the measurement and reduction has been carried out under the supervision of J. K. E. HALM, to whom also we owe the discussion of the results contained in the Introduction (pages I—LXV). The examination of discordances in the magnitudes derived from the measured diameters as well as the reductions and the general checking of the results has been carried out under the control of A. J. WILKIN.

The main object of the Cape investigation was to find the relationship between the measured diameters and the corresponding photographic magnitudes based on a fundamental photometric system. Before this problem could be solved it was necessary to establish, in some specially selected region, a sequence of photographic magnitudes. For this purpose the region within one degree of the South Pole was chosen.

A general statement of the photometric problem is given in section A of the Introduction. Use has been made of the general equations

$$f(D) = I\,t/\psi\,,$$
$$\psi = \tfrac{1}{2}(i^{\alpha} + i^{-\alpha})\,,$$
$$i = I/I_0, \qquad \alpha = 0{,}25\,,$$

where D is either diameter or opacity, I intensity of the incident light in the interval of time, t, during which the star is exposed, and I_0 is the intensity at which the iso-actinic curve $f(D) =$ const. attains a minimum (optimal intensity).

Experiments may be so arranged that one of the three variables, D, I, and t, is kept constant.

1. By exposing the stars of a region for a certain interval of time, t_c, the relation between D and I for all star-images on the plate is expressed by

$$f(D) = \frac{I}{\psi}\,t_c\,.$$

The curve (D, I) is known as the "intensity scale".

2. A series of images is obtained from every star by exposing it successively for different intervals of time. The images of one and the same star refer to the same intensity, I_c, and the relation between D and t is therefore expressed by:

$$f(D) = \frac{I_c}{\psi_c}\,t\,.$$

The curve (D, t) is termed the "time scale".

For images of the same opacity D in both the series we find:

$$\frac{I}{I_c}\frac{t_c}{t} = \frac{\psi}{\psi_c}.$$

Suppose we are dealing with intensities very much smaller than the optimal intensity I_0; we may then write:

$$\frac{\psi}{\psi_c} = \left(\frac{i}{i_c}\right)^{-\alpha} = \left(\frac{I}{I_c}\right)^{-\alpha}$$

and

$$\left(\frac{I}{I_c}\right)^{1+\alpha} = \frac{t}{t_c}.$$

Expressing I in magnitudes and adopting $\alpha = 0{,}25$:

$$m - m_c = 2{,}0 \log t_c/t.$$

Since t_c and t are known from the observed data, the only unknown quantity is m_c. Obviously it can be determined if the magnitude of at least one star on the plate is known from other sources.

In establishing the South Polar Sequence, the magnitudes of the Pleiades according to the determinations by HERTZSPRUNG[1] and by Miss LEAVITT[2] have been used.

With regard to the accuracy of this method it has been found that the mean difference irrespective of sign between Cape and HERTZSPRUNG is $\pm$ 0m,07 and that between HERTZSPRUNG and Harvard $\pm$0m,06. The Cape determinations, therefore, nearly reach the standard of accuracy of the best photometric determinations.

The following small table will give an idea of the use of this method and of the accuracy obtained by it:

Groups	$\log t_c/t$	m (HERTZSPRUNG)	$m - 2\log t_c/t$	Obs.—Comp.
I	0m,00	3m,88	3m,88	+0m,04
II	+0 ,77	5 ,36	3 ,82	−0 ,02
III	+1 ,41	6 ,64	3 ,82	−0 ,02
IV	+1 ,67	7 ,14	3 ,80	−0 ,04
V	+1 ,86	7 ,65	3 ,93	+0 ,09
VI	+2 ,07	8 ,05	3 ,91	+0 ,07
VII	+2 ,35	8 ,53	3 ,83	−0 ,01
VIII	+2 ,59	8 ,96	3 ,78	−0 ,06
IX	+2 ,76	9 ,33	3 ,81	−0 ,03

The method of observing was as follows: On each of six plates six exposures of 1000, 500, 250, 125, 62, and 31 seconds duration were made. The plates used were lantern plates of low sensibility (6 units in the H and D system).

Then faster plates were used, and from these it was found that photometrically correct magnitudes can be obtained throughout the whole range of measurable diameters or opacities, if the investigation is carried out by steps proceeding from slow to fast plates. Apart from a constant, the slow plate furnishes a correct scale of magnitudes of the brighter stars, the faintest of which determine the zero-point of the faster plate. The latter, in its turn, furnishes the scale of still fainter magnitudes, so that eventually a sequence of magnitudes through the whole range of measurable actinic effects is obtained, based on a photometrically correct uniform scale, which is independent of extraneous observational data, the only external evidence required being the magnitude of one star on the plate.

[1] A N 186, p. 171 (1910).
[2] Harv Ann 71, No. 3, Table XLVI (1911).

The correctness of the results was tested by comparison with 65 stars in the Selected Area 101 according to the improved values of SEARES.

The following group comparison is of interest:

Group	Mt. Wilson – Cape	n
$8^m - 10^m$	$0^m,00$	6
10 – 11	–0 ,04	11
11 – 12	–0 ,05	12
12 – 13	–0 ,01	12
13 – 14	+0 ,03	16
14	+0 ,06	8

The provisional photographic magnitudes for the South Polar Sequence were then used for a standardization of the magnitudes of the Cape Astrographic Zones. For this purpose 24 nearly equidistant areas situated along the parallel $\delta = -45°$ were selected. Exposures of 20 and 6 minutes duration of both the polar and one of the zonal areas were made on the same plate, the observations being taken at approximately the same zenith distances. For each plate a smoothed curve was then constructed, the coordinates being the measured diameters D and the polar magnitudes m. From this graph the magnitude of any star of the zonal sequence was obtained by using its observed diameter as argument.

The graphs (D, m), which contain the intensity-scales of the individual plates, shew appreciable differences from plate to plate. From a detailed discussion of the material it was found that the intensity-scales of the 24 graphs can be brought into fair coincidence by adding appropriate constants to the magnitudes m. That means that the gradation of the scale is very approximately the same on all plates and that only the zero-point of the m-coordinate shows appreciable differences.

An interesting result is that when m is grouped according to the epochs of observation (ultimate values 1911,1–1914,6) there is a pronounced "run" in the m's from positive to negative. As the same brand of plates was used it seems that this may depend on a linear decrease having occurred in the sensibility of the plates.

The difference between the tabular magnitudes obtained with good and bad definition is rather constant throughout the entire range of magnitudes and in the mean $= +0^m,34$, in the sense that the plates with good definition show larger tabular magnitudes than the plates with bad definition. It seems that definition has no appreciable effect on the course of the scale, but considerably influences the value of the zero-point.

By using the sequences of magnitudes in the 24 Selected Areas of the Cape Zone, a mean intensity-scale (D, m) was established in the following way by means of the diameters published in the Cape Astrographic Zones:

a) The conversion of the diameters into magnitudes can be based on the intensity scale, provided that the correction to the zero-points, Δm, can be determined from one or more known magnitudes.

b) Known visual magnitudes and colour indices from the R H P (Harv Ann 50), as well as photographic magnitudes for faint stars (Harv Ann 71, No. 2), were also used in order to establish the connection between diameter and magnitude.

c) Lastly, an intensity scale can be derived from the time scale provided that the optimal intensity and one magnitude are known. A time scale $(D, \log t)$ was derived from a number of plates on which the diameters were measured in accordance with the system in the Astrographic Zones. From the investigations of the South Polar Sequence and the Pleiades the optimal intensity was found to correspond to $m_0 = 2^m,93$. By accepting that $D = 4{,}0$ and $m = 6{,}60$, it follows that $m - m_0 = 3^m,67$ and $\log t_c/t$ — const $= + 1^m,61$. We have generally:

$$\log \tau = C - \log t.$$

The numerical value of C is 3,51.

The results are collected in the following table:

Diameter (D)	Magnitudes (m)			Mean.
	a	b	c	
8,0	—	2^m,87	—	2^m,87
7,5	—	3 ,28	—	3 ,28
7,0	—	3 ,70	—	3 ,70
6,5	—	4 ,15	—	4 ,15
6,0	—	4 ,61	4^m,65	4 ,63
5,5	—	5 ,09	5 ,10	5 ,10
5,0	—	5 ,59	5 ,56	5 ,58
4,5	—	6 ,10	6 ,07	6 ,08
4,0	6^m,60	6 ,63	6 ,60	6 ,61
3,5	7 ,17	7 ,19	7 ,22	7 ,19
3,0	7 ,77	7 ,81	7 ,87	7 ,82
3,5	8 ,45	8 ,50	8 ,57	8 ,51
2.0	9 ,23	9 ,27	9 ,35	9 ,28
1,8	9 ,60	9 ,60	9 ,66	9 , 62
1,6	9 ,99	9 ,99	10 ,01	10 ,00
1,4	10 ,40	10 ,39	10 ,42	10 ,40
1,2	10 ,86	10 ,82	10 ,87	10 ,85
1,0	11 ,38	—	11 ,37	11 ,38
0,8	12 ,00	—	11 ,98	11 ,99
— 1	—	—	12 ,7	12 ,7
— 5	—	—	13 ,7	13 ,7

As for the relation between diameter and magnitude of the 24 Selected Areas investigated, we refer to Table XXI in the Introduction to the photographic work of the Cape.

Now the question arises whether the variations in gradation of the CPD plates are also sufficiently small and accidental to disappear almost completely, as is the case with the plates of the intensity scale. To test this, altogether 8 Selected Areas were examined. It was found that in these cases the tabular magnitudes were approximately correct in scale.

The following general conclusion can be drawn:

In a catalogue of stars the diameters of which have been observed on a small number of plates (three or more) the establishment of a mean intensity scale (D, m) is sufficient for the determination of systematically correct magnitudes for all the stars, provided that each plate contains at least one star whose magnitude is known from extraneous sources.

In the Cape Astrographic Zones at least two diameters for every star are available from overlapping plates. In addition, the CPD diameters can be constructed for the stars brighter than 11,5, so that every magnitude depends on the mean of the results obtained from three different plates.

The work of deriving necessary zero-point corrections for the Cape Astrographic and CPD plates in order to convert the tabular magnitudes into correct photographic magnitudes was greatly facilitated by the thorough comparison of the CPD plates with the Harvard Visual System (Harv Ann 80, No. 13). On every plate a number of CPD stars and their diameters were collected. The CPD magnitudes were then converted into photographic magnitudes, and the tabular magnitudes were taken from the table. The difference between these and the corrected CPD magnitudes is the required plate correction, Δm.

A Table BC gives the corrections to tabular magnitudes for every plate of the Cape Astrographic zones, 1512 in number.

The following statement in the Introduction is very important: "It is a matter of serious consideration whether the standardization of the magnitudes of all the Astrographic Catalogues might not be undertaken on the basis of this simple method, which would require comparatively little work beyond what has been done already".

The above method of deriving photographic magnitudes has been applied to the diameters of the 20843 stars contained in the Cape Zone Catalogue. The magnitude of every star is based on three, and sometimes four, determinations. By comparing the single results with the mean, the mean error of a single magnitude based on the observation of one diameter was found to be $\pm 0^m,28$. This error is due to a variety of causes, the most prominent of which are the accidental errors in the measurements of the diameters, errors arising from the irregularities in the images, especially near the edges of the plates, and from local irregularities in the sensitiveness of the film, and errors due to variation in the gradation of the intensity-scale.

With regard to the first-mentioned type of error, the investigation of the South Polar Sequence has shown the mean error from a single diameter to be $\pm 0^m.22$. Doubtless the accuracy of the estimation of the diameters of the Zone Catalogue is inferior to that of the more precise method adopted in the observations of the South Polar Sequence, so that the increase in the mean error from $0^m,22$ to $0^m,28$ is due, probably to a large extent, to the larger accidental discrepancies in the measurements of the diameters. There is thus good reason for assuming that the errors arising from the other above mentioned sources are of a comparatively low order.

An investigation of the size of the error of the zero-points gives the result that the m. e. of a single CPD magnitude would be $\pm 0^m,11$.

If the full amount of this error is attributed to the Cape system, the m. e. of a single catalogue magnitude based on the measurement of three diameters would be $\pm 0^m,20$.

It is found that the scale of the Cape visual system, i. e. (Cape photographic — colour index), is in good agreement with the scale of the Mount Wilson photovisual system. In order to reduce the Cape system to the Mount Wilson system, a zero-point correction of $+0^m,04$ is required.

If we directly compare the photographic results at the Cape and Mount Wilson and use the Harvard Selected Regions (Harv Ann 71, No. 4) as auxiliary standards, the same value is found for the zero-point correction.

Lastly, a comparison with Mount Wilson has been made by using the star numbers for a certain limiting magnitude according to Publ. of Astronomical Laboratory at Groningen No. 27. The mean difference, Mount Wilson — Cape, for stars between $6^m - 11^m$ is $+0^m,06$, which is thus essentially the same as has been found from using two other methods of comparison.

59. SCHWARZSCHILD's Extra-focal Method. A classical paper concerning photographic photometry is K. SCHWARZSCHILD's: Die Bestimmung von Sternhelligkeiten aus extrafokalen photographischen Aufnahmen[1]. He took up the suggestion of using extra-focal images of the stars for photometric purposes.

[1] Publ d. KUFFNERschen Sternwarte Wien 5 (1900).

Much care was given to several questions, for instance, whether extra-focal or intra-focal images should be used. The degree of blackness of the images was measured, a scale obtained by exposure according to the formula being used.

The relation between blackness, time of exposure and intensity was discussed, and SCHWARZSCHILD introduced in this work the constant p that has been named after him. The photographic magnitudes of 34 stars in the Pleiades are the practical result of this very important methodical investigation.

Extensive use of the extra-focal method has been made by E. S. KING at Harvard. Some of his results are given in connection with the discussion of the colours.

60. PARKHURST, Yerkes Actinometry. This work[1] is not only remarkable because of its high value from the point of view of photographic photometry; it also gives the photovisual magnitudes of stars down to $7^m,5$ in the zone between $+73°$ and $+90°$ declination. The choice of zero-point on the scale is of fundamental importance for a catalogue giving stellar magnitudes. During the time the work was being performed PICKERING suggested an international system for photometry. This important suggestion, which was adopted by PARKHURST, was as follows:

For the stars of spectral type A0 of the Harvard classification and of magnitude $5^m,5$ to $6^m,5$ the photographic and visual magnitudes shall be assumed equal, and visual magnitudes shall be reckoned on the Harvard scale.

The way of obtaining an absolute photometric scale is described in a paper by PARKHURST and JORDAN[2]. The magnitudes were also checked by the plates taken of the Pleiades.

The catalogue gives the photographic and photovisual magnitudes of stars together with the spectral classes as classified at Yerkes Observatory.

The photographic magnitudes are on an average dependent on measurements of 4,7 plates. The photovisual are averages from 3,4 plates. The mean errors on a plate are $\pm 0^m,068$ for the photographic and $+0^m,046$ for the visual magnitudes. The errors, classified according to magnitudes, are shown in the following small table:

Limiting magnitudes	Photographic		Photovisual	
	Mean error	n	Mean error	n
Brighter than $6^m,0$	$\pm 0^m,049$	33	$\pm 0^m,058$	58
6,00 to 6,49	30	28	49	61
6,50 to 6,99	31	66	58	104
7,00 to 7,24	36	30	58	87
7,25 to 7,49	34	54	56	107
7,50 to 7,74	36	66	47	73
7,75 to 7,99	44	74	59	61
Fainter than $8^m,0$	62	256	70	44
Mean of all	$\pm 0^m,047$	601	$\pm 0^m,053$	702

As regards the photographic magnitudes the stars brighter than $6^m,0$ and fainter than $8^m,0$ have large mean errors. For the photovisual magnitudes a dependence on brightness is not very evident.

A comparison can also be made for 151 stars in common with the list of polar stars in the Göttinger Aktinometrie. The agreement is good except for stars of mean magnitude $7^m,78$, where the coloured stars show deviations. The differences are dependent on the colours, and in fact we have established here the existence of a colour equation.

[1] Ap J 36, p. 169 (1912). [2] Ap J 26, p. 244 (1907).

There were 598 stars in common with the Potsdam Durchmusterung. The comparison shows that the scale of the Yerkes Actinometry is nearly identical with that of the PD. This again shows that PARKHURST had succeeded in establishing a photovisual photometry, i. e., a scale that closely agrees with that established visually by the Potsdam observers. The scale curves showing the course of the differences in different magnitudes are given on p. 221 of PARKHURST's paper.

Lastly a comparison was made with the Harvard visual magnitudes. It showed that the scale was very nearly the same for the white stars and that the coloured stars are measured slightly brighter at Yerkes than at Harvard.

The photographic magnitudes were also compared with the Greenwich magnitudes in M N 72, p. 695 (1912), based directly on the Harvard Polar Sequence.

61. Special Catalogues and Researches. a) R. S. DUGAN has measured the photographic magnitudes of 359 stars in the Pleiades[1]. Four plates taken with the 16-inch Bruce telescope were used. The whole cluster was not measured, but only a zone some 27′ broad in right ascension and 60′ in declination. As standards of magnitudes for the first plate 24 stars between magnitudes $8^m,19$ and $12^m,04$ from GAULTIER's catalogue were used, and 8 stars between magnitudes $7^m,6$ and $10^m,4$ as determined by WOLF. The curves for the different plates connecting diameters and magnitudes, which were hyperbolas according to the author, were extrapolated in order to derive the fainter magnitudes.

The catalogue gives magnitudes between $11^m,70$—$15^m,35$ and a chart accompanies the paper.

From stars measured on three plates the mean error, $\pm 0^m,114$, of one determination was derived.

b) SCHILLER. For this investigation[2], which was also carried out at the Observatory at Heidelberg, 12 plates were used. The plates were reduced in practically the same way as in DUGAN's measurements. A comparison was also made between the measurements of SCHILLER and those of DUGAN. 27 stars in common gave the formula:

$$m_S = m_D + 0,23\,(m_D - 10) - 0,01\,.$$

The stars in SCHILLER's catalogue have magnitudes between the values 8,06 and 16,06. The region measured is situated in $3^h37^m20^s$—$3^h39^m20^s$ and $23°5'$—$24°5'$. A chart of the measured region is given.

c) GAULTIER. Four plates taken in Algiers between 1895 and 1899 served as a basis for the photographic magnitudes of 300 stars[3]. The published catalogue is arranged for groups of magnitudes between $3^m,9$ and $12^m,3$. Nine stars are suspected of variability. A chart with the magnitudes of the author is attached to the paper, which also gives a collection of all magnitudes estimated between 1650 and 1882.

A second catalogue of 30 stars in the Pleiades was published later on[4]. Moreover, improved values for the earlier stars are given on the basis of a measured new plate.

d) HNATEK[5]. The extra-focal method of SCHWARZSCHILD was applied to 71 stars in the Pleiades. The agreement with SCHWARZSCHILD's results is very good. There is a dependence on the distance from the centre in the sense that at the centre the magnitudes are brighter by $0^m,04$, and at a distance of 5 cm fainter by $0^m,10$.

[1] Publ Astrophys Inst Heidelberg 2, Nr 2 (1906).
[2] Publ Astrophys Inst Heidelb 2, Nr 10 (1906). [3] Bull Soc Astr de France 14, S. 441 (1900).
[4] Bull Soc Astr de France 15, S. 491 (1901). [5] A N 186, p. 129 (1910).

e) Coma Berenices. HNATEK used the extra-focal method and impressed an absolute scale on the plates by the aid of a "hole" photometer[1]. As zero-point that of the International Scale was used. Photographic magnitudes were derived for 104 stars. A detailed comparison was made with the results of PICKERING[2], HEINRICH[3], BELIAWSKY[4], v. PRITTWITZ[5], and Potsdam[6]. A variation in the difference $m_{\text{Hnatek}} - m_{\text{Pick.}}$ was found, which is explained by the fact that PICKERING's magnitudes are based on a portion of the spectrum at λ 4320. The differences in the comparisons with HEINRICH and BELIAWSKI are more or less constant and are due to the fact that in the former case the magnitudes are based on PD and Mrs. VON PRITTWITZ's determinations and that in the second case SCHWARZSCHILD's scale lies some $0^m,4$ to $0^m,5$ lower than the Harvard scale.

f) Photometric Sequence for Praesepe. The agreement between magnitudes for stars in this cluster as given in Harv Ann 45, 54, 70 and 84 was not very good. By the aid of a wedge and ZÖLLNER photometer GRAFF undertook the determination of the magnitudes of 40 stars between $6^m,3$ and $11^m,5$. The agreement with the Harvard magnitudes is in the mean very good. If we assume the mean error to be equal in both series we find its value to be $\pm 0^m,10$.

g) Scale of FA-Plates. The limiting photometric magnitude of the FRANKLIN-ADAMS plates has been derived by GYLLENBERG[7]. He makes use of the number of stars in the photographic CdC and assumes a law for $N(m)$ as function of m as derived from the investigations of CHARLIER[8]. The comparison stars in Harv Ann 37, II are also used. The limiting magnitude found is $14^m,98 \pm 0^m,20$.

62. The Greenwich Catalogues. Photographic magnitudes of the stars brighter than $9^m,0$ have been determined at Greenwich in the Astrographic zone $+65°$ to $+90°$ (London 1913). Each field has been photographed when at the same altitude as the pole, and the polar area containing the sequence was photographed immediately before or immediately after the exposure on the same plate with exposures of equal length. Also two exposures were given to the region. It was necessary to extend the polar sequence to include all stars in the polar region brighter than $9^m,0$. Thus the magnitudes of 100 stars were determined, which were then used as standards. The diameters of the stars were measured and for their conversion into magnitudes the formula,

$$m = C - k\sqrt{d}$$

where d is the measured diameter of the image, was used. The constants were determined from the measurements of the standard stars, divided into three groups.

By the aid of the (revised) secondary standards the error in the measured magnitudes due to the position of the star on the plate was determined. The means of the residuals of the magnitudes of the polar standards obtained from the measurements of images in these areas were equated to expressions of the form:

$$ax + by + cx^2 + dy^2 + exy + f.$$

The mean differences between the magnitudes derived from the diameter formula and the adopted magnitudes of the polar stars range from $-0^m,151$ to $+0^m,096$.

---

1 A N 193, p. 313 (1913).
2 Harv Ann 26, pt. II, p. 264 (1897).
3 A N 183, p. 299 (1910).
4 Pulkova Mitt 4, p. 85 (1911).
5 A N 161, p. 1 (1903).
6 P D Catalogues.
7 Lund Medd Sér. I, No. 87 (1918).
8 Lund Medd Sér. II, No. 8 (1912).

As the standards have been taken from Harv Circ 170, the magnitudes of the Greenwich catalogue are referred to this system.

The mean error of a single determination of a magnitude on an average plate has been found to be $\pm 0^m{,}083$, which was obtained by comparing the residuals of polar standards with one another. A comparison was also made between the two observations of one star in every twenty stars of the catalogue, from which the mean error was found to be $\pm 0^m{,}094$. This value is to be preferred, because it includes the error caused by the possible changes in the sky when the region and the pole are compared.

The number of objects in the first catalogue is 2329.

The second catalogue gives the photographic magnitudes of 5514 stars, situated between $+65°$ and $+75°$ (Edinburgh 1914). During the course of the work the magnitudes of a considerable number of stars situated within $5°$ of the pole were determined with sufficient accuracy to allow their use as secondary standards. It has thus become possible to utilize measurements of images up to a distance of $5°$ from the centre of the field, whereas in the first catalogue only distances of $3°$ radius could be used. As the plates showed stars considerably fainter than $9^m{,}0$, these were added to the original programme in order to expedite the magnitude determinations required for the Astrographic Catalogue.

In the first and second catalogues the diameters were measured by two observers. For the fainter stars a comparison scale, which was an actual photograph with the same instrument, consisting of a number of star images with graded exposures differing by about $0^m{,}25$, was put side by side with the plate and viewed simultaneously. The polar standards gave the scale equivalents for each plate.

The magnitudes are based on the standards in Harv Circ 170.

The large field of $10°$ made it necessary to consider the different effects of atmospheric absorption in different parts of the field. The size of this effect is $0^m{,}01$ per degree at the zenith distances in question. The extreme error is found to be $\pm 0^m{,}027$ when the y-coordinate is $\pm 5°$.

The discordances between the two observations of the stars in the zone $+65°$, which were taken as fairly representative of the whole, were grouped in order of magnitude. The mean discordances give the mean errors of a determination from a single plate as follows:

Magnitude	Mean discordance between two observations	Mean error of one determination
$7^m{,}0$— $8^m{,}0$	$\pm 0^m{,}133$	$\pm 0^m{,}118$
8 ,0— 9 ,0	±0 ,115	±0 ,102
9 ,0— 9 ,5	±0 ,106	±0 ,095
9 ,5—10 ,0	±0 ,144	±0 ,127
10 ,0—10 ,5	±0 ,148	±0 ,132
fainter than 10,5	±0 ,153	±0 ,136

63. Schwarzschild, Göttinger Aktinometrie[1]. This work is one of the most important in the branch of photographic photometry. It embraces the stars brighter than $7^m{,}5$ within the zone $+0°$ to $+20°$ decl. and gives the photographic magnitudes of 3522 stars within this zone and of 167 polar stars. The intention was originally to survey the northern sky and to obtain a photographic pendant to the Potsdam Durchmusterung. As far as we know, the work is to be

[1] Aktinometrie der Sterne der BD bis zur Größe 7,5 in der Zone 0° bis + 20° Deklination. Teil A. Astr. Mitt. d. Kgl. Sternw. Göttingen 14 (1910). — Teil B. Abhandl. d. Kgl. Gesellsch. d. Wiss. Göttingen, Math Phys Kl, Neue Folge 8 Nr 4 (1912).

considered as finished by now. From a methodical point of view it is of considerable interest because SCHWARZSCHILD's ingenious invention, the "Schraffierkassette" (moving plate-holder), was used. This apparatus is somewhat complicated and its use laborious and troublesome, but its great advantage is that the irregularities of the optics which in the extra-focal photometry give rise to sources of error are avoided in this case. (At present E. HERTZSPRUNG is using the moving plate-holder in his extensive studies of variable stars at Leiden.)

Three exposures were taken on each plate and an approximate reduction to magnitudes was performed by using the interpolation methods developed by SCHWARZSCHILD in an important paper[1].

In order to reduce the provisional magnitudes to an absolute scale, a photographic comparison was made, the Pleiades being used. Use was also made of the circumstance that the plates partly overlapped.

The accuracy of the results was investigated by SCHWARZSCHILD. Using the average deviations of the results of one plate from the mean of four or five plates, on which the same star appeared, he found the following result after using two hundred and fifty stars:

Magnitude	Average deviations from mean	Mean error of the mean
$2^m,0-5^m,5$	$0^m,041$	$\pm 0^m,033$
5 ,5 – 6 ,5	0 ,029	0 ,022
6 ,5 – 7 ,5	0 ,032	0 ,024
7 ,5 – 8 ,5	0 ,053	0 ,039
fainter than 8,5	0 ,064	$\pm$0 ,054

The increase of the deviations from brighter to the fainter stars is due to the defect that only in the case of intermediate brightness do all the three exposures give measurable densities. From this table the average error is found to be $\pm 0^m,034$, which is considerably lower than in the case of the Potsdam Durchmusterung.

In the second part, published in 1912, the magnitudes of the first catalogue were reduced to an absolute scale. Care was taken to eliminate the influence of the systematic error depending on the right ascension and declination. The zero point of the scale was brought into accordance with the proposal of PICKERING that has been mentioned earlier. The comparison between the final magnitudes and the Potsdam and Harvard ones gave colour indices for practically all the stars. The following system of formulae connects the different magnitude scales (P = Potsdam; H = Harv Ann 45; HR = Harvard Revised Photometry):

$$m_G = m_P - 1^m,03 - 0^m,09\,(m_P - 6,5) + 0^m,63\,Sp$$
$$m_G = m_H + 0,54 - 0,031\,(m_H - 6,5) + 0,49\,Sp$$
$$m_H = m_{HR} - 0,04 + 0,02\,Sp$$
$$m_G = m_{HR} - 0,58 - 0,031\,(m_{HR} - 6,5) + 0,51\,Sp$$
$$m_P = m_{HR} + 0,52 + 0,086\,(m_{HR} - 6,5) - 0,13\,Sp.$$

64. Pulkovo Measurements of Northern BD Stars. Recently a publication issued in 1915 has been distributed from Pulkovo, giving the photographic magnitudes of 2747 BD stars down to BD $9^m,0$ between $+75°$ and $+90°$ declination. This zone was observed at Simeis by S. BELIAWSKY who used the astrographic camera of that observatory[2]. The Pleiades and Coma Berenices were used as standards, in such a way that equal exposures were given to the sequence

[1] A N 172, p. 65 (1906). [2] Pulkovo Bull 6, p. 263 (1915).

and the region in question. An intermediary scale of the same kind as was used in SHAPLEY's and v. ZEIPEL's work gives the comparison between the standard stars and those in the region. Because of the Pleiades not being at the same altitude as the regions the differential extinction had to be applied.

The probable error of a magnitude of unit weight between $6^m,2$ and $10^m,0$ is $\pm 0^m,059$.

A comparison has been made with 168 stars in the vicinity of the pole, determined by SCHWARZSCHILD. It has been found that the stars of the spectral classes B5—A9 can be represented by the expression:

$$m_B - m_S = -0^m,005 + 0,050\,(m - 6,0).$$

For the other spectral classes the expression is quite different and takes the parabolic form:

$$a\,(m - 7,25)^2.$$

The number of stars is too small to permit numerical evaluation. A comparison was also made with the Yerkes Actinometry for 382 stars. The same phenomena were found as in the former comparison. Thus the stars of types B5 to A9 are expressed by the formulae:

$$m_B - m_Y = -0^m,035 + 0,030\,(m - 6,0)$$

while for the other classes a quadratic expression of the form:

$$a\,(m - 8,0)^2$$

is suggested.

An analogous discussion was also undertaken for the magnitudes in common with the Greenwich catalogue. The following formulae were found:

		n
B5—A9	$m_B - m_G = -0^m,02 + 0,13\,(m - 6,0)$	153
F0—F9	$= +0\,,13 + 0,04\,(m - 6,0)$	55
G0—M	$= +0\,,26 - 0,05\,(m - 6,0)$	175

Lastly 18 stars were compared with the North Polar Sequence. The expression $m_B - m_N = -0^m,01 + 0,04\,(m - 6,0)$ shows that the agreement is good.

64a. Further Photometric Catalogues of the Pleiades. Many special catalogues of photographic magnitudes have been issued. Most of these contain determinations of magnitudes in clusters such as the Pleiades, Hyades, Praesepe, and Coma Berenices. Especially the first cluster has been used many times as a test object when new photometric methods have been applied or new photometers have been tested out. There exist at least some twenty catalogues of magnitudes of stars in or near the Pleiades (cf. ciph. 61). No doubt it would be a very profitable task to reduce the existing measures of m_{ph} and m_v to the International Scale, still more so because of the existence of colour indices of the said stars, independent of the two magnitude scales (e. g. the determination of effective wave lengths by HERTZSPRUNG). When such a determination of final magnitudes of Pleiades stars is undertaken we have in the Pleiades another sequence of magnitudes and colour indices of high accuracy. Of the determinations of magnitudes in the Pleiades we only mention that of HERTZSPRUNG[1] and that of SHAPLEY and Miss RICHMOND[2]. The former adopted as zeropoint that of the Göttinger Aktinometrie and the latter derived photographic and photovisual magnitudes and colour indices by polar comparisons with the Mount Wilson standards[3].

[1] A N 199, p. 247 (1914). [2] Mount Wilson Contr 218 (1921).
[3] Mount Wilson Contr 97 (1915).

For some 450 stars between $11^{m},5$ and $14^{m},5$ common to the two lists SEARES[1] has found for the photographic magnitudes:

$$m_{\mathrm{MW}} - m_{\mathrm{Hertz.}} = +0^{m},02.$$

Small corrections for zero point and colour equation must be applied to reduce the differences to a homogenous system. If C is the Mount Wilson co-

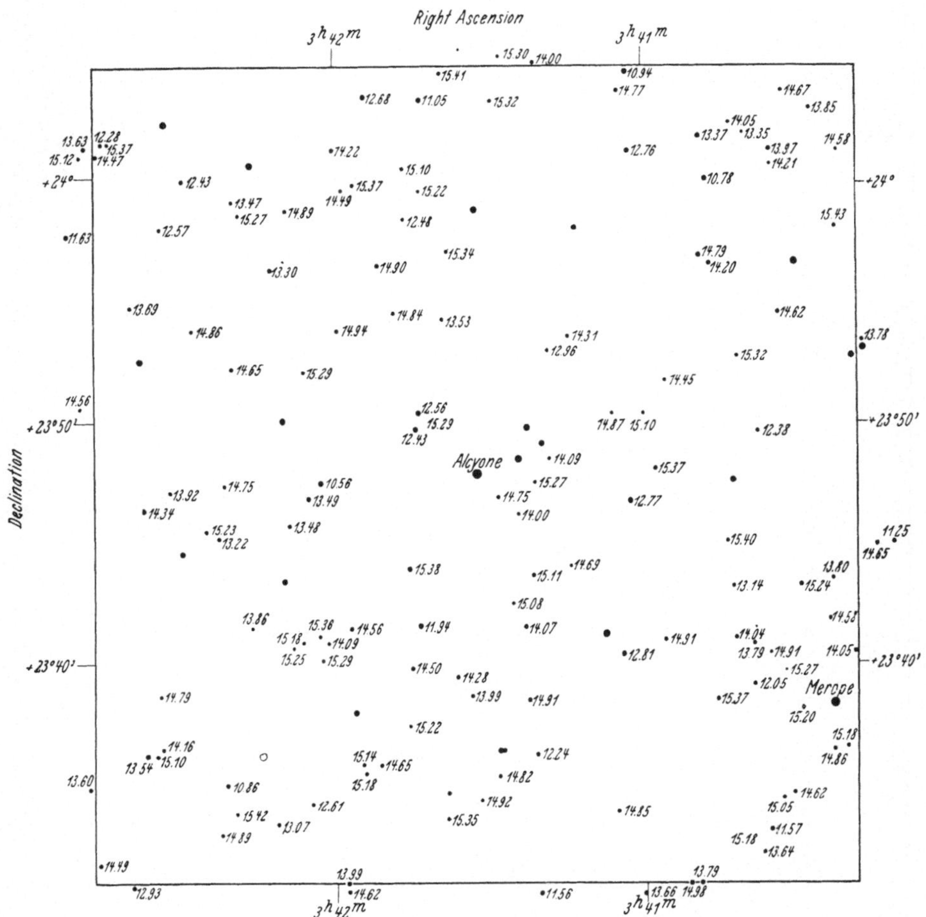

Fig. 78. Chart of the central part of the Pleiades giving the magnitudes of fainter stars according to the determination of HERTZSPRUNG.

lour index, HERTZSPRUNG'S photograpic magnitudes must be reduced to the Mt. Wilson system by adding $+0^{m},05 - 0,10C = -0^{m},02$ in the mean.

65. Selected Areas. The aim of this great undertaking by J. C. KAPTEYN was to bring together all the elements which at the present time must seem most necessary for a successful attack on the problem of the structure of the sidereal world. It was necessary for the plan to be restricted to embracing only a certain number of selected regions or areas distributed over the whole of the sky. The Selected Areas are 206 in number, regularly distributed over the sky.

[1] Trans Internat Union I, p. 81 (1922).

Besides, several particularly interesting regions (the special plan), to the number of 46, are added. The details concerning these are to be found in KAPTEYN's paper: Plan of Selected Areas (also reprinted by VAN RHIJN).

The S A were intended to give visual and photographic magnitudes for some 200000 stars. The three catalogues published in Harv Ann 101—103 give

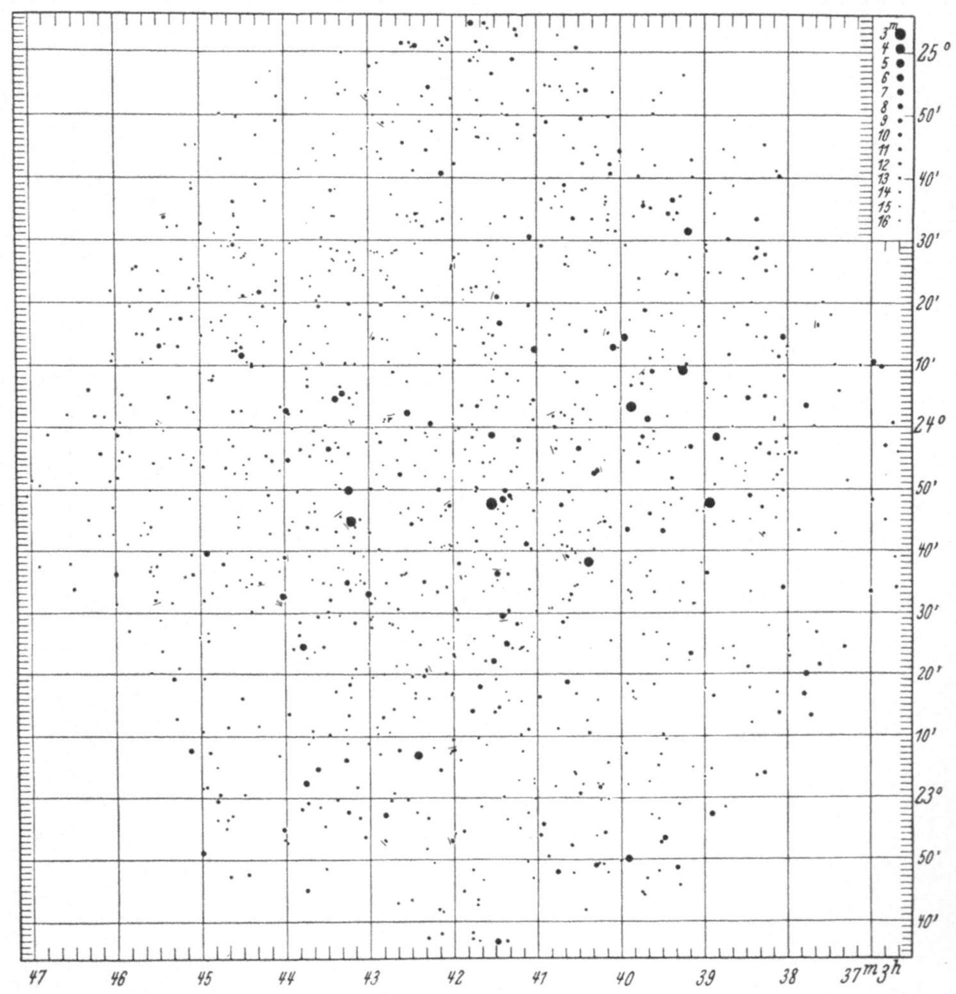

Fig. 79. Chart of the Pleiades cluster (HERTZSPRUNG).

the photographic magnitudes for the stars brighter than 16^m of the systematic plan according to the scheme:

	Decl.	n	Area
Harv Ann 101	0° to +90°	84002	127,9 □°
Harv Ann 102	−15 ,, −30	65763	97,2 ,,
Harv Ann 103	−45 ,, −90	82226	41,8 ,,
		231991	

The work has been performed at the Harvard and Groningen Observatories, the former taking the plates and determining the magnitude of a sequence in each region, the latter measuring and reducing the plates. Since the areas were distributed all over the sky, the Harvard Observatories at Cambridge and Arequipa were made use of for the execution of the observations. Three different telescopes with apertures of 16, 24 and 8 inches were used. The scale used was that in Harv Ann 71. A comparison with the pole was not made for each plate, but only for the first, sixth, and twelfth of each box. Generally one of the Harvard standard regions was included. Exactly 10 min exposure was given to each plate. The mean of the three polar plates gave the connection between magnitude and scale readings.

The reduction of the estimated diameters to the International System depends entirely on the photographic magnitude-standards determined at the Harvard Observatory both with regard to scale and zero-point.

The error was derived in two ways for the northern part: a) from the differences between the catalogue magnitudes and those of the standard region; b) from a comparison of five rejected plates with the results of the repeated plates.

In the first case, on the assumption that the uncertainty in the two sources is equal, the mean error was found to be $\pm 0^m,14$. In the second case, from a material of 578 stars, the mean error was determined as $\pm 0^m,18$. The value $\pm 0^m,16$ was adopted as a compromise. A summary investigation did not show any marked change of the m. e. with the magnitude. KAPTEYN points out the agreement of the m. e. with that of the Cape Durchmusterung.

For the southern part of the work VAN RHIJN derives practically the same value of the mean error.

A fourth volume is intended to contain the special plan. As regards the present state of the whole work an account is given in the Reports published by KAPTEYN and by VAN RHIJN.

PANNEKOEK[1] has tried to find the systematic errors of the estimates on the scale in the Selected Areas by supposing that in each area the increase in the numbers of stars follows the same simple law. He finds corrections varying between $+ 0^m,07$ and $- 0^m,06$. SEARES, MARY C. JOYNER and MYRTLE L. RICHMOND[2] have given the following formula for the reduction of the selected area magnitudes m_{SA} to the international scale:

$$m - m_{SA} = +0,06\,(m_{SA} - 6) + 0,21\,C\,, \qquad (m \leqq 10)\,,$$

where C is the colour index.

66. Yerkes Photometry of the Selected Area Zones. A very important contribution to the photometry and colours of faint stars was finished by PARKHURST[3] before his lamentable death. It consists of photovisual and photographic magnitudes of stars in the Selected Areas of KAPTEYN with $\delta = +45°$. The work was carried out with three instruments, the 6-inch UV doublet, the 24-inch reflector, and the 40-inch refractor of the Yerkes Observatory. The object was to find m_{pv} and m_p for all the stars whose parallaxes and proper motions had been determined with the 40-inch instrument and to test the capability of the instruments for photometric work. Magnitudes were derived from 472 plates for 1550 stars. The scale was based on diffraction images obtained with objective gratings. For the determination of the zero-point the Polar Sequence was used as well as white stars in P D and Harv Ann 45, reduced to the International System. The agreement was found to be very good in the two cases.

The accuracy has been investigated by PARKHURST using residuals from the reflector plates and 40-inch plates:

$\varepsilon(m_{pv})$	n	$\varepsilon(m_{pv})$	n
$\pm 0^m,06$	180	$\pm 0^m,05$	449
$\pm 0\;,06$	176		

[1] B A N 1, p. 53 (1922). [2] Mount Wilson Contr No. 289 (1925). = Ap J 61, 303 (1925).
[3] Yerkes Publ 4, part 6 (1927). The preparation for publication was completed after the author's death by Alice Hall Farnsworth.

It was found that too small grating constants for two partial gratings had been assumed (11 and 13 percent), which makes the correction to the 40-inch scale = 1,02 times the correction to the reflector scale. A direct comparison in 18 areas gave a contradictory result, viz., the corr. to the reflector scale = 1,02 times the corr. to the 40-inch scale. It was decided to apply a correction of 12 per cent to the mean of both scales.

The colour equation between the photovisual magnitudes and the reflector magnitudes was found to be of the form:

$$m_{\mathrm{pv}} + 0{,}10 \text{ Colour index} = m_{\mathrm{Refl.}}$$

The colour difference between the reflector and the 40-inch telescope was found to be in the mean $-0^{\mathrm{m}},11$ and of such constancy that there was no evidence of any systematic colour equation between the two instruments.

The agreement with the Mount Wilson improved values of the Selected Areas is good and the mean difference $m_{\mathrm{MW}} - m_{\mathrm{Yerkes}}$ amounts only to $+0^{\mathrm{m}},06$. There is also a good agreement between PARKHURST's magnitudes and those in Harvard Standard Regions.

67. HERTZSPRUNG's Leiden Catalogue. Photographic magnitudes of 658 stars from plates taken mainly by W. H. VAN DEN BOS with the 33-cm Leiden refractor have been derived by E. HERTZSPRUNG[1]. Grating and intra-focal photographic images were used. There were some difficulties as regards the determination of the scale. HERTZSPRUNG suggests an ingenious method for establishing an absolute magnitude scale. Take two stars for which the provisional m's have been determined. If we were able to find the provisional value of m of their combined light, it is clear that the deviation of this provisional m from the m calculated as if the adopted provisional magnitude scale were correct would determine the correction factor to that provisional scale. This method is related to the differential method of SCHWARZSCHILD and it might be of much interest to apply both methods in practice[2].

The method will be applied to double stars whose distance between the components is equal to the distance of the spectra of the first order from the central image. Experiments on two double stars gave a correction factor of $k = 1{,}023$, but HERTZSPRUNG considers this value as too uncertain to be used.

The connection between HERTZSPRUNG's magnitudes and those of KING in Harv Ann 76 no. 6 is given through the formula:

$$m_K - m_H = +0^{\mathrm{m}},081 + 0^{\mathrm{m}},034\,(m_K + m_H - 7) - 0^{\mathrm{m}},038\,(c_2/T - 2)$$

where c_2/T is the colour. The formula is reduced to:

$$m_K = -0^{\mathrm{m}},099 + 1{,}038\,m_H - 0^{\mathrm{m}},038\,(c_2/T - 2)\,.$$

A comparison with the Göttinger Aktinometrie showed that HERTZSPRUNG's scale is intermediate between that of Göttingen and that of KING.

The mean error of one magn. from one plate is $\pm 0^{\mathrm{m}},08$ and the average mean error of the magnitudes is $\pm 0^{\mathrm{m}},05$.

---

[1] B A N 1, p. 201 (1923).

[2] Let m_A and m_B be the magnitudes in the correct system and m'_A, m'_B the magnitudes in the provisional system, the grating constant being l and l' respectively and the scale factor k; then we have the equations:

$$m_A = k\,m'_A\,,$$
$$m_B = k\,m'_B\,,$$
$$l = k\,l',$$
$$10^{-k m'_A} + 10^{-k m'_B} - (10^{-k m_A - l'} + 10^{-k m_B - l'}) = l'$$

which, if m'_A, m'_B, and l' are given, determine k, l, m_A, and m_B.

68. Photographic Catalogue of Lund. K. G. MALMQUIST[1] has determined the photographic magnitudes of 3700 stars. This work was undertaken in connection with his determinations of colour indices, which are to be reviewed later on. The colour indices have been used to derive the photovisual magnitudes.

The instrument used was the Zeiss-Astrograph of the Lund Observatory with an aperture of 18 cm and a focal length of 90 cm. As scale the Harvard Standard Region $12^h + 31°30'$ was used, which was impressed on each plate of the first series. In order to reduce the magnitudes to the International System the corrections given in Harvard Bulletin 781 have been applied. An intermediary photographic scale has been used, which was brought into contact with the film side of a field plate.

The derivation of the corrections to the centre of the plate was, on account of the instrument showing a rapid deterioration of the images with increasing distance from the optical axis, according to the author, "a very difficult task".

For this purpose stars in the Paris and Oxford zones of the C d C were used. The magnitudes derived for stars within 20 mm of the optical axis were used as standard stars, and the magnitudes in other parts of the plates were reduced to the same scale, and from the mean magnitude the centre corrections was derived.

An examination of a Polar Sequence plate convinced the author that the colour equation of the instrument is so small that it may be neglected.

From the differences between individual magnitude determinations for each star the following mean errors have been found by MALMQUIST for the magnitudes given in the catalogue:

Magnitude	Mean error
$8^m,00$ — $8^m,99$	$\pm 0^m,151$
9 ,00 — 9 ,99	0 ,122
10 ,00 — 10 ,99	0 ,125
11 ,00 — 11 ,99	0 ,128
12 ,00 — 12 ,99	0 ,115
13 ,00 — 13 ,99	0 ,094
14 ,00 and fainter	0 ,132

In the catalogue the general accuracy of the photographic magnitudes is noted in such a way that the sign : means a mean error between $0^m,20$ and $0^m,30$ and :: a mean error larger than $0^m,30$.

69. The Mount Wilson Catalogue of Photographic Magnitudes. This work has its origin in a request of J. C. KAPTEYN in 1909 for assistance in determining the magnitudes of stars in Selected Areas. The investigations have been carried out under the supervision of F. H. SEARES, who has also carried out considerable parts of the work for the catalogue in connection with his photometric work at Mount Wilson. The catalogue, which is now in the press, will appear under the names of SEARES, KAPTEYN, and VAN RHIJN with Miss JOYNER and Miss RICHMOND as assistants. The catalogue[2] gives the photographic magnitudes, number of individual observations, and approximate coordinates of 67941 stars in 139 Selected Areas between the North Pole and declination $-15°$. The results are arranged in the serial order of the Areas, those for each area being divided between a main list, which gives the mean magnitude of all the stars in the list adopted at Groningen, and a supplementary list including stars outside the Groningen fields, which are therefore measured only at Mount Wilson. The

[1] Lund Medd Ser II, No. 37 (1927).

[2] Mount Wilson Catalogue of Photographic Magnitudes in Selected Areas 1—139 by F. H. SEARES, J. C. KAPTEYN and P. J. VAN RHIJN assisted by MARY C. JOYNER and MYRTLE RICHMOND (= Wash Carnegie Inst Publ No. 402) (1930).

fields of the main list are $15' \times 15'$ for galactic latitudes less than 40° and $20' \times 20'$ for higher latitudes.

The derivation of the standards required an extended investigation of methods that might be used to establish a scale of magnitudes over a wide interval of brightness as well as the determination of the magnitudes of the North Polar Sequence, which have played an important part in the reduction of the measurements of the Mount Wilson Catalogue.

Standard magnitudes have also been determined in each Area. These have been based on four plates containing exposures with full aperture, with diaphragms of various sizes, and with a wire gauze screen. The derivation of the distance-correction necessitated a troublesome investigation on account of the difficulties involved. It was concluded that the results of the whole undertaking could be much strengthened by determining the magnitudes of all the stars visible on these plates instead of only those of a limited number to be used as standards. Since the fields measured at Groningen and at Mount Wilson were not entirely consistent, the enlarged programme has also increased the number of stars available for statistical investigations.

A very important item in the preparation of the catalogue was the standardization of the photometric zero-points of intercomparison photographs. The stars in each Area were connected with those in the adjacent Areas in the same zone by means of multiple exposures on the same plate. Each zone was similarly connected at six equidistant points with the Polar Sequence. The magnitude scales have been independently derived for each Area with the aid of the diaphragm and screen methods. The resulting mean scale is identical with that of the Polar Sequence and hence with the International Photographic Scale adopted at the Rome meeting of the International Astronomical Union (1922). This is shown by the intercomparison photographs of the 37 Areas connected directly with the Pole.

Magnitude	Polar scale – S. A. scale	n
$11^m,5$	$0^m,00$	58
12 ,5	−0 ,01	181
13 ,5	0 ,00	398
14 ,5	0 ,00	528
15 ,5	0 ,00	400
16	−0 ,01	52

The measurements and the reduction of the plates of the original plan, embracing one exposure of 60^m and one of 3^m, were made at Groningen, and the other plates (diaphragm and intercomparison) were measured and reduced at Mount Wilson. The combination of the two series of measurements and the preparation of the manuscript for the catalogue were also made at Mount Wilson.

The limiting magnitude averages $18^m,5$. Even the brightest stars within the limits of the adopted fields have been measured. The stars brighter than 13^m are comparatively few on account of the smallness of the field.

The accuracy of the magnitudes varies between rather wide limits and is closely related to the number of observations for individual stars. For stars of intermediate brightness with measurable images for all of the exposures the number of separate determinations is 16. The very faintest stars are visible on only one plate. The average mean errors including the uncertainty in the determination of the zero-point are as follows:

Number of observations	Mean error
1	$\pm 0^m,234$
3	0 ,128
5	0 ,104
7	0 ,093
9	0 ,087
11	0 ,085
13	0 ,082
15	0 ,080

The catalogue provides numerous standards well distributed over the sky north of $-15°$ and of high value in determining the magnitudes of faint stars in neighbouring regions. The data included in the catalogue form a statistical population of sufficient extent to permit a discussion of the general features of stellar distribution. Such a discussion has also been undertaken by SEARES and some of his results will be summed up briefly later on.

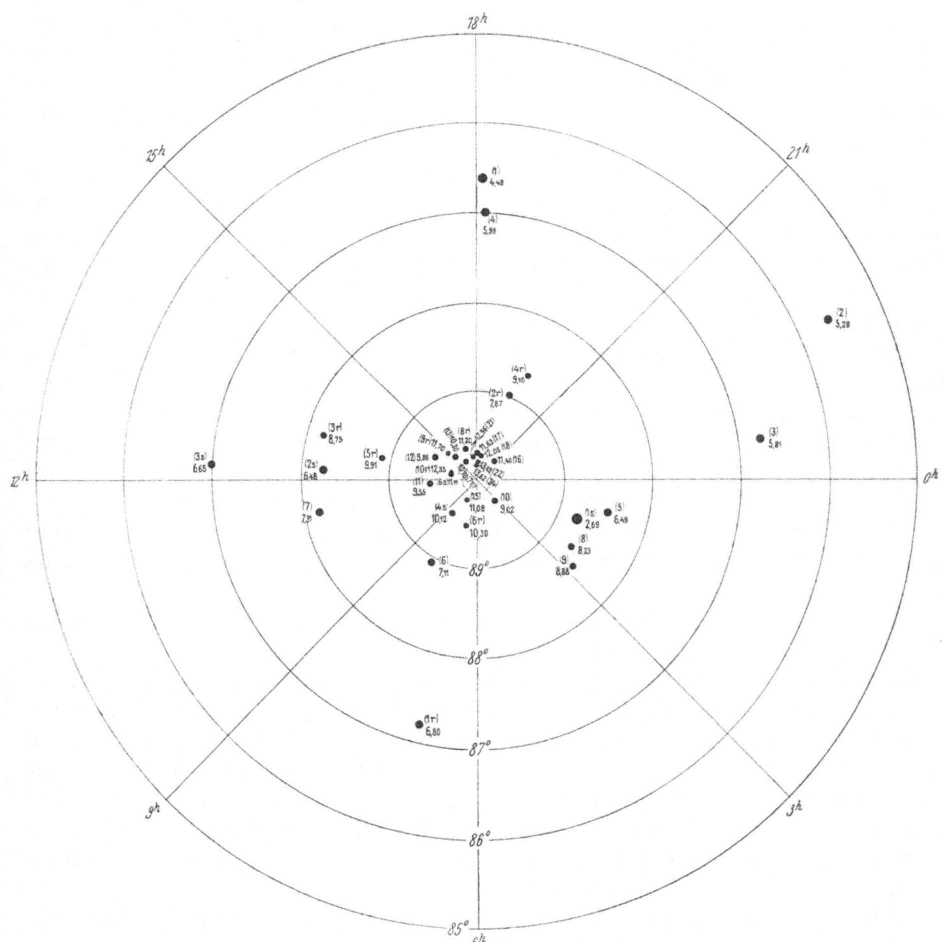

Fig. 80. Chart showing the brighter stars in the North Polar Sequence according to Miss LEAVITT's determination in Harv Ann 71, part 3. The magnitudes are photographic. The numbers and letters within the parentheses are designations used by Miss LEAVITT.

70. The Polar Sequence. E. C. PICKERING[1] had the ingenious idea of establishing a photometric scale in the sky, which could be used as a standard for the photographic magnitudes.

The magnificent work of determining the magnitudes in the North Polar Sequence was performed by Miss HENRIETTA LEAVITT[2] and the results are given at length in her paper Harv Ann 71, which also contains valuable methodical

[1] Harv Circ 108 (1906).

[2] Harv Circ 125 (1907), 150 (1909), 160 (1910), 170 (1912), and Harv Ann 71, p. 36 (1917).

results. A total number of 96 stars was measured. The fact that the colour index varies for different telescopes and under changing photographic conditions, makes it of importance that the stars should have the same colour, if possible. In the sequence adopted the stars brighter than $11^m,4$ are A stars, except for one which is an F star. Between $11^m—16^m$ the stars are relatively white. For stars fainter than 16^m the colour was not known at the time Miss LEAVITT published her results. Twelve stars between $6^m,80$ and $13^m,48$ form a special sequence.

Altogether 13 different telescopes were used, whose apertures varied from 1,2 to 150 cm. The magnitudes are based on 299 photographs. Large errors are introduced by the effect of the distance from the centre of the plates, and also by changes of focus. The corrections in question were carefully determined and applied.

In accordance with PICKERING's suggestion the zero-point is the Harvard magnitude of A stars between $5^m,5$ and $6^m,5$. A detailed comparison between the scales and zero-points underlying the determination led to a small change in the magnitudes given in Harv Circ 170, so that the scale was altered by $+0^m,048$ per magnitude for stars brighter than 17^m, while the zero-point was changed by $0^m,08$.

The considerable variation of the colour equation of different scales of photographic magnitudes led Miss LEAVITT to extend the definition of the zero-point to read as follows: "Photographic magnitudes coincide with photometric magnitudes of Class A0 between the magnitudes $5^m,5$ and $6^m,5$, and are fainter than the photometric magnitudes by $1^m,00$ for stars having spectra of Class K0 between the same limits".

Miss LEAVITT asked for investigations of the standard scale by different observers using different methods, and foresaw the appearance of discrepancies in the results.

A sequence of stars is also given in Harv Ann 71 part 3 for 36 stars in the Pleiades and 18 stars in the Praesepe.

The principal advantage is certainly that all stars in the scale are situated at a given place at practically the same altitude during long intervals of time. Another advantage is that all photometric measurements can be reduced and referred to a uniform scale, and thus determinations at different places are made comparable with each other. It is also valuable to have the stars in the same region of the sky, and to avoid having to photograph a number of objects in different regions, which has been considered necessary when determining colours or colour equivalents by means of standards.

There are also certain disadvantages, but it will be possible to overcome most of them even with our present resources.

The instrument must be very accurately adjusted. If the true pole and the instrumental pole do not coincide, the stars on the plate at some distance from the optical axis will not appear round but as trails. The method described by J. SCHEINER[1] is very convenient for reducing the influence of the errors of the mounting to a minimum. The troubles caused by the influence of differential extinction will in some cases be considerable. In low latitudes the Pole is near the horizon and the extinction must be carefully determined. Besides the extinction seems to be dependent on the azimuth at least in observatories close to large cities.

Another trouble is the colour equation of the instrument because of the impossibility of using instruments with the same sensibility curve for different wave lengths.

[1] Die Photographie der Gestirne p. 100, Leipzig (1897).

To avoid changes in transparency the starting and ending of the exposure with the sequence is to be recommended. But when an instrument with restricted light capacity is being used and work is being done in rapidly changing climates, this precaution will not always be possible.

Even if the instrument is accurately adjusted, the character of the images may be systematically different for the field and the sequence, principally on

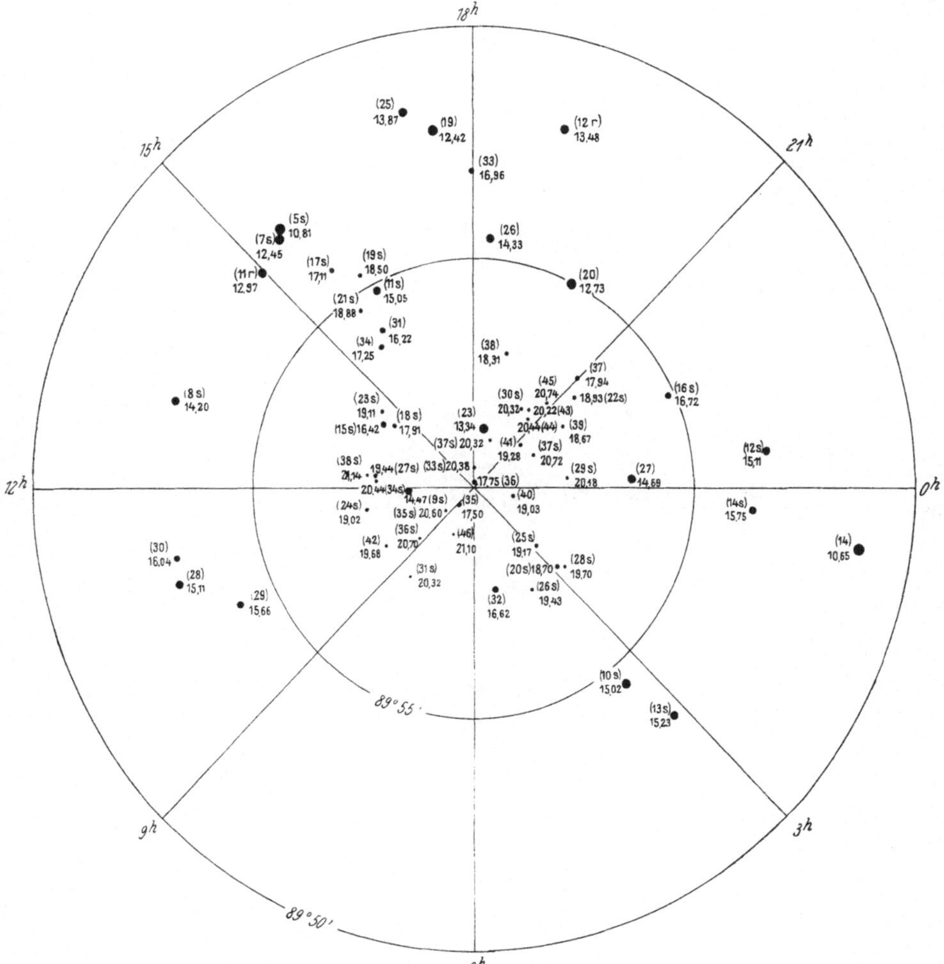

Fig. 81. Chart of the North Polar Sequence according to Miss LEAVITT's determination in Harv Ann 71 part 3. The magnitudes are photographic magnitudes. The numbers and letters within parentheses are the designations used by Miss LEAVITT.

account of varying atmospheric conditions, such as humidity, and on account of changes in the optical system.

In spite of the difficulties the use of the sequence is necessary in all accurate photometric work. From what has been said it is evident that the establishment of secondary sequences is a very important task. Work has also been done in this respect at Harvard and other observatories, and the best secondary sequence is certainly that in the Pleiades, but it does not include as yet as many faint stars as the North Polar Sequence does, and the existing determinations have not been reduced as yet to the International Scale.

71. Mount Wilson Sequence and International Scale. F. H. SEARES, the eminent authority on stellar photometry, has submitted an excellent report to the International Astronomical Union concerning the determination of stellar magnitudes[1]. In the following review of the work concerning the scales we are greatly indebted to this report as well as to other papers by SEARES.

The aim of SEARES's extensive work for the construction of the International Scale was to form a series of standards on the normal scale of POGSON. The results

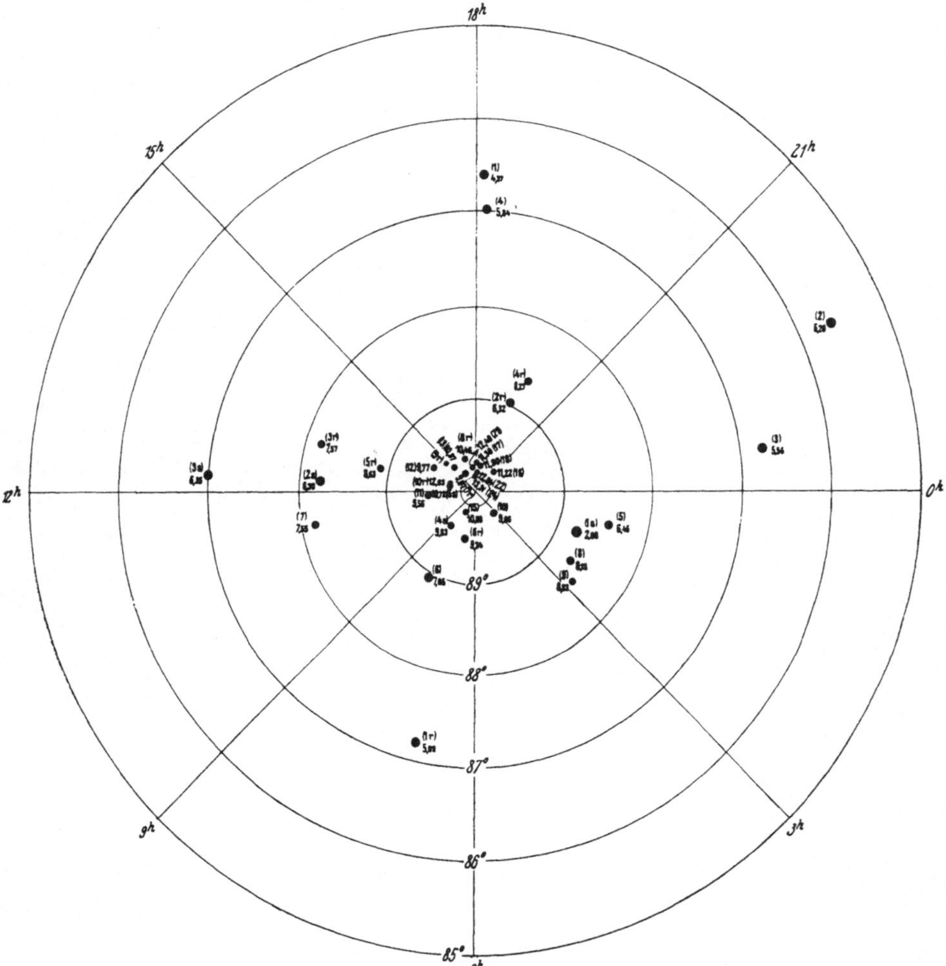

Fig. 82. Photographic magnitudes of brighter stars in the North Polar Sequence according to Mount Wilson Contribution No. 235.

now obtained reach a magnitude of 20^{m} for the photographic region, and $17^{m},5$ for the visual region. Results from different observatories are in such good accordance for the photographic magnitudes that one can safely conclude that the scale reaches such an accuracy for stars brighter than 16^{m} that the inclusion of future data will not materially change the present results. SEARES has found that the average deviation of a magnitude for any of the observatories from the adopted means, including all systematic differences in scales between $2^{m},6$ and $10^{m},0$, is $+0^{m},024$, corresponding to a mean error of $\pm 0^{m},030$.

[1] Trans Internat Astr Union 1, p. 69 (1922); Mount Wilson Contr No. 288 (1924).

As the magnitudes of the faintest stars are dependent on the Mount Wilson measurements alone, it is important that the measurements should be tested in some independent way. SEARES found a possibility for a test in the method of exposure ratios which he detected. He has compared his colour indices, derived without using the photographic or photovisual scales, with the indices resulting from the scales themselves. It is evident that the scales must be mutually consistent, and it is very unlikely that they should both be affected by the same

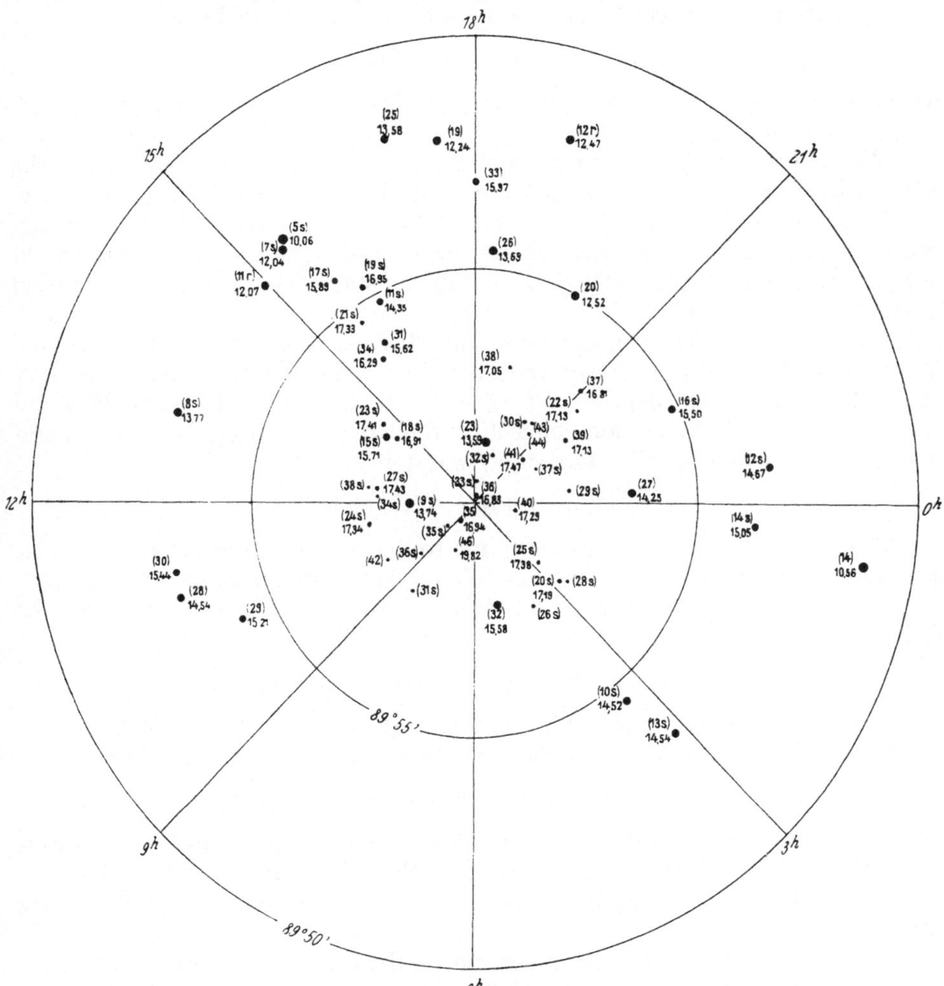

Fig. 83. Photographic magnitudes of fainter stars in the North Polar Sequence according to Mount Wilson Contr No. 235.

systematic errors of equal size. The only magnitudes which should be taken with reservation, until new determinations have been presented, are those of the very faintest stars. In the case of these magnitudes there are possibilities of systematic errors having crept in owing to the stars being near the limit of visibility in the plates. But, on the other hand, these magnitudes are outside the range of intensity covered in most of the problems attacked so far.

When photometric results obtained with different instruments are compared with each other, it is necessary to take the colour into account, as the differences in the colour perception and the differences in colour sensibility for different telescopes and photographic plates are more or less of the same order of magnitude, so that they cannot be neglected. A standard system of colours ought to be adopted, to which the magnitudes should be referred. SEARES points out that this is not feasible at present, because observational data cannot be reduced from a certain system to any other, unless the colours of the individual stars are known, which is not the case in general.

The colours of the polar stars are known, and SEARES has reduced the material to a homogeneous system, equations of the following form being used:

$$m = m_c + a + bC$$

where a and b are constants, m_c the magnitudes of a certain catalogue, and C is the colour index.

We are here giving a brief review ot the results reached by SEARES as to the relation between different scales of the magnitudes in the North Polar Sequence.

Photographic scale. Göttingen. The results obtained in SCHWARZSCHILD's Actinometry have been already mentioned. The colour equation could not be derived on account of the small number of stars. The zero-point correction is $+0^m{,}05$.

Greenwich I. These magnitudes have been determined by CHAPMAN and MELOTTE[1], who used the 26-inch reflector. The standards are taken from Harv Circ 170 and the scale depends on 17 stars of magnitudes between $8^m{,}9$ and $14^m{,}2$. The following equation connecting the Greenwich magnitudes (m_{Gr}) and the magnitudes in Harv Circ 170 (m_{170}) was found:

$$m_{\mathrm{Gr\,I}} = m_{170} + m_{\mathrm{red}} + 0^m{,}21 + 0{,}23\ C.$$

The term m_{red} expresses the reduction to the Harvard homogeneous system; it is 0 for 9^m and increases to a maximum of $0^m{,}03$ for stars between 12^m and 16^m.

Greenwich II. This scale has been established by S. JONES[2] who used the 13-inch equatorial. JONES has found the formula:

$$m_{97} = m_{\mathrm{Gr\,II}} + 0^m{,}01 + 0{,}001\ (m - 10) - 0{,}04\ C,$$

where m_{97} refers to the magnitudes in the scale of Mount Wilson Contr 97.

SEARES has found by means of the revised Mount Wilson results[3]:

$$m_{60} = m_{\mathrm{Gr\,II}} - 0^m{,}00 - 0{,}01\ C,$$

where m_{60} refers to the improved Mount Wilson Scale as given in Mount Wilson Contr No. 235 and depending upon the 60-inch and 10-inch instruments.

Thus the correction may be neglected and the magnitudes used as given by JONES.

Harvard I. This designation denotes all data for stars between 10^m—16^m in Harv Ann 71 part 3. Harvard II relates to stars brighter than 9^m. The original magnitudes of stars fainter than 16^m have to be excluded as being of a provisional nature on account of partial neglect of the distance correction when the material was reduced.

Miss LEAVITT has reduced the results of the North Polar Sequence to the colour system of the one-inch (Cooke) and the 8-inch (Draper) telescopes. SEARES has found that most of the fainter stars whose colours were unknown at the time of the appearance of Harv Ann 71, are red or reddish and thus appreciable corrections are needed to reduce them to a homogeneous system. SEARES has reduced the original magnitudes to the Harvard homogeneous scale defined

[1] M N 74, p. 40 (1913). [2] M N 82, p. 21 (1921).
[3] Mt Wilson Contr 235 (1922).

in Mount Wilson Contr No. 98, where the following general relation is given:

$$m_{\mathrm{MW}} - m_{\mathrm{HH}} = +0{,}061\,(m_{\mathrm{HH}} - 6{,}0) + 0{,}06\,C\,,$$

where m_{MW} stands for the Mount Wilson magnitudes and m_{HH} for the magnitudes in the Harvard homogeneous scale. The changes are generally smaller than $0^{\mathrm{m}},06$, but should nevertheless be applied.

Reducing these results to the system of the reflector and the zero-point of the International Scale, SEARES has found

$$m_{\mathrm{H\,I}} = m_{\mathrm{H\,71}} + m_{\mathrm{red}} + 0^{\mathrm{m}},24 + 0{,}06\,C\,.$$

Harvard II. Together with the results for brighter stars in Harv Ann 71 (Group 1, 2 and 5) the results of KING with the 24-inch reflector in Harv Ann 76 form the magnitude system of Harvard II. SEARES finds the corrections for the zero-point to be: Group I (extra-focal observations): $-0^{\mathrm{m}},06$; Group II (measures of four stars made with a plate of Iceland Spar): $-0{,}01$, Group V (measures with the 1-inch Cooke anastigmat + screen): $+0^{\mathrm{m}},01$; measures with the 24-inch reflector: $-0^{\mathrm{m}},09$. To group V is also added the colour correction $+0{,}04\,C$.

Mount Wilson 60. Details are given in Mt Wilson Contr 97 and 235. For stars between 11^{m} and 16^{m} exposures on the Pole in order to establish the zero-point for the Selected Areas could be used.

Mount Wilson 10. SEARES and HUMASON have used a screen on the 10-inch Cooke refractor for the determination of stars in the Polar Sequence of magnitudes. The screen constant was determined from photographic magnitudes of the Mount Wilson 60 series. It is found that the instrument has an unusually large colour correction of $+0^{\mathrm{m}},19\,C$. For further details Mt Wilson Contr No. 234 and 235 should be consulted.

Potsdam. Using the 80 cm refractor and applying the "halfscreen method" (Halbgittermethode) of SCHWARZSCHILD, DZIEWULSKI[1] has determined the magnitudes of 222 polar stars. The following formula:

$$m_{\mathrm{Pots}} = m_{\mathrm{Dz}} + 0^{\mathrm{m}},33 - 0{,}10\,C$$

reduces his measurement to the Potsdam system.

Yerkes. The magnitudes in the Yerkes Actinometry[2] give

$$m = +0^{\mathrm{m}},46 + 0{,}94\,m_{\mathrm{y}} - 0{,}07\,C$$

where m is the reduced magnitude and m_{y} the magnitude given in Yerkes Actinometry.

The value of the colour coefficient is not very accurately known.

The Mount Wilson Polar Sequence. The photographic Harvard scale of the North Polar Sequence has been independently confirmed through the extensive work by SEARES, who has also determined the photovisual scale of the stars in the sequence.

The work was started in 1910 and the results for the photographic magnitudes between $10^{\mathrm{m}},5$ and $17^{\mathrm{m}},5$ were ready in 1912[3].

The methods are described in Mt Wilson Contr No. 80. A number of diaphragms and wire-gauze screens were used, and the relation between scale readings and magnitudes was found graphically by applying the differential method of SCHWARZSCHILD[4]. For an account of the different effects and sources of errors we refer to the last-mentioned papers. Later on the relation between the Mount Wilson scale and that of the Harvard homogenous scale (m_{HH}) was given as above; it is valid for stars brighter than 16^{m}.

[1] A N 198, p. 65 (1914).
[2] Ap J 36, p. 169 (1912).
[3] Mt Wilson Contr 70 (1913).
[4] A N 172, p. 65 (1906).

Various tests, which are described in the papers by SEARES, showed that the work of Miss LEAVITT and SEARES has resulted in a photographic mean scale that is but little affected by the errors in the reduction constants. For the bright stars there is a marked difference, which is nearly linear in character and is approximately represented by the formula:

$$m_{MW} - m_H = -0{,}070(11{,}0 - m_{MW,H}) \qquad 2^m > m_{MW,H} > 9^m .$$

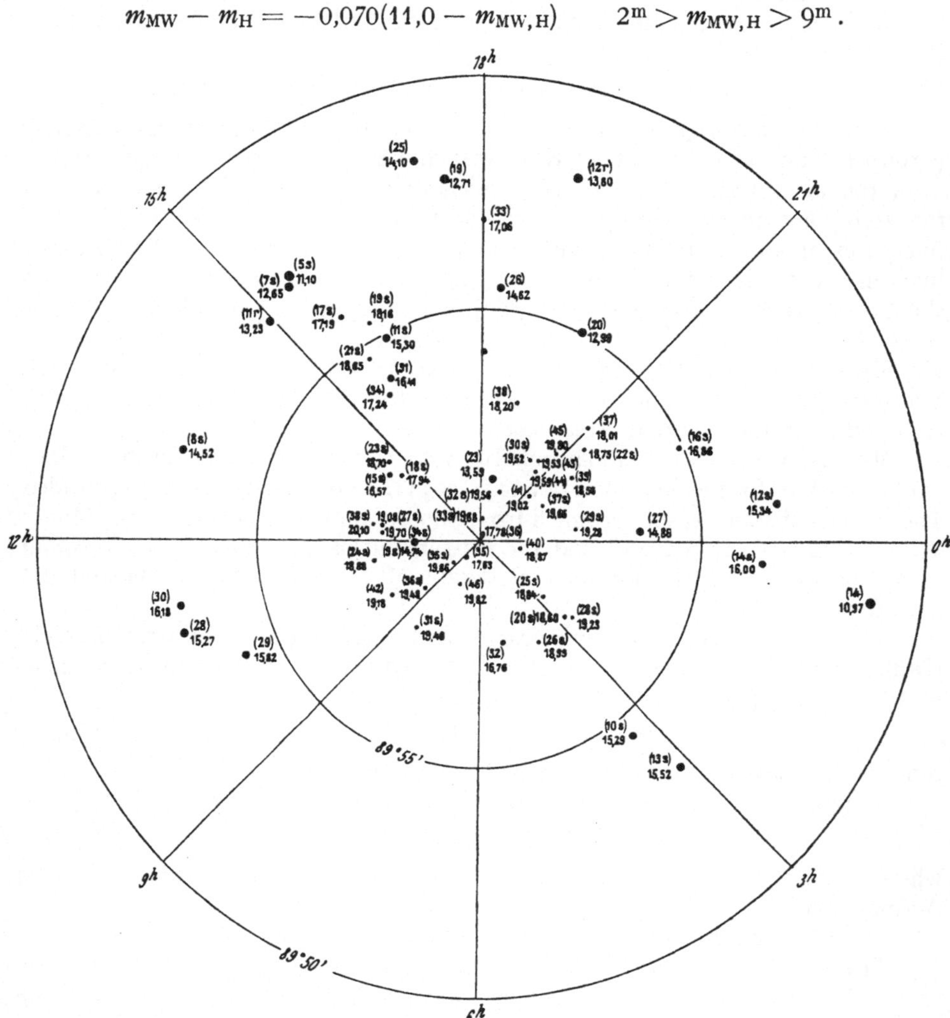

Fig. 84. Photovisual magnitudes of fainter stars in the North Polar Sequence according to Mount Wilson Contr No. 235.

Here m_{MW} means the magnitudes in the provisional Mount Wilson Scale, m_H the magnitudes in the Harvard Polar Sequence, and $m_{MW,H}$ the combined magnitudes. When the faint stars are included and the results referred to the international zero-point, the differences through the range $2^m - 15^m{,}5$ are expressed by: $m_{MW} - m_H = +0^m{,}37 - 0{,}070\ (11{,}0 - m_{MW,H})$ [1],

in which the second term is disregarded for $m_{MW,H} > 10$.

The photovisual scale was determined in precisely the same manner as the photographic scale. The visual scale of Harvard (Harv Circ 170) and the photo-

[1] Mt Wilson Contr 70 (1913).

visual scale of SEARES coincide at 6^m and 12^m, but on account of the colour equation certain differences of importance appear.

The catalogue given in the Mount Wilson Polar Sequence includes 645 objects, of which 617 have been investigated with regard to m_{ph}, and 339 with regard to m_{pv}. Of these objects 311 appear in both groups. Comparisons are made with the results at Harvard, Greenwich, and Potsdam. The m_{ph} embrace the interval

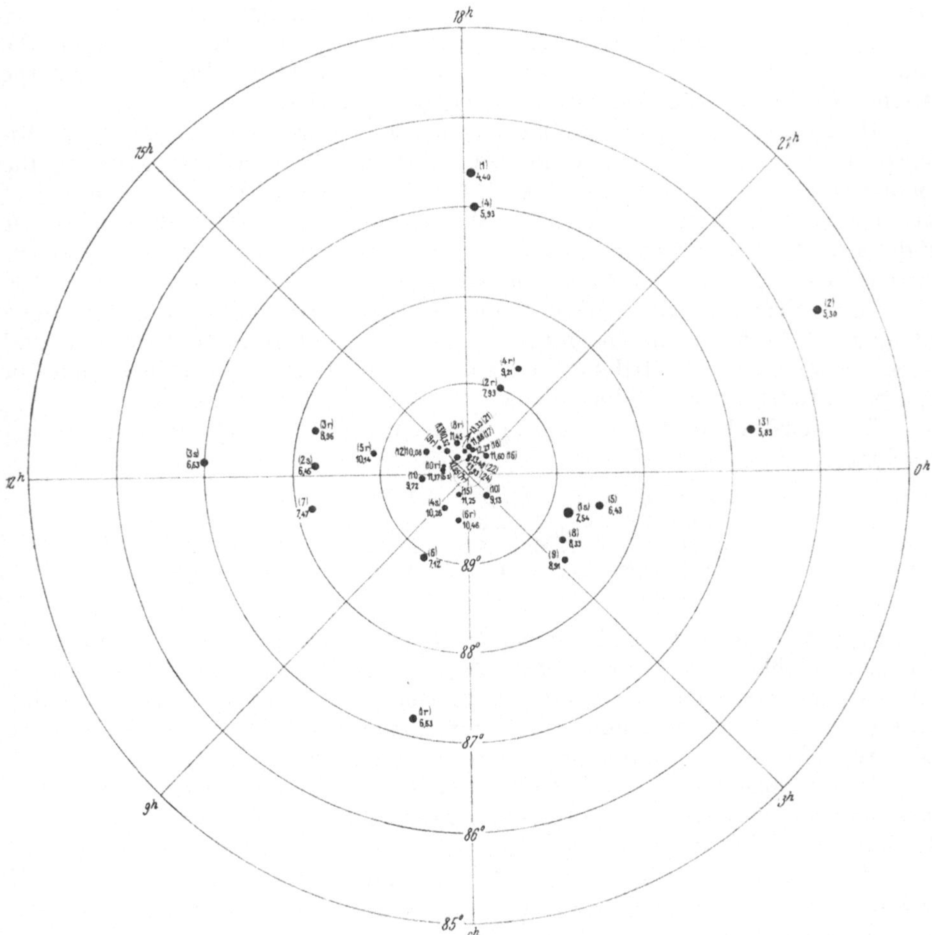

Fig. 85. Photovisual magnitudes and positions of brighter stars in the North Polar Sequence according to Mount Wilson Contr No. 235.

$2^m,0-20^m,0$ and the m_{pv} the interval $2^m,0-17^m,5$. The Mount Wilson determination is also invaluable for a knowledge of the colours of faint stars. Attention should be called to the useful table in G. EBERHARD's contribution to this Handbook (Band II, zweite Hälfte, Tabelle 22, p. 495, and 23, p. 496—500) giving the m_{pv} and m_{ph} as well as the coordinates of all stars included in the North Polar Sequence.

A revision of the Polar Sequence has been published jointly by SEARES and M. L. HUMASON[1], which includes additional data secured with the 10-inch

---

[1] Mt Wilson Contr No. 234.

Cooke refractor. The use of that instrument represents a radical change in conditions as compared with the use of the 60-inch reflector. The modest aperture of the former implies an entirely different relation between the diffraction pattern and the resulting photographic image of a star. Its short focal length of 45 inches means that the influence of atmospheric tremor is diminished. The large field results in small corrections for the distance error. Lastly the large coefficient of the colour equation (= 0,19 C) presents an extreme example of the difference in colour sensibility that may affect the instrumental apparatus.

The final investigation concerning the photographic scale including the material in Mount Wilson 97 combined with the material obtained using the 10-inch Cooke refractor led to the following conclusions.

If the combined photographic magnitudes are assumed to be correct between 10^m and 15^m, the magnitudes of the 10-inch Cooke refractor confirm the photographic scale between 6^m and 12^m of the revision for the reduction of accidental error discussed in Mount Wilson Contr 235. The values of colour indices found from exposure ratios show that the two scales are mutually consistent to 17^m photovisual magnitude with a maximum difference of $0^m,09$ at 10^m.

Table VIII in Mount Wilson Contr 235 summarizes all of the work done at Mount Wilson on the magnitudes and colours of stars near the Pole and should always be consulted by the student when he needs individual values of the magnitudes in the "International Standard Magnitude Scale".

The weighted mean deviation of a star in the Polar Sequence is according to SEARES:

Göttingen	$0^m,038$	M W 60	$0^m,022$
Gr. I	0 ,028	M W 10	0 ,025
Gr. II	0 ,025	Potsdam	0 ,041
H I (LEAVITT) . .	0 ,019	Yerkes	0 ,047
H II (KING) . . .	0 ,029		

On account of the small number of objects small differences in zero-point are unavoidable. For example the Göttinger Aktinometrie is referred to the international zero-point, but the comparison of SEARES gives a zero-point difference of $0^m,05$. The number of stars being small, this difference cannot be a reliable test as regards either agreement or disagreement.

SEARES concludes that a combination of the Mount Wilson photovisual measurements with data from other observatories is not to be recommended. An appreciable difference in scale between Mount Wilson and Harvard is evident.

The following relation connects the Harvard visual measurements m_{Hv} of magnitudes and the photovisual magnitudes in Harv Ann 71, part 3 with the m_{pv} reduced to the Mount Wilson colour system:

$$m_{\mathrm{Hv}} = m_{\mathrm{Harv\ 71}} - 0{,}14\,\mathrm{C}.$$

The two scales are in agreement at 6^m and 12^m, but between these values there is an irregularity that cannot be disregarded.

The discussion of the colour indices revealed that this irregularity must certainly be ascribed to an irregularity in the Harvard visual scale.

In Harv Bull 781 revised values of the stars brighter than $9^m,5$ are given. The correction of the magnitude scales for the brighter stars necessitates an adjustment of the whole sequence as is shown in the report of SEARES. In future work the Harvard photographic magnitudes will be based on the International Scale of standards. To correct the various photographic sequences in Harv Ann 71 and elsewhere the following values should be applied to all Harvard photographic magnitudes, published prior to May 1922:

Mag.	Corr.	Mag.	Corr.	Mag.	Corr.
2,5	$-0^m,14$	8,0	$+0^m,07$	13,5	$+0^m,26$
3,0	$-0\ ,12$	8,5	$+0\ ,10$	14,0	$+0\ ,25$
3,5	$-0\ ,10$	9,0	$+0\ ,14$	14,5	$+0\ ,25$
4,0	$-0\ ,08$	9,5	$+0\ ,18$	15,0	$+0\ ,24$
4,5	$-0\ ,08$	10,0	$+0\ ,21$	15,5	$+0\ ,22$
5,0	$-0\ ,06$	10,5	$+0\ ,22$	16,0	$+0\ ,19$
5,5	$-0\ ,05$	11,0	$+0\ ,24$	16,5	$+0\ ,16$
6,0	$-0\ ,04$	11,5	$+0\ ,24$	17,0	$+0\ ,12$
6,5	$-0\ ,02$	12,0	$+0\ ,24$	17,5	$+0\ ,06$
7,0	$0\ ,00$	12,5	$+0\ ,25$	18,0	$-0\ ,01$
7,5	$+0\ ,03$	13,0	$+0\ ,26$	18,5	$-0\ ,11$

The revision of the scale for brighter stars was made chiefly by Miss Leavitt. After her death the work has been completed by Miss Walkey and H. Shapley.

72. Harvard Standard Regions. According to a plan by E. C. Pickering, which was outlined in 1884, the sky has been divided into 48 equal parts. The brighter stars in these parts are to be determined photometrically by means of visual and photographic methods, and their spectra classified. The fainter stars are to be determined by means of a sequence situated in the centre of each region. In Harv Ann 71, part 4 Miss Leavitt gives the results of her determination of the standard magnitudes in the 48 regions, which thus serves as a system of secondary standards. Altogether 1995 stars were measured for photographic magnitudes in the said regions and 49 stars around the Southern Pole with a mean error that can be estimated as averaging $\pm 0^m,11$.

73. The Greenwich Polar Sequence. This work was perfomed by Chapman and Melotte[1]. The instrument used was the 30-inch reflector with a diffraction grating placed in front of the telescope. The measurements were made with a scale of numbered and graded comparison stars, which were viewed simultaneously together with the stars to be measured in the eyepiece of a comparator by means of an arrangement of reflecting prisms. For standardization, magnitudes of the pole field are used. The magnitudes determined are 262. From the residuals the mean error of an estimate on a single long exposure plate is estimated to be $\pm 0^m,114$ with regard to stars of 12^m-15^m. On the short exposure plates the estimates seem to have a mean error of $\pm 0^m,118$.

Chapman has noted that a slight systematic difference exists between the calculated and measured magnitude intervals of the gratings, viz.:

	Grating I	Grating II
Calculated	$2^m,66$	$4^m,04$
Measured	2 ,72	4 ,18

The differences are not large, but their systematic character deserves attention, and they cannot be attributed either to accidental errors in the measurements of the dimension of the grating or to the determinations of the photographic magnitudes. Chapman suggests that the fact that the images are not truly round produces a small systematic error, or else that there is a small systematic error in the Harvard scale. He is inclined to accept the latter explanation, but wishes to await further material before making definite conclusions.

74. Potsdam Polar Standards. W. Dziewulski[2] has determined at Potsdam the magnitudes of 222 polar stars of magnitudes 9^m-16^m. The half-grating method of Schwarzschild was used. A colour equation was found relative to

[1] M N 74, p. 40 (1913).
[2] A N 198, p. 65 (1914).

the systems of Harvard and Greenwich, as is evident from the relations:

$$m_P = m_H + 0{,}20\ C$$
$$m_P = m_G + 0{,}28\ C - 0^m{,}02$$

whereas the zero-point correction was insignificant in both cases. As is seen from the comparison of SEARES, the colour equation becomes smaller by 0,10 C when a reduction is made to the International Scale, but the zero-point correction increases to $+0^m{,}33$.

DZIEWULSKI[1] finds that the mean error of the standards varies between $\pm 0^m{,}014$ for stars of $12^m{,}0-13^m{,}0$, and $\pm 0^m{,}053$ for stars of $15^m{,}7-16^m{,}1$, and averages $\pm 0^m{,}025$.

75. Some General Conclusions with Regard to the Derivation of Apparent Magnitudes. In connection with the estimated magnitudes a general phenomenon is found to prevail, viz., the scales show a curvature in the sense that the fainter magnitudes have too small intervals. What are called $9^m{,}5$ in B D are really stars between 10^m and 11^m. The faintest stars cannot be discerned any longer as regards their real magnitudes. In the oldest estimates the faintest assigned magnitude cover too narrow an interval on the correct scale, whereas in modern estimates the converse is true. The scale of ARGELANDER's Uranometria Nova illustrates this remarkable change in the behaviour of the estimated magnitudes.

Another general phenomenon of great interest is the dependence of the scale on the star density of the background. This is evident already in the ancient estimates of magnitudes and also in such modern series as B D and A G.

One explanation might be that the faintest stars in the Milky Way are not seen. But the faintest stars noted in these regions are thought to be of the same magnitude as the faintest stars from the outside regions. On the other hand the phenomenon also appears in other magnitudes than the very faintest; thus it seems more reasonable to assume that the background excercises some influence on the estimates of the magnitudes. In favour of the first hypothesis is the fact that, as VAN DE LINDE[2] has shown, HEIS, HOUZEAU, and ARGELANDER have observed a smaller number of the visible stars within the Milky Way zone than outside it. Comparing the numbers in the three uranometries with the catalogue of BACKHOUSE[3], VAN DE LINDE has found:

	HEIS/BACKHOUSE	HOUZEAU/BACKHOUSE	ARGELANDER/BACKHOUSE
Northern region . .	0,88	0,54	0,46
Milky Way	0,62	0,48	0,38
Southern region . .	0,77	0,66	0,49

a_4) Special Problems of Stellar Photometry.

76. General Remarks concerning Photometric Problems. The eclipsing binaries are not variable stars; they have generally been included by tradition in the chapters in text books dealing with the binary stars.

In addition to the photometric studies of eclipses there are a number of problems that can be successfully attacked by the aid of stellar photometry. Such are the determination of the figures of stars, that deviate considerably from the spherical shape. In the case of double stars, there is the orbital eccentricity which produces changes in the interaction of the radiation of the components and which produces tidal phenomena. Also the effect of resisting media may have to be included.

---

[1] A N 198, p. 71 (1914).
[2] De Verdeeling der heldere Sterren. Dissertation. Rotterdam (1921).
[3] See paragraph 35.

The question whether the Sun is a variable star is also of much importance. The results reviewed elsewhere concerning the magnitude of the Sun are unable to reveal whether it is variable, and measurements of its light are too complicated and uncertain to make it possible to decide about the variation. In direct measurements of the Sun's radiation a serious difficulty is involved on account of the absorption in the atmosphere. The reflected light of planets may be used. The first experiments by MÜLLER[1] and KING[2] did not give positive results on account of the small variations used. GUTHNICK and PRAGER[3] have measured the light of Saturn during its opposition in 1914 to 15, in 1917, and in 1921. In 1921 Jupiter was also measured. W. E. BERNHEIMER[4] has discussed these observations and finds a certain correlation between the light of Saturn, or rather the difference Saturn $-\alpha$ Leonis, and the solar constant. On the other hand there is no connection between the light of Jupiter and the solar constant. The evidence given by the planets is not in favour of the theory that the solar radiation undergoes small variations in short periods.

The possible rotation of small planets and the deviation in their figure from the spherical form are problems that can be solved by the methods of stellar photometry. The same applies to the important question of the determination of the total brightness of surface objects, such as globular clusters and anagalactic nebulae.

77. The Density or Background Effect. In all the photometric estimates and measurements so far investigated there seems to be an effect in the magnitudes depending on the intensity of the background. I think that the explanation of this phenomenon might be that the background acts in the same manner as the "veil" on a photographic plate[5]. Suppose that μ is the observed magnitude and μ_0 is the limiting magnitude or the magnitude that is equal to the intensity of the background and that the corresponding intensities are I and I_0. If m is the magnitude free from the influence of the background, we have:

$$\mu = -2{,}5 \log (I + I_0),$$

$$\mu_0 = -2{,}5 \log I_0,$$

$$m = -2{,}5 \log I.$$

Thus:

$$I_0 = I \cdot 10^{0{,}4(m-\mu_0)}$$

and:

$$\mu - \mu_0 = m - \mu_0 - 2{,}5 \log \left(1 + \frac{I}{I_0}\right),$$

$$\mu - \mu_0 = m - \mu_0 - 2{,}5 \log (1 + 10^{0{,}4(m-\mu_0)}).$$

The following table shows the size of the influence of the background:

$\mu-\mu_0$	$\Delta\mu=m-\mu$	$\mu-\mu_0$	$\Delta\mu=m-\mu$
$6^m{,}0$	$0^m{,}00$	$1^m{,}0$	$0^m{,}55$
5 ,0	0 ,01	0 ,8	0 ,70
4 ,0	0 ,03	0 ,6	0 ,93
3 ,0	0 ,07	0 ,4	1 ,28
2 ,0	0 ,19	0 ,3	1 ,55
1 ,5	0 ,31	0 ,2	1 ,95

The fact that the mean deviations from the photometric measures of the estimated magnitudes discussed in Harv Ann 14 are arranged in the systematic

[1] Potsdam Publ 8 (1893). [2] Harv Ann 59, p. 262 (1912); 81, p. 208 (1923).
[3] Berlin Babelsberg Veröff I (1915); II (1919).
[4] Probleme der Astronomie. Festschrift für HUGO V. SEELIGER (1924).
[5] A N 193, p. 81 (1912).

way as is shown in the following table where a correlation between the course of the Milky Way and the value of the mean residual is clearly indicated, speaks in favour of the opinion that the light of the Milky Way acts as a veil. A final investigation into that question is highly desirable but scarcely feasible for the present on account of our limited knowledge as well of the intensity of the light of the Milky Way as of the number of stars in different part of the sky.

Influence of Milky Way in photometric estimates.

δ \ α	0^h	1	2	3	4	5	6	7	8	9	10	11	12	13	14	15	16	17	18	19	20	21	22	2
80°																								
70											−1	−1	−1	−1	−1	−1	−1							
60										−1	−1	−1	−1	−1	−1	−1	−1	−1						
50										−1	−1	−1	−1	−1	−1	−1	−1	−1						
40											−1	−1	−1	−1	−1	−1	−1	−1						
30					+1	+1	+1	+1				−1	−1	−1	−1	−1	−1							
20	−1	−1			+1	+1	+1	+1				−1	−1	−1	−1	−1							−1	−
10	−1	−1	−1		+1	+1	+2	+1	+1			−1	−1	−1	−1							−1	−1	−
0	−1	−1	−1	−1		+1	+2	+2	+1			−1	−1	−1				+1				−1	−1	−
−10	−1	−1	−1	−1		+1	+2	+2	+1		−1	−1	−1	−1		+1	+1	+1	+1		−1	−1	−1	−
−20	−1	−1	−1	−1	−1	+1	+2	+2	+2		−1	−1	−1			+1	+1	+1			−1	−1	−1	−
−30°	−1	−1	−1	−1	−1		+1	+2	+2	+1	−1	−1	−1		+1	+1	+1	+1			−1	−1	−1	−

Unit is $0^m,1$. Data are taken from Harv Ann 14.

78. The Influence of Multitude and Incompleteness on Photometric Catalogue Differences. When two photometric catalogues are compared the method generally used is to form mean values of groups of differences between objects in common. Owing to the influence of the multitude of the stars and the incompleteness of the catalogues such mean differences will not give correct values for establishing the scale reduction. I think that CHARLIER[1] was the first to point out this phenomenon and to deduce a formula for correcting its influence. Later on PANNEKOEK[2] has investigated the question in detail and derived formulae suitable for the numerical calculation of the amount of the influence excercised by multitude and incompleteness in different cases in practice. We give here a summary of his formulae worked out in the way preferred by CHARLIER and other students of mathematical statistics.

The common assumption in the different cases is that the distribution of the errors of each catalogue follows a Gaussian law. The catalogue that is compared by combining all the stars of a certain magnitude μ will be called the first; the one to which it is compared will be called the second. The observed magnitudes are called μ; m stands for the true magnitudes and σ for the dispersion into m; σ_2 for the dispersion into μ.

Case 1. Both catalogues are complete; the second gives the true magnitudes. It is assumed that the number of stars of magnitude $m \pm \frac{1}{2} dm$ is:

$$a(m) = A_0 e^{bm} = A_1 e^{b(m-\mu)}$$

(SEELIGER's hypothesis; cf. Lund Medd 100, p. 12),

further that the errors of the first catalogue, $m - \mu$, have the frequency:

$$\frac{1}{\sigma\sqrt{2\pi}} e^{-\frac{(\mu - m)^2}{2\sigma^2}} .$$

---

[1] Lund Medd Sér II No. 8, p. 25 (1912).

[2] Publ Astr Inst Univ Amsterdam No. 1, p. 9 (1924).

Then the average difference $\overline{m - \mu}$ is given by:

$$\overline{m-\mu} = \frac{\frac{1}{\sigma\sqrt{2\pi}}\int\limits_{-\infty}^{+\infty} A_1 (m-\mu)\, e^{b(m-\mu)} e^{-\frac{(\mu-m)^2}{2\sigma^2}}\, dm}{\frac{1}{\sigma\sqrt{2\pi}}\int\limits_{-\infty}^{+\infty} A_1 e^{b(m-\mu)} e^{-\frac{(\mu-m)^2}{2\sigma^2}}\, dm},$$

or

$$\overline{m-\mu} = \frac{A_1 b\sigma^2 e^{\frac{b^2\sigma^2}{2}}}{A_1 e^{\frac{b^2\sigma^2}{2}}} = b\sigma^2$$

and

$$\Delta m = \frac{b\sigma^2}{2}.$$

Case 2. Both catalogues are complete and both have accidental errors.

The number of stars of magnitude m, observed in the second as μ_2 and in the first as μ_1 is:

$$\frac{1}{\sigma_1\sigma_2 2\pi}\int\limits_{-\infty}^{+\infty} d\mu_2 \int\limits_{-\infty}^{+\infty} dm\, A_0 e^{bm} e^{-\frac{(\mu_1-m)^2}{2\sigma^2}} e^{-\frac{(\mu_2-m)^2}{2\sigma_2^2}}.$$

The average difference is given by:

$$\overline{\mu_1-\mu_2} = \frac{\int\limits_{-\infty}^{+\infty} d\mu_2 \int\limits_{-\infty}^{+\infty} dm\, (\mu_1-\mu_2)\, A_1 e^{b(\mu_1-m)} e^{-\frac{(\mu_1-m)^2}{2\sigma^2}} e^{-\frac{(\mu_2-m)^2}{2\sigma_2^2}}}{\int\limits_{-\infty}^{+\infty} d\mu_2 \int\limits_{-\infty}^{+\infty} dm\, A_1 e^{b(\mu_1-m)} e^{-\frac{(\mu_1-m)^2}{2\sigma^2}} e^{-\frac{(\mu_2-m)^2}{2\sigma_2^2}}},$$

which can be reduced to:

$$\overline{\mu_1-\mu_2} = b\sigma^2.$$

Case 3. The second catalogue stops at magnitude l. Then the integration gives:

$$\overline{\mu_1-\mu_2} = b\sigma^2 - \frac{\alpha}{\sqrt{2\pi}}\cdot\frac{e^{-K^2}}{\psi(K)}$$

where:

$$\psi(K) = \frac{1}{\sqrt{\pi}}\int\limits_{-\infty}^{K} e^{-x^2} dx, \qquad K = \frac{1}{\alpha\sqrt{2}}(l - b\sigma^2),$$

$$\alpha^2 = \sigma^2 + \sigma_2^2.$$

This formula is a special case of the more general formula:

$$\overline{\mu_1-\mu_2} = b\sigma^2 - \frac{\frac{A\alpha}{\sqrt{2\pi}} e^{-K^2} + \frac{\beta\alpha'}{\sqrt{2\pi}} e^{-K'^2}}{A\psi(K) + B\psi(K') + C},$$

which is valid when a catalogue consists of two parts with different limiting magnitudes.

If the two limits l and l' are not too nearly coincident PANNEKOEK finds that we may substitute for this expression a series of partial corrections, each computed for a single limit occurring alone.

Case 4. The first catalogue is cut off at $\mu_1 + l$. In this case the formulae are the same as in case 2.

Case 5. The second catalogue is incomplete; all stars are omitted that fall below a certain limit in a third catalogue.

The number of stars of magnitude μ_1 with observed magnitudes μ_2 and with μ_3 in the two other catalogues and with true magnitudes m is given by:

$$\frac{1}{\sigma\sigma_2\sigma_3\sqrt{8\pi}^3} d\mu_2 d\mu_3 \int A(m)\, e^{-\frac{(\mu_1-m)^2}{2\sigma^2} - \frac{(\mu_2-m)^2}{2\sigma_2^2} - \frac{(\mu_3-m)^2}{2\sigma_3^2}}\, dm.$$

The average difference is here:

$$\overline{\mu_1 - \mu_2} = b\sigma^2 - \frac{\sigma^2}{\beta\sqrt{2\pi}} \cdot \frac{e^{-K^2}}{\psi(K)}$$

where:

$$\beta^2 = \frac{1}{\sigma^2 + \sigma_3^2}, \qquad K = \frac{l - b\sigma^2}{\beta\sqrt{2\pi}}.$$

Case 6. The first catalogue is incomplete; at a certain magnitude the real number A of stars in the sky should decrease according to this law:

$$A = A_0 \frac{1}{\sigma_3\sqrt{2\pi}} e^{bm} \int_{-\infty}^{l} e^{-\frac{(\mu-m)^2}{2\sigma_3^2}}\, d\mu = A_1 \frac{1}{\sigma_3\sqrt{2\pi}} e^{b(\mu_1-m)} \int_{-\infty}^{l} \frac{e - (z - [\mu_1 - m])^2}{2\sigma_3^2}\, dz,$$

where $z = \mu - \mu_1$.

The correction will then be written:

$$\sigma^2\left(b - \frac{1}{\beta\sqrt{2\pi}} \cdot \frac{e^{-K^2}}{\psi(K)}\right) = b'\sigma^2.$$

Case 7. The second catalogue is incomplete; from a certain magnitude, l, the number of stars decreases according to the Gaussian law.

In this case the number of stars in the second catalogue must be multiplied by:

$$\frac{1}{\sigma_2\sqrt{2\pi}} \int_{-\infty}^{l} e^{-\frac{(\mu-m_2)^2}{2\sigma_3^2}}\, d\mu.$$

Thus the number of stars with a magnitude μ_1 in the first and μ_2 in the second catalogue is:

$$A_1 \frac{1}{\sigma\sigma_2\sigma_3} d\mu_2 \int_{-\infty}^{l} e^{-\frac{[z-(\mu_2-\mu_1)]^2}{2\sigma_3^2}}\, dz \int_{-\infty}^{+\infty} e^{b(\mu-m) - \frac{(\mu_1-m)^2}{2\sigma^2} - \frac{(\mu_2-m)^2}{2\sigma_2^2}}\, dm.$$

Putting: $\sigma^2 + \sigma_2^2 + \sigma_3^2 = \gamma^2$ and $z = \mu - \mu_1$ we get after two integrations:

$$\overline{\mu_1 - \mu_2} = \frac{\int_{-\infty}^{l} \frac{2b\sigma^2\sigma_3^2 + 2\alpha^2 z}{\gamma^2} e^{-\frac{(z-b\sigma^2)^2}{2\gamma^2} + \frac{b^2\sigma^2}{2}}\, dz}{\int_{-\infty}^{l} e^{-\frac{(z-b\sigma^2)^2}{2\gamma^2} + \frac{b^2\sigma^2}{2}}\, dz}$$

and putting:

$$\frac{l - b\sigma^2}{\gamma\sqrt{2}} = K,$$

we get:

$$\overline{\mu_1 - \mu_2} = b\sigma^2 - \frac{\alpha^2}{\gamma\sqrt{2\pi}} \cdot \frac{e^{-K^2}}{\psi(K)}.$$

In order to illustrate the numerical size of the corrections we take the following example from the paper of PANNEKOEK.

A comparison was made between Harv Ann 24 and Harv Ann 45 by MÜLLER and KEMPF in the introduction to their P D. Besides a dependence on A R which they think may be caused by a different treatment of the extinctions, they find a marked variation of the difference with the magnitude (argument μ in Harv Ann 45). The number of stars common to both catalogues increases strongly towards $6^m,5$ and then decreases as rapidly. The corrections for multitude computed according to the corresponding formula make the variation found by MÜLLER and KEMPF disappear.

m	Harv Ann 24 — Harv Ann 25	n	Correction for multitude	Harv Ann 24 — Harv Ann 45 corrected
$4^m,5-5^m,0$	$+0^m,17$	21	$-0^m,05$	$+0^m,12$
5 ,0 — 5 ,5	+0 ,02	66	−0 ,04	−0 ,02
5 ,5 — 6 ,0	+0 ,07	182	−0 ,05	+0 ,02
6 ,0 — 6 ,5	+0 ,06	539	−0 ,06	0 ,00
6 ,5 — 7 ,0	+0 ,01	896	+0 ,02	+0 ,03
7 ,0 — 7 ,5	−0 ,05	214	+0 ,08	+0 ,03

79. The Reflecting Effect in Eclipsing Variables. In several cases of eclipsing variables the components are so close that the reflection by the faint star of light from the bright star can be seen, when the light curve is examined, on account of the fact that the hemispheres turned towards and away from the bright component have unequai luminosity. The light is thus not constant between the eclipses and if the stars are not spherical the constant phase will also be affected. EDDINGTON[1] has considered the physical theory of the reflection effect. He supposed that the translation of the light curve into a heat curve was performed and investigated the amount of energy reflected. Then the problem is simple because a star necessarily re-emits completely the radiation falling on it. This means that the heat-albedo could be taken as 1.

A more detailed explanation is given in the original paper. The meaning of the statement is that the internal state of a star cannot be appreciably altered by light falling on its surface. The stream of energy will continue to pour out from it at a normal rate. Since it is the net outward flow that is thus described, the gross outward flow is greater by the amount of the incident radiation. When the additional radiation is referred to reflected, it is only a convention, as the actual process is absorption followed by re-emission.

J. STEBBINS[2] has investigated the reflection of light in a close binary system. For diffuse reflection from the companion, expressions for the albedo have been derived from the full phase on the assumption of the validity of LAMBERT'S law and the absorption law. It has been found that the reflection in the partial phases will mask to some extent the effect due to ellipsoidal form. Since the radiation effect usually amounts to only a few hundredths of a magnitude, STEBBINS concludes that there is no prospect of obtaining anything but a rough indication of the amount of the reflection effect.

Galactic latitude	Amount of light per square degree (Unit is one magnitude in the Harvard visual scale)
40°	0,012
50	0,011
60	0,009
70	0,009
80	0,008
90	0,008

[1] M N 86, p. 320 (1926).

[2] Probleme der Astronomie. Festschrift für Hugo v. Seeliger (1924).

80. The Total Amount of Starlight. From the number of stars within the different magnitudes the total amount of the starlight can be summed up. From Groningen Publications No. 27 for stars fainter than $5^m,5$ the results shown in the table (page 339 below) have been found[1].

Now the brightness of the sky can, also, be estimated or measured by direct methods. The following values have been obtained:

Authority	Brightness of the sky per □°	Source
NEWCOMB . .	0,029	Ap J 14 p. 297 (1901)
BURNS . . .	0,050	Ap J 16 p. 166 (1902)
ABBOT . . .	0,075	Ap J 27 p. 20 (1911)
YNTEMA . . .	0,140	Gron. Publ No. 22 (1909)
VAN RHIJN .	(0,130)	Mount Wilson Contr. No. 173 (1919)
VAN RHIJN .	0,165	Groningen Publ No. 31 (1921)

Unit in the above and in following tables of this paragraphs is a star of $1^m,00$ in the Harvard visual scale.

G. BURNS compares the brightness of the sky with the extra-focal images of a star of known magnitude. The sky is observed with one eye and the telescopic extra-focal image with the other.

ABBOT, YNTEMA, and VAN RHIJN have used photometers. A mean of their values thus has probably a higher chance of being correct than those of G. BURNS and NEWCOMB.

Now, the brightness of the sky is not due exclusively to the stars. An additional light impression is caused by earthlight. The light of the stars scattered by the atmosphere is not contained in the earthlight. VAN RHIJN finds that even during the same night the earthlight may change by as much as 20 per cent, but there seems to be no systematic tendency in these variations. A dependence on azimuth is clearly shown which is explained by a dependence of the earthlight on the position of the measured area relative to the ecliptic. If the measurements are arranged according to latitudes (ecliptical coordinates) it is found that the amount increases when the latitude decreases.

The earthlight must then be connected with the zodiacal light. The absolute amount of this source cannot be computed but only its excess over a mean value. VAN RHIJN computes this excess, basing his computation partly on FESSENKOFF's[2] observations and giving special attention to the brightness of the counterglow, and finds that his material does not contradict SEELIGER's theory[3] of the zodiacal light. Uncertainty is involved on account of our imperfect knowledge of the law of phases.

The light for galactic latitudes $>40°$ is practically independent of the starlight and can be used for a separation of starlight and other sources of skylight.

As the brightness of the sky also varies with the zenith distance it seems possible that part of the earthlight is due to a persistent aurora. This conclusion is strengthened by the fact that SLIPHER[4] has photographed the green auroral line in the spectrum of the background of the sky.

According to VAN RHIJN[5] the light of the sky is, supposing average conditions prevail, a sum of the light of various sources as follows:

[1] P. J. VAN RHIJN: On the Brightness of the Sky at Night and the Total Amount of the Starlight, Mount Wilson Contr No. 173 (1919).

[2] La Lumière Zodiacale. Thèse. Paris 1914.

[3] Über kosmische Staubmassen und das Zodiakallicht. Sitzber math phys Kl Bayer Akad Wiss 31, Heft 3 (1900).

[4] Lowell Bull 3, No. 1 (1916). [5] Groningen Publ No. 31 (1921).

Source	Amount
Zodiacal light	0,071
Scattered earthlight	0,032
Aurora Borealis	0,024
Direct starlight	0,029
Scattered starlight	0,009

K. GRAFF[1] is the first who has compared different parts of the Milky Way by aid of photometric means. For that purpose a universal photometer of his own construction was used and the relative intensity of 47 small areas determined which were used for deriving isophotes (= lines of equal intensity). Later on J. HOPMANN[2] used a wedge photometer constructed by GRAFF and measured the relative intensity of 13 small Milky Way areas. These intensities were then used as standards for reduction of 256 estimates. The measures as well as the estimates were performed at Christmas Island in 1922. Including estimates and measures over the Northern Milky Way an isophotic chart was constructed.

HOPMANN[3] has made an extensive discussion of his measures and reduced his photometric system to that of VAN RHIJN. Assuming the intensities of HOPMANN, H, to be connected with those of VAN RHIJN, R, in the following manner:

$$aH = R + C,$$

where a and C are constants, HOPMANN found, comparing such parts of the sky where the two observers had observed principally the zodiacal light, the earthlight, and the Aurora Borealis, the following formula:

$$R = -0{,}1183 + 0{,}1540\,H.$$

As to the distribution of the starlight in the Milky Way HOPMANN derives:

Galactic lat.	Amount	Galactic lat.	Amount
2°,5	0,077	20°,0	0,046
10 ,0	0,066	27 ,5	0,037

Of special interest is the derivation of the intensity of a number of condensations in the Milky Way.

The following comparison between the total amount of starlight as observed by VAN RHIJN and YNTEMA on the one hand and computed from data in the Groningen Publications 27 and 18 on the other may be of interest:

Galactic lat.	VAN RHIJN	YNTEMA	Gron. 27	Gron. 18
0°	0,080	0,088	0,055	0,186
10	,065	,055	,043	,118
20	,044	,033	,027	,055
30	,026	,024	,020	,032
40	,015	,020	,015	,022
50	,014	,018	,014	,018
60	,012	,015	,012	,015
70	,011	,013	,011	,013
80	,010	,013	,010	,012
90	0,010	0,012	0,010	0,012

This table gives an interesting confirmation of the conclusion derived from other evidences, namely that the magnitude scale in Groningen 18 must be considerably in error.

How valuable the investigations here mentioned may be, the interesting field of photometry applied to the nocturnal sky is certainly by no means exhausted.

[1] Bergedorf Astr Abh 2, No. 5 (1920). [2] A N 219, p. 189 (1923).
[3] A N 222, p. 81 (1924).

Of recent estimates of intensities in the Milky Way the extensive ones of A. PANNEKOEK[1] should be mentioned.

81. The Limiting Magnitude of Star Catalogues. For what magnitude intervals the magnitude catalogues are complete is a very important question. Even in case of stars visible to the unaided eye it will be difficult to find a source including them all. In fact it has been shown by VAN DE LINDE[2] that there does not exist any Uranometry giving the stars visible to the unaided eye at a certain place during average conditions.

E. C. PICKERING[3] has introduced the term limiting magnitude, which measures the magnitude in a certain catalogue at which the number of brighter stars omitted is equal to the number of fainter stars included. As it is impossible to determine directly the number of bright stars omitted, the determination will be effected by counts of the number of faint stars included.

If m_f denotes the magnitude of the faintest stars in a certain catalogue, m_l the limiting magnitude, m_c the magnitude for which all stars brighter than and of this magnitude have been included, and $N(m)$ is the actual number of stars, brighter than magnitude m, and $N'(m)$ the corresponding number in the catalogue considered, and m is expressed in a photometrically correct scale, then we have according to the definition the proportionality:

$$\frac{N(m_l)}{N(m_c)} = \frac{N'(m_f)}{N'(m_c)}.$$

Further $N(m_c) = N'(m_c)$. In case of magnitudes numerically lower than m_c, $N(m)$ is larger than $N'(m)$, because of the incompleteness of individual catalogues owing to doubtful identifications, stars far to south, selection among faint stars, doubles, variables, etc.

We can write the above equation:

$$\log N(m_l) = \log N'(m_f) - \log N'(m_c) + \log N'(m_c).$$

For sake of smoothing out the observed numbers of N', the N' values of m_{c+a} and m_{c-a} are introduced. In older catalogues the quantity a can be taken as corresponding to the smallest interval used when estimating m (e. g. in PTOLEMAIOS's catalogue some $\frac{1}{3}^{\mathrm{m}}$) and in modern catalogues it can properly be taken as $0^{\mathrm{m}},5$.

We have then:

$$\log N(m_l) = \log N'(m_f) - \log N'(m_{c-a}) + \log[N'(m_{c+a}) - N'(m_{c-a})]$$

or:

$$\log N(m_l) = A - B + C.$$

The three terms A, B, and C can thus be formed from the data in the catalogue and a table giving $N(m)$ as function of m furnishes a means of reading m_l when $A - B + C$ is known.

The limiting magnitude is dependent on the degree of precision of the selection, because great carefulness in the selection will determine the limiting magnitude very precisely. As a measurement of the precision PICKERING introduced the uncertainty. This quantity is the change which would take place in the limiting magnitude if the supernumerary faint stars were omitted.

---

[1] Die nördliche Milchstraße, Leiden Ann Deel XI, Derde Stuk (1920); Die südliche Milchstraße, Lembang Ann 2, No. 1 (1928).

[2] De verdeeling der heldere sterren. Thesis. Utrecht (1921).

[3] Limiting Magnitudes and Uncertainty of Star Catalogues. Harv Ann 76, No. 9 (1916).

In this case the total number of stars in the catalogue will be $N'(m_l)$ instead of $N'(m_f)$. If the limiting magnitude in this case is $m_{l'}$, then we have:

$$\log N(m_{l'}) - \log N(m_c) = \log N'(m_l) - \log N'(m_c).$$

If $\log N'(m_l)$ is denoted by D then we have as above:

$$\log N(m_{l'}) = D - B + C$$

and the change:

$$\Delta m_l = m_l - m_{l'}$$

then defines the uncertainty.

For establishing the function $N(m)$, Pickering has made use of the results from Harv Ann 48, No. 5 as well as of those from Groningen and Greenwich. The Harvard law of distribution is assumed to be the correct one. Accordingly the Groningen results are reduced to those of Harvard. The following short table gives the connection between m_l and $A - B + C$.

m_l	$A-B+C$	m_l	$A-B+C$
$4^m,0$	2,66	$9^m,0$	5,13
5 ,0	3,17	10 ,0	5,57
6 ,0	3,68	11 ,0	5,98
7 ,0	4,18	12 ,0	6,35
8 ,0	4,66	13 ,0	6,68

The method is very convenient, especially if it is applied with the aid of graphs. A more extensive table of each tenth of the magnitudes and the logarithms $N(m)$ is given in table II in Pickering's paper and can be selected as a basis for graphical methods. The ratios $\frac{N(m_1+m_2)}{N(m_1)}$, where m_2 varies from 0,0 to 2,0 and m_1 from 4,0 to 10,0 and can be used for deriving the difference $m_l - m_c$, are tabulated in table III of the same paper whereby it was tacitly assumed that the ratios are equal for the whole sky or for any part of it. As an example we mention the determination of the limiting magnitude of the catalogue of Ptolemaios. The total number of stars discussed in Harv Ann 14, p. 343 is 757. It seems that few stars as bright as 4^m have been omitted and this magnitude is assumed as a basis. The photometrically corrected magnitude of Ptolemaios' 4^m is $4^m,35$. Of the 299 stars estimated as being of 4^m by Ptolemaios we may assume that half are brighter and half are fainter than their mean magnitude $4^m,35$. All the 263 stars estimated as brighter than 4^m are assumed to be actually brighter. Hence $150 + 263 = 413$ and $757/413 = 1,83$. From the table we find corresponding to $4^m,35$ and 1,83 the difference $m_l - m_c = 0^m,51$. Thus $4^m,35 + 0^m,51 = 4^m,86$ is the limiting magnitude. To derive the uncertainty we must determine the number of stars fainter than $4^m,86$. There are 576 stars brighter than $5^m,4$ and 5 of magnitude $5^m,4$. As the mean magnitude of the latter is $4^m,66$ we may assume that it corresponds to a total number of $576 + {}^1/_2 \cdot 5 = 578$. In like manner the magnitude $4^m,86$ corresponds to 618 stars. The number of stars omitted, which are brighter than $4^m,86$, is therefore $757 - 618 = 139$. The ratio 618/413 is 1,50 and the corresponding correction from the table is $0^m,34$. The difference $4^m,86 - (4^m,35 + 0^m,34) = 0^m,17$ is the uncertainty of the catalogue.

In like manner the limiting magnitude and the uncertainty were investigated for a number of catalogues by Pickering. On account of the great interest connected with these results we give them in the following table:

Catalogue	$N'(m_f)$	m_c In the scale of catal.	m_c Reduced to RHP	$N'(m_{c-a})$	$\frac{N'(m_f)}{N'(m_{c-a})}$	$m_l - m_c$	m_l	$N'(m_l)$	$\frac{N'(m_l)}{N'(m_{c-a})}$	$m_{l'} - m_c$	Δn
Ptolemaios	757	4	4^m,35	413	1,83	0,51	4^m,86	618	1,50	0^m,34	0^m,
Al Sûfi	830	4	4 ,24	372	2,23	0,68	4 ,92	709	1,91	0 ,55	0 ,
Uranometria Nova (ARGELANDER). .	3188	5	4 ,98	930	3,43	1,06	6 ,04	2623	2,82	0 ,89	0 ,
Heis.	5354	5	4 ,95	902	5,94	1,53	6 ,48	—	—	—	—
Bonn. Durchmusterung	314925	8,5	8 ,64	43352	7,26	1,96	10 ,60	259290	5,98	1 ,76	0 ,
Uranometria Argentina	7685	6,0	5 ,78	1970	3,89	1,18	6 ,96	—	—	—	—
Uranometria Argentina	7685	6,5	6 ,27	3464	2,22	0,70	6 ,97	—	—	—	—
Harvard Photometry	4193	5,24	5 ,24	1391	3,01	0,94	6 ,18	3442	2,47	0 ,80	0 ,
Revised Harvard Photometry . . .	8761	5,24	5 ,24	1893	4,63	1,32	6 ,56	7747	4,09	1 ,25	0 ,
Revised Harvard Photometry . . .	8761	5,74	5 ,74	3453	2,54	0,80	6 ,54	7642	2,21	0 ,69	0 ,
Revised Harvard Photometry . . .	8761	6,24	6 ,24	6106	1,43	0,31	6 ,55	7695	1,26	0 ,20	0 ,

82. Zero-Point for the Photometric Measures. It would be very convenient to select the magnitude of the Sun as a zero-point for the magnitudes of the stars. This involves extending the magnitude-scale back to -27^m.

A photometric comparison between the Sun and the stars is a difficult problem as it involves an enormous weakening of the light of the Sun. The first comparison seems to have been made by C. HUYGENS, who found that Sirius has a light amounting to $^1/_{50} \cdot 10^6$ of the Sun's, which corresponds to -22^m,1 for the Sun's magnitude.

The next determination was made by WOLLASTON, who used a flame as intermediary. The image of the Sun was reflected into a tube by the aid of a thermometer and compared with the flame. The weakness of the method is that the loss of light of the Sun owing to the reflection could not be computed with any certainty. WOLLASTON assumed an arbitrary loss of 50 per cent and found the ratio Sun/Sirius $= \frac{1}{2} \cdot 10^{10}$, which corresponds to a magnitude of the Sun of -25^m,7.

H. N. RUSSELL[1] has discussed the value of the magnitude of the Sun. He states that of the older determinations we can only use that by ZÖLLNER[2]. He compared with his photometer the starlike image of the Sun, formed by a suitable combination of lenses and shade-glasses, and the direct image of Capella, formed by a larger objective, the same artificial star. The light of the lamp had to be assumed to be absolutely constant throughout the interval of four months covered by the observations. The Sun was observed on six days and Capella on eleven nights and the average deviation of a single observation by day or night from the general mean was only 0^m,05. ZÖLLNER found the difference between Capella and the Sun to be $-26^m,87 \pm 0^m,05$. As the Harvard magnitude of Capella is 0^m,21, this gives the magnitude of the Sun as equal to $= -26^m$,66.

The next determination is due to C. FABRY[3], who compared the light of a standard electric lamp, shining through a blue filter, with that of the Sun weakened

[1] Ap J 42, p. 103 (1916).
[2] Photometrische Untersuchungen p. 107, 124—125. Leipzig (1865).
[3] C R 137, p. 973 and 1242 (1903).

by divergence through a lens of short focal length, by means of a LUMMER-BRODHUN prism. Then the same standard lamp, reduced to a stellar image, was compared by means of a microscope objective and reflection from a glass plate with Vega seen directly with the naked eye. FABRY finds that the Sun at the zenith and at mean distance gives light equal to 100000 bougies décimales at 1 m, and that Vega in the zenith is equal to a candle at 780 m. Reduced to the Harvard scale the magnitude of the Sun is found to be $-26^m,80$.

CERASKI[1] has used Venus, and has compared that planet, near elongation and in full daylight, with the Sun, reflected from the surface of a small lens at a distance of 152 m. Auxiliary comparisons were also made between Venus and various bright stars. On the Harvard scale the value of the Sun's magnitude is $-26^m,60$ according to CERASKI.

In the determination of W. H. PICKERING[2] at Jamaica in 1901, a 12-inch telescope of 135 feet focal length was used in the following way. By means of a shadow photometer the brightness of the centre of the focal image of the Sun was seen shining through a small aperture about one mm in diameter in front of the objective, and compared with that of the extra-focal images of stars two to three mm in diameter. As a comparison light a pentane lamp was used, which shone through a cell filled with a blue solution. Ten independent determinations on as many days and nights were carried out and gave as the Sun's magnitude $-26^m,83$.

Collecting all the results used by RUSSELL we have:

ZÖLLNER	$-26^m,66$
FABRY	$-26\ ,80$
CERASKI	$-26\ ,60$
PICKERING	$-26\ ,83$
Mean value	$-26^m,72 \pm 0^m,04$

The agreement is remarkably good when the difficulties of determining accurately the amount of the enormous weakening of the Sun's light necessary in all the methods are considered. The fact that the processes are so radically different makes it highly improbable that appreciable systematic errors have affected the mean value.

The photometry of the stars thus embraces a range of 48 magnitudes. The ratio between the light of the Sun and that of the faintest stars for which magnitudes have been determined is 1000000000000 to 1.

RUSSELL also mentions two determinations of the photographic magnitude of the Sun.

KING[3] has compared the Sun with Polaris, Arcturus, and Capella by means of his extra-focal method and finds a mean value of $-25^m,83$ with an average deviation of only $0^m,07$ for the 11 determinations. RUSSELL assumes that the probable error is somewhat larger than this agreement would indicate on account of the uncertainty of the measurements of the absorption of the filters employed.

O. BIRCK[4] has used a method essentially similar to that of PICKERING's visual magnitudes. The differences between the Sun and Vega, Arcturus, and Capella were determined. By using KING's photographic magnitudes the following table was obtained:

---

[1] Ann de l'Obs Moscou 2nd Ser. 5, p. 1 (1911).

[2] Harv Ann 61, p. 56 (1912).

[3] Harv Ann 59, p. 248 (1908).

[4] Das photographische Helligkeitsverhältnis der Sonne zu den Fixsternen. Dissert Göttingen (1909).

	Diff. in magnitude	Colour index		Photographic magnitude	Photographic magnitude of Sun	
		BIRCK	KING		Uncorrected	Corrected
Vega	$25^m,95$	$0^m,00$	$0^m,00$	$0^m,14$	$-25^m,81$	$-26^m,15$
Capella	27 ,04	1 ,02	0 ,82	1 ,03	−26 ,01	−26 ,01
Arcturus	27 ,76	1 ,71	1 ,09	1 ,33	−26 ,43	−26 ,20

The corrected results were found by RUSSELL from the following considerations. The photographic differences in magnitude between Vega and Capella, and between Vega and Arcturus can be derived from BIRCK's observations. The Harvard visual magnitudes give in comparison with BIRCK's values the colour indices, which turn out considerably larger than those of KING. If BIRCK's results are approximately reduced to the Harvard photographic scale the corrected values are obtained. These agree very well, and their mean value is $-26^m,12$.

RUSSELL gives double weight to KING's value and finds the mean for the photographic magnitude of the Sun on the Harvard scale to be $-25^m,93$.

Thus we have:

$C_\odot$ = colour index of the Sun $= -25^m,93 - (-26^m,72) = +0^m,79$, which is in very close agreement with the mean value for G0 stars, $+0^m,72$ according to KING.

83. Skewness and Excess in the Frequency of the Apparent Magnitudes. This problem has been discussed by MALMQUIST[1], who starts from the assumption that the distribution functions of apparent and absolute magnitude $a(m)$ and $F(M)$ have the form of a frequency function of type A. Thus:

$$a(m) = a_0 \varphi(m) + a_3 \varphi^{III}(m) + a_4 \varphi^{IV}(m) + \cdots$$
$$F(M) = f(M) + A_3 f^{III}(M) + A_4 f^{IV}(M) + \cdots,$$

where $a_0, a_3, a_4, \ldots, A_3$ and $A_4, \ldots$ are constants, $\varphi(m)$ and $f(M)$ are Gaussian curves, φ^{III} and f^{III} the third derivatives of φ and f, respectively, and so on. Further:

$$D(r) = D_0 e^{-r^2/2\varrho^2},$$

where D_0 is the density around the Sun, r the distance, and ϱ the dispersion in r.

Using the fundamental equation of stellar statistics CHARLIER[2] has derived the relations:

$$m_0 = M_0 + 5 \log \varrho + 0{,}7922,$$
$$\alpha^2 = \sigma^2 + 1{,}1020,$$

where σ is the dispersion around the mean value M_0 of M and α the dispersion around m_0. MALMQUIST then derives the relation between skewness in the apparent magnitudes ($s = a_3/\alpha^3$), and in the absolute magnitudes ($S = A_3/\sigma^3$), and the relation between excess in the apparent magnitudes ($e = a_4/\alpha^4$) and in the absolute magnitudes ($E = A_4/\sigma^4$) and finds:

$$s = \frac{1}{\alpha^3}(\sigma^3 S + 0{,}5304),$$

$$e = \frac{1}{\alpha^4}(\sigma^4 E + 0{,}2448).$$

Thus the skewness s and excess e in the distribution of the apparent magnitudes are mainly dependent on the values of the corresponding quantities in the absolute magnitudes. The coefficients σ^3/α^3 and σ^4/α^4 will not deviate very much from

[1] Lund Medd Ser. I, No. 96 (1920).
[2] Lund Medd Ser. II, No. 14 (1916).

one, and s and e will then be small quantities which can be neglected in the first approximation. The normal frequency function:

$$a(m) = \frac{a_0}{\alpha\sqrt{2\pi}}\, e^{\frac{-(m-m_0)^2}{2\alpha^2}}$$

thus will generally express the distribution of the apparent magnitudes with sufficient accuracy.

84. The Number of Stars within Certain Limits of Magnitude. The number of stars within a certain magnitude interval dm, or the number $N(m)$ of stars brighter than the magnitude m, plays an important rôle in stellar statistics. In connection with the luminosity law $N(m)$ gives a means of computing the density law $D(r)$. Although the treatment of our knowledge concerning the empirical function $N(m)$ may not entirely fall within this chapter, we are going to treat it shortly because it must be founded entirely on the results of the photometric work. As soon as accurate tables of the course of $N(m)$ according to galactic latitude and longitude or $N(m, \lambda, \beta)$ have been established, many problems in stellar photometry will be reduced to counting the stars of a certain magnitude, from which counts the corresponding magnitude is simply and accurately derived from a mere interpolation in the said table.

Instead of $N(m)$ many times $A(m)$, or the number of stars of magnitude $m: N(m+1) - N(m)$ or $N(m) - N(m-1)$ are used.

The value of B D for deriving $N(m)$, provided its magnitudes could be reduced to an accurate scale, was early recognized. H. v. SEELIGER[1] has undertaken the extensive work of classifying the stars in that catalogue according to the magnitudes. The investigations of PICKERING[2] were used in order to reduce the B D scale to the Harvard one. In 1898 he published an extensive paper in which he derives his well-known laws expressing the distribution of the B D stars.

1. The numbers $A(6{,}0)$ to $A(9{,}0)$ increase much more slowly in proportion to the numerical increase of the magnitude than would be the case if the distribution of these objects was uniform in space and their luminosity the same.

2. The number $A(m)$ increases more and more in proportion to m as we approach the Milky Way.

As KOBOLD[3] had found that the second law was not confirmed in the case of stars brighter than $6^m{,}0$, SEELIGER also has investigated the brighter stars. Irregularities were found in the B D photometric scale, so it was not very well possible to use it without reduction. Instead, the investigation was based on the H S. In accordance with earlier results of KOBOLD it was found that $\log A' = \log A(m) - \log A(m-1)$ *decreases* for bright stars as we approach the Milky Way.

[1] Die Verteilung der Sterne auf der nördlichen Halbkugel nach der Bonner Durchmusterung. München Sitzber d K Akad d Wiss Math-phys Kl 14, p. 521 (1884); Über die Verteilung der Sterne auf der südlichen Halbkugel nach Schönfeld's Durchmusterung. München Sitzber d K Bayer Akad d Wiss Math-phys Kl 16, p. 220 (1886); Die Verteilung der in beiden Bonner Durchmusterungen enthaltenen Sterne am Himmel. München Ann d Sternw 2 C, p. 1 (1888); Über die Größenklassen der teleskopischen Sterne der Bonner Durchmusterungen. München Sitzber d K Bayer Akad d Wiss Math-phys Kl 28, p. 147 (1898); Betrachtungen über die räumliche Verteilung der Fixsterne. München Abh d K Bayer Akad d Wiss II Kl 19, p. 564 (1898); Betrachtungen über die räumliche Verteilung der Fixsterne. München Abh d K Bayer Akad d Wiss Math-phys Kl 25, p. 569 (1909); Zur Verteilung der Fixsterne am Himmel. München Sitzber d K Bayer Akad d Wiss 29, p. 363 (1899); Über die räumliche Verteilung der Sterne im schematischen Sternsystem. München Sitzber d K Bayer Akad d Wiss (1911), p. 413; Über die räumliche und scheinbare Verteilung der Sterne. Sitzber d K Bayer Akad d Wiss (1912), p. 451.

[2] Harv Ann 48, p. 149 (1903). [3] V J S 34, p. 192 (1899).

CHAPMAN and MELOTTE[1] have made extensive star counts on the FRANKLIN-ADAMS plates. The number of stars between 12^m and 17^m was derived on a basis of photometric estimates in 750 fields reduced to the Polar scale (H S). These counts were combined with a number of sources that gave $N(m)$ for the bright stars. The following formulae:

$$\log N(m) = -0{,}18 + 0{,}720\,(m - 11) - 0{,}0160\,(m - 11)^2,$$
$$\log N(m) = +0{,}09 + 0{,}660\,(m - 11) - 0{,}0130\,(m - 11)^2$$

show the limits within which the constants can be varied without the values of $\log N(m)$ coming into conflict with the observations.

Extensive star counts on the FRANKLIN-ADAMS charts and plates have been performed by CHARLIER and his pupils and preliminary results have been published[2]. The final reductions to a photometric scale will be carried out by CHARLIER.

The principal results with regard to the number of stars brighter than a certain magnitude can be found in SEARES's[3] papers and in CHAPMAN's and MELOTTE's papers. An excellent summary of VON SEELIGER's work has been given by G. DEUTSCHLAND[4]. The following synopsis gives a selected number of the most important results recently published by SEARES.

The determination of the spatial density is very restricted as regards its possibilities as long as we use the individual determinations of M. Somewhat more general results can be expected from the photometric data regarding the number of stars brighter than a certain magnitude. On account of the uncertainty about the course of $\varphi(M)$, and because of our complete lack of knowledge in what way this function varies in different parts of the sky all results based on $N(m)$ are subject to great uncertainty.

In the Mount Wilson derivation of the mean distribution of stars, differences between the two galactic hemispheres as well as systematic deviations in longitude were neglected in the first approximation. In Mt Wilson Contr 346, SEARES and MARY C. JOYNER have studied in detail the deviations in $N(m)$ in different parts of the sky from the mean symmetrical distribution. It was found that the principal irregularity is of the form:

$$\log N(m) - \overline{\log N(m)} = \Delta = a + b\cos(\lambda - \lambda_0)\,.$$

The values of a are small; b decreases with increasing galactic latitude, but is approximately the same for equal latitudes north or south. For a certain m the longitude λ_0 is much the same for all latitudes. The theoretical deviations for an eccentric location of the Sun within a stellar system having spheroidal symmetry have the following form, if we neglect higher harmonics:

$$s \pm G + F\cos(\lambda - \lambda_1) \mp k\cos(\lambda - \lambda_2)\,,$$

where s is a systematic correction to the mean distribution tables, F and G are parameters, depending upon the latitude, the spatial density and the coordinates of the centre of the system relative to the Sun.

The upper sign refers to northern, the lower to southern latitudes, λ_1 is the longitude of the centre and λ_2 the longitude of the true galactic pole relative to the adopted pole.

[1] Mem R A S 60, p. 145 (1915).
[2] Lund Medd Ser. II, No. 31 (1923).
[3] Intern Astr Union Transact 2, p. 692 (1925) (together with VAN RHIJN); Mt Wilson Contr 301 (together with P. J. VAN RHIJN, MARY C. JOYNER, and MYRTLE L. RICHMOND) (1925).
[4] V J S 54, p. 25 (1919); Sirius 52, p. 173 (1919).

On the basis of the observed data the positions of the centre and the pole have been derived. These form a sequence which begins with values for the local cluster of bright helium stars and terminates with the centre for the globular clusters and the pole of the Milky Way. The variation of the longitude of the centre is from 244° to 325°. The galactic pole changes from 170° and 83° to 350° and 87°.

The sequences indicate that the bright early B stars are the nucleus of a much larger local system than what is generally meant with this term. The extended local system should have a diameter of at least 20000 light years, and includes stars brighter than 16^{m}. The local cluster stars are responsible for three fourths of the total spatial density around the Sun and for one half of the mean total density in the galactic plane at distances of about 2500 light years.

The attempts to locate the centre of the stellar system by neglecting the local cluster and by supposing that spherical symmetry can be assumed lead to contradictory results. It seems that the influence of the local cluster cannot be ignored. The dominant rôle of the local system prevents the location of the centre of the larger system. The ratios of the number of stars in directions centre (ξ) and anticentre (ζ) have been computed and lead to the ratio of total light in the said directions $B_{\xi}/B_{\zeta} = \frac{3{,}04}{1{,}29}$ which are equal to the ratios of the integrated spatial densities in the same directions.

The method of SCHWARZSCHILD was applied for a derivation of the spatial densities in the same directions. Stars of low M do not occur in a sufficient number in the assumed forms of $\varphi(M)$. The effect is that the densities derived decrease too slowly with increasing distance. Their relative values in different directions should be little affected. The steady decline in $D(r)$ with increasing r does not exclude the existence of large fluctuations, such as would be produced by a number of successive star clouds distributed in the direction of the Sun to centre direction. The distribution in the directions centre and anticentre is very asymmetrical. The curves $D(r)$ may be resolved into a symmetrical or nearly symmetrical curve, representing the larger cluster, and an asymmetrical curve, representing the larger system. The analysis cannot lead to a unique value, but it seems clear that a resolution into two symmetrical systems is impossible.

The density of the larger system increases toward the centre and reaches a maximum value at a distance of 3000 light years or more. This maximum has been formerly identified with the centre of the larger system. The courses of the lines $D(r) = \text{const.}$ agree in indicating that the geometrical centre is at a much greater distance. The asymmetry of the system is rather large, but it does not seem much larger than in the case of the small Magellanic cloud and other anagalactic objects. SEARES is inclined to attribute the asymmetry to an influence of obscuration, analogous to that seen in highly inclined spiral nebulae.

The diameter of the system is 300000 light years, if it may be regarded as co-extensive with the system of globular clusters, and at least 260000 light years if the absolute magnitude of normal B0—B2 stars may be assigned to the faint blue stars in the Milky Way. Since it seems probable that stars in the direction anticentre occur at distances as great as 30000 light years, the excentric stellar system will have an extension of at least 200000 light years.

The Sun is situated almost exactly in the galactic plane as defined by faint stars, but a little to the north of that determined by Cepheids, Novae, faint B stars, and other special classes of objects. The distance of the Sun from the central condensation cannot be determined with any accuracy and SEARES adopts with some hesitation the value of SHAPLEY of 68000 light years. The centre of the local system is only 330 light years distant from the Sun. It scentre is in $\lambda = 230°$, some 150 light years south of the galactic plane.

The agreement of the first harmonics derived from different zones of galactic latitude does not exclude the existence of higher harmonics, which are, of course, involved in the residuals given by the above theoretical formula. The average deviation from spheroidal symmetry for the limiting magnitudes 9^m—18^m averages 25—30 per cent of the number of stars for low latitudes, and decreases to 8—10 per cent for latitudes higher than 40°. The largest negative residuals are associated with obscured areas or dark nebulae.

The results as to the number of stars above a certain limiting magnitude is given in the following table according to a number of modern investigations.

Number of stars above a certain limiting magnitude.

m	Number of stars	Authority	
$6^m,75$	$0,11004 \cdot 10^6$	E. C. PICKERING	Counts in the DRAPER Catalogue.
11 ,0	$1,0133 \cdot 10^6$	HENIE	Counts on the copy of the Harv. Sky Map in possession of the Observatory at Lund.
11 ,0	$1,0752 \cdot 10^6$	NORT	Counts on a copy of the Harv Sky Map in possession of the Utrecht Observatory at reduction of HENIE's counts.
11 ,5	$2,676 \cdot 10^6$	STROOBANT	According to counts in the Astrographic Catalogues.
13 ,5	$9,854 \cdot 10^6$	STROOBANT	According to counts on the Astrographic Chart.
15 ,7	$20, \cdot 10^6$	W. and J. HERSCHEL . . .	Stars counted in the star-gauges of the HERSCHEL's.
15 ,7	$27, \cdot 10^6$	VON SEELIGER	HERSCHEL's stars as reduced by VON SEELIGER.
16	$32, \cdot 10^6$	VAN RHIJN	According to KAPTEYN's counts.
17	$54,9 \cdot 10^6$	CHAPMAN and MELOTTE .	According to counts on the FRANKLIN-ADAMS plates.
17 —18	$64, \cdot 10^6$	GORE	Astronomical Essays
18	$200, \cdot 10^6$	SEARES	According to the Mount Wilson Catalogue of Selected Areas.

Having access to the function $N(m)$ and supposing it to be of the form:

$$\log N(m) = a + bm + cm^2$$

where a, b, and c are constants the corresponding function $A(m)$ can easily be formed. Then the number between the magnitudes $-\infty$ and $+\infty$ can be computed by integration and thus the total number of stars in the sky derived. In the adjoining table the best existing determinations of $N(m)$, including all magnitudes, have been collected.

Estimated total number of stars in the Milky Way system.

Number of stars	Authority
$10 \cdot 10^9$	VON SEELIGER
$3,4 \cdot 10^9$	VAN RHIJN
$2 \cdot 10^9$	CHAPMAN and MELOTTE
$7 \cdot 10^9$	HERTZSPRUNG
$30 \cdot 10^9$	SEARES

85. WILLIAM HERSCHEL's Determination of the Space-penetrating Power. Sir WILLIAM HERSCHEL[1] introduced the idea that all the stars have the same absolute brightness and that their apparent magnitudes only depended upon their distance from us. By a series of experiments he determined the "space-penetrating power" of each of his telescopes and used his results for an equaliz-

[1] Collected Scientific Papers 2, p. 31 (1912).

ation of the star-light. Two telescopes exactly equal in every respect were chosen and placed side by side. Pairs of stars possessing equal brightness were selected by the simultaneous use of both telescopes. By diminishing the aperture of one telescope directed to a bright star, and by keeping the other telescope unchanged and by directing it to a fainter star, the two objects could be equalized in light. From the relative light and the relative size of the apertures, the relative light of this pair of stars could be accurately computed. He thus found that the space-penetrating power of his largest telescope was 61,18, which means that an object appears as bright in this instrument at a distance of 61,18 as with the unaided eye at unit distance.

On the other hand using the same telescope HERSCHEL has counted in the mean 82 stars in a field of 15′ 4″. As there are 1535 stars in unit space brighter than 6^m it follows that the faintest stars seen by HERSCHEL ought to be 22 times as distant as the stars of 6^m. According to the determination of space-penetrating power the said telescopic stars ought to be 61,18 times as distant as the sixth magnitude stars. This contradiction escaped the notice of HERSCHEL and was first pointed out by F. G. W. STRUVE[1]. He thought that the only explanation was a general extinction in space, which affected the gauges. If the coefficient of extinction per magnitude is λ, the apparent magnitude of the faintest stars seen is $\frac{H}{(22)^2}\lambda^{21}$, which is equal to $\frac{H}{(61,18)^2}$. By solving the equation, the value $\lambda = 0,907$ is found. Later on, STRUVE found a considerably smaller value of the extinction corresponding to λ, 0,990651, which corresponds to a change of $0^m,9$ for six magnitudes and which is, of course, impossible in the light of modern data. On the other hand, a general extinction may be present in our stellar system.

86. The Relative Distribution of $N(m)$. The distribution of the values of $N(m)$ on the sphere will give us some idea of the Milky Way structure. It cannot be expected that even for m fainter than 15^m $N(m)$ will give a detailed representation of the Milky Way, because of the influence of nebulae and clusters, and because of the influence of certain optical and physiological phenomena.

The question of the Milky Way structure is treated in another chapter and here we shall only mention that the method of harmonic analysis is suitable when the distribution of the $N(m)$ on the celestial sphere is being discussed.

S. OPPENHEIM[2] has applied the method when discussing the star counts performed by Dr. H. NORT[3] on the Harvard Sky Map.

S. OPPENHEIM has developed the $N(11)$ as determined by NORT by using harmonic analysis, and has found the following general expression:

$$\begin{aligned} N(11) = 21,40\,[&+0,2241\cos\ (\lambda - 307^\circ,9)\\ &+0,1352\cos 2\,(\lambda - \ \ 70,7)\\ &+0,1631\cos 3\,(\lambda - 283,4)\\ &+0,0814\cos 4\,(\lambda - 266,6)\\ &+0,0294\cos 5\,(\lambda - 281,7)\\ &+0,0120\cos 6\,(\lambda - 270,0) + \cdots] \end{aligned}$$

where λ is the galactic longitude.

Then expressions were derived for nine different zones of galactic latitudes.

A further analysis showed that the values $N(11)$ can be represented by means of an ellipsoid with its centre at:

$$A = 15^h,3;\ D = -40^\circ$$

[1] Etudes d'Astronomie Stellaire. St. Pétersbourg (1847).

[2] Sitzber Akad Wiss Wien Math Nat Kl Abt IIa, Bd. 130, Heft 6 (1924).

[3] Recherches astr. de l'Obs. d'Utrecht 7 (1917), also Dissert Utrecht.

which gives the direction of the centre of the stellar (local) system as defined from the objects included in the counts used. One axis of the ellipsoid is directed towards the Milky Way Pole, another coincides with the direction of the vertex of the stellar motions, and the third makes a right angle with the two others.

The ellipsoid has such a form that it seems, in fact, to be composed of two elementary ellipsoids which have two axes coinciding with each other, the third axis making an angle of 90° with each of the others.

OPPENHEIM has also reached the same results when analyzing the stellar motions and their distribution.

87. TUCKER's Definition of an Absolute Scale[1]. The attempts to grade the stars according to their brightness have been based upon the adoption of an arbitrary scale, the basis of which may be said to be the order of magnitudes first given to the brighter stars by the earliest naked-eye observers. The expression for the summation of the number of stars of any order of brightness is usually put in the form:

$$N(m) = a b^m .$$

In this expression, m is the limiting magnitude, a the number of stars for unit radius, and b depends upon the assumption concerning the distribution of stars in space. If the distribution is uniform b will equal to 3,98.

The law expressed by the above formula is, as has been pointed out by R. H. TUCKER, that of the absolute scale of magnitudes. As counts and estimates of the actual numbers of faint stars become available, the value of b can be determined at various points in the scale, if we treat it as a variable quantity, in order to conform to the deviation of fact (real distribution) from theory (uniform distribution) as was originally assumed.

From Harv Ann 48 the value of b has been determined by a process of smoothing out the observed numbers. The value of this constant ranges from 3,34 at 0^m to 2,95 at $13^m,0$. A number of interesting conclusions are derived with regard to the behaviour of different scales; of these we quote the photometric magnitudes of the D M scales, as derived by TUCKER from star counts, supposing the above quoted law gives a correct expression for the dependence between m and $N(m)$.

Magnitude in catalogue	Photometric values of			
	ARGELANDER	Cordoba	SCHÖNFELD	Cape
$7^m,0$	$7^m,06$	$7^m,17$		
8 ,0	8 ,09	8 ,30		
9 ,0	9 ,39	9 ,69	$9^m,34$	$9^m,58$
9 ,1	9 ,54	9 ,86		
9 ,2	9 ,69	10 ,03		
9 ,3	9 ,90	10 ,18		
9 ,4	10 ,09	10 ,34		
9 ,5	(10 ,53)	10 ,71		
9 ,6		10 ,78		
9 ,7		11 ,03		
9 ,8	10 ,5	11 ,26		
9 ,9		11 ,43		
10 ,0		(12 ,02)	10 ,70	
10 ,1				11 ,1
10 ,2				
10 ,3				11 ,5
10 ,4		12 ,0		

[1] Publ A S P 25, p. 144 (1913); Lick Bull 9, p. 162 (1918).

Magnitude in catalogue	Photometric values of ARGELANDER I	II	III	Adopted mean
$7^m,0$	$6^m,9$	$7^m,1$	$7^m,1$	$7^m,0$
8 ,0	7 ,9	8 ,1	8 ,1	8 ,0
9 ,0	9 ,2	9 ,4	9 ,3	9 ,3
9 ,1	9 ,4	9 ,5	9 ,5	9 ,5
9 ,2	9 ,6	9 ,8	9 ,6	9 ,7
9 ,3	9 ,9	10 ,0	9 ,8	9 ,9
9 ,4	10 ,2	10 ,0	10 ,0	10 ,1
(9 ,5)	10 ,6	10 ,7	10 ,3	10 ,5

Magnitude in catalogue	SCHÖNFELD photometric values	Magnitude in catalogue	Photometric values of Cordoba I	II	III	Adopted mean
$7^m,0$	$7^m,0$	$7^m,0$	$7^m,0$	$7^m,0$	$6^m,9$	$7^m,0$
8 ,0	8 ,1	8 ,0	8 ,1	8 ,1	7 ,9	8 ,0
9 ,0	9 ,3	9 ,0	9 ,7	9 ,5	9 ,1	9 ,6
9 ,1	9 ,6	9 ,1	9 ,9	9 ,7	9 ,2	9 ,8
9 ,2	9 ,8	9 ,2	10 ,2	9 ,9	9 ,3	10 ,0
9 ,3	10 ,0	9 ,3	10 ,4	10 ,1	9 ,5	10 ,2
9 ,4	10 ,2	9 ,4	10 ,6	10 ,3	9 ,7	10 ,4
9 ,5	10 ,5	9 ,5	10 ,9	10 ,5	9 ,9	10 ,7
9 ,6	10 ,6	9 ,6	11 ,1	10 ,8	10 ,1	10 ,9
9 ,7	10 ,7	9 ,7	11 ,4	11 ,0	10 ,3	11 ,2
9 ,8	10 ,9	9 ,8	11 ,6	11 ,2	10 ,6	11 ,4
9 ,9	11 ,0	9 ,9	11 ,8	11 ,5	11 ,0	11 ,6
10 ,0	11 ,1	(10 ,0)	12 ,0	11 ,9	11 ,6	12 ,0

88. KAPTEYN's Method. A scheme has been worked out by J. C. KAPTEYN which shows how stars of different magnitudes are distributed in space according to their luminosity. Let i_m be the apparent brightness of a star of magnitude m and take $i_0 = 1$. Then $i_{m+1}/i_m = 1/\delta = \frac{1}{2{,}512}$ and $i_m = \delta^{-m}$.

An (infinite) number of spheres are imagined around the Sun as the centre in such a way that the ratio of the radii of two successive spheres is $r_{j+1}/r_j = \delta^{\frac{1}{2}}$. A star which is situated successively on the surfaces of the spheres changes its apparent brightness by one magnitude for each change.

From the table given below the apparent magnitude of a star of determined M, that is successively placed in the middle of the various shells can be read. The spherical shell, the mean parallax of which is 0″,1, is called a_0, and the density is d_0. We suppose that there are c stars of $M = -5{,}5$, d of $M = -4{,}5$, e of $M = -3{,}5$, &c. If in shell a_1 the density is d_1, the number of stars with $M = -4{,}5$ is cd_1, etc. The quantity ε is equal to the ratio of the volumes of two succesive spheres. As $\varepsilon = \left(\frac{r_{j+1}}{r_j}\right)^3$ and r are selected so that $\log(r+1) - \log r = 0{,}2$, $\log \varepsilon$ will be 0,6 and $\varepsilon = 3{,}98$. The diagonals in the table will according to our definitions mark the lines for which $M =$ constant.

Shell	m / π	−7,5	−6,5	−5,5	−4,5	−3,5	
a_{-2}	0″,251	$\varepsilon^{-2}cd_{-2}$	ε^{-2}—dd_{-2}	$\varepsilon^{-2}ed_{-2}$	$\varepsilon^{-2}fd_{-2}$	$\varepsilon^{-2}gd_{-2}$	$M = -1{,}5$
a_{-1}	0 ,158		$\varepsilon^{-1}cd_{-1}$	$\varepsilon^{-1}dd_{-1}$	$\varepsilon^{-1}ed_{-1}$	$\varepsilon^{-1}fd_{-2}$	$M = -2{,}5$
a_0	0 ,100			cd_0	dd_0	ed_0	$M = -3{,}5$
a_1	0 ,063				εcd_1	εdd_1	$M = -4{,}5$
a_2	0 ,040					$\varepsilon^2 cd_2$	$M = -5{,}5$
a_3	0 ,025						

In order to apply this method we may take the simple case of density being constant. Then all d's are equal in the scheme, and $N(m+1)/N(m) = \varepsilon = \delta^{\frac{3}{2}}$; and therefore $N(m) = C\delta^{\frac{3}{2}m} = C i_m^{-\frac{3}{2}}$, and $A(m) = c\,\delta^{\frac{3}{2}m}$; and also $\pi_m/\pi_{m+1} = \delta^{\frac{1}{2}}$ as was found by SCHIAPARELLI. A number of other assumptions with regard to the form of the density law or luminosity law can be made. The advantage of the method is that no assumption has to be made with regard to the mathematical form of the density-function, which is always a drawback. Further the use of the scheme of KAPTEYN is to be preferred to using certain solutions of the fundamental integral-equation in stellar statistics, because the former treatment pays some attention to the fact that the numbers in successive shells do not correspond to variables but to variates (the frequencies are not of a continuous nature but given as discontinuous numbers), a circumstance that has often been overlooked in work within this branch, but which is of certain importance. The equations in stellar statistics do not express the relations between the finite numbers in space, but are built on the supposition that the characteristics of stars are continuous functions. Future work concerning the theory of stellar statistics will certainly involve the development of fundamental relations based on the theory of discontinuous functions.

89. The Limits of Unaided Vision. The 6^m stars are generally considered as the boundary between stars visible and invisible to the unaided eye. Nevertheless, fainter stars can be seen, especially when the observer knows the direction in which the object lies and when the light of the sky is screened down.

In order to find the limit of unaided vision during such favourable circumstances, H. D. CURTIS has made some experiments at the Lick Observatory[1]. Two screens were attached to the 12-inch telescope and with the declination set the instrument was swept slowly in A R.

The faintest stars seen by the aid of this method are of the Harvard magnitude $8^m,2$.

To this may be added the fact that the present writer is able to see with his unaided eye the object M33, having a total magnitude of $6^m,8$. His eyes are astigmatic but in other respects normal.

P. MEESTERS[2] has compared the Harvard charts of variable stars with the sky. He finds the limit of his unaided vision to be $6^m,76$ for white stars and $6^m,91$ for yellow stars. It seems that MEESTERS has even keener eyesight than HEIS had.

90. Shadows Cast by Starlight. The question if the planets and stars can cast shadows is of a certain general interest from the point of view of stellar photometry. It is well known that Venus and Jupiter cast distinct shadows, but it does not seem to have been recognized that, provided proper precautions are taken to exclude extraneous light, the shadows cast by a number of bright stars can be seen without difficulty. H. N. RUSSELL has investigated this question[3]. A darkened room with an open window is used and the aperture of the window is screened down to an area of a square foot or less. The image of the opening will then appear on a large piece of paper held in the path of the star rays as soon as the eyes of the observer have become rested from the glare of the ordinary lights. The shadow of a convenient object may then be made to fall on the screen. By this method RUSSELL has succeeded in distinguishing the shadows of 29 stars, the faintest of which is 15 Argus = $2^m,9$. The Pleiades group and the sword of Orion also cast perceptible shadows.

---

[1] Lick Bull 2, p. 67 (1901).

[2] De in Nederland met het bloote oog zichtbare Sterren. Amsterdam (1921).

[3] Pop Astr 10, p. 242 (1900).

It follows from photometric considerations that the illumination of the paper by Sirius is about 1/40 as bright, area for area, as the general background of the sky, outside of the Milky Way. No perceptible colour is shown by this faint light.

The shadow cast by Sirius has been photographed by E. TOUCHET[1].

91. Length of Exposures for Reaching a Certain Magnitude. It is very difficult to give any general rules for the proper exposure time in order to reach a certain magnitude with a certain instrument. The data in the following table are very rough and only intended to give the reader a general idea of the order of magnitude of the exposure time necessary to obtain traces of a certain magnitude. When measurable images are wanted, the exposures have to be doubled at least. The numbers refer to high-speed commercial plates and the sensibility in the HD system ought to equal or exceed 300. Among developers which "take out" very faint impressions on a plate we should like to mention that of Ross and those having agents such as hydroquinone, pyro, and metol in proper combinations.

Exposure time in minutes.

Limiting magn.	Harv Cook 1-inch lense	Upsala Twin 6-inch telescope	Upsala 13-inch refractor	Lick Crossley reflector	Mount Wilson	
					60-inch reflector	100-inch reflector
10^m						
11	6 min.	1 min.				
12	20	2				
13	60	6				
14	200	20	1 min.			
15	600	63	4	1 min.		
16		190	12	2	1 min.	
17		(630)	40	6	2	1 min.
18			130	20	6	2
19			(400)	63	20	6
20				200	63	20
21				630	200	63
22					630	200
23						630

For exposures around 200 minutes the faintest stars that can be reached with the instrument are generally recorded. The gain when a longer exposure is taken is generally small. The limiting magnitudes for the above instruments are roughly: $14^m,0$, $16^m,6$, $18^m,8$, $20^m,8$, $22^m,2$, and $23^m,2$, respectively. It is generally considered that finely grained plates such as Seed 23 are not able to reach as faint objects, *ceteris paribus*, as the more coarsely grained plates. This is true to a certain extent, but F. E. Ross has found that the limit of efficiency is reached much later for Seed 23 than for the faster plates. If a very long exposure is taken with the former the gain is still considerable and thus it is possible to reach as faint, if not fainter, magnitudes with the slower plates as with the faster.

92. Accuracy Attainable with Photographic Methods. We cannot treat this question in any detail. Of the many investigations concerning the subject we mention that of A. HNATEK[2], who finds that in the most favourable cases the minimum error will be 3 to 4 per cent, or, expressed in magnitudes, $\pm 0^m,03$ to $\pm 0^m,04$. In the case of lantern slides the error is slightly lower. Another investigation in the same direction is that of Mademoiselle CLAVIER[3] concerning

[1] B S A F 17, p. 446 (1903).
[2] Z f wiss Photogr 16, p. 323 (1916); 18, p. 198 (1919).
[3] B A Mem et Var 3, p. 341 (1923).

the influences of uniform errors on the photographic plates. A number of surfaces were impressed on the plates in such a way that the opacity law could be derived and surfaces corresponding to equal light could be measured on different parts of the plate. The mean error of ordinary plates was found to be $\pm 0^m,02$ or $\pm 0^m,03$. In the case of old plates, which are not sensitized and have a rather strong chemical veil round their borders, their error is somewhat larger and varies between $\pm 0^m,02$ and $\pm 0^m,04$. In the case of sensitized plates the error is considerably larger and varies between $\pm 0^m,03$ and $\pm 0^m,07$. These mean errors correspond to the effects in the plates themselves, and give a limiting value of the highest accuracy attainable when photographic plates are measured.

With regard to the measurement of diameters not much seems to be known, but the evidence is against the conclusion that the error of the plate itself affects the diameter or astrogamma[1] quite as much as the density.

A photometer of SCHILT's type permits the measurement of the density or of the integrated light on the plate with such accuracy that the maximum accuracy can be reached. But there also have to be added the errors arising from the imperfectly known relation between the density or diameter and the magnitudes, further the errors due to imperfect guiding, the errors arising from variable atmospheric conditions during exposure, the errors due to imperfect knowledge of the reduction to the centre, etc. It seems that the average mean error can be estimated as nearly $\pm 0^m,10$ for one determination. The minimum external mean error attainable scarcely seems to be below $\pm 0^m,06$, which is about the same accuracy as can be reached with visual methods (Potsdam). Many astronomers believe in the superiority of the photographic methods. As far as accuracy is concerned there is certainly no superiority, but the photographic methods have certain advantages which are not present in the visual methods.

93. Secular Changes in the Light of the Stars. Much has been written about this interesting question, but I do not think a review would be of much value, on account of the uncertain conclusions reached. I only wish to mention the investigations of ZINNER and a small contribution of my own.

ZINNER has used the reduced magnitudes in his catalogue and computes the yearly changes Δm_i in the magnitude according to the formula:

$$\Delta m_i = \frac{m_0 - m_i}{t_i}; \qquad \varepsilon_i = \frac{\sqrt{\varepsilon_0^2 + \varepsilon_i^2}}{t_i},$$

$$\Delta m = \frac{\Sigma \Delta m_i p_i}{\Sigma p_i}; \qquad \varepsilon = \sqrt{\Sigma (\varepsilon_i p_i)^2}.$$

where m_0 is the magnitude in the catalogue of ZINNER and m_i the magnitude in a certain catalogue, ε and ε_i are the corresponding mean errors, t_i is the epochal difference with 1894 as zero-point, and p_i is the weight.

For 668 stars observed by W. and J. HERSCHEL, ZÖLLNER, and SEIDEL it was possible to derive a value of Δm_i. Nine of these values refer to bright objects, or their change is difficult to establish for other reasons. Of the rest, 435 had to be excluded as they had a mean error equal to or larger than Δm_i. There are thus 224 stars left, which vary between the limits $0^m,0007$ and $0^m,0059$ per year and have mean errors between $0^m,0003$ and $0^m,0024$ with an average value of $\pm 0^m,0013$.

---

[1] The astrogamma is introduced by ROSS and defined as Γ in the equation:

$$d = a + \Gamma \log i$$

where d is the diameter and i the brightness. The astrogamma is thus equal to $2,5/b$ in SCHEINER's equation (see ciph. 52).

The distribution as regards the spectral classes is as follows:

Spectral class	Positive	Negative	Mean change in magnitude
B	31	35	$-0^m,00023 \pm 0^m,00027$
A	11	23	$-0\ ,00077 \pm 0\ ,00031$
F	20	11	$+0\ ,00075 \pm 0\ ,00040$
G	16	6	$+0\ ,00101 \pm 0\ ,00050$
K	25	35	$-0\ ,00027 \pm 0\ ,00025$
M + N	4	5	$-0\ ,00046 \pm 0\ ,00071$
All	107	115	$-0\ ,00007 \pm 0\ ,00035$

Thus the B and K stars do not vary outside the limits of their mean errors. The A, F, and G stars seem to be liable to some slight variation but, of course, it is difficult to decide whether the computed changes should be considered as real. This determination depends on the modern measurements, but the change is confirmed by the oldest catalogues. The estimates are not accurate enough in individual cases, and, besides, the older observers were certainly not free from bias.

In order to test the large changes derived from Sûfi, Zinner has selected 20 stars in order to see whether the observations from the 16th and 17th Centuries confirmed the changes or not. For 10 stars these observations were not available. In eight cases the old observations (mainly Kirch's) did not confirm the changes derived from Sûfi. There are only two cases, α Ursae Minoris and α Ophiuchi, where it is possible that the change can be real.

Finally, the mean changes were derived from Ptolemaios's as well as from Brahe's data.

Authority	Positive changes Δm	Positive changes $\varepsilon_{\Delta m}$	n	Negative changes Δm	Negative changes $\varepsilon_{\Delta m}$	n
Al Sûfi	$+0^m,00070$	$\pm 0^m,00003$	16	$-0^m,00071$	$\pm 0^m,00002$	53
Ptolemaios	$+0\ ,00042$	$\pm 0\ ,00003$	15	$-0\ ,00036$	$\pm 0\ ,00007$	47
Brahe	$+0\ ,00144$	$\pm 0\ ,00042$	13	$-0\ ,00153$	$\pm 0\ ,00036$	46
19th Century	$+0\ ,00066$	$\pm 0\ ,00035$	16	$-0\ ,00050$	$\pm 0\ ,00050$	53

The progressive change from Ptolemaios to Brahe is certainly due to the fact that Brahe and Al Sûfi have not established their scale independently of Ptolemaios.

It is clear that the changes must be small. It might seem premature to look for such changes, but we may remember that we know of many cases of very rapid light-variation with high amplitude. There is nothing to prevent also having to deal with changes of long periods as well as with real secular changes. It will certainly be one of the most important tasks of future photometry to analyse the observations for changes of this kind.

The method employed by Zinner is applicable to a possible general increase or decrease in the starlight. If the A stars systematically change their light relatively to the K stars, this only means a change in the colour equation.

From an investigation of my own[1], the following mean systematic difference between some of the old catalogues and Harvard was formed:

	Old magnitudes — Harv.
Ptolemaios	$-0^m,18$
Al Sûfi	$+0\ ,02$
Tycho Brahe.	$+0\ ,18$
Tycho Brahe.	$+0\ ,04$

[1] V J S 61, p. 230 (1926).

The moderate size of the systematic difference may only show that the scales agree with regard to the zero-point. But even then it seems possible to use the systematic differences for an investigation of a possible secular change. We suppose we had a sufficient number of star catalogues, and also had reason to assume that their systematic errors had a Gaussian distribution, then it ought to be possible to find a general change in m. The above differences have been represented by the formula:

$$\Delta m = +0^m{,}1272 + 0^m{,}000146\,(t - 1900)$$
$$\pm 0\;{,}146 \quad \pm 0\;{,}000112$$

where t is the epoch and the mean errors are indicated below the constants.

This implies that the light of the stars decreases with an amount of the order of 1^m per 10000 years. Certainly the real change is smaller, but the above value may be regarded as an upper limit. A secular change can originate from two causes. The absolute magnitude may undergo changes, or the motions of the stars may be such that the apparent magnitude changes from time to time. A third cause may arise if we have to pass such a region of space that part of the starlight has gradually to undergo local extinction or scattering. As regards a general change in absolute magnitude we do not know anything at present. With regard to the effect of our motion through space the largest change should arise if the K-effect in the radial velocities of the giants were real, as most of our objects are giants. It is easy to show that the decrease in the magnitudes, due to such a systematic motion, does not suffice to explain the above change. The third alternative is possible, although not very plausible.

Anyhow, it seems that the stock of photometric observations in our present possession shows that the secular change of the starlight in general, as well as the changes in individual cases, must be smaller than 1^m during 10000 years. When the observations have been extended for several thousand years in the future, we shall have in the photometric catalogues a treasury of data concerning the nature of the variation in the starlight which will be invaluable.

The changes based on AL SÛFI magnitudes and on those of the 19th Century show a correlation:

AL SÛFI	19th Century	n
$-0^m{,}0005$	$+0^m{,}0002$	29
$-0\;{,}0005$	$-0\;{,}0007$	54
$-0\;{,}0004$	$-0\;{,}0011$	16
$+0\;{,}0005$	$+0\;{,}0007$	16
$+0\;{,}0007$	$+0\;{,}0007$	16
$+0\;{,}0002$	$+0\;{,}0012$	10

I have computed the correlation from 120 individual cases and find:

$$r = 0{,}298 \pm 0{,}083\,.$$

It seems that there is a real correlation, but that the SÛFI change is much smaller than the modern change. On account of the smallness of r it would be premature to draw any conclusion until more material has been collected.

When discussing the results of the numerous measurements of the photometric standards used at the Harvard Observatory, PICKERING also touched upon the question of a secular change[1]. The observations in question covered an interval of 23 years and were compared with the corresponding ones of PTOLEMAIOS, Al SÛFI, and WILLIAM HERSCHEL. It was found that the values of the variations were insignificant. The changes suggested by the Harvard observations did not

[1] Harv Ann 64, p. 217 and ff. (1912).

obtain any support from the oldest observations. PICKERING said that he was going to take up a discussion of the secular variation as derived on the basis of the catalogue in the Almagest. It is likely that his lamented death prevented him from carrying through this plan.

94. The Total Magnitude of Star Agglomerations. It is possible to estimate the total magnitudes of most of the agglomerations of stars, such as open and globular clusters and anagalactic nebulae.

Remarks concerning the magnitude of the individual objects are to be found scattered through the literature. The first systematic series estimates of nebulae and clusters was HOLETSCHEK's. By using the unaided eye or the smallest possible instrumental means in order to make the objects starlike or at least as concentrated as possible, he estimated the total magnitudes of 676 objects. These observations have been published[1] at length and contain many valuable observational details.

Other series of a similar kind have been performed by STONE[2], KRITZINGER[3], E. C. PICKERING[4], and by KOBOLD and WIRTZ[5].

HOPMANN[6] has measured photometrically a number of the objects in the list of HOLETSCHEK, which can thus be reduced to a photometric system. The following relation between the scales has been found:

HOLETSCHEK	HOPMANN	n	HOLETSCHEK	HOPMANN	n
$7^m,64$	$8^m,10$	7	$9^m,16$	$9^m,91$	12
8 ,30	8 ,49	7	9 ,37	10 ,51	9
8 ,64	9 ,03	8	9 ,55	10 ,77	11
8 ,84	9 ,12	7	9 ,74	10 ,56	8
9 ,00	9 ,63	7	10 ,30	11 ,44	6

The mean errors in the estimates of HOLETSCHEK and HOPMANN are:

Total magnitudes	ε_m	Total magnitudes	ε_m
$8^m,80$ — $8^m,99$	$\pm 0^m,27$	<11,00	$\pm 0^m,25$
9 ,00 — 9 ,99	±0 ,30	All	±0 ,28
10 ,00 — 10 ,99	±0 ,30		

In 1924 Miss ADELAIDE AMES of the Harvard Observatory published some preliminary results of her estimates of the total magnitudes of nebulae seen on the Harvard plates[7]. During a stay at the Greenwich Observatory the present writer found that the FRANKLIN-ADAMS plates could be used advantageously for estimates of the total photographic magnitudes of nebulae and clusters. Time did not permit me to go over all the plates available for the purpose. The results for a few selected regions giving the total photographic magnitudes of 215 objects, have so far been published[8].

In order to obtain photometric standards several regions of the Selected Areas were identified. Then a number of starlike nebulae were compared with the stars. These nebulae were used as a kind of secondary standard in order to obtain the magnitudes of the nebular objects with a more appreciably extended surface. As a rule the estimates were in rather good agreement. In some cases uncertainties are involved because of the comparison of objects situated in different parts of the plates.

---

1 Wien Ann d K K Sternw 20, p. 39 (1907).
2 Leander McCormick Publ 1, p. 175 (1893).
3 Sirius 48, p. 111 (1915).
4 Harv Ann 33, p. 135 (1900).
5 Strassburg Ann 4, p. 79 (1911).
6 A N 214, p. 425 (1921).
7 Harv Circ 257 (1923); 294 (1926).
8 Ark Mat Astr Fys 20 A, No. 13 (1927); Upsala Medd No. 21.

SHAPLEY and Miss AMES have published estimates of altogether 103 objects[1].

Recently estimates of 2775 objects have been performed at the Harvard Observatory by Miss AMES[2]. From this work we have overtaken the relation between angular diameter and total magnitude as given in Fig. 86.

If we compare common objects and derive the relation-line between different observers and then square and add the residuals, the following mean errors are found:

	$\varepsilon(m)$
SHAPLEY and AMES	$\pm 0^m,15$
LUNDMARK	$\pm 0\ ,29$
HOLETSCHEK	$\pm 0\ ,34$
Strassburg Observers	$\pm 0\ ,50$

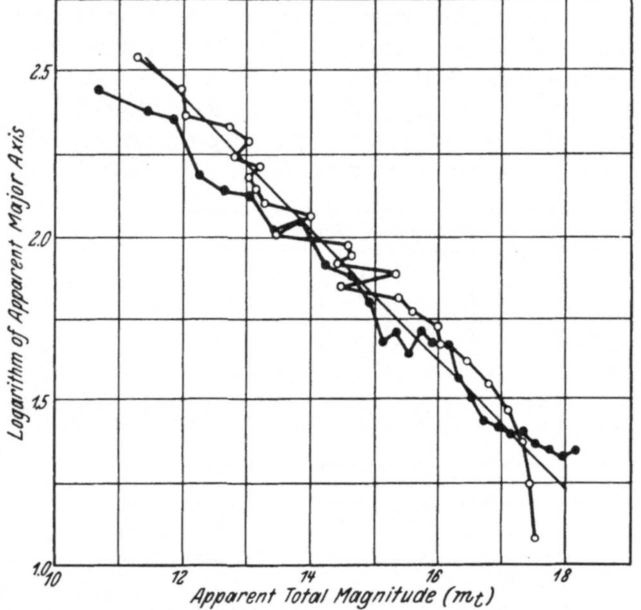

Fig. 86. Relation between the total magnitudes of 2775 galaxies and the logarithms of the apparent major axes. The fact that the relation is very nearly linear speaks against the existence of an appreciable general absorption in space.

The first value may be too low. The lower accuracy of the Strassburg magnitudes is explained from the fact that the estimates are by-products in the measurements of the coordinates of the nebulae.

Reducing my values to the system of SHAPLEY and Miss AMES and reducing HOLETSCHEK's observations to the Harvard magnitudes by applying the corrections to the BD stars as derived by PICKERING and others, we find:

$$m_{\text{Lundm.}} - m_{\text{Hol.}} = +0^m,6$$

which agrees well with our present conception of the mean colour index of anagalactic nebulae.

We give here the relation between the different earlier observers as it has been determined by WIRTZ[3] (Abbreviations: Hol. = HOLETSCHEK; Str. = Strassburg; L. Mc. = Leander McCormick; Kr. = KRITZINGER).

$$m_{\text{Hol.}} = -6{,}60 + 1{,}613\ m\,,$$

$$m_{\text{Hol.}} = 10{,}011 + 1{,}102\,(m_{\text{Str.}} - 10) - 0{,}233\,(m_{\text{Str.}} - 10)^2\,,$$

$$m_{\text{Str.}} = -0{,}69 + 1{,}103\ m_{\text{Hol.}}\,,$$

$$m_{\text{L. Mc.}} = -14{,}57 + 2{,}450\ m_{\text{Str.}}\,,$$

$$m_{\text{Str.}} = -13{,}91 + 1{,}89\ m_{\text{L. Mc.}}\,,$$

$$m_{\text{Hol.}} = 1{,}94 + 0{,}808\ m_{\text{Kr.}}\,,$$

$$m_{\text{Kr.}} = 2{,}05 + 0{,}763\ m_{\text{Hol.}}\,,$$

$$m_{\text{Hol.}} = 1{,}75 + 0{,}831\ m_{\text{Pi.}}\,,$$

$$m_{\text{Pi.}} = 2{,}04 + 0{,}747\ m_{\text{Hol.}}\,.$$

[1] Harv Circ 294 (1926). [2] Harv Bull 864 (1922). [3] A N 204 p. 190 (1917).

The following table gives the reduction of the different series to each other:

Str.	Hol.	Diff.	Kritz.	Diff.	Pick.	Diff.
7	$6^m,5$		$6^m,0$		$6^m,2$	
		$1^m,1$		$1^m,2$		$1^m,0$
8	7 ,6		7 ,2		7 ,2	
		1 ,1		1 ,2		1 ,0
9	8 ,7		8 ,4		8 ,2	
		1 ,1		1 ,2		1 ,1
10	9 ,8		9 ,6		9 ,3	
		1 ,1		1 ,2		1 ,1
11	10 ,9		10 ,8		10 ,4	
		1 ,1		1 ,2		1 ,1
12	12 ,0		12 ,0		11 ,5	

95. The Convergence of the Sum of Stellar Light. From a scrutiny of the empirical data it seems that the total amount of stellar light has a finite value. This will be caused by three circumstances or by a combination of two of them, viz.

1. the stars are finite in number,
2. there is an absorption or extinction in space,
3. the stars are infinite in number, but distributed in such a way that the sum of their light becomes finite.

We know that the Milky Way system is limited in space and we know that the probable number of its members is $3 \cdot 10^{10}$. Although the luminosity law is not very well known, it is nevertheless possible to compute with a fair degree of accuracy the total absolute magnitude, M_{tot}, of our stellar system, because of the fact that we know the frequency of the absolutely brighter stars fairly well which contribute most to the sum of the magnitudes. The result is:

$$M_{tot} = -16^M,5 .$$

It is difficult to estimate the mean error but it should not exceed $\pm 1^M$.

It seems to be fairly well established that the anagalactic nebulae are stellar systems of the same kind as our Galaxy. Novae and Cepheids have been observed in five cases in sufficient number to give a fair parallax. In some 25 cases the brightest ordinary stars of the Galaxy in question have been observed, and their apparent magnitudes roughly determined. It is a fair assumption that the mean apparent magnitude of, say, the twenty brightest stars corresponds to the mean absolute magnitude of the twenty absolutely brightest stars in our stellar system. This quantity is taken to be $-6^M,8$ and thus we have:

$$M_{tot} = m_{tot} - \bar{m}_{max} - 6{,}8 ,$$

where m_{tot} is given by observations and $\bar{m}_{max}$, the mean apparent magnitude of the twenty brightest stars, is also found from observations.

Further a few M_{tot} can be derived from the data of double anagalactic nebulae. About 200 pairs have been found that are apparently nearer each other than would be the case with a haphazard distribution in space. In the case of physically connected nebulae we have:

$$M^A_{tot} - M^B_{tot} = m^A_{tot} - m^B_{tot}$$

or

$$\Delta M_{tot} = \Delta m_{tot} ,$$

where A and B stand for the primary and the companion respectively.

The absolute magnitudes of some 30 galaxies have thus been derived without making any assumptions on the constancy of M_{tot} or the absolute dimensions:

Elliptical objects	$-13^M,4$
Spirals	$-17\ ,5$
Magellanic Clouds	$-15\ ,3$

The dispersion averages $\pm 1^M,12$.

The total apparent magnitude can be used, within the different subgroups, as a fair measure of the total absolute magnitude.

The problem that concerns us here is the total amount of the light of the anagalactic objects. As far as our present data go, it seems that the galaxies form a system of a higher order, the metagalactic system, which has a finite extension in space. It seems that the number of galaxies should not exceed 10^7. If all the galaxies are assumed to have an absolute total magnitude of $-16^M,5$ (which is probably too high), the total absolute magnitude of the metagalactic system will be

$$M_{\text{total}}^{\text{metagal.}} = -34^M$$

thus about 600 times brighter than the Sun appears in the sky. The modest value of the total magnitude of the metagalactic system might be taken to suggest the presence of a general absorption. As has been shown by SHAPLEY and the present writer, absorption in the metagalactic space must be very small[1].

Now it is quite possible that the stars are infinite in number, but distributed in such a way that the sum of their light has a finite value. The stars would have to be arranged, as has been shown by CHARLIER[2] and J. SELETY[3], in cluster formations, and certain criteria as to the relation between the dimensions of two clusters of successive orders and the number of members in a cluster of a certain order would have to be fulfilled.

The derivation of the sum of the total light is easy if we assume that N_1 stars build up a spherical galaxy of the first order, that N_2 galaxies of the first order form a galaxy of the second order, and so on to infinity.

We have:

$$M_i = N_i M_{i-1} = M_0 \prod_{i=1}^{\infty} N_i .$$

The observer is supposed to be situated at the common centre of galaxies of different orders.

The sum of the light of a spherical galaxy is found by the application of the well-known theorem in Celestial Mechanics of the sum of attraction of n points forming a sphere:

$$\text{Total light} = \sum_{i=1}^{\infty} \frac{h_i}{R_i^2} .$$

The series converges if:

$$\frac{h_i}{R_i^2} > \frac{h_{i-1}}{R_{i-1}^2} \quad \text{or} \quad \frac{h_i}{h_{i-1}} > \frac{R_i^2}{R_{i-1}^2} > N_i .$$

Thus the convergence criterion is:

$$\frac{R_i}{R_{i-1}} > N_i^{\frac{1}{2}} .$$

If the special theory of relativity holds true, it can be shown[4] that:

$$N_i^{\frac{1}{2}} > \frac{R_i^{\frac{1}{2}}}{R_{i-1}^{\frac{1}{2}}} .$$

Thus the complete criterion is:

$$N_i > \frac{R_i}{R_{i-1}} > N_i^{\frac{1}{2}} .$$

As far as our present data go the criterion is fulfilled. This does not prove that the Universe is built up infinitely, but it shows that this is quite possible.

[1] For a thourough account of modern results as to the absorption in space see the paper of C. SCHALÉN: Zur Frage einer allgemeinen Absorption des Lichtes im Weltraum. A N 236, p. 249 (1929); Upsala Medd No. 45.

[2] Lund Medd Ser. I, No. 98 (1922).

[3] Ann d Phys 4. Folge, 68, p. 281 (1922); 72, p. 58 (1923).

[4] LUNDMARK, Studies of Anagalactic Nebulae, First Paper, Upsala Medd No. 30 (1927).

b) Stellar Colours[1].

b_1) Direct Estimates of Colour.

96. HAGEN's Discussion of Scales. The colour of a star can only be found as a relative quantity and it has to be referred to some standard system in somewhat the same way as the classifications of spectra are referred to the system of the New HENRY DRAPER Catalogue or the new Mount Wilson system.

Many difficulties will certainly be met with when attempts are made to establish a colour sequence by verbal descriptions or by assigning to the different objects numerical values in accordance with a certain scale. Still, the difficulties should be overcome and a colour sequence should be established.

I have a notion that astronomical opinion nowadays is not in favour of visual observations. The enthusiasm of the early days of colour equivalents, when it was thought possible "to measure the colour by objective methods", has perhaps faded somewhat. A more sceptical conception of the colour problem certainly prevails now. But there is still a religious faith in photographic methods. In cases of colour estimates there is no doubt that an experienced observer, using a large reflector, can more rapidly reach results as accurate as when using photography. The estimates by K. GRAFF of colour in a number of sequences undoubtedly show the correctness of this statement. The reply will be made: "But we have the photographs and can always remeasure them". — Yes, they can be remeasured, but this will very seldom happen in practice, since the impressions on the plates can be measured with such an accuracy that nothing can be gained by remeasuring the plate-material again. The main trouble is to establish the real connection between the measured quantities and the photometric quantities to be derived.

I think that for the determination of colours visual methods are of much value for another reason: they afford more possibilities for contributions from skilful amateurs than the photographic methods do. On the other hand modern astronomy concerns itself with measurable quantities and thus it will also be of much importance to determine colour equivalents.

When direct estimates of colours are being made, three elements have to be distiguished, namely brightness, tone, and shade. The brightness or its logarithmic equivalent, the magnitude, has been dealt with separately, but enters as a parameter into the determination of colours.

In order to obtain a clear conception of the nomenclature we shall make use of a very important paper by Father J. G. HAGEN[2] dealing with the different scales for estimates of colour.

In most dictionaries the terms tone, hue, shade, and tint are given as synonyms. To remove this ambiguity HAGEN has used tone and shade in a technical sense as the qualitative and quantitative elements respectively. The quality of colour depends on the limits of the corresponding wave lengths in the visual spectrum, and the quantity on its mixture with the two auxiliary "colours", white and black. The colours of the spectrum are often compared with the gamut in music. If then the spectrum is a chromatic gamut, its colours are the tones. Thus the quantity or shade can be called the nuance. Some writers use the expressions tone and nuance (shade) in exactly the reverse sense.

---

[1] The intimate relationship between the magnitude and the effects of the colours, such as the colour equation or the colour index, has made it quite impossible to treat the magnitudes and the colours independently. Several results concerning the colours of the stars have already been mentioned and anticipated. Some other results will be given in connection with the treatment of stellar diameters or stellar masses.

[2] Ap J 34, p. 261 (1911).

In the measurements of colours the two elements, tone and shade, must be referred to certain scales. The tone scale consists of a number of fundamental colours, but here we meet very different opinions. In the art of painting, red, yellow, and blue are generally considered the primary colours. To these green was added by LEONARDO DA VINCI, but later on it has been considered a mixed colour. NEWTON distinguished in his Opticks between seven colours, but his distinction between blue and indigo has not been kept. Whether blue or violet is really a primary colour has been very much discussed. Experiments on colour blindness led HELMHOLTZ and MAXWELL to select violet and to disregard blue.

HAGEN points out that astronomers must base the colour scales on the colours actually shown by the stars, when they are observed through our atmosphere. That was also the point of view taken in ARGELANDER's famous paper[1], "Aufforderung an Freunde der Astronomie".

NEWTON[2] represented the shade-scale graphically by arranging the colours along the circumference of a circle, whose centre represented white. He called the shade of the colour the saturation, and its quantity is thus proportional to the length of the radius. Black is not contained in the scale, although NEWTON also considered the mixture of colours with black as a quantitive element of colour.

From NEWTON's mode of representing the colours it follows that the solar spectrum is the normal one, with maximum shade or saturation. The colours diminish with regard to shade as we go towards white and towards black. The paler colours are thus the under-saturated and the darker colours the over-saturated.

LAMBERT[3] has pointed out the lack of the pole for black in NEWTON's shade-scale and has supplied it by means of a cone erected over the circle as base. The vertex represented black.

HAGEN represents the two-dimensional scale of colours by means of a sphere, which is certainly the most natural way. The poles represent white and black and the spectral colours lie around the equator. Thus the half meridian circles are loci for constant tone, and the parallel circles loci for constant shade. Each diameter designates two complementary colours that make grey together, and thus white and black also are complementary.

The next question is to find proper symbols for the intervals in the two-dimensional scale. Four kinds of symbols have been used in this connection, viz., technical words, coloured diagrams, numbers, and letters.

With regard to words, they can be used for defining the tones by reference to the solar spectrum or to other natural objects of a certain colour. When words are used for defining the shades, the difficulties will be many on account of the vagueness of every attempt to describe the degree of black and white in a certain colour.

The first diagram of chromatic tones and shades was published in 1686 by WALLER. Artistic representations of colours were given by CHEVREUL[4] for industrial purposes. This work gives 10 "cercles chromatiques" and 13 colour scales and a solar spectrum. A simplification of CHEVREUL's system has been adopted

---

[1] Jahrbuch für 1844 herausgegeben von H. C. Schumacher (1844).

[2] Opticks etc. London (1704).

[3] Beschreibung einer mit Calau'schem Wachse ausgemalten Farbenpyramide etc. Augsburg (1772).

[4] Exposé d'un moyen de définir et de nommer les couleurs d'après une méthode précise et expérimentale, avec l'application de ce moyen à la définition et à la dénomination des couleurs d'un grand nombre de corps naturels et de produits artificiels. Mém de l'Acad des Sciences et de l'Inst Imp 33, with Atlas: Cercles Chromatiques (1861).

in P. KLINCKSIECK's and TH. VALETTE's work: Code des Couleurs, Paris 1908, a system that seems to be sufficient for most practical purposes. The translation of different names for colours in the French language is also of value.

HAGEN expresses the opinion that the diagrams of LACOUTURE[1] are unsurpassed in beauty and correctness. I think this opinion can be sustained.

W. OSTWALD[2] has published a extensive atlas giving the colours. This work has not been accessible to me.

W. H. SMYTH[3] has published, in 1864, a diagram specially designed for star colours, which gives 6 different tones and 4 degrees of shading. Another diagram was published by W. H. PICKERING[4], which gives altogether 12 tones and shades of blue, yellow, and red.

A general drawback with such diagrams, when they are used for estimating star colours, is that they have to be used in artificial light. Care must then be taken to use the same kind of articifial light when the diagram is used in practice. On the other hand, they are certainly of value, especially to the inexperienced observer, as an intermediary or reference scale.

Numbers were first used by H. J. KLEIN[5] to represent the intervals in a chromatic scale with the explicit purpose of imitating ARGELANDER's magnitude scale. He dealt only with strongly coloured stars and divided the interval, yellow to deep fiery red, into five steps designated from 0 to 5.

J. SCHMIDT has extended the scale to embrace all the stars and practically kept the steps of KLEIN, but moved the numbers back towards white and ended with 10 = red. SAFARIK seems to have used twenty degrees on this scale and KÖHL used a scale containing 11 steps from 1^c = deep red to 12^c = white. INNES introduced a scale beginning with 1 = red and through passing over 8 = white to 14 = reddish violet.

Very little uniformity prevails in the numerical designation of shades. CHEVREUL has used twenty degrees from white to black, symmetrically distributed with regard to the normal curve. LACOUTURE has reduced the ten grades on each side to six. P. SMYTH has four degrees from deep saturation to white, and puts the numbers in the form of exponents over the words designating colours and the diagrams, and the same notation has been followed by ESPIN. FRANKS and BACKHOUSE have used the exponents in an opposite direction, so that 1 is white-tinged and 4 the deepest tints; KRÜGER has used only three exponents but in the same direction as FRANKS.

Letters were made use of first by H. C. VOGEL. He started from the three primary or simple colours:

W(eiß)	G(elb)	R(ot)
W(hite)	G(Yellow)	R(ed)

and by means of combinations and permutations interpolated four mixed colours between them:

W, GW, WG, G, RG, GR, R.

White was thus considered one of the principal sidereal colours. As has been pointed out by HAGEN, one of the intervals, W—G or G—R, is too large, because it is not clear whether G means yellow or orange. VOGEL did not use symbols for the quantitative elements.

---

[1] Répertoire Chromatique. Solution raisonnée et pratique des problèmes les plus usuels dans l'étude et l'emploi des couleurs. Paris (1890).

[2] Der Farbenatlas, Ausgabe A (in 13 Kästen), Ausgabe B (auf 103 Tafeln). Leipzig u. Jena (1918 ff.).

[3] Sideral Chromatics. London 1864.

[4] Pop Astr 25, p. 419 (1917).

[5] A N 70, p. 105 (1868).

In his extensive estimates of star colours FRANKS[1] used the following colours:

R, O, Y, G, B, P, W, L.

The first five are the principal colours of the spectrum and the three last letters stand for purple, white and lilac respectively. Combinations of letters are avoided, but the shading of the colours is expressed in three degress, Capital Roman, small Roman and small Italic types, the order being from strong to faint saturation.

Later on FRANKS modified his scale, the six colours were kept, and white designated by O as in NEWTON's chromatic circle. This necessitated the expression of orange by or. The designation P was changed into V. The shadings were to be indicated by numerical exponents according to the practice of SMYTH.

KRÜGER interpolated the colour symbol O between VOGEL's G and R.

A designation for black is met for the first time in LACOUTURE's work. White and black are used to indicate the shadings. Mixed pale or dark colours are thus expressed by three letter symbols.

In English B cannot be used for black and HAGEN recommends the letter S as the initial of sombre (sub-umbra), which is also in accordance with the French sombre, the German schwarz, and the Italian scuro.

HAGEN then compares the various scales; they can be divided into one- and two-dimensional scales. The former mainly use numerical and the latter literal symbols.

The two-dimensional scales (NEWTON, SMYTH, FRANKS, and BACKHOUSE) choose the six fundamental colours: Red, orange, yellow, green, blue, violet (or purple), and four degrees of shading, in which the arrangement of tones and shades differs in the case of SMYTH from that of FRANKS and BACKHOUSE.

It might seem at first that the symbol for darkness, N, is not quite necessary in the case of astronomical colours. But HAGEN points out that for variable stars in particular the darker shadings should be contained in the scale. Thus the chromatic sphere seems to give a better two-dimensional scale than the circle or the cone.

The linear scale seems to originate from ARGELANDER. He suggested the use of the scale:

purpur, rot, orange, gelb, weiß, bläulich,

with two or three gradations between them. White replaces the green in the spectral sequence. The scale is empirical and intends to represent what ARGELANDER actually saw in the stars. HAGEN assumes that by purpur ARGELANDER meant the colour of the variable stars i. e. when they decline to their minimum brightness, or a darker shade of red. Bläulich cannot be a fundamental colour either. It is the faintly bluish shade which white stars sometimes appear to have.

SCHMIDT translated the scale of ARGELANDER into numerical symbols.

0^c =	White	6^c =	Orange
1 =		7 =	
2 =		8 =	
3 =		9 =	
4 =	Yellow	10 =	Red
5 =			

[1] Harv Ann 14 (1884).

The four primary colours of ARGELANDER have been adopted by most observers in Europe and America. They have been used by SECCHI, by DUNÉR, by SAFARIK (10—20 subdivisions), by CHANDLER and YENDELL, and by KRÜGER and OSTHOFF.

The scale of OSTHOFF[1], as it has been used extensively, is given in detail:

- 0^c = White
- 1 = Yellow-whitish (white dominates)
- 2 = White yellow (white and yellow in equal proportions)
- 3 = Bright or pale yellow
- 4 = Pure yellow
- 5 = Dark yellow
- 6 = Reddish yellow (yellow dominates)
- 7 = Red yellow (red and yellow in equal proportions)
- 8 = Yellowish red (red dominates)
- 9 = Red with minute proportions of yellow
- 10 = Red.

The relation between the scale of Father HAGEN and the scales of other observers is seen from the adjoined table.

It was possible for FR. KRÜGER[2] to compare 3162 stars, which had been observed on the two-dimensional scale by FRANKS as well as on the one-dimensional scale by KRÜGER and OSTHOFF. Means were apparently taken of the observations of the latter without any reduction to a common scale. (See table on top of page 368).

Later on FRANKS and HAGEN[3] made a comparison using 4067 stars whereby the mean values changed but little.

In order to examine the effect of saturation, HAGEN has collected the letter symbols that nearly corre-

	BW −1	W 0	YW 1	WY 2	Y 3	OY 4	YO 5	O 6	RO 7	OR 8	R 9	SR 10
SCHMIDT		weiß				reines helles Gelb		intensiv goldgelb				rot ohne gelb
SAFARIK		rein weiß		gelblich weiß	strohgelb	messing-gelb schwefel-gelb	zitronen-gelb	goldgelb orange	rötlich orange	feuer-farbig	blutrot	karminrot tiefstes Rot
CHANDLER		white	yellow white	yellow	yellow orange	orange						intense red
KRÜGER		rein weiß	bläulich weißgelb	gelblich weiß	gelblich	rein gelb	strohgelb	orange	goldgelb	rötlich	kupferrot	rein rot
OSTHOFF	bläulich-weiß	weiß	gelblich-weiß	weißgelb	blaßgelb	rein gelb	dunkel-gelb	rötlich-gelb	rotgelb	gelblich rot	rot mit Spur von gelb	rot
INNES		white	yellowish	yellow	deep yellow	orange yellow	orange	orange red	reddish	red	very red	deep red

[1] A N 153, p. 141 (1900); 192, p. 85 (1912).
[2] Specola Astronomica Vaticana IX, p. XXI (1917).
[3] Ibid. XV, p. X (1923).

Yellowish colours			Orange colours			Peculiar colours		
FRANKS	KRÜGER and OSTHOFF	n	FRANKS	KRÜGER and OSTHOFF	n	FRANKS	KRÜGER and OSTHOFF	n
W	$1^{c},99$	553	$\overline{OY}^1$	$4^{c},23$	4	B^1	$2^{c},01$	7
Y^1	3 ,09	799	$\overline{OY}^2$	5 ,86	78	G^1	2 ,39	13
Y^2	4 ,96	653	$\overline{OY}^3$	6 ,42	72	$\overline{YG}^1$	2 ,82	55
Y^3	5 ,88	493	O^1	3 ,30	15	$\overline{YG}^2$	3 ,57	3
			O^2	6 ,35	253	R^1	3 ,40	6
			O^3	6 ,70	137	$\overline{OR}^1$	3 ,35	2
			$\overline{OR}^3$	7 ,71	17	V^2	1 ,50	2

spond to the same number and combined them into mean values. In this way the following table is obtained.

KRÜGER and OSTHOFF . .	$2^{c},0$	$3^{c},2$	$4^{c},6$	$5^{c},9$	$6^{c},4$	$6^{c},7$	$7^{c},7$
Normal colours	W	Y^1	Y^2	$\overline{OY}^2$	O^2		
Shifted colours	—	O^1	$\overline{OY}^1$	Y^3	$\overline{OY}^3$	O^3	$\overline{OR}^3$

As symbol O of FRANKS and the corresponding symbol of the Potsdam scale are nearly in agreement as regards their numerical value, this means that FRANKS and Potsdam nearly agree with regard to their zero-point.

The normal colours of FRANKS have the index 3. This shows that the four degrees of shading introduced by SMITH have only a theoretical value.

The shifted colours are displaced towards white. The fact that the shifted symbols contain orange, which is a tone and not a shading, shows that what the two-dimensional scale expresses as a shading is a tone in the one-dimensional scale. The symbols in the lowest line of the above table have certainly only a theoretical meaning, like the exponent 4, and the symbols in the vertical rows in the table are practically identical and therefore of little value.

The range of the letter symbols is far larger than the range of the normal colours in the table.

As regards the peculiar colours, they are situated outside the numerical scale. This fact might seem to be in favour of a letter scale, but KRÜGER and OSTHOFF maintain that these peculiar colours are only seen in double stars.

The letters may be of a certain advantage in connection with the peculiar type, and there is nothing to prevent us keeping the letter symbols in such a case.

It must not be forgotten that when a two-dimensional scale is being used it is not allowable to take the means without any reduction. In this case the colours can be represented by vectors. Half the sum of two vectors is, as is well known, a third vector drawn to the middle point of the line which joins their extremities.

The arithmetical operations are obvious in the linear scales, but it should be noted that the numbers to be operated on must not differ by many units. For if such is the case the results are affected by some uncertainty, because we do not know if the scale is linear from a physical point of view. It is a mere assumption that the intervals are really comparable, and it is likely that the intervals when reduced to the same physical unit, e. g., colour index or temperature, would present several deviations from the average.

Independently of the Aufforderung by ARGELANDER B. SESTINI undertook to observe the colours of the stars[1]. His observations in Rome during the years 1844—1846, originally published in different volumes of Memorie del Collegio Romano, have been re-edited by HAGEN[2] who has also appended modern observations of colour to those by SESTINI. LINDEMANN has used SESTINI's

[1] Mem del Collegio Romano (1843).

[2] Specola Astronomica Vaticana III (1911).

description of the colours. VOGEL has omitted one of the colours between W and R and has reduced SCHMIDT's scale from 10 to 6 steps, in order not to have too many subdivisions. During the course of the Potsdam Durchmusterung it was found that the subdivisions were insufficient. Thus each interval was divided into three, which gave a scale of 18 degrees between W and R with an average value of about two thirds of SCHMIDT's degrees:

White	W−	W	W+
Yellowish-White	GW−	GW	GW+
Whitish-Yellow	WG−	WG	WG+
Yellow	G−	G	G+
Redish-Yellow	RG−	RG	RG+
Red	R−	R	R+

HAGEN says that if some mathematical correlation could be established between the literal and the numerical symbols, the result would be not only a comparison between the scales of one and two dimensions, but also an easy transition from one to the other without any ambiguity. He has made the correlation by changing the four irregularly distributed numerals of SCHMIDT, viz. 0, 4, 6, and 10, into the arithmetical progression:

$$0^c = \text{white},\ 3^c = \text{yellow},\ 6^c = \text{orange},\ 9^c = \text{red}.$$

Between these four cardinal points the combinations and permutations of the four letters will exactly fill the numerical intervals as is shown in the following table:

White			Yellow			Orange			Red		
BW	W	YW	WY	Y	OY	YO	O	RO	OR	R	SR
−1	0	1	2	3	4	5	6	7	8	9	10

The two limiting shades, −1 and 10, are presumed to correspond to ARGELANDER's bläulich = bluish-white, and purpur = sombre-red. As the scale divides the interval W−R into 9 steps it may be called in accordance with HAGEN's proposal the chromatic scale of nine grades.

In order to discriminate between SCHMIDT's and OSTHOFF's steps, generally designated 1^c—10^c, and those of HAGEN, it seems appropriate to give the denomination 1^H to the intervals of the chromatic scale.

The main difference between the two scales is that the latter puts the yellow a degree nearer the white. The difference in red is of very little practical importance on account of the scarcity of stars that ought to have been assigned the symbol R. In VOGEL's scale one of the two middle sections of the table is cut out. CHANDLER has orange = 4. The scale of SAFARIK agrees pretty well with that of HAGEN:

3 = straw-yellow, 6 = orange, 9 = blood-red.

Between KLEIN's numbers and HAGEN's there is the relation:

$$n^H - n^K = 5\,.$$

HAGEN then suggests a similar change in the two-dimensional scale. He reduces the four degrees of saturation to three in order to coordinate them with the three shadings,

YW, WY, Y.

For this change the chromatic sphere is used and the matter is illustrated by a large diagram printed in two colours, representing the linear scale within the two-dimensional scale, and given in Vol. III (facing p. XXVIII) of the Publications of the Specola Vaticana.

A numerical evaluation by ARMELLINI-CONTI[1] of the four principal colour scales has given:

HAGEN	OSTHOFF	KRÜGER	SESTINI	HAGEN	OSTHOFF	KRÜGER	SESTINI
1^H	1^c,7			5^H	4^c,8	5^c,4	5^c,3
2	2 ,1	1^c,8	2^c,2	6	6 ,0	5 ,5	5 ,7
3	3 ,0	3 ,1	2 ,2	7	6 ,8	6 ,6	7 ,1
4	3 ,9	5 ,1	4 ,1	8	7 ,6	7 ,7	7 ,6

97. KRÜGER. The first catalogue of KRÜGER[2] published in 1893 contains the colours of 2153 + 144 stars. The work was taken up again in Altenburg in 1898,

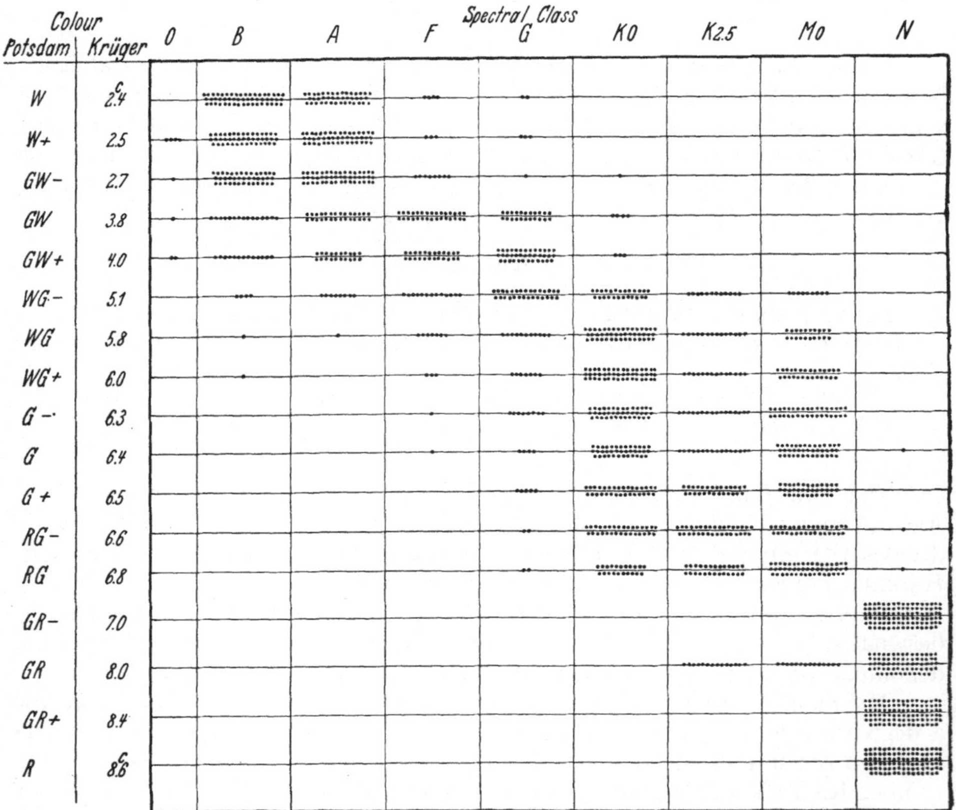

Fig. 87. Relation between spectral class and star colour according to estimates at Potsdam and by KRÜGER.

a refractor of 17,2 cm (aperture $f/17$) being used, which was moved to Aarhus in Denmark in 1910 and used there until the death of KRÜGER.

The second catalogue[3] contains the objects in the first edition together with more than 3000 other stars. The total number of objects is 5915. As a rule coloured objects were observed, but in order to obtain the relation to other scales, also a number of white stars have been included.

The magnitudes for stars fainter than $7^m,0$ were also estimated. The colours are expressed in the 9-grade scale of HAGEN. At least three different observations form the foundation for the catalogue colours.

[1] G. ARMELLINI-CONTI, Sul confronto dei catalogi colorimetrici di HAGEN, SESTINI, OSTHOFF e KRÜGER. Rend R Acc Naz dei Lincei 5, p. 70 (Roma 1927).

[2] Kiel Publ d Sternw VIII (1893).

[3] Specola Astronomica Vaticana VII (1914).

The connection between the spectral classes, the Potsdam colours, and the colours of OSTHOFF is illustrated in fig. 87.

The phenomenon that the colour increases for the fainter apparent magnitudes within the different spectral classes is very clearly present in the following summary:

Magn.	B	*n*	A	*n*	F	*n*	G	*n*	K	*n*	K2—K5	*n*	M	*n*
1^m	$1^c,5$	5	$2^c,0$	5	$3^c,0$	1	$3^c,4$	1	$4^c,6$	1	$6^c,5$	1		
2	2,2	9	2,4	9	3,7	4	4,8	1	5,5	6	6,0	4	$7^c,2$	1
3	2,5	20	2,8	20	3,9	15	4,6	17	5,5	46	6,3	7	6,8	7
4	2,6	76	2,9	81	4,1	38	5,1	49	5,7	113	6,5	20	7,1	26
5	2,4	86	2,7	78	4,3	42	5,2	60	5,7	239	6,4	77	7,1	92
6	2,5	5	2,6	24	4,6	20	5,2	54	5,8	312	6,5	123	7,2	190
7	2,5	1	2,8	19	4,4	30	5,2	56	5,8	294	6,5	171	7,2	465
8			3,0	12	4,6	60	5,3	14	6,0	88	6,7	67	7,4	1378
9											6,9	6	7,6	106
	2,4	202	2,8	248	4,3	156	5,1	252	5,7	1099	6,4	476	7,2	2265

The general increase in colour per magnitude $\frac{\Delta c}{\Delta m}$ is about $0^c,1$ within each spectral class.

The Index Catalogue[1] is a catalogue mainly worked out by KRÜGER, giving an index of the stars for which colours have been determined by SESTINI, KRÜGER, and OSTHOFF. Besides, the colours according to KRÜGER's estimates are given. The number of estimated stars is altogether $5915 + 2316 = 8231$. The estimates by FRANKS are also given. KRÜGER himself had planned to use the catalogue for a number of statistical researches. Certainly there is much more work to be done regarding the visual colours and their relation to other elements of stellar astronomy.

98. OSTHOFF and WIRTZ. Sources of Error. We are now going to consider a number of sources of error influencing the direct estimates of colours.

Colour sensitiveness as a function of time during the same night[2]. The sensitivity curve has been derived by OSTHOFF from a large amount of material. It shows that it takes a certain time before the eye is adapted for colour observations. This time, which varies on an average between 30^m and 45^m, differs from observer to observer.

As a foundation for the curve only bright stars were used, and for faint stars the conditions are somewhat different.

Adaptation to darkness having once been reached, a short interruption does no matter. Gazing at the faintly illuminated chart affects colour by a mean amount of only $0^c,1$.

The best adaptation for colours is after a restful night. In order to illustrate the effect we quote the following observations of mean colour by OSTHOFF.

Condition of observer	Light from street lanterns		Lights in windows of near houses	
	near	distant	yellow	red light
Immediately after leaving bed . .	$2^c,96$	$3^c,42$	$4^c,80$	$7^c,17$
$10^m - 15^m$ later	1,51	2,54	4,33	5,33
Half an hour later.	1,67	2,33	4,00	5,00

Change in sensitiveness from one day to another. This varies from day to day between limits that can be estimated as $0^c,5$ for an experienced observer. The change is largest for red stars, smaller for yellow, and has not been verified for white stars. There seems to be a general correlation between the change for yellow and red stars.

[1] Specola Astronomica Vaticana IX (1917). [2] A N 153, p. 141 (1900); 205, p. 1 (1917).

Fatigue. If the observer becomes tired for a short time it does not seem to have any influence on colour observations. When the observer is fatigued from the beginning of the observations, the colours are noted as uncommonly reinforced. There are very few investigations with regard to this interesting matter.

Influence of moonshine. The light of the moon introduces false colours. OSTHOFF finds the influence to be $-0^c{,}14$ and $-0^c{,}16$, respectively, for his two instruments. The change on account of moonshine is, of course, dependent on the angular distance from the moon. When colour observations are under discussion care should be taken to eliminate the influence of a bright background.

Influence of atmospheric conditions. When the air is hazy or cloudy the colour increases in the numerical scale. OSTHOFF computes the influence to be of the order $+0^c{,}3$ and $+0^c{,}4$ for the two instruments employed. Among different kinds of clouds the cirri increase the colour most. Disturbances arise also from smoke from the chimneys of factories.

The unsteadiness of the air does not seem to have any influence on the estimated colours. Also in cases of strong scintillation it is possible to determine the main colour of an object with considerable accuracy.

Influence of altitude. An increase in the numerical scale is shown, when the altitude decreases. Of course, observations below an altitude of 20° should not be made at all.

Contrast. In the case of close doubles there is a considerable effect from the nearness of the two objects. As soon as the angular distance increases, there seems to be little or no influence of the colour of one object, if the two objects are estimated in quick succession. OSTHOFF has investigated close doubles and finds the following results:

Difference in magn.	Increase in colour for the red component	Decrease in colour for the white component	n
White comp ftr than red, Δm to $-0^m{,}6$	$+0^c{,}56$	$-0^c{,}05$	11
Both comp equal, Δm from $-0^m{,}5$ to $+0^m{,}4$ . . .	+0 ,45	−0 ,38	13
White comp brtr than red, Δm from $+0^m{,}5$ . . .	+0 ,55	−0 ,75	5

An influence of contrast taken in a wide sense is noted when a bright star and a faint star of the same colour are observed in succession. In such a case the eye is not sensible to the colour of the faint star.

The influence of brightness on the colour. When the light decreases the colour increases. The numerical value can be found from observations of the same objects by means of instruments of different optical power. OSTHOFF has found in this way:

Colour	Difference per 1^m	
	Finder minus 4 inch	Refractor minus 4 inch
White	$-0^c{,}18$	$+0^c{,}50$
Yellow	+0 ,25	+0 ,32
Orange	+0 ,27	+0 ,20

A very thorough review of the catalogue of OSTHOFF has been made by WIRTZ[1]. Because of the completeness of the catalogue with regard to the northern sky — it can safely be assumed to contain all stars brighter than $5^m{,}5$ — it is a valuable source for some general considerations.

[1] V J S 54, p. 2 (1919).

We give the correlation-surface of OSTHOFF's colours with regard to the magnitude.

Apparent magnitude \ Colour C	$0^c,5$	$1^c,5$	$2^c,5$	$3^c,5$	$4^c,5$	$5^c,5$	$6^c,5$	$7^c,5$	$8^c,5$	Σ
$0^m,0 - 0^m,4$		1		1	1					3
0 ,5 − 0 ,9			2				1			3
1 ,0 − 1 ,4		1	1		1		1			4
1 ,5 − 1 ,9		5		1						6
2 ,0 − 2 ,4		5	4	2	2	2	3			18
2 ,5 − 2 ,9		2	4	4	2	6	5			23
3 ,0 − 3 ,4		2	14	7	5	10	4			42
3 ,5 − 3 ,9			30	18	14	18	12	1		93
4 ,0 − 4 ,4		1	34	36	20	28	14	2		135
4 ,5 − 4 ,9		7	78	51	31	44	51	19		281
5 ,0 − 5 ,4	3	64	146	51	55	66	70	39	2	496
	3	88	313	171	131	174	161	61	2	1104

The deficiency of stars between the degrees 4^c and 5^c is evident and the cause, as pointed out by SCHWARZSCHILD, is that these stars are on an average absolutely faint.

Next WIRTZ divides the celestial sphere, with regard to the Galaxy, into zones, each of 20° in β, and computes the relative densities D or numbers n of stars in each zone, taking the number in the galactic zone as unity.

Galactic zone	$0^c,0$—$2^c,9$		$3^c,0$—$4^c,9$		$5^c,0$—$6^c,9$		$7^c,0$—$8^c,9$	
	D	n	D	n	D	n	D	n
+70° to +90°	0,36	20	0,62	19	0,62	20	0,00	0
+50 +70	0,46	51	0,70	42	0,70	45	0,72	8
+30 +50	0,35	51	0,61	49	0,53	45	1,58	23
+10 +30	0,53	76	0,53	41	0,85	70	0,70	10
−10 +10	1,00	142	1,00	78	1,00	82	1,00	14
−30 −10	0,85	102	1,02	67	0,85	59	1,18	14
−50 −30	0,42	35	0,44	20	0,58	28	0,60	5
−70 −50	0,21	6	0,34	5	0,71	11	1,50	4

The table shows that the concentration in the Milky Way becomes smaller when the colour increases, but the number of objects is rather small, which compels one to be careful before arriving at conclusions. WIRTZ forms the ratio: numbers in zones +30° to +90° to numbers in zones 0° to 30° and finds:

Colour group	$0^c,0-2^c,9$	$3^c,0-4^c,9$	$5^c,0-6^c,9$	$7^c,0-8^c,9$
Galactic concentration . .	2,59	1,72	1,65	1,05

The red stars are concentrated but little, if in fact at all, towards the Galaxy.

Next the B stars have been subtracted, which resulted in the following distribution:

Galactic zone	Proportion of stars			
	$0^c,0$—$2^c,9$	n	$3^c,0$—$4^c,9$	n
+70° to +90°	0,47	20	0,70	19
+50 +70	0,57	48	0,76	41
+30 +50	0,44	49	0,67	48
+10 +30	0,63	69	0,58	40
−10 +10	1,00	108	1,00	69
−30 −10	0,72	66	1,01	59
−50 −30	0,40	25	0,49	20
−70 −50	0,24	6	0,38	5
Galactic concentration	2,20		1,55	

WIRTZ compares the colour estimates of OSTHOFF and those he made with the 49 cm refractor at Strassburg in connection with double star work. The comparison is also of interest as in the case of WIRTZ the data have been obtained with the aid of a large refractor.

Colour observed by OSTHOFF and by WIRTZ.

OSTHOFF	WIRTZ	n	WIRTZ	OSTHOFF	n
0c,38	1c,05	4	0c,37	2c,03	13
1 ,40	1 ,84	17	1 ,42	2 ,27	22
2 ,43	1 ,94	37	2 ,35	3 ,41	53
3 ,41	2 ,41	24	3 ,31	4 ,13	32
4 ,37	2 ,76	23	4 ,36	5 ,17	13
5 ,38	3 ,64	25	5 ,11	6 ,10	9
6 ,59	4 ,13	12	6 ,75	6 ,90	2
7 ,15	6 ,25	2			

The relation curve is derived graphically and is expressed by means of the following small table:

OSTHOFF	Relation line (WIRTZ)	OSTHOFF	Relation line (WIRTZ)
0c	0c,6	5c	3c,7
1	1 ,1	6	4 ,6
2	1 ,6	7	6 ,2
3	2 ,2	8	8 ,0
4	2 ,9		

WIRTZ's observations are a by-product in his observations of double stars and are, naturally enough, not of the same accuracy as the estimates of OSTHOFF.

With regard to the B stars it is of interest to test whether the colours afford support for the conclusion reached by CHARLIER that the class B0 is later than B1. The mean colours are:

Spectral class	Mean colour	n	Spectral class	Mean colour	n
Oe5	2c,9	5	B3	1c,9	24
B0	2 ,1	81	B5	2 ,4	18
B1	2 ,8	8	B8	2 ,3	19
B2	2 ,2	16	B9	4 ,0	3

It is an interesting fact that the B3 stars are the whitest of all stars.

Then comparisons are made between colours and the effective wave lengths:

OSTHOFF ($\bar{c}$) and BERGSTRAND and HERTZSPRUNG.

BERGSTRAND						HERTZSPRUNG					
$\bar{c}$	$\lambda_{\text{eff.}}$	n	$\lambda_{\text{eff.}}$	$\bar{c}$	n	$\bar{c}$	$\lambda_{\text{eff.}}$	n	$\lambda_{\text{eff.}}$	$\bar{c}$	n
0c,5	420,9	4	417,8	1c,7	5	1c,3	415,5	5	413,9	2c,4	8
1 ,1	418,9	2	422,5	1 ,3	5	2 ,6	417,8	13	418,5	2 ,8	17
2 ,0	421,6	2	432,4	5 ,2	3	3 ,2	418,3	9	423,3	3 ,9	9
3 ,4	422,3	3	438,4	5 ,9	3	4 ,4	427,3	7	430,5	5 ,4	6
5 ,4	438,1	4	445,9	6 ,6	6	5 ,4	427,9	6	437,7	6 ,3	3
6 ,7	445,7	9	451,6	6 ,8	5	6 ,4	439,0	6	445,0	5 ,9	3
7 ,1	450,7	3	465,5	8 ,4	2						
8 ,4	465,6	2									

The relation curves are expressed in the following smoothed values:

Colour	$\lambda_{\text{eff.}}$		Colour	$\lambda_{\text{eff.}}$	
$\bar{c}$	BERGSTRAND	HERTZSPRUNG	$\bar{c}$	BERGSTRAND	HERTZSPRUNG
$0^c,5$	4190		$4^c,5$	4310	4260
1 ,0	4200	4140	5 ,0	4340	4290
1 ,5	4210	4150	5 ,5	4370	4320
2 ,0	4220	4160	6 ,0	4410	4360
2 ,5	4230	4170	6 ,5	4450	4400
3 ,0	4240	4190	7 ,0	4500	
3 ,5	4260	4210	7 ,5	4550	
4 ,0	4290	4240	8 ,0	4600	
4 ,5	4310	4260	8 ,5	4660	

The comparison of different instruments is not an appropriate method for deriving $\frac{\Delta c}{\Delta m}$ on account of the factors that give rise to the colour equation of the instrument. Two instruments of equal construction but different power would give better results, but also in this case there are a number of difficulties. Another method is to use rotating sectors for diminishing the light of the observed objects in a known proportion. The observations of OSTHOFF gave the value of $\frac{\Delta c}{\Delta m}$ as $+0^c,7$ for stars between 1^m and 2^m, but the value is uncertain.

Experiments have also been made by OSTHOFF with the aid of an artificial red star. He finds from them $\frac{\Delta c}{\Delta m} = 0^c,55$. No differences between white and dark backgrounds could be found.

More definite experiments have given the values:

Epoch	White	Orange
1901	$+0^c,86$	$+1^c,20$
1902—1903	+0 ,52	0 ,70

The value of $\frac{\Delta c}{\Delta m}$ is certainly variable with the time.

The variable stars change their colours in the same direction with decreasing magnitude. It cannot be said without special investigation in what proportion this is due to changes in the atmosphere of the stars and to changes in the eye of the observer. Anyhow, it is an important phenomenon, which deserves careful study.

99. Changes in the Colour of the Stars. Much has been written about supposed changes in the colours of the stars. In many instances the observations are due to observers who do not seem to have had the necessary experience to be able to make the delicate and difficult observations of colour changes. The most accurate observations of such changes are undoubtedly those of OSTHOFF[1].

It is well known that the Cepheids and Long-period Variables as well as the Novae show changes in their colour. It is more difficult to say whether also ordinary stars suddenly change their colour. The question cannot very well be settled, because in those cases where real changes of colour have been observed it is quite possible that the light has also undergone changes.

100. LAU. During the years 1914—16 LAU[2] determined the visual colours of 744 stars with the aid of a refractor of 95 mm aperture and with a magnification of 170. The usual precautions were taken and LAU also wrote down his notes

[1] AN 153, p. 241 (1900). See also: Specola Astronomica Vaticana VIII, p. XXVIII (1916).
[2] AN 205, p. 49 (1917).

in the dark, and he surveyed a given region alternately from east to west and vice versa. The stars were observed in the meridian or at a constant altitude, in order to avoid the effect of the atmospheric dispersion.

LAU also estimated the proportions of the three fundamental colours, white, yellow, and red. If the proportions of white and yellow were equal, the notation 1 W 1 G was written down, if the yellow appeared to be twice as great as the white 1 W 2 G, and so on. In the reduction the following scale was used:

$$\begin{aligned} \text{White} &= 0^c \\ \text{Yellow} &= 5^c \\ \text{Red} &= 10^c \end{aligned}$$

The author remarks that no other colours except those that appear in the cooling of metals are to be seen among the stars. The stars called "flushed white" by FRANKS were found to be yellowish-white. Mixed colours with white and red as components were not observed at all.

From the stars observed three times the following mean error of one observation has been derived:

Spectral class	Apparent magnitude		
	1^m—3^m	3^m—4^m	4^m—5^m
B	$\pm 0^c,71$	$\pm 0^c,57$	$\pm 0^c,79$
A	0 ,54	0 ,72	0 ,71
F	0 ,61	0 ,53	0 ,53
G	0 ,20	0 ,53	0,37
K	0 ,43	0 ,30	0 ,37
M	0 ,15	0 ,25	0 ,43

The errors do not vary sensibly with the apparent magnitude, but they depend on the spectral class. The B and A stars have on an average a mean error of $\pm 0^c,74$ and the F and G stars $\pm 0^c,66$ and $\pm 0^c,41$ respectively. Lastly the K and M stars can be estimated with such accuracy that the same quantity only amounts to $\pm 0^c,37$.

On account of the slightly different scale employed, when comparing the above errors with those of OSTHOFF, they must be multiplied by 0,8 for whitish yellow stars, and by 1,2 for red stars. The mean error of all stars and classes is thus reduced to $0^c,44$, which agrees very well with the values of OSTHOFF.

Although the atmospheric dispersion was avoided, some observations were made in order to determine its influence. If we denote with C_z the apparent colour, and with C_0 the colour at the zenith, and with a_z the absorption in magnitudes, the equation:

$$C_z = C_0 + 0^c,028\, a_z$$

is found.

The increasing "deepness" or depth of the colour with decreasing brightness was at first investigated by the same means as those used by OSTHOFF. Better results were found when a grating absorbing three magnitudes was employed. No certain dependence on the spectral class could be traced. The observations gave the formula:

$$C = C_0 + 0{,}90\, m - 0{,}07\, m^2.$$

OSTHOFF's observations gave:

$$C = C_0 + 0{,}84\, m - 0{,}07\, m^2.$$

The deepness for different magnitudes can be computed from the expression $0^c,49 + 0^c,42\, m$ in the two cases.

The relation between magnitude and colour for each class in the scheme of Miss MAURY was investigated and gave equations of the form:

$$C_m = C_s + \alpha (m - m_s),$$

where C_m and m are the colour and magnitude, and C_s and m_s the mean values of colour and magnitude of stars in a certain spectral class, and α a constant. The value of α is $+0^c,46$ for all spectral classes, whereas the observations with the grating give $+0^c,48$. It is clear that the increase of colour with the magnitude is caused by physiological effects. Certainly no appreciable absorption in space exists among the stars in Miss MAURY's catalogue.

The stars belonging to MAURY's divisions c and ac have, in general, a deeper (darker) colour than the corresponding groups in the a and b divisions. The degree of deepness grows smaller as the index of the spectral class grows larger:

IIIc—Vc	VIc—VIIIc	XIIc—XIIIc
$+0^c,99$	$+0^c,62$	$+0^c,31$

LAU draws the conclusion which is remarkable for that time (1914—1916), namely that the abnormally low density is a characteristic feature of the c stars.

LAU has made an important comparison between his estimates and those of OSTHOFF, HAGEN, and KRÜGER. All observations have been reduced to $3^m,5$.

Spectral class Miss MAURY	Colour according to				
	OSTHOFF	HAGEN	KRÜGER	LAU	Mean
I	$2^c,8$	$2^c,9$	$2^c,5$	$2^c,5$	$2^c,8$
II	2 ,6	2 ,6	2 ,4	2 ,8	2 ,6
III	2 ,6	2 ,5	2 ,6	3 ,0	2 ,7
IV	2 ,1	2 ,2	2 ,2	2 ,1	2 ,2
IV^2	2 ,1	2 ,1	2 ,2	2 ,2	2 ,2
V	2 ,7	2 ,7	2 ,5	2 ,6	2 ,6
VI	2 ,3	2 ,7	2 ,5	2 ,2	2 ,4
VII	2 ,5	2 ,4	2 ,5	2 ,7	2 ,5
VIII	2 ,6	2 ,7	2 ,5	2 ,8	2 ,7
IX	3 ,0	2 ,9	2 ,9	3 ,1	3 ,0
X	3 ,2	3 ,3	3 ,3	3 ,3	3 ,3
XI	3 ,4	3 ,4	3 ,4	3 ,9	3 ,5
XII	3 ,9	3 ,8	3 ,9	3 ,7	3 ,8
XIII	4 ,4	4 ,6	4 ,4	4 ,4	4 ,4
XIV	4 ,6	4 ,5	4 ,6	4 ,6	4 ,6
XIV^2	5 ,4	5 ,4	5 ,3	5 ,4	5 ,4
XV	5 ,6	5 ,6	5 ,6	5 ,6	5 ,6
XV^2	6 ,4	6 ,6	6 ,3	6 ,4	6 ,4
XVI	6 ,8	6 ,6	6 ,6	6 ,4	6 ,6
XVII	6 ,9	6 ,6	6 ,8	6 ,8	6 ,8
XVIII	6 ,7	6 ,4	6 ,9	6 ,7	6 ,7
XIX-XX	6 ,7	7 ,3	7 ,1	6 ,9	7 ,0

The systematic errors between the scales are so small that they can safely be ignored. The mean error averages $\pm 0^c,39$ for OSTHOFF, $\pm 0^c,52$ for HAGEN, $\pm 0^c,33$ for KRÜGER, and $\pm 0^c,40$ for LAU. The catalogues of OSTHOFF, KRÜGER, and LAU have more or less the same accuracy. If all except HAGEN are assigned weight one, the weight for the latter amounts to a half.

A comparison was made between the catalogues of OSTHOFF and LAU. The mean error of the differences OSTHOFF—LAU was computed from the differences themselves, as well as from the mean errors of the catalogues. The good accordance shows that the individual colour within a certain group in Miss MAURY's spectral catalogue must vary comparatively little. The dispersion cannot very well exceed $0^c,13$.

The values of $\Delta c/\Delta m$ show rather good agreement in the cases of the different observers.

Group	OSTHOFF	HAGEN	KRÜGER	LAU	MÖLLER
Orion stars.	$+0^c,32$	$+0^c,30$	$+0^c,18$	$+0^c,43$	$+0^c,35$
I. Type (SECCHI) . .	+0 ,31	+0 ,29	+0 ,24	+0 ,48	+0 ,32
II. ,, (,,) . .	+0 ,27	+0 ,25	+0 ,16	+0 ,46	+0 ,08
III. ,, (,,) . .	+0 ,20	+0 ,21	+0 ,16	+0 ,39	—

In the mean:
$$\frac{\Delta c}{\Delta m} = 0^c,28 \pm 0^c,08.$$

LAU finds that his observations with a grating give almost the same value for $\Delta c/\Delta m$ as the observations themselves, or $0^c,48$ and $0^c,44$. OSTHOFF finds $0^c,42$ by means of diaphragms, whereas the observations give $0^c,28$. LAU's investigation shows that OSTHOFF systematically overestimates the amount of the "deepness", which will explain the high value in the former case.

LAU had planned a general catalogue of star colours, and for that purpose he also compared the estimates of SESTINI in the scale of OSTHOFF. For stars of yellow colours the result is:

Colour SESTINI	Mean colour	Reduced to $3^m,5$	Mean magn.	n
Gialla chiara, assai chiara . . .	$3^c,17$	$3^c,05$	$4^m,37$	21
Gialla	4 ,06	3 ,95	4 ,32	165
Gialla chiara b or bb	4 ,68	4 ,65	3 ,68	8
Gialla b.	5 ,43	5 ,32	4 ,27	23
Gialla cupa	5 ,00	4 ,92	4 ,10	2

The stars called O by HAGEN give:

Colour	$\bar{c}$	Reduced to $3^m,5$	$\bar{m}$	n
Arancia chiara	$5^c,47$	$5^c,33$	$4^m,50$	11
Oro chiaro	5 ,50	5 ,50	3 ,50	2
Arancia.	5 ,85	5 ,68	4 ,74	11
Oro	5 ,81	5 ,65	4 ,67	9
Arancia-oro	5 ,96	5 ,90	4 ,16	14
Arancia-rame	6 ,06	5 ,90	4 ,62	5
Oro b and bb.	6 ,08	6 ,00	4 ,05	6
Arancia assai b	6 ,45	6 ,41	3 ,80	4
Arancia-oro bb	6 ,64	6 ,56	4 ,08	5
Arancia b	7 ,10	6 ,95	4 ,75	3

The mean error of the differences SESTINI — OSTHOFF averages $\pm 0^c,11$, which corresponds to a mean error of $\pm 0^c,9$ for SESTINI. If OSTHOFF is given a weight of one, the weight of SESTINI should be 0,25.

The evaluation of FRANKS's symbols gives [Harv Ann 14, p. 94 (1884)] the following results:

Symbol	Mean colour	Reduced to $3^m,5$	$\frac{\Delta c}{\Delta m}$	n
b	$2^c,47$	$2^c,45 \pm 0^c,51$	$+0^c,30$	6
W	2 ,62	2 ,52 ± 0 ,49	0 ,21	190
w	2 ,74	2 ,46 ± 0 ,46	0 ,08	58
g	2 ,95	2 ,82 ± 0 ,71	0 ,09	41
y	4 ,10	3 ,87 ± 1 ,06	0 ,25	317
Y	5 ,84	5 ,69 ± 0 ,67	0 ,24	176
o	6 ,46	6 ,21 ± 0 ,49	0 ,21	57
O	6 ,82	6 ,60 ± 0 ,35	0 ,29	64
o	7 ,01	6 ,67 ± 0 ,32	0 ,36	9

In the new edition of the catalogue of SESTINI, HAGEN[1] expresses the colours by means of combinations of capital letters. The most important have been evaluated by LAU as follows:

Symbol	Mean colour	$\bar{c}$ reduced to $3^m,5$	Mean magnitude	$\frac{\Delta c}{\Delta m}$	n
B	$2^c,20$	$1^c,98 \pm 0^c,26$	$5^m,03$	—	6
BW	3 ,15	2 ,99 ± 1 ,46	4 ,63	—	4
W	2 ,67	2 ,61 ± 0 ,49	3 ,92	$+0^c,28$	121
YW	3 ,02	2 ,95 ± 0 ,66	4 ,01	−0 ,09	124
WY	3 ,12	3 ,03 ± 0 ,61	3 ,17	−0 ,35	26
Y	4 ,15	4 ,04 ± 1 ,06	4 ,30	−0 ,23	219
OY	5 ,66	5 ,55 ± 1 ,03	4 ,27	+0 ,20	112
YO	5 ,70	5 ,57 ± 1 ,11	4 ,43	—	7
O	5 ,95	5 ,80 ± 0 ,91	4 ,52	+0 ,34	126
RO.	6 ,86	6 ,71 ± 0 ,62	4 ,53	+0 ,12	27
R	7 ,75	7 ,75 ± —	Var	—	2

Besides, VB is found to be $1^c,8$, VY = $2^c,5$, YB = $2^c,7$, and BY = $2^c,5$. All these combinations also represent white colours.

101. J. MÖLLERS's Estimates. On board a vessel in the tropical parts of the Atlantic and Pacific Oceans this author estimated with the unaided eye or an opera-glass the colours of 173, mainly southern, stars and of four planets[2]. He built up his own scale and the relation between his scale and that of OSTHOFF is given here:

Symbol		Corresponding colour
MÖLLER	OSTHOFF	
1^c	$0^c,0$	(pure white)
2	1 ,0	(yellowish white)
3	2 ,0	(yellow and white in equal proportions)
4	3 ,0	(whitish yellow)
5	4 ,0	(pure yellow)
6	5 ,4	(reddish yellow)
7	7 ,0	(red and yellow in equal proportions)
8	8 ,4	(yellowish red)
9	10 ,0	(pure red)

The following mean differences were found:

OSTHOFF—MÖLLER . . . . . $+0^c,05$ (OSTHOFF's 4-inch)
OSTHOFF—MÖLLER . . . . . + 0,19 (OSTHOFF's refractor)

102. The Relation of the Stars' Colours to the Milky Way. The catalogue of MÜLLER and KEMPF is the most extensive catalogue of direct colour estimates we have. The general relation between the colours and the magnitudes and the galactic coordinates has been investigated by these authors[3].

The distribution of the 14172 objects in percentages is as follows:

Magnitude	Colour symbol			
	W	GW	WG	G etc.
$0^m,0 - 3^m,9$	16,4%	34,0%	19,5%	30,2%
4 ,0 − 4 ,4	11,5	39,2	22,5	26,9
4 ,5 − 4 ,9	9,9	37,9	17,3	35,0
5 ,0 − 5 ,4	9,6	34,5	21,7	34,1
5 ,5 − 5 ,9	8,9	39,4	24,6	27,1
6 ,0 − 6 ,4	12,6	34,7	26,4	26,3
6 ,5 − 6 ,9	11,3	37,5	31,6	19,6
7 ,0 − 7 ,4	13,1	43,8	31,2	11,9
7 ,5 − 7 ,9	17,3	50,9	25,7	6,1
8 ,0 − 8 ,4	21,7	59,7	16,1	2,4
8 ,5 − 8 ,9	26,0	62,0	10,8	1,2
All	14,4	44,6	26,5	14,4

[1] Specola Astronomica Vaticana III (1911).
[2] A N 166, p. 305 (1905). [3] A N 180, p. 249 (1909).

It is important to see whether systematic differences in the estimates might have any influence on the results. Between the two main instruments, the refractors constructed by STEINHEIL and SCHRÖDER, the adjoined differences are found:

Colours	STEINHEIL–SCHRÖDER
W	$+1^c,8$
GW	$-0,2$
WG	$-0,1$
G	$-1,3$
RG etc.	$-3,0$

If the classes W and GW are combined into one group and WG and G into another the following relative frequencies are obtained:

Apparent magnitude	Percentage		Apparent magnitude	Percentage	
	White	Yellow		White	Yellow
$0^m,0-3^m,9$	50,3	49,7	$6^m,5-6^m,9$	48,8	51,2
4 ,0–4 ,4	50,8	49,2	7 ,0–7 ,4	57,0	43,0
4 ,5–4 ,9	47,7	52,3	7 ,5–7 ,9	68,2	31,8
5 ,0–5 ,4	44,1	55,9	8 ,0–8 ,4	81,4	18,6
5 ,5–5 ,9	48,3	51,7	8 ,5–8 ,9	88,0	12,0
6 ,0–6 ,4	47,3	52,7			

The decrease in yellow stars for magnitudes below $7^m,4$ cannot be real and is at least partly explained by the fact that the PD cannot aim at completeness below $6^m,5$.

Finally the following synopsis shows the influence of the galactic latitude in the distribution of the star colours:

Galactic latitude	Brighter than $6^m,0$			$6^m,00-6^m,99$			$7^m,00$ and fainter		
	W+GW	WG	G	W+GW	WG	G	W+GW	WG	G
+90° to +70°	42,2	31,3	26,5	49,5	37,2	13,3	66,5	27,7	5,8
+69 +50	50,2	23,2	26,5	46,1	31,9	22,1	67,2	27,6	5,2
+49 +30	39,5	26,8	33,7	44,7	34,9	20,4	63,8	30,5	5,7
+29 +10	41,1	22,1	36,8	45,7	32,1	22,2	66,7	26,7	6,6
+ 9 −10	48,2	22,3	29,4	52,8	24,8	22,4	70,3	20,8	8,9
−11 −30	60,8	16,9	22,3	51,3	29,4	19,3	65,4	25,0	9,5
−31 −50	51,0	20,3	28,7	48,6	29,9	21,5	61,6	28,1	10,3
All	47,8	22,5	29,7	48,9	30,0	21,1	66,6	25,7	7,7

The so-called phenomenon of KAPTEYN, i.e. that the blue stars are more numerous in the Milky Way, is not present in the distribution of the brighter stars. For these the maximum frequency occurs in a zone between $b = -11°$ and $-30°$. Besides a minimum in the frequency of the white stars does not present itself in the Milky Way Pole regions, as is supposed to be the case in the earlier investigations of KAPTEYN[1].

103. BELL's Study of the Double Star Colours. This investigation[2] aimed at discovering the relation of the visible colours of the stars to known facts regarding their spectra.

The author points out that no extraordinary colours exist among the stars. He quotes the analysis of the Potsdam colours by PICKERING. The chances are 15 to 1 that an A star is either white or yellowish white, and the chances are 150 to 1 that it will not be called a full yellow.

The following list, taken from WEBB's Celestial Objects, exhibits the vagaries of nature or of vision in double star chromatics: Indigo, grayish white, olive bluish gray, bluish red, violet, tawny, ruddy or dusky, brownish, pale rose, fawn colour, olive blue, greenish blue, pale green, yellowish green, green, rosy, gray, lilac, mauve, bluish green, ashy yellow, bright green, greenish blue, pale

[1] Cape Photographic Durchmusterung, Cape Ann III, Introduction (1896).
[2] Ap J 31, p. 234 (1910).

tawny, sombre, cool gray green, greenish red, ruddy olive, dusky red, garnet. A complete list of the epithets used for the colours of the double stars would run to nearly a hundred.

BELL discusses in detail the various results concerning the colours of the double stars and finds the masked influence of a personal equation in such estimates. He compares the percentages of various colours for stars brighter than $6^m,5$ according to FRANKS[1] and to PICKERING[2] (see the adjoined table).

FRANKS's colours	FRANKS	Potsdam observers
White	31 %	13,2%
Yellowish white	20,8	40,2
Pale or whitish yellow . . .	22,4	23,0
Yellow	11,8	20,6
Pale orange (reddish yellow) .	7,7	2,98
Orange (yellowish red) . . .	5,6	0,02
Orange red.	0,7	—

The infrequency of white or whitish stars and the frequency of those strongly yellowish in the lists of coloured doubles are very striking.

The relation of colour to difference in magnitude in double stars is striking. STRUVE has found Δm to be $0^m,5$ for stars having the same colour and $2^m,0$ or more for those showing distinctly different colours. HOLDEN[3] has found:

Δm	
$0^m,53$	same colour
2 ,44	different colours.

The striking colours of many comites of double stars have been thought to be real as they persist, when the primary is hidden by an occulting bar. Besides, STRUVE has pointed out that if such colours are due to contrast they must be complementary, which, as a rule, they are not. But colours due to subjective causes do not, in fact, disappear through the use of an occulting bar and the argument of STRUVE is also fallacious. Simultaneous contrast, which involves the shifting of contrast colours towards complementary effects, is quite probably mainly a physiological phenomenon, inasmuch as it may occur, to a certain extent, in very brief glimpses, as is shown by MAYER, and once introduced it sometime persists, as HELMHOLTZ has noted, until the eye gets rid of its bias and makes a fresh start. An important source of subjective coloration is fatigue colour. In applying the colour diagram of MAXWELL to the double stars one will find the true complementary produced only in the absence of any subjective colour in the comes.

A third cause that is tending to produce subjective colour is the PURKINJE-effect. When any very faint objects are observed the tendency is to see them, irrespective of their real colours, as faded blue-green merging into a whitish hue (anagalactic objects), or at least to rob them in very large measure of red and yellow tints, which marks the transition stage between cone and rod vision[4]. As a powerful and very interesting modifying influence in observed colours one must consider the "dazzle tints", i. e., the subjective colorations corresponding to positive after-images.

The whole subject of fatigue colours and dazzle tints is set forth in a masterly paper by BURCH[5].

These four causes of subjective coloration operating simultaneously will explain why truly complementary hues are likely to be the exception. These

[1] M N 68, p. 672 (1908). [2] Harv Ann 64, p. 125 (1912).

[3] Amer Journal of Science 19, p. 467 (1880).

[4] As most of the authors who have commented upon the PURKINJE-effect BELL was not aware of the fact that this effect had only been proved in case of surfaces. The observations of GALLISSOT and DANJON have shown that for luminous points a reversed PURKINJE-effect takes place (see ciph. 47 and 48). [5] Phil Trans B 191, p. 1 (1900).

causes will also have another result, namely, when double stars are observed, brilliant colours will normally appear only when the difference in magnitude between the components is considerable, so that the primary can impose its fatigue and dazzle tints upon the faded light of the comes.

BELL investigated these phenomena by means of artificial stars. The results of the experiments showed among other facts that merely cutting down the intensity of one of the elements of the artificial double invariably shifted its colour progressively from the initial colour through a variety of transition hues, sometimes complicated by the appropriate dazzle tints, to a distinctly greenish, bluish-green, or bluish colour, depending on the value of the intensity of the illumination. Shifting the observation from the tired eye to the fresh one changed the hue somewhat, but did not abolish the subjective colours.

Experiments were also made with artificial clusters and BELL concludes that in the case of multiple stars and of clusters subjective colours play the chief, if not the only rôle in determining the apparent tints observed. This does not seem to be the case with the estimates of GRAFF[1]. Comparisons between these and several colour equivalents have been made by COLLINDER[2], who finds a very pronounced correlation.

104. FRANKS's Determination of the Relation between Star Colours and Spectra. This investigation[3] includes a comparison of 4175 stars estimated for colour by W. S. FRANKS with most of their spectra as given in the Revised Harvard Photometry. The adjoining synopsis give the colours used and their frequency.

	Colour	Symbol	Percentage	n
1	White	O	31,0	1083
2	Yellowish white	Y^1	20,8	729
3	Pale yellow	Y^2	22,4	782
4	Yellow	Y^3	11,8	413
5	Pale orange	Or^2	7,7	271
6	Orange	Or^3	5,6	195
7	Orange red	$Or\,R^2$	0,7	24

Summary of comparison between colour and spectra:

Colour	Spectral Class											
	Od—Oe	Oe5	B—B5	B8—A4	A5—F2	F5	F7—G8	K	K2—K5	Ma—Md	N	n
White		5	191	793	78	7	9					1083
Yellowish white		4	50	338	204	47	69	17				729
Pale yellowish	2	1	1	59	143	51	242	260	12	11		782
Yellow					9	4	68	274	32	26		413
Pale orange						2	41	155	29	44		271
Orange							13	84	36	59	3	195
Orange red							2	4		10	8	24
n	2	10	242	1190	434	111	444	794	109	150	11	3497

The following regression-line deductions are of interest:

Spectral class		Conclusion
O	Maximum	in white gradually decreasing to pale yellow.
B	„	„ „ The whitest stars.
A	„	white but less white than B.
F	„	yellowish white, but the pale yellow not far behind.
F5	„	between yellowish white and pale yellow-the latter slightly in excess.
G	„	in pale yellow—very pronounced.
K	„	„ „ yellow and yellow—nearly equal.
K2—K5	„	pretty evenly distributed between yellow, pale orange and orange.
M	„	in orange, but pale orange not far behind.
N	„	orange red—very pronounced.

[1] A N 219, p. 297 (1923).
[2] Stockholm Ark Mat Astr Fys. 19 B, No. 11 (1929). — Upsala Medd No. 9 (1926).
[3] M N 67, p. 539 (1907).

105. Suspected Old Observations of Colour. In his paper "Rubra canicula" G. SCHIAPARELLI[1] also commented on the fact that PTOLEMAIOS, in his work Tetrabiblos[2], attributes one of the planets to the different stars. He also suggests that the reason for such comparisons may be that there was agreement as regard to colour, but that also other unknown criteria may have been applied.

Later on, this question has been discussed in an extensive paper by F. BOLL[3]. Without knowing SCHIAPARELLI's work BOLL has found during his studies of the documents from the antique astronomy that the old astronomers very likely described the colours of the stars in such a way that one or two planets of more or less the same colour are mentioned together with a certain star. The reason for noting the colours has certainly been of an entirely astrological nature: it has probably been thought that the stars of a certain colour had a certain influence on certain temperaments.

BOLL has edited the famous astrological handbook of PTOLEMAIOS, Tetrabiblos, which earlier only existed in two unsatisfactory editions from 1535 and 1553. He also discovered a hitherto unknown text called Pseudo-PTOLEMAIOS, which gives information about the colours of the brighter stars. Also in a manuscript of an unknown author from 379 AD (Anonymus) and in several other manuscripts, dependent on that source, a number of bright stars are mentioned together with planets. Such is also the case in BAYER's Uranometria. Nothing is explained there about why planets are attributed to stars, but in the older literature there are clear statements concerning the agreement as their "temper" between certain stars and certain planets.

But what is fundamental? Have the astrologers distributed the "tempers" of the planets among the stars on a basis of actual comparisons with regard to colour (and light), or has some other principle prevailed?

If actual comparisons have been made we should have an invaluable wealth of colour observations hidden in the astrological literature. Thanks to the careful discussion of the different texts undertaken by BOLL we are now in a position to test the question by means of a translation of the supposed observations into a modern colour scale.

BOLL himself made extensive comparisons and concluded that the agreement was satisfactory on an average, although individual deviations occurred. In a critical review C. WIRTZ[4] has denied that there is any agreement at all between the colours derived from the attribution of planets to stars and the modern colour scale.

The planets have the following colours according to different observers:

Planet	OSTHOFF	MÖLLER	WIRTZ	Mean.
Mercury	—	—	5^c,6	5^c,6
Venus	3^c,4	3^c,6	—	3 ,5
Mars	7 ,0	7 ,9	8 ,0	7 ,6
Jupiter	3 ,6	3 ,6	3 ,5	3 ,6
Saturn	—	5 ,3	4 ,2	4 ,8

The scale of the astrologers was evidently very narrow, as the whitest colour was 3^c,5 and the reddest 7^c,6.

---

[1] Atti dell' I R Acc di Sci Lett ed Arte d Agiati Ser. III, Vol. II, Fasc II. Rovereto (1896).

[2] For an edition of Tetrabiblos in English see J. M. Ashmand: Ptolemaios' Tetrabiblos. London (1917).

[3] Antike Beobachtungen farbiger Sterne. Mit einem Beitrag von C. BEZOLD. Abh d Bayr Ak d Wiss Philos-philol u hist Kl. 30. Bd, 1. Abh. München (1918).

[4] V S J 55, p. 27 (1920).

WIRTZ has derived a colour catalogue for 91 stars from the material in BOLL's memoir. Since the colour cannot be estimated with any accuracy for stars fainter than $2^m,5$, all fainter stars were excluded. From the Sterntafeln of NEUGEBAUER the declination for the year 1000 B.C. was taken, in order to see if any influence of the atmosphere can be traced. When two planets were attributed to a star the colour is the mean of the colours of the planets. If "in smaller degree" was added to a certain planet in the old sources a weight of $^1/_2$ was assigned. On account of the dependence of Anonymus and Pseudo-PTOLEMAIOS on Tetrabiblos the former were given half weight.

If there should really prove to be old observations of colours hidden in the astrological texts, it will, of course, be of much importance to know from what age they originate. This question has been discussed by BOLL and by C. BEZOLD in an appendix to the paper of the former. There seems to be no doubt that the comparison of planets and stars with regard to the "temper" originated in Babylon. PTOLEMAIOS also often refers in his Syntaxis to the Babylonian observations.

The astrological observations certainly date from an epoch at least 1000 years before our era.

The distribution table which is formed from WIRTZ's catalogue shows that there is little or no correlation between the presumed colour observations and the modern values of colour.

OSTHOFF \ PTOLEMAIOS	$3^c,0$—$3^c,9$	$4^c,0$—$4^c,9$	$5^c,0$—$5^c,9$	$6^c,0$—$6^c,9$	$7^c,0$—$7^c,9$	Sum
$0^c,0$—$0^c,9$		1	1			2
1 ,0—1 ,9	3	8	2	4	6	23
2 ,0—2 ,9	3	12	2	5		22
3 ,0—3 ,9	2	6	2	1		11
4 ,0—4 ,9		5	2	2	2	11
5 ,0—5 ,9	2	3	1	2		8
6 ,0—6 ,9	2	3	2	4	1	12
7 ,0—7 ,9		1		1		2
Sum	12	39	12	19	9	91

The coefficient of correlation is found to be $r = -0{,}003 \pm 0{,}105$. Thus there is no relation between the two systems of colours.

Even the modest expectation of being able to trace the two main colour classes in the material cannot be realized, as is seen from the following table:

OSTHOFF	PTOLEMAIOS	n
$2^c,0$	$5^c,2$	47
5 ,0	5 ,1	44

Even if the observations of the astrologers were made 4000 years ago this time is a moment in the life of the stars. It is impossible that in the meantime the bright stars have changed their colour, i. e. their spectral class, in such a degree as is indicated above.

Thus there is only the possibility left that the astrologers classified the stars in some other way and not only according to their colour. There must have been other points of view too; or have the original observations been forgotten and the stars divided up partly according to new principles?

WIRTZ admits that his conclusion with regard to the absence of a relation between the two colour scales will not be seconded by many readers of BOLL's work. He adds sarcastically that it is not impossible that the thesis will be advanced that the crowd of visible stars was divided by the antique observers into two groups partly covering each other, the white stars and the red stars.

It is, of course, very difficult to know whether the absence of any correlation really excludes the possibility of colour observations in the astrological documents. Even the careful editions of BOLL cannot avoid corrupted sources. Besides the original sources may have been lost. A comparison with the modern colour scale may not be quite justified on account of the possibility that the seven planets were distributed among the stars, not only with regard to colour, but with regard to general characters. Besides the modern colour scale has not been tested in Mesopotamia, where the condition for colour observations may be slightly different. It has to be remembered that observations of effective wave lengths have given colour indices for Jupiter and Saturn quite as large as that for Mars. There may be observers who also note these planets as red by visual inspection.

In order to see how large the dispersion may be in the astrologers' colours we have used the catalogue of the bright stars given by BOLL and translated the planetary symbols into colours. In the case of two planets we have given double weight to the colours of the first one.

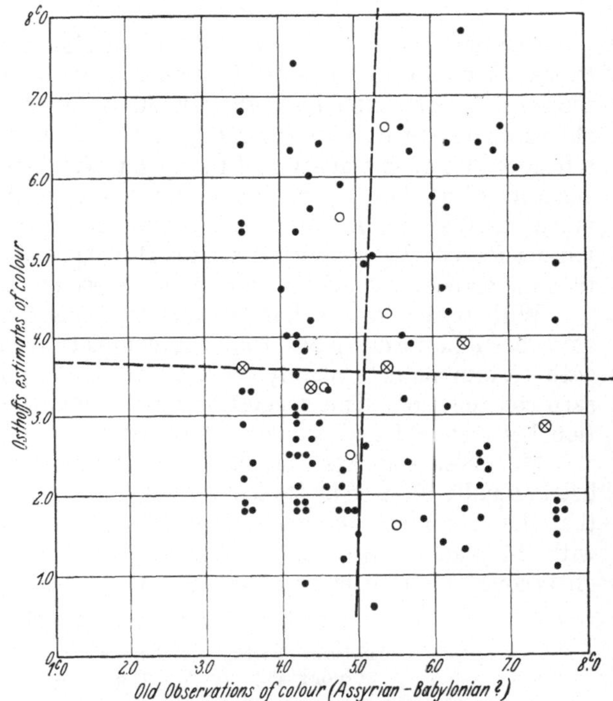

Fig. 88. Comparison between old colour observations and modern determinations of colour. The two regression lines are nearly at right angles to each other, which shows the total absence of any correlation. The material used is the catalogue compiled by WIRTZ.

The observations are not independent of each other, but it is clear that the dispersion is considerable. Even if the astrologers compared planets and stars with regard to colour accurate results cannot be expected. Also an influence of $\Delta c/\Delta m$ may be present. Still, it is doubtful what the astrological attribution of planets to fixed stars really expressed.

If we select from the list the stars attributed to planets in which only one planet has been mentioned we find the following mean values with their mean errors:

Planet	Mean colour	Mean error	*n*	Planet	Mean colour	Mean error	*n*
☿	$2^c,4$	$\pm 0^c,46$	8	♃	$2^c,4$	$\pm 0^c,25$	8
♂	5 ,6	±0 ,73	7	♄	3 ,3	±0 ,25	6

As it has been shown from the work of WIRTZ that the old observers did not arrange the stars according to their colours alone when they compared them with the planets, it seems justifiable to conclude that some systematic errors have crept into the observations.

O. MEISSNER[1] has pointed out that except for Sirius it is striking that the fainter stars show considerable differences in colour in both directions and that in general the old colours seem to be 2^c redder than the modern ones.

If the 27 brightest stars apart from Sirius are selected one receives quite another impression from the material than if one uses the 89 stars in the catalogue deduced by WIRTZ.

For these 27 bright stars the following relation is found:

Colour Babylonian-Assyrian	OSTHOFF	Colour difference	
$5^c,12$	$2^c,82$	$+2^c,30$	$\pm 0^c,31$

The mean error in one estimate of one star is about $\pm 1^c,63$, which according to the view of LUDENDORFF "Certainly is not very small, but is not altogether too large". MEISSNER finds the coefficient of correlation between modern and old estimates to be $r = +0,423 \pm 0,173$, which value seems to have real significance. As an explanation of the systematic difference found above, MEISSNER suggests that the stars have most frequently been observed when rising and setting. This seems reasonable on account of the use of the observations for astrological purposes. It is also expressly stated by the old astronomers that the planets when rising and setting showed a red colour (BOLL, loc. cit. p. 144).

With regard to the fainter stars it seems that no individual observations have been performed, but that the complete constellations have been ranged with certain planets mainly on account of the colour of the principal star of each constellation. This suggestion might explain the entire absence of a correlation between old and modern observations.

MEISSNER suggests that also the colour of several of the fainter stars has been actually observed. He has excluded all stars where there was no likelihood that the colour was the underlying principle for coupling the object in question with the planets. Although this selection cannot be free from bias it is quoted on account of its considerable interest:

Stars of magnitude 2^m—3^m.

Babylonian-Assyrian	OSTHOFF	Difference		n
Colour $4^c,7$	$3^c,6$	$+1^c,1$	$\pm 0^c,25$	36

Mean error of one observation: $\pm 1^c,48$. Coefficient of correlation between old and modern observations: $+0,278 \pm 0,160$.

106. Suspected Changes in the Colour of Sirius. The question of the colour of Sirius in olden times presents a difficult problem, because PTOLEMAIOS mentions the star as red. In Megiste Syntaxis the star is called ὑπόκιρρος or reddish. The same word is used for Aldebaran, Antares, Arcturus, Betelgeuze, and Pollux. SCHIAPARELLI[2] thought that the word originated from an error in the Almagest, but such an explanation does not seem tenable. The question has been recently discussed at length by T. J. J. SEE[3], who has published an investigation he made at Berlin some 30 years ago. All the Greek and Latin sources have been examined, and the evidence for classifying Sirius as a red star seems indeed overwhelming.

Of the 1022 stars in the catalogue of PTOLEMAIOS[4] only six are distinguished by colour. The silence with regard to many of the bright stars of spectra A—G shows that the stars noted as reddish must have shown a conspicuous colour.

[1] A N 231, p. 391 (1927).

[2] Rubra Canicula. Considerazioni sulla mutazione di colore che si dice avvenuta in Sirio. Atti d I R Accad di Scienze. Ser. III, Vol. II, Fasc. II. Rovereto (1896).

[3] A N p. 229, 245 (1927).

[4] See f. i. edition by Peters and Knobel. Washington Carnegie Inst Publ No. 86 (1915).

Among the other testimonies that of SENECA[1] is remarkable. He says: "in the heavens also there does not appear one colour of things, but the redness of the Dog Star is deeper, that of Mars milder, that of Jupiter nothing at all, the splendour being turned into pure light".

We cannot follow the interesting discussion by SEE in detail. To sum up it appears that the highest authorities of classic antiquity attributed to Sirius a ruddy colour. No authority can be found expressly saying that the star was white.

Also an astrological reason put forward by SEE seems worth considering. The ancient world ascribed the intense heat of the "Dog days" to the influence of the burning Dog star. According to Astrology evil influences were usually supposed to proceed from bodies presenting a "fiery", "violent", or "angry" appearance, and salutary influences from the great multitude that shine with a clear, brilliant light.

In a critical review of SEE's work, H. OSTHOFF[2], the eminent authority on star colours, reverses the last argument and thinks that the conception of the evil influence of Sirius made the observers in the antique world believe that the star simply must be red. OSTHOFF suggests that the scintillation near the horizon may have given the coloured impression of the star. He also points out that the many Roman authors quoted by SEE were not astronomers by profession. On the other hand general knowledge of astronomy was more common among educated men in ancient times than is the case nowadays.

Sirius is not mentioned as regards its colour by AL SÛFI, who calls Algol red. He has related an Arabic table according to which Sirius and Canopus are spoken of as sisters. This implies that these stars could not very well have had conspicuously different colours in the 10th century.

Most astronomers do not seem to be willing to accept a change in the colour of Sirius, which, if the observations of PTOLEMAIOS and SENECA are correct, must have taken place between 137 AD and 980 AD. But do we really know enough of the physics of the stars to be able to decide that such a change is quite impossible? We know for certain that novae change their colour. Most of the stars must have passed the nova stage and especially the A stars seem to be physically adapted for to reach that stage[3]. It is by no means impossible that Sirius has passed the nova stage long ago and that some changes in the later stages occurred some two thousand years ago. There are also possibilities that changes of the same kind may have occurred in its companion. If we suppose that this star increased in magnitude by 10^m, then the combined magnitude of Sirius and its companion would have been $-2^m,3$ instead of $-1^m,6$. Such a difference in magnitude would scarcely have been noted on account of the inaccurate magnitude scale, even if PTOLEMAIOS had witnessed the outburst, but the change in colour would certainly have been significant. J. PLASSMANN[4] has also suggested this explanation and he relates that in the AVESTA[5], the Holy Scripture of the Persians, it is said: "We honour the magnificent Tischtrya-star (Sirius), the white shines with his bright unparalleled rays". This may be an interesting contribution to the possibility that the star was white in the 5th century B.C., but it does not disqualify the nova theory. We cannot expect to find any evidence by investigating the now exhibited spectrum of the companion of Sirius, because the nova spectrum does not show any peculiar features

[1] Quest. Nat. Lib. 1, cap. 1, 6.
[2] Die Himmelswelt 37, p. 167 (1927), also A N 229, p. 443 (1927).
[3] Publ A S P 35, p. 95 (1923). [4] Himmelswelt 37, p. 165 (1927).
[5] Avesta, 8th sacrificial song.

a comparatively short time after the outburst. Perhaps every star has been a pre-nova. In any case it is very difficult to locate older novae by means of classifying their spectra some decades after the outburst.

With regard to the supposed change in the colour of Sirius MEISSNER presents the following suggestion: On account of the large amount of the extinction at the horizon no star except Canopus is actually seen to rise or set. The astronomers of Mesopotamia thought that the colour shown at the horizon was the real colour of the star and that the white colour at a higher altitude was due to some paling process.

MEISSNER also points out that there is no reason to assume that changes in colour have been established through the old observations. It is strange that the variability of Mira Ceti was not noted by old observers. It is on the other hand possible that the Arabian observers noted the variability of Algol.

A. STENZEL[1] has given an interesting contribution to the question of the colour of Sirius. This star played a very important rôle in Egypt. It was the heliacal rising of Sirius which inaugurated the new astronomical year. The name of Sirius, Soped, or Sati (Sothis) is very common on a number of Egyptian monuments. The hieroglyph designating Sirius contains a triangle, and this is given in the inscriptions with a red surface. The colours are not selected at random, but are given according to their symbolical significance.

The principal place for the cult of Sothis was the temple at Dendera (Tantari). In the inscriptions of this temple the Isis is named the golden and it is also said: "Beautiful is your sight, you Goddess Tumit, Lady of the Orient and Occident, Goddess of the red veil which loves the bright colour."

From a number of rather definite statements concerning the colour it seems justifiable to conclude that the old Egyptians saw Sirius as a red or golden star.

b_2) Measured Star Colours.

107. Pioneer Work for Measuring Star Colours. One of the earliest attempts to use a scale in the determination of the colours of the stars was that of A. SECCHI[2]. His method consisted of comparing the colours of stars with those of the lines of a spark spectrum. Later on B. KINCAID[3] suggested the use of a metrochrome for accurate estimates of star colours. The apparatus consisted of a fine platinum wire, mounted in the focal plane and rendered incandescent by an electric current, together with a rotating drum with six radial openings. The second, fourth, and sixth of these transmitted the normal light of a lantern, the other three contained flatsided bottles containing chemical solutions of different colours. By varying the rotational speed of the drum the different star colours could be produced. No observations based on the application of this apparatus seem to have been published.

W. H. M. CHRISTIE[4] has made measurements of colours with the aid of an apparatus constructed on the principle of CLERK MAXWELL's colour-box, in which the red, green, and blue parts of three spectra were made to overlap. Through variations of the width of the slits any mixed colour could be produced.

The good agreement between the colours of 34 stars as measured by F. ZÖLLNER[5] in 1868 and the values c_2/T according to modern determinations prove the possibility of using his colorimeter for actual determinations of colours. It is to be regretted that the readings of the colorimeter have not been used for derivation of c_2/T.

[1] A N 231, p. 387 (1927).
[2] Mem Obs Coll Rom 1852—55, p. 135.
[3] M N 27, p. 265 (1867).
[4] M N 34, p. 111 (1874).
[5] A N 71, p. 357 (1868).

In his paper "On the Colours of the Variable Stars" C. CHANDLER[1] made the very important suggestion of interposing a shade of coloured glass, and then estimating the magnitudes e. g. according to the step-degree-method. By comparing such estimates with ordinary visual ones colour indices were in fact obtained. CHANDLER seems in reality to have been the inventor of the colour-index method. Estimates of the relative diminution were performed for 77 red variable stars. Comparisons were made between the two scales, the relative diminution estimates and the directly estimated colours.

108. Colour Equivalents. The colour can strictly speaking be determined only by direct estimates with the naked eye or by the aid of an instrument and is as such not very well suited for measurement purposes.

As the colour equivalent of a star there may be taken a very well measurable quantity that depends in a simple way on the colour itself. Among the principal equivalents the following may be mentioned:

I. The colour index as already defined in the introduction as the difference in magnitude between two different parts of the spectrum.

II. The exposure-ratio method.

III. The method of using a central diaphragm and comparing the outside ring with the central dot.

IV. The method of covering the central part of the objective with one filter and the other part with a filter of another colour.

V. Effective wave lengths and minimum wave lengths.

VI. The method of attaching an objective prism in front of the central part of the objective and comparing the image of the uncovered part with the prismatic image.

Ia, b. In practice the colour index proper has generally been used, the difference photographic—visual magnitude or $C = C_{pv} = m_p - m_v$. Since the photovisual scale very nearly coincides with the visual scale the colour index $C_{ppv} = m_p - m_{pv}$ will give practically the same results.

Ic, d. Other colour indices of importance are the photoelectric colour index and the heat index. The former is the difference between the photoelectric magnitudes obtained with and without a filter and the latter the difference between the radiative magnitudes measured in the same way: $C_{rv} = m_r - m_v$.

Ie. The radiative magnitudes also permit the use of $C_{rp} = m_r - m_p$ which gives a wider range to the colour index.

If. The ultra violet magnitudes as introduced by SEARES make a number of new combinations possible. Extensive measurements of m_{uv} have not, so far, been published.

Ig. The colour index obtained by the use of heterochrome photometers (e. g. of NORDMANN's construction) is closely related to the ordinary colour index C.

The determination of colour indices depends on the fundamental scales for all the different magnitudes used. For the photographic magnitudes there is the well-established North Polar Sequence as determined and confirmed by the cooperation of Harvard, Mount Wilson, Potsdam, and Greenwich. For the brighter visual magnitudes we have the Harvard and Potsdam system, which seems fairly well established although there are small systematic differences which remain unexplained. A North Polar scale of photovisual magnitudes for fainter stars has been determined by SEARES, but this can only be regarded as an extension of the visual series if no systematic difference between the two exists. Coincidences between both scales take place at 6^m and 12^m. The difference

[1] A J 8, p. 137 (1888).

seems to be dependent on the colour. What this finally comes to is that in the colour index systematic errors are involved which depend on the colour, in other words, what we actually determine is a function of the colour and some unknown parameters. Fortunately these unknown quantities do not play an important part, and do not sensibly diminish the value of the colour index determined in this way.

The colour indices I c, d, e are of much importance, but their information is restricted at present to the brightest stars. The same applies to I g.

Among the disadvantages of the colour index determinations the following may be mentioned:

Dying the plates may introduce systematic errors or possibly a non-uniform sensitivity of the plates. The dying also decreases the general accuracy attainable.

Corrections have to be applied corresponding to the distance from the centre and these may contain uncertainties.

If an auxiliary scale is used and magnitudes are determined by estimates, as in the ARGELANDER method, trouble will arise from the definition of the images on the plates. This can be largely overcome by using different scales, corresponding to varying definitions. There also seems to be a change in the personal equation with time.

A great advantage of the exposure-ratio method is that every part of the field can be used, as the shape of the images will not greatly influence the accuracy, and that measurements are not necessary; estimates will do, especially as a large number of images are secured on the same plate. Among the disadvantages it may be mentioned that the connection between exposure ratios and colour is not known a priori, and has to be determined in the course of the work.

It has been tacitly assumed that the filters and the colour sensitivity of the plates are invariable. This applies, of course, to all methods using dyed plates and colour filters.

The colour is dependent on the intensity of the image and the reduction to standard intensity is sometimes troublesome.

In more unfavourable climates it seems difficult to avoid the influence of changing transparency, which is of course a drawback in every photometric method that uses intermediate or long exposures in such climates.

As has already been mentioned, the method, being independent of the standard sequences, has afforded a valuable check on the colour indices obtained by SEARES in the North Polar Sequence.

109. KAPTEYN's Phenomenon. In 1890 J. C. KAPTEYN found[1] that the plates of the C P D contained more stars than SCHÖNFELD's part of B D near the Milky Way and a progressively smaller number towards the poles of the Milky Way. Analogous results were found when the C P D was compared with the C D M.

KAPTEYN suggested, in order to see whether the stars of the Milky Way were bluer than other stars, that photographs might be taken of the same regions both with ordinary and with isochromatic plates in different galactic latitudes. A number of such plates were taken at the Cape and investigated by W. DE SITTER[2]. As no colour screen was used the isochromatic plates were not sufficiently different from the others to show the expected systematic differences of colour. Besides, the magnitudes of the C P D were not true photographic magnitudes either.

The problem was later taken up again by DE SITTER. From a comparison of the Potsdam C d C plates with B D, SCHEINER[3] had found that the number of stars in the former source in excess of those in B D increased 3,6 times from the Milky Way pole to the Milky Way regions. During a stay at the Cape DE SITTER

[1] Cape Photographic Durchmusterung (1890).
[2] Groningen Publ No. 2 (1900). [3] A N 147, p. 1 (1898), and 149, p. 165 (1899).

secured[1] the visual magnitudes of 791 stars through photometric work with a ZÖLLNER photometer, and through visual estimates by R. T. A. INNES. The mean error of the final visual magnitude of a star was found to be $\pm 0^m,15$. Also 29 plates were taken and photographic magnitudes derived for the stars determined visually.

From a detailed discussion DE SITTER found that there is a want of homogeneity in the average colour of the stars of the galactic areas, the colours of the polar areas being on the other hand practically homogeneous. The mean photographic difference between a Milky Way star and a star of equal visual brightness at the Milky Way pole varies between $+0^m,68$ and $-0^m,32$ for the different areas. The author concludes that it is impossible to say whether the variation has any relation to the general structure of the Galaxy, or whether it is caused by real differences in the relative intensity of different spectral parts, or by a general absorption in one part, whether the stars of the same spectral intensities are grouped together in space or whether the effect is due to selective absorption by intervening cosmical clouds or nebulous masses.

110. Colour Indices from the Göttinger Aktinometrie[2]. The distribution of the colour indices with regard to their value, both in and outside the Milky Way regions, is found to be as follows:

Colour index	Inside	Outside	All	Colour index	Inside	Outside	All
$-0^m,75$	6	0	6	$+0^m,95$	51	35	86
0 ,65	8	2	10	1 ,05	50	27	77
0 ,55	31	17	48	1 ,15	35	28	63
0 ,45	69	34	103	1 ,25	31	28	59
0 ,35	89	34	123	1 ,35	28	27	55
0 ,25	72	34	106	1 ,45	47	25	72
0 ,15	55	44	99	1 ,55	31	26	57
−0 ,05	54	24	78	1 ,65	26	13	39
+0 ,05	58	42	100	1 ,75	19	7	26
0 ,15	42	24	66	1 ,85	5	1	6
0 ,25	29	14	43	1 ,95	4	0	4
0 ,35	32	9	41	2 ,05	2	0	2
0 ,45	18	7	25	2 ,15	0	0	0
0 ,55	20	28	48	2 ,25	1	0	1
0 ,65	25	28	53	2 ,35	0	0	0
0 ,75	41	49	90	2 ,45	1	1	2
0 ,85	59	42	101				

The low frequency of colour indices around the values $0^m,4$ and $0^m,5$ is very remarkable. This fact seems to have been pointed out for the first time by Ö. BERGSTRAND[3].

When the colour indices are grouped according to A R, the following distribution is found:

Limits of right ascension	Mean colour index	Limits of right ascension	Mean colour index
$23^h,6-1^h,5$	$+0^m,55$	$11^h,6-13^h,5$	$+0^m,56$
1 ,6− 3 ,5	+0 ,35	13 ,6−15 ,5	+0 ,59
3 ,6− 5 ,5	+0 ,27	15 ,6−17 ,5	+0 ,56
5 ,6− 7 ,5	+0 ,33	17 ,6−19 ,5	+0 ,43
7 ,6− 9 ,5	+0 ,49	19 ,6−21 ,5	+0 ,64
9 ,6−11 ,5	+0 ,52	21 ,6−23 ,5	+0 ,50

[1] Groningen Publ No. 12 (1904).

[2] Aktinometrie der Sterne der B. D. bis zur Größe 7,5 in der Zone 0° bis +20° Deklination. Teil A: Unter Mitwirkung von BR. MEYERMANN, A. KOHLSCHÜTTER und O. BIRCK. Göttingen (1910). Teil B: Unter Mitwirkung von BR. MEYERMANN, A. KOHLSCHÜTTER, O. BIRCK und W. DZIEWULSKI. Berlin (1912).

[3] Upsala, Nova Acta R Soc Ser IV, vol 2 (1909).

The so-called phenomenon of KAPTEYN is shown in the sense that the stars between $4^m,5$ and $7^m,0$ in the Milky Way have a bluer colour than the outside stars. But SCHWARZSCHILD has pointed out that there is, so far, no such phenomenon if the latter means, that stars of the same spectral class should have a more blue colour in the Milky Way than outside it. He thinks that KAPTEYN's phenomenon is the same as that discovered by E. C. PICKERING, according to which the Milky Way contains more white stars than the outside regions. The phenomenon is illustrated in this fourfold distribution:

	White stars	Yellow stars
Near Milky Way	545	492
Far from Milky Way . .	278	391

PICKERING is of opinion that the phenomenon is very pronounced. In fact the correlation Milky Way/Colour is rather small. Computing the Bernoullian coefficient of correlation, r_β, I have found:

$$r_\beta = +0{,}123 \pm 0{,}024$$

thus a low value which propably not is real.

The relation between the colour indices at Göttingen and the colours estimated at Potsdam is found to be:

PD	Mean colour	$\bar{C}$	PD	Mean colour	$\bar{C}$
W	$2^c,4$	$-0^m,27$	WG	$5^c,2$	$+0^m,75$
W+	2 ,5	−0 ,25	WG+	5 ,5	0 ,88
GW−	2 ,8	−0 ,14	G−	6 ,1	1 ,20
GW	3 ,3	+0 ,01	G	6 ,4	1 ,37
GW+	3 ,8	+0 ,18	G+	6 ,5	1 ,43
WG−	4 ,1	+0 ,22	RG−	6 ,7	1 ,55
			RG	6 ,9	1 ,72

Also here the jump between WG− and WG is very evident.

The relation between the colours of OSTHOFF and the colour indices is given by means of the formula:

$$\text{Colour of OSTHOFF} = 3{,}25 + 3{,}01\, C - 0{,}522\, C^2.$$

SCHWARZSCHILD estimates the mean error in the directly estimated colour to be $= 0^m,14$, which proves that the estimates of OSTHOFF are of considerable value.

111. The Absolute Colour Index Scale from the Göttinger Aktinometrie. If the eye were sensible only to the wave length $\lambda_1 = 5700$ and the photographic plates only to $\lambda_2 = 4250$, the two intensities would be[1]:

$$J_1 = C\lambda_1^{-5}\left(e^{\frac{c_2}{\lambda_1 T}} - 1\right)^{-1}; \qquad J_2 = C\lambda_2^{-5}\left(e^{\frac{c_2}{\lambda_2 T}} - 1\right)^{-1}$$

and the corresponding intensities in magnitude:

$$m_v = +2{,}5 \log \lambda_1 \left(e^{\frac{c_2}{\lambda_1 T}} - 1\right) + K_1,$$

$$m_{ph} = +2{,}5 \log \lambda_2 \left(e^{\frac{c}{\lambda_2 T}} - 1\right) + K_2,$$

and the colour index C:

$$C = m_{ph} - m_v = 2{,}5 \log \frac{\lambda_2}{\lambda_1} \frac{e^{\frac{c_2}{\lambda_2 T}} - 1}{e^{\frac{c_2}{\lambda_1 T}} - 1} + K.$$

The scales can be defined in such a way that $K = 0$. Thus:

$$C_{\text{Abs.}} = 2{,}5 \log \frac{J_1}{J_2} + 10 \log \frac{\lambda_1}{\lambda_2}.$$

[1] Göttinger Aktinometrie Teil II, p. 29 (1912).

The value $C_{Abs.}$ or the absolute colour index has the property that its value is zero for a body having infinitely high temperature.

c_2/T	T	$C_{Abs.}$	$C_{Harv.}$	$C_{Pots.}$	Mean Harv., Potsd.
0,0	∞	$0^m,00$	$-0^m,05$	$-0^m,14$	$-0^m,10$
0,5	28400°	0 ,18	+0 ,32	−0 ,04	+0 ,14
1,0	14200	0 ,43	0 ,57	+0 ,29	0 ,43
1,5	9500	0 ,70	0 ,88	0 ,56	0 ,72
2,0	7100	1 ,00	1 ,14	0 ,86	1 ,00
2,5	5700	1 ,31	1 ,45	1 ,37	1 ,41
3,0	4700	1 ,63	1 ,77	1 ,49	1 ,63
3,5	4100	1 ,95	2 ,09	1 ,81	1 ,95
4,0	3600	2 ,27	2 ,41	2 ,13	2 ,27
4,5	3200	2 ,59	2 ,73	2 ,45	2 ,59
5,0	2800	2 ,91	3 ,05	2 ,87	2 ,96
5,5	2600	3 ,23	3 ,37	3 ,09	3 ,23
6,0	2400	3 ,55	3 ,69	3 ,41	3 ,55

The zero point of the observed colour indices has been shifted in order to fit the absolute colour index of the Sun.

112. The Exposure-Ratio Method. This method was outlined by F. H. SEARES in 1916[1]. It consists in determining the ratio of exposure times necessary to produce photographic or photovisual images, or blue and yellow images, of the same size. The both images should be on the same plate in order to reach the highest possible precision. An isochromatic plate exposed behind a yellow filter registers the yellow image as usual. The same plate used without filter gives the blue image. To some extent also longer wave lengths will be included in this image, but owing to the relatively small sensitiveness of the isochromatic plate to the yellow and orange rays, the shorter wave lengths will still be of predominating influence.

In practice a number of blue images with the exposure time increasing in geometrical progression have to be impressed on the plate together with at least one yellow image. The diameters of the blue images, or their scale readings, are plotted against the logarithms of exposure time and from the nearly linear curve thus derived can be read the exposure time for a blue image of the same size as the yellow image considered. The ratio of the interpolated exposure to that which produced the yellow image, or the exposure-ratio compared with similar ratios for stars of known colour, gives then the colour index.

Considerations should be given to the following points. First, if there are differences of gradation between the blue and yellow images, the exposure-ratios of bright and faint stars of the same colour will be different or in other words they will depend upon the size of the images. The matter is investigated by observing the same stars with different apertures and exposure times. It is advisable to adjust exposure and aperture so, that the resulting yellow image is of a certain standard size. Small deviations in size of the standard image are of no consequence.

Second, the atmospheric extinction reduces the intensity of blue light more than that of yellow and thus modifies the exposure-ratio. The effect is differential but still appreciable if the zenith distance is not very small. Tables of corrections can be derived by determining the exposure-ratio for the same star at different zenith distances.

Third, care should be taken to bring the plates into equilibrium with the atmospheric conditions before beginning the exposures. Otherwise the absorption

[1] Mt Wilson Comm No. 33 (1916).

of moisture during the exposure may introduce variations of sensitiveness and gradation which are capable of influencing seriously the results. This precaution, of course applies to any photographic method.

SEARES states in a second communication[1] that the method is expeditious, and under favourable conditions precise. It is entirely independent of stellar magnitudes and hence avoids the systematic errors which so easily enter as a result of uncertainties in the magnitude scale or in their zero points.

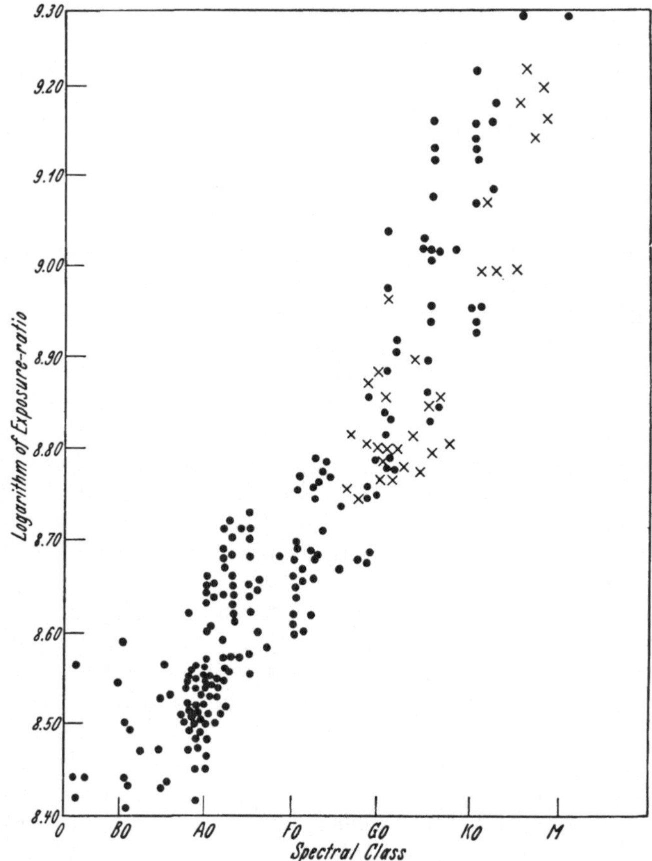

Fig. 89. Relation between logarithm of exposure-ratio, and spectral class according to SHAJN's measures of binaries.

The method has been used at Mount Wilson for derivation of the colours of stars in the North Polar Sequence[2]. Some of the results found at this work have been briefly mentioned in an earlier cipher.

One of the most interesting consequences of the determinations of the colours of stars near the Pole is to be found in the confirmation of an earlier result by SEARES, namely, that there are no faint stars in this region with negative or small positive colour indices. The lower limit of the colour index gradually increases as the fainter stars are approached and at the 16th photographic magnitude its value is of the order of 0,5 magnitude. The absence of faint white stars is known to be a characteristic of other regions as well but, on

[1] Mt Wilson Comm No. 38 (1916).
[2] Mt Wilson Contr No. 235 (1922) = Ap J 56, p. 97.

the other hand, SEARES has also examined certain Milky Way regions where stars of the 14th or 15th magnitude are nearly normal in colour and thus must include a considerable number of objects that are white. — SEARES's exposure-ratio method has been applied by G. SHAJN[1] for the determination of the colour indices of double stars. For that purpose the normal astrograph at the Pulkova Observatory was used. At first it was tried to determine λ_{eff} but this work was soon given up, especially on account of the wellknown absence of variation in λ_{eff} between spectral classes B and F5, and instead the method of exposure-ratio was adopted. The precautions advised by SEARES were taken.

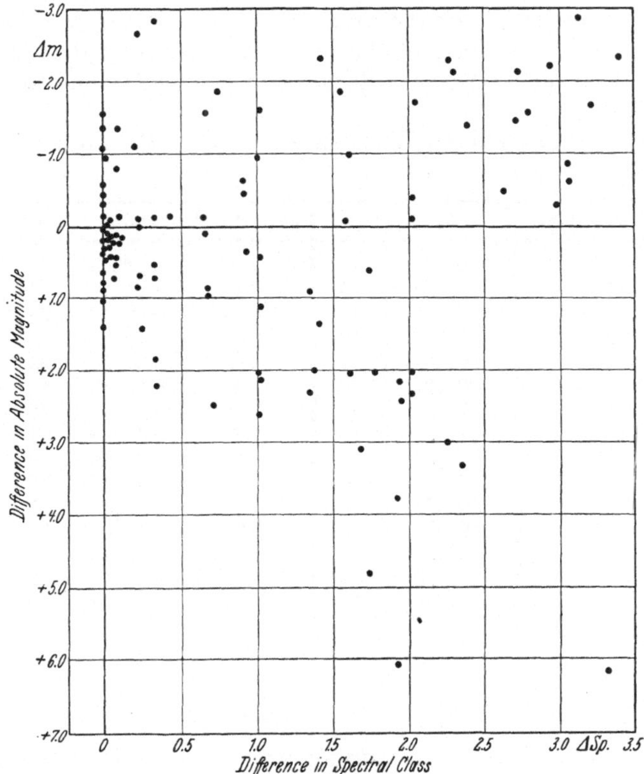

Fig. 90. The difference in apparent or absolute magnitude as a function of the difference in spectral index or spectral class (G. SHAJN).

As a normal for the colour index, the system of KING was selected. The following formula converts the logarithm of exposure-ratio, E, into C_{King}:

$$C_{\text{King}} = -0{,}30 + 2{,}109\,(\log E - 8{,}42)\,.$$

Altogether 123 pairs were observed for E. The relation between exposure-ratio and spectrum is shown in the fig. 89. SHAJN is of opinion that the dispersion is quite real, even within the separate groups of giants and dwarfs. He suggests as explanation the individual physical properties of stars. The said dispersion illustrates the difficulties involved in translating colours equivalents into spectra and vice versa.

---

[1] Results of Observations of the Double Stars and their Relation to the Giants and Dwarfs Theory, Pulk Bull X, p. 276 (1925).

The variation of Δm or ΔM with ΔS as given in the following table and illustrated in the figure 90 is of interest:

$\Delta \log E$	ΔM	ΔS	n	$\Delta \log E$	ΔM	ΔS	n
−0,40 to −0,76	1^M,29	2^S,70	15	0,00—0,05	0^M,53	0^S,07	20
−0,30 „ −0,40	1 ,19	1 ,80	4	0,05—0,10	1 ,27	0 ,77	12
−0,20 „ −0,30	0 ,94	0 ,77	4	0,10—0,20	1 ,54	1 ,17	9
−0,10 „ −0,20	0 ,81	0 ,47	3	0,20—0,30	2 ,87	1 ,73	7
−0,05 „ −0,10	0 ,75	0 ,25	3	0,30—0,40	2 ,94	1 ,80	2
−0,00 „ −0,05	0 ,81	0 ,14	18	0,40—0,60	6 ,30	3 ,40	1

This table shows that for the same ΔS, ΔM is larger for dwarf systems than for giant ones. The form of the curve also suggests that an integration for finding M as a function of S will not give as result a one to one relation. (Compare ciph. 139.)

113. Yerkes Actinometry[1]. The relation between PARKHURST's spectra and colour index is as follows:

Spectral class	C	n	Spectral class	C	n
B2—B6. . .	-0^m,36	5	F2	$+0^m$,43	11
B7. . . .	−0 ,23	3	F3	+0 ,46	4
B8. . . .	−0 ,16	12	F4—F6	+0 ,60	24
B9. . . .	−0 ,14	10	F8—F9	+0 ,70	6
A0. . . .	−0 ,01	55	G0	+0 ,90	22
A1. . . .	+0 ,02	22	G1	+0 ,88	11
A2. . . .	+0 ,12	30	G2	+0 ,95	25
A3. . . .	+0 ,12	23	G3	+1 ,08	13
A4. . . .	+0 ,22	17	G4	+1 ,06	9
A5. . . .	+0 ,22	20	G5	+1 ,12	25
A6. . . .	+0 ,27	12	G6—G9	+1 ,24	7
A7. . . .	+0 ,31	7	K0	+1 ,29	16
A8. . . .	+0 ,31	9	K1—K3	+1 ,40	13
A9. . . .	+0 ,39	1	K4—K6	+1 ,59	32
F0. . . .	+0 ,33	20	K8	+1 ,64	1
F1. . . .	+0 ,33	9	M	+1 ,73	5
			N	+2 ,74	1

The straight line:

$$S + 1{,}11 = 2{,}22\, C$$

represents the colour indices of PARKHURST very closely; S is the spectral index, C = the colour index and $S = 0$ for spectral class B0, $S = 5{,}0$ for spectral class M.

The colour-index scale of the three principal systems is given here:

Spectral class	KING	PARKHURST	SCHWARZSCHILD	Mean colour index
B0	-0^m,31	-0^m,6	-0^m,41	-0^m,42
B5	−0 ,17	−0 ,28	−0 ,27	−0 ,21
A0	0 ,00	−0 ,06	−0 ,07	0 ,00
A5	+0 ,18	+0 ,18	+0 ,13	+0 ,21
F0	+0 ,32	+0 ,40	+0 ,33	+0 ,42
F5	+0 ,52	+0 ,63	+0 ,53	+0 ,63
G0	+0 ,71	+0 ,86	+0 ,77	+0 ,84
G5	+0 ,90	+1 ,08	+1 ,03	+1 ,05
K0	+1 ,16	+1 ,32	+1 ,28	+1 ,26
K5	+1 ,42	+1 ,55	+1 ,73	+1 ,47
M.	+1 ,67	+1 ,77	+2 ,1	+1 ,68

[1] Ap J 36, p. 169 (1912).

114. MALMQUIST's Colour Catalogue. In order to determine for a sufficiently large area around the north pole of the Galaxy the colour indices of all stars down to 14^m K. G. MALMQUIST[1] has applied the method of F. H. SEARES and determined the colour indices for 3700 stars. The instrument used was the reflector of the Bergedorf Observatory. The exposure ratios were translated into colour indices by the aid of the North Polar Sequence, by means of which corrections were applied for the influence of the magnitude of the stars (or the effect of gradation). From a comparison with SEARES the mean errors were found to vary between $\pm 0^m,089$ and $\pm 0^m,175$ for one estimate according to the quality of the image. From the deviation from the mean values of C the probable error was found to vary between the limits $\pm 0^m,090$ and $\pm 0^m,184$. It was also found that no systematic difference was present between the colour indices of SEARES and those of MALMQUIST, neither when they are grouped according to their angular distance from the optical axis nor according to the apparent photovisual magnitude or the colour index. A small correction has to be applied on account of the fact that the assumption of a linear connection between scale reading and colour index was not strictly valid. After this correction has been applied the accordance with the system of SEARES is perfect.

The magnitudes and the colour indices have been used for a number of stellar-statistical investigations. The stars for which $0^m,00 \leqq C \leqq 0^m,24$ were separated and their distribution in space investigated. Only 39 stars belonging to the material, when stars fainter than 13^m were omitted, were found in the direction of the Galactic Pole. The density in space was derived. Within 800 light-years the spatial density is sensibly constant, but then decreases rapidly and at a distance of 3300 light-years is only 0,01 of the value around the Sun. The boundary of the stellar system for stars of the colour class in question seems to have been reached.

At Harvard[2] about five hundred stars within the region investigated by MALMQUIST have been classified by Miss CANNON. If the spectra are translated into colour indices it seems that a considerable systematic difference exists, inasmuch as $C_{\text{Lund}} - C_{\text{Harv.}}$ is on an average $= -0^m,497$ (502 stars). If KING's table for converting spectra into colour indices is used instead of MALMQUIST's the residual is reduced to about half the value. It is not possible at present to explain the discrepancy, which is an object for further investigations at Lund as well as at Harvard. It seems that the connection between faint spectra and colours is not as solid as in the case of bright stars. Part of the discrepancy may perhaps be ascribed to the effects of luminosity, which might be considerable in the case of absolutely faint dwarfs.

115. The Spectral Index. The spectral index S is defined as the values in a numerical scale of the spectral classes of Harvard. The most common notation supposes the scale to be linear and selects the zero point at A0 = 0. Thus F0 = 1 and M0 = 4. (See also ciph. 112.)

C. V. L. CHARLIER[3] has computed the correlation between S and C (colour index) and has found the very high value of $r = +0,958 \pm 0,011$, from which the following regression lines have been deduced:

---

[1] Lund Medd Ser. II, No. 37 (1927).

[2] Harv Bull No. 859 (1928).

[3] Lund Medd Ser. II, No. 19 (1918). See also Introduction to Stellar Statistics. Lund (1921).

Spectral class	s	c	c	s	Spectral class
B0	$-1^s,0$	$-0^m,46$	$-0^m,4$	$-0^s,70$	B3
B5	$-0,5$	$-0,28$	$-0,2$	$-0,30$	B7
A0	$0,0$	$0,00$	$0,0$	$+0,10$	A1
A5	$+0,5$	$+0,23$	$+0,2$	$+0,50$	A5
F0	$+1,0$	$+0,46$	$+0,4$	$+0,90$	A9
F5	$+1,5$	$+0,69$	$+0,6$	$+1,30$	F3
G0	$+2,0$	$+0,92$	$+0,8$	$+1,70$	F7
G5	$+2,5$	$+1,15$	$+1,0$	$+2,10$	G1
K0	$+3,0$	$+1,38$	$+1,2$	$+2,50$	G5
K5	$+3,5$	$+1,61$	$+1,4$	$+2,90$	G9
M0	$+4,0$	$+1,84$	$+1,6$	$+3,30$	K3
			$+1,8$	$+3,70$	K7
			$+2,0$	$+4,10$	M1

The frequencies of the spectral classes as given earlier in this chapter can be used for the derivation of a frequency table of the distribution of the colours in N H D.

The spectral classes in the New Draper Catalogue thus can be translated to colour indices. Through the relation in Harv Ann 70 the colour index has been computed for the stars in the New Draper Catalogue and from that the photographic magnitude has been given.

Another source for obtaining colour indices en masse would be a comparison of the Cape Photographic Durchmusterung with the Cordoba Durchmusterung, or of the Cape Astrographic Zones (the corrected photographic scale) with the Cordoba Durchmusterung. In neither case can any individual accuracy be expected, but the mean colour indices ought to be statistically correct.

116. STERNBERCK's Investigations[1]. This investigation was started in order to test the exposure-ratio method suggested by SEARES[2]. At first the relation between the exposure-ratio method and the laws of photographic photometry is discussed. From the fact that KRON's results, which were reached by the use of light sources in the laboratory, have been confirmed when applied to stellar photometry by KRON himself, STERNBERCK concludes that KRON's density law: $S = \varphi(s)$, $s = t\psi(J)$ (where S is the opacity of the point considered, J its intensity and t the exposure time) is valid for photographic as well as for photovisual rays (the constants will, of course, be different). Further it is known that:

$$\psi(J) = J^{q(\lambda)}.$$

Denoting with t_{pv} and t_p the exposure times necessary to impress photovisual and photographic images of a certain density we have:

$$t_{pv}\,\psi_1(J_{pv}) = K_1,$$
$$t_p\,\psi_2(J_p) = K_2.$$

The ratio:

$$\frac{J_{pv}}{J_p} = c$$

is a characteristic of the star colour. From these equations we obtain:

$$\log\frac{t_p}{t_{pv}} = \log\varepsilon = \Psi(c, J_p).$$

Thus $\log\varepsilon$ contains a magnitude equation, which is partly dependent on the variation of $p = 1/q(\lambda)$ and which is called the p-effect.

---

[1] Berlin-Babelsberg Veröffentlichungen. Bd. V, H. 2 (1924).

[2] Mt Wilson Comm No. 33 (1916), 38 (1917), and 59 (1919); Mt Wilson Contr No. 235 (1922).

Methods are derived in order to eliminate the influence of the magnitude equation. STERNBERCK finds that the p-effect curve can be used for all plates of the same brand (emulsion).

The author points out the considerable difficulties met with when a photographic refractor is used and photovisual images are needed. Thus the atmospheric conditions ("sight") seem to exercise a larger influence on the photographic images than on the photovisual ones. When the extrafocal method was used (the extra-focal photographic images had to be compared with intra-focal photovisual images) the disks of the stars were not uniform on account of the zonal errors, but could be measured with accuracy.

As this method involved the use of filters two other methods were also tested. In 1916, in a paper written in Russian, TICHOV[1] had suggested covering the central region of the objective with a round disk. Later on N. TAMM[2] independently suggested the same method for determining colour equivalents. An extra-focal plate will give a central image of the yellow light surrounded by a ring representing the blue rays. The intensity ratio of the ring to the point

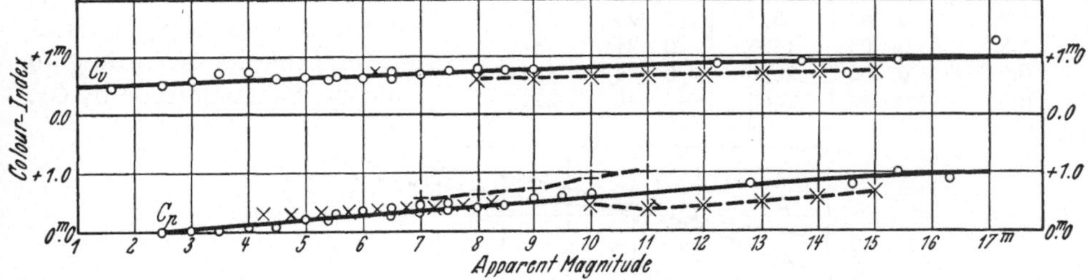

Fig. 91. The colour index as a function of apparent magnitude (SEARES). The curves give the function according to visual grouping (C_v), and photographic grouping (C_p). Circles are data from Revised Harvard Photometry and for faint stars from Mount Wilson measures at the Pole; crosses on full-line curve are data from Göttinger Aktinometrie. The dotted curves are based on KREIKEN's determinations of colour indices in Milky Way fields and show the influence of the concentration of blue stars in this region.

gives a measurement of the colour. STERNBERCK modified the method in such a way that the photographic images were measured by the density of the rings, and the photovisual by the diameter of the central images. The ratio of the exposure times, necessary to give a certain density and a certain diameter, gave the colour equivalent. Lastly the central screen was taken away and stars photographed near the yellow focus. In the central parts the blue rays dominate. The mixture of the different sorts of light results in a decrease of the sensibility, especially for white stars. The time necessary for these exposures becomes comparatively long.

The observations were performed in 1922—23, the Berlin-Babelsberg refractor being used. On 98 plates some 2000 exposures were impressed, and colour equivalents were derived for 182 stars.

E. ÖPIK[3] has applied G. TICHOV's method to the determination of the colour indices of 104 stars. The circular diaphragm in front of the objective makes an extra-focal photograph appear as a ring (radiation for which the system is achromatized) surrounding a point (radiation with its focus in the plane of the plate). In order to eliminate the PURKINJE-effect scales were so constructed that rings were compared with rings and points with points.

[1] Bull de l'Acad Russe des Sciences 1916. The paper has been published in English A N 218, p. 145 (1923). [2] V J S 56, p. 206 (1921), and A N 216, p. 331 (1922).

[3] Dorpat (Tartu) Publ 26, No. 3 (1925).

The mean error of a colour index averages $\pm 0^m,09$. An interesting result of this investigation is the rather pronounced luminosity effect in the colour indices observed.

117. The Mean Colour Index as a Function of m. A knowledge of the mean colour is of much importance in statistical discussions of stellar distribution. The material available is affected by some uncertainties, but none of these is of serious consequence. Some of the magnitude scales require small corrections in order to be reduced to a homogeneous system. Since the change in mean colour with magnitude is small, the influence of scale error is small.

F. H. SEARES[1] has discussed the dependence of C on m. The following sources have been used:

Revised Harvard Photometry. The mean colour indices for grouping according to visual magnitude are denoted by C_v, and the corresponding quantities for grouping according to photographic magnitude by C_p.

Apparent magnitude	C_v	n	C_p	n
1,6	$+0^m,49$	55		
2,5	0 ,53	36	$+0^m,00$	23
3,0	0 ,61	88	0 ,03	45
3,5	0 ,73	130	0 ,08	84
4,0	0 ,75	249	0 ,09	148
4,5	0 ,64	458	0 ,17	299
5,0	0 ,68	876	0 ,24	500
5,5	0 ,63	1563	0 ,31	826
6,0	0 ,66	2658	0 ,34	1344
6,5	+0 ,72	2669	+0 ,40	2217

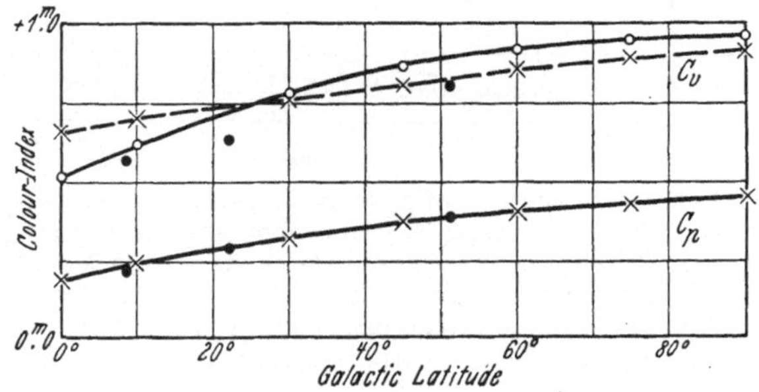

Fig. 92. The dependence of colour index on galactic latitude according to visual grouping (C_v), and photographic grouping (C_p).
Points = Revised Harvard Photometry, circles = HENRY DRAPER Catalogue, crosses = Göttinger Aktinometrie.

The mean colour index from the HENRY DRAPER Catalogue of 98675 stars brighter than $8^m,75$ was used according to the following synopsis:

Apparent magnitude	C_p	C_v	Apparent magnitude	C_p	C_v
5,4	$0^m,20$	$0^m,61$	8,5	$0^m,42$	$0^m,76$
6,5	0 ,29	0 ,66	9,0	0 ,50	0 ,74
7,0	0 ,33	0 ,68	9,5	0 ,58	
7,5	0 ,39	0 ,75	10,0	0 ,68	
8,0	0 ,38	0 ,78			

Here all galactic latitudes were taken together. If the stars brighter than $8^m,25$ were considered to a number of 58508, SEARES found:

Galactic latitude	C	Galactic latitude	C
0°	0,51	60°	0,94
10	0,62	75	0,94
30	0,79	90	0,93
45	0,86		

[1] Mt Wilson Contr No. 287 (1925).

Further the data of the Göttinger Aktinometrie and Yerkes Actinometry were used as well as KREIKEN's C_λ (colour indices derived from effective wave lengths) and the colour indices of the polar stars and of stars in the Harvard Standard regions.

From a combination of these data the following regression lines resulted for all latitudes together:

$$C_v = 0^m{,}50 + 0{,}029\, m_v \text{ (visual grouping)}$$
$$C_p = -0^m{,}18 + 0{,}071\, m_p \text{ (photographic grouping).}$$

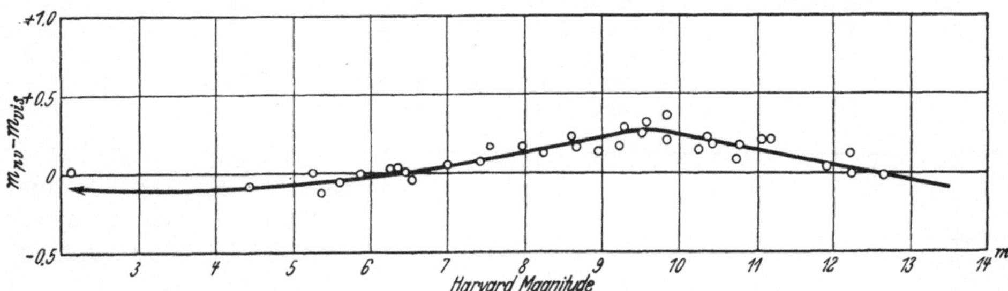

Fig. 93. Reduction of Harvard visual scale to Mount Wilson photovisual system. The ordinates give the difference $m_{pv} - m_{vis}$ as reduced to the photovisual colour system. The cause of the considerable maximum at $9^m{,}5$ is as yet unexplained.

A variation with galactic latitude was also indicated, but could not be established with much accuracy, as most of the data bearing on this change refer to stars between $5^m{,}7$ and $7^m{,}4$.

118. Yerkes Photometry of Selected Areas[1]. The mean colour index according to different galactic latitudes and m is as follows:

Apparent magnitude	$b = 7°$		$b = 20°$		$b = 55°$	
	$\bar{C}$	n	$\bar{C}$	n	$\bar{C}$	n
9^m	$+0^m{,}55$	28	$+0^m{,}56$	24	$+0^m{,}45$	17
10	0 ,53	64	0 ,76	45	0 ,40	28
11	0 ,70	112	0 ,55	102	0 ,36	29
12	0 ,70	306	0 ,46	183	0 ,26	73
13	0 ,73	147	0 ,52	171	0 ,24	72

The distribution of the spectral types in galactic and non-galactic regions of 1470 stars is in percentages:

Spectral class	Galactic regions $b = -18°$ to $+28°$	Non-Galactic regions $b = +39°$ to $+72°$
B	7	12
A	8	10
F	15	24
G	27	33
K	31	20
M	11	2

The deficiency of K and M stars outside the galactic regions is an interesting feature.

119. Visual Effective Wave Lengths. The effective wave length was first used as a colour equivalent by G. C. COMSTOCK[2], who measured 51 stars visually with a mean error of ± 50 A for a single determination. The range in λ_{eff} is comparatively small in the visual region and on that account not much has been

[1] Yerkes Publ IV, Pt VI (1927). [2] Ap J 5, p. 26 (1897).

done since. A contribution was made by H. E. LAU[1], who measured 68 stars. The method has recently been improved by H. GRAMATZKI[2]. A cylindrical lens is used so that the spectra of the first order become lines instead of points. This permits a higher accuracy in the measurements. Results have been published for 80 stars.

Spectral class	GRAMATZKI	LAU	COMSTOCK
B	5451 ± 9 A	5653 A	5630 A
A (white) . . .	5449 ± 9		
A (yellow) . . .	5539 ± 8	5659	5640
F	5510 ± 13	5708	5682
G	5551 ± 20		
K	5567 ± 13	5768	5738
M	5692 ± 34	5847	

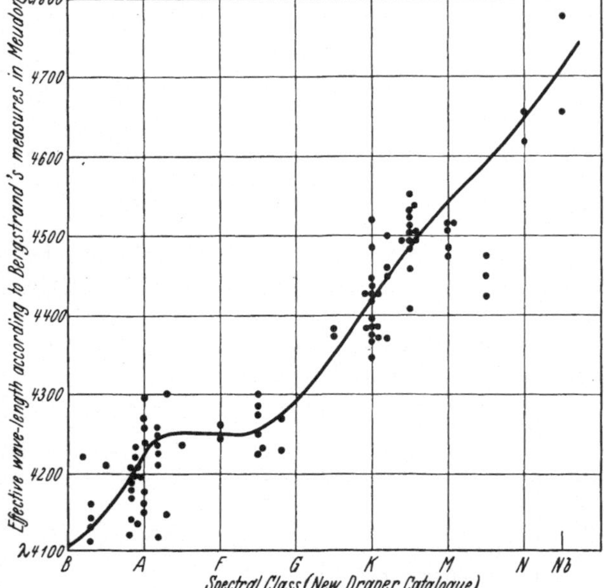

Fig. 94. Relation between the measures of λ_{eff} of BERGSTRAND at Meudon (1909) and modern determinations of spectral class.

The larger range in the case of GRAMATZKI is due to the better achromasy of a reflector as compared with a refractor.

120. BERGSTRAND's Work in Meudon. Ö. BERGSTRAND[3] was the first to apply the idea of HERTZSPRUNG of using the effective wave lengths as colour equivalents. With the aid of the reflector at the Observatory of Meudon, in front of which a grating with parallel wires was placed, a number of exposures were secured of 92 stars, having different colours according to the direct estimates. The measurements were reduced to a common value of the diameter. The following relation was found between the colours in P D and the effective wave lengths:

Colour in Potsdam Durchmusterung	λ_{eff}	n	Colour in Potsdam Durchmusterung	λ_{eff}	n
W	4199 A	10	WG+	4434 A	8
W+	4207	7	G−	4440	6
GW−	4185	8	G	4484	6
GW	4218	12	G+	4451	5
GW+	4244	4	RG−	4485	7
WG−	4334	4	RG	4496	5
WG	4377	6			

The λ_{eff} does not vary much for the colours W to GW+. The brusque transition from GW+ to WG− does not depend on any irregularity in the scale of P D. The same phenomenon is also found when the scale of OSTHOFF is used.

[1] A N 173, p. 81 (1906). [2] A N 222, p. 145 (1924).
[3] Upsala Nova Acta R Soc, Ser. IV, Vol. 2 (1909).

The relation between spectral type and λ_{eff} is the following:

Spectral class	λ_{eff}	n	Spectral class	λ_{eff}	n
B5 . . .	4165	5	K1 . . .	4436	16
A1 . . .	4230	10	M . . .	4497	7
F8 . . .	4380	3	N . . .	4655	2

In the curve in fig. 94 we have used the spectral types in the New DRAPER Catalogue.

Also some experiments were made in which a colour screen was used together with the grating; it was possible to reach a greater accuracy in these cases, but on the other hand the amplitude of the λ_{eff} is very small, as will be seen from the following:

Class of stars	λ_{eff}
White stars	5492
Yellow stars	5535
Reddish stars	5550

The values of λ_{eff} do not seem to depend on the length of the exposure.

The combination of screen and grating will no doubt be of much importance for the solution of several other problems besides the determination of the colour equivalents.

121. HERTZSPRUNG's Studies. The general accuracy of the method of effective wave lengths and their possibility as colour equivalents were investigated by E. HERTZSPRUNG in a comprehensive paper in 1911[1]. The material consisted of numerous plates and exposures secured at the Urania Observatory at Copenhagen. Altogether the effective wave lengths of 276 objects, which were situated in the open clusters: Pleiades, Hyades, Praesepe, and Coma Berenices, or were specially selected stars, were determined. Other evidence concerning the colour of the stars within the mentioned clusters was also collected and the colours were compared with the effective wave lengths.

HERTZSPRUNG points out that the spectrum must be a function of the temperature as well as of the absolute magnitude and emphasizes the importance of the spectral classification of Miss MAURY, as it gives a criterion for a classification of absolute magnitudes into two subdivisions according to the c- and the a- and b-characteristics.

The general accuracy of the λ_{eff} corresponds to a mean error of ± 20 A, which is equal to a mean error in the colour index of $\pm 0^m,10$. Thus it is evident that the method can compete with other schemes for measuring colour equivalents. The change of λ_{eff} with the apparent intensity of the (central) image on the plate is a troublesome correction, and covers a spectral interval of 125 A. It is dependent on the spectral class, although the dependence could not be determined with sufficient accuracy in this case. Furthermore λ_{eff} is dependent on the thickness of the atmospheric layer traversed by the light of the object.

The effective wave lengths of the Pleiades show a steady increase towards red with decreasing (absolute) magnitude. Between 6^m and 11^m the change $\frac{\Delta\lambda}{\Delta m}$ is roughly 13 A. In the case of the Hyades such a change cannot be found in the λ_{eff}, but is present in the colour index. HERTZSPRUNG classified the spectra of the Hyades stars by estimating the line-ratio $\frac{K}{K + \text{hydrogen lines}}$. The results were in good agreement with those of SCHWARZSCHILD. Plotting the line-ratios against the magnitudes, a progressive rate was found which corresponds

[1] Potsdam Publ No. 63 (1911).

to $\frac{\Delta C}{\Delta m} = 0^m,12$. In the Hyades group there are certain stars of high luminosity but red colour. These correspond to the red giants according to our present terminology.

The Sun falls into the colour-luminosity diagram of the Hyades group.

The Praesepe stars show the same behaviour as the stars in the Hyades cluster. In Coma Berenices it is not possible to distinguish the physical members from the background stars. The absolutely faint stars with measured parallaxes show a marked increase in the effective wave length.

During his stay at Mount Wilson HERTZSPRUNG measured 184 stars in the open cluster N G C 1647 for effective wave lengths[1]. The following connection between the colour index C and λ_{eff} was found:

$$C = 0{,}005\, \lambda_{\text{eff}} - 21{,}17.$$

A steady increase in λ_{eff} with decreasing brightness was found:

$$\lambda_{\text{eff}} = 4340 + 30{,}5\,(m - 12{,}76)$$

hence $\frac{\Delta\lambda}{\Delta m} = 30{,}5$ A.

Furthermore 42 stars of low absolute magnitude were measured[2]. Plotting the λ_{eff} as a function of the absolute magnitudes HERTZSPRUNG found that, on the supposition that the stars are black-body radiators, the relation between absolute brightness and colour as established from stars of absolute magnitudes between -2^M to $+8^M$ deviates from the same theoretical relation in the same way and to about the same extent as we should expect from the known increase in density and decrease in mass with decreasing absolute brightness. Between $+8^M$ and $+13^M$ the deviations from the black-body radiation are too large to be explained by facts already known.

The change $\frac{\Delta\lambda}{\Delta m}$ was found to be 26 A, which is in good agreement with the results from the investigation of N G C 1647.

SEARES[3] has determined the colour indices of 47 stars in this cluster and found the relations:

$$163\, C_{\text{HERTZSPRUNG}} = \lambda_{\text{eff}} - 4260$$

$$C_{\text{HERTZSPRUNG}} = C_{\text{SEARES}} - 0^m{,}16\,.$$

He also concluded that as regards the question of accuracy there seems to be little choice between the two methods.

122. Determinations of Effective Wave Lengths at Upsala and Lund. Ö. BERGSTRAND and B. LINDBLAD[4] measured 44 stars in order to see if the Zeiss twin-astrograph at Upsala Observatory could be advantageously used for the λ_{eff}. It was found that the amplitude of the λ_{eff} was somewhat smaller than in the case of a reflector, but still large enough for a fair determination of the λ_{eff}. Although the material is small the authors find a sudden change in λ_{eff} between the Potsdam colours WG— and WG (corresponding to λ 4243 and λ 4348 respectively).

A case where the amplitude of λ_{eff} is rather too small for accurate measurements of the effective wave lengths has been investigated by K. G. MALMQUIST and J. OHLSSON[5] with the aid of the astrographic camera at Lund ($a = 18$ cm, $F = 90$ cm). They have found the following relation between spectral class and λ_{eff}:

Spectral class	Effective wave length	Spectral class	Effective wave length
A5	4296	G5	4369
F0	4293	K0	4347

[1] Ap J 42, p. 92 (1915). [2] Ap J 42, p. 111 (1915). [3] Ap J 42, p. 120 (1915).
[4] Ark Mat Astr Fys 11, No. 17 (1916). [5] Lund Medd Ser. I, 101 (1921).

Thus it will be difficult to deduce the spectral classes from the λ_{eff} except for the very red stars.

123. Lindblad's Investigations. An extensive methodical investigation concerning the use of the λ_{eff} as colour equivalents is made by LINDBLAD in a paper of 1917[1]. The instrument used was the Upsala twin-astrograph and altogether 79 stars were measured. The curves giving the correlation for the relative size of the diameter of the central image were derived by means of successive approximations and had a rather complicated course. A new determination of the value $\Delta\lambda/\Delta_{\text{atm. layer}}$ was made and the value 16,1 was found. The following determination of the mean error of the λ_{eff} shows the general accuracy of the method:

Diameter value	$\varepsilon(\lambda)$	ε_D
0,045 mm	±30	
0,070 ,,	±21	±0,004 mm
0,090 ,,	±17	

The following formula connects the diameter D, exposure t, and magnitude m:

$$m = 14{,}30 - 4{,}08 \log D + 2 \log t.$$

Eight hours is the longest exposure time in practice with this instrument. The faintest stars that can be measured for λ_{eff} by means of the instrument in question are thus of magnitude $13^{\text{m}},6$.

The relation between spectral class and λ_{eff} is:

Spectral class	λ_{eff}	n	Spectral class	λ_{eff}	n
Oe5	4123	2	F5—F8 . .	4236	5
B0—B3 . .	4118	6	G0	4281	3
B5	4112	4	G5	4296	4
B8	4159	5	K0	4326	9
A0	4179	9	K5	4420	3
A2	4192	6	Ma	4434	4
F0	4229	4			

The relation between the colour index in the Göttinger Aktinometrie and λ_{eff} is:

C	λ_{eff}	C	λ_{eff}
$-0^{\text{m}},59$	4118	$+0^{\text{m}},30$	4236
$-0\ ,45$	4112	$+0\ ,74$	4281
$-0\ ,39$	4159	$+0\ ,83$	4296
$-0\ ,27$	4179	$+1\ ,06$	4326
$-0\ ,22$	4192	$+1\ ,50$	4420
$+0\ ,13$	4219	$+1\ ,60$	4434

For computing the absolute temperature the following formula was derived:

$$\lambda_{\text{eff}} = 4034 + 114 \frac{10^4}{T}.$$

The following table evaluates the Potsdam colours in λ_{eff}:

Potsdam colour	λ_{eff}	n	Potsdam colour	λ_{eff}	n
W	4160	5	WG—	4240	1
W+	4159	18	WG	4283	9
GW—	4165	19	G—	4385	7
GW	4222	17	G	4333	3
GW+	4208	3	G+	4397	2
			RG— to RG	4417	2

[1] Ark Mat Astr Fys 13, No. 26 (1917).

The following relation between OSTHOFF's colours and λ_{eff} was found:

OSTHOFF's colours	λ_{eff}	n	OSTHOFF's colours	λ_{eff}	n
2^c,16	4152	10	4^c,60	4289	4
2 ,71	4165	13	5 ,11	4294	7
3 ,19	4184	7	5 ,60	4362	1
3 ,63	4227	3	6 ,23	4411	3
4 ,18	4259	5	6 ,61	4413	7

Using the first spectrographic parallaxes of ADAMS and JOY (500-star list), LINDBLAD derived for K stars

App. magn.	$\Delta C/\Delta M$	n
2^m-3^m	−0,10	16
3 −4	−0,07	15
4 −5	−0,25	8

In the mean $\frac{\Delta C}{\Delta M} = -0^c,12$. M can be computed from the following formula:

$$M = 0^M,75 + 8^M,3\,(5,8 - C).$$

The agreement is shown here:

$\bar{\pi}$	$\bar{M}_{\text{obs.}}$	$\bar{M}_{\text{comp.}}$	n	m
0″,009	−2,49	−2,07	10	2,95
0 ,029	+0,18	+0,43	10	3,24
0 ,059	+2,34	+2,74	10	3,50
0 ,157	+4,88	+3,74	9	3,77

A better agreement was obtained when parallaxes smaller than 0″,020 were excluded.

For 39 stars belonging to the spectral classes F5 to G, LINDBLAD found on an average $\frac{\Delta C}{\Delta M} = -0^c,13$; he discussed if this could be the effect of selective absorption in space, but concluded that even the comparatively large coefficient of VAN RHIJN could by no means give an explanation. He solved the equations of VAN RHIJN anew and found that the facts considered by the latter to be in favour of the existence of selective absorption in space could be just as well explained by the assumption that there is a change in the absolute magnitude with colour. VAN RHIJN's data gave $\frac{\Delta C}{\Delta M} = -0^M,037 \pm 0^M,014$, whereas the mean of LINDBLAD's values is $-0^M,053$.

Lastly he found that there was a difference in the effective wave lengths for stars of the same spectral class having different M. Collecting the G8—K5 stars in his table 27 we have:

$\bar{M}$	λ_{eff}	n
0^M,46	4368	10
5 ,58	4319	8

In a subsequent paper by LINDBLAD[1] the minimum effective wave length, $\lambda_{\min}$, was introduced as a colour equivalent. This quantity is defined as the distance in λ between the inner limits of the grating spectra of a certain order. The $\lambda_{\min}$ depends on the instrument, atmosphere, and chemical properties of the plate in about the same way as does λ_{eff}, but it is more sensitive to absorption than the ordinary wave length. The investigations were continued and new results given in LINDBLAD's thesis, where also a method was developed for determining M from measurements of λ_{eff} and $\lambda_{\min}$ without a knowledge of the spectral class or other equivalents of the colour.

[1] Ap J 49, p. 289 (1919).

124. Effective Wave Lengths of the Pleiades. On plates taken with the Mount Wilson 60-inch reflector HERTZSPRUNG[1] has impressed images of the Pleiades with exposures of 1800, 570, 180, 57, 18, 6, and 2 seconds. The area to be investigated was about 2° × 2°, while the 60-inch reflector with full aperture has a field of only about 20′ diameter for the determinations of λ_{eff}. In order to increase the size of the field, the reflector was diaphragmed down to 40 inches. The measurements are based on 68 plates and altogether 1246 stars were measured. Also spectra of the second and third order were used.

As normal colour indices the C_H values of the Göttinger Aktinometrie were chosen. Instead of giving the λ_{eff} themselves HERTZSPRUNG has reduced his

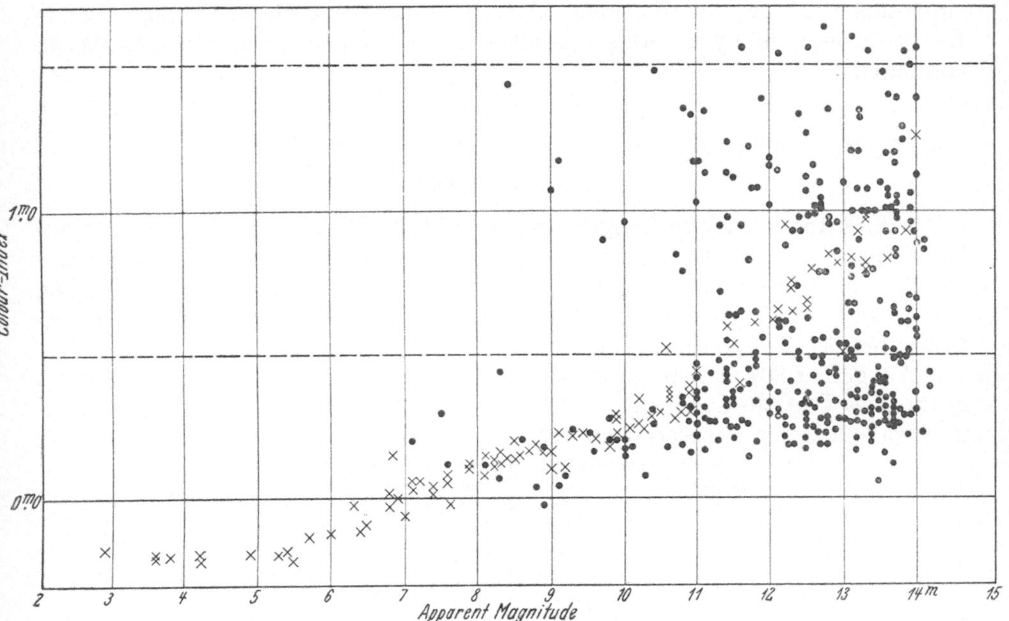

Fig. 95. The colour indices as a function of apparent magnitudes in the Pleiades group (according to HERTZSPRUNG's measures). The effective wave lengths have been transformed to colour indices.

measurements to the C_H-system (SCHWARZSCHILD's photographic magnitudes—Harvard visual magnitudes). The λ_{eff} and the C_H are connected by means of the formula:

$$\lambda_{\text{eff}} = 200\, C_H + 4216{,}4.$$

The accuracy of the results can be judged from the following table:

Diameter of central image	Mean error $\varepsilon(C_H)$	n	Diameter of central image	Mean error $\varepsilon(C_H)$	n
0,05	±0,58	82	0,11	±0,16	165
0,06	0,41	125	0,12	0,13	177
0,07	0,33	222	0,13	0,13	66
0,08	0,23	254	0,14—0,15	0,15	80
0,09	0,19	246	0,16—0,17	0,16	121
0,10	0,15	211	0,18—0,19	0,12	70
			0,20—0,22	0,11	27

[1] Copenhague Mém de l'Acad R des Sciences et des Lettres, Sect des Sc, 8me Sér. IV, No. 4 (1923).

The estimates of the diameter d of the central star-image were converted into provisional magnitudes and then reduced to the scale of HERTZSPRUNG.

A remarkable fact concerning the physical members of the Pleiades is the very regular change in colour with the magnitude. A least-square solution gave:

$$C_\lambda = -0{,}662 + 0{,}0945\, m_{ph}.$$

For the Hyades HERTZSPRUNG had found earlier[1]:

$$C_\lambda = -0{,}253 + 0{,}0935\, m_{ph}.$$

Thus the coefficient $\Delta C/\Delta M$ is practically the same in the two cases. The difference between stars with the same colour in the two clusters is $2^m{,}675$. By supposing that stars of the same colour have the same absolute magnitude in the two groups and by adopting a parallax of $0'',027$ for the Hyades HERTZSPRUNG computed:

$$\pi_{\text{Plej.}} = 0'',0079.$$

An analogous comparison with the Ursa Major cluster gave:

$$\pi_{\text{Plej.}} = 0'',0088.$$

125. Effective Wave Lengths of Faint Milky Way Stars. Ö. BERGSTRAND has communicated[2] the general results of an investigation of effective wave lengths within the areas $\alpha = 19^h 49^m$ to $19^h 59^m$; $\delta = +35°10'$ to $+37°10'$. The plates taken by the Upsala observers with the aid of the Zeiss twin-triplet have exposures of 483^m, 60^m, and 10^m.

Altogether 803 stars brighter than $13^m{,}4$ have been measured. From 43 stars in the New DRAPER Catalogue the following provisional evaluation of the effective wave lengths in spectral classes is found:

Spectral class	λ_{eff}	Spectral class	λ_{eff}
O	4110	(G)	4230
B	4160	G5	4250
A	4190	K	4290
(F)	4210	M	4390

The decrease in the relative number of white stars when faint magnitudes are reached seems to be illustrated from the table:

Mean magnitude	$\overline{\lambda_{eff}}$ All stars	n	$\overline{\lambda_{eff}}$ White stars	n	$\overline{\lambda_{eff}}$ Yellow stars	n
7,5	4211	43	4178	25	4257	18
9,5	4213	66	4182	45	4280	21
10,4	4214	70	4182	41	4260	29
11,4	4210	134	4185	81	4248	53
12,2	4223	139	4184	61	4253	78
12,8	4222	141	4184	73	4262	68
13,4	4222	180	4184	86	4257	94

Although the differences are small in the λ_{eff} for all stars, the author thinks that the tendency of faint stars to increase their colour towards red, as has been found by HERTZSPRUNG[3], SEARES[4], and others is also shown in the above table. The fact that the relative frequency of A stars suddenly undergoes a change is explained by the hypothesis that the stars of higher M that are near the boundary of the "local system" are of apparent magnitude 12^m. If the $\varphi(M)$ is equal

[1] A N 209, p. 120 (1919). [2] Jubiläumsnummer der A N, p. 3 (1921).
[3] Potsd Publ 22, p. 5 (1911) and numerous subsequent papers.
[4] Mt Wilson Contr No. 81 = Ap J 39, p. 361 (1914).

in all parts of the universe then the apparent magnitude m depends on M and r. When fainter and fainter stars are included it will happen that at a certain magnitude the absolutely brightest stars are comprised. In the following magnitudes the relative number of absolute faint stars will increase. As the yellow stars are absolutely fainter than the white, BERGSTRAND concludes that the above-mentioned change can be explained.

He estimates the absolutely brightest stars to have $M = 0$ which gives the distance as 2500 parsecs, and adds that the star-cloud between β and γ Cygni is probably situated farther away.

126. The Determinations of λ_{eff} at Greenwich[1]. The 30-inch reflector was diaphragmed down to 20 inches in order to increase the field of good definition. The total number of measured stars is 4472, of which 3733 are contained in the B D. This is thus the most extensive determination of λ_{eff} hitherto published.

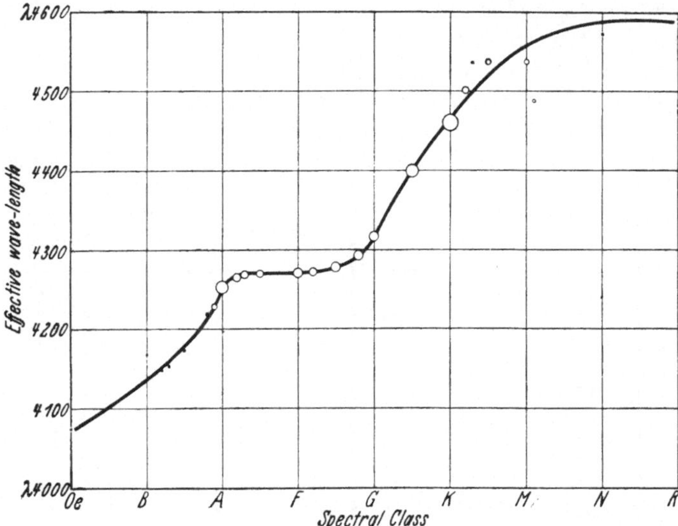

Fig. 96. Relation between effective wave length and spectral class according to the measures at Greenwich. The areas of the dots and circles are proportional to the number of stars in each group.

The area investigated lies between 80° northern declination and the pole. The photographs have been measured and discussed by DAVIDSON and MARTIN.

If we group the stars according to magnitude and the value of λ_{eff}, there is little evidence of any change in the percentage of stars of different colours. The number of stars whose λ_{eff} is less than 4200 (B0—B5 stars) is about 0,5 per cent of the total number. About 8 per cent have λ_{eff} less than 4250 (early A stars). A large number, 43 per cent, have λ_{eff} between 4250—4300 (A—G). The number of very red stars, λ_{eff} around 4500, is 6 per cent.

A comparison of λ_{eff} with spectral type for stars whose parallaxes have been determined was made in order to see whether λ_{eff} enables a discrimination to be made between giants and dwarfs of the same spectral type. It was found that in all cases where the effective wave length is greater than the average for that spectral class the star is one of great luminosity, and in the opposite case one of small luminosity.

[1] Determinations of Effective Wave Lengths of Stars Made at the R. Observatory, Greenwich, in the Years 1920 to 1925 under the Direction of Sir FRANK W. DYSON, London (1926).

The mean parallaxes of different groups of stars were obtained from the proper motions in the Astrographic Catalogue Vol. IV. The results are:

λ_{eff}	m_{ph}	$\bar{\pi}$	Mean absolute magn.	
			M_{ph}	M_{vis}
>4500	9,88	0″,00145	$-0^m,3$	$-1^m,3$
4450–4500	10,03	0 ,00262	+2 ,0	+1 ,1
4400–4450	9,91	0 ,00275	+2 ,1	+1 ,4
<4250	9,62	0 ,00226	+1 ,4	+1 ,4

The first group contains no dwarfs. The second group contains a small percentage of dwarfs, but there does not appear to be any sharp distinction. Between λ 4250 and λ 4400 there is a much larger percentage of faint stars than in the first group. Thus of 40 G5 stars 8 have $\overline{M} = 5^m,3$, while the remainder have $\overline{M} = 2^m,2$ photographic. The group $\lambda < 4250$ seems to be very homogeneous.

The relation between the Greenwich system of λ_{eff} and the spectral classification of Upsala has been investigated by H. PETERSSON[1]. At first the mean λ_{eff} was determined for intervals of spectral class and apparent magnitude. It was found that the Greenwich λ_{eff} as well as the spectral classification of PETERSSON are reliable for faint apparent magnitudes. Next the following correlation was determined:

Harvard spectral class	λ_{eff}	Mean error	n	Harvard spectral class	λ_{eff}	Mean error	n
Oe5 . . .	4074		1	F2 . . .	4272	±16	32
B	4167		1	F5 . . .	4279	14	59
B2 . . .	4148		2	F8 . . .	4292	24	62
B3 . . .	4153		5	G0 . . .	4315	36	59
B5 . . .	4174		4	G5 . . .	4401	49	114
B8 . . .	4219	±41	5	K0 . . .	4461	55	163
B9 . . .	4227	25	13	K2 . . .	4501	44	38
A0 . . .	4252	18	95	K3 . . .	4531		1
A2 . . .	4265	14	57	K5 . . .	4539	35	15
A3 . . .	4268	13	26	Ma . . .	4539	18	12
A5 . . .	4269	14	32	Mb . . .	4488	50	6
F0 . . .	4270	17	60	N	4572		1
				R5 . . .	4591		1

Using his own spectral classes PETERSSON finds for late stars:

Spectral class	λ_{eff}	n	Spectral class	λ_{eff}	n
d G0. . .	4330	22	g K0 . .	4475	23
g G5. . .	4422	22	d K0 . .	4425	25
d G5. . .	4372	21	g K0 . .	4515	22
g G8. . .	4452	25	d K0 . .	4462	9
d G8. . .	4409	26	g K5 . .	4555	21

There seems to be a decided difference in λ_{eff} for late giants and dwarfs, as was first found by LINDBLAD[2] and later on confirmed by means of a number of investigations. The absolute magnitude can undoubtedly be approximated to a certain extent from the (accurate) spectral class and an (accurate) value of a colour equivalent.

The effective wave length is practically constant from A2 to F5 and cannot be used within this interval for a discrimination of the spectral class. DYSON

[1] The Distribution of Distances and Velocities of Stars in the Carrington Zone on the Basis of Spectrophotometric Analysis. Thesis, Upsala 1927 = Upsala Medd No. 29.

[2] Ark Mat Astr Fys 13, No. 26 (1918).

and MARTIN have shown that by the use of Ilford Panchromatic plates and a screen that absorbs the blue rays it is possible to establish a sensitive relationship between spectral class and λ_{eff}. The principal disadvantage is, of course, that the exposures have to be considerably increased (some 10 times).

The relation between the λ_{eff} and the Upsala colour equivalent, E_g, being a spectrophotometric measurement of the break in the continuous spectrum at the band G is as follows:

Spectral class	λ_{eff}	$\log E_g$	n
F8 . .	4303	0,06	98
d G0. . .	4331	0,24	56
d G5. . .	4371	0,45	19
d G8. . .	4405	0,62	10
d K0 . .	4409	0,74	5
d K2 . .	4462	1,16	3
g G5. . .	4423	0,62	45
g G8. . .	4448	0,82	52
g K0 . .	4480	1,07	26
g K2 . .	4519	1,53	14
g K5 . .	4552	1,76	15
g M . . .	4528	1,63	9

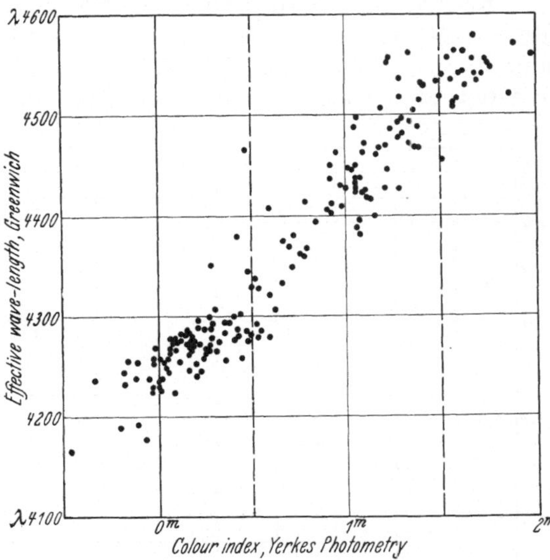

Fig. 97. The colour indices of the Yerkes Actinometry as compared with the effective wave lengths measured at Greenwich. Investigating the non-linear relation between the two quantities HERTZSPRUNG finds a mean error in the λ_{eff} translated into magnitudes of $\pm 0^{\text{m}},041$.

127. VON KLÜBER's Modification of the λ_{eff} Method. This method[1] makes use of a round prism with an angle of 1°, which is attached to a plane glass, placed in front of an objective, which is larger than the round prism. The annular part of the objective, which is free from the prism, gives a central image of the star, and the prism produces a spectrum, which is very little elongated on account of the small dispersion.

If n_λ is the index of refraction for the wave length λ and if the angle i between the optical axis and the rays falling on the objective, is small, then we may write with sufficient accuracy:

$$\delta_\lambda = (n_\lambda - 1)\, i$$

where δ_λ is the deviation.

If f is the focal length and s is the measured distance between the centre of the central image and the optical centre of the spectrum we have the relation:

$$\operatorname{tg} \delta = \frac{s}{f}\,.$$

From the formulae the wave length can be found, when s and the constants of the prism are known. A determination of the constants involves several difficulties, the principal one having its origin in the fact that different parts of the objective are used for producing the central image and the spectrum. The value of s, being a colour equivalent, can be used without reduction, and the relation between s and stars with a known colour can be derived.

The principal advantage of VON KLÜBER's method is that with a given instrument fainter objects can be investigated than when a grating is used. The loss of light in the second case amounts to $2^{\text{m}},5$, but in the former case it can

[1] Astr Abh Ergh zu den A N, Bd. 5 No. 1 (1925).

be reduced to $0^m,75$. A gain of $1^m,75$ is very important when faint stars or nebulae are being investigated.

The correction for the apparent intensity of the image shows the same general behaviour in this case as in that of λ_{eff}.

A rough classification can be made from the appearance of the spectra in both cases.

Results have been derived for the 25 stars selected by BERGSTRAND and ROSENBERG as a sequence for λ_{eff}. The effective wave lengths were also measured and it seems that VAN KLÜBER's central-objective prism method is slightly superior to the λ_{eff} method as regards accuracy.

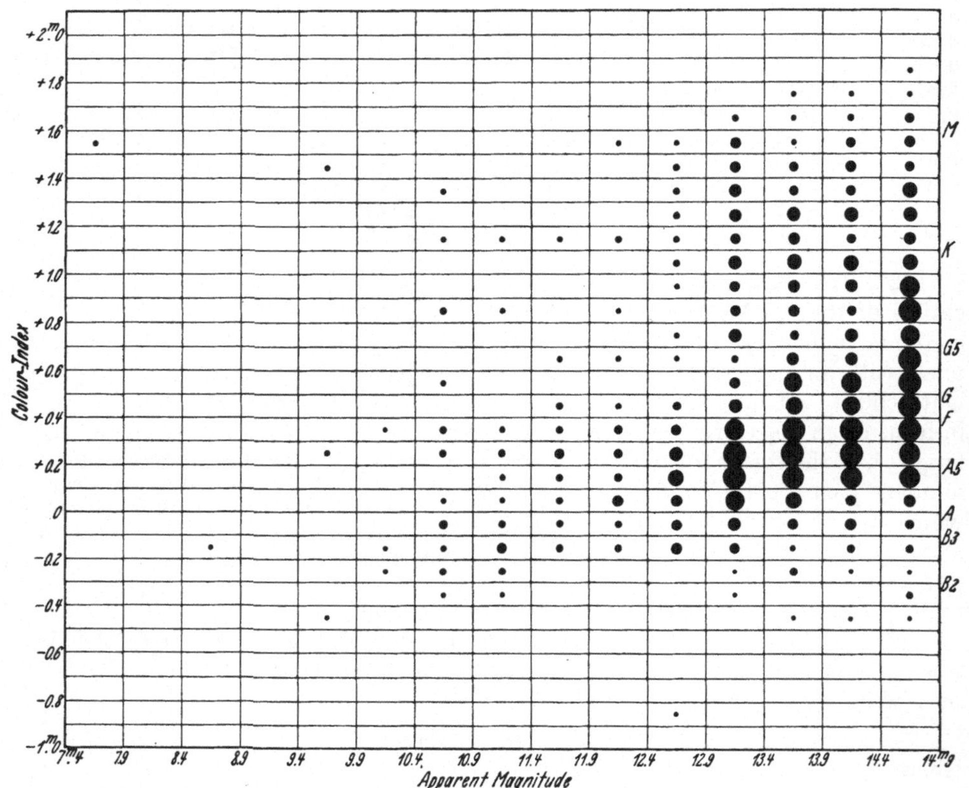

Fig. 98. Correlation surface showing the distribution of the material measured by KREIKEN for λ_{eff} and m with regard to these quantities. The areas of the black circles are proportional to the number of stars within each group.

128. KREIKEN's Researches. In the extensive work of determining colour indices on the basis of HERTZSPRUNG's Mount Wilson material by E. A. KREIKEN[1] the photographic magnitudes have also been determined. It has been assumed either that, except for the zero-point correction, the diameter-magnitude curve for the plates was the same as that for the Pleiades as measured by HERTZSPRUNG or that, when diameters are plotted as ordinates against magnitudes as abscissae on the Mount Wilson plates, the curve will be shifted by a constant magnitude along the axis of abscissae as compared to HERTZSPRUNG's curve.

[1] On the Colour of the Faint Stars in the Milky-Way and the Distance of the Scutum-Group. Groningen 1923.

As this curve is based on plates taken with the same instrument the assumption seems to be justifiable.

A comparison was made between the two observers as regards their diameter estimates.

The mean error of one single colour determination of a star with diameter 0,10 mm was found to be $= \pm 0^m{,}17$, which is in excellent agreement with the corresponding value found by HERTZSPRUNG.

According to HERTZSPRUNG a difference in the colour index of 1 magnitude corresponds to a difference in λ_{eff} of 200 A or in the case of KREIKEN to 0,319 revolutions of the screw. Thus the equation:

$$C_\lambda = m_p - m_v = -6{,}489 + \frac{1}{0{,}319} B_\lambda ,$$

where B_λ is the effective wave length expressed in revolutions of the screw, will transform the measurements directly into colour indices.

A selection of the material as regards limiting magnitude will be unavoidable on account of the fact that the red stars have somewhat shorter spectra than the blue stars and can therefore be measured down to a somewhat fainter magnitude than the latter. In order to have a uniform limiting magnitude all stars with central images less than 0,08 mm have been rejected.

The frequencies of C_λ have been used for deriving the distance of the star clouds on the basis of KAPTEYN's formula for the mean magnitude:

$$\overline{m} = \frac{\int\limits_{-\infty}^{m_0} m\, e^{-h^2 (m - M_0 + 5 \log \pi)^2}\, dm}{\int\limits_{-\infty}^{m_0} e^{-h^2 (m - M_0 + 5 \log \pi)^2}\, dm}$$

where m_0 is the limiting magnitude, $h = 1/\sigma\sqrt{2}$ or the measure of precision, and M_0 the zero-point in the normal distribution of the absolute magnitudes, viz.

$$e^{-h^2 (M - M_0)^2} .$$

Graphical solutions of the integral equation above have been introduced by the present writer[1].

The equation assumes that the stars are all at the same distance. If this is not the case it can be assumed that a_i stars are at a distance r_i and so on. We then have:

$$\overline{r} = \frac{\sum^n a_i r_i}{\sum^n a_i} ,$$

$$\overline{m} = \frac{\sum a_i m_i}{\sum a_i} .$$

In order to avoid the difficulty of not knowing a_i KAPTEYN[2] has introduced $\varphi(m)$, chosen in such a way that $\overline{\varphi(m)}$ is a linear function of r. We then have:

$$\overline{\varphi(m)} = \frac{\int\limits_{-\infty}^{m_0} \varphi(m)\, e^{-h^2 (m - 5 \log r - M_0)^2}\, dm}{\int\limits_{-\infty}^{m_0} e^{-h^2 (m - 5 \log r - M_0)^2}\, dm} .$$

---

1 Stockholm Acad Handl 60, No. 8 (1920). Also Thesis Upsala (1920).
2 Mt Wilson Contr No. 147 = Ap J 47, p. 146 (1918).

KREIKEN applies this formula to his C_λ in the Scutum Cloud and finds its distance to be 1500 parsees. The range in depth in this cloud is comparatively small. The range is so small that it cannot perceptibly influence the distribution of the apparent magnitudes, when stars of all spectral classes are taken together. Thus:

$$\varphi(m) = \varphi(M) + \text{constant}.$$

The observed apparent curves, $\varphi(m)$, are found to agree with the curve of KAPTEYN and VAN RHIJN up to $-0^M,5$.

The change in colour index with apparent magnitude for stars in or near the galactic plane has been investigated by E. A. KREIKEN[1]. The material consisted mainly of colour indices derived on basis of measured effective wave lengths, viz. HERTZSPRUNG's[2] Mount Wilson measurements of the Pleiades (1246 stars), KREIKEN's measurements of HERTZSPRUNG's material in 8 selected galactic regions (about 5000 stars), J. H. PETERSSON's frequencies of C, based on 1020 stars in Aquila[3], and 313 colour indices in the galactic cloud near M11 as determined by SHAPLEY[4] were used too. The following representations of the colour index were found:

		n
Scutum, Aquila	$C = -1,120 + 0,105\, m_{ph}$	2044
	$C = +0,661 + 0,006\, m_{ph}$	80
Cepheus	$C = -0,587 + 0,068\, m_{ph}$	428
Cygnus, Pleiades, Lyra	$C = +0,170 + 0,025\, m_{ph}$	1812
Pleiades Cluster[4]	$C = -0,662 + 0,094\, m_{ph}$	$4^m,5$—$11^m,0$

The frequencies of the colour indices in the different regions are given and it seems that the increased frequency of blue stars is rather sudden. None of the curves represent a normal distribution.

The total amount of light of a certain magnitude is given by $A(m)\,e^{-0,4m}\,dm$ where $A(m) = e^{a+bm-cm^2}\,dm$ (apparent luminosity curve), and the mean colour of the sky is found by evaluating the integral:

$$\overline{C} = \frac{\int\limits_{-\infty}^{+\infty} e^{a+bm-cm^2-0,4m}(A+Bm)\,dm}{\int\limits_{-\infty}^{+\infty} e^{a+bm-cm^2}\,dm}$$

if we use the linear formula $C = A + Bm$ for the colour index.

The fraction of light emitted by stars fainter than magnitude 17,3 is negligible.

The principal results are:

	$\overline{C}$
Scutum	$+0^m,36$
Cepheus	+0 ,36
Cygnus, Lyra, Pleiades	+0 ,49

A. PANNEKOEK[5] has derived the value $0^m,43$ as mean colour index of the Scutum Cloud by comparing the visual brightness of the star cloud with the photographic brightness. FATH[6] has found that the integrated spectrum of some Milky Way clouds (Sagittarius-, Scutum-, and Cygnus Cloud) is approximately of the solar type that corresponds to a value of C of about $+0^m,7$.

---

[1] M N 87, p. 196 (1927).

[2] E. HERTZSPRUNG: Effective Wave Lengths of Stars in the Pleiades. København Acad Mém, tome IV, N. 4 (1923).

[3] Ark Mat Astr Fys Bd. 19 A, Nr. 3 (1925).

[4] Mt Wilson Contr No. 126 (1917) = Ap J 45, p. 164 (1917).

[5] B A N 2, p. 23 (1923). [6] Mt Wilson Contr No. 63 (1912) = Ap J 36, p. 362 (1912).

129. Standardization of λ_{eff}. In 1922 Ö. BERGSTRAND and H. ROSENBERG[1] proposed that a sequence of effective wave lengths should be established. Twenty-five stars north of $+79°$ decl. of magnitudes $4^m,57$—$6^m,25$, the colour indices of which vary between $-0^m,46$ and $+1^m,87$, were selected for the purpose. Results have been published so far from the observatories of Greenwich[2] (DAVIDSON and MARTIN) and Innsbruck[3] (OBERGUGGENBERGER) and from Berlin-Charlottenburg (VON KLÜBER)[4].

The principal drawback with the sequence suggested is that the stars are not within the same field. The use of a large instrument for securing exposures of a sufficient number of sequence stars takes such a long time, that the assumption of uniform transparency cannot very well be made. Besides, the stars are rather bright for a large instrument.

E. HERTZSPRUNG[5] published in 1922 effective wave lengths of 129 stars within $0°,5$ of the North Pole. The measurements are based on photographs taken with the 60-inch Mount Wilson reflector. The following relation between the colour index, C_λ, and the Mount Wilson magnitudes (International Scale) resulted:

$$C_\lambda = 0{,}632 + 0{,}888\,(m_p - 13{,}533) - 0{,}902\,(m_{pv} - 12{,}723).$$

The magnitudes of the HERTZSPRUNG stars vary between $9^m,48$ and $16^m,1$ photographic. It seems that if some intermediate and brighter stars were to be measured in the polar region it would be possible to obtain pretty soon a sequence of λ_{eff} meeting all requirements.

The absolute term in the above equation represents the value of the zero-points of the two systems. Its value should be zero. The fact that this is not the case does not prove that the Mount Wilson colour indices for faint stars are too large as compared with those of stars near 6^m, because the zero-points have been based upon very different material.

SEARES has compared the λ_{eff} as determined at Greenwich with the colour indices of Mount Wilson. If the mean differences between the two series are made equal to zero for white stars of 6^m the transformation formula reads:

$$178\,C_\lambda = 4250 - \lambda_{\text{eff}}.$$

The values of C_λ agree better with the colour indices, C_m, than they do with those from the exposure-ratios C_e. Especially around 9^m is the agreement better. This suggests that the systematic deviation in $C_m - C_e$ is to be attributed to the exposure-ratios. Thus the photographic and photovisual scales have been confirmed from the data of λ_{eff}.

130. Discussion of the Method of λ_{eff}. If λ is the effective wave length, i the angle of the rays of a star with the optical axis, σ the total deviation of the rays after passing the grating, then we have:

$$n\lambda = c\sin\sigma - 2\,c\sin\sigma\sin^2\frac{i}{2} + 2c\sin i\sin^2\frac{\sigma}{2}\,.$$

The second and third term on the right hand side can be dropped, as i is small, and then we can write

$$\lambda = \frac{c}{n}\sin\sigma\,,$$

---

[1] A N 215, p. 447 (1922). [2] M N 84, p. 425 (1924).

[3] Wien Akad d Wiss Sitzber Abt. IIa 137, p. 275 (1928).

[4] Astr Abh Ergh Bd. 5, No. 1 (1925).

[5] Mt Wilson Contr Nr. 231 = Ap J 55, p. 370 (1922).

where n is the order of spectrum considered. If s is the distance of the spectra of the first order and f is the focal length we have:

$$\tfrac{1}{2}s = f \sin \sigma$$

or:

$$\lambda = \frac{c}{2}\frac{s}{f} = ks.$$

Thus the effective wave lengths are proportional to the distances of two spectra of the same order.

According to what has been said earlier (see ciph. 3) about the definition of the colour index we have the definition:

$$\lambda_{\text{eff}} = \frac{\int\limits_{\lambda_1}^{\lambda_2} \lambda \varphi(\lambda)\, e(\lambda)\, d\lambda}{\int\limits_{\lambda_1}^{\lambda_2} \varphi(\lambda)\, e(\lambda)\, d\lambda}.$$

As has been pointed out on an earlier occasion, no clear definition exists in the literature on the subject as to whether λ_{eff} in the practical measurements is the value found by bisecting the blackened area or by setting on the "centre of gravity" which then gives λ_{eff} as defined above. It is evident that in practice the last way has been followed, and although measurements of this kind may seem only vaguely defined, accuracy in measuring can certainly be obtained after some experience.

The construction of a coarse grating does not introduce such delicate technical problems as are met with in the manufacture of colour screens. Besides, an objective grating may be more easily kept in order and controlled than a colour screen.

If it is always used in the same position relative to the objective, no trouble will arise, even if the grating is slightly tilted relatively to the optical axis, as this introduces only a small systematic error. As is the case with SEARES's, KING's, TICHOV's and NORDMANN's methods, work can be done for λ_{eff} independently of a photometric scale. Nor is any dyeing of the plates needed with its accompanying uncertainties, and almost any kind of high-speed commercial plates can be used to advantage. If we leave out of consideration the correction of λ_{eff} for the relative strength of the image, the reduction of the measurements is simple and does not involve much labour.

One of the principal advantages with the λ_{eff} method is perhaps its usefulness for objects of measurable extension on the plate, such as nebulae, planets, and comets. The use of other methods, except SEARES's (exposure-ratio) and perhaps GUTHNICK's (photo-electric), seems to present difficulties in these cases.

On the other hand, it must be admitted that the method is of rather idiosyncratic behaviour, and, unless carefully used, certainly possesses several disadvantages.

Every observer of λ_{eff} has met with the so-called photographic PURKINJE effect, i. e. the dependence of λ_{eff} on the intensity of the images on the plate. (This follows immediately from the formula above.) In order to overcome this, the material has generally been reduced to an arbitrarily selected value of the diameter of the central image. Now the effect depends also on the spectral class, and the correction can thus be obtained only by successive approximations, as λ_{eff} is unknown. The occasionally complicated forms of the empirical reduction curves used in corrections for this effect show how much care should be taken in order to get an accurate correction for this important source of error.

For some instruments the effect has been found to be so large that the measurements cannot properly be used for obtaining knowledge of λ_{eff}. On the other hand, this effect appears to be very small, if perceptible at all, in the determinations of λ_{eff} for surface objects with the Upsala 6-inch twin telescope, at least as long as large intensities are not used.

The effective wave lengths as measured are in a high degree dependent on the optical properties of the instruments used. With regard to refractors nothing can be said before a thorough test has been made. The system of λ_{eff} depends entirely on achromatism. Thus it was found at Upsala that changes in the optical system (the middle lenses of the triplets were slightly readjusted in order to obtain a good focus over as large a field as possible) changed the λ_{eff} so much that another scale had to be established. Reflectors ought to be well suited for work of this kind. Their behaviour has, however, not quite come up to theoretical expectations. A comparison between the results of BERGSTRAND, WOLF, and Greenwich is, in this respect, very instructive[1]. The apparent relation of the system of λ_{eff} to a re-silvering of the mirror is rather obscure, as was found by WOLF[2].

A connection between λ_{eff} and changes in focal length, arising from temperature, has been suspected by ROSENBERG[3] who has investigated this problem and found, by using a grating of high precision, that changes in the focal length as small as $\frac{1}{1680}$ could be detected in the measured effective wave lengths. Of still greater importance is his result that small focal changes influence very considerably the shape of the reduction curves for the photographic PURKINJE-effect.

Other optical effects are the distortion of the images by the distance from the centre and by atmospheric dispersion. These can be made less serious by reducing the aperture of the objective, by avoiding too long exposures, or by working at small zenith distances. The Greenwich observers have measured their λ_{eff} for stars with distorted images, as they found these not less suitable for accurate measurement. They do not state, however, whether there exists any systematic difference between the λ_{eff} for those images and for images near the centre of the plate.

In an instrument used by LINDBLAD and LUNDMARK a rotational error was found, i. e. λ_{eff} changes when the grating is rotated in front of the objective.

Sometimes a chromatic irregularity of the lens is evident, which acts in much the same way as atmospheric dispersion. The most natural explanation is that the lens acts as a prism, and that in the resulting spectra there is dispersion in two directions.

The state of the atmosphere is of some importance for work of this kind. VALLIN and LUNDMARK[4] thus found a fog effect for λ_{eff} at Upsala. When λ_{eff} obtained through a clear sky and low humidity were compared with those obtained through a foggy sky and high humidity, a mean difference of 30 A was found in the sense that a fog increases λ_{eff}. DAVIDSON and MARTIN[5] also mention an influence of the same sort. They do not state, however, whether this effect is systematic and in the same direction. The fog effect has recently been investigated by ROBB[6].

The effect of the thickness of the plates on the λ_{eff} has been investigated by EBERHARD[7].

---

[1] M N 82, p. 495 (1922). [2] A N 213, p. 49 (1921). [3] A N 213, p. 329 (1921).
[4] Unpublished results. [5] M N 82, p. 65 (1921). [6] Ap J 65, p. 315 (1927).
[7] Probleme der Astronomie (Seeliger-Festschrift), p. 115 (1924).

The influence of chemical fog or other photographic effects has not yet been investigated; nor has any comparison been made between different kinds of plates. Preliminary work such as is reported in LINDBLAD's thesis[1], shows the importance of a proper selection of plates. In the measurement of the plates a personal element is introduced, as the short spectra are generally not symmetrical and their appearance varies with the spectral class. When measuring λ_{eff} of spirals and clusters LINDBLAD and LUNDMARK[2] have found very good agreement, if we take into account the difficulty of measuring objects of this kind. From a great number of measurements there resulted a mean difference between the two observers of 10 A, which is quite negligible. A comparison of their measurements of stars has shown very little systematic difference between these two observers for all spectral classes except B, where LINDBLAD's measurements of λ_{eff} are decidedly smaller.

Among the disadvantages of the λ_{eff} method the fact may be mentioned that in dense regions the position of the short spectra can be influenced by photographic effects; this is sometimes called photographic repulsion and has been investigated by F. E. ROSS[3]. This applies to the other colour equivalent methods, but in a less degree, because, when dealing with effective wave lengths, we have at least three times as many images on the plate. Compared with the colour index method, the λ_{eff} method is relatively somewhat inefficient; for it has been found that when the grating is used, exposures must be increased at least ten to fifteen times, whereas the colour screen increases the exposures only six times. Moreover, in the latter case underexposed images are still measurable, which is not true for the short spectra.

It is doubtful whether λ_{eff} has any real significance for classifying purposes when applied to objects with bright emission lines. Thus the extraordinarily low value of 3940 A for the Ring Nebula, as found by LINDBLAD and LUNDMARK and later on confirmed by KREIKEN[4], must be ascribed to the presence of the bright doublet at 3727 A. The spirals gave wave lengths corresponding to the spectral classes G—K, which is in close accordance with the results from actual spectral classification. For the planetaries a mean effective wave length of 4130 A was found, corresponding to the spectral classes B—F. It thus seems possible with this method to separate the spirals from the planetaries, which it is sometimes impossible to do from considerations of structure alone. In a recent paper BALANOVSKY[5] has found for 16 objects, probably spirals, a mean effective wave length of 4420 A (in his system corresponding to G), which is in good agreement with LINDBLAD and LUNDMARK. On the other hand, 4 of his planetaries give in the mean 4760 A, which abnormal value seems to be difficult to explain at present, even if we may assume that N_1 and N_2 have contributed to the recorded spectrum. Although he thus gets λ_{eff} higher for planetaries than for spirals, which is contrary to the Upsala results, his system can nevertheless be used for distinguishing between these two kinds of nebulae.

The close dependence of λ_{eff} on the optical system and the several other effects mentioned are the reason why no two results of different observers are directly comparable. It is thus to be regretted that no standard sequence has been established. It may be suggested that this can be performed by laboratory experiments with monochromatic light. Until this has been accomplished, however, astronomers should be careful in their interpretation of observations of effective wave lengths, and should not use them without reduction, as has

[1] Upsala Universitets Årsskrift (1920). Also thesis Upsala.

[2] Ap J 46, p. 206 (1917) and Ap J 50, p. 376 (1919).

[3] Ap J 53, p. 349 (1921). [4] Thesis Groningen 1923. [5] A N 215, p. 233 (1922).

been done in some cases, for the derivation of colour equivalents. Likewise I am of the opinion that it is not justifiable to use the value of the effective wave lengths for a certain spectral class as determined by another observer with another instrument and other plates as a point of reference for one's own measurements, which procedure has been followed by several investigators.

In spite, however, of all the systematic errors to which a determination of λ_{eff} is subject, there is, unquestionably, a relation between λ_{eff} and spectral class. With regard to the main features of this relation there is general agreement among observers. It appears that the increase of λ_{eff} is rather rapid between B and A, is slow (for some observers almost zero) from A to F, and large again from F onward, till at M it goes down towards Mb, Mc, and up towards N.

DAVIDSON and MARTIN[1] publish in their paper an extensive diagram giving the relation between their individual λ_{eff} and spectral classes, from which it will be seen that at G5, e. g., the measured effective wave lengths cover the entire range of the mean λ_{eff} from A to K5. When it comes to discussing the possible causes of this unexpected effect, we are now able to arrive at definite conclusions respecting the error in spectral classification. For, apart from the Harvard spectra used by DAVIDSON and MARTIN, we have another determination of the spectral classes by G. H. TEN BRUGGEN CATE[2] for the stars in this part of the sky. As his paper deals with a comparatively small number of stars and great care was taken in the classification, we can give this determination a weight at least equal to that of the New HENRY DRAPER Catalogue. A comparison between the two different classifications will then give us an idea of the error to be expected in a single determination of spectral class. If A0 is denoted by 0 and K0 by $3^{\text{s}},0$ then the relation between Harvard H and TEN BRUGGEN CATE can be written

$$B = 0{,}89\,H + 0^{\text{s}}{,}4.$$

The deviations from this straight line lead to the following mean errors in the classification (unit: one spectral sub-division); for:

A0 . . .	$\pm 0^{\text{s}},7$	G0 . . .	$\pm 3^{\text{s}},4$
F0 . . .	2 ,2	G5 . . .	2 ,8
F5 . . .	3 ,1	K0 . . .	2 ,2

It is clear from this that errors in spectral classification alone cannot be responsible for the large discordance in λ_{eff} for G5—K, as the mean errors decrease where the discordances increase. This conclusion is supported from a discussion[3] by LUNDMARK and LUYTEN where the Greenwich λ_{eff} are plotted against TEN BRUGGEN CATE's spectra.

Another estimate of the allowable error in spectral classification is found from a comparison between Harvard and Mount Wilson spectra for 489 stars (between 0 and 8 hours of R A and between F0 and M, taken from Mount Wilson Contribution, No. 199).

The spectral indices were again used and a rigorous solution by least squares was made in the following way: Let each star be represented by a point whose x equals the Harvard spectrum, and whose y equals the Mount Wilson spectrum; then the problem is to lay a straight line, the relation-line, through this cluster of points in such a way that the sum of the squares of the perpendiculars of these points from the line is a minimum. Write the equation of the line as follows:

[1] M N 82, p. 65 (1921).

[2] Determinations and Discussion of the Spectral Classes of 700 Stars, mostly Near the North Pole. Dissertation Groningen 1920.

[3] M N 82, p. 504 (1922).

$L = -x \sin\alpha + y \cos\alpha - l = 0$ (where α is the angle between the line and the axis of x); then the perpendicular distance $D(x_1, y_1)$ from a point (x_1, y_1) to the line $= L(x_1, y_1)$; accordingly $\Sigma L^2 (x_1, y_1) = \text{Min.}$ After differentiation and substitution we find for α and l:

$$\tan 2\alpha = \frac{2\Sigma xy - \frac{2}{n}\Sigma x \Sigma y}{\Sigma x^2 - \Sigma y^2 + \frac{1}{n}\Sigma y \Sigma y - \frac{1}{n}\Sigma x \Sigma x}$$

and

$$l = \frac{\cos\alpha\, \Sigma y - \sin\alpha\, \Sigma x}{n}$$

where n is the number of stars. This derivation, however, only holds when the mean errors in the two co-ordinates are equal; in order to obtain this the ordinates were increased in a proportion which gives double weight to Mount Wilson.

A solution of the equations then yields:

$$W \doteq 1^{S},15\ (\pm 0^{S},17)\ H - 1^{S},6\ (\pm 2^{S},0)$$

on the spectral index scale, where F $= 0^{S}$ and M0 $= 3^{S},0$.

The mean errors in spectral classification then turn out to be:

$\pm\, 2^{S},0$ for Harvard, $\pm 1^{S},4$ for Mount Wilson,

these being of the same order of magnitude as was previously found.

Another useful parameter for indicating errors in spectral classification is the correlation constant. In the case Harvard—Mount Wilson:

$$r = 0{,}932 \pm 0{,}006.$$

This is relatively low, for CHARLIER has found as high a value as 0,958 for the correlation constant between Harvard spectra and PARKHURST's colour indices.

The difference Mount Wilson—Harvard has been investigated by ADAMS and JOY[1], who found $-1^{S},6$. On the basis of a more comprehensive inquiry[2] they derived later on a difference of opposite sign, varying irregularly with the spectral class itself. It may also be mentioned that they found a pronounced difference in the values H—W for giants and dwarfs, in the sense that Mount Wilson classifies the dwarfs comparatively redder than it does the giants (by approximately two spectral sub-divisions).

If we now consider another possible cause of the discrepancy in the λ_{eff} for late-type stars, a cause bearing on the irregularities in the spectra (and accordingly in the λ_{eff}) of the stars, and depending on the giant or dwarf character (as pointed out by SEARES and by LINDBLAD), we come to the conclusion that this cannot produce such a marked effect on the range. In agreement with present data we assume that among the K0 stars the dwarfs outnumber the giants in space by 100 to 1; further, that their respective mean absolute magnitudes are $+6^{m},0$ and $+0^{m},8$, both having a Gaussian distribution. After making use of KAPTEYN's and VAN RHIJN's density law we find that at apparent magnitude $8^{m},5$ (which is lower than the mean of the Greenwich stars) the dwarfs form only 30 per cent of the total number. This would accordingly cause a spread on the b l u e side of the maximum, whereas DAVIDSON and MARTIN's figures show that most of the dispersion is on the r e d side.

As the evidence we have regarding the errors in the measurements of effective wave length (for one instrument only) does not account for discrepancies

[1] Mt Wilson Contr No. 142 = Ap J 46, p. 316 (1917).

[2] Pop Astr 29, p. 143 (1921) = Publ Amer Astr Soc IV. p. 202 (1923).

of the size of those indicated in our earlier work[1], we come to the conclusion that at present individual effective wave lengths cannot be used to predict spectral class with high certainty.

Whether this is due to the various causes mentioned above or to real differences in λ_{eff} even among giants of the same spectral class, we cannot yet say, but the facts seem to point largely towards the latter possibility.

It seems that between A and G λ_{eff} is nearly independent of the spectral class, whereas between G and M the uncertainty is too large, or the curve reverses (WOLF)[2], and the only safe predictions that can be made are for the classes O and N. We can therefore say that the relation between spectral class and effective wave length is not of the nature of a one to one correspondence. The most that can be said is that upon the spectral class a rough estimate of λ_{eff} may be based, but that the direction of this inference may not be reversed.

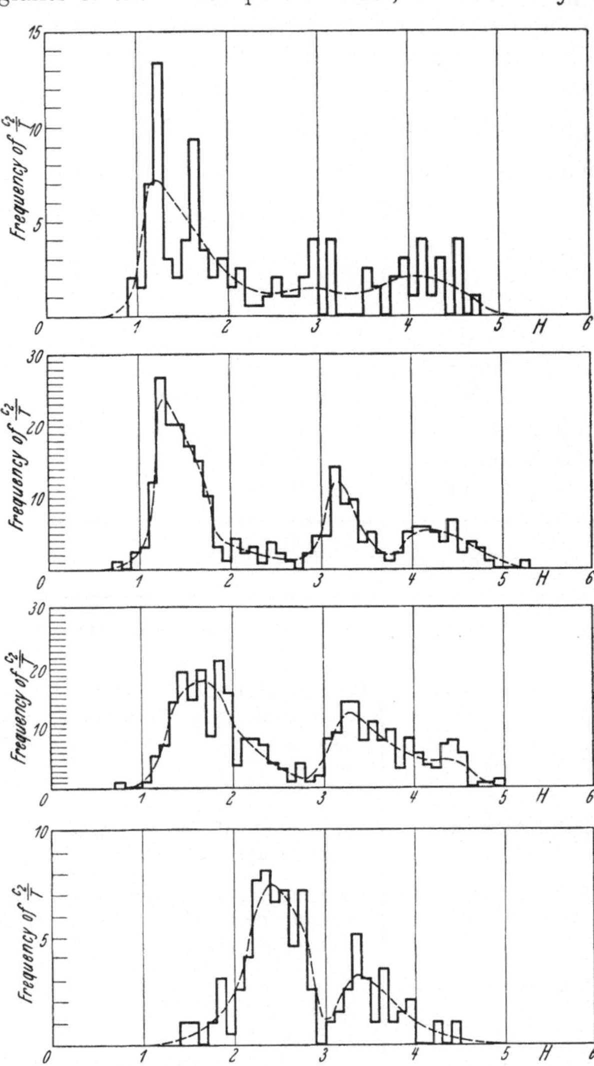

Fig. 99. Frequency of c_2/T for giant and dwarf stars. The lowest curve is based on 80 stars, for which $m + 5 \log \mu$ is situated between the limits $+1$ and $+4$; next curve is based on 293 stars, for which the said limits are -2 $+1$; the next curve is based on 245 stars, for which the said limits are -5 and -2; finally, the top curve is based on 94 stars, having their $m + 5 \log \mu$ between the limits -8 and -5.

131. Reduction of Colours to a Standard System. A number of the most important sources for colours or colour equivalents have been reduced to a common system by HERTZSPRUNG[3], and mean values have been formed according to the derived weights of the sources. As a standard scale for the colour equivalents that of the reciprocal temperatures or c_2/T was chosen, where c_2 is the constant of PLANCK's formula and is assumed to have the numerical value 14600.

[1] M N 82, p. 495 (1922). See especially fig. 2.
[2] A N 213, p. 49 (1921).
[3] Mean Colour Equivalents and Hypothetical Angular Semi-Diameters of 134 Stars Brighter than the Fifth Magnitude and within 95° of the North Pole. Leiden Sterrewacht Ann XIV, St. 1 (1922).

The method used by HERTZSPRUNG to reduce the different series to c_2/T is in principle the correlation method. Suppose that the two values to be compared are x and y, and that they have mean errors ε_x and ε_y. Then the variables are first reduced to equal accuracy by multiplying the values of y by $\sqrt{\varepsilon_x}/\sqrt{\varepsilon_y}$. Thus $y' = y\sqrt{\varepsilon_x}/\sqrt{\varepsilon_y}$. x and y' are taken as coordinates of a point in the xy'-plane and the axes are rotated through an angle ϑ. The new coordinates are: $x'' = x\cos\vartheta + y'\sin\vartheta$ and $y'' = -x\sin\vartheta + y'\cos\vartheta$. If the x'' axis is the line sought as a representation of the linear relation between x and y' then the least square solution: $\frac{\Sigma x'' y''}{\Sigma x''^2}$ must give the x'' axis in which case we have the condition: $\Sigma x'' y'' = 0$, which determines the angle from the relation:

$$\Sigma(x\cos\vartheta + y'\sin\vartheta)(-x\sin\vartheta + y'\cos\vartheta) = 0$$

or:

$$\mathrm{tg}^2\vartheta + \mathrm{tg}\vartheta\frac{\Sigma x^2 - \Sigma y'^2}{\Sigma x y'^2} = 1\,.$$

Thus the line of symmetry in the terminology of the theory of correlation, or the major axis of the ellipse of BRAVAIS, is determined by this method.

The effective temperatures of 199 bright stars, as measured in Potsdam by SCHEINER and WILSING, have been used unchanged. The small weight of these determinations is astonishing.

The following sources were compared.

Author		Source	Abbr.	n	Weight of one obs.
J. WILSING	Effective Temperaturen von 199 helleren Sternen	Potsd Publ No. 74	W	199	9
H. ROSENBERG . .	Photogr. Unters. d. Intensitätsverteilung in Sternspektren	Nova Acta Abh K. Leop Carol Deutschen Akad Bd. 101, No. 2, p. 110 (1914)	R	95	20
E. S. KING	Comb. out of focus results from several instruments	Harv Ann 76, No. 6 (1915)	K	159	34
P. GUTHNICK and P. HÜGELER	Beobacht. d. Helligkeit, des Farbenindex und des Spektrums der Nova Aquilae 3	A N 210, p. 345 (1920)	G	60	28
B. LINDBLAD . . .	Die photogr. effekt. Wellenlänge als Farbenäquivalent d. Sterne	Ark Mat Astr Fys 13, No. 26 (1918)	Lb_1 Lb_2	58 12	13 9
	On the Use of Grating Spectra	Ap J 49, p. 297 (1919)	Lb_3	18	35
	On the Distribution of Intensity in the Continuous Spectra	Upsala Dissertation 1920	Lb_4	29	9
H. OSTHOFF	Die Farben der Fixsterne	A N 153, p. 173 (1900)	O_4	615	22
	Farbenkatalog. Beobacht. mit terr. Refraktor		Ot	201	23
F. KRÜGER	Neuer Katalog farbiger Sterne	Specola Vaticana 7 (1914)	Kr	609	16
H. E. LAU	Unters. über die Farben der Fixsterne	A N 205, p. 65 (1918)	Lau	672	5

(Continued.)

Author		Source	Abbr.	n	Weight of one obs.
G. MÜLLER and P. KEMPF	Photometrische Durchmusterung	Potsd Publ 17 (1907)	P	461	7
HERTZSPRUNG	Combination of m_{ph} of the DRAPER Catalogue and m_{vis} of R H P	Leiden Ann 14, 1. Stuk	D H	573	12
K. SCHWARZSCHILD and coworkers	Combination of m_g of the Göttinger Aktinometrie and the m_p of Potsdam Durchmusterung	Göttinger Aktinometrie (1912)	G P	164	20
J. G. HAGEN . . .	Colori stellari.	Specola Vaticana 3 (1911)	Ha	596	7
J. WILSING	Messungen der Farben, der Helligkeiten u. der Durchmesser der Sterne	Potsd Publ 24, p. 12 (1920)	Wk	38	25

The relation-lines according to HERTZSPRUNG are collected below and may be used to reduce the measurements in a certain system to those of another.

Linear regressions

$W - 1{,}243 = 0{,}5272\,R$	59 (+69)	stars
$W - 2{,}440 = 1{,}6045\,(K - 0{,}489)$	108	,,
$W - 2{,}498 = 7{,}1206\,(G + 0{,}7789)$	42 (+41)	,,
$W - 2{,}000 = 0{,}0101\,(Lb_2 - 4190)$		
$W - 0{,}9906 = 0{,}0164\,(Lb_3 - 4100) - 0{,}0000196\,(Lb_3 - 4100)^2$		
$W + 72{,}56 = 0{,}018\,Lb_4$		
$W + 0{,}9152 = 0{,}7002\,Ot$		
$W - 2{,}722 = 2{,}346\,(D\,H - 5{,}122)$	162	,,
$W - 1{,}628 = 1{,}535\,m_g - 1{,}436\,m_p$	31	,,

Non-linear regressions

Lb_1 λ	4157	4175	4189	4225	4287	4330	4429	Relation-line
W	1,31	1,52	1,62	2,13	3,09	3,64	4,34	

O_4	2,35	2,68	3,10	3,82	4,93	5,20	5,59	6,53	Relation-line
W	1,24	1,54	1,79	2,39	3,05	3,25	3,61	4,49	

Argument	1,0	1,5	2,0	2,5	3,0	3,5	4,0	4,5	5,0	5,5	6,0	6,5	7,0	7,5	8,0
(Arg. OSTHOFF)	0,16	0,56	0,96	1,36	1,76	2,13	2,46	2,76	3,08	3,50	3,97	4,47	4,97		
(Arg. KRÜGER)	0,44	0,82	1,19	1,57	1,91	2,19	2,45	2,74	3,10	3,52	3,96	4,41	4,88	5,37	5,87
(Arg. LAU)	—	—	0,80	1,00	1,20	1,41	1,67	1,99	2,35	2,80	3,32	4,05	4,93	5,83	6,73

P	W	W+	GW−	GW	GW+	WG−	WG	WG+	G−	G	G+	RG+	RG
W	1,20	1,51	1,82	2,11	2,33	2,55	2,79	3,05	3,35	3,66	4,01	4,36	4,71

The question of a possible magnitude equation in c_2/T was tested by the values of this quantity being grouped according to magnitude and spectra of the classes A0—A5. From 195 objects the following table was derived:

Magn.	$2^m{,}69$	$3^m{,}98$	$4^m{,}42$	$4^m{,}69$	$4^m{,}92$
c_2/T	1 ,66	1 ,60	1 ,64	1 ,67	1 ,50

The last value in the second row is too small on account of the different proportions of the spectral sub-groups. After correcting for the influence of this error HERTZSPRUNG has obtained computed values, which show the presence of a magnitude equation.

The mean values of c_2/T for different spectra in the Harvard classification are as follows:

Spectral class	c_2/T	n	Spectral class	c_2/T	n	Spectral class	c_2/T	n
Oe5 . . .	1,32	4	A0 . . .	1,48	82	G0 . . .	2,87	31
B0 . . .	1,48	11	A2 . . .	1,59	58	G5 . . .	3,14	29
B1 . . .	1,65	9	A3 . . .	1,80	20	K0 . . .	3,53	154
B2 . . .	1,63	14	A5 . . .	1,89	35	K2 . . .	4,07	10
B3 . . .	1,25	47	Fo . . .	2,20	37	K5 . . .	4,32	29
B5 . . .	1,36	27	F2 . . .	2,47	1	Ma . . .	4,53	29
B8 . . .	1,44	21	F5 . . .	2,40	32	Mb . . .	4,35	5
B9 . . .	1,50	8	F8 . . .	2,67	17	Comp. . .	3,22	4

For the Mount Wilson spectral classification HERTZSPRUNG has found the following mean values of c_2/T:

Spectral class	c_2/T	n	Spectral class	c_2/T	n	Spectral class	c_2/T	n	Spectral class	c_2/T	n
A5 . .	2,00	1	F6 . .	2,51	6	F6 . .	2,30	1	K0 . .	3,50	18
A6 . .	2,16	1	F7 . .	2,45	6	F9 . .	2,83	5	K1 . .	3,54	14
A7 . .	2,08	4	F8 . .	2,71	5	G0 . .	2,92	3	K2 . .	3,83	24
A8 . .	2,26	3	F9 . .	2,62	3	G1 . .	2,92	3	K3 . .	3,86	12
A9 . .	2,30	4	G0 . .	2,68	4	G2 . .	3,15	6	K4 . .	4,14	15
F0 . .	2,11	9	G1 . .	2,64	5	G3 . .	3,09	3	K5 . .	4,33	15
F1 . .	2,30	3	G5 . .	2,76	2	G4 . .	3,15	8	K6 . .	4,36	9
F2 . .	2,29	7	G8 . .	3,03	2	G5 . .	3,25	14	K8 . .	4,35	1
F3 . .	2,34	6	G9 . .	3,11	1	G6 . .	3,20	14	Ma . .	4,49	19
F4 . .	2,36	8	K1 . .	3,05	1	G7 . .	3,29	16	Mb . .	4,45	8
F5 . .	2,48	10				G8 . .	3,39	15	Mc . .	3,93	1
						G9 . .	3,48	10			

132. The Existence of Preferential Colour Indices. In the introduction to the Cape Zone Catalogue[1] J. HALM raised a theory concerning the existence of preferential colour indices. If there was a very close relation between colour index and spectrum, the frequency curve of colour indices of a selected spectral type would be a normal error curve with its maximum at the mean colour index of the type. Now it is known that the difference in mean colour index between giants and dwarfs can be considerable.

A comparison of the photographic magnitudes of the Cape Zone Catalogue with Harvard visual magnitudes gives colour indices for about 17600 stars in the $-45°$ zone of declination. The distribution of the colour indices within the different types suggests that the frequency curves are components of two or three normal curves. The general conclusion from a preliminary investigation may be stated as follows: There appear to exist at least six (probably seven) distinct colour indices, round which the stars are preferentially grouped in accordance with the Gaussian law of errors. The maxima of the error curves are situated at regular intervals of $0^m,45$ and have been located at $-0^m,9$, $-0^m,50$, $-0^m,05$, $+0^m,40$, $+0^m,85$, $+1^m,30$, $+1^m,8$. Each type contains at least two of these curves which are supposed to have the same dispersion. The only variable quantities are the amplitudes of the error curves, which depend on type, galactic position, and apparent brightness.

The material was divided into altogether 25 groups with regard to position in the sky and spectral classes.

[1] Magnitudes of Stars Contained in the Cape Zone Catalogue of 20843 Stars for Equinox 1900, Zones $-40°$ to $-52°$. Reduced and Prepared for Press under the Direction of H. SPENCER JONES. London (1927).

When we consider the highly complex character of the observed distribution curves, which vary both with spectral class and galactic position, it is, as the author remarks, "doubtless remarkable that a set of six error curves kept at rigidly fixed intervals should represent so closely 26 curves solely varying in their amplitudes".

The following are the values of the preferential colour indices.

Designation	$m \leqq 6{,}0$	$m > 6{,}0$	Mean	All stars
0	−0,9	−0,8	−0,85	−0,95
I	−0,48	−0,42	−0,45	−0,50
II	−0,08	−0,02	−0,05	−0,05
III	+0,37	+0,40	+0,38	+0,40
IV	+0,83	+0,85	+0,84	+0,85
V	+1,32	+1,35	+1,33	+1,30
VI	+1,72	—	+1,72	+1,75

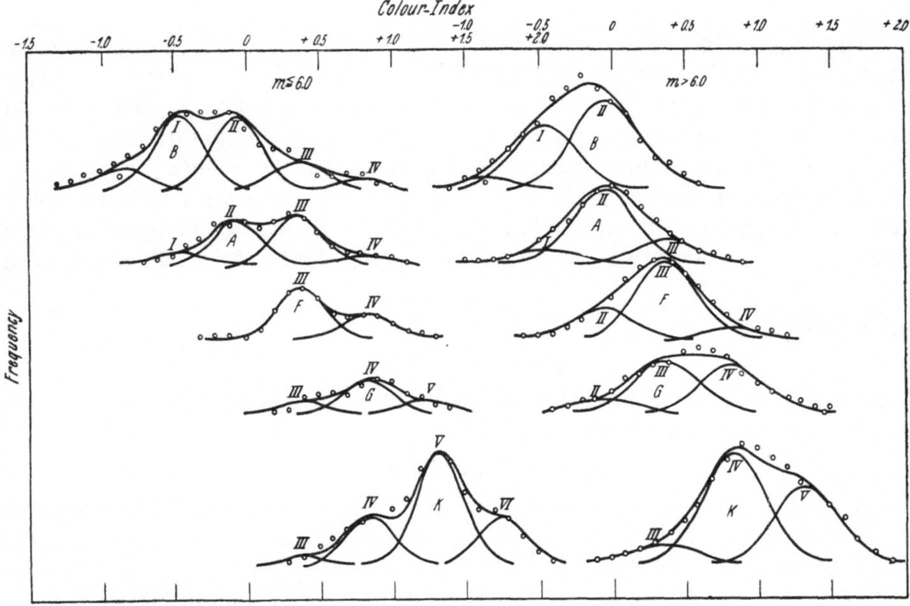

Fig. 100. Dissection of the frequencies of the colour indices of the different spectral classes into components, each supposed to have a normal or Gaussian distribution and the same modulus or dispersion. The Roman numbers refer to the separate components. The ordinates are the number of the stars within each group.

The dispersion of the error curves must be attributed to two causes: the error of observation in the determinations of C and the actual dispersion in colour index around the preferential values. For the photographic magnitudes in the Cape Catalogue the mean error is $\pm 0^m{,}19$. The mean error of the Harvard magnitudes is estimated to be $\pm 0^m{,}20$. Consequently the mean error of a single colour index is $\pm 0^m{,}28$.

On the other hand the modulus of the error curves was found to be 2,5, corresponding to a mean error of $\pm 0^m{,}28$. Halm thus finds the remarkable result that the dispersion of these curves can be entirely accounted for by the accidental deviations of the individual colour indices from the preferential colour indices.

This result was accepted with some hesitation and further tests were made, the Harvard Selected Regions in H A 71, No. 4 being used. The material consisted of 579 stars and it was found that the preferential values of colour index were situated at:

$$-0^m{,}04, \quad +0^m{,}38, \quad +0^m{,}84, \quad +1^m{,}30.$$

Also values of c_2/T as introduced by WILSING and by HERTZSPRUNG were used. The following maxima were found:

Class	Cape c_2/T	Harvard	WILSING and HERTZSPRUNG	Mean c_2/T	n	0,788 n	T_n
0	0,0	—	—	0,0	0	0,00	∞
I	0,78	—	0,8	0,79	1	0,79	18150
II	1,55	1,57	1,53	1,55	2	1,58	9075
III	2,34	2,32	2,32	2,33	3	2,36	6050
IV	3,14	3,14	3,20	3,16	4	3,15	4538
V	4,00	3,97	3,97	3,98	5	3,94	3630
VI	4,74	—	4,63	4,69	6	4,73	3025

The results included in the table seem to suggest that the stars cannot assume all effective temperatures with equal facility. The stars are highly selective in their choice of the temperatures of their radiating layers. The stars seem to "jump" rather abruptly from one preferential temperature to another and to avoid temperatures lying between these selected values.

This singular phenomenon is of universal character and independent of distance and position in space. The values in the above table suggest a simple relation. The preferential temperatures T_n are found from the expression:

$$c_2/T_n = 0{,}788\, n,$$

and hence:

$$n T_n = 1850$$

where n is a one-integer number between 0 and 6. The preferential colour indices are expressed by the relation:

$$C = 0{,}45\, n - 0{,}92.$$

According to a preliminary notice[1] B. STICKER has discussed the question of the preferential colour indices. The sources used were the catalogue of BRILL (134 objects), the colour indices in the Göttinger Aktinometrie, and the colour indices of 1550 stars in zone +45° of Selected Areas. This material has been reduced together with that of HALM to the scale of the Göttinger Aktinometrie. The existence of preferential values is confirmed, as will be seen from the following summary.

Group	BRILL	Göttingen	Cape	Yerkes	Mean	c_2/T	T
I	$(-0^m{,}50)$		$(-0^m{,}50)$	—	$-0^m{,}50$	0,7	20800
II	$-0\ ,31$	$-0^m{,}30$	$-0\ ,35$	$-0^m{,}34$	$-0\ ,32$	1,25	11480
III	$+0\ ,18$	$+0\ ,12$	$+0\ ,11$	$+0\ ,24$	$+0\ ,16$	2,07	6930
IV	$+0\ ,79$	$+0\ ,80$	$+0\ ,75$	$+0\ ,78$	$+0\ ,78$	2,94	4480
V	$+1\ ,48$	$+1\ ,42$	$+1\ ,43$	$+1\ ,48$	$+1\ ,45$	3,93	3660
VI	—	—	$(+1\ ,90)$	$(+1\ ,99)$	$+1\ ,95$	4,7	3050
	0^m-5^m	6^m-8^m	8^m-10^m	10^m-14^m			

STICKER says that it is not necessary to interpret the phenomenon as meaning that stars can only reach certain distinct values of T. The possible explanation is that we observe several groups of stars of the same age. The different frequencies of colours in clusters suggest such an explanation.

[1] Die Naturwissenschaften 17, p. 542 (1929).

133. Colour Equation of Star Catalogues. From what has been said it is clear that if the magnitudes of the stars in a given catalogue whose spectra are of class A are reduced to a certain standard scale (RHP or PD) it will generally be found that the stars of class K if similarly reduced will be systematically a little brighter or fainter than in the standard scale. This difference is known as the colour equation of the catalogue. This quantity depends as to its amount upon the wave lengths included in a certain case of measuring or estimating magnitudes. The colour equation thus can be said to

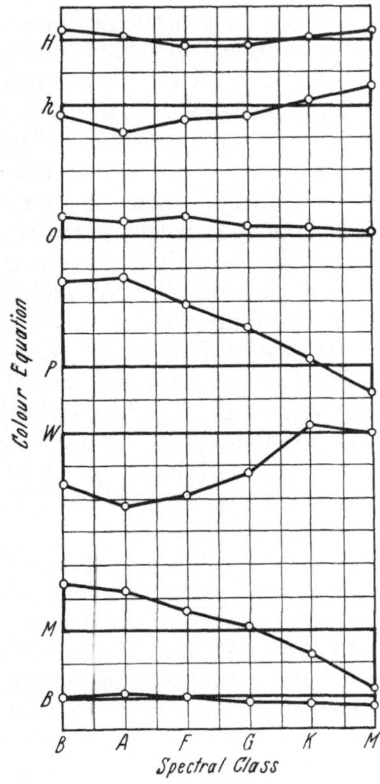

Fig. 101. Colour equations of star catalogues (H = William Herschel's estimates); (h = John Herschel's estimates); (O = Uranometria Oxoniensis); (P = Potsdam Durchmusterung); (W=Williams, Catalogue of 1081 Stars); (M = Bailey's observations in Harv Ann 46, pt. 2); (B = Uranometry by Bailey included in Harv Ann 50). Unit for the ordinates is $0^{m},1$.

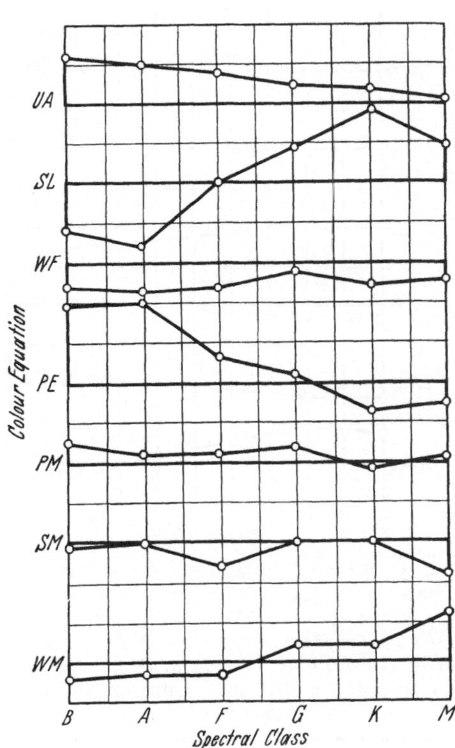

Fig. 102. Colour equations of star catalogues (UA = Uranometria Argentina); (SL = Seidels photometric catalogue); (WF = Wolff's photometric catalogue); (PE = Peirce's photometric researches, Harv Ann 9); (PM = Pickering's photometric measurements, Harv Ann 14); (SM = Searle's photometric measurements, Harv Ann 14); (WM = Wendell's photometric measurements, Harv Ann 14). Unit for the ordinates is $0^{m},1$.

be a function of the sensitiveness of the eye of the observer to light of different colours, the emulsion of the photographic plate, the screen, the selective absorption of the lenses, and the special construction or type of the photometer.

Every catalogue of estimated magnitudes containing so many objects that the mean reduced difference in magnitude between the K and A stars can be accurately derived furnishes an excellent test of the sensitiveness of the eyes of the observer to rays of different colours. If the catalogue is based on photometric measures the colour equation of the instrument is combined with that of the observer.

E. C. PICKERING[1] has derived the colour equation of a number of wellknown photometric catalogues. His main results are illustrated in the figures 101—105. At first it was tried to group the catalogues according to apparent magnitude and spectral class. It appeared that the number of stars in each group was so small that it was decided to combine those differing only in magnitude. In the case of the early catalogues those stars were combined in regions north of the Milky Way (denoted N), south of the Milky Way (S), partly in the Milky Way (P), and in the Milky Way (P). In combining the groups according to magnitude their algebraic mean was taken. If means of the residuals of the individual stars had been taken this would give undue weight to the fainter stars.

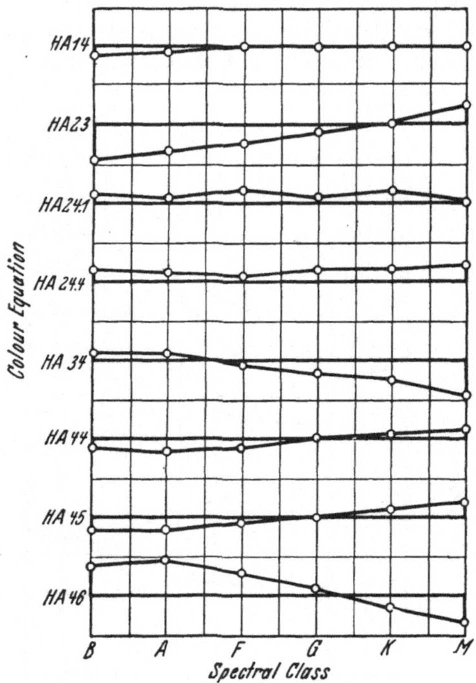

Fig. 103. Colour equations of the different Harvard Catalogues: Harv Ann 14 (= HA 14); Harv Ann 23 (= HA 23); Harv Ann 24, pt. 1 (= HA 24, 1); Harv Ann 34 (= HA 34); Harv Ann 44 (= HA 44); Harv Ann 45 (= HA 45); Harv Ann 46 (= HA 46). Unit for the ordinates is $0^m,1$.

Of the results found those relating to Ptolemaios's Catalogue are of a certain interest because they render it probable that no great change has taken place in the human eye or in the relative colours of the stars during the last two thousand years.

This conclusion gains support through an investigation of mine concerning the estimates of magnitudes by PTOLEMAIOS, AL SÛFI and TYCHO BRAHE[2]. The systematic deviation from the mean scale was grouped according to colour index as follows:

Colour Index	Systematic deviation from the mean scale					
	PTOLEMAIOS		AL SÛFI		TYCHO BRAHE	
		n		n		n
$-0^m,4$ to $-0^m,2$	$+0^m,16$	163	$+0^m,10$	162	$-0^m,01$	137
+0 ,1 „ +0 ,2	−0 ,06	368	−0 ,09	364	+0 ,02	354
+0 ,3 „ +0 ,5	+0 ,04	103	+0 ,03	101	0 ,00	100
+0 ,6 „ +0 ,8	+0 ,08	80	+0 ,10	82	−0 ,02	75
+0 ,9 „ +1 ,4	−0 ,01	269	0 ,00	264	−0 ,04	252
+1 ,5 „ +1 ,8	−0 ,01	35	+0 ,06	35	+0 ,22	34

If the colour equation is derived practically the same results will be found as those given by PICKERING in case of PTOLEMAIOS and AL SÛFI.

The Chinese catalogue mentioned in ciph. 17 furnishes a case where a grouping according to magnitude and spectral class is possible. The results found by me are given in the following table.

[1] Harv Ann 76, p. 11 (1916). See also PICKERING's discussion of the Revised Harvard Photometry in Harv Ann 64, p. 91 (1912).

[2] Upsala Medd No. 15 (1926).

Chinese magnitude / Spectral class	2^m	n	3^m	n	4^m	n	5^m	n	6^m	n
O — B3	$1^m,96 \pm 0^m,16$	16	$3^m,10 \pm 0^m,14$	29	$4^m,06 \pm 0^m,09$	59	$4^m,72 \pm 0^m,07$	55	$5^m,30 \pm 0^m,06$	70
B5 — B9	$2\ ,37 \pm 0\ ,18$	5	$3\ ,64 \pm 0\ ,12$	19	$4\ ,42 \pm 0\ ,08$	42	$4\ ,88 \pm 0\ ,08$	82	$5\ ,62 \pm 0\ ,04$	116
A0 — A5	$2\ ,31 \pm 0\ ,10$	21	$3\ ,51 \pm 0\ ,11$	47	$4\ ,35 \pm 0\ ,06$	110	$4\ ,95 \pm 0\ ,05$	165	$5\ ,68 \pm 0\ ,03$	345
F0 — F8.	$2\ ,06 \pm 0\ ,06$	5	$3\ ,67 \pm 0\ ,13$	27	$4\ ,21 \pm 0\ ,07$	58	$5\ ,08 \pm 0\ ,07$	91	$5\ ,84 \pm 0\ ,05$	183
G0 — G8	—	—	$3\ ,34 \pm 0\ ,15$	22	$4\ ,46 \pm 0\ ,09$	38	$4\ ,97 \pm 0\ ,08$	59	$5\ ,70 \pm 0\ ,05$	152
K0 — K2	$2,\ 20 \pm 0\ ,15$	9	$3\ ,47 \pm 0\ ,08$	47	$4\ ,23 \pm 0\ ,05$	102	$4\ ,90 \pm 0\ ,05$	150	$5\ ,74 \pm 0\ ,04$	305
K5 — Ma	$2,\ 51 \pm 0\ ,21$	8	$3\ ,47 \pm 0\ ,16$	13	$4\ ,25 \pm 0\ ,11$	35	$5\ ,01 \pm 0\ ,08$	55	$5\ ,68 \pm 0\ ,06$	101

The colour equation is found as follows:

Magnitude	Colour equation	Mean error
2^m	$-0^m,11$	$\pm 0^m,18$
3	$-0\ ,12$	$\pm 0\ ,14$
4	$-0\ ,06$	$\pm 0\ ,07$
5	$-0\ ,04$	$\pm 0\ ,07$
6	$+0\ ,09$	$\pm 0\ ,05$

Fig. 104. Colour equation of Cape Photographic Durchmusterung (C P D) and the DRAPER Catalogue in Harv Ann 27 (D C).

+,30	SL
+,25	
+,20	W
+,15	Sf
+,10	Hp Bn WM CE 23S JH
+,05	Arg EE HE PE 44
,00	Hs WF PM SM 14 23N 24,1 24,4 45 WH UO
—,05	BD UA 34 B
—,10	46,1
—,15	
—,20	PE 46,2
—,25	PD

Fig. 105. Distribution of different values of the colour equation. Abbreviations not explained in the legends of the figures 101—104: Sf = AL SÛFI's estimates; Hp = HIPPARCHOS's est.; Bn = BEHRMANN's est.; Hs = HEIS's est. in Atlas Coelestis Novus; Arg = ARGELANDER's est. in Uranometria Nova; BD = Bonner Durchmusterung; CE, EE, HE, and PE estimates by CHANDLER, EDMANS, HOWARD, PICKERING in H. A. 14. Numbers refer to different volumes of Harv Ann; PD = Potsdam Durchmusterung.

As could be expected in beforehand the colour equation is a function of m. Although the uncertainty is considerable it seems to be out of question that the value of the colour equation in the Peking catalogue changes its sign between 5^m and 6^m. It seems that the white stars when the apparent magnitude increases become slightly redder as should be the case if the GALLISSOT effect is present. A final decision can not be made because of the difficulty to say if the red stars grow whiter or redder when their apparent intensity decreases. The course of the reduction curves speaks in favour of the first mentioned explanation.

c) Absolute Magnitudes and Some Other Problems.

134. Definition and Units. Parallax Methods. The absolute magnitude is the apparent magnitude reduced to a certain unit distance. There are many suggestions as regards a proper stellar unit. The following are the most important:

Stellar unit	In parallax	Astronomical units	In km	In light years
Light-year	$\frac{3{,}257}{\pi}$	63311	$9{,}5 \cdot 10^{12}$	1
Parsec	$\frac{1}{\pi}$	206264,8	$3{,}08 \cdot 10^{13}$	0,316
Sternweite	$\frac{1}{\pi}$	206264,8	$3{,}08 \cdot 10^{13}$	0,316
Astron	$\frac{1}{\pi}$	206264,8	$3{,}08 \cdot 10^{13}$	0,316
Metron	$\frac{1}{\pi}$	206264,8	$3{,}08 \cdot 10^{13}$	0,316
Siriusweite	$\frac{1}{5\pi}$	1031324	$1{,}541 \cdot 10^{15}$	0,063
Siriometer	$\frac{0{,}206265}{\pi}$	1000000	$1{,}495 \cdot 10^{15}$	0,065
HERSCHEL	$\frac{3{,}443}{\pi}$	66890	10^{13}	1,058

Of these the light-year is the one most used. The parsec is also comparatively much used and would be more so also if it were not for its awful name. The Siriometer has also been used in a number of earlier investigations at the Observatory of Lund.

It seems that the light-year is the unit that can be defined most accurately. The velocity of light has recently been very accurately determined by A. A. MICHELSON. Further, this velocity is a universal constant of enormous importance for our conception of the physical world.

Although the pro's and contra's of the different units will not be discussed here, I think it should be mentioned that the best unit of length will certainly be the centimetre. The intimate relation between modern physics and modern astronomy makes it very desirable that the C G S system should be accepted in astronomy.

In the computation of the absolute magnitudes three units for the standard distance have mainly been used, namely: one parsec, ten parsecs, and one siriometer. The use of ten parsecs has been recommended by the International Astronomical Union and has been generally accepted. The absolute magnitude M (or M_{01}) thus defined is given by the formula:

$$M = m + 5 + 5\log\pi,$$

if one parsec is used by:

$$M_1 = m + 5\log\pi\,,$$

and if one siriometer is used, by:

$$M_{02} = m + 3{,}428 + 5\log\pi\,.$$

The conception of absolute magnitude is about as old as the first successful determinations of stellar parallax about 1838. Very soon after that time it was clear that the absolute magnitude was by no means the same for all stars. The considerable uncertainty in the parallaxes made conclusions with regard to the absolute magnitudes very uncertain until the methods of measuring parallaxes had gained in accuracy and the material included several hundred stars.

The rate of development of our knowledge of stellar parallaxes is found from the following synopsis:

Epoch	Number of measured parallaxes	Source
1838	3	
1882	34	HOUZEAU, Vademecum de l'Astronomie (1882).
1892	87	PRITCHARD, Oxford Publ No. 4.
1910	365	KAPTEYN, Groningen Publ No. 24.
1917	625	WALKEY, Catalogue J. B. A. A. Appendix (1917).
1924	1870	SCHLESINGER's Catalogue (1924).
1929	2300	Manuscript catalogue kept at the Observatory of Lund.
1932	2600	

The absolute magnitude of stars is determined with the aid of measured or derived annual parallaxes or on the basis of certain effects in spectra, depending in some measurable way on the luminosity. The principal methods for deriving parallaxes are the following:

a) Trigonometric methods. The measurements are nowadays mainly performed on photographs. Comparatively few stars have a measurable parallax. It can thus safely be assumed that a parallax star measured relatively to five or six other stars well distributed on a plate will give a tolerable value of the relative parallax. The mean parallax of the comparison stars can be approximated from investigations based on proper motions and radial velocities or with the aid of other methods. The most dangerous systematic error in parallax measurements are the hour-angle errors which include a differential flexure effect, and the effect of the atmospheric dispersion, which has a yearly period and behaves in the same way as the annual parallax. When the hour-angles are kept within $\pm 30^{m}$ the effects of the hour-angle errors will be inappreciable. It seems to be of more advantage to use a visual refractor with filter than a photographically corrected system. But the disadvantage is that in the former case very faint stars cannot very well be reached.

If the parallax stars and the comparison stars are not of the same intensity on the plates a magnitude equation will arise. This equation is eliminated by different devices, of which the most common is a rotating sector in the pencil of the parallax star, which diminishes its light to the same order of magnitude as that of the comparison stars.

b. The Dynamical Parallax Method. In the case of a binary star we have the relation:

$$\pi_d = \frac{a}{P^{\frac{2}{3}}(\mathfrak{M}_1 + \mathfrak{M}_2)^{\frac{1}{3}}}\,.$$

Inasmuch as the masses of the stars do not vary considerably, the dynamical parallax π_d can be approximated. If we assume the mass-sum to be known in the cases where the orbital elements are not known, but the relative motion is established, it will also be possible to approximate π_d. One method given by H. N. RUSSELL[1] and further developed by JACKSON and FURNER[2] is to compute the most probable values of the parameters involved, viz the secant of the inclination, and the mean value of the third root from the sum of masses.

The formula adopted is then:

$$\pi_d = \operatorname{cosec} i \, (\mathfrak{M}_1 + \mathfrak{M}_2)^{-\frac{1}{3}} \left[\left(\frac{\Delta\theta}{\Delta t}\right)^2 - \frac{1}{4\pi_c \varrho} \frac{\Delta^2 \varrho}{\Delta t^2}\right]^{\frac{1}{3}}.$$

The other method developed by HERTZSPRUNG[3] is to compute the minimum dynamical parallax, $\pi_{m,d}$ or the parallax corresponding to a parabolic motion of the companion relative to the principal star:

$$\pi^3_{m,d} = \frac{r^2}{8\pi^2} \frac{\Delta\theta}{\Delta t}.$$

When deriving this formula HERTZSPRUNG assumes $\mathfrak{M}_1 + \mathfrak{M}_2 = \odot$. The dynamical parallax is then found from the empirical relation:

$$\pi_d = 1{,}86\, \pi_{m,d}.$$

It seems to be a matter of taste which method is to be prefered.

The scheme of deriving dynamical parallaxes also in such cases where the observed relative motion does not suffice to give orbital elements was given, as has been pointed out by J. JACKSON[4], in F. G. W. STRUVE's Mensurae Micrometricae of 1837[5]. It has also been used by J. H. MÄDLER, but seems since then to have been practically forgotten until H. N. RUSSELL and E. HERTZSPRUNG in 1911 independently presented their methods. Extensive computations of dynamical parallaxes have been performed by RUSSELL[1], JACKSON and FURNER[2], G. VAN BIESBROECK[6], and the present writer[7].

For a long time the dynamical parallaxes were termed hypothetical parallaxes. This name is not very suitable nowadays when we have a number of hypothetical methods.

A special case of the dynamical parallaxes are those derived from the absolute dimensions of an orbit of a double star, as computed from observed differences in radial velocity between the components and the elements derived from visual observations. There are only a dozen cases where such a parallax has been derived as yet[8].

c. Moving Cluster Parallaxes. In cases where the members of a certain group of stars have a strictly parallel motion through space the parallaxes can be derived by using the proper motions and the radial velocities for deriving the elements of the parallel motion stream velocity and the convergence point from which the individual parallaxes are easily derived. The group motion parallaxes π_g are of extreme value, as they are independent of the trigonometric measurements and thus apt to reveal the existence of systematic errors in the directly measured parallaxes. On the other hand, it is very often very difficult to decide if a star is a physical member of a certain moving cluster or not. Strictly speaking it should be necessary to know the parallax beforehand in order to know if it is a cluster member. Only two of the moving clusters are well established, the Hyades cluster and the Ursa Major cluster. But as the members

---

[1] Publ American Astr Soc 2, p. 50 (1911).
[2] M N 81, p. 3 (1920).
[3] A N 190, p. 113 (1911).
[4] Obs 45, p. 341 (1922).
[5] p. CLXII ff.
[6] Publ Yerkes Obs 5, p. 4 (1927).
[7] Parallaxes of Southern Binaries, unpublished.
[8] LUNDMARK, Gravitational Parallaxes. Upsala Medd No. 35 (1927).

seem to be scattered all over the sky it is a profitable task to search for new members of these clusters and thus to increase the number of available moving cluster parallaxes.

Extensive researches of the moving clusters have been made by KAPTEYN, who has specially studied the large Scorpio-Centaurus cluster[1]. The parallaxes of several hundred B stars have been derived. In special cases the values are uncertain, but the papers of KAPTEYN are of much interest on account of their masterly discussion of a difficult problem.

N. RASMUSON[2] has discussed earlier work and made new determinations of the elements of different moving clusters. His work has been much consulted by workers in the field of stellar parallaxes. Of recent work we may mention that of LEDERSTEGER[3], who has extended the Ursa Major cluster with 53 new members.

d. O. STRUVE's Method. This method has recently been invented and determines the distance from its effect on the absorption of detached calcium[4].

e. The Spectral Proper Motion Parallax. W. J. LUYTEN and the present writer have shown[5] that the mean absolute magnitude $\overline{M}$, taken within a narrow interval of spectral subdivision, in combination with $H = m + 5 + 5\log\mu$ will give the empirical relation:

$$\overline{M} = a + (m + 5 + 5\log\mu)b + (m + 5 + 5\log\mu)^2 c,$$

or:

$$\overline{M} = a + bH + cH^2.$$

As I think the aim of this method has sometimes been misunderstood, it might be stated that it was not, of course, started in order to compete with any direct or indirect method that is able to give individual values. The $\pi_{\mu s}$ parallaxes are only to be used in cases where no other approximation of π is obtainable. At the time the method was started there existed comparatively few methods for deriving the absolute magnitude of stars of spectral classes earlier than F. In order to use the proper motion material that we now possess (some 30000 stars) efficiently, the derivation of $\pi_{\mu s}$ may be of some importance.

The method has much in common with the method of deriving mean parallaxes of stars developed by KAPTEYN, VAN RHIJN, and others. The essential difference is that this method does not use narrow intervals of spectral subdivisions, but derives a general relation between the logarithms for $\pi_{\mu s}$ and for μ. The mean parallax formula takes into account the variation of the parallax with galactic latitude, this influence is ignored when deriving $\pi_{\mu s}$. An improvement of this method, taking into consideration the said variation, and to work out special reduction formulae for the stars which belong to one of the three main groups, according to their direction, viz: the extended Ursa Major cluster, the Taurus cluster, and the central group, would probably prove to be of value.

The method depends on the assumption that the tangential motion of the members of a certain (narrow spectral) group is fairly constant. The rather surprising fact that the method breaks down as seldom as it actually does in practice seems to be explained by the selection of the material. Most of the stars we now know anything about have a measurable proper motion. The group of stars with their space motions practically in the line of sight does not

---

[1] Ap J 40, p. 43; Mt Wilson Contr 82 (1914); Ap J 40, p. 104, 146, 255; Mt Wilson Contr 147 (1918); Ap J 47, p. 189.

[2] Lund Medd Ser. II, No. 26 (1921).

[3] A N 224, p. 153 (1925). [4] M N 89, p. 567 (1929).

[5] Lick Bull 10, p. 153 (1922); Publ A S P 35, p. 209 (1923).

enter our present material, and thus the dispersion in tangential velocity is small within a certain spectral sub-division.

f. „Spectroscopic“ (Spectrographic) Parallax. Next we have a considerable number of methods that determine M on the basis of spectral criteria and give the spectrographic parallax as a by-product. The method of ADAMS and KOHLSCHÜTTER determines the absolute magnitude from the variation in the intensity of certain enhanced lines. This method has been adopted with certain variations and modifications in the work at the observatories of Victoria, Harvard, Sidmouth, Yerkes, and Arcetri. At Upsala spectra of very small dispersion have been used and variations in the intensity of broader parts of the continuous spectrum, as discovered by LINDBLAD, have been studied.

For certain classes of stars, e. g. the Cepheids, the absolute magnitude can be determined according to the period-luminosity relation, the dependence having been discovered by Miss LEAVITT[1] and later on worked out by E. HERTZSPRUNG[2], H. SHAPLEY[3], and many others. It might also be possible to establish a period-luminosity relation in the case of long-period variables, as has been indicated from recent work at Harvard.

The eclipsing variables present another case where special methods of parallax determination can be worked out[4].

In the case of distant clusters and anagalactic nebulae several other methods are now being developed. The most important seems to be the determination of the mean absolute magnitude of a certain group of stars (Novae, Cepheids, O stars, globular clusters, the most luminous stars in a galaxy etc.) or the shifting of the luminosity curve $\varphi(M)$ of a certain class of stars in our system until it fits the curve $\varphi(m)$ of a distant system. The distribution of the apparent magnitudes being practically the same in a distant system of stars as the distribution $\varphi(M)$, the distance of the system is thus obtained, when $\varphi(m)$ is compared with the curve $\varphi(M)$ in our galaxy for the corresponding population.

g. PLUMMER's Method. By assuming that the B and A stars have a motion in space parallel to the Milky Way plane H. C. PLUMMER[5] has determined hypothetical parallaxes, the constants being derived from proper motions and radial velocities. The method is certainly interesting and has had its value, but it seems that nowadays quite as much is to be obtained from the methods of comparing the elements of the apparent and linear motion in the form in which they have been worked out by KAPTEYN[6], VAN RHIJN[7], or LUYTEN and the present writer[8], R. E. WILSON[9], and others. In the latter case the computed parallaxes will always be positive, whereas in PLUMMER's method considerable negative values sometimes occur.

135. The Trigonometric Parallaxes. Thanks to the energy and skilfulness of a number of mainly American astronomers our knowledge of trigonometric parallaxes has advanced wonderfully since 1912. Of very high importance is the work begun at the Allegheny Observatory by FRANK SCHLESINGER, the eminent authority on astronomy of position. Under his supervision some 700 parallaxes have been derived, and also a number of questions relating to the

---

[1] Harv Circ 173 (1912). [2] A N 192, p. 261 (1912).

[3] Mt Wilson Contr (1918), 151, 152, 153, and numerous subsequent papers. For a bibliography see H. SHAPLEY, Stellar Clusters (1930).

[4] SHAPLEY, Princeton Contrib No. 3 (1915).

[5] Lick Bull 7, p. 30 (1912).

[6] A N 146, p. 97 (1898); Ap J 40, p. 43 (1914); 47, p. 104, 146, 155 (1918).

[7] Groningen Publ 34 (1923).

[8] Publ A S P 35, p. 209 (1923); also unpublished results.

[9] A J 36, p. 49 (1925).

methods have been treated. With regard to these investigations we refer to the papers in Allegheny Publications and to the summary by SCHLESINGER[1].

Another of the leading parallax-observatories is the Leander McCormick. Its director, S. A. MITCHELL, has done an immense piece of work conjointly with a number of skilful colleagues and assistants. The determination of trigonometric parallaxes at Mount Wilson is a very large piece of work and has been carried out by A. VAN MAANEN alone.

Important contributions within this sphere have also been made by the observers at Yerkes, Sproul, and Dearborn.

In Europa Greenwich is the leading parallax-observatory; among other observatories smaller contributions have been communicated from Pulkova, Upsala, and Berlin-Babelsberg.

At most of the observatories engaged in parallax work a number of investigations of a general nature have also been performed. Special care has been given to the question of the size and nature of the systematic errors.

The contributions up to 1929 from the different observatories are shown in the following synopsis:

Institute	Number of measured parallaxes
Allegheny Observatory	800
Leander McCormick Observatory	1000
Yerkes Observatory	250
Mount Wilson Observatory	300
Royal Greenwich Observatory	300
Sproul Observatory	250
Dearborn Observatory	100

In order to see how the modern parallax-measurements are distributed in the sky and with regard to their distances we have only some results from an investigation by the late G. SCHNAUDER[2]. His material embraced 1380 objects.

Single objects . . . . 57 % of measured parallaxes
Double objects . . . . 21,5 ,, ,, ,, ,,
Objects with 3 components . . . . . 8 ,, ,, ,, ,,
Multiple objects . . . . 13,5 ,, ,, ,, ,,

For some 200 stars there only exist observations not confirmed by photographic ones.

As regards their distribution in the sky 87% of all parallaxes relate to stars in the northern hemisphere. If the northern sphere is divided into three equal zones, SCHNAUDER finds:

Polar zone . . . . 28,5% of measured parallaxes
Intermediate zone . . . . 33,5 ,, ,, ,, ,,
Galactic zone. . . . . 38,5 ,, ,, ,, ,,

The distribution as regards the right ascension of the measured parallaxes is according to SCHNAUDER:

Right ascension	0^h—2^h	3^h—5^h	6^h—8^h	9^h—11^h	12^h—14^h	15^h—17^h	18^h—20^h	21^h—23^h
Observed percentage	12,5	12,5	13	10	9	14	17	12
Expected percentage	13	13,5	13	11	11	12	13,5	13

A preference is given to the region Cygnus-Aquila.

[1] Probleme der Astronomie (Seeliger Festschrift), (1924), p. 422.

[2] Ergebn d exakt Naturw 2, p. 19 (1923).

The distribution as regards the declination is as follows:

	Polar zone	Intermediate zone	Equator
Observed percentage . . .	44	28,5	27,5
Expected percentage . . .	34	33,5	32,5

The polar zone is given preference in the observations as the zone of zenith for most of the observatories is situated within it. Besides, the Greenwich programme is purposely restricted just to the polar zone.

As regards the distribution within different groups of the distances SCHNAUDER finds:

Parsecs	Percentages	Parsecs	Percentages
0—10	7,5	>30	49
10—20	14,5	negative	
20—30	14	parallaxes	15

136. Systematic Errors in the Stellar Parallaxes. The existence of systematic errors in stellar parallaxes that are a function of the right ascension was pointed out by A. FLINT in 1916[1]. Using the much more extensive material existing in 1918 A. VAN MAANEN[2] confirmed the existence of such errors (compare the table on the following page) and pointed out that an error varying with A. R. must have a seasonal origin. He mentions a number of causes that singly or in combination with one another may contribute to the effect. We quote the following excellent treatment of these causes[3]:

"To begin with, an observer will in general be more tired at the time of the morning exposures than during the evening exposures, and the fatigue will be different in summer from that in the winter season with its long cold nights. No instrument is perfect, and the guiding will be less efficient when the observer is tired. The effect will be different for observatories where the same observer works the whole night and for those where a change in observer is made about midnight. The driving clock is likely to require slightly different corrections in the morning and in the evening, which again will be different in summer and winter. The hour-angles at which the plates are taken, too, will in general be different for morning and evening exposures, even when, as in modern parallax work, all hour-angles are kept fairly small; there is, further, the possibility of differences in seeing between morning and evening, and both these two factors will vary for different times of the year; there is, finally, the effect of galactic latitude; at those times of the year when the fields observed are in or near the Milky Way we are able to choose comparison stars much closer to the central stars and, accordingly, less likely to be affected by any of the influences discussed."

To this the suggestion could be added that the systematic difference in mean absolute magnitude between the Milky Way and outside regions, on account of the high proportion of early spectral classes in the former, might enter into the parallax determinations and introduce a periodic error. If so, there ought to be a general correlation between the different sets of systematic errors.

We have computed the correlation between VAN MAANEN's results for the principal observatories, but the values of r are very low and probably not real except.

[1] A J 29, p. 199 (1916).
[2] Mt Wilson Contr 189 (1920).
[3] Mt Wilson Contr 357 (1928).

Systematic correction in absolute magnitude for the value $\pi = 0'',010$. (According to VAN MAANEN's investigations.)

Right Ascension	MountWilson	Allegheny	McCormick	Yerkes	Sproul	Greenwich	Dearborn
0^h	$-0^M,07$	$+0^M,48$	$-0^M,28$	$+1^M,93$	$-0^M,91$	$+3^M,19$	$+2^M,13$
1	−0 ,52	+0 ,28	−0 ,33	+2 ,11	−1 ,22	+2 ,93	+2 ,04
2	−0 ,91	+0 ,22	−0 ,11	+2 ,04	−1 ,41	+2 ,24	+0 ,83
3	−1 ,11	+0 ,24	+0 ,22	+1 ,67	−1 ,59	+1 ,32	0 ,06
4	−1 ,11	+0 ,28	+0 ,67	+1 ,04	−1 ,74	+0 ,33	−3 ,06
5	−0 ,83	+0 ,30	+1 ,15	+0 ,24	−1 ,87	−0 ,54	−4 ,60
6	−0 ,46	+0 ,26	+1 ,54	−0 ,61	−2 ,11	−1 ,13	−5 ,25
7	−0 ,09	+0 ,13	+1 ,76	−1 ,32	−2 ,43	−1 ,37	−4 ,86
8	+0 ,24	−0 ,02	+1 ,72	−1 ,82	−2 ,69	−1 ,24	−3 ,54
9	+0 ,37	−0 ,24	+1 ,52	−1 ,98	−2 ,89	−0 ,89	−1 ,76
10	+2 ,26	−0 ,39	+1 ,15	−1 ,82	−2 ,93	−0 ,50	−0 ,04
11	−0 ,09	−0 ,41	+0 ,72	−1 ,37	−2 ,78	−0 ,20	+1 ,04
12	−0 ,50	−0 ,30	+0 ,33	−0 ,76	−2 ,35	−0 ,07	+1 ,09
13	−0 ,91	−0 ,04	+0 ,07	−0 ,15	−1 ,69	−0 ,20	0 ,00
14	−1 ,26	+0 ,35	−0 ,02	+0 ,39	−0 ,93	−0 ,54	−2 ,08
15	−1 ,37	+0 ,85	+0 ,04	+0 ,72	−0 ,10	−0 ,93	−4 ,63
16	−1 ,24	+1 ,32	+0 ,24	+0 ,87	+0 ,61	−1 ,24	−7 ,06
17	−0 ,87	+1 ,69	+0 ,50	+0 ,85	+1 ,13	−1 ,32	−8 ,73
18	−0 ,37	+1 ,91	+0 ,67	+0 ,74	+1 ,37	−1 ,04	−9 ,25
19	+0 ,13	+1 ,95	+0 ,76	+0 ,67	+1 ,35	−0 ,41	−8 ,47
20	+0 ,54	+1 ,80	+0 ,67	+0 ,69	+1 ,04	+0 ,50	−6 ,49
21	+0 ,72	+1 ,50	+0 ,48	+0 ,89	+0 ,59	+1 ,45	−3 ,84
22	+0 ,69	+1 ,13	+0 ,20	+1 ,22	+0 ,07	+2 ,37	−1 ,13
23	+0 ,39	+0 ,76	−0 ,11	+1 ,59	−0 ,48	+3 ,02	+1 ,00

For other values of parallaxes use: $\frac{\text{tabular value}}{\pi} \times 0'',010$.

137. HERTZSPRUNG's Discovery of Giants and Dwarfs. The contents of HERTZSPRUNG's two papers, Zur Strahlung der Sterne[1], are very remarkable. Several ideas and results in modern astronomy can trace their origin back to those papers of 1905 and 1907. As they are contained in a journal not generally devoted to astronomical researches, and are also, owing to other reasons, not very well known, a review of their contents will be given here.

The starting point was the difference concerning the sharpness of lines, classified by Miss MAURY[2] by the addition of the subscript a, b, and c to the spectral classes. Miss MAURY has subsequently found that there is no decided difference between the a and b characters, or the fuzzy lines. The c character is distinguished by the strongly defined sharpness of the spectral lines. These are uncommonly sharp and narrow, and the relative intensities of the lines are not comparable with the corresponding ratios in the spectrum of the Sun. Several lines not identifiable in the solar spectrum also appear.

Miss MAURY has suggested that the a and b stars, on the one hand, and the c stars on the other, represented different lines of evolution. In order to test this question HERTZSPRUNG has used the proper motions in such a way that μ was reduced to its value for $m=0$. Then the central values of $\mu_{m=0}$ were used, i. e. the value of μ for which there are an equal number of values both over and below it. The corresponding values of m reduced to $\mu = 0'',01$, or H_μ, were then computed. These values correspond to the values of $m + 5 + 5 \log \mu$ used later, and are generally closely related to the M_μ of LUYTEN, LUNDMARK, and others.

The radial velocities for some 60 a and b stars have a normal distribution around zero, with a mean error of ± 20 km. Thus the projections of the absolute

[1] Z f wiss Photogr 3, p. 429 (1905); 5, p. 86 (1907).
[2] Harv Ann 28; pt. 1 (1897).

motions must be normally distributed on a line, say, in the tangential plane. In an inquiry concerning the mean error in M, assuming that all stars in a group have the same absolute magnitude, HERTZSPRUNG finds that at least the A stars must have a rather constant absolute magnitude. He used the μ's and found the mean for 102 A stars to be $-0'',048 \pm 0'',0448$. The mean error corresponds to a linear value of ± 20 km/sec and thus gives the mean parallax $= 0'',0112$ for $\overline{m} = 4^m,84$ which corresponds to a mean absolute magnitude of

$$\overline{M} = 1,16.$$

A colour equivalent was derived from a comparison of the visual Harvard magnitudes m_H with the photographic magnitudes, m_D, derived from the intensity of the light at the G-line (λ 4308). The difference $m_H - m_D$ was formed, and it was assumed that a linear relation existed between the two quantities and the computed values of m_D which corresponded to $m_H = 4^H,5$. The following extract from table I in HERTZSPRUNG's paper may be of interest:

Spectral class	H	n	Mean colour index	n	$\frac{m_H}{m_D}$
B3	$-0,63 \pm 0,41$	38	0,13	66	1,30
B5—B9	$+2,25 \pm 0,29$	21	0,06	41	1,31
A	$+3,05 \pm 0,21$	47	0,11	81	1,31
A5—F	$+4,06 \pm 0,33$	34	0,29	57	1,26
F5	$+5,23 \pm 0,40$	30	0,53	45	1,38
G—G5	$+2,93 \pm 0,31$	24	0,76	37	1,45
K	$+4,38 \pm 0,22$	74	0,94	102	1,51
K5	$+2,77 \pm 0,56$	21	1,19	28	1,75
Ma	$+3,28 \pm 0,30$	15	1,33	31	1,43

The ratios m_H/m_D should be equal to 1. The increase from white to red is caused, according to HERTZSPRUNG's opinion, by a photographic PURKINJE effect, which has its probable origin in a different gradation in the photographic film of the spectra and the lamp used for comparison.

The stars in the above table are b and a stars. Another table is constructed for the c and ac stars. Only proper motions of the fundamental stars have been used and the data ought to be comparatively reliable. HERTZSPRUNG points out that there is an increase in the reduced proper motion with increasing apparent magnitude. He says that this could be interpreted as an extinction of the light and points out that the fainter c stars also show a more advanced colour. Anyhow, it is evident that the reduced μ's of the c stars are very small, only amounting to a few hundredths of a second of arc. He concluded that the c stars must be as bright intrinsically as the Orion stars. Another interesting conclusion is that with increasing number in the spectral sequence of Miss MAURY the c character decreases and ends just where the bright K stars set in.

Next a table is compiled giving the measured absolute parallaxes for 53 stars and the values of the parallaxes reduced to the common magnitude zero. A comparison between the mean values of these parallaxes and the means of the yearly proper motions is given in the following table:

Spectral class	$\overline{\pi}_{m=0}$	$\overline{\mu}_{m=0}$	$\overline{\mu}/\overline{\pi}$	n
B—B3	0'',0255	0'',075	2,9	6
B5—B9	0 ,106	0 ,282	2,7	5
A—F	0 ,166	0 ,498	3,0	16
F—G5	0 ,525	0 ,777	1,5	7
K—K5	0 ,156	0 ,645	4,1	11
Ma	0 ,115	0 ,453	3,9	3

From a discussion of the parallaxes HERTZSPRUNG concludes that the parallaxes, reduced to equal magnitude, show a maximum around the G class. α Carinae and α Aurigae, which evidently possess high luminosity, are exceptions.

HERTZSPRUNG gives the following formula for reducing the magnitude M_r of a double star to unit values of the mass and the parallax. The quantity M_r then obtained is:

$$M_r = m_H + 5 \log \pi + 5/3 \cdot \log \mathfrak{M},$$

where $\mathfrak{M}$ is the mass ($\odot = 1$). M_r can be derived from double stars if the period P and the major axis a are known, as:

$$\pi^3 \mathfrak{M} = a^3/P^3 .$$

The last formula but one can then be written:

$$M_r = m_H + 5 \log a - 10/3 \log P.$$

HERTZSPRUNG uses the 53 known orbits with accurate elements and computes M_r and gives the following formula connecting m_H and m_D and M_r:

$$M_r = -1{,}23 - 2{,}23\,(m_H - 4{,}99) + 3{,}583\,(m_D - 5{,}39) .$$

If we consider the stars γ Leonis and 70 Ophiuchi which do not differ very much with regard to their masses and which belong to the same spectral class XVa, it seems that their absolute magnitudes differ by nearly 6 magnitudes. HERTZSPRUNG points out that if the surface intensity is equal, the density of γ Leonis is 1/3000 of that of 70 Ophiuchi.

The Sun represents a somewhat "earlier" type than 70 Ophiuchi and it seems very unlikely that it should, when it becomes cooler, be a γ Leonis star. This object must then represent an earlier evolution stage or a collateral series. The hypothesis of the existence of two such collateral series is apt to explain why the stars for which $m_H < 5$, the c- and the Orion stars, are the absolutely brightest and the yellow, not the red stars, are the absolutely faintest. The series are represented by the two sequences of objects given in the table:

Spectral class	Sequence I	Modern data		Spectral class	Sequence II	Modern data	
		M	Diameter			M	Diameter
	Nebulae				Nebulae		
Oe5	S Monocerotis	−2,3	11	B2	χ Orionis	−1,4	24
B0	ε Orionis	−3,7	12	B8p	β Orionis	−5,8	
B2	γ Orionis	−0,2	9	Aop	α Cygni	−4,1	
B8	α Leonis	+0,3	3,8	F8p	δ Can Majoris	−3,0	
A0	α Can Majoris	+1,3		K0	α Bootis	+0,5	30
A5	α Aquilae	+2,0	2,8	Ma	α Orionis	−3,7	300
F5	α Can Minoris	+3,0	2,0		Vogel's type IV		
G0	Sun	+4,8	1,0		(M giants)	0,0 to 3,0	
K0	70 Ophiuchi	+5,6	0,7				
K5	61 Cygni	+8,0					
B9	o^2 Eridani BC	+10,3					
	"dark" stars						

In his second paper HERTZSPRUNG pointed out the importance of testing the influence of the selection in a given material, when questions relating to the evolution of the stars are being dealt with. He considered the stars which were then known to have a parallax larger than 0″,100. The 95 stars of this group did not show a division of the absolute magnitudes of the red stars into two distinct groups. The mean values were as follows:

Spectral class	$\overline{M} \pm \varepsilon(\overline{M})$	n	Potsdam colour	$\overline{M} \pm \varepsilon(M)$	n
A	$5^{M},45 \pm 0^{M},97$	11	W	$3^{M},34 \pm 0^{M},34$	6
E, F, G	$5\ ,37 \pm 0\ ,20$	34	GW	$6\ ,07 \pm 0\ ,29$	20
H, J, K	$6\ ,85 \pm 0\ ,39$	24	WG	$6\ ,14 \pm 0\ ,23$	23
Ma	$9\ ,94 \pm (0\ ,47)$	3	G	$8\ ,88 \pm 0\ ,67$	4
			RG	$2\ ,40 \pm$	1

Some 90 per cent of the stars within 10 parsecs are absolutely fainter than our Sun. Finding 90 per cent of the stars within 5 parsecs fainter than our Sun, HERTZSPRUNG predicted correctly that future determinations of parallaxes will increase the said proportion.

The fact that practically none of the absolutely bright red stars (giants according to our recent terminology) are found within the space investigated shows that these are very rare in space as compared with stars of the normal solar series (dwarfs). It is possible that the former pass comparatively quickly through the red stage, or the giant stars may form a collateral branch in the evolution. The author left it undecided whether there is a gap between the two stages or intermediate ones existed.

The author reduced the proper motions of the classes XIIIa and XIVa in the classification of Miss MAURY to equal apparent magnitude. The disparity in the reduced proper motion was considerable in the latter class and HERTZSPRUNG concluded that that was also the case for the absolute magnitude. He also found from the remarks of Miss MAURY that the intensity of the nowadays well-known line λ 4077 varied with the reduced proper motion and hence with the absolute magnitude. Here, as far as I am aware, the possibility of the spectrographic method has been suggested for the first time, because HERTZSPRUNG says that whatever the causes of the large differences in absolute magnitude may be, there must still exist some spectral criterion which will indicate the variation of M.

Further evidence for the existence of the two groups of red stars with regard to the distribution of their absolute magnitudes was specially collected from a study of binary stars. Using the pairs with known trigonometric parallax, dynamical parallax, and proper motion, HERTZSPRUNG derived the relation:

$$\frac{\Delta(\log\mu - \log\pi_d)}{\Delta M} = 0{,}10 \pm 0{,}024\,.$$

Roughly speaking this formula says that the variation in the logarithm of the tangential velocity amounts to a tenth of the variation in absolute magnitude. The most plausible explanation according to HERTZSPRUNG is that the double stars being absolutely faint also have small masses. Among other possible explanations he also mentions that the double stars have been formed at the same time and those that are moving faster have been cooling more rapidly than the ones having slower motion. The absolutely faint stars are the oldest and their velocity increases with their age through acceleration analogous with the formation and the fall of rain-drops.

For the binaries belonging to the type of Capella HERTZSPRUNG found $M = -3$. The difference $m_A - m_B$ or the difference in magnitude between the principal star and the companion ought to be smaller in the photographic light than in the visual light. For the binaries of the solar type the converse inequality ought to take place, i. e., one can expect that the intrinsically fainter component is the redder of the two stars. The data at hand showed this to be the case.

Since the physical members of the Pleiades are at the same distance the fainter stars ought to be redder than the brighter ones. HERTZSPRUNG has found:

Spectral class	m Potsdam	$\overline{m}$ Harvard	n
V	$4^m,44 \pm 0^m,30$	$4^m,22$	8
VI	$6\ ,25 \pm 0\ ,17$	$5\ ,90$	5
VII	$7\ ,03 \pm 0\ ,09$	$6\ ,58$	5
X	$6\ ,72$	$6\ ,57$	1

Thus the spectral absolute magnitude relation is confirmed. HERTZSPRUNG derived as the parallax of the cluster: $0'',0085 \pm 0'',0020$, which is in excellent agreement with recent results.

There are a number of interesting suggestions in this paper, of which we can only mention a few. HERTZSPRUNG points out the possibility that the Sun has originated in the Orion nebula.

The giant nature of the N stars is made plausible from their small proper motions and strong galactic concentration.

For the three Cepheids, δ Cephei, ζ Geminorum, and η Aquilae, the order of magnitude of their parallaxes was fixed at $0'',005$.

The c and ac stars of a special group have high absolute magnitudes. This is especially the case with the former class.

The relation between spectral class and colour was investigated on the basis of OSTHOFF's first catalogue. Corrections were applied for the change of the colour c with apparent magnitude m and for the change of c produced by the thickness of the atmospheric layers passed, so that the colours were finally reduced to equal m and 1,3 atmospheres. The relation between c and m is not quite linear but can be assumed to be so at a first approximation. The relation between the colours as estimated by OSTHOFF and by the Potsdam Observers was also investigated.

138. Luminosity of Different Spectral Classes. In 1906 A. PANNEKOEK[1] published an investigation concerning the luminosity of different spectral classes. Parallaxes were not available and the proper motion had to be used as an indicator of the distance. A better measure than the (reduced) proper motion is the parallactic motion. The τ- and υ-components were taken from KAPTEYN's papers concerning the BRADLEY stars. The mean values of q (parallactic motion) and τ were also reduced to 4^m through multiplication with the photometric factor $10^{0,2(m-4)}$.

The spectral classification of Miss MAURY was used as the basis. Comparing the values $\overline{\tau}$ and $\overline{q}$ PANNEKOEK found that the mean linear velocity is not constant for all classes, but increases as further stages of development in the spectral series are reached, which is in agreement with the earlier conclusions of CAMPBELL. As regards the absolute brightness it is found that it constantly decreases when we proceed from the Orion stars to the solar stars. For the stars of the same type as Arcturus the absolute magnitude increases again and PANNEKOEK concludes that these stars possess, on an average, a much larger surface and volume than the other second type stars of the classes F and G. This result was at variance with the theory of stellar evolution then accepted.

In order to decide whether the high luminosity of the Arcturian stars was caused by high mass or low density PANNEKOEK made use of the values of the mass-function of spectroscopic binaries. The material was not extensive enough

[1] Proc Amsterd Ac 9, p. 134 (1906).

to give a decision. The higher mass of the Sirian stars as compared with the solar stars was shown to be probable and the low values of the masses of the c stars, if the mass-function is considered as an equivalent of the mass, was also noted.

139. RUSSELL's Investigations on Stellar Development. At the meeting of the American Astronomical Society in 1914 H. N. RUSSELL delivered his

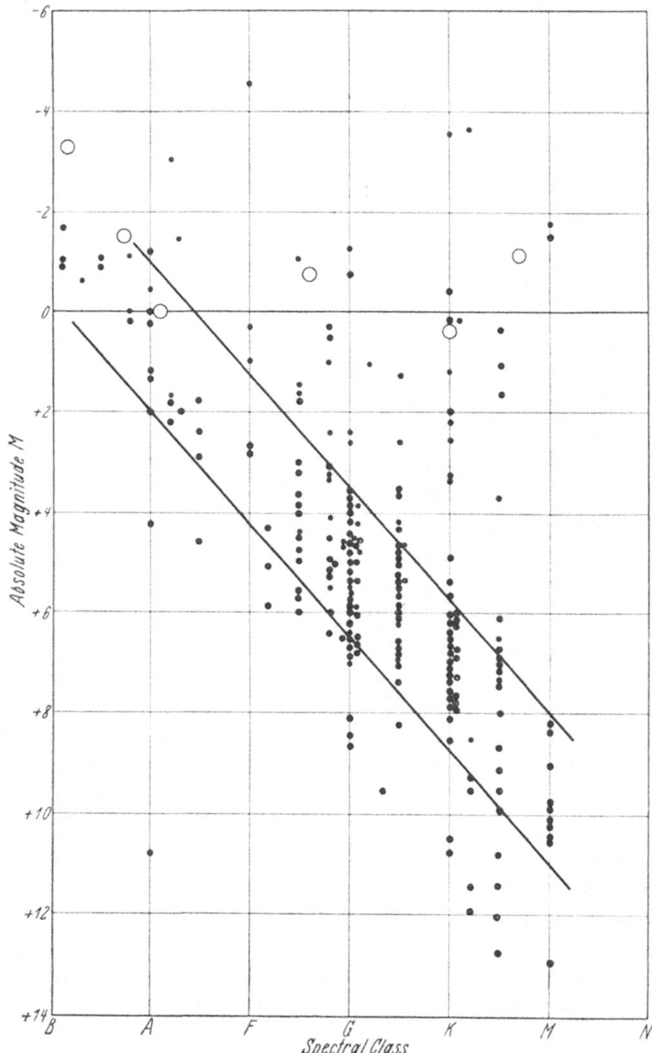

Fig. 106. The original RUSSELL-diagram derived on basis of all the direct measures of parallax available in the spring of 1913 when the diagram was constructed. The small dots represent results derived from the poor parallaxes and the larger dots good determinations. The open circles represent mean results for numerous bright stars of small proper motion.

famous address[1] concerning the relations between the spectra and other characteristics of the stars. The author points out that many of the results exhibited are not new, but the paper also contains a number of new results and ideas, and

[1] Pop Astr 22, p. 275, 331 (1914); Publ Amer Astr Soc III, p. 22.

is of a certain interest because of its views on stellar evolution, which originated from a discussion of the distribution of absolute magnitudes.

The measurements of stellar parallaxes had advanced since HERTZSPRUNG made the first attempt to analyse the curve $M = \psi(S)$. The diagram plotted from the existing parallax measurements suggested that there are two great

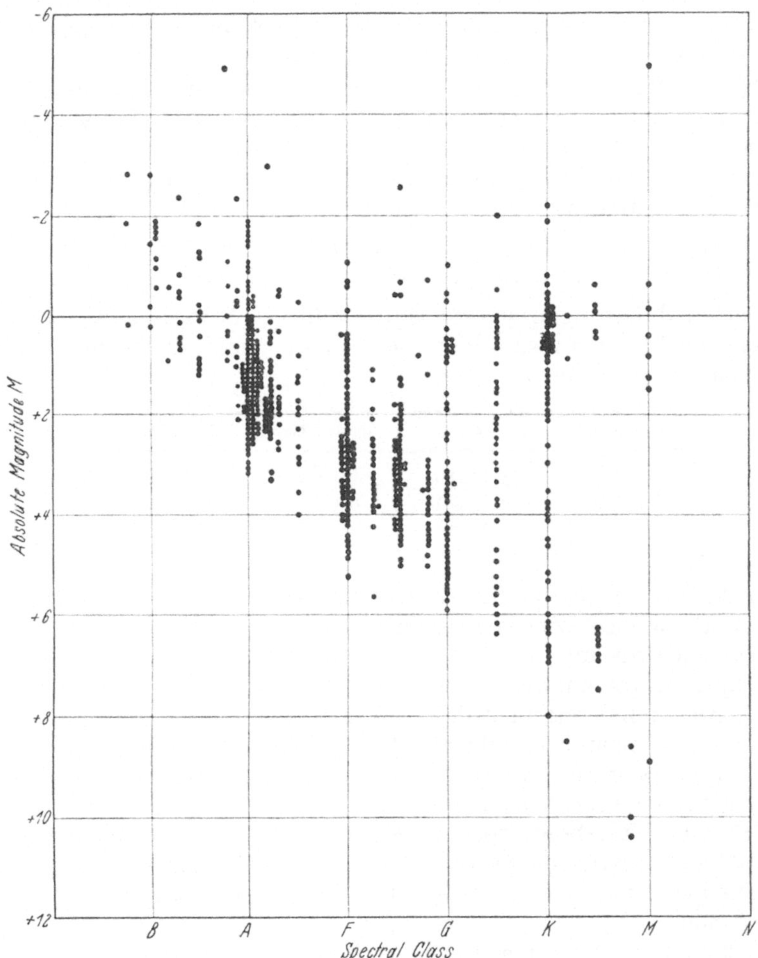

Fig. 107. The distribution of absolute magnitudes M with regard to spectral class as derived by RUSSELL in 1914 on basis of some 550 stars for which he had derived dynamical parallaxes. Part of the dispersion in M is caused by the inaccurate assumption of a constant value of the masses when computing the parallaxes. It is remarkable that the above diagram shows most of the principal features of the diagrams of the same kind based upon the most modern data. Compare for instance figures 109, and 110.

classes of stars: one of great brightness averaging hundreds of times the brightness of the Sun and varying very little in brightness from one class of spectrum to another; the other of less brightness, which falls off very rapidly with increasing redness.

RUSSELL says that HERTZSPRUNG has applied to these classes the "excellent names of giant and dwarf stars". These names do not originate from HERTZSPRUNG; I have tried to find where they first occur, but so far without success.

RUSSELL also discussed the evidence from the moving clusters concerning $M = \psi(S)$ and found the conclusions drawn from the parallax measurements corroborated by the new and independent data. He also pointed out the equal form of $M = \psi(S)$ for the different clusters.

Somewhat earlier HERTZSPRUNG and RUSSELL had rediscovered the STRUVE-method for deriving "the dynamical parallax". RUSSELL was able to derive the parallax for more than 550 pairs showing a relative change. The diagram showed the same division of the stars into two main series.

Lastly 80 eclipsing variables, whose parallaxes were determined on the basis of assumptions concerning the surface brightness, were discussed and used for the diagram.

RUSSELL has derived the following formulae for computing the mean absolute magnitude of the dwarfs:

$M = 1^{\mathrm{M}},4 + 2^{\mathrm{M}},1\ (Sp - \mathrm{A})$ Stars of directly measured parallax.
$M = 0\ ,6 + 2\ ,1\ (Sp - \mathrm{A})$ Stars in moving clusters.

He concluded that the mean error of the values of M for the different spectral classes was $\pm 0^{\mathrm{M}},39$ in the mean. For one absolute magnitude he found the following size of the mean error:

Spectral class	Parallax stars	Spectral class	Cluster stars
A—F8	$\pm 1^{\mathrm{M}},10$	B0—B9	$\pm 0^{\mathrm{M}},88$
G—G5	$\pm 1\ ,28$	A —A5	$\pm 0\ ,84$
K—M	$\pm 1\ ,44$	F —F8	$\pm 0\ ,71$
		G —G5	$\pm 1\ ,00$

RUSSELL discusses the question of the general size of the stellar masses, and points out that the most conspicuous thing about those stellar masses that have been determined with any approach to accuracy is their remarkable similarity. While the range in the known luminosities of the stars will exceed a million-fold and that in the well-determined densities is nearly as great, the range in the masses so far investigated is only about fifty-fold. The author also concludes that "there appears, from the rather scanty evidence at present available, to be some correlation between mass and luminosity".

LUDENDORFF had shown conclusively in 1911[1] that the average mass of the spectroscopic binaries of spectral class B is three times as great as that of the spectroscopic binaries of other spectral classes and may exceed ten times that of the Sun.

A determination of the mean masses of stars of various spectral types was undertaken by RUSSELL. He compared the M as determined from the relative motion in double stars and thus containing the effect of the mass, with the M as measured directly, or derived from the elements of the motion in moving clusters, and thus free from this effect. In fact the same method was used later on by SEARES.

Spectral class	M (Observed)	M (Reduced to $\mathfrak{M}=1$)	Mass
B2	—1,4	—0,6	3,0 ☉
A0	+0,5	+1,1	2,3
A5 dwarfs	+1,5	+1,6	1,2
F0	+2,4	+2,8	1,7
F3	+3,3	+3,1	0,8
F8 and G0	+4,6	+4,0	0,5
G5 dwarfs	+5,1	+4,2	0,3
K0 dwarfs	+6,4	+5,5	0,3
K5 and M	+8,9	+7,7	0,2
G	—0,2	+0,6	3,0
K0	+0,2	+0,5	1,5
K5 and M	—0,3	0,0	1,5

[1] A N 189, p. 145 (1911).

Among the other main results reached by RUSSELL in his address we mention the following:

The differences in brightness between the stars of different spectral classes and between the giants and dwarfs of the same spectral sub-division do not arise from differences in mass. The mean masses of the various groups of stars are very similar.

The surface brightness or intensity of the stars diminishes rapidly with increasing redness, changing by about three times the difference in colour index or rather more than one magnitude from each class to the next.

The mean density of the B and A stars is a little more than one-tenth of the density of the Sun. The densities of dwarf stars increase with increasing redness from this value through that of the Sun to a limit that cannot at present be exactly defined. This increase in density together with the diminution in surface brightness accounts for the rapid fall in luminosity with increasing redness among these stars.

The mean densities of the giant stars diminish rapidly with increasing redness from 1/10 ⊙ for A-stars to less then 1/20000 for M stars. This counteracts the change in surface brightness and explains the approximate equality in luminosity of all these stars.

The actual existence of stars of spectra G and K whose densities are of the order derived here is proved by several examples among the eclipsing variables, all of which are far less dense than any of the more numerous eclipsing stars of "early" spectral type with the sole exception of β Lyrae.

These facts have a decided bearing on the problem of stellar evolution and they constituted the platform from which RUSSELL built up his well-known cosmological scheme.

Of the propositions which may be made regarding stellar evolution there is one that has been generally accepted for long time that as a star grows older it contracts. Since contraction converts potential energy of gravitation into heat which is transferred by radiation to cooler bodies it appears from thermodynamic principles that the general trend of change must in the long run be in this direction. It is conceivable that in some particular epoch during the lifetime of a star there might be so rapid an evolution of energy that it temporarily surpasses the loss by radiation and leds to an expansion against gravity. It is clear that this must be a passing stage in the career of the star and it will still be true in the long run that the order of increasing density is the order of advancing evolution.

If the stars studied by aid of the data mentioned above are arranged in order of increasing density the starting point is a giant star of class M. Then the giant series has to be followed in the reverse order (M, K, G, F, A) from that in which the spectra are usually placed and then with density still increasing, though at a slower rate, the series of dwarf stars in the usual order of the spectral classes (B, A, F, G, K, M) past the Sun to the absolutely faintest stars known.

There is no doubt that this is the order of increasing density but is it also the order of advancing age? If so, we are led back to the hypothesis of NORMAN LOCKYER[1] that a star is hottest near the middle of its history and that the redder stars fall into two groups, one of rising and the other of falling temperature. Accepting this hypothesis it follows that the giant stars then represent succesive stages in the heating up of a body and must be more primitive the redder they

[1] Phil Trans 184, p. 688 (1902); London R S Proc (1899), p. 186.

are. The dwarf stars represent successive stages in its later cooling and the reddest of these are the farthest advanced. We have no longer two separate series of stars to deal with but a single one beginning and ending with class M and having class B near the middle.

The terms "earlier" and "late" generally applied to the corresponding spectral classes before the appearence of RUSSELL's theory and still many times used are actually misleading if the view on stellar evolution as suggested by RUSSELL is correct.

Next, the question is dealt with if the scheme of increasing density is physically possible and it is found that the scheme is in conspicuous agreement with the conclusions which may be reached directly from elementary and very probable physical considerations.

The stars in general are certainly masses of gas and the great majority of them are at any given moment very approximately in stable internal equilibrium under the influence of their own gravitation and very nearly in a steady state as regards the production and radiation of heat, but they are slowly contracting on account of their loss of energy. The behaviour of such a mass of gas has been investigated by LANE, RITTER, EMDEN[1] and others. As long as the density of the gaseous mass remains so low that the so called gas laws represent its behaviour with tolerable accuracy and as long as the gaseous mass remains built up on the same model[2] the central temperature at geometrically homologous points varies inversely as the radius (LANE's law). If after the contraction the star is built only approximately on the same model as before this law will be approximately but not exactly true.

The temperature of the layers emitting the bulk of radiation will also rise as the star contracts but more slowly since the increase in density will make the gas effectively opaque in layers whose thickness is an ever decreasing fraction of the radius. The temperature of the outer nearly transparent layers in which the FRAUNHOFER absorption takes place will be determined almost entirely by the energy density of the flux of radiation through them from the layers below.

When the gaseous mass slowly looses energy and contracts, its effective temperature will increase and the mass will grow whiter. Meanwhile the dimensions and surface of the mass will diminish and this will at least partially counteract the influence of the increased surface brightness or even overbalance it. This process will go on until the gas reaches such a density that the departures of its behaviour from the laws which hold true for a perfect gas become important. This critical density will, of course, first be reached at the centre of the mass. The principal departure from the simple gas laws will be that the gas becomes more difficultly compressible. Thus a smaller rise in temperature than that demanded by the elementary theory will suffice to preserve equilibrium after further contraction. The rise in temperature will therefore slacken and finally cease first at the centre and later in the outer layers. Further contraction will only be possible when it is accompanied by a decrease in temperature, and the heat expended in warming the mass during earlier stages of evolution will now be gradually transmitted to the surface and liberated by radiation. The contraction also generates radiation.

During this stage the behaviour of the mass will resemble roughly that of a cooling solid body, except that the rate of decrease of temperature is far

---

[1] For bibliography see EMDEN: Gaskugeln (1907).

[2] This means that the density and temperature at geometrically homologous points vary proportionally to the central density or temperature.

slower. The dimensions and surface brightness diminish now both and the luminosity of the body will fall off very rapidly as the light of the mass grows

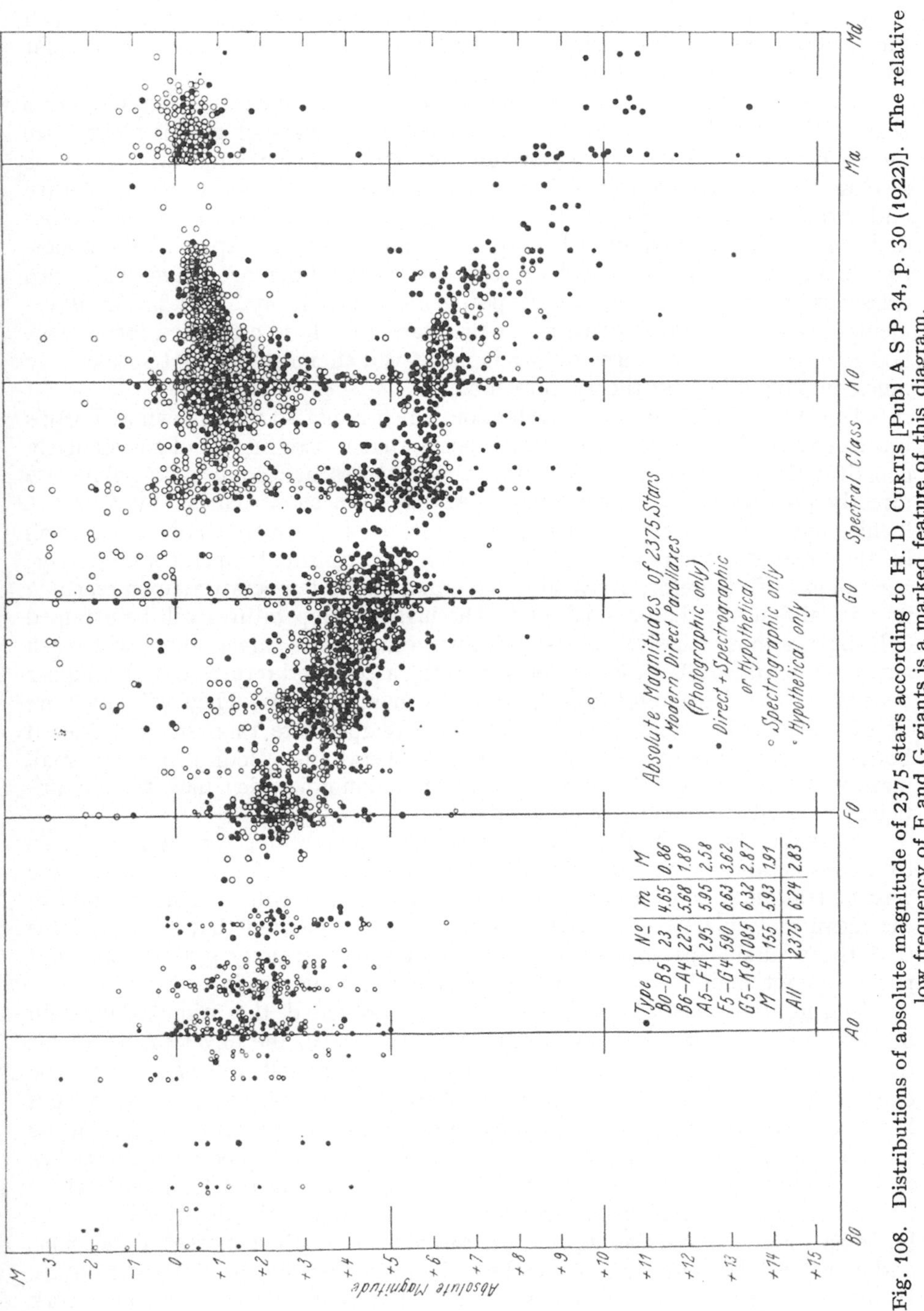

Fig. 108. Distributions of absolute magnitude of 2375 stars according to H. D. Curtis [Publ A S P 34, p. 30 (1922)]. The relative low frequency of F and G giants is a marked feature of this diagram.

redder. Sooner or later the mass muss liquefy and then solidify and the former star will be cold and dark. These changes should not begin in general until the surface temperature has fallen far below that of the stars of class M (around 3000°).

It will not be possible to estimate the critical density at which the rise of temperature will cease. So much can be said that it certainly will be much greater than that of ordinary air and at least for substances of moderate molecular weight considerably less than that of water.

RUSSELL is of opinion that the resemblance between the characteristics that might thus be theoretically anticipated in a mass of gas of stellar size during the course of its contraction, and the series of giant and dwarf stars "is so close that it might fairly be described as identity". The variations in density and surface brightness during the giant stage will nearly compensate each other and keep the said stars more or less equal in luminosity. The rapid fall of luminosity among the dwarf stars and the ever increasing difference between the two classes is exactly what might be expected from the theory of gases. The agreement between the actual densities of the stars and those estimated for gaseous bodies in the different stages of development is a circumstance that speaks very much in favour of the theory presented.

RUSSELL points out that a deduction which is not so well known as LANE's law can be made from the elementary theory that in two masses of gas similarly constituted and of equal masses the temperatures at homologous points are directly proportional to their masses. As has been shown earlier by RUSSELL in his address the effective surface temperature of the more massive body will be the greater though to a less degree than the central temperature. A large mass of gas will therefore arrive at a higher maximum temperature upon reaching its critical density, than a small one. The highest temperatures will be attained only by the most massive bodies and, all through their career, these will reach any given temperature at lower density on the ascent and return to it at a higher density on the descending scale than a less massive body. They will therefore be of much higher luminosity for the same temperature than small masses if both are rising towards their maximum temperatures. Bodies of very small mass will reach only a low temperature at maximum which may not be sufficient to enable them to shine at all.

RUSSELL also pointed out that at stellar temperatures the gases must be to a considerable extent dissociated and ionized. This fact will not change the way of reasoning here, because the principal influence will be a diminution of the mean molecular weight of the stars. He also suggested that it is probable that the available potential energy of a star is not entirely gravitational but partly, if not mainly, of radioactive or similar atomic origin.

During the time passed since RUSSELL presented the stimulating train of ideas and suggestions, only briefly indicated above, the development of his theory concerning the stellar evolution has principally followed two ways. The one is that of theoretical investigations, where the principles of thermodynamics have been applied in order to explain the general development of a star during its life-time as radiating body and the RUSSELL-diagram has been scrutinized as to its meaning for stellar evolution. The second is that of practical work where it has been tried to find from direct or indirect evidence the distribution of the luminosities of the stars as well as the distribution of their masses, dimensions and densities. The most notable workers along the first line are SCHWARZSCHILD, EDDINGTON, JEANS, VON ZEIPEL, VOGT, MILNE and ROSSELAND and their work is reviewed in other parts of this Handbook. As to the work along the second line the following parts of this chapter will mention the principal contributions without aiming at completeness. In as much as a complete review of the work concerning the stellar evolution is nearly impossible on account of the enormous number of individual contributions we will here restrict our task to giving a short sketch of the development during the period since RUSSELL presented his address.

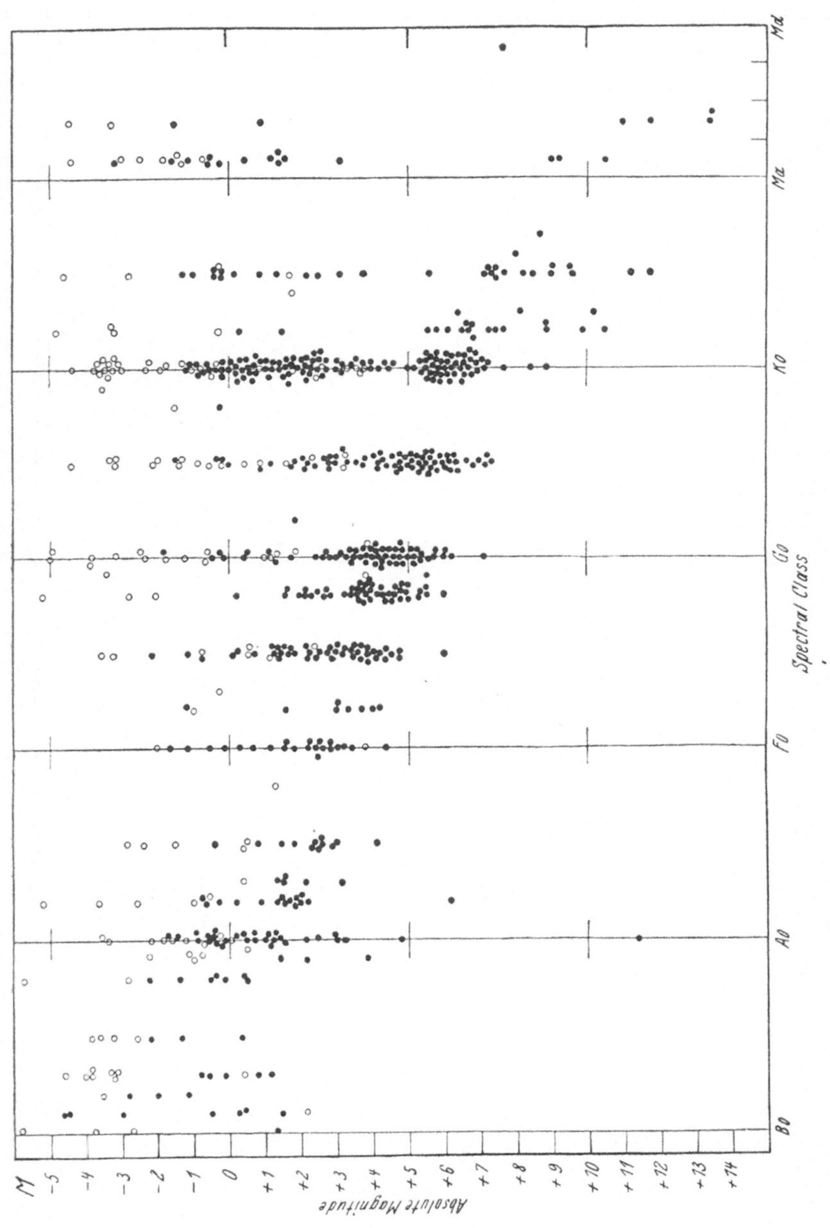

Fig. 109. RUSSELL diagram as derived from the absolute magnitudes based upon 675 parallaxes measured at the Leander McCormick Observatory. Dots are positive parallaxes. Circles are negative parallaxes assumed to mean a relative parallax of 0″,000. The accurate way the distribution of M is represented in this diagram speaks highly in favour of the accuracy in the Leander McCormick measures and the absence of systematic error in them.

In the years 1916—1917 EDDINGTON[1] published his famous theory of the interior of the stars, based on the hypothesis of radiative equilibrium introduced by K. SCHWARZSCHILD[2]. In its first form the theory was in good agreement with the features of the RUSSELL diagram and this diagram began to be generally

[1] M N 77, p. 16, 596; Ap J 48, p. 205 (1918).

[2] Göttinger Nachr 1906, p. 41.

thought of as a proof of the continued development of the stars from the giant stage. But in 1924 the idea of evolution was quite changed. EDDINGTON[1] reached then the conclusion, when confronting his theory with the mass-luminosity

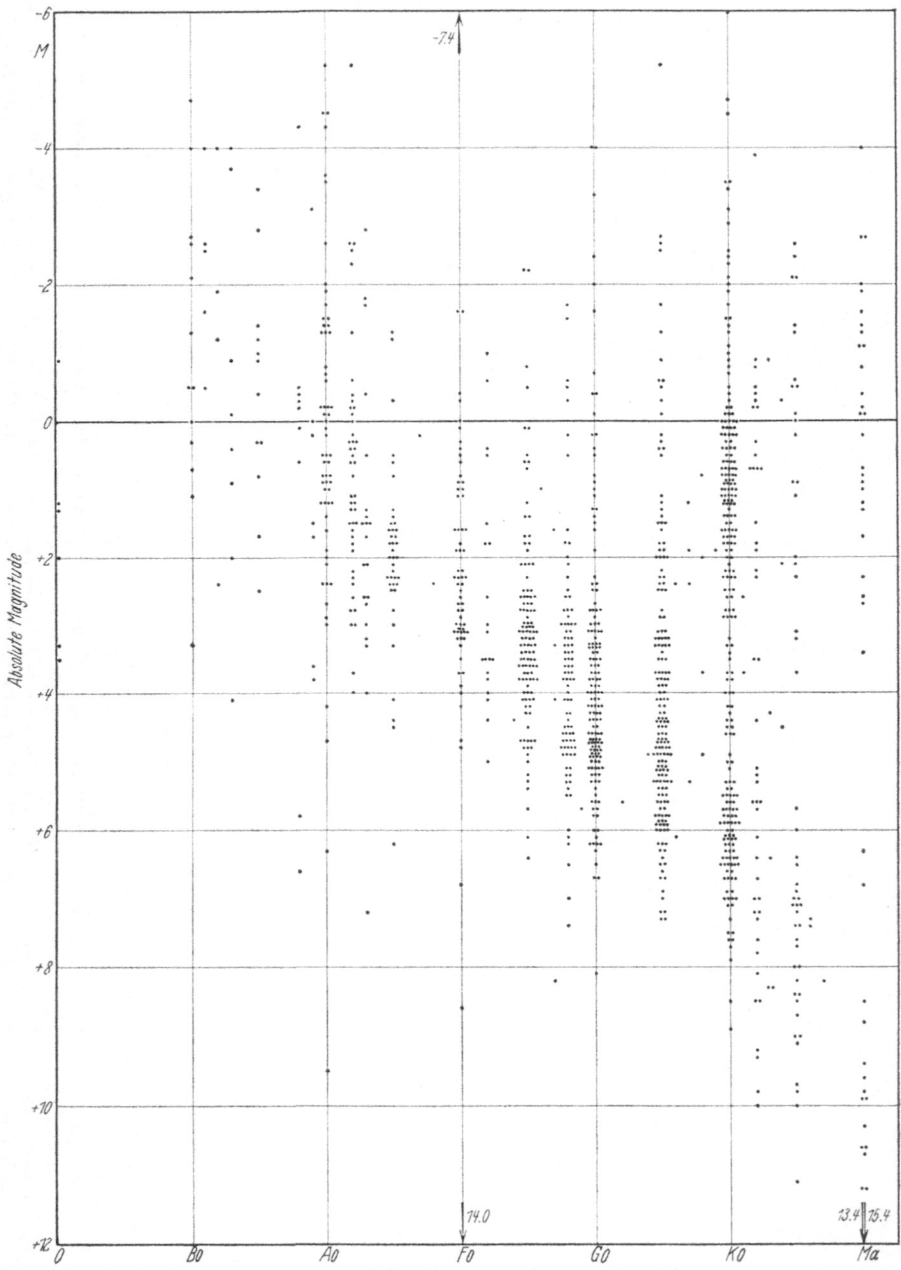

Fig. 110. RUSSELL diagram based upon data in SCHLESINGER's Catalogue of Parallaxes. Only positive values of the relative parallaxes have been used.

[1] M N 84, p. 308 (1924)

relation, that whether the stars are giants or dwarfs their luminosity is fully determined by their mass and only by their mass. The reducing effect supposed to be exercised by the high density of the dwarf stars was shown not to exist. The effect seems to be founded upon the modifications in the gas laws at the high density in question. This part of the RUSSELL theory had therefore to be abandoned. On the other hand theory and observation agreed about that the star gases seem to be in an ideal condition, that is to say, obey the gas laws. The high correlation between mass and luminosity shows that a star can not develop from giant to dwarf without its mass diminishing to a small fraction of the original mass. It is the difference in mass and not in density that causes the great difference in the luminosity of the giants and dwarfs.

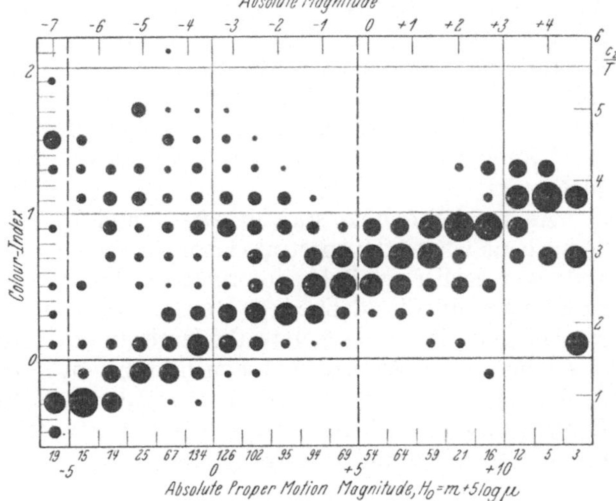

Fig. 111. Correlation surface of colour indices, or c_2/T, and absolute proper motion magnitude H_0, according to HERTZSPRUNG (see ciph. 131). The material consists of stars near the Sun. The colour indices are those of the Göttinger Aktinometrie. For a number of faint stars of which no colour index was available the colour index equivalent to the Mount Wilson spectrum has been adopted. The diagram gives the relative numbers of stars between the limits $-0{,}6$ to $-0{,}4$, $-0{,}4$ to $-0{,}2$ of colour index for each interval of one of $m + 5 \log \mu$, expressed in percentages of the total number in the same interval. The areas of the circles are proportional to these percentages. The total numbers for each interval are given at the bottom of the diagram [B A N 1, p. 91 (1922)].

The leading problem since 1924 has been: assuming the way of evolution to be that suggested by RUSSELL (increasing density) how to explain the mass reduction? This reduction is to day, of course, easily explainable without we are obliged to think that stars throw off matter into space. The star radiates energy and as energy is equivalent to mass, the stellar mass will be diminished by radiation if no other factors work in the other direction.

The problem is then reduced to the question of determining the cosmical time scale. For a discussion of this question we refer to ciph. 232 and 233.

It seems peculiar that little or no attention has been given to the possibility of increasing the energy of the stars from their capture of meteors. Recent work concerning the frequency of meteors points in direction that these are not a local phenomenon but a universal one. Furthermore, the dark nebulae may be explained as agglomerations of meteors. Even restricting us to the humble density of meteoric matter around the Sun we find that the mass of the Sun will be nearly doubled during a time of 10^{13} years. Thus the reduction of mass may at least partly be compensated from the energy given to the Sun by the meteors encountered during its lifetime.

Although the increasing density of stars is the natural way of evolution it seems that we also have to deal with such episods in the life of the stars where their bodies suddenly expand to such a degree that the density will diminish a million-fold. Recent work concerning Novae points definitely into that direction[1]. It also seems from several cases where the magnitudes of Novae before outburst are known that these originate as well from giants as from dwarfs[2]. The frequency of Novae as compared with the probable number of stars in our stellar system shows that the Nova outbursts occur so often that even if we assume all stars capable to become Novae, the phenomenon still will be repeated many times during the probable life-time of the stars[3]. The Nova-stage is a phenomenon certainly not restricted to special classes of stars and by no means a rare occurrence in the life of most of the stars. A complete theory of stellar evolution thus has also to take account to the Nova phenomenon.

As to the second line of research, that concerning the form of the RUSSELL diagram, it was pretty soon clear that the two main lines in this diagram have not the form of ⦣ but rather of \|. The relative scarcity of F and G stars is illustrated already in one of RUSSELL's own diagrams (Fig. 107) and substantiated through the work by CURTIS and by LUYTEN. The former used all the information with regard to absolute magnitudes available in 1922[4] and the latter[5] used instead of M its equivalent $H = m + 5 + 5\log\mu$ and thus could base his diagram on 4446 stars.

As to the double stars it was found by F. C. LEONARD[6], who classified 80 systems where the spectral classes of the components were not known before and used earlier available information (158 systems), that the objects investigated undoubtedly defined a RUSSELL diagram. He also concluded that the smaller mass is always before the larger in development, if the RUSSELL scheme of evolution is assumed. Subsequent work by SHAJN[7], where colour indices of double stars were measured, confirmed these conclusions. LUNDMARK and LUYTEN[8] used 260 double stars where both spectra were known and constructed the RUSSELLdiagram, integrating the relation:

$$dm = dM = \int_{S_1}^{S_2} \varphi(S)\, dS$$

where S is the spectral class or spectral index. The constant of integration was determined from 4 well-known stellar distances (obtained by special methods) and so the diagram is constructed without using any trigonometric parallaxes. The agreement between the absolute magnitudes as derived directly from measures of parallax and from differences in magnitudes of double stars speaks for the smallness of the systematic errors of the modern trigonometric parallaxes.

Although it can be said that the double stars form no exceptions as regards the general distribution of their masses it seems that the RUSSELL diagram in this case has the appearance of $<$ instead of ⦣. On the other hand, it seems not impossible that the RUSSELL diagram of single stars also will prove to have the same form as that of the double stars.

Many of the RUSSELL diagrams prepared are but of restricted value as they are founded on little extensive material and no such reduction is performed that the diagrams represent the actual conditions in space. This item is nearer discussed in ciph. 159.

[1] Some years ago I suggested on basis of investigations concernings Nova Aquilae 1918 the Nova outburst to be caused by an expansion or rather an explosion of the star [Publ A S P 34, p. 207 (1922)]. At this time I was not aware of EVERSHED's investigation in M N 79, p. 482 (1919), where practically the same theory was advanced, probably for the first time.

[2] Publ A S P 35, p. 95 (1923).

[3] C. LÖNNQUIST, Lund Obs Circ No. 7 (1932).

[4] Publ A S P 35, p. 33 (1922).

[5] Lick Bull 10, p. 135 (1922).

[6] Lick Bull 10, p. 169 (1923).

[7] Pulkovo Bull X, p. 276 (1925).

[8] A J 35, p. 93 (1923).

Our knowledge of the real form of the RUSSELL diagram is still imperfect. The dwarf series, nowadays generally termed the main series[1], is comparatively well established but even there differences in opinion still exist as to the actual course of the series. Y. ÖHMAN[2] has suggested on account of a colour excess in A3—A5 over F0 stars found by him that there exists a discontinuity in the main series between A5—F0, and he obtains also some support from a RUSSELL diagram prepared by LINDBLAD[3]. I think that also more recent material will give some support to this conclusion.

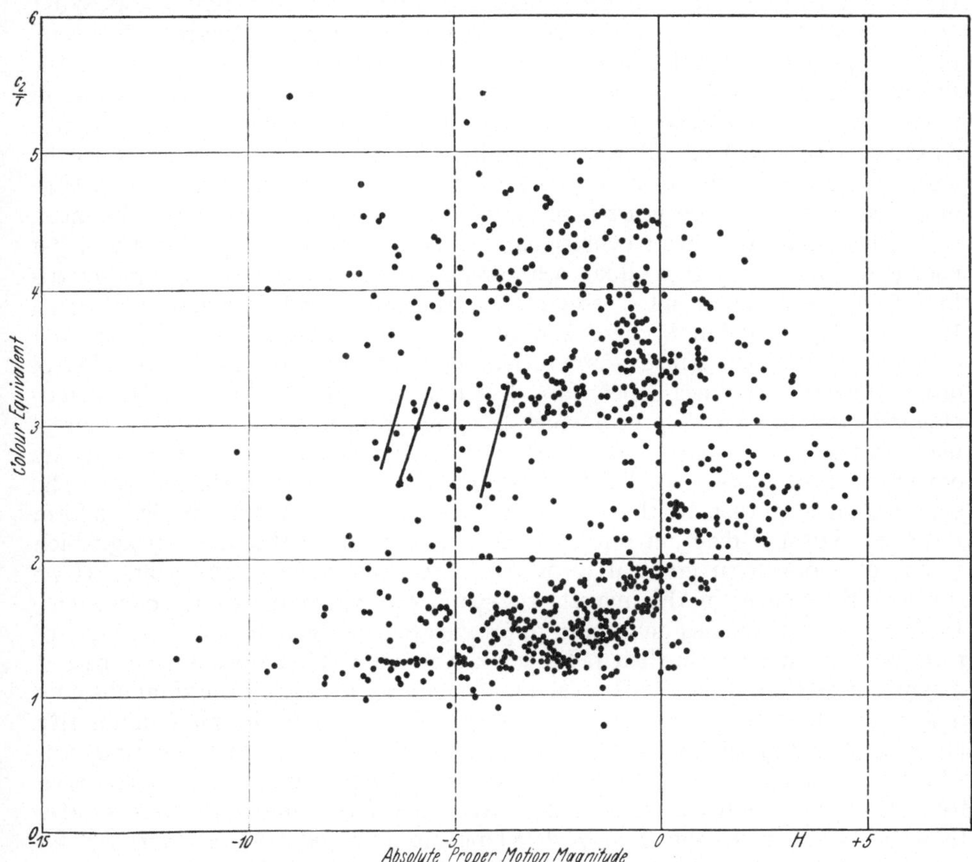

Fig. 112. Distribution of all the 734 stars in HERTZSPRUNG's colour catalogue with regard to absolute proper motion magnitude, $H' = m + 5\log\mu$, and colour equivalent c_2/T.

Many astronomers do not consider the RUSSELL diagram as indicating the course of the stellar evolution. EDDINGTON has pointed out that it may be instead of a line of evolution a locus of equilibrium points. If we do not admit that the mass of an evolving star changes, than the lines of the diagram are no indicators of the stellar evolution stage.

Theories assuming a discontinuous course of stellar evolution have not so far been worked out, but it seems that the time is ripe for such a kind of quantization idea applied to stellar evolution. The work of STRÖMBERG[4] recently

[1] The main series includes also the B and A stars of ordinary brightness.

[2] Upsala Medd No. 48 (1930).

[3] Pop Ast Tidskr 1922, p. 33.

[4] Mt Wilson Contr 385 (1930); 396 (1930); 411 (1930), 430 (1931).

concluded showing the existence of distinct maxima of frequencies in the absolute magnitudes will probably give rise to an application of the ideas of quantum-mechanics in macrocosmos.

The increasing knowledge of anagalactic systems seems not to favorise the idea of a stellar evolution, or, it suggests the time-scale is immense long. We do not only know that the integrated spectra or colour indices of anagalactic systems are very uniform but it seems also that these characteristics do not vary very much during some hundred million of years. In the metagalactic cluster in direction of the Northern Milky Way Pole first discovered by MAX WOLF ($\alpha = 12^h\,55^m$; $\delta = +28°$) the spectra and effective wave lengths of several objects, presumably belonging to the cluster, have been investigated[1]. The distance of this cluster cannot be determined very accurately on account of present difficulties of establishing a very accurate metagalactic distance scale. Anyhow, I believe that all workers within this field agree in assigning to the cluster a minimum distance of $1{,}6 \cdot 10^7$ light-years. Now, it seems that a time of 10^9 years or even 10^{10} years is of very little consequence as regards the mean spectral class or the mean evolutionary stage of two anagalactic systems. Of course, one might say that equal mean spectral class does not prove equal distribution of the different sub-classes of spectrum and absolute magnitude (equal RUSSELL diagrams) but if there were unusual proportions of K and M stars in a certain system, then unusual proportions of B and A stars also are needed in order to balance the mean spectrum. But if so, we should expect to find blend-effects in the integrated spectra which does not seem to be the case. The agreement between the integrated spectra and colour indices among anagalactic objects then suggests a similarity as regards the properties of the spectra. The explanation may then be that either the stars in a certain system always form the same RUSSELL diagram during their development or the time during which a star develops from giant M to dwarf G must last at least 10^{12} years. If we consider the masses of the anagalactic systems we reach the same conclusions. In five cases the masses have been approximated for such systems[2] on basis of rotational components in their radial velocity and found to be of the same order of magnitude as that of our stellar system. The fair agreement of the total absolute magnitudes of different anagalactic systems suggests that the mass-luminosity law is a universal phenomenon. If we arrange the galaxies along our time axis which at least has a length of $1{,}5 \cdot 10^9$ years, we cannot find such a decrease in total absolute magnitude as would be the case if the time scale were shorter than 10^{12} years as to the time of development from giant M to dwarf G[3].

140. ADAMS and KOHLSCHÜTTER's Investigations in 1914. The problem whether there are any spectral peculiarities within the two groups of stars that evidently differ radically with regard to the size of their absolute magnitude M was raised by W. S. ADAMS and A. KOHLSCHÜTTER in 1914[4], and criteria were sought for that would enable M to be determined from observations of the behaviour of the spectra. The principal differences in the spectra of the two groups are small. The continuous spectrum of the small proper motion stars is relatively fainter in its violet, as compared with its red, than the spectrum

[1] LUNDMARK, Studies of Anagalactic Objects. I. Upsala Medd No. 30 (1927); HUMASON, Mt Wilson Contr 426; Ap J 74, p. 35 (1930); HUBBLE and HUMASON, Mt Wilson Contr 427; Ap J 74, p. 43 (1930).

[2] Upsala Medd No. 40 (1928); Populär Astron Tidskrift 11, p. 85 (1930).

[3] For deriving the relative distances we do, of course, not use m_{tot}, but base the distances on the dissolution of anagalactic objects into separate stars, the apparent dimensions and redshift velocity relation.

[4] Ap J 40, p. 385 (1914).

of the large p.m. stars (KAPTEYN and others thought at the time that this proved the existence of a selective absorption in space).

The hydrogen lines are abnormally strong in a considerable number of the small p.m. stars and weak in the spectra of the large p. m. stars, which show the titanium oxide bands.

Certain separate lines are weak in the large p. m. stars and strong in the small p. m. stars and conversely.

The results from the investigation of the intensity of the continuous spectrum are seen in the following table:

Spectral class	Mean p. m.	Average class	Density	
			at λ 4220	at λ 4955
A0—A9	0″,020	A2	0,30	0,32
	0 ,13	A3	0,29	0,32
F0—F9	0 ,012	F4	0,25	0,37
	0 ,66	F6	0,32	0,37
G0—G4	0 ,009	G3	0,22	0,41
	0 ,64	G2	0,33	0,41
G5—G9	0 ,011	G7	0,25	0,48
	0 ,64	G7	0,40	0,48
K0—K4	0 ,011	K2	0,16	0,44
	0 ,70	K1	0,31	0,44

If the proper motions are taken as a measure of the distance, it seems that the different groups in the interval F—K have nearly the same ratio of distances. Thus it appears that at least a part of the absorption in the violet part of the spectrum of distant stars must be ascribed not to scattering of light in space but to conditions in the stellar atmospheres. Similar conclusions were reached by MONCK a little later.

The abnormal strength of the hydrogen lines in the spectra of certain small p. m. stars was more closely investigated. It was found that 98 per cent of the hydrogen absorption must occur in the stellar atmospheres and that but two per cent can possibly be due to hydrogen gas in space. Systematic differences of intensity for certain lines between stars of large and stars of small proper motion soon became evident during the course of these investigations. Pairs of lines were selected not differing much as to their wave lenghts and of as nearly as possible the same intensity and character.

The intensities of the spectral lines were then estimated by the aid of a step-degree method (Stufenschätzungsmethode) and the material uniformly reduced. The first object in view was to investigate the changes of the estimated intensity differences with the spectral type and to form a classification system depending on certain well defined criteria. The second object was to investigate changes with absolute magnitude.

The following pairs showed the largest and most definite changes with spectral class:

F8—G6: $\lambda\,4227/H\gamma$; $\lambda\,4326/H\gamma$; $H\gamma/\lambda\,4352$; $H\gamma/\lambda\,4405$; $H\gamma/\lambda\,4384$,

G6—K9: $\lambda\,4326/H\gamma$; $H\gamma/\lambda\,4352$; $H\gamma/\lambda\,4405$; $H\beta/\lambda\,4872$; $H\beta/\lambda\,4958$.

These lines have been used to determine the class of each individual star. The mean error of one determination depending on three plates is 0,4 of a subdivision of the Harvard scale.

The M were computed from measured parallaxes when the latter were available. In other cases the M were computed from the following formula:

$$M = +0{,}975\, m + 4{,}3 \log \mu .$$

The most prominent cases of lines in which systematic differences are seen to exist between the stars of high and low luminosity are the following:

High luminosity			
Strong	Weak	Strong	Weak
4216 Sr 4395 Ti, V, Zr	4325 Sc 4435 Ca	4408 V, Fe	4456 Ca 4535 Ti

The following five pairs of lines were selected as the basis for the determination of M:

$\lambda\,4216/\lambda\,4250$; $\lambda\,4395/\lambda\,4415$; $\lambda\,4408/\lambda\,4415$; $\lambda\,4456/\lambda\,4462$; $\lambda\,4456/\lambda\,4495$.

If the difference, value of line intensity—normal value, is denoted by D, the following formulae were found:

$$\text{F8—G6}: \; M = +5{,}6 - 1{,}6 D,$$
$$\text{G6—K9}: \; M = +6{,}8 - 1{,}8 D.$$

For the very faintest stars the linear relation does not seem to hold strictly. The average difference between the two sets of M is:

F8—G6 stars: $1^{M},6$,
G6—K9 stars: $1\;\;,5$.

A table giving M for 71 stars of the classes F8—G6 and 91 stars of the classes G6—K9 ends the paper.

141. Subsequent Mount Wilson Investigations. On account of the war KOHLSCHÜTTER had to leave Mount Wilson and thus could not take part in the further investigations. The next advance was made in 1917, when ADAMS jointly with A. JOY published "spectroscopic parallaxes" of 500 stars[1]. Experiments to measure the relative intensities of the absolute magnitude lines did not give satisfactory results. It was found that it was better to estimate the difference in the intensity. A new system for spectral classification was established on the basis of the estimated intensities of certain lines that varied with spectral class alone. The connection between the absolute magnitude and the ratio of the line intensities could not be established with such accuracy as has been possible later on. A main result of special interest was the gap between giant and dwarf stars, a gap that increased more and more when later spectral types were reached.

Of course, the method does not give any parallaxes. These are a by-product of the determination of the absolute magnitudes. The determination of the latter is the salient point of the method. In this chapter we are not concerned with the parallaxes themselves, which only serve as a means of computing the absolute magnitude.

Trigonometric parallaxes could be used, to the number of 360, for a comparison with the spectroscopic values in the list of 500 stars. The average difference, without regard to sign, was found to be 0″,026.

The comparison as regards spectral class and absolute magnitude gives the following synopsis:

[1] Mt Wilson Contr No. 142 (1917); Ap J 46, p. 316 (1917).

M / Spectral class	$-3^M,0$ to $+1^M,9$ $\pi_s-\pi_{tr}$	n	$2^M,0$ to $4^M,9$ $\pi_s-\pi_{tr}$	n	$+5^M,0$ to $7^M,9$ $\pi_s-\pi_{tr}$	n	$8^M,0$ to $13^M,3$ $\pi_s-\pi_{tr}$	n	All $\pi_s-\pi_{tr}$	n
F0—F8	+0″,003	12	+0″,004	62	+0″,018	14			+0″,006	88
F9—G7	+0 ,005	22	+0 ,007	35	+0 ,007	34			+0 ,007	91
G8—K2	−0 ,004	38	−0 ,011	21	+0 ,002	35	−0 ,021	2	−0 ,004	96
K3—K9	+0 ,011	16	−0 ,108	1	+0 ,014	23	+0 ,005	17	+0 ,008	57
Ma—Md	+0 ,004	14	−0 ,049	1			+0 ,008	13	+0 ,004	28
Mean	+0 ,002	102	+0 ,001	120	+0 ,008	106	+0 ,004	32	+0 ,0037	360

One of the principal results of this list was the confirmation of the giant and dwarf division as earlier suggested by HERTZSPRUNG, RUSSELL and others. The phenomenon is illustrated by the frequency curves given in the paper. Among others results the following may be mentioned.

Abnormal intensity of the hydrogen lines is a characteristic of the highly luminous stars of the later spectral classes. This phenomenon is certainly related to the giant and dwarf division among these stars.

The lines used for determinations of absolute magnitude are lines which in laboratory sources show marked variations with vapor-density and temperature. The lines which are strong in the highly luminous stars are strong in the electric spark, and those strong in the absolutely faint stars are strong in the electric furnace.

The stars of high luminosity are characterized by spectra in which both the hydrogen lines and the enhanced lines are abnormally strong. In the spectra of the absolutely fainter stars all of the low temperature lines are strong. The lines used for the determinations of magnitude form but a part of this more general spectral distinction.

The next catalogue of spectrographic parallaxes from Mount Wilson was issued in 1921 and contains results for 1646 stars[1]. W. S. ADAMS, A. H. JOY, G. STRÖMBERG, and CORA G. BURWELL took part in the work. The material of the first catalogue is included in this new list. Since the publication of the first list numerous developments and changes had taken place in the methods used. Besides, the number of trigonometric parallaxes had increased rapidly and added greatly to the accuracy of the constants used as the basis of the method.

In order to make the spectroscopic parallaxes directly comparable with the trigonometric ones the following deductions were made.

The absolute magnitude as given by the provisional tables is called M_s, and the finally derived absolute magnitude as given in the catalogue M_S, the most probable absolute magnitude M_p, and the real absolute magnitude M. The corresponding parallaxes are π_s, π_S, π_p and π respectively. Further ΔM_s is the systematic correction to M_s for reducing it to M_p, and π_{tr} the trigonometric parallax, ε_π its real error and ε_M the accidental error of $M_s + \Delta M_s$.

Then we can write:

$$M = M_s + \Delta M_s + \varepsilon_M = m + 5 + 5\log(\pi_{tr} + \varepsilon_\pi)$$

and:

$$M_s = m + 5 + 5\log\pi_s.$$

Hence:

$$\Delta M_s + \varepsilon_M = 5\log(\pi_{tr} + \varepsilon_\pi) - 5\log\pi_s$$

or:

$$\pi = \pi_{tr} + \varepsilon_\pi = \pi_s \cdot 10^{0,2(\Delta M_s + \varepsilon_M)} = \pi_s \cdot 10^{0,2\Delta M_s} \cdot 10^{0,2\varepsilon_M}.$$

[1] Mt Wilson Contr No. 199 (1921); Ap J 53, p. 13.

Putting:

$$\sigma = 10^{0,2\Delta M_s}, \text{ and } \tau = 10^{0,2\varepsilon_M},$$

we have:

$$\pi = \pi_s \sigma \tau .$$

If τ is developed into series and terms of higher order than the second one are dropped we have:

$$\tau = 1 + \frac{0,2\,\varepsilon_M}{0,4343} + \frac{1}{2}\left(\frac{0,2\,\varepsilon_M}{0,4343}\right)^2$$

or:

$$\tau = 1 + 0,4605\,\varepsilon_M + 0,1060\,\varepsilon_M^2 .$$

From the equations above we obtain:

$$\pi_s \sigma = \frac{\pi_{\text{tr}} + \varepsilon_{\text{tr}}}{\tau} = (\pi_{\text{tr}} + \varepsilon_\pi)\,(1 - 0,4605\,\varepsilon_{\text{tr}} + 0,1060\,\varepsilon_M^2)$$

or:

$$\pi_s \sigma - \pi_{\text{tr}} A = \varepsilon_\pi - \pi \cdot 0,4605\,\varepsilon_M$$

where:

$$A = 1 + 0,1060\,\varepsilon_M^2 .$$

STRÖMBERG then makes use of the approximate formula:

$$\overline{xy} = \bar{x} \cdot \bar{y} ,$$

valid if the two variables are independent of each other. The quantity π_s is dependent upon ε_M and thus upon τ. A separation has been made above and taking the arithmetical means of a number of equations of the form of the next second last above we obtain:

$$\overline{\pi_s \sigma} - \overline{\pi_{\text{tr}}} \cdot \bar{A} = 0 ,$$

or:

$$\frac{\overline{\pi_s \sigma}}{\bar{A}} = \overline{\pi_{\text{tr}}} = \pi .$$

The quantity $\pi_s \sigma$ is the parallax which corresponds to the most probable magnitude M_p. The mean of a series of these parallaxes will not be exactly equal to the mean of the corresponding trigonometric parallaxes but is systematically larger. The explanation to this are the unequal effects produced in the spectrographic parallax by a positive and a negative error of the same size in the estimation of M.

In order to make the spectrographic parallaxes directly comparable with the trigonometrical ones the quantity $\pi_s \sigma / \bar{A}$ is defined as the resulting spectrographic parallax π_S.

We have the relations:

$$M_S = m + 5 + 5 \log \pi_s \sigma - 5 \log \bar{A} ,$$

$$M_S = m + 5 + 5 \log \pi_S ,$$

$$M_p = M_S + 5 \log \bar{A} .$$

In the determination of $\sigma / \bar{A}$ the following equations were used:

$$\pi_s \sigma - \pi_{\text{tr}} A = \varepsilon_{\text{tr}} - \pi \cdot 0,46\,\varepsilon_M$$

If the range in the values of π_s is small, we find by forming means and multiplying by the relative weights p of the trigonometric parallaxes:

$$\frac{\sigma}{\bar{A}} = \frac{\Sigma p \pi_{\text{tr}}}{\Sigma p \pi_s} .$$

If the range in π_s is considerable, the equation does not utilize the higher weight of the larger parallaxes to their full extent. A least-square solution might give erroneous results on account of the effect of large values in the ε_{tr} and ε_M. A compromise between the two modes is accordingly used. Dividing the results into a few groups, according to the size of π_s, we have:

$$\frac{\sigma}{A}\Sigma p\pi_s - \Sigma p\pi_{tr} = 0.$$

$$\text{Weight} = \frac{1}{k^2\Sigma p + \frac{\eta^2(\Sigma p\pi_0)^2}{n}},$$

where k is the mean error in the π_{tr} of unit weight, n the number of the stars, and:

$$\eta^2 = \frac{0{,}04\,\varepsilon_M^2}{(\text{Mod})^2}.$$

The numerical values of k and η used are: $\pm 0'',015$, and $\pm 0'',205$, which correspond to a mean error in the corrected M of $\pm 0^M,3$.

For the derivation of reduction curves the parallactic motion was used and also the peculiar motion.

142. The A Stars and B Stars. In 1922 ADAMS and JOY[1] published a spectrographic method of determining the absolute magnitudes of stars of spectral class A.

The spectrographic method used in the determination of the absolute magnitude of stars of the later spectral classes cannot in general be applied to those of the spectral class A. As a rule these stars do not show the great variations in the intensities of the lines that are known in the later classes. On the other hand, there are a number of differences in the appearance of their lines. In Sirius the lines are sharp and well-defined and in others, such as Altair, very diffuse and vague. This fact was used as a criterion dividing the A stars into two groups.

The first step and the most important one was an accurate determination of spectral class. This was based on the metallic arc lines. Spectra in which the helium lines do not appear and where such prominent arc lines as λ 4326 are not present are called A0. With increasing intensity of the arc lines the stars are classified as A1, A2 etc. When the helium lines, λ 4026 and λ 4471, just appear, the spectrum is classified as B9, and when λ 4471 and λ 4481 are equal, as B8. This system is based on the classification made by KOHLSCHÜTTER in the years 1912—1914.

In addition to the ordinary symbols the letters "n" for nebulous and "s" for sharp lines have been appended in order to indicate the general characters of the spectral lines. Also the c characteristics noted by Miss MAURY are used in the classification at Mount Wilson.

As preliminary standards for the absolute magnitudes 36 parallaxes of stars in Taurus and Ursa Major taken from RASMUSON's thesis on moving clusters were used.

For the various spectral subdivisions the following mean values of M have been obtained:

Spectral class	M	n	Spectral class	M	n
A0n	$+0^M,9$	6	A2s	$+0^M,6$	3
A1n	1 ,4	4	A3s	1 ,0	2
A2n	2 ,0	3	A4s	1 ,2	3
A3n	1 ,9	5	A6s	2 ,1	1
A4n	2 ,2	4	A7s	2 ,2	3
A5n	2 ,9	2	A8s	2 ,7	4
A6n	3 ,0	2	F1s	3 ,0	3
A7n	2 ,8	5	F2s	2 ,8	1
A8n	2 ,8	3			
F0n	3 ,2	2			

[1] Mt Wilson Contr No. 244 (1922); Ap J 46, p. 242.

The results for the Taurus and Ursa Major group were extended to include 82 stars with absolute magnitudes from moving cluster parallaxes and 104 stars with reliable trigonometric parallaxes. The number of objects used was altogether 148.

The stars were classified independently by ADAMS and by JOY. The parallaxes have been used for a comparison with the proper motion and a satisfactory correlation was found. The curve derived gives for $\mu = 0$ $\pi = +0'',009$ which is in good agreement with an earlier result of STRÖMBERG[1].

The same method has also been applied by the same authors to the B stars[2]. The first step was the accurate determination of the spectral class. The Harvard system has been rather accurately followed, except that the systems differ as regards the classes B5—B8. The spectral intervals B2—B5 and B8—A0 are greater than B5—B8, which is also confirmed by the fact that the Harvard observers have not classified any stars in this last interval. With regard to the O stars the nomenclature suggested by H. H. PLASKETT[3], who used the notation O5, O6 etc., has been adopted. Also the division into nebulous and sharp lines has been kept.

At first the trigonometric and moving cluster parallaxes alone were used, but finally the statistical values of PLUMMER, CHARLIER, KAPTEYN, and others, and the mean magnitude of sub-groups of the B stars, have also been used.

The final smoothed out values are shown in the following table:

Spectral class	Nebulous lines	Sharp lines	Spectral class	Nebulous lines	Sharp lines
O5—O9	$-2^M,5$	$-2^M,5$	B6	$-0^M,3$	$-0^M,9$
B0	$-3\ ,1$	$-3\ ,1$	B7	$-0\ ,1$	$-0\ ,8$
B1	$-2\ ,4$	$-2\ ,6$	B8	$+0\ ,1$	$-0\ ,6$
B2	$-1\ ,5$	$-2\ ,0$	B9	$+0\ ,5$	$-0\ ,2$
B3	$-0\ ,9$	$-1\ ,5$	A0	$+0\ ,9$	$+0\ ,2$
B4	$-0\ ,6$	$-1\ ,2$	A1	$+1\ ,3$	$+0\ ,6$
B5	$-0\ ,5$	$-1\ ,1$	A2	$+1\ ,7$	$+0\ ,9$

These values have been used in the determination of the absolute magnitudes of 300 stars of the O and B classes for which spectrograms were available at Mount Wilson.

A comparison of the results M_S given by the spectrographic method with those derived from moving clusters M_{Cl} is of interest:

Group	M_{Cl}	M_S	n
Pleiades	$-0^M,1$	$+0^M,1$	8
Perseus	$0\ ,0$	$-0\ ,5$	20
Orion	$-2\ ,0$	$-2\ ,0$	17
Scorpio-Centaurus . .	$-0\ ,5$	$-1\ ,0$	11

The relationship of M to the total proper motion has been investigated. The authors say that when a wide range in apparent magnitude is involved, the best procedure is to use the value of $H' = 0,2\,m + \log\mu$ instead of the proper motion itself, and to compare with M, as has been done by LUNDMARK and LUYTEN[4]. The following relation derived by the latter authors is of interest:

[1] Mt Wilson Contr No. 170 (1919).
[2] Mt Wilson Contr No. 262 (1923); Ap J 57, p. 294.
[3] Publ Astrophys Obs Victoria 1, p. 325 (1922).
[4] Lick Bull 10, p. 153 (1922).

M	H	Limits of $H' = 0{,}2\,m + \log\mu$	n
$-2^{M},01$	$-4^{H},05$	$<8{,}50$	12
-1 ,49	-1 ,75	8,50–8,74	11
-1 ,37	-0 ,50	8,75–8,99	23
-1 ,18	-0 ,35	9,00–9,14	23
-0 ,80	$+1$,10	9,15–9,29	43
-0 ,47	$+1$,70	9,30–9,39	38
-0 ,48	$+2$,25	9,40–9,49	27
-0 ,25	$+2$,75	9,50–9,59	32
-0 ,15	$+3$,25	9,60–9,69	32
$+0$,13	$+3$,70	9,70–9,79	29
$+0$,21	$+4$,30	9,80–9,94	18
$+0$,61	$+5$,35	$>9{,}75$	11

It is clear from the close relation between $H = m + 5 + 5\log\mu$ and M that our knowledge of the proper motions can also in the case of early-type stars be used to give mean magnitudes. The difference between the method of ADAMS and JOY on the one hand and that suggested by LUNDMARK and LUYTEN on the other is essentially that in the latter case it will probably not be possible to decide from the proper motion alone if an object is n or s. For the late B stars it is essential to know to which spectral sub-class a star belongs. For the early classes the difference is rather small and the use of H will in many cases give a fair approximation of M.

143. Further Researches on the Mt. Wilson Observatory. The division into giants and dwarfs according to the list of 1646 stars has been investigated by ADAMS and JOY[1] and the results are contained in the following table giving the absolute frequencies:

Absolute magnitude	Spectral class						Sum
	A	F0–F9	G0–G9	K0–K3	K4–K9	M	
$-3^{M},0$ to $-2^{M},6$			9	1		1	11
		3	9	2		1	15
-2 ,5 „ -2 ,1		3	3			1	7
-2 ,0 „ -1 ,6		5	8				13
-1 ,5 „ -1 ,1		8	11			2	21
-1 ,0 „ -0 ,6		6	20	1		9	36
-0 ,5 „ -0 ,1	1	5	34	9	1	20	70
0 ,0 „ $+0$,4	2	5	50	35	38	55	185
$+0$,5 „ $+0$,9	2	11	60	90	88	30	281
1 ,0 „ 1 ,4	8	10	38	71	8	7	142
1 ,5 „ 1 ,9	20	21	30	36		1	108
2 ,0 „ 2 ,4	9	35	14	26			84
2 ,5 „ 2 ,9	3	46	12	7			68
3 ,0 „ 3 ,4		61	9	7			77
3 ,5 „ 3 ,9		93	12	3			108
4 ,0 „ 4 ,4		40	32	5			77
4 ,5 „ 4 ,9		5	45	5			55
5 ,0 „ 5 ,4		1	50	2			53
5 ,5 „ 5 ,9		2	50	21	4		77
6 ,0 „ 6 ,4			11	46	19		76
6 ,5 „ 6 ,9				11	19		30
7 ,0 „ 7, 4				1	4		5
7 ,5 „ 7 ,9					11		11
8 ,0 „ 8 ,4					8	2	10
8 ,5 „ 8 ,9					3	2	5
9 ,0 „ 9 ,4					1	5	6
9 ,5 „ 9 ,9						6	6
10 ,0 „ 10 ,4						6	6
10 ,5 „ 10 ,9							
	45	360	507	379	204	148	1643

The results from this list are in excellent agreement with those derived from the list of 500 stars of 1917.

[1] Publ Amer Astr Soc 4, p. 201 (1920).

The frequencies are apparent and should be reduced to equal volumes of space. A determination of the necessary reduction involves the use of the luminosity law, of which our knowledge is rather insufficient at present.

An interesting result was derived from a comparison of the Harvard and Mount Wilson estimates of spectral types. Investigating the first 12 hours of right ascension it was found that the systematic spectral difference between Harvard and Mount Wilson is not the same for giants and dwarfs in the sense that the latter are bluer. This is undoubtedly an effect of the different temperature scale for giants and dwarfs. The correlation coefficient between the two classifications has been computed as +0,932, which is comparatively low and suggests the presence of some other parameter than the spectral class alone (see ciph. 130).

A special paper[1] by ADAMS, JOY, and HUMASON has been assigned to the study of the absolute magnitude of the M stars. The stars were classified according to the system adopted by the committee on the spectral classification of stars. The main basis are the intensities of the bands of Ti-oxide, but minor consideration is also given to some other spectral characteristics. The classes are M0, M1, M2, ... in order of increasing intensities of the bands. The majority of the M stars belongs to the M0—M4 classes. The long-period variables belong to the more advanced classes. In the case of dwarfs it seems probable that those later than M5 are too faint, apparently, to be detected and observed.

The following lines have been used for the derivation of the absolute magnitudes:

Giants		Dwarfs		Giants		Dwarfs	
λ 4077	Sr^+	λ 4318	Ca	λ 4389	Fe	λ 4586	Ca
4207	Fe	4435	Ca	4489	Fe	4607	Sr
4215	Sr^+	4454	Ca	H_γ			
4258	Fe	4535	Ti-blend	H_β			

The mean absolute magnitude of M giants is $-0^M,2$ with individual values ranging between $-4^M,5$ and $+0^M,7$. The brightest of the M dwarfs is $+7^M,0$. Thus the "gap" embraces more than 6^M. There is a regular decrease in $\overline{M}$ for the various sub-classes as is shown in the table:

Spectral class	$\overline{M}$	Spectral class	$\overline{M}$	Spectral class	$\overline{M}$	Spectral class	$\overline{M}$
M0	$+8^M,2$	M2	$+9^M,2$	M4	$+10^M,4$	M6	$+12^M,4$
M1	+8 ,7	M3	+9 ,8	M5	+11 ,1		

The absolute magnitude of ordinary giants has a small dispersion. This phenomenon is certainly real because, if it were caused by failure of the spectral criteria, one might expect to find a small dispersion among the supergiants as well, which is not the case. Comparisons of $\overline{M}$ with $\overline{H}$ show a fairly regular, though small, increase in the former with decrease in the latter. H is dependent on M and on the linear cross-motion. The value of H may be accounted for by a considerable value for the dispersion in the linear cross-motion and a small value in M. The fact that the dispersion in radial velocity in the normal M giants is larger than for any other class of giant stars supports the theory of a small dispersion in M. The giant stars in common with the list of 1646 stars are $0^M,4$ brighter on an average, but $0^M,8$ fainter than those in common with YOUNG and HARPER. The agreement with these authors is much better for the early than for the later sub-divisions. The differences become progressively greater and

[1] Mt Wilson Contr No. 319 (1926); Ap J 44, p. 225 (1926).

amount to $1^M,5$ for the stars of class M5 and M6. The Mount Wilson values are also some $0^M,7$ fainter on an average than the absolute magnitudes of RIMMER.

144. Yerkes Work Concerning A Stars. Miss A. V. DOUGLAS[1] has made a study of the A stars, based on the one-prism slit spectrograms taken at the Yerkes Observatory. She was "guided less by theory than by appearances" in selecting the ratios λ 4215/λ 4227, λ 4233/λ 4227, 4535/4481 for the determination of M. In each case relations have been found to exist between these ratios and the absolute magnitude.

As a second line of attack the width of the lines was used. The author states that there are strong theoretical reasons, which chiefly involve RAYLEIGH scattering, that lead to an anticipation of a relation between the width of the spectral lines and M. In the case of λ 4481, $H\delta$ and the K-line of calcium, such relations have been established by Miss DOUGLAS.

For the reduction curves 80 stars with known trigonometric parallaxes or theoretical parallaxes derived for members of the Ursa Major or Taurus clusters were used. The Mount Wilson scheme of dividing B and A stars into s and n according to the line character was also adopted. In the case of H and M a strong correlation was found between width and M for the s group, but such was not the case with the n group. For the other criteria both n and s groups yielded correlations.

Magnitudes and parallaxes have been derived for 200 stars of the classes B9—F0. The mean error of M was estimated as $\pm 0^M,7$. Comparison with 108 stars in common with the determinations of absolute magnitude at Mount Wilson gave:

$$\overline{M}_{\mathrm{MW}} - \overline{M}_{\mathrm{Ye}} = +0^M,09\,.$$

Twelve physical pairs existed in the material. They gave an average of:

$$\overline{\Delta M} - \overline{\Delta m} = \pm 0^M,34\,,$$

where theoretically it should be:

$$\overline{\Delta M} = \overline{\Delta m}.$$

The small size of the mean deviation is a good test of the high accuracy of the method.

145. STRÖMBERG's Method for the Systematic Errors[2]. The comparison of spectrographic parallaxes enables us, theoretically at least, to determine the true systematic correction for each observer and for the spectrographic system as well. The fact that the systematic (and accidental) errors in π_s are proportional to the parallaxes themselves, while the errors in π_{tr} are independent of the size of the parallax, gives the equation:

$$k_1 \pi_S = \pi_{\mathrm{tr}} + k_2$$

from which the systematic correction k_2 to the trigonometric parallaxes, and the correction factor k_1 to the absolute magnitude, that is the systematic correction to the spectrographic absolute magnitude, can be determined. A more detailed derivation of the above equation is given at the beginning of STRÖMBERG's paper.

It is known from VAN MAANEN's investigation that the parallaxes determined at Sproul and at Yale (heliometer) have a considerable magnitude equation.

[1] Ap J 64, p. 262 (1926).

[2] Mt Wilson Contr No. 220 (1922); Ap J 55, p. 11.

As such an error will appreciably affect the determination of k_1, the results of a final solution were adopted where the said parallaxes had been omitted.

Simultaneous solution	k_2 STRÖMBERG		k_2 VAN MAANEN
McCormick	+0″,0003	±0″,0021	−0″,0016
Allegheny	+0 ,0026	±0 ,0017	+0 ,0027
Yerkes	+0 ,0034	±0 ,0022	+0 ,0036
Mt Wilson.	−0 ,0030	±0 ,0014	−0 ,0020

Spectral class	Absolute magnitude	k_1	ΔM_S
A6—F	$1^M,0-5^M,5$	1,041 ± 0,045	$+0^M,09 \pm 0^M,09$
gF	0,0	1,090 ± 0,254	−0 ,19 ± 0 ,52
gG	3,0	1,135 ± 0,092	+0 ,27 ± 0 ,18
dG	3,1	0,987 ± 0,037	−0 ,03 ± 0 ,08
gK	4,0	1,012 ± 0,086	+0 ,03 ± 0 ,18
dK	4,1	0,972 ± 0,034	−0 ,26 ± 0 ,41
gM	3,0	0,888 ± 0,167	+0 ,14 ± 0 ,11
dM	7,0	1,068 ± 0,052	

The mean parallax of the comparison stars was taken into account and the following reductions from relative to absolute parallax were found:

	Red. to absolute parallax
Allegheny	+0″,008 ± 0″,002
McCormick.	+0 ,005 ± 0 ,002
Mount Wilson	−0 ,001 ± 0 ,001
Yerkes	+0 ,008 ± 0 ,002

146. Victoria Spectrographic Parallaxes. A very important contribution concerning the absolute magnitudes of the stars has been made at the Victoria Observatory by R. K. YOUNG and W. E. HARPER[1]. The first years of the work of this observatory were devoted to the determination of the radial velocities of stars, mainly from the Boss catalogue. The spectra have also been utilized for the determination of absolute magnitudes. A search was made for new pairs of spectral lines that show variations with the absolute magnitude. The following table gives the lines:

Ratio λ_1/λ_2	Used in the spectral classes		Ratio λ_1/λ_2	Used in the spectral classes	
	YOUNG	HARPER		YOUNG	HARPER
λ4071,9/λ4077,9	A2—M6	A2—M6	λ4455,1/λ4462,0	F8—K9	F8—K9
λ4161,6/λ4167,6	A2—K2	A2—K9	λ4451,1/λ4494,7	G2—K9	G0—K8
λ4202,2/λ4207,0	G5—M6	G4—M6	λ4455,1/λ4482,4	G6—M6	G6—M6
λ4215,7/λ4250,6	A2—K2	A2—K2	λ4455,1/λ4489,4	G5—M6	G6—M6
λ4247,2/λ4250,6	A2—K2	A2—K9	λ4482,4/λ4494,7	M0—M6	G8—M6
λ4258,5/λ4250,6	F2—M6	A5—M6	λ4489,4/λ4494,7	G5—M6	G8—M6
λ4271,7/λ4290,1	A2—G4	A2—G4	λ4494,7/λ4496,5	G5—M6	G8—M6

Besides these, there are a number of other lines that show variations with the absolute magnitude, among which the authors mention

λ 4233, 4305, 4348, 4395, 4474, 4522, and 4531.

The paper of R. K. YOUNG and W. E. HARPER contains a very detailed account of the methods used by them. Their numerous comparisons with other data are of high value for obtaining a more intimate knowledge of the behaviour of the spectrographic method. The catalogue gives the results concerning the absolute magnitudes of each measured pair of lines as well as the results of YOUNG

[1] Publ Dom Astrophys Obs Victoria 3, p. 3 (1924).

and HARPER separately. The material given could certainly be used with much advantage for a detailed study of the errors in the method as well as for studies of the behaviour of the results from individual pairs.

The greatest divergence from the results at Mount Wilson was found for giants of the classes K3 to K9. The following table is quoted as it also gives

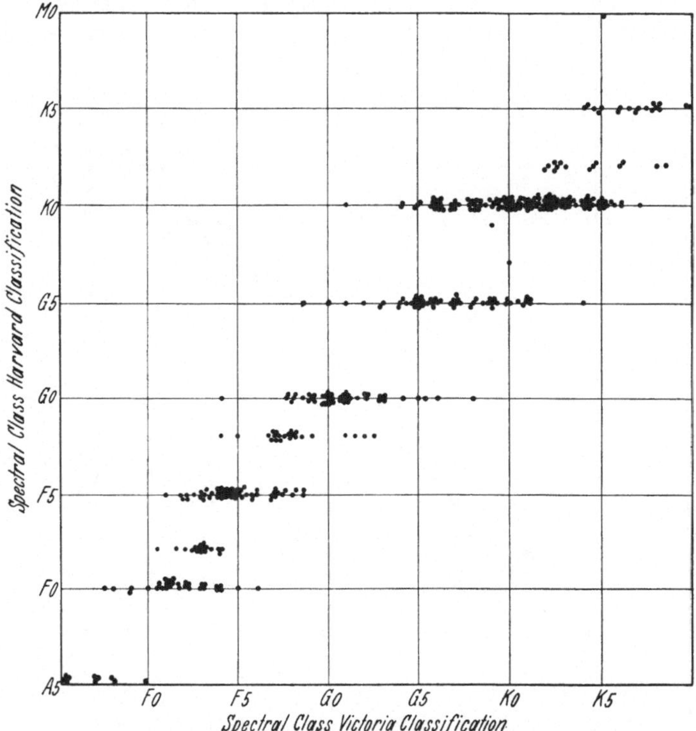

Fig. 113. Comparison between the Victoria and Harvard classification of stellar spectra in the first twelve hours of right ascension. The "resting lines" parallel with the x axis show that the Harvard classification does not give such a detailed description of the stellar spectra as the Victoria classification.

an illustration of the determination of M from the line intensities and of the difference between the results of the two observers at Victoria.

Spectral class	Mean absolute magnitude M				
	YOUNG	HARPER	Mt. Wilson	From total proper motion	n
K2	$-3^M,9$	$-3^M,3$	$-2^M,3$	$-2^M,7$	3
K1	$-1\ ,0$	$-0\ ,7$	$-0\ ,1$	$-1\ ,3$	5
K1	$0\ ,0$	$0\ ,0$	$+0\ ,7$	$-0\ ,1$	7
K1	$+1\ ,6$	$+1\ ,5$	$+1\ ,0$	$+1\ ,4$	16
K1	$+1\ ,0$	$+0\ ,9$	$+1\ ,3$	$+1\ ,0$	9
K1	$+3\ ,1$	$+2\ ,8$	$+2\ ,7$	$+3\ ,4$	8
K4	$-1\ ,7$	$-1\ ,2$	$0\ ,0$	$-2\ ,5$	8
K4	$-0\ ,5$	$0\ ,0$	$+0\ ,5$	$-1\ ,6$	9
K4	$+1\ ,5$	$+1\ ,4$	$+0\ ,5$	$+0\ ,6$	9
K7	$+0\ ,4$	$+0\ ,4$	$+0\ ,5$	$+0\ ,2$	10
M2	$-3\ ,2$	$-2\ ,9$	$-1\ ,5$	$-3\ ,1$	7
M2	$+0\ ,6$	$-1\ ,2$	$+0\ ,2$	$-0\ ,7$	24
M2	$-1\ ,4$	$-0\ ,6$	$+0\ ,2$	$-0\ ,3$	12
M1	$+0\ ,3$	$-0\ ,2$	$+0\ ,3$	$+0\ ,2$	11

The mean error of the absolute magnitudes of YOUNG and HARPER is found from the residuals in the different pairs to be $\pm 0^{M},25$ for F stars and is continually decreasing until it reaches the M stars for which it is estimated to be $\pm 0^{M},18$. This is the internal error. The external error can be estimated from a comparison with the Mount Wilson magnitudes, and the trigonometric values of M have given as a rough value $\pm 0^{M},68$ or very nearly the same value as results from various determinations of the mean error of the Mount Wilson magnitudes.

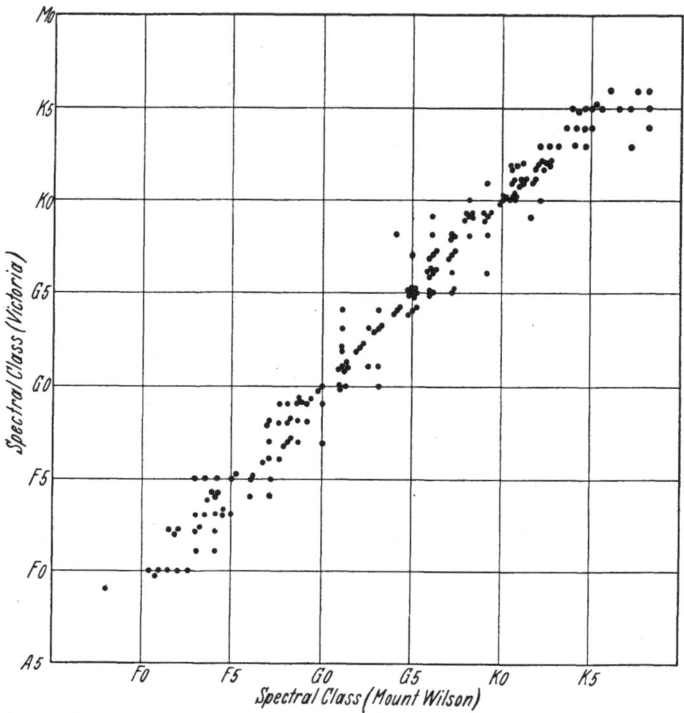

Fig. 114. Comparison between the Mount Wilson and Victoria classification in the first twelve hours of right ascension. The small scattering and the crowding around the diagonal line shows the high accuracy and freedom from systematics in the two independent systems of classification.

147. Work of Norman Lockyer Observatory. The determination of the spectral types F0 to M has been performed by RIMMER[1] and the determination of the earlier classes by EDWARDS[2] and WOODS[3]. Altogether absolute magnitudes for 1600 stars have hitherto been determined. The material used was obtained with the McCLEAN 12-inch prismatic camera of the Norman Lockyer Observatory. The dispersion on the plate is such that the distance $H\beta$ to $H\delta$ is 21,4 mm. The intensities of the lines have been measured with the aid of a graduated neutral tinted wedge cutting. A homogeneously exposed Kodak film has been cut into graduated-sized strips and a stepped wedge has been constructed by the strips being placed one over the other. The wedge has been tested in various ways and the results of different observers have been investigated. The sources of error to be eliminated are principally the following:

[1] M N 83, p. 47 (1922); 84, p. 366 (1924), (100 stars); 85, p. 439 (1925), (100 stars); 87, p. 364 (1928), (gives a catalogue of all 300 parallaxes determined to that date); 88, p. 175 (1929), (125 stars); 90, p. 523 (1930), (175 stars).

[2] London Mem R A S 62, p. 113 (1923), (500 stars); 64, p. 1 (1927), (525 stars).

[3] M N 87, p. 387 (1927), (300 stars).

The variation in the illumination. The influence was found to be negligible.

The personal equation. This was found to be rather constant between the different observers and of moderate size.

The variations caused by varying qualities of the negatives themselves with regard to the density of the continuous spectrum and the definition of the lines, and variations due to unequal conditions of developing and fixing, the presence of fog, etc. The error produced by these variations does not exceed the error of measurement. As two or more negatives of each star were measured these sources of error seem to have been reduced to a minimum influence. Too dense or too thin negatives were rejected, although the results obtained were not very different from those that had been normally exposed.

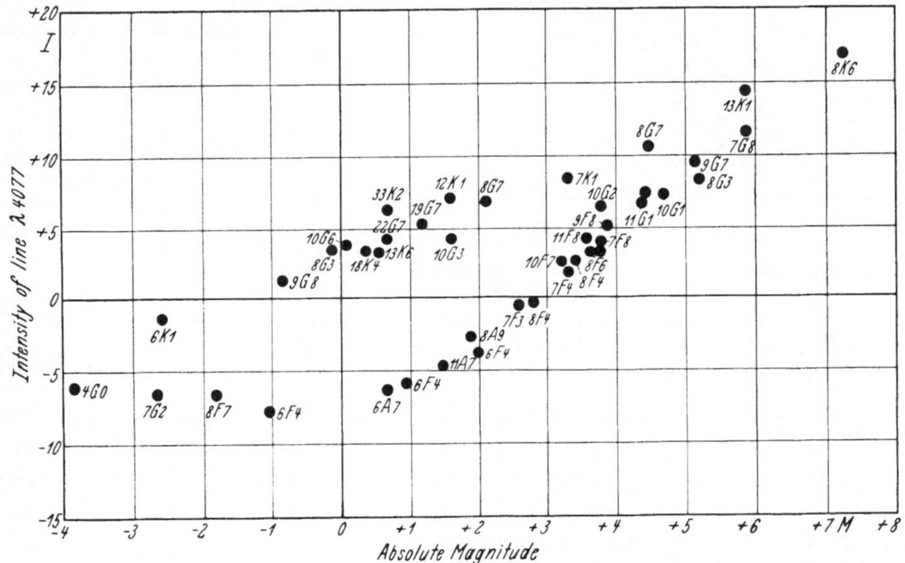

Fig. 115. Preliminary relation of line intensity (λ 4071 : λ 4077) to absolute magnitude. The numbers adjoined to the spectral classes give the number of stars forming the mean value which each dot represents (Young and Harper).

Stars of the same absolute magnitude and spectral type may be subject to different physical conditions that affect some of the spectral lines. It has been found that the sharpness or diffusedness of a line affects to some extent the wedge measurements.

Errors in the assumed spectral class of a star. This source of error has been reduced as much as possible by the use, for absolute magnitude determinations, of only those pairs that belong to such spectral classes that their relative intensities are not greatly affected.

The pairs used for the absolute magnitude determinations are:

Line-pair	Range of spectral class	Line-pair	
λ 4072/λ 4078	F0—G8	λ 4371/λ 4179	Very luminous spectral classes; F and G stars and very luminous stars of class M
λ 4216/λ 4227	F0—G4	λ 4384/λ 4395	
λ 4216/λ 4250	F0—Ma	λ 4384/λ 4400	
λ 4272/λ 4290	F0—F6	λ 4384/λ 4418	
λ 4444/λ 4455	F6—Ma		
λ 4455/λ 4462	G5—K6		
λ 4326/Hγ	M stars of very high luminosity.		
λ 4352/Hγ			
λ 4375/λ 4384			

For deriving the standard reduction curves mean trigonometric parallaxes were alone used. The correction adopted for converting relative values to absolute ones was +0″,005 for Allegheny, Greenwich, McCormick, and Sproul, and +0″,002 for Mount Wilson. All determinations have been assigned equal weights.

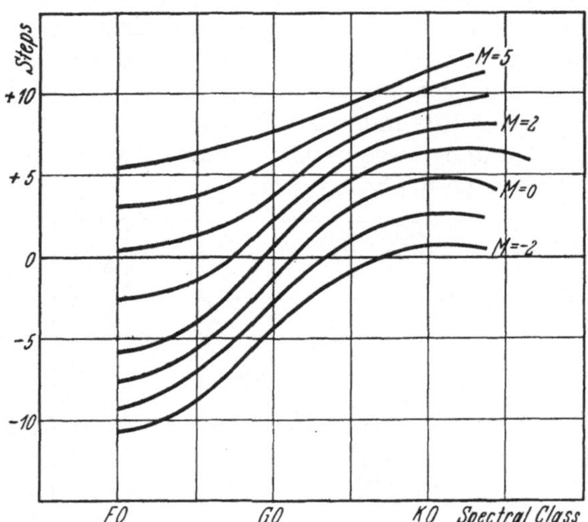

Fig. 116. Relation between estimated line intensity λ 4071 : λ 4077 and spectral class for different values of absolute magnitude (YOUNG and HARPER).

The comparison between the results of RIMMER and Mount Wilson gave a good agreement, as is shown from the following table mainly taken from RIMMER's catalogue:

Spectral class	$M_{MW}-M_{NL}$	$\varepsilon(M)$	n
A8—A9	$-0^M,52$	$\pm 0^M,69$	18
F0—F1	−0 ,43	0 ,54	13
F2—F3	−0 ,34	0 ,44	15
F4—F5	−0 ,09	0 ,33	30
F6—F7	+0 ,10	0 ,34	40
F8—F9	−0 ,05	0 ,25	14
G0—G1	0 ,00	0 ,24	28
G2—G3	−0 ,10	0 ,38	21
G4—G5	−0 ,19	0 ,34	38
G6—G7	−0 ,12	0 ,38	73
G8—G9	+0 ,02	0 ,31	74
K0—K1	−0 ,09	0 ,30	65
K2—K3	−0 ,10	0 ,31	24
Ma—Mb	−0 ,11	0 ,30	47

The excellent agreement with Mount Wilson for all classes later than F4 also shows that the trigonometric parallaxes can be used alone for all accurate derivations of absolute magnitudes from line intensities. It seems scarcely necessary to apply the somewhat elaborate scheme used in Mt Wilson Contr 199 when deriving spectrographic magnitudes.

The second catalogue of RIMMER[1] is an extension of the first catalogue. Spectra have been classified in accordance with Mt Wilson Contr 199, but the results are not communicated. For the standard stars use was also made of moving cluster parallaxes (F stars), parallactic motions (Pseudo-Cepheids), and a combination of trigonometric parallaxes and proper motion data (late K and M stars).

For the M stars some very large differences between the Mount Wilson and Norman Lockyer Observatory results occur, in the same way as was noted by YOUNG and HARPER. The agreement between the Victoria results and those of RIMMER is very good for the M stars as well as for the K stars. The comparison with Mount Wilson is given in detail.

Spectral class	$\overline{M}_{NL}-\overline{M}_{MW}$	n	Spectral class	$\overline{M}_{NL}-\overline{M}_{MW}$	n
F0	$+0^m,10$	31	G5	$-0^m,22$	14
F0	−0 ,05	15	K0	+0 ,24	32
F5	+0 ,04	24	K2	+0 ,28	24
Pseudo-Cepheids	−0 ,21	20	K5	+0 ,37	25
G0	−0 ,12	11	Ma, Mb	+1 ,23	19

The considerable difference for the M stars may be attributed to the different absolute magnitude criteria adopted in each case and to the different methods of treating the material.

[1] London Mem R A S 64, p. 1 (1927).

47 of RIMMER's objects are in common with Harvard. The systematic differences in the sense Harv — N.L.O. are very small for 6 G5 and 30 K0 stars, viz. $-0^{M},03$ and $+0^{M},05$. For 8 K2 and K5 stars the systematic difference is $+0^{M},26$ or very close to the value given above for Mount Wilson.

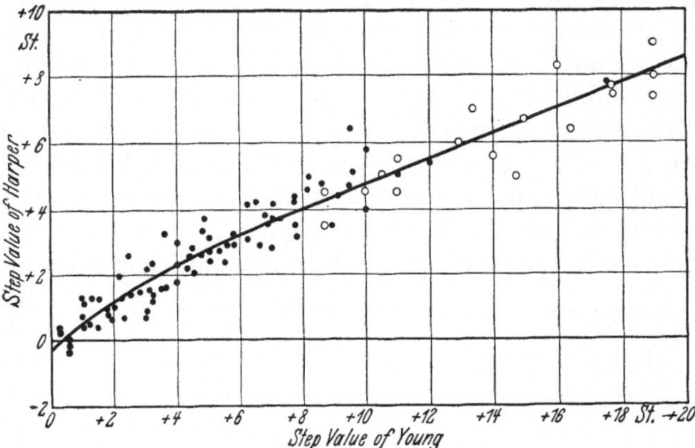

Fig. 117. Comparison between the measured step values by YOUNG and HARPER each using his own scale.

A comparison was also made between the trigonometric parallaxes in SCHLESINGER's catalogue of 1924 and those of the author according to different spectral classes. The differences are small. The results for the K2 and K5 stars ($\pi_{tr} - \pi_{s} = -0'',006$) suggest that these objects are even brighter than as found by RIMMER.

Some of the F stars presented difficulties in the estimates. The exceptions are discussed in detail by RIMMER. It seems to be difficult to decide if these objects are to be placed among the giants.

The work at the Norman Lockyer Observatory in order to determine the parallaxes of earlier types was inaugurated by EDWARDS[1]. About 200 negatives were picked out in order to discover which lines were subject to variation with change of absolute magnitude. In each of the negatives the intensities of all of the prominent lines were measured together with those of any fainter lines that might be suggestive of variation. Then the intensities were reduced to a common standard and as such the mean intensities of $H\beta$, $H\gamma$, $H\delta$ and $H\varepsilon$ were used. The various lines reduced to this standard were then plotted individually against the spectral types and against the absolute magnitudes when the latter were known. Any relation between line intensity and type or absolute magnitude could be detected in this way. In the cases where a variation was evident the lines were carefully re-measured and compared with lines situated comparatively near. More accurate curves were thus obtained in a second approximation.

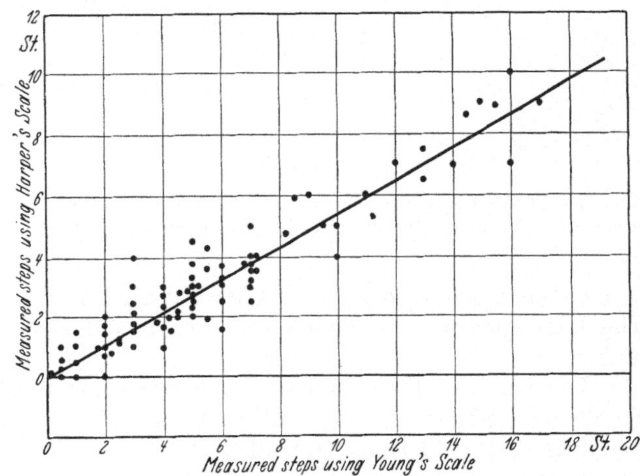

Fig. 118. Curve showing measured steps when both scales were used by one observer.

---

[1] M N 83, p. 47 (1922).

The wedge method was also used in the case of early spectral classes. The system of spectral classification used at Harvard was not entirely satisfactory for the stars considered. In a search for lines varying with the spectral type it was found that the best lines were those of the diffuse series of the helium doublets λ 4472 and 4026. The Mg^+ line 4481 also gave fairly good results when it was measurable, but a maximum intensity was reached at B5, which rendered classification work in that neighbourhood less accurate. The principal classification work is thus done by the aid of the helium lines λ 4472 and λ 4026. The mean results are generally not very different from the Harvard ones.

In determinations of the absolute magnitudes of the B stars there is considerable difficulty owing to the lack of a sufficient number of accurate trigonometric parallaxes upon which to found the reduction curves. It is true that valuable results have been reached, but the principal difficulty is that many B stars are so remote and their parallax obtained by direct measurements consequently so small that they are within the limits of their mean errors. EDWARDS[1] has used theoretical values as well as moving cluster parallaxes of KAPTEYN and dynamical parallaxes derived by JACKSON and FURNER and in some cases by RUSSELL. Also the parallaxes of PLUMMER were used.

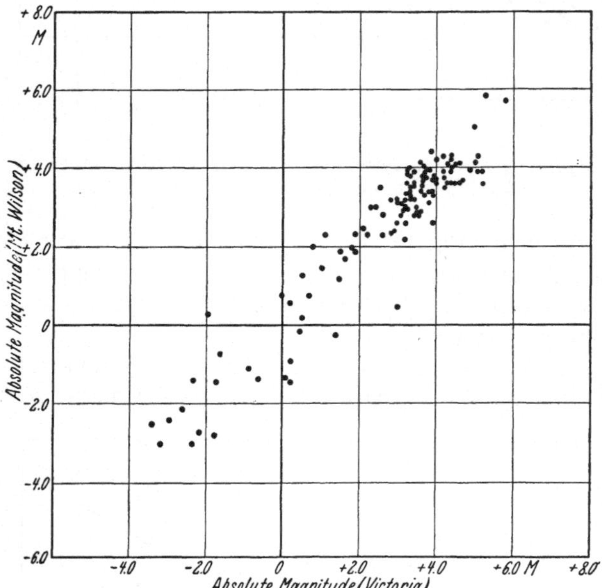

Fig. 119. Comparison of Victoria and Mount Wilson determinations of absolute magnitude for spectral classes F0—F9.

It was found that the helium lines of the diffuse singlet series were likely to give results. The doublet series also appeared to vary with M, but a good curve could not be established. The lines chosen were λ 4388 and λ 4144, and these were compared with $H\gamma$ and $H\delta$.

The inclusion of the parallaxes of PLUMMER is not commendable. On account of the theory used he finds the parallaxes 0",044, 0",017, and 0",011 for three physical members of the Pleiades that all must have the parallax very near to 0",007. The parallaxes of EDWARDS are 0",044, 0",016, and 0",015, respectively.

The next paper[2] is partly a revision of the foregoing one. For the final curves only trigonometric results of photographic methods and moving cluster parallaxes were used. The same corrections as used by RIMMER were applied in this and the following lists of B stars. The line pair $H\gamma/\lambda$ 4471 was also used in this paper, and the reduction curves for this pair from trigonometric values and group motions are given in fig. 1 of the paper.

Table I of the paper gives the spectrographic parallaxes of 100 B stars. Most of the stars in the earlier list have been included.

[1] M N 83, p. 47 (1922).
[2] M N 84, p. 366 (1924).

Comparisons were made with various kinds of parallaxes. The comparison between 14 dynamical and spectrographic parallaxes is of considerable interest. The mean difference, $M_{\text{spec}} - M_{\text{dyn}} = +0{,}5$, suggests an error in the average mass of the B stars, but the material is not sufficient to be decisive.

The Mount Wilson division into sharp and diffuse line stars is supported by the results of EDWARDS. The mean absolute magnitude deduced for 15 stars of the Mount Wilson class s is $-2^{\text{M}},3$ and for 21 stars of class n is $-1^{\text{M}},8$, a result that agrees very well with that found at Mount Wilson.

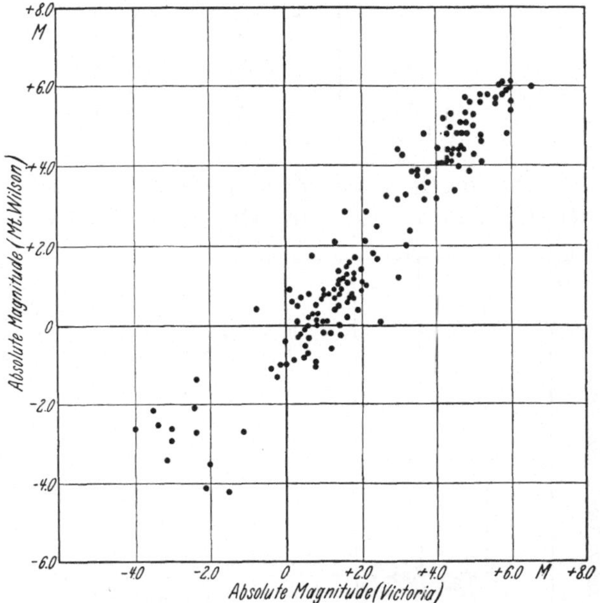

Fig. 120. Comparison of Victoria and Mount Wilson determinations of absolute magnitude for spectral classes G0–G9.

The next[1] list is a continuation of the former and does not give any essentially new methods. It has been found that the reduction curves can be based on stars of a single subdivision of spectral type alone. Thus B3 stars showed variations in the strength of the helium lines, which depended on the absolute magnitude to such a degree that a reduction curve could be established closely approaching the reduction curves deduced from all types. This means that the measurements give something more than a simple classification of spectral type according to the Harvard system. Other quantities are apparently included in the measurements, among which an effect of the sharpness or other line characters is certainly one of the most important.

For certain types of stars the wedge method is not so well adapted. In these stars, the considerable breadth and diffusedness of the hydrogen lines reduce the apparent intensities of the latter in comparison with the helium lines, with the result that the measured magnitudes are usually too bright. Sometimes the reverse is the case, the helium lines being more affected than the hydrogen lines.

The author reaches the conclusion that the very nebulous line stars together with the majority of the late B stars require to be treated by a different method. He says that the method of ADAMS and JOY is the best spectrographic method at present available in this case.

The systematic differences between the results of ADAMS and JOY and those of the Norman Lockyer Observatory have been investigated and found to be of the same order of magnitude as those in the first list. The following comparison from the latter may be quoted:

[1] MN 85, p. 439 (1925).

Spectral class Mt Wilson	Mean absolute magnitude			Spectral class Harvard
	NLO	MtW	Harv	
B0n	$-3^M,1$	$-3^M,1$	$-3^M,2$	B0
B1	−3 ,3	−2 ,6		
B1n	−3 ,3	−2 ,4	−3 ,1	B1
B2s	−2 ,4	−2 ,0		
B2n	−1 ,3	−1 ,5	−2 ,6	B2
B3s	−2 ,3	−1 ,5		
B3n	−1 ,9	−0 ,9	−1 ,9	B3
B4n	−1 ,6	−0 ,6		
B5s	+0 ,1	−1 ,1	−0 ,7	B5
B5n	−0 ,9	−0 ,5		
B7n	−0 ,9	−0 ,1	−0 ,3	B8
B8n	−0 ,2	+0 ,1		
B9s	−0 ,4	−0 ,2	−0 ,2	B9

The fourth paper[1] gives the collected results for 300 B stars and contains detailed comparisons with other results in the same line of work.

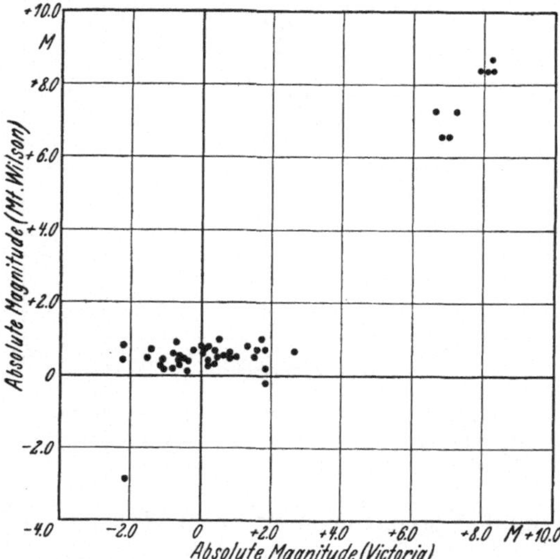

Fig. 121. Comparison of Victoria and Mount Wilson determinations of absolute magnitude of spectral classes K3—K9. There is a strong systematic deviation between the two systems for giants of the said classes.

The lines adopted for the classification were λ 4121, 4144, and 4472 due to helium, the Mg^+ line λ 4481, and the silicon double λ 4128, 4131.

The Harvard system was followed as closely as possible. In the following enumeration the additional remarks are put within parentheses. B0 to B2 are identical with Harvard.

Class B3. H lines about 0,5 as intense as in α Canis Maj. He lines more prominent than in B2 owing to the extreme faintness of some O, Si, and C lines (λ 4472 much stronger than λ 4481); λ 4121 stronger than Si double.

Class B4. He lines less prominent than in class B3; λ 4472 stronger than λ 4481; Si double approximately equal to λ 4121 and fainter than λ 4144.

Class B5. Increased intensity in *K* line and in Si double, which is stronger than λ 4121, but fainter than λ 4144. Line λ 4481 0,7 times as intense as λ 4472 [Ca line *K* usually just visible].

Class B6. Si double much stronger than λ 4121, nearly equal to λ 4144; λ 4472 slightly stronger than λ 4481; hydrogen lines slightly increased in intensity, helium lines less prominent.

Class B7. Si double equal to λ 4144; λ 4121 scarcely visible; λ 4481 equal to λ 4472; λ 4026 and λ 4472 easily seen, other He lines faint or invisible; *K* much fainter than λ 4026.

Class B8. Helium lines λ 4026 and λ 4472 are present together with several lines prominent in class A0; *K* is less intense than λ 4472 and slightly fainter than λ 4481; Si double stronger than λ 4144.

[1] M N 87, p. 364 (1928).

A comparison with the Harvard classes gives the summary:

Revised class	B0	B1	B2	B3	B4	B5	B6	B7	B8	B9
HENRY DRAPER (mean class)	B0	B0,8	B1,9	B3,0	B3,5	B4,7	B6,1	B7,8	B8,2	B8,9

The negatives were also examined from the point of view of line characters. Great differences exist between different lines in the B stars as regards their degree of sharpness, but these differences appear to exhibit a continuous gradation from very sharp to very nebulous types. Still it was difficult to measure the sharpness or to establish a numerical scale, and thus the spectra were divided into five groups according to the sharpness of the lines. These groups have been designated

ss, s, ns, n, nn,

in order of decreasing sharpness, where the extreme groups are exceptional and the middle group, ns, represents the average character. There was also some correlation between the degree of sharpness and the spectral class, in the sense that the average sharpness appeared to decrease from B0 to B9 stars.

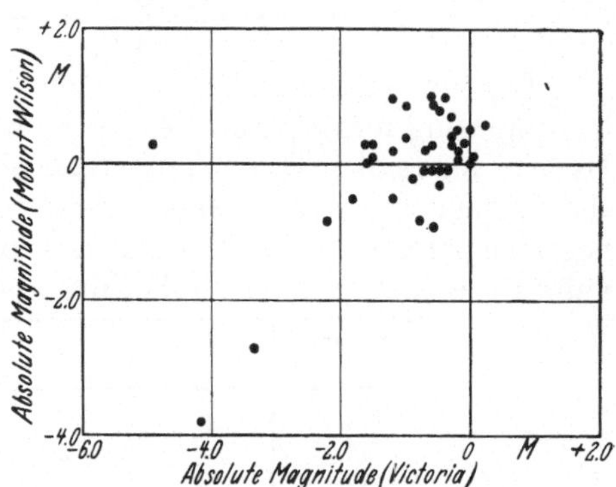

Fig. 122. Comparison of Victoria and Mount Wilson absolute magnitudes for spectral class M. The very discordant star is μ Cephei.

Curves were deduced separately for each line-character group and spectral subdivision. The chief difficulty lay in the small number of stars present in some of the groups.

The parallactic motions (the adopted elements for the Sun's motion were: $A = 270°$; $D = +34°$ and $V_0 = 21$ km/sec) were used. Radial velocities and peculiar motions were not used for obtaining mean parallaxes, since it has been shown by RUSSELL that for B stars the parallactic motions should give better results than the peculiar motions. From the mean parallax, $\overline{\pi}$, the mean absolute magnitude, $\overline{M}$, of each group was deduced from the formula:

$$\overline{M} = 5 \log \left\{ \frac{n}{\sum 10^{-0,2\,m}} \right\} + 5 + 5 \log \overline{\pi},$$

n being the number of stars in the group and m the apparent magnitudes.

The results are shown in the following summary:

Mean absolute magnitude deduced from parallactic motions.

Spectral class	ss	s	ns	n	nn
B1	$-3^M,7$	$-3^M,5$	$-2^M,6$		
B2	$-2\ ,9$	$-2\ ,7$	$-2\ ,3$	$-1^M,9$	
B3	$-2\ ,8$	$-2\ ,3$	$-1\ ,8$	$-1\ ,5$	$-1^M,2$
B4		$-1\ ,5$	$-1\ ,2$	$-0\ ,6$	
B5	$-2\ ,0$	$-1\ ,6$	$-1\ ,1$	$-0\ ,8$	
B6		$-1\ ,7$	$-0\ ,8$		
B7		$-1\ ,5$	$-0\ ,7$		
B8		$-1\ ,0$	$-0\ ,7$	$-0\ ,2$	
B9		$-0\ ,6$	$-0\ ,1$	$+0\ ,1$	

From the moving cluster data the following mean magnitudes were derived:

ss	s	ns	n	nn
$-2^M,6$	$-1^M,8$	$-1^M,4$	$-0^M,7$	$-0^M,6$

and if the spectral classes were taken into account:

	ss—ns	s—n
B0—B3	$-0^M,9$	—
B2—B4	—	$-0^M,7$

and for the B3 stars separately:

ss	s	ns	n	nn
$-2^M,3$	$-1^M,7$	$-1^M,4$	$-0^M,8$	$-0^M,7$

Trigonometric parallaxes are not sufficiently numerous to be of any value in computing mean parallaxes for groups as small as those under consideration. Use was made of F. DYSON's[1] method of correcting trigonometric parallaxes from the frequency curve of the observed parallaxes, by means of which method negative parallaxes are given the most probable positive values, and large positive values are reduced. The following results were finally adopted:

Spectral class	ss	s	ns	n	nn
B2,1	$-2^M,5$				
B2,3		$-2^M,1$			
B3,4			$-1^M,6$		
B3,5				$-1^M,0$	$-0^M,3$

The assembly of the curves finally adopted is shown in fig. 1 in the fourth paper and the results are given in the following table:

Spectral class	$\overline{M}$				
	ss	s	ns	s	nn
B0	$-3^M,9$	$-3^M,7$	$-3^M,4$	$(-3^M,2)$	
B1	−3 ,2	−2 ,9	−2 ,6	−2 ,3	
B2	−2 ,7	−2 ,3	−1 ,9	−1 ,6	
B3	−2 ,3	−1 ,8	−1 ,4	−1 ,1	$-0^M,8$
B4	−2 ,0	−1 ,3	−1 ,0	−0 ,7	−0 ,5
B5	−1 ,6	−1 ,0	−0 ,7	−0 ,5	−0 ,3
B6		−0 ,8	−0 ,5	−0 ,3	−0 ,1
B7		−0 ,6	−0 ,2	0 ,0	
B8		−0 ,5	−0 ,1	+0 ,2	
B9		−0 ,3	0 ,0	+0 ,3	

The results were compared with those of Mount Wilson, Arcetri (ABETTI), and Upsala (SCHALÉN)[2]. We form the fourfold table:

Norman Lockyer	Mount Wilson		
	s	n	Sum
ss + s	22	12	34
ns + n + nn	14	63	77
Sum	36	75	111

The Bernoullian coefficient r_B is found to be $+0,458 \pm 0,049$, which shows that there is a decided correlation between Mount Wilson and Norman Lockyer results. The objects in common with ABETTI are too few to allow any conclusions to be drawn. With regard to the groups of SCHALÉN the following fourfold table is obtained:

Norman Lockyer	Upsala		
	$\sigma\sigma, +\sigma\sigma, -\sigma\sigma$	$\sigma+, +\sigma$	Sum
ss + s	11	5	16
ns + n + nn	9	16	25
Sum	20	21	41

[1] M N 86, p. 686 (1926).

[2] Upsala Medd No. 10 (1926).

The coefficient of correlation, $r_B = +0{,}332 \pm 0{,}284$, scarcely suggests a real correlation but the few objects in common prevent the drawing of any definite conclusion. In this case the correlation will be lower, since SCHALÉN was unable to trace any resemblance between his groups and those of ADAMS and JOY. EDWARDS suggests that this lack of agreement may possibly be attributed to the very much smaller dispersion used by SCHALÉN causing the wings surrounding the hydrogen lines to become more prominent. It has frequently been noticed, he adds, that spectral lines occur that are sharp and narrow, but superimposed on a comparatively broad, hazy, though much fainter, band. The description of the line is thus a function of the instrument that is used.

The parallaxes in common with other observers gave:

Mount Wilson — EDWARDS = +0″,0015
ABETTI — EDWARDS = +0″,0036
SCHALÉN — EDWARDS = +0″,0015

The absolute magnitudes of the Norman Lockyer Observatory are thus the brightest given. The general agreement with the other methods is good, except in the case of SCHALÉN's $\sigma +$ group.

The comparison between the parallaxes of EDWARDS and those of the moving cluster stars is as follows:

Group	π_{spec}	π_{group}	Difference	M_{spec}	M_{group}	ΔM	n
Pleiades	0″,0138	0″,0138	0″,0000	$-0^M,45$	$-0^M,45$	$0^M,00$	5
Perseus	0 ,0085	0 ,0100	−0 ,0015	−0 ,75	−0 ,23	−0 ,52	16
Orion	0 ,0067	0 ,0059	+0 ,0008	−2 ,56	−2 ,89	+0 ,33	14
Scorpio-Centaurus .	0 ,0113	0 ,0119	−0 ,0006	−1 ,79	−1 ,65	−0 ,14	29
All groups	0 ,0099	0 ,0104	−0 ,0005	−1 ,61	−1 ,47	−0 ,14	64

The parallaxes of 300 A stars of spectral class A0—A5 derived from plates taken at the Norman Lockyer Observatory have been derived in a paper by H. C. WOODS[1]. A preliminary solution, for the mean absolute magnitudes of groups of stars, showed plainly the dependence of the absolute magnitude on both line character and spectral class.

As standards for the reduction curves trigonometric parallaxes were used, corrected according to F. DYSON's[2] method. The spectra were classified in the six groups A0 to A5. The metallic lines are not shown sufficiently clearly to enable them to be measured, but their intensity could be estimated. The character of the hydrogen lines was estimated, and the stars arranged in grades on this basis.

The results given in the following table also indicate the system of classification of the lines according to their degree of sharpness:

Spectral class	s	ns+	ns	+ns	n	n+	nn
B9	$-0^M,3$	$-0^M,2$	$-0^M,1$	$0^M,1$	$0^M,3$	$0^M,5$	$0^M,8$
A0	−0 ,1	0 ,1	0 ,2	0 ,4	0 ,6	0 ,8	1 ,1
A1	0 ,2	0 ,4	0 ,6	0 ,8	1 ,0	1 ,2	1 ,4
A2	0 ,5	0 ,7	0 ,9	1 ,1	1 ,3	1 ,5	1 ,7
A3	0 ,8	1 ,0	1 ,2	1 ,4	1 ,6	1 ,8	2 ,0
A4	1 ,1	1 ,3	1 ,5	1 ,7	1 ,9	2 ,1	2 ,3
A5	1 ,4	1 ,7	1 ,9	2 ,1	2 ,2	2 ,4	2 ,6

WOODS mentions the correction necessary on account of many of the stars being spectroscopic binaries. When both spectra are seen a correction of $+0^m,5$ is applied to the apparent magnitude.

[1] M N 87, p. 387 (1927). [2] M N 86, p. 686 (1926).

The following results were found, comparing the parallaxes with other determinations:

Source	Norman Lockyer	Mean of sources	Dispersion	n
Trigonometric parallaxes	0″,0322	0″,0312	±0″,0156	107
Moving cluster ,,	0 ,0455	0 ,0450	±0 ,0098	23
Dynamical ,,	0 ,0221	0 ,0269	±0 ,0156	26
Adams a. co-workers ,,	0 ,0284	0 ,0298	±0 ,0069	143
DOUGLAS ,,	0 ,0256	0 ,0280	±0 ,0081	124
ABETTI I ,,	0 ,0252	0 ,0247	±0 ,0056	62
ABETTI II ,,	0 ,0224	0 ,0237	±0 ,0060	85

Among the errors that may affect determinations of absolute magnitudes WOODS mentions those arising from inaccuracy in the classification of the spectra. Poor negatives will usually lead to the assignment of too nebulous a line grade and too early a spectral type.

148. Arcetri Determinations of *M*. G. ABETTI tried in 1919[1] to classify the stars of spectral classes B0—G8 numerically by using the intensity of certain metallic lines in a way analogous to that of ADAMS and KOHLSCHÜTTER. He reached the important conclusion that the width of the hydrogen lines and of the calcium bands was not only dependent on the spectral class, but also on the absolute magnitude of the stars. The observations referred to were made with the small equatoreal of the Collegio Romano and the objective prism of SECCHI, and were taken up later on at the observatory at Arcetri.

The first catalogue of ABETTI[2] deals with the derivation of the absolute magnitudes for the first type of SECCHI (B8—F6). The instrument is the 30 cm refractor of the observatory (f/7,6). The distance between $H\xi - H\alpha$ is 26 mm. For the classification of the spectra the width of the line K and the difference of width of the lines $H\varepsilon + H$ and K were used.

Three sub-groups concerning the degree of sharpness of the spectral lines were introduced, viz.

n, sn, and s.

The spectroscopic parallaxes for 159 stars were derived. A comparison between the M in common with Mount Wilson showed that they were in good accordance, which indicates a definite correlation. The systematic difference was very small:

$$M_{\mathrm{MW}} - M_{\mathrm{A}} = +0^{\mathrm{M}},2 .$$

The agreement as regards the parallaxes is very good:

Spectral class	MW—A	n
B7—A1	+0″,003 ± 0″,0015	29
A2—A4	+0 ,001 ± 0 ,0030	24
A5—F2	+0 ,003 ± 0 ,0032	25

ABETTI also investigated the relation between M and $H = m + 5 + 5 \log \mu$ and found systematic differences between M_H and M as computed from his parallaxes, which call for a revision of the curves connecting M and H.

The second catalogue[3] contains 275 stars of the same spectral classes. The earlier results have been re-discussed and are included in the new catalogue. New curves were derived giving the relation between M and the spectral class for the three groups n, s, and sn.

---

[1] Mem S A It (3) 1, p. 149 (1920).
[2] Arcetri Publ Fasc 41 (1924).
[3] Arcetri Publ Fasc No. 42 (1925).

The mean parallaxes were also computed from the formula of VAN RHIJN. The following comparison is of interest:

Spectral class	μ	M_A	$\overline{M}_\mu$	n
A0,7s	0″,035	$+0^M,6$	$+0^M,2$	40
A5,8s	0 ,046	+1 ,6	+0 ,7	19
A0,6sn	0 ,046	+0 ,8	+0 ,3	42
A6,1sn	0 ,060	+1 ,7	+1 ,0	19
A3,3sn	0 ,111	+2 ,3	+1 ,0	18
B9,7n	0 ,051	+0 ,7	+0 ,5	46
A6,0n	0 ,103	+2 ,0	+1 ,5	31
F3,8n	0 ,186	+2 ,9	+2 ,5	23

The mean difference, $\overline{M_A - M_\mu}$, is $+0^M,52$ and it will be of much interest to look for an explanation of the discrepancy when more material has been collected. If the difference between the mean parallaxes is computed, the relation is as follows:

$$\bar{\pi}_A - \bar{\pi}_\mu = +0'',004 .$$

This is in substantial agreement with VAN RHIJN's result of comparing the mean parallaxes with the trigonometric.

149. Harvard Determinations of *M*. In 1916 H. SHAPLEY suggested the possibility of using small dispersion spectra for the determination of the luminosities and distances of stars[1]. He found that the dispersion used for Harvard photographs of the brighter stars is sufficient for the application of the Mount Wilson criteria. Moreover several other characteristics have been found to be of use as indices of absolute brightness. The first Harvard paper on the determination of the absolute magnitudes contained the distances of 50 stars between G7—K3 determined by H. SHAPLEY and B. LINDBLAD[2].

The same luminosity criteria were used as those employed at Mount Wilson together with the "cyanogen" absorption bands and the lines of hydrogen, calcium, and manganese.

The mean systematic difference between the M in the sense LINDBLAD—SHAPLEY was found to be $+0^M,31$ and was slightly dependent on M. The average deviation without any correction for the systematic difference was found to be $0^M,32$.

The following comparison shows the agreement between the Harvard and Mount Wilson results:

	H—MW	Sh—MW	Ld—MW
Average	$0^M,29$	$0^M,37$	$0^M,34$
Systematic	−0 ,09	−0 ,09	−0 ,11

In the next paper[3] H. SHAPLEY and Mrs. MARTHA B. SHAPLEY have determined the absolute magnitudes of 87 bright stars of classes G7—K3. The systematic difference between the results of the two observers is H S — M B S is $+0^M,06$. The average difference in spectral classification between Harvard and Mount Wilson is one spectrum unit for 94 stars compared.

In a third paper[4] SHAPLEY and Miss ADELAIDE AMES determined the magnitudes of 233 southern stars.

The relation of line intensity to M has been calibrated by means of the proper motions in the BOSS catalogue.

[1] Publ Amer Astr Soc 3, p. 222 (1916).
[2] Harv Circ 228 (1921).
[3] Harv Circ 232 (1922).
[4] Harv Circ 243 (1923).

The view expressed earlier that the parallax system of Mt Wilson Contr 199 makes the K giants on an average intrinsically somewhat too faint has found support in the Harvard investigations as is seen in the following synopsis.

Source	Systematic difference	Average difference	N
M W 199 — Harv Circ 243	$+0^M,53$	$\pm 0^M,51$	23
Harv Circ 228, 232 — Harv Circ 243 . .	$+0\ ,39$	$\pm 0\ ,46$	43

The $\overline{M}$ of the 233 stars is $+0^M,67$, which is in agreement with an unpublished mean value derived by LUYTEN from the trigonometric parallaxes of northern K0 stars.

SHAPLEY points out that the intensity of the spectral line λ 4215, on which a large part of the determinations is based, is actually a direct indicator of ionization conditions and only indirectly an index of total luminosity. Any irregularities in the behaviour of ionized strontium in stellar atmospheres will invalidate the spectroscopic parallaxes.

The fourth paper[1] gives the M for 116 southern G5—K5 stars. The variation of the intensity of the Sr-lines for giants of classes K2—K5 is much less pronounced than for giants of classes G0—K0. For the K2 giants, in particular, there is not much gained by deriving individual absolute magnitudes except for a few of the brightest and faintest stars, since for the remainder it is nearly as satisfactory to adopt a mean value of $+0^m,5$.

150. Yü's Method for the Absolute Magnitudes of A Stars[2]. The starting point is that the temperature of the stars must be taken into account, if a simple relation is to be found between the intensity difference of adjacent regions in the spectrum and the absolute magnitude. A careful search was made for an intensity-magnitude relation among the A stars. A plot of R, defined as the intensity of the continuous absorption at the limit of the BALMER series, and the M gave no correlation. Next it was found that the following relation existed:

$$M = 11,5\,R - \frac{T}{1600} + 3,25$$

where T is the absolute temperature. Investigations were carried out that made it very probable that the correlation underlying the above equation has a real significance.

The constants have been determined from 39 stars and then the absolute magnitude was derived for 24 stars.

Among the advantages of the method the principal one seems to be that spectral classification is not involved in the data. The intensity R is founded on a direct photometric basis, as has been shown by YÜ in an earlier paper[3]. The absorption R is continuous and unlike the line intensity it is independent of the slit and dispersion as well as of seeing, focus, and imperfect guiding.

In the case of objective prism observations it is also independent of "seeing", imperfect guiding, and inaccurate focus.

The most serious limitation of the method is that the region of continuous hydrogen absorption lies in the ultra-violet and is therefore not readily accessible with the instruments now in use.

151. Absolute Magnitudes from Variations in Small Dispersion Spectra. In connection with work concerning the determination of effective wave lengths

[1] Harv Circ 246 (1923).
[2] Lick Bull 12, p. 155 (1926).
[3] Lick Bull 12, p. 104 (1926).

LINDBLAD[1] tried to determine the absolute magnitude from a combination of two colour indices. When working at Mount Wilson he used the 10-inch Cooke refractor with a 6° objective prism for the development of a spectro-photometric method for determining absolute magnitude. A comparison between the spectra of Sirius and o^2 Eridani ($M = +10{,}8$) revealed a number of differences. The most conspicuous feature was the very considerable widening of the wings of the hydrogen lines in the later stars, which was connected with a decided weakening of the high-numbered members of the BALMER series. The distribution of energy, between $H\varepsilon$ and $H\zeta$ in the fainter stars, was different on account of a considerable decrease in intensity in the region λ 3889 — λ 3907 as compared with the region λ 3907 — λ 3935. This may be partly caused by the expansion of the wings of $H\zeta$, but also a suspected widening of λ 3906 Si may contribute.

In order to measure the degrees of intensity of different portions of the spectrum a time scale was used. Series of exposures with the exposure time varying in the ratio $\sqrt{2}$ were taken on the same plate. A certain region, a, is compared with the region b in such a way that two scale images where the intensities in both regions are most nearly equal are compared, and by the interpolation of a tenth of the interval the corresponding exposure time was estimated. $\log E = t_a/t_b$ is then a measure of the change of energy from region a to region b. The process is analogous to that used by SEARES in determining the colour of a star by the method of exposure-ratios. In the case of bright stars the $\log E$ had to be estimated directly, a method which gave results consistent with the case where comparisons between different exposures could be made.

The juxtaposition of the two spectral regions makes the influence of slight changes in the sensibility curve of the plate negligible. The same is the case for a variation in zenith distance within reasonable limits from one object to another, and for a difference in gradation. If the law of density in the form used by SCHWARZSCHILD is $S = \varphi(I\,t^p)$ and I_a, I_b are the acting intensities for the first and second regions, and if we have equal intensities ($S_a = S_b$), then:

$$\log\frac{I_a}{I_b} = p\log\frac{t_a}{t_b} = p\log E\,.$$

If the same brand of plates is used and the photographic conditions are kept as constant as possible, there is reason for assuming p to be tolerably constant, though experience shows that large deviations may occur.

According to KRON's results there cannot be any change in $\log E$ with the density of the images on account of the parallelism of the curves $\log It$ and $\log I$ for different degrees of intensity. A slight change of $\log E$ may occur with the absolute value of I or with m due to the bending of the curves and corresponding to a change of p with I.

When images of stars with accurate values for m occur on the plates the value of p can be determined. We can write $p = \frac{\Delta \log I}{\Delta \log t}$ and thus the interpolation of the density of successive images of one star into the scale defined by the images of the second star gives p. From the Mount Wilson measurements a decrease of p with m was found:

$$\frac{\Delta p}{\Delta m} = -0^m{,}026 \pm 0^m{,}003 \text{ for } m = 6^m{,}5\,.$$

As a mean for the first seventeen plates measured it was assumed that

$$p = 0{,}97 - 0{,}026\,(m - 5)\,.$$

[1] Mt Wilson Contr 228 (1922); Ap J 55, p. 85.

The mean error of $\log E$ was estimated from a series of four images on a single plate as $\pm 0^m,037$.

The method was first applied to the Hyades, Pleiades and Ursa Major stream, Praesepe, and some Orion stars. Altogether 71 new absolute magnitudes were determined from the groups. It may be remarked that the value of π used for the Hyades group is undoubtedly too small.

The increase in $\log E$ for increasing M can be used for determining the absolute magnitudes for stars of the classes B8—A3. The relation curves cannot give very trustworthy results for the later classes.

Other criteria had to be used for stars between G and M. Miss MAURY has remarked that the sudden decrease of intensity in the spectrum from λ 3889 to the ultraviolet in her classes XIV and XV as well as the degree of absorption in the regions λ 4055—λ 4078 and λ 4144 — λ 4216 is developed to a different degree for different stars within these types. The variation of intensity in these regions is further thought to be correlated to some extent with the strength of the lines λ 4215, λ 4227 and the compound λ 4076,8, λ 4077,9.

The absorption in ultra-violet was shown by KAPTEYN to be negatively correlated with the proper motion and this fact led him to assume the presence of a considerable selective absorption in inter-stellar space. This influence is not due to absorption, but is certainly connected with the absolute magnitude as found by LINDBLAD with the aid of very short objective-grating spectra. The colour effect of the absolute magnitude which is to be explained elsewhere, was evidently somewhat more strongly developed in the minimum wave length than in the effective wave length. LINDBLAD suspected that the effect could be due to an increase of absorption with the M in the heavily winged lines of iron in the region λ 3870—λ 3890.

Later on it was found that the effect was due to a variation in strength with M in the strong "cyanogen" band with its first head at λ 3883. Using the Crossley reflector LINDBLAD found that three cyanogen bands with first heads at λ 4216, λ 3883, and λ 3590 showed the same behaviour. The effect is strongest in the band at λ 3883 and the region most sensitive to changes in M seems to be the region around and between the heads λ 3871 and λ 3883.

By the use of essentially the same method as that for A stars the absorption in the band at λ 3883 was measured and compared with a region adjacent to the absorbing region. The maximum effect is found in G8—K0 stars and a rapid decrease takes place from G0 to G5. The effect is strongest for giants of G5—K5 and much weaker for dwarfs. The Cepheids and Pseudo-Cepheids show very little change. All stars with $\log E > 0{,}17$ are giants.

A decided disadvantage arises from the fact that the region in question being situated in violet necessitates the making of rather long exposures for the redder stars. On that account an investigation of the favourably situated band λ 4216 was made. The regions to be compared were λ 4144—λ 4184, λ 4184 to λ 4227, and λ 4227—λ 4272.

A scheme for classifying the objective-prism spectra was developed by the use of types determined by ADAMS and his co-workers. The classification is based on the appearance and relative strength of the hydrogen lines, the Ca-line λ 4227, the TiO bands, and the general distribution of energy in the spectrum.

Parallaxes and absolute magnitudes were derived for 78 stars. The open cluster Messier 11 was also investigated and the spectra were classified for 85 stars.

The work of B. LINDBLAD[1] at Upsala has been based on plates taken with a 6-inch telescope and with the aid of a prism having an angle of 9°,7, giving the dispersion 1,4 mm between $H\gamma$ and $H\varepsilon$. The measurement of the exposure ratio of λ 3895—3907 to λ 3907—3925 proved to be rather difficult, when it was attempted at Upsala. The difference of density between the two regions was subject to considerable disturbance, probably due for the most part to unsteadiness of the air during the exposures. Instead of these regions the varying sharpness of the hydrogen lines was made use of and a general classification of B8—A3 stars was undertaken.

The general scheme used at Upsala is a modification of the ordinary Harvard classification (PICKERING-CANNON). Besides this, the giants and dwarfs are separated mainly by the cyanogen criteria. Another addition of importance is the possibility of obtaining the colours of the stars for the classes later than G0 by estimating the degree of contrast between the two regions of the spectrum immediately adjacent to the band G, and, for the early classes, by comparing the spectral region of the red side of $H\gamma$ with the region $H\varepsilon - H\zeta$.

The B0—B5 stars can be separated from other classes by the extreme narrowness of the hydrogen lines, the low colour, and traces of the most prominent helium lines, such as λ 4026. The spectra cannot be used for a discrimination between the sub-classes of the early B stars.

The stars of classes B8—A0 are classified according to the following scheme:

Class	
$\sigma\sigma$	The hydrogen lines are very sharp and scarcely visible with the dispersion used.
σ++	The hydrogen lines are very sharp but appear somewhat wider than in the foregoing class.
σ+	The hydrogen lines do not appear wider in the weak images than in the strong ones, though the lines are wider than in the spectrum of f. i. α Cygni.
σ	The width of the hydrogen lines increases only a little from strong to weak images.
σ—	The sharpness of the lines is smaller than in preceding classes, but is still noticeable. The colour estimated by comparing the region of the red side of $H\gamma$ with the region $H\varepsilon - H\zeta$ is low.
ϱ	The wings of the hydrogen lines are rather strong, but the colour is as low as in the preceding class.
μ	The wings of the hydrogen lines are very strong. The colour is the same as in the preceding classes or slightly more advanced.
μ—	The colour is somewhat more advanced than in the preceding class.
$\varkappa$—	The colour is considerably more advanced than in the preceding classes.
$\varkappa$	An absorption at λ 3906 probably due to some arc-lines of Si and Fe is often slightly marked.
$\varkappa$+	The arc-line absorption at λ 3906 is comparatively well marked.

Groups with strongly winged lines, especially those where the latter are combined with advanced colour, such as $\varkappa$ —, $\varkappa$, and $\varkappa$ +, cannot occur among the B8 stars. Classes σ — and ϱ are not used for A2 and A3 stars.

The methods have been applied to stars within the Greenwich Zone. The general catalogue gives data for 1066 stars within that zone and for 200 stars also classified in the Mount Wilson list of 1646 stars. A third table gives the results of a classification of stars in the Pleiades group, for which a parallax value of 0″,0079 is derived.

[1] Upsala Nova Acta Ser. IV, 6, No. 5 (1925). First paper. — Volumen extra ordinem editum 1927 = Upsala Medd No. 11. Second paper. Further the following papers by LINDBLAD should be consulted: Summary of Results Concerning the Determination of Absolute Magnitudes by the Cyanogen Criterion. Upsala Medd No. 18 (1927); On the Absolute Magnitudes and Parallaxes of Bright Stars Determined by the Cyanogen Criterion. Upsala Medd No. 28 (1927); B. LINDBLAD and C. SCHALÉN, The Luminosities, Individual Parallaxes and Motions of B and A Type Stars. Upsala Medd No. 17 (1927).

The relation between cyanogen absorption and absolute magnitude has been derived for G and K stars with the aid of the data from the proper motions of the Boss and Cincinnati stars and of the Greenwich zone. Absolute magnitudes as determined by ADAMS and his co-workers have also been used. The tables of VAN RHIJN were used for converting proper motions into mean parallaxes.

The sub-groups of the A stars show an interesting correlation with the reduced proper motion:

Spectral class	6m,0	n	Spectral class	6m,0	n
$\sigma+$, σ, $\sigma-$	0″,028	9	μ	0″,046	10
ϱ	0 ,027	8	$\varkappa$	0 ,069	7

The A stars show relatively sharp hydrogen lines and σ stars must be of comparatively high luminosity, since it has been shown that general sharpness of the absorption lines, or Miss MAURY's c character, means an extraordinarily high value of the absolute brightness. LINDBLAD points out that this is also in harmony with modern views on stellar constitution. If T is the effective temperature, R the radius, $\mathfrak{M}$ the mass, M the bolometric magnitude, and G the gravitation constant, we shall have:

$$\frac{dP}{\varrho\, dr} = \frac{G\mathfrak{M}}{R^2}.$$

$$M = -2{,}5 \log R^2 T^4 + \text{Const.}$$

If EDDINGTON's mass-luminosity relation is written $\log \mathfrak{M} = f(M)$, we have:

$$\log \frac{dP}{\varrho\, dr} = f(M) + 0{,}4 M + 4 \log T.$$

The two terms $f(M)$ and 0,4 M vary in opposite directions, but the variation in the latter case is much larger, so that $\log \frac{dP}{dr}$ decreases with decreasing M, that is with increasing luminosity. If the general opacity of the atmospheric layers is more or less independent of the luminosity, the gradient of the gas-pressure will decrease with increasing luminosity in still greater proportion.

LINDBLAD has investigated the general accuracy of the determinations of M at Upsala and concludes that the mean error of one determination is $\pm 0^{M},8$.

In another paper LINDBLAD has investigated further the relation between absorption and M and extended the analysis of the special distribution of the stars in the Greenwich zone to a part of the zone situated as far as possible from the Milky Way.

A revised diagram was derived, which gives the spectral type and absolute magnitude as a function of the cyanogen absorption in the region λ 4144—4184 and the colour effect at the band G. The material consisted of proper motion stars from different regions of the sky and spectrographically determined M. In order to avoid the considerable systematic differences between different lists the Mount Wilson values alone have been used.

The comparison between the mean absolute magnitudes as determined at Mount Wilson ($\overline{M}_W$) and as computed from VAN RHIJN's tables ($\overline{M}_G$) with the determinations of Upsala ($\overline{M}$) is quoted from LINDBLAD in the table on p. 483.

From these comparisons the mean error of the estimates of M at Upsala is found to be $\pm 1^{M},0$ in both cases. It is also found that the accuracy of VAN RHIJN's method for the spectral classes G and K is certainly higher than he thought.

A comparison is also made between the cyanogen absorption ratios N' (λ 3840 — 3883; λ 3883 — 3933) and N (λ 4144 — 4184; λ 4227—4272); the relation:

$$N' = 2{,}26\, N + 0{,}69$$

represents the interval $\log E_g < 1,3$ or spectral classes earlier than gK5.

152. CARRINGTON Circumpolar Zone. J. H. PETERSSON[1] has determined the spectral classes for about 2000 CARRINGTON stars and also the absolute magnitudes for a great number of them. The four regions investigated are centred at $+85°$ and 0^h, 6^h, 12^h, and 18^h respectively. The scheme of spectral classification is that of LINDBLAD. The luminosities were determined from the cyanogen bands λ 3883 and λ 4216. The last-mentioned band, although weaker than the first, especially proved to be convenient. The cyanogen absorption, named $\log E_n$, was determined from comparing images which had been impressed with the exposure times in the ratio $\sqrt{2}/1$. The break in the continuous spectrum at the G band which was introduced by LINDBLAD and called $\log E_g$, was estimated in a way analogous to that of $\log E_n$. The spectral regions compared in the first case are λ 4144—λ 4184 and λ 4227 to λ 4272 and in the second λ 4227—G and G—λ 4383.

A comparison was made between the values of $\log E$ determined by LINDBLAD and PETERSSON and a slight systematic difference was found. The measurements were reduced to the system of the former. From a nomogram giving the curves spectral class = const and M = const with $\log E_n$ and $\log E_g$ as variables, the values of M are read.

log E_g																	
0,0—0,3		n	0,3—0,5		n	0,5—0,8		n	0,8—1,0		n	1,0—1,3		n	>1,3		n
$\overline{M}_W$	$\overline{M}$		$\overline{M}_W$	$\overline{M}$		$\overline{M}_W$	$\overline{M}$		$\overline{M}_W$	$\overline{M}$		$\overline{M}_W$	$\overline{M}$		$\overline{M}_W$	$\overline{M}$	
$-0^M,2$	-3^M	1	$-0^M,9$	$-0^M,1$	1	$-4^M,2$	$-1^M,5$	1	$+0^M,3$	$+0^M,3$	8	$+0^M,5$	$+0^M,4$	15	$+0^M,6$	$+0^M,4$	7
+4 ,2	+3 ,9	12	+0 ,7	+1 ,9	1	−0 ,7	−0 ,4	5	+1 ,9	+1 ,1	7	+4 ,3	+1 ,7	1	+9 ,2	+9 ,1	3
			+1 ,7	+2 ,6	2	+1 ,1	+0 ,5	5	+4 ,8	+4 ,7	5	+6 ,5	+6 ,1	6			
			5 ,5	+5 ,3	4	+5 ,5	+5 ,6	6									

log E_g																	
0,0—0,3		n	0,3—0,5		n	0,5—0,8		n	0,8—1,0		n	1,0—1,3		n	>1,3		n
$\overline{M}_G$	$\overline{M}$		$\overline{M}_G$	$\overline{M}$		$\overline{M}_G$	$\overline{M}$		$\overline{M}_G$	$\overline{M}$		$\overline{M}_G$	$\overline{M}$		$\overline{M}_G$	$\overline{M}$	
$-2^M,3$	-3^M	1	$-0^M,1$	$+0^M,3$	5	$-0^M,5$	$-0^M,3$	5	$+0^M,1$	$+0^M,3$	16	$-0^M,2$	$+0^M,6$	16	$-0^M,5$	$+0^M,2$	9
+2 ,1	+3 ,2	5	+2 ,0	+5 ,0	2	+1 ,4	+1 ,3	12	+1 ,5	+0 ,4	7	+1 ,5	+0 ,9	15	+1 ,6	+2 ,0	2
+4 ,0	+4 ,0	26	+4 ,0	+4 ,9	11	+3 ,9	+4 ,1	13	+4 ,1	+4 ,6	13	+3 ,8	+5 ,2	6	+3 ,5	+3 ,6	1
+5 ,0	+2 ,8	4	+5 ,5	+5 ,4	3	+5 ,2	+4 ,9	8	+5 ,6	+5 ,6	3	+5 ,7	+6 '2	4	+6 ,0	+8 ,5	3

In this table E_g is the logarithmic exposure ratio for the regions G—λ 4383 and λ 4227—G. E_g thus measures the break of the continuous spectrum at the band G and plays a part in the Upsala classification scheme for the "late" classes.

[1] The Distribution of Distances and Velocities of Stars in the Carrington Zone on Basis of Spectrophotometric Analysis. Upsala Medd No. 29, also Thesis Upsala (1927).

The apparent magnitudes measured at Greenwich in the scale of Harv Circ 180 have been reduced to the International Scale. The $\log E_g$ were reduced to the scale of PARKHURST and further to the scale of Greenwich corrected to Harvard. Then the following mean photographic magnitudes were found (see adjoining Table).

Spectral class	$\overline{M}_{ph}$	Spectral class	$\overline{M}_{ph}$
dG0	4,9	gG0	(1,0)
dG5	5,3	gG5	1,6
dG8	5,5	gG8	1,8
dK0	6,0	gK0	1,7
dK2	6,7	gK2	1,5
dK5	(7,5)	gK5	1,7
dM	(8,0)	gM	1,7

The rarity of the gG0 stars is a very important result. Among the 182 G0 stars in the general catalogue 174 are dwarfs and only 8 giants. With regard to fainter G stars it can safely be assumed that they are all giants without any sensible error being introduced into the (statistic) conclusions based on this assumption. Distances and velocities of the stars were computed by means of the photographic magnitudes and proper motions published from Greenwich, and the constants of the velocity-ellipsoids were derived according to the method of CHARLIER. The agreement with the values of the Boss stars and the agreement with similar results of LINDBLAD shows that the Upsala magnitudes used for calculating the linear velocities must be nearly correct. Further it is found by PETERSSON that the effect of the galactic rotation in the region investigated for stars nearer than 417 parsecs is -5 km/sec. The parallactic motion is computed to be 16,6 km/sec in the same region and thus a pure parallactic motion of 11,6 km/sec in the direction of galactic longitude is to be expected. The parallactic motion really observed in longitude is 9,8 km/sec.

The number of stars investigated is 2003. Most of these are of apparent magnitudes $8^m,5-11^m,6$. The material can be considered complete down to $m_{ph}=11^m,0$. The relative frequency of the different spectral classes is as follows:

Spectral class	Relative frequency	Spectral class	Relative frequency
A0	0,047	gG0	0,004
A2—A5	0,056	gG5	0,070
F0—F2	0,092	gG8	0,082
F5—F8	0,348	gK0	0,057
dG0	0,087	gK2	0,039
dG5—DG8	0,049	GK5	0,032
dK0—dK2	0,017	gM	0,018
dM	0,002		

The spatial distribution of stars of different classes was also investigated. Of special interest is the relation between the heights above the galactic plane as expressed in parsecs and the corresponding different values of the absolute density $D(r)$.

$D(r)$	B8—A3 Height		$D(r)$	gG0—K2 Height	
	LINDBLAD	PETERSSON		LINDBLAD	PETERSSON
0,35	72	64	0,16	180	124
0,30	84	69	0,15	192	138
0,25	104	80	0,14	206	148
0,20	125	101	0,13	218	158
0,15	139	115	0,12	230	170
0,10	156	132	0,11	254	186
0,08	167	150	0,10	290	200
0,055	185	185			
0,04	204	226			
0,02	254	280			

This method should be used for a determination of the galactic "dip".

The form of the luminosity curve of the A stars was examined, the material being divided into two groups, namely the stars nearer than 240 parsecs and the stars farther away than 240 parsecs. The number of stars between certain limits of M were derived and reduced to equal volumes and corrected for the star density. Then the curves showing the course of $\varphi(M)$ were formed. The maximum frequency is attained at $M = +2{,}3$ for stars with distance smaller than 240 parsecs and at $M = +2{,}6$ for stars more distant than the said limit. A secondary maximum at $+0^M{,}6$ is also indicated. To this group of stars correspond the A stars with narrow hydrogen lines. The displacement of the principal maximum is small, but neither does the form of the two curves coincide. It seems that the distribution of absolute magnitude is probably not quite independent of the distance as is generally assumed in statistical investigations.

In the near future the remaing part of the CARRINGTON zone will be surveyed spectroscopically by PETERSSON.

153. Photometric Measurements of Absolute Magnitude Regions. Y. ÖHMAN[1] has used a SIEGBAHN photometer and a SCHILT photometer in order to measure by photometric methods the effects in stellar spectra used as criteria for the determination of M in the work at Upsala. In order to determine the density curves of the plates the prism-grating method of HERTZSPRUNG was used so that the magnitudes of the corresponding wave lengths in the spectrum of the central image and those of the first order differed by a constant amount of $2^m{,}486$. The difference in magnitudes G_m between the regions $\lambda\,4232$—$\lambda\,4272$ and $\lambda\,4326$—$\lambda\,4368$ and the difference C_m between the regions $\lambda\,4138$—$\lambda\,4173$ and $\lambda\,4232$—$\lambda\,4972$ were measured for 77 late stars, previously investigated by LINDBLAD. The first quantity corresponds to the colour effect at the band G and the other to the cyanogen absorption. The correlation r between the estimated values of G_m and the measured values was high; thus r is equal to $+0{,}946 \pm 0{,}012$. The correlation between measured and estimated values of C_m is slightly lower, viz. $r = +0{,}930 \pm 0{,}016$. No correlation exists between the two quantities G_m and C_m. The mean error in LINDBLAD's estimates is determined by ÖHMAN as $\pm 0^m{,}032$ for the colour effect and as $\pm 0^m{,}034$ for the cyanogen absorption.

The quantities G_m and C_m were also determined for 64 bright G and K stars. The method of exposure-ratios is not so suitable for bright stars as for fainter ones, because the exposure-times become very short. The spectra of these stars were analysed with the SIEGBAHN photometer as well as with the SCHILT photometer. A comparison could be made between the absolute magnitudes of 56 stars determined at Upsala as well as at Mount Wilson. The stars were divided into two groups, corresponding to colour effect $G_m < 0{,}8$ and $> 0{,}8$ (in LINDBLAD's units) and the mean errors were found to be $\pm 1^M{,}1$ and $\pm 0^M{,}6$ respectively, the mean error of the Mount Wilson determinations being taken as $\pm 0^M{,}6$.

Finally 71 B and A stars were investigated. The luminosity classes of LINDBLAD are based on the degree of sharpness of the hydrogen lines. The total contours of the spectral lines are considered and not the central parts preferably studied by others. ÖHMAN measured the sharpness of the spectral lines by the aid of the SCHILT photometer and the measurements proved a good correlation with LINDBLAD's classes. Finally one diaphragm was used in the measurements, which corresponded to a spectral region of some 20 A in width. The measurements also gave the total or integral value of the spectral region considered. ÖHMAN

[1] Photometric Studies of Effects of Luminosity and Colour in Short Stellar Spectra. Upsala Medd No. 33 (1927).

finds that the measured quantity corresponds to the harmonic mean intensity of the spectral region. It is of great interest that the measured contrast of the line to the continuous spectrum is independent of the density of the image for the straight line portion of the characteristic curve. For faint densities the measurements correspond to the mean intensity logarithm of the spectral region.

The spectral lines $H\gamma$, $H\delta$, and K were measured as colour equivalents and compared with the difference in magnitude between the wave lengths λ 3912 and λ 4415. The following mean errors have been found for $H\gamma$ and $H\delta$: $\pm 0^m,056$ (49 stars) and $\pm 0^m,058$ (48 stars) respectively. The correlation between the magnitude difference derived from the ratio of intensity λ 4415/λ 3912 and the strength of the K line is not very good, which indicates that the intensity of the K line may be a vague criterion of colour for A stars. The correlation is good between λ 4415/λ 3912 and HERTZSPRUNG's colour indices of Pleiades stars. The classes of LINDBLAD, based on the sharpness of the hydrogen lines and on the colour of the stars, are very satisfactorily correlated with the measured intensities of the $H\gamma$ line. Analogous correlations are found for the $H\delta$ line. If the Upsala classes are plotted as a function of the strength of $H\gamma$ or $H\delta$ and the colour equivalent λ 4415/λ 3912, a certain similarity is found between the resulting diagrams and the luminosity diagram given by CHING-SUNG YÜ[1]. The colour equivalent λ 4415/λ 3912 would thus correspond to the temperature in his diagram. YÜ has, however, used a higher dispersion and only measured the central parts of the lines, which probably explains some differences between the diagrams. When the strengths of $H\gamma$ and $H\delta$ are compared an influence of temperature and luminosity is revealed in the sense that for absolutely bright stars $H\delta$ is stronger than $H\gamma$, but for absolutely faint stars $H\gamma$ is about as strong as $H\delta$. This fact may probably be used as a criterion of absolute magnitude for A stars showing faint hydrogen lines.

154. Stonyhurst Work. Rev. H. MACKLIN[2] has started work at Stonyhurst with regard to the determination of spectrographic parallaxes. On account of the comparative smallness of the dispersion (14,26 mm or 10,71 mm from $H\beta$ to $H\varepsilon$) the close pairs of lines used at Mount Wilson could not be employed. The following lines were used instead: λ 4227 Ca, λ 4396 Fe, λ 4215 Sr^+, λ 4290 Ti^+, λ 4353 V^+, λ 4387 Ti, λ 4444 Ti^+, of which the two first decrease in intensity with increase of absolute magnitude and the other five increase under the same conditions. The line-intensities were measured by the aid of a photographic wedge fitted to a HILGER measuring machine.

Results are given for 30 stars all previously determined at Mount Wilson and at Norman Lockyer Observatory. The agreement between the three sources is very good indeed.

155. Influence of Spectroscopic Binaries. A source of error of a certain importance is the high frequency of spectroscopic binaries in any photometric material of brighter stars. If the magnitudes are not properly corrected, the spectrographic parallaxes will be affected systematically. D. L. EDWARDS[3] has investigated this question and pointed out that two sources of error are introduced if the correction is ignored. The consideration of a number of visual binaries gives a mean value of $m_A - m_B = 0{,}9$ and the correction required to reduce total magnitude to the magnitude of the primary is $0^m,4$. Thus:

$$m_A = m_{AB} + 0{,}4.$$

[1] Lick Bull 12, p. 135 (1926). [2] M N 85, p. 444 (1925).

[3] M N 88, p. 695 (1928). The errors have also been pointed out and described during my Upsala lectures.

The corresponding correction applied by WOODS is $0^M,5$, and the mean value of the correction used at Mount Wilson is $0^M,37$, whereas YOUNG and HARPER have used $0^M,54$.

The first error introduced is, of course, that spectroscopic binaries are entered with a wrong m. The second error arises in the formation of the standard reduction curves connecting M with the criteria used. Suppose that there are n spectroscopic binaries of N stars used for the derivation of the curves. M is then correct for $N - n$ stars and 0,4 too bright for n stars. The adopted curve will then be displaced 0,4 n/N along the M-axis (provided that the frequencies of spectroscopic binaries are independent of M). The true value of M is then $M + 0{,}4\,\frac{n}{N}$. This correction applies to all parallaxes which have to be corrected with the factor $1{,}2^{\frac{n}{N}}$. In the case of spectroscopic binaries both sources of error act and give the correction factor $1{,}2^{\frac{n}{N}-1}$.

If no correction from m_A to m_{AB} has been used for the derivation of the reductional curves the π_{tr} are rendered directly comparable with the π_{sp} by determining $k = \overline{\pi_{tr}/\pi_{sp}}$. The quantity $k\pi_{sp}$ will then differ from π_{tr} only on account of accidental errors and of the spectroscopic binary errors. If the spectroscopic binaries are considered, we find the relation:

$$\overline{\pi_{tr}/k\pi_{sp}} = 1{,}2^{\frac{n}{N}-1},$$

and if the single stars are considered:

$$\overline{\pi_{tr}/k\pi_{sp}} = 1{,}2^{\frac{n}{N}}.$$

The existence of several unknown binaries among the supposed single stars will introduce certain (small) changes in the figures derived below:

Source	N	k	$1{,}2^{\frac{n}{N}-1}$	$1{,}2^{\frac{n}{N}}$
Mt Wilson 1646 stars	648	1,05	0,85	1,02
A stars	184	1,03	0,89	1,02
B stars	81	1,22	—	—
YOUNG and HARPER	353	0,97	0,85	1,02
RIMMER	364	1,04	0,87	1,04
Miss DOUGLAS	74	1,01	0,92	1,10
Abetti	93	1,03	0,90	1,06
WOODS (uncorrected stars)	131	0,96	0,88	1,06

156. Necessity of a Standard System. There are at present some 7000 stars that have been investigated for absolute magnitude or spectrographic parallax. Owing to a number of circumstances the magnitudes are not comparable without suitable reduction. Thus there are considerable differences between the different series with respect to the dispersion of the spectrograms (Mount Wilson, Upsala) or the method employed (Mount Wilson, Upsala). Further the spectral criteria employed are sometimes quite different. The methods of reduction have also varied considerably. There are a number of other questions to be considered. No criticism of any of the valuable results concerning the absolute magnitudes is implied in the proposal that a discussion of the different series will prove to be a desirable and very remunerative task. Such a reduction is, in fact, started at the Observatory of Lund. A card catalogue has been written out containing the available material and it is intended to use all new direct evidence concerning trigonometric parallaxes and parallaxes derived by the aid of indirect methods, and indirect evidence based on proper motions and radial velocities.

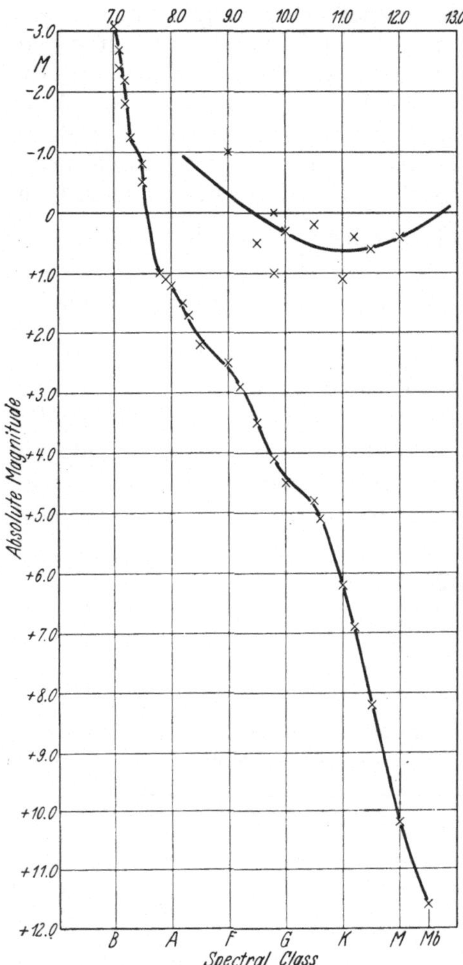

Fig. 123. Mean values of absolute magnitude for different spectral classes as derived from some 1500 trigonometric parallaxes (the material available in 1923).

157. Absolute Magnitude from Spectral Class and Proper Motion. R. E. WILSON[1] has introduced a slight modification in the method used by LUNDMARK and LUYTEN for computing the absolute magnitude of stars from their spectrum and proper motion. The combination of the mean absolute magnitude with the KAPTEYN formula modified by STRÖMBERG: $\log\pi = \alpha + \beta\overline{m} + \gamma\log(\overline{\mu} + k)$, gives:

$$\overline{M} = a + b\overline{m} + c\log(\overline{\mu} + k)$$

where a, b, c, and k are constants. For b the expression tang ψ was introduced. The parallaxes of more than 1300 stars of classes Oa—F0 for which determinations existed, were computed by means of a formula of the above type and a comparison was made with different sets of parallaxes. The following numerical values of the above formula were adopted:

Spectral class	Mean magnitude formula
O	$\overline{M} = 9{,}0 + 0{,}88m + 1{,}65\log(\mu + 0'',004)$
B0	$= 5{,}3 + 0{,}60m +$,,
B1	$= 5{,}0 + 0{,}55m +$,,
B2	$= 4{,}6 + 0{,}50m +$,,
B3	$= 4{,}0 + 0{,}45m +$,,
B5	$= 3{,}0 + 0{,}35m +$,,
B8	$= 2{,}7 + 0{,}31m +$,,
B9	$= 2{,}3 + 0{,}30m +$,,
A0	$= 2{,}0 + 0{,}28m +$,,
A2	$= 1{,}7 + 0{,}26m +$,,
A3	$= 1{,}1 + 0{,}25m +$,,
A5	$= 0{,}9 + 0{,}23m +$,,
F0	$= 0{,}4 + 0{,}20m +$,,

By the aid of the formulae for computing the difference between n and s stars, according to the Mount Wilson classification, a mean of $+0^{M},18$ is found, whereas the Mount Wilson value is $+0^{M},37$. The mean difference is thus only half the value expected. Still, it is very interesting to note the same systematic trend proving the correlation between the n and s division and H.

As long as no extensive discussion exists in which all the present material is used for derivation of M from spectral class and proper motion, it seems to me that the curves derived in 1922—23 and given in figures 124—126 can be used for deriving a preliminary value of M or π when no individual determination is available.

[1] A J 36, p. 49 (1925).

158. The Distance Method of OTTO STRUVE[1]. The intensities of the detached Ca^+ line, *K*, in stars of earlier classes than B5 were found by O. STRUVE[2] to be strongly correlated with the distances of the stars. The method based on that discovery is of special interest, because it gives a means of determining the distance itself. The method does not involve any assumption concerning the

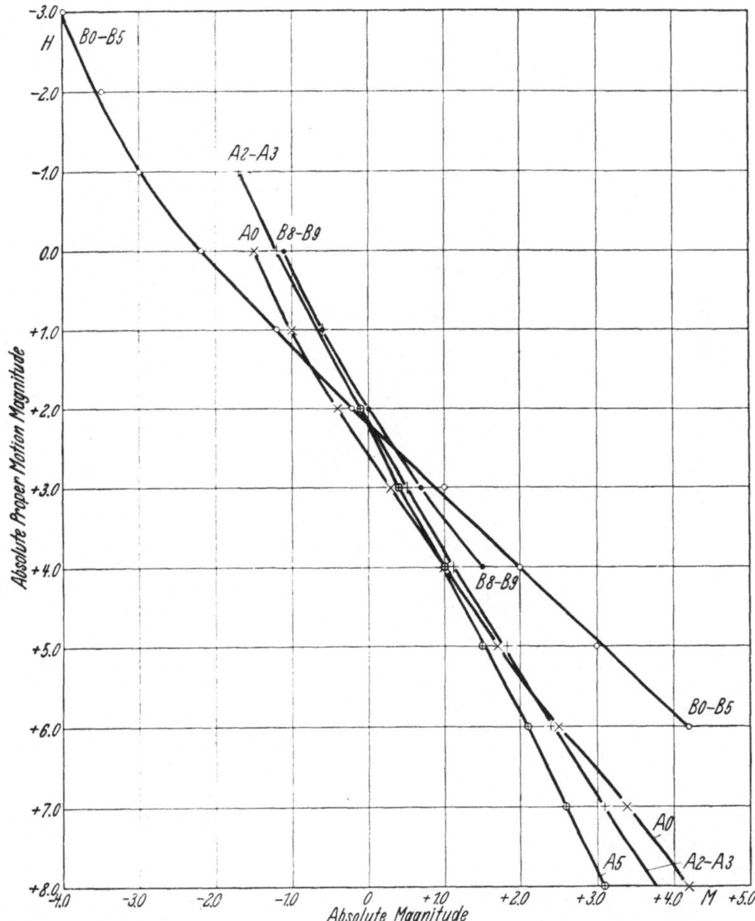

Fig. 124. Relation between $H = m + 5 + 5 \log\mu$ (ordinates) and M (abscissae) for spectral classes B0—A5. Although the material available at present might modify the curves slightly, still this and the two following figures can be used for approximating absolute magnitudes or the parallaxes when only proper motions and spectral classes are known.

masses of the stars. The only objection to the method may be that we do not know, as yet, if the density of Ca^+ in space varies uniformly with distance. But when our knowledge of the spatial distribution of O—B5 stars has advanced, the method of STRUVE may prove to be of much value for a study of the properties of the Ca^+ "Universe". Besides, on account of its foundation on the distances the method will be of much assistance in the determination of systematic errors in parallaxes or absolute magnitudes, etc.

The intensities of the detached *K* line were expressed in an arbitrary scale, defined in such a way that the central intensity of this line in Vega is 10. A

[1] M N 89, p. 567 (1929). [2] Ap J 67, p. 353 (1929).

crude calibration has shown that one unit of STRUVE's scale corresponds to a step of $0^{m},1$ in the differences in intensity between continuous spectrum and the

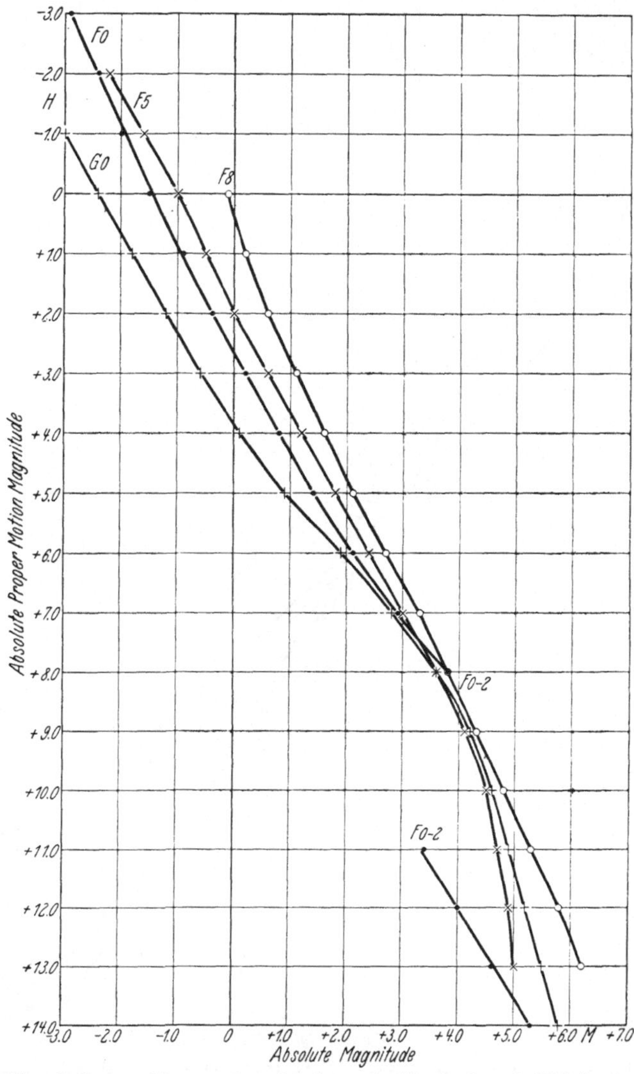

Fig. 125. Relation between $H = m + 5 + 5 \log \mu$ (ordinates) and M (abscissae) for spectral classes F0—G0. The discontinuity in the F0—F2 curve is an interesting feature of the diagram.

centre of the line. The curve connecting intensity i of the K line and distance D was derived from the following data:

Group of stars	$\bar{i}$ of the K line	D in parsecs
Stars of class B5	0,0	100
EDDINGTON's moving cluster in Perseus	1,7	150
Orion nebula group	2,1	180
Cluster in Cygnus	3,9	1100
Clusters in Perseus	4,0	2300
Oe5 stars	4,2	1450
Oe stars	4,8	1730
Three galactic novae	5,5	1000
P Cygni stars.	7,5	6500

It was also found that i increases linearly with the apparent magnitude. The above data are satisfactorily represented by the equation:

$$i = 0{,}35\,m + 0{,}82.$$

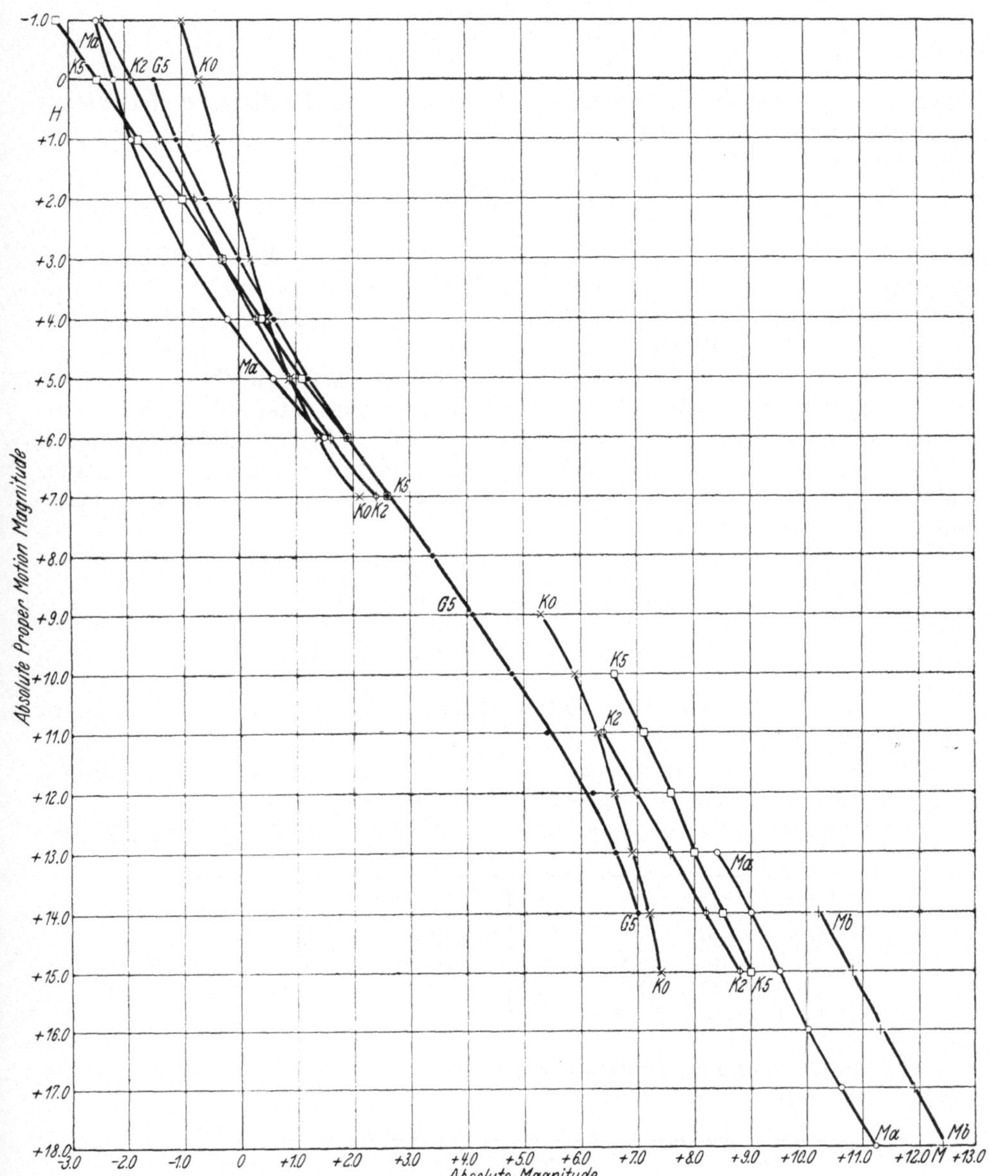

Fig. 126. Relation between $H = m + 5 + 5\log\mu$ (ordinates) and M (abscissae) for spectral classes K0—Mb. The discontinuity of all curves except the G5 one is an interesting feature of this diagram and another confirmation of the giant and dwarf theory.

It has been shown by R. E. WILSON[1], C. V. L. CHARLIER[2], B. P. GERASIMOVIČ[3], and O. STRUVE that M is itself a function of m. From GERASIMOVIČ's work the following formula was adopted:

$$M = 0{,}48\,m - 4{,}20.$$

Thus:

$$i = 3{,}36\log D - 5{,}38.$$

[1] A J 36, p. 49 (1925). [2] Lund Medd 2, No. 34 (1926). [3] V J S 61, p. 219 (1926).

From this equation new independent values of i and D can be derived. Besides, a number of additional points were obtained from the Ca^+ residual radial velocities. These may be explained by J. H. OORT's galactic rotation:

$$V = K + DA \sin 2(l - l_0),$$

where K is the red shift assumed to be constant, D is the distance of the stars considered, and A is equal to:

$$\frac{V}{4R}\left(1 - \frac{R}{K_1}\frac{\partial K_1}{\partial R}\right),$$

where R is the distance of the Sun from the centre, V the circular velocity near the Sun, and $K_1 = V^2/R$. Finally l_0 is the galactic longitude of the centre of the stellar system. For the Ca^+ lines $K = 0$. All radial velocities together give $DA = +5{,}3 \pm 1{,}0$ km/sec.

GREAVES, DAVIDSON, and MARTIN[1] and also GERASIMOVIČ[2] have confirmed the conclusion of O. STRUVE[3] that stars with strong detached Ca^+ lines are usually more yellowish than stars with faint lines. E. HERTZSPRUNG[4] has noted that the yellow B stars are strongly concentrated towards the galactic plane, while no such concentration is observed in the white stars. The following intensities were found:

Group of stars	$\bar{i}$
14 yellow stars	$4{,}5 \pm 0{,}4$
4 white stars	$2{,}1 \pm 0{,}3$

The determination of the relative distances of the two groups presents considerable difficulties.

The derived curve can also be shown to be consistent with the established fact that the Ca^+ intensity is greatest in the earliest classes.

A correlation between proper motion and line intensity is also present:

Proper motion	$\bar{i}$	n	Proper motion	$\bar{i}$	n
0″,000—0″,009	3,2	53	0″,040—0″,049	2,2	11
0 ,010—0 ,019	3,1	43	0 ,050—0 ,079	1,3	3
0 ,020—0 ,029	3,0	22	≧0″,080	0,5	2
0 ,030—0 ,039	1,9	18			

The larger trigonometric parallaxes are associated with fainter lines than the smaller parallaxes:

$\bar{\pi}$	$\bar{i}$	n	$\bar{\pi}$	$\bar{i}$	n
−0″,011	3,1	14	+0″,011	2,7	15

Finally it has been shown that the components of physical double stars have identical Ca^+ intensities. Consequently the condition:

$$\Delta M = \Delta m,$$

[1] M N 89, p. 125 (1928). [2] Harv Bull 864 (1929).
[3] A N 227, p. 377 (1926); Obs 52, p. 52 (1929).
[4] B A N 1, p. 204 (1923).

is fulfilled. The calibrations of the scales of line intensity and of line width give the results:

Intensities expressed			Line width.	
in astronomical units	in magnitude	in percentage of absorption	in steps	in AE.
1	$0^m,1$	9	1	1,2 A
2	0 ,2	17	2	1,4
3	0 ,3	24	3	1,7
4	0 ,4	31	4	1,9
5	0 ,5	37	5	2,1
6	0 ,6	42	6	2,3
7	0 ,7	47	7	2,5
8	0 ,8	52	8	2,8
9	0 ,9	56	9	3,0
10	1 ,0	60	10	3,2

The distances of 92 stars of spectral classes O—B3 have been derived. 9α Camelopardalis $= -10^M,5$ (B0) has the highest absolute magnitude and f^1 Cygni $= +1^M,1$ (B0) the lowest. The most distant of the investigated stars is χ^2 Orionis (9200 parsecs) and the nearest α Virginis (30 parsecs).

The Mount Wilson criteria for the determination of M of the B stars allow for a certain amount of dispersion within each spectral sub-division by the classification according to the s and n characteristics. STRUVE has shown that only a part of the observed dispersion in line width is due to differences in M. It is known that axial rotation must and does have an appreciable effect upon line width. A grouping according to the line width will not tell us anything concerning the relation of the corresponding mean absolute magnitudes to each other without making certain simplifying assumptions.

The comparison between STRUVE's results and those obtained at the Mount Wilson and Norman Lockyer Observatories showed that there was a certain correlation between the absolute magnitudes. The mean values of STRUVE are, as a rule, higher than the earlier determinations. STRUVE has further tested the correlation between his absolute magnitudes and those of ADAMS and JOY and of EDWARDS by comparing the values of his M with the estimated width of line He λ 4472. The differences in M between stars of classes n and s are somewhat larger than those of ADAMS and JOY, but of the same order of magnitude as the values obtained by EDWARDS. STRUVE concludes that line width is rather strongly correlated with absolute magnitude, but obviously there is no one-to-one relationship, since line width also depends upon rotation and probably also upon other factors not related to absolute magnitude.

O. STRUVE has also investigated the theoretical line contour of the detached lines of Ca^+ on the basis of the classical electron theory. The computed theoretical curve does not agree so well with the empirical one as might have been expected. The disagreement is probably to be attributed to the uncertainty of the calibration of intensities. This calibration is, of course, of no importance whatever so long as only the empirical curve is considered.

159. Distribution of Absolute Magnitude and Spectral Class. The so-called HERTZSPRUNG-RUSSELL diagrams are based on the numbers of stars observed within different spectral classes. They are not of much value as a foundation for cosmological theories and deductions owing to the fact that they represent a selection of the stars. If we include all stars down to apparent magnitude $5^m,0$, the maximum distance is about 1000 parsecs. The number is 1500, but the total number of stars within a sphere having a radius of 1000 parsecs is, according to KAPTEYN's and VAN RHIJN's investigations, $30 \cdot 10^6$. It is clear that even if we possessed knowledge of the accurate absolute magnitudes of all stars brighter than $5^m,0$, which is far from being the case, we could not base much on knowledge of merely a 1/20000 part of the material.

It is necessary to know the distribution in spectral classes and absolute magnitudes of all stars within a given space.

We do not know completely the stars situated even within 10 parsecs. The author has found the number to be some 150 and it can be computed that some 80 stars are missing. The RUSSELL diagram for these stars shows some features of interest. Firstly there does not seem to be any actual separation between K giants and K dwarfs, secondly the "white dwarfs" must be comparatively numerous, and thirdly the B stars are scarce. The last-mentioned fact is also confirmed from a statistical investigation of the number of B stars in the DRAPER Catalogue. It is found that $N(m)$ reaches a maximum at $10^m,0$ according to the formula:

$$\log N(m) = a + bm + cm^2.$$

This is certainly not an effect of incomplete knowledge of B stars between magnitudes 8^m and 9^m. Even if we exclude all stars fainter than $6^m,75$, the same conclusion is derived:

$$\log N(m) = -0{,}358 + 0{,}742\,\mathrm{m} - 0{,}0362\,\mathrm{m}^2.$$

Thus there should scarcely be any B stars fainther than 17^m—18^m, and the total number can be estimated to be between 6000 and 8000.

The existence of the bridge joining giants and dwarfs of class K was made probable one of the first times from an investigation by W. J. LUYTEN[1] of the distribution of M for stars nearer than 25,3 parsecs.

There are two ways that enable us to deduce the absolute RUSSELL diagram from the apparent one.

R. HESS[2] has derived the frequency surfaces $\psi(M, S)$ of the spectral class S and M. The material used are the spectrographic parallaxes in the Mount Wilson catalogues, the Harvard catalogues, and the Norman Lockyer catalogues. The vitiation of the distribution of stars occasioned by the selection in the material is neutralised by the weights or frequencies being calculated according to the formula:

$$p = \frac{N(m, \mu, s)}{n(m, \mu, s)}$$

where $N(m, \mu, s)$ means the true number of stars having magnitudes, total proper motions, and spectral classes within the intervals dm, $d\mu$, and ds, and $n(m, \mu, s)$ are the corresponding numbers within the same intervals in the material used. The inverse value of p thus expresses the probability of a certain star being included in a certain catalogue.

The frequency surface has been derived in such a way, and the projections of the level-lines are shown in fig. 127.

Another way has been followed by K. G. MALMQUIST[3]. He has proved that if the apparent $\psi(M, S)$, including stars to a certain limiting magnitude m_0, follows a normal error curve, this holds good, too, for the distribution in space of all stars belonging to the same class, provided the value of m_0 is not too high. For stars brighter than $6^m,0$ he finds that the F and M stars have a normal distribution of $\psi(M, S)$, whereas the G and K stars can be considered to be

[1] Ap J 62, p. 8 (1925). [2] SEELIGER-Festschrift p. 265 (1924).
[3] Lund Medd, Ser II, No. 32 (1924).

composed of two normal distributions. As no M dwarfs are of $6^m,0$, their mean M and dispersion can be calculated in various ways. Among stars $<9^m,5$ and north

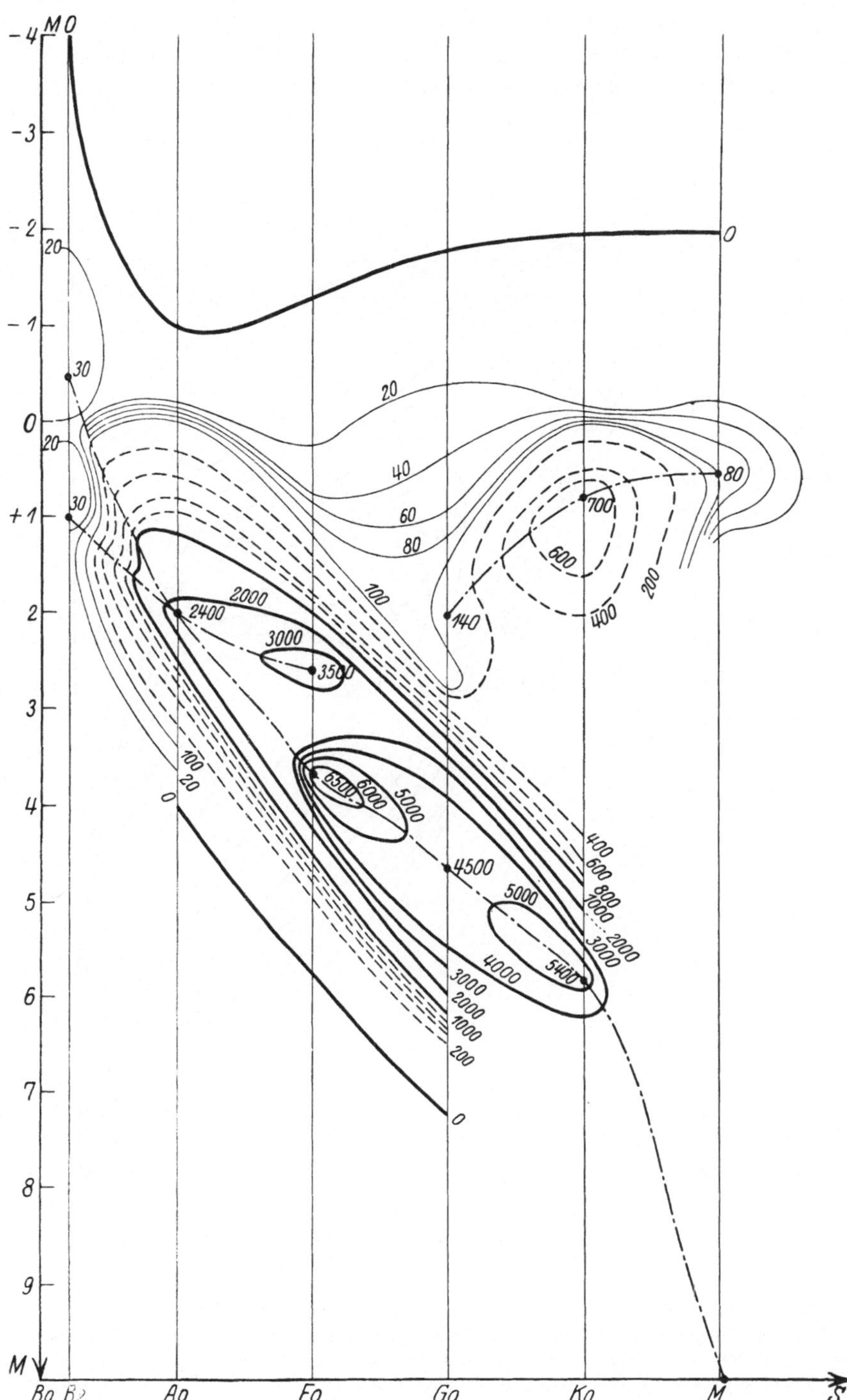

Fig. 127. The HESS diagram showing the correlation surface $\psi(M, S)$ between spectral class S and absolute magnitude M. [From R. HESS, SEELIGER-Festschrift p. 272 (1924).]

of $-30°$ Declination 18 M dwarfs are known. Computing the real distribution of the stars in space MALMQUIST finds M and S for the various groups.

C. LÖNNQUIST[1] has made use of MALMQUIST's results and computed the frequency surface $\psi(M, S)$ for stars within 100 parsecs. The results given in fig. 128 are not, without reduction, comparable with those of HESS, because the spectral division employed is somewhat different, and because HESS does not go

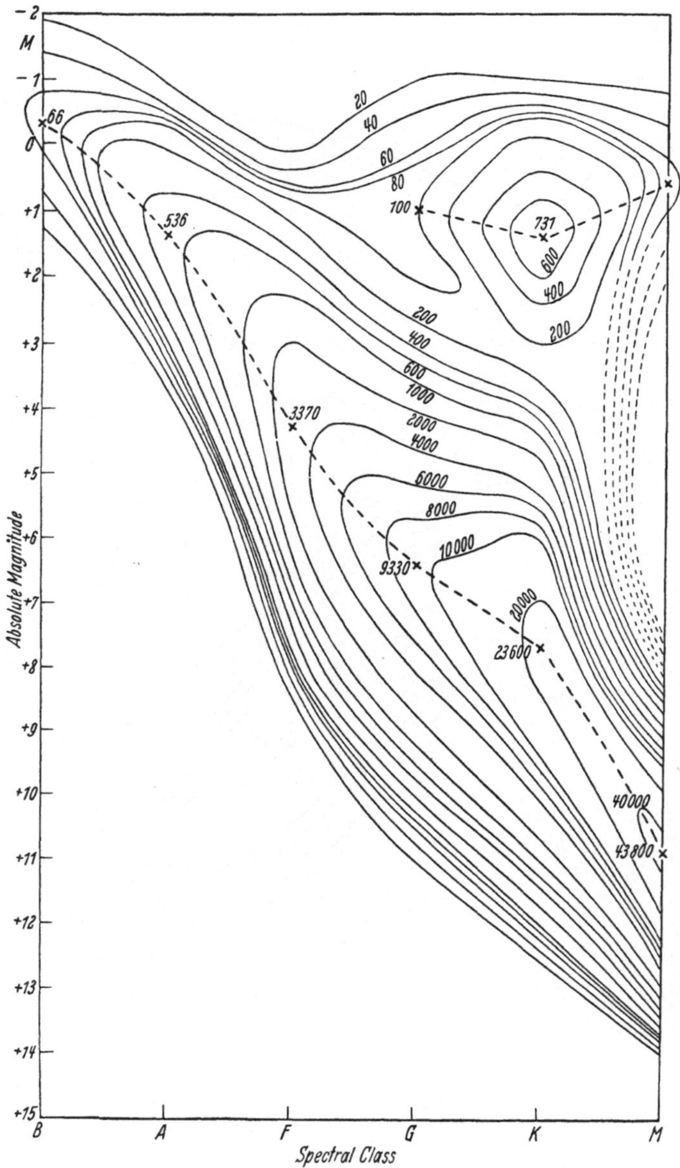

Fig. 128. HESS diagram as constructed by C. LÖNNQUIST (Upsala Medd No. 25, 1927). A detailed comparison between this diagram and the original one by HESS is very interesting as showing, in spite of the different material employed, a good agreement on the whole but also several interesting local deviations, especially in the dwarf mountain range.

[1] On the Evolution of the Stars with Mass Reduction. Ark f Mat, Astr, Fys 20 A, No. 21 (1927). Also Thesis Upsala = Upsala Medd No. 25.

farther down than to $6^m,5$. This restriction has probably influenced the K dwarf maximum, which only reaches 5400 as compared with 23600 obtained by LÖNNQUIST.

In both figures the stars of the giant group lie like a limited hillock besides the enormous mountain ridge of the main series. The limit is not clearly pronounced. The intermediate territory near the K line has, according to HESS, a level of between 200—400 on an average with a hill whose highest point is 700. LÖNNQUIST finds nearly the same height for the hillock (731), but the pass separating it from the mountain-ridge has a height of only 160.

The division into giant M and dwarf M is very pronounced, and whereas the giant hill for the M class only reaches 65, the ridge of the dwarf mountain for the M class reaches the immense height of 43800.

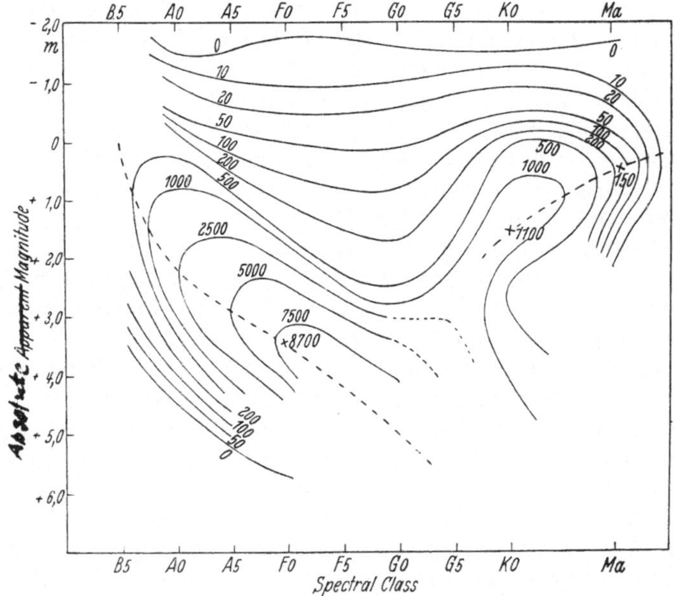

Fig. 129. HESS diagram constructed by C. LÖNNQUIST [Upsala Medd No. 25 (1927)] on basis of stars in globular clusters. There are certain deviations from the curves in fig. 128 which might be real.

Several astronomers have urged that the giant and dwarf division is not real, but only caused by the selection. The foremost holder of this view is P. J. VAN RHIJN[1], who, using mainly the same method as HESS, has investigated the distribution of M stars and comes to the conclusion that there is no valley between the M giants and M dwarfs. LÖNNQUIST is of opinion that VAN RHIJN's figures and illustration show the contrary. At that point of the diagram where the frequency is lowest its logarithm must be at least 0,8 less than the corresponding value for the most frequent giant magnitude. In the depression between the giants and dwarfs the frequency is thus at the most 1/6 of the maximum frequency of the giants.

When we remember that our knowledge of the M dwarfs and the distribution in space of the stars is imperfect, it seems very possible that future data will elevate the dwarf mountain much more and thus practically level the cleft

[1] Seeliger-Festschr p. 247 (1923).

between giants and dwarfs. But this division will certainly be for a long time a catalogue phenomenon occurring in our sources, even if it is not very pronounced in space.

The evidence for the distribution of M and S in star clusters has been investigated by TEN BRUGGENCATE[1] and C. PARVULESCO[2]. The globular clusters show quite a different luminosity diagram. The former author has applied HESS's method to the globular cluster M3 and to the open clusters M11, M37, and N G C 1647. The evidence is not accurate enough to permit any conclusions concerning the possible division into giants and dwarfs.

It may be of interest to note that the open clusters seem to be very lacking in dwarf stars, and that faint dwarfs do not occur at all in many of them. In such clusters as the Pleiades and M34 there are no stars absolutelyf ainter thanthe Sun.

160. The General Distribution of Absolute Magnitudes. The distribution of M is a very important and interesting chapter in stellar astronomy. The most natural method is, of course, to group the M obtained from a number of representative determinations of parallaxes and to investigate the general form of the frequencies $\varphi(M)$. Provided we have a sufficient number of parallaxes it will be possible to determine $\varphi(M)$ for different values of the spectral class, or of the colour indices, or for different values of the apparent magnitude.

Serious troubles arise when the method is applied, because it is necessary to have a typical selection of parallaxes from a unit of space. Even with the aid of the indirect methods it is very difficult to survey a volume of space completely or in such a way that the results concerning M give the right form of $\varphi(M)$. It cannot be said that we know the distances of all the stars nearer than 10 parsecs and even within 5 parsecs our knowledge is imperfect.

By the aid of the proper motions an extension of our knowledge of π and thus of $\varphi(M)$ can be obtained, but also in this case unavoidable difficulties appear.

Another of the chief methods is to compute the $\varphi(M)$ from other data. It is well known that there exist two integral equations connecting $\varphi(M)$, π_m, $N(m)$, and $D(r)$, where $N(m)$ is the number of stars of magnitude m, π_m the mean parallax for a certain magnitude, and $D(r)$ the density of stars within the space-unit. A knowledge of $N(m)$ and π_m gives a means of computing $\varphi(M)$ and $D(r)$. The integral equations in question are of a complicated nature, and it is necessary to have access to accurate series of values of $N(m)$ and π_m. It has only recently been possible to obtain accurate knowledge of the number of stars within given limits of magnitude on account of the lack of an accurate photometric scale. The mean parallax formula cannot be established without extensive and perfect knowledge of stellar parallaxes. Thus very high accurateness cannot be expected either when the latter method is applied.

The first who tried to establish a luminosity law seems to have been COMSTOCK[3], who derived the form:

$$\varphi(i) = \Gamma \frac{H}{i} \log\left(\frac{H}{i}\right)$$

where H is the upper limit for the luminosity corresponding to $M = -0{,}6$, i the luminosity and Γ a constant. This expression was later[4] on changed into

$$\varphi(i) = \Gamma \frac{H}{i} \log\left(\mu + \frac{H}{i}\right),$$

where μ is a new parameter, subject to the limitation $0 < \mu < +1$.

[1] Z f Phys 24, p. 48 (1924); Seeliger-Festschr p. 50 (1914).
[2] Bull Astr 5, p. 5 (1925). Also Thesis Paris.
[3] A J 24, p. 139 (1904); 25, p. 169 (1907).
[4] A N 185, p. 297 (1910).

In SEELIGER's work much attention was given to the form of the luminosity law. He deduced in his memoir of 1911[1] the general form:

$$\varphi(i) = \Gamma\left\{\left(\frac{H}{i}\right)^{\nu} e^{-k^2\left[\left(\log\frac{i}{H}\right)^2 + b\log\frac{i}{H} + c\right]} - \frac{i}{H}\, e^{-k^2 c}\right\}.$$

The constants of the formula have been chosen in such a way that $\varphi(H) = 0$. The following numerical representation was found:

$$\varphi(i) = e^{5,261\,x - 0,345\,x^2}\{1 + 2,959x - 0,691\,x^2\}$$

where $x = -\log\frac{i}{H}$.

This formula, being founded on the mean parallaxes of KAPTEYN and the values of $N(m)$ derived in Groningen Publ No. 18, cannot be considered very accurate to-day. A second law was derived by SEELIGER of the form:

$$\varphi(i) = e^{-5,384\,x - 0,230\,x^2} - e^{1/0,434}\,x$$

which agrees very well with the former expression.

KAPTEYN's important and extensive work concerning the formation of the luminosity curve has its origin in a paper[2] of A N 3487/88, where he derived, from the parallactic motions of the stars in AUWERS-BRADLEY, π_m for stars between $2^m,0$ and $9^m,0$. In Groningen Publ No. 8 the work was extended, the trigonometric parallaxes of 58 stars (mainly measured by FLINT) being used in combination with the proper motions in AUWERS-BRADLEY. The constants for the π_m were determined separately for the two main types of stars (VOGEL I and II) and for all stars together. KAPTEYN introduced in this paper the assumption that the quantities $z = \log\pi - \log\pi_0$, where π is the real and π_0 the probable parallax, are distributed according to a Gaussian frequency curve, and that the "probable error of this curve is $0,19\,\pi_0$".

The derivation of the luminosity law was made in Groningen Publ No. 11. From SEELIGER's numbers for $N(m)$ the numbers $N(m, \mu)$ were derived by counting the number of proper motions between intervals of m and μ in proper motion catalogues. The space was divided into a number of shells with radii r selected in such a way that $\log r$ increases by 0,4 for each shell. The mean parallax was determined for each of the numbers $N(m, \mu)$ according to the formula for $\pi_{m\mu}$. By the aid of the value of the probable deviation of the error curve, $\log\pi - \log\pi_0$, calculations were made that determined which of the numbers found occur in each shell.

KAPTEYN has also varied the values of the constants of $\pi_{m\mu}$ considerably in order to test to what degree the luminosity curve changed. It appeared that the different solutions gave results that varied but little.

In A J 24, p. 115 (1904) KAPTEYN has derived an analytical expression for the luminosity curve, and has discussed in a critical way the views of COMSTOCK concerning the absorption of light in space.

The first results of KAPTEYN were based on SEELIGER's counts of faint stars (the B D material). When KAPTEYN had derived the number of stars for given limits of magnitude (Groningen Publ No. 18) he undertook, later on, a new derivation of the luminosity law. He started from the analytical formulae for $\varphi(i)$ and $D(r)$ previously given. These were substituted in the integral formula for $N(m)$ and the numbers of stars calculated and compared with the observed

[1] München, K Bayer Akad Sitzber 41, p. 413 (1911).
[2] A N 146, p. 96 (1898).

numbers. The constants for $D(r)$ were changed, but the luminosity law remained unchanged.

For an empirical derivation of the luminosity curve the star clusters can be used as well as the anagalactic systems dissolved into separate stars. In both cases we have collections of stars at such a distance that $N(m)$ directly gives the form of $\varphi(M)$, provided that the background and foreground stars have been eliminated.

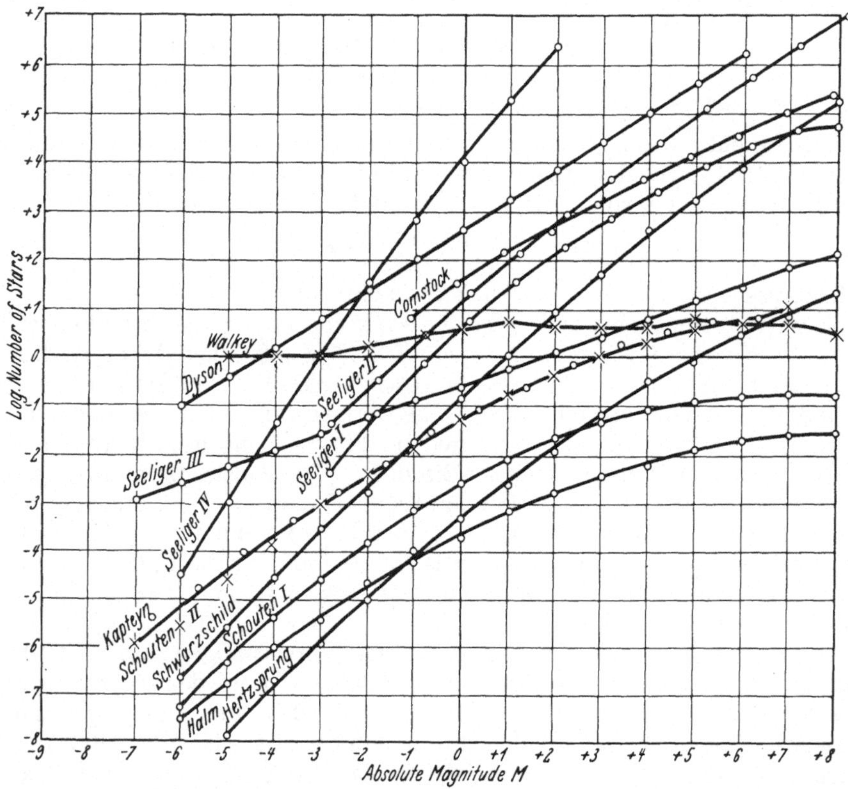

Fig. 130. Different determinations of the luminosity curve as collected by SCHOUTEN in his dissertation (Groningen 1918). The considerable differences between the course of the curves illustrates the difficulties to establish an accurate luminosity curve.

With regard to the clusters the principal objection is that we do not know that the luminosity has the same general distribution within them as within the stellar system. R. TRÜMPLER[1] finds that this is not the case, and warns us against employing cluster data or deriving parallaxes from cluster objects on the assumption that the luminosity law in clusters has the same general form as in the Milky Way system. VON ZEIPEL[2] finds that for M3 the luminosity law is of another form than KAPTEYN's. It has been pointed out by H. v. ZEIPEL that the probability that in a cluster as many stars are within the interval $M + dM$ is not the same. Analyzing the form of $N(m)$, or the function giving the number of the stars brighter than m, he finds the following expression for the luminosity law:

$$\varphi(M) = e^{\sqrt{a+km}} + C$$

where a, k and C are constants.

[1] Allegh Publ 6, p. 45 (1929).

[2] Arkiv f Mat, Astr Fys Bd. 11, No. 22 (1916).

This law, if revised with more modern data from magnitudes of cluster stars, could be useful for deriving the relative parallaxes of clusters.

As regards the open clusters it seems that different luminosity laws hold good in different cases, but that if the investigation is restricted to a comparatively narrow spectral interval (e. g. the A stars) it can be assumed that the stars within

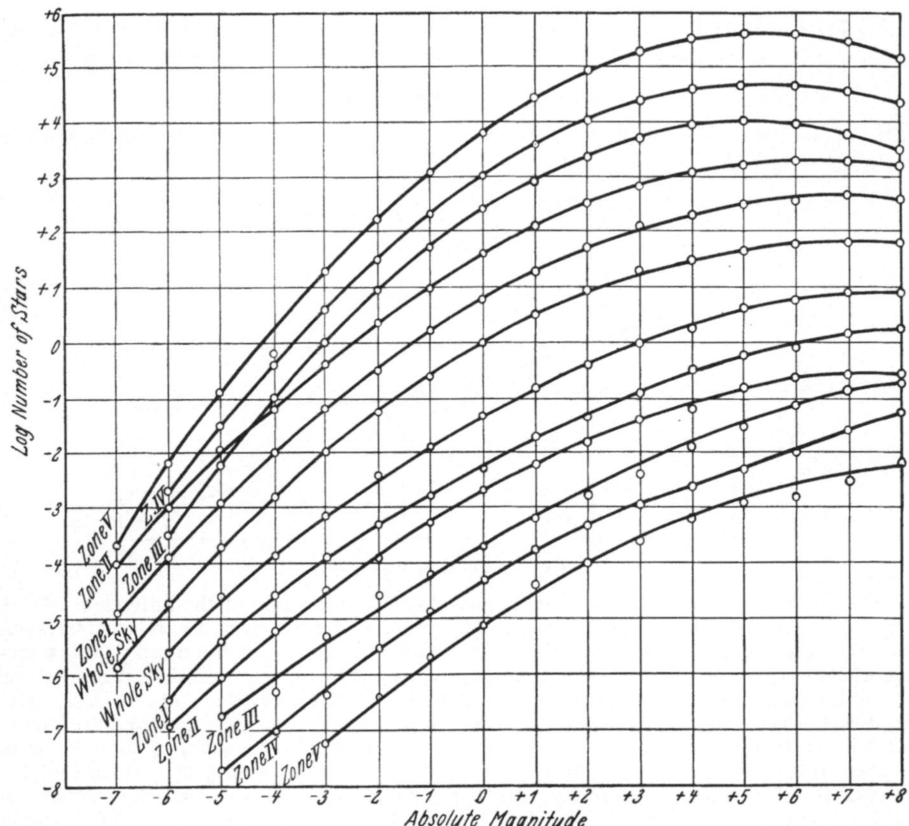

Fig. 131. The luminosity curve for different zones of galactic latitude according to SCHOUTEN's investigations. The zones are defined as follows:

Zone I	is the part of the sky between gal. lat. $b \pm 10°$		
II	,, ,, ,, ,, ,, ,, with b from	$-10°$ to $-30°$ and	$+10°$ to $+30°$
III	,, ,, ,, ,, ,, ,,	$-30°$,, $-50°$	$+30°$,, $+50°$
IV	,, ,, ,, ,, ,, ,,	$-50°$,, $-70°$	$+50°$, $+70°$
V	,, ,, ,, ,, ,, ,,	$-70°$,, $-90°$	$+70°$,, $+90°$

The six curves of the upper part of the figure relate to the determination based on the method of SCHWARZSCHILD, and the six curves of the lower part to the determination based upon the method of KAPTEYN. It appears that the luminosity curve does not vary sensibly with the galactic latitude. The investigations of SCHOUTEN also illustrate the difficulties present when determining the maximum point of the general luminosity curve.

such an interval have a normal distribution with regard to their absolute magnitude. H. SHAPLEY[1], S. RAAB[2], and P. COLLINDER[3] have used this assumption and derived parallaxes for a number of open clusters.

---

[1] Mt Wilson Comm No. 62 (1919). [2] Lund Medd Ser. II, No. 28 (1922).
[3] Lund Ann No. 2 (1931).

In a series of papers F. H. SEARES[1] has given valuable contributions to the luminosity law. A combination of the luminosity and density functions of KAPTEYN and VAN RHIJN leads to the apparent luminosity function according to the expression:

$$\psi(M, m) = 4\pi\,\varrho^2 d\varrho\, \Delta(\varrho)\, \Phi(M)$$

which gives the numerical values:

$$\log\psi_1(M, m) = \log\Phi(M) - 0{,}584 + 0{,}6\,(m - M)\,, \quad (m - M = -\infty \text{ to } +8)$$

$$\log\psi_2(M, m) = \log\Phi(M) - 2{,}840 + 1{,}076\,(m - M) - 0{,}0262\,(m - M)^2$$

$$(m - M = +8 \text{ to } +\infty)\,.$$

$\psi(M, m)$ is the total number of stars in the sky with magnitudes between the limits $M \pm \frac{1}{2}\,dM$ and $m \pm \frac{1}{2}\,dm$.

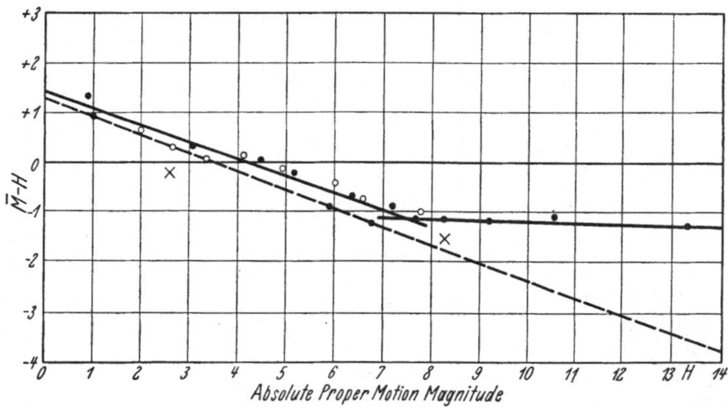

Fig. 132. The absolute proper motion magnitude H as a function of the difference $\overline{M}-H$ according to SEARES. The relation given in the graph is identical with the relation between mean absolute magnitude and H. The dotted line corresponds to the mean parallax formula of KAPTEYN and VAN RHIJN and is in agreement with the theoretical relation derived from a luminosity curve and a distribution curve of the logarithms of the tangential velocities, both assumed to be Gaussian curves. The circles represent 181 stars from the Mount Wilson list of 1646 stars with proper motions between the limits 0″,3 and 0″,5 and the points 325 stars from LUYTEN's list of stars with proper motions exceeding 0″,5 [Lick Bull 11, p. 1 (1923)]. The wide departure for intrinsically faint stars indicates that the frequencies of the absolute magnitudes of these stars are not distributed in accordance with a simple Gaussian curve.

For comparison with the formula stars of $m > 5{,}0$ were considered. A constant density and a normal error curve for $\varphi(M)$ can be assumed, which reduces $\Phi(M)$ to another error curve with the same dispersion as $\Phi(M)$, but with a maximum for which M is increased by the amount $0{,}3/(\mathrm{Mod}\,h)^2$. The agreement between the theoretical curve and the observations is "superficially bad" according to SEARES, but can be accounted for by the assumption of uncertainties in the theory. LUYTEN[2] points out that the frequency of $H = m + 5 + 5\log\mu$ for stars brighter than $5^{\mathrm{m}}{,}0$ accords fairly well with a normal error curve with $H = 2{,}8$ and $\sigma_H = \pm 2^{\mathrm{H}}{,}5$. If we take $M = -2 + 0{,}65\,H$ it is found that $\Phi(\dot{M})$ is an error curve with $M_0 = 0$ and $\sigma_M = \pm 2^{\mathrm{M}}{,}2$. The agreement between the curve and the theoretical one is as good as can be expected. LUYTEN also points out that the value of M for A and K giants used by SEARES is comparatively low. The theoretical excess of very bright stars, which is a conspicuous feature in SEARES's diagram, is probably real.

[1] Mt Wilson Contr No: 271; Ap J 59, p. 11 (1924).
[2] Harv Circ 262 (1924); Ap J 62, p. 8 (1925).

SEARES has found in the distribution of all absolute magnitudes very little trace of the giant and dwarf sub-divisions.

The parallactic motions are used in the derivation of the luminosity law on the assumption that there is no dependence between the amount of the Sun's motion and the mean absolute magnitude of the group considered. STRÖMBERG's results show the existence of such a relation and this calls for a revision of the mean parallax formula, and hence of the luminosity law and the density law.

SEARES finds that in the part of the sky covered by observations most of the stars of low luminosity that are apparently brighter than $10^{m},0$ have been observed. The maximum M_0 or at least the dispersion in M of the luminosity function is too small. According to SEARES, LUYTEN's conclusion that M_0 must be increased is invalid owing to selection in the data discussed.

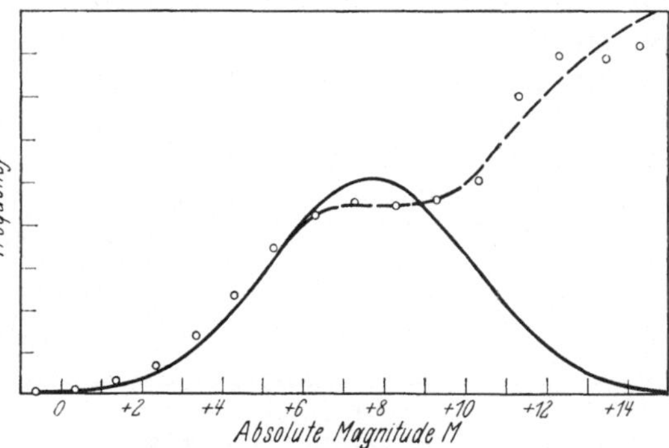

Fig. 133. The luminosity curve according to SEARES (circles and dotted line). The full-drawn Gaussian curve corresponds to the luminosity curve derived by KAPTEYN and VAN RHIJN. The material used for the construction of the dotted line is the same as that in fig. 132. The sudden transition from one linear relation for $\overline{M}$ to the other in this figure is partly responsible, at least, for the peculiar distribution of frequencies at $M = +8$.

The second paper[1] of SEARES deals with the distribution function for stellar velocity and serves as a foundation for the third paper[2], which deals with the form of the luminosity function. The fact that few stars of low luminosity are known introduces an uncertainty in the mean parallaxes of stars. If we make use of the quantity H the relation:

$$\overline{M} = +1,55 + 0,645 H - 5(\log\overline{\pi} - \log\underline{\pi})$$

is easily derived.

A theoretical formula for mean parallax can be derived from the distribution functions for density, tangential velocity, and luminosity. The first two of these are known with little uncertainty. When the error curve of KAPTEYN and VAN RHIJN is adopted for the distribution of M, the theoretical formula for the mean parallax differs from the empirical relation only by the presence of a small term in m, which for stars near the Sun is too small to be detected. If the luminosity function is an error curve, the mean parallaxes of the stars of low luminosity should be represented by a linear formula of H with coefficients agreeing with those of the formula of KAPTEYN and VAN RHIJN.

The results for $H < 12,5$ and $H > 13,0$ are entirely different. SEARES finds:

$$\log\underline{\pi} - \log\mu = -0,72 - 0,066 H,$$
$$\log\underline{\pi} - \log\mu = -1,17 - 0,006 H,$$

respectively.

Thus the agreement is good in the first case, but in the second there does not seem to be any dependence between H and $\log\overline{\pi} - \log\mu$. The lines intersect at $H = 12^{m},5$, which corresponds to $M = +6^{M},4$. SEARES concludes that the luminosity function of KAPTEYN and VAN RHIJN is applicable as long as we are dealing with stars whose magnitudes fall on the ascending branch of the curve. If the absolute magnitudes of the fainter stars are distributed normally the constants must differ from those of the function of KAPTEYN and VAN RHIJN.

Thus the luminosity function *does not seem to be a symmetrical curve*.

[1] Mt Wilson Contr No. 272; Ap J 59, p. 274 (1924).

[2] Mt Wilson Contr No. 273; Ap J 59, p. 310 (1924).

This is also the principal result of an investigation of E. ÖPIK[1]. The frequency of proper motions of stars was derived from the Johannesburg and Helsingfors Blinkmicroscope observations on the basis of the method of coefficients of perception. A table giving the distribution of m and μ is derived for stars $6{,}44 > m > 13$ and μ between $1'',945$ and $0'',040$. This table is then used for determining the luminosity and density functions. The mean parallax formula is derived on the basis of an investigation of GORAKH PRASAD[2] which gives the relation between M and H for some 2200 stars of different spectral classes. The curve can be derived as far as to the point $M = 10{,}5$ and $H = 16$ but beyond this point there is an extrapolation on account of the lack of data for absolutely very faint stars.

The luminosity and density laws were computed from the data, the probable error in $\log \pi$ as determined from m being assumed to be $\pm 0{,}16$. The luminosity curve shows a maximum at $6^{M},5$ and a steady increase from 12^{M}, which suggests a combination of two normal curves. A comparison with KAPTEYN's, VAN RHIJN's[3] and SEARES's[4] results is given:

Number per 10^6 parsecs of unit spatial density.

M	ÖPIK	KAPTEYN & VAN RHIJN	SEARES
$+1^{M},5$	920	380	550
+2 ,5	2350	910	1270
+3 ,5	5070	1830	2410
+4 ,5	7140	3200	4050
+5 ,5	8090	4790	5700
+6 ,5	8290	6100	6970
+7 ,5	7620	6860	7400
+8 ,5	6990	6430	7200
+9 ,5	6900	5290	7700
+10 ,5	7690	3960	8800
+11 ,5	9980	2410	11700
+12 ,5	14000	1510	12900

161. The Luminosity Curve from Differences in Magnitude of Double Stars. An interesting and important method suggested and used by E. ÖPIK[5] should be mentioned in this connection. Let M_A and M_B be the absolute magnitudes of the components of a double star and let $\psi(M_A)$, $\varphi(M_B)$ and $\chi(M_B, \Delta m)$ be the frequencies of M_A, M_B and $M_B - M_A = \Delta M = \Delta m$ respectively; then we have the equation:

$$\varphi(M_B) = \int_a^{\infty} \psi(M_A)\,\chi(M_B, \Delta m) \cdot d(\Delta m).$$

The lower limit a can be taken as zero when M_B is the magnitude of the secondary so that $M_B > M_A$. If the cases $M_B > M_A$ and $M_B < M_A$ are likely to occur with equal chance as is the case in the third component of certain pairs then $-\infty$ can be taken for a.

As to the form of $\chi(M_B, \Delta m)$ there are two extreme cases to be considered. The one is that $\partial\chi/\partial M_B = 0$, or that the frequency function of the difference ΔM only depends on this difference. This case involves a strong correlation between M_A and M_B. In such a case the distribution of M_A cannot be derived without knowledge of the distribution of M_B or vice versa. The observations furnish the form of the function χ which is not influenced by the selection in

[1] Tartu (Dorpat) Publ 26, No. 4 (1925).
[2] M N 85, p. 157 (1924).
[3] Mt Wilson Contr 188 (1920).
[4] Mt Wilson Contr 273 (1924).
[5] Publ Obs Tartu (Dorpat) 25, No. 5 and 6 (1923, 1924).

the magnitudes M_A. The second case is that $\partial\chi/\partial(\Delta m) = 0$, which means that the observed distribution of Δm depends only on M_B. The components are then entirely independent of one another and the equation becomes:

$$\varphi(M_B) = \int_{-\infty}^{\infty} \psi(M_B - \Delta m)\,\chi(M_B)\,d(M_B - M_A)$$

or:

$$\varphi(M_B) = \chi(M_B) \int_{-\infty}^{+\infty} \psi(M_A)\,dM_A = c\,\chi(M_B).$$

Thus the true distribution of the absolute magnitudes of the secondaries will be identical with the observed distribution, multiplied with a certain magnifying or diminishing factor c.

Also in this case, of course, there will be no influence due to a possible selection in M_A.

At first 645 close pairs were discussed, and their m's were adopted according to BURNHAM's General Catalogue. The scale being based on direct estimates is, of course, uncertain. As accurate measurements were available only for a small number (Harvard), it was not possible to reduce the m's to the Harvard scale. The data are strongly influenced by the selection in m and Δm because of the difficulties met with in discovering faint components of close pairs and distant pairs (low values of m_{AB}). The observed data were corrected for the influence of the selection according to a method described by ÖPIK.

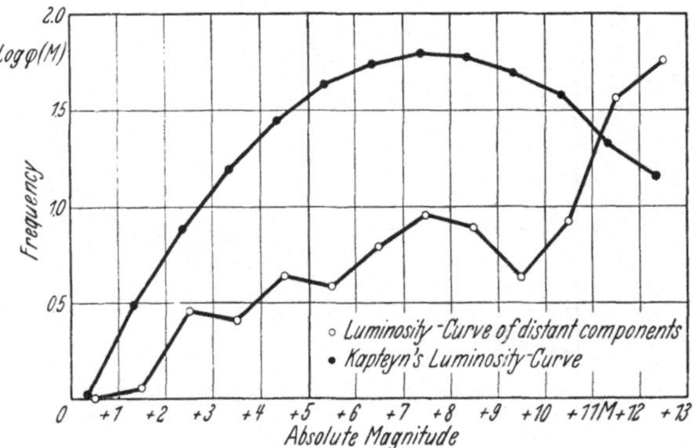

Fig. 134. Luminosity curve as derived from distant companions in binary systems (ÖPIK) and compared with KAPTEYN's luminosity curve.

It was found that the form of $\varphi(M_B)$ is practically the same for the different spectral classes except B0—B9. This indicates that the first of the above-mentioned cases is satisfied. The curve derived for all spectra has no resemblance to a (Gaussian) normal distribution. It seems that components of close doubles cannot be regarded as representative of single stars. The probability of a component of a close pair having a certain luminosity is not independent of this luminosity, but depends on the luminosity of the other component. This has some bearing on the problem of deriving the luminosity curve of stars in our immediate neighbourhood on account of the high percentage of double and multiple stars.

In his second paper ÖPIK calibrated the magnitude scale of the different observers of double stars with the aid of the photometric measurements at Harvard. His tables 4, 5, and 6 furnish a means for converting estimates of m made by different observers into photometric magnitudes. The tables are based on pairs where the primary is brighter than 6^m, but for a first approximation they may be used for fainter pairs also. The final correction of m could be represented by the equation:

$$\Delta m' = k(\varrho)\,\Delta m + \Delta m_0$$

where $k(\varrho)$ is a coefficient depending on the distance ϱ and Δm_0 a residual correction, which is generally small.

The selection of the list has been discussed and numerical values for the completeness of the data have been determined.

From 1246 companions considered to be physical the distribution of the parallaxes π and relative magnitudes m have been derived, the different spectral classes being treated separately. The projected density of the companions per unit of the projected area is found to vary inversely as the square of the projected (absolute) distance D. This means that the spatial density of components of double stars should vary inversely as the cube of the distance. Such a distribution will lead to an infinite total number of companions. The law will be changed on

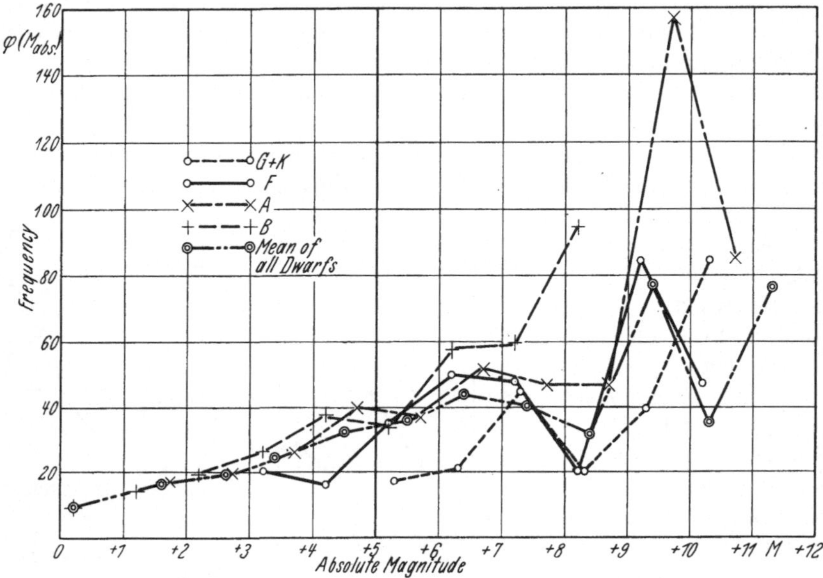

Fig. 135. The luminosity curve as derived by E. ÖPIK, using distant companions in binary systems.

account of the existence of an upper limit of distance where external forces put a boundary to the sphere of action of a single star. There seem to exist some (small) real deviations of the projected density from the inverse square law. That these deviations are of a similar character for all spectral classes and absolute magnitudes, super-giants excluded, implies the existence of a general law governing the origin of companions. The distribution of the distances is such that similar maxima correspond to greater distances for stars of greater mass. This is quite contrary to what should be expected if the spectral sequence presented an evolutionary course in the direction earlier to later with a corresponding mass reduction. This fact will be made clearer from the following small table:

Spectral class	Displacement in $\log D$	D/D_0	Mass	Theoretical value of D/D_0
dK	−0,33	0,5	0,70	8
dG	−0,60	0,25	1,0	5
dF	−0,09	0,8	2,5	2,2
dA	−0,06	0,9	6	0,9
dB	+0,03	1,1	9	0,6
gG	+0,39	2,5		
gK	+0,21	1,6		
gM	+0,60	4,0		

Instead of changing according to the law $D \sim 1/\mathfrak{M}$ the ratios D/D_0, i. e. the proportions in which the distances D_0 of the mean curve must be changed in order to make this curve as similar as possible to the curve for a certain class, appear to change in the same direction as the masses, although not so rapidly. The ratios D/D_0 are best represented by $c\mathfrak{M}^{\frac{1}{3}}$, but if that formula was extended to the giant branch, the resulting mean mass would be 130 ⊙ (G stars), 35 ⊙ (K stars), 500 ⊙ (M stars), which seems very high, if not altogether impossible.

The distribution of Δm of close companions having D smaller than 220 astronomical units or of Δm of distant companions with D larger than 220 astronomical units shows a remarkable similarity for stars of the dwarf series, and hence a similarity of origin may be suspected. ÖPIK says that the close companions are likely to have been originated through fission, whereas the distant companions may represent apparently independent nuclei in the "nebula" from which the system may have developed. The capture theory is highly improbable in the case of wide systems and seems to be impossible in the case of close binaries.

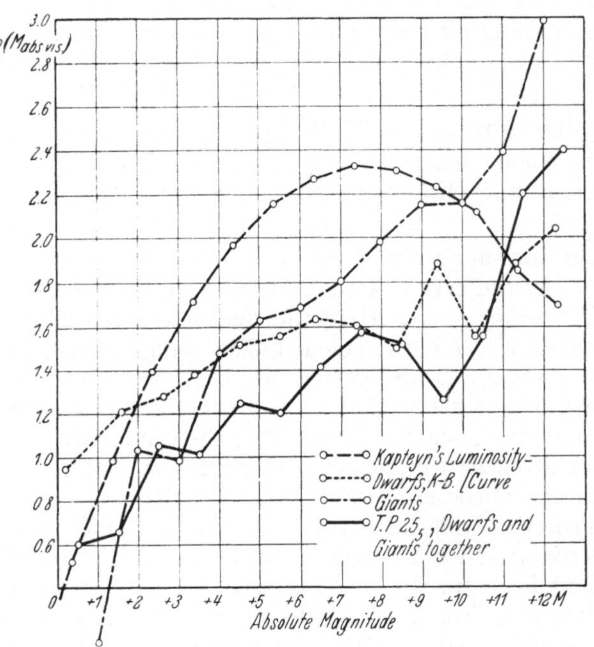

Fig. 136. Luminosity curve as derived on basis of distant companions in binary systems (giants and dwarfs) and compared with the luminosity curve of KAPTEYN.

ÖPIK thinks that a stationary state of the stars (no evolution) is not altogether impossible, but leans towards a weak faith in stellar evolution. The evolutionary change of a star brighter than $M=6$ cannot exceed 1^{m} since its origin, if gravitational contraction, or radio-active cooling are considered as the main cause of evolution. It seems that almost the same variety in the absolute magnitudes and spectra as is observed at present has existed since the origin of the stars. This variety must very probably be chiefly attributed to the initial conditions, such as differences in mass. A slight increase of age with advancing spectral class is suggested for the dwarf sequence by the relative displacement of the luminosity curves of close companions.

The best agreement with observed facts is found on the basis of a hypothesis of periodical evolution with a radio-active law of cooling, and on the assumption that catastrophes (Novae) occur independently for the companions of a binary system.

The luminosity curve of giant companions differs considerably from the luminosity curves of dwarfs.

The second paper of ÖPIK also contains a catalogue of some 1850 double stars which gives the magnitudes of the components, the spectral class, and other useful information. This catalogue is the best source we have at present as regards the magnitudes of the components of double stars.

162. Determination of Absolute Magnitudes of Faint Stars. We have already mentioned some of the results showing that there is a dependence upon absolute magnitude in the colour of stars. The first who pointed out that the spectral distribution or the colour is related to intrinsic brightness seems to have been W. S. ADAMS[1] in 1914. Through the subsequent work of SEARES, HERTZSPRUNG, LINDBLAD, BALANOVSKY, and others it is clear that there is a dependence between the colour or some of its equivalents and the absolute magnitude. It is certainly possible to determine M, at least roughly, when the colour and spectrum are known. Such determinations do not possess much interest, because, as soon as the spectrum can be observed, M can also be determined from a knowledge of the relation:

$$C = a + bS + cM + dM^2.$$

I. BALANOVSKY and V. HASE[2] have found the following general relation between the colour index in the system of the Göttinger Aktinometrie and the absolute magnitude (Mount Wilson system):

$$C = 1{,}11 - 0{,}147\,M + 0{,}010\,M^2.$$

The formula is valid for G and K stars.

In the case of faint stars it is of much importance to obtain even a rough knowledge of M. Here the question arises whether it is possible to use two values of C, i. e. two colour indices, and, through a combination of the equations, to eliminate S and derive M. LINDBLAD[3] has introduced the minimum wave length as a new colour equivalent and has derived the absolute magnitude in a number of cases. Work along the same lines by SEARES and BOTTLINGER has not given positive results. It seems that not every colour equivalent is adapted for the purpose. It is to be hoped that new work will be started both on the theoretical dependence of C, M, and S and with regard to the possibility of deriving M for faint stars. In the case of dwarfs of late classes it will be sufficient for a number of purposes to possess a criterion which tells us if a star is a giant or a dwarf. A knowledge of that fact gives a fair approximation of the parallax as soon as the object is further away than 50 light-years.

163. Distribution of the Spectrographic Parallaxes with Regard to m and Spectral Class. B. ŠČIGOLEV[4] has studied the correlation between $-\log\pi$ and m of 2716 stars, for which the spectrographic parallaxes have been reduced to the Mount Wilson system.

For giant stars the correlation coefficient ranges between $+0{,}331 \pm 0{,}116$ (G0—G4 stars) and $+0{,}840 \pm 0{,}030$ (K5—K9 stars) and for dwarfs between $+0{,}410 \pm 0{,}294$ (Ma stars) and $+0{,}896 \pm 0{,}016$ (F5—F9 stars). We quote some of the values of r together with the corresponding regression-lines:

Spectral class	Coefficient of correlation r	Regression-lines
B2	$+0{,}826 \pm 0{,}023$	$-\log\pi = 1{,}1720 + 0{,}1858m$
A1	$+0{,}768 \pm 0{,}019$	$-\log\pi = 0{,}9184 + 0{,}1718m$
gK1	$+0{,}772 \pm 0{,}017$	$-\log\pi = 0{,}9276 + 0{,}1830m$
gK2	$+0{,}840 \pm 0{,}030$	$-\log\pi = 0{,}9836 + 0{,}1935m$
gMa	$+0{,}782 \pm 0{,}039$	$-\log\pi = 1{,}1758 + 0{,}1700m$
gMb	$+0{,}834 \pm 0{,}050$	$-\log\pi = 1{,}2767 + 0{,}1627m$
gK	$+0{,}772 \pm 0{,}016$	$-\log\pi = 0{,}9327 + 0{,}1855m$
gM	$+0{,}781 \pm 0{,}033$	$-\log\pi = 1{,}2018 + 0{,}1682m$
dF1	$+0{,}815 \pm 0{,}093$	$-\log\pi = 0{,}5080 + 0{,}1682m$
dF2	$+0{,}896 \pm 0{,}016$	$-\log\pi = 0{,}3800 + 0{,}1798m$
dG1	$+0{,}795 \pm 0{,}035$	$-\log\pi = 0{,}3178 + 0{,}1655m$

[1] Mt Wilson Contr. 78; Ap J 39, p. 89 (1914). [2] Bull Inst Astr Leningrad No. 18 (1927).
[3] Upsala Universitets Årsskrift (1920), also Thesis Upsala.
[4] R A J 5, p. 236 (1929).

In these cases a knowledge of the spectral class is sufficient to give a rather accurate value of the parallax with the aid of the regression-lines.

164. Relation between Absolute Magnitude and Radial Velocity. As was discovered by ADAMS and STRÖMBERG[1] there is a linear relation between the absolute magnitude and radial velocity. In the mean it has been found that:

$$\frac{\Delta V}{\Delta M} = 1{,}5 \text{ km/sec.}$$

The following table constructed by WIRTZ[2] shows the general relation between V and M for different spectral classes:

M	F	G	K	M
− 2	11,9 km/sec	12,9 km/sec	13,9 km/sec	16,9 km/sec
0	13,4	15,6	16,2	19,2
+ 2	15,0	18,8	19,0	21,8
+ 4	17,0	22,7	22,1	
+ 6		27,5	25,8	
+ 8			30,2	32,3
+10				36,8

The present writer, and probably also several others, have tried to use the relation between V and M in order to compute the absolute magnitude from given values of the radial velocities, but so far without success. The correlation is so low that the results will be very uncertain, and they can by no means compete with the results obtained by the aid of the proper motions.

The best results as regards hypothetical parallaxes and hypothetical absolute magnitudes will undoubtedly be obtained when the two characteristics of the motions, the proper motion μ and the radial velocity v, are combined. Except for the attempts by HERTZSPRUNG[3] there seem to be no methods developed where v and μ are utilized for a determination of M. In this connection the problem of the correlation between the proper motion and the radial velocity arises. In a joint paper W. A. GYLLENBERG and K. G. MALMQUIST have discussed this question at length both from the theoretical and the practical point of view[4].

165. CHARLIER's Determinations of the Distribution of Helium Stars[5]. If r is the distance of a star, we have:

$$r = R \cdot 10^{0{,}2\,m}.$$

If d is the radius of the star, and T its temperature, we can write

$$R = d f(T).$$

For a narrow interval of spectral class $f(T)$ does not vary considerably, and thus the parameter R, the distance for which the apparent magnitude of the star equals zero, has a rather constant value, or at least only a small dispersion. The absolute magnitude is derived from the formula:

$$M = m + 5 - 5 \log R - 1{,}57.$$

The computation of R is performed in the following way.

Let U, V, W denote the components of the velocity of a star in a system of rectilinear coordinates, that has its axis of z (W) directed along the radius vector, its axis of X in the direction of increasing α, and its axis of Y in the direction of increasing δ.

[1] Mt Wilson Contr No. 131 (1917); Ap J 45, p. 293.

[2] MÜLLER-POUILLET, Lehrbuch der Physik, 11. Aufl., Bd. V, II. Hälfte. Braunschweig 1928.

[3] A N 185, p. 92 (1910).

[4] Lund Medd Ser I, No. 108 (1925).

[5] Lund Medd Ser II, No. 14 (1916).

Let further U'', V'', W'' denote the components of the velocity of the same star referred to the common astronomical equatorial system. Then if the direction-cosines are introduced it is found that:

$$\begin{aligned} U &= -U'' \sin\alpha + V'' \cos\alpha\,, \\ V &= -U'' \sin\delta \cos\alpha - V'' \sin\delta \sin\alpha + W'' \cos\delta\,, \\ W &= U'' \cos\delta \cos\alpha + V'' \cos\delta \sin\alpha + W'' \sin\delta\,. \end{aligned}$$

If $u = \mu_\alpha \cos\delta$, $v = \mu_\delta$ denote the proper motions of a star we have:

$$U = ru; \quad V = rv$$

or introducing the value of the distance r and using the abbreviations:

$$u' = 10^{0,2m}\, u; \; v' = 10^{0,2m}\, v$$

we have:

$$U = Ru'; \; V = Rv'.$$

Applying the method of least squares we are now able to determine from each of the three above equations the mean values of U'', V'', W'', that is the mean velocity of the B stars referred to the Sun. The third equation gives the values of U'', V'', W'' expressed in linear measure. The first two contain as unknown the parameter R. A comparison with the former results gives us the value of R.

If R cannot be considered as a constant, the two first of the above equations give us the mean values of:

$$\frac{1}{R}\,U'', \quad \frac{1}{R}\,V'', \quad \frac{1}{R}\,W''$$

and a comparison with the results of the radial velocity equation gives the mean value of $1/R$ or $\overline{1/R}$.

The solution of the equation system is simplified by the use of determinants.

Considering the group of stars in question (e. g. the B stars) to be at rest and denoting the velocity of the Sun with Ω_0, we have:

$$\begin{aligned} \Omega_0 \cos D \cos A &= -U_0'', \\ \Omega_0 \cos D \sin A &= -V_0'', \\ \Omega_0 \sin D &= -W_0'', \end{aligned}$$

where A and D are the coordinates of the apex.

CHARLIER has pointed out on several occasions the difference between the mean value x of a quantity and the inverse value of the mean value of $1/x$. Especially on account of dispersion there is a great difference between $\bar{\pi}$ and $1/\bar{r}$, where π and r denote the parallax and the distance of a star.

The necessary condition for using the equation:

$$\bar{\pi} = \frac{1}{\bar{r}}$$

is that the frequency distribution of the r-values does not contain values of r that are small, that moreover this frequency distribution is nearly a normal example of type A, and that the dispersion is small.

These conditions are, evidently, fulfilled for the values of R, at least as far as the B stars are concerned. Hence we are entitled to put:

$$\overline{\left(\frac{1}{R}\right)} = \frac{1}{\bar{R}} = \frac{1}{R_0}\,.$$

The mean errors in R are difficult to determine; CHARLIER estimates them to be about 8 or 10 per cent.

It seems that a correlation between R and m exists that implies that the more distant B stars have a fainter luminosity than the nearer ones:

Apparent magnitude	R in light-years
$<4^m,0$	43,73
$<5\ ,0$	36,45
$5^m,0-6\ ,0$	26,27

If all B stars possess the same temperature and the same radius, then the value of R ought to be independent of m. If R varies it is because either the temperature or radius or both are dependent on m. Variability of the radius would imply that the larger stars are more condensed near the centre of the Milky Way, a conclusion that is by no means improbable.

It is necessary to inspect more closely the hypothesis of equal temperature of the B stars. We must consider the sub-classes of spectral class B, that is:

B0, B1, B2, B3, B5.

The values of R and M for these classes have been determined by CHARLIER given in the adjoined table.

The considerable decrease of R with increasing M is no longer exhibited. Using the results from all stars brighter than 6^m we have:

Spectral class	R in light-years	M
B0	20,45	$-0^M,56$
B1	56,36	$-2\ ,76$
B2	56,36	$-2\ ,76$
B3	26,19	$-1\ ,10$
B5	28,56	$-1\ ,29$

The variation of R with m is thus explained by the fact that of the brighter stars 27 per cent belong to the classes B1 and B2, whereas there are only 11 per cent of such stars among the stars between 5^m-6^m.

The sequence of evolution ought to be B1, B2, B3, B5, B0, which also finds some support from the descriptions of the spectra in Harvard Annals.

The first paper of CHARLIER contains investigations concerning the form of the system of the B stars and the coordinates of its centre. A catalogue of the distances and space coordinates of 805 stars finishes the paper.

In accordance with E. C. PICKERING's views on the subject CHARLIER thought that the Harv Ann 56 had exhausted the B stars. The question of the completeness of the B stars in our catalogues and of the size or extent of the system of the B stars has no direct bearing on the problems considered in this chapter.

In a subsequent paper CHARLIER[1] has investigated the Galaxy of B stars composed by the 3674 stars of

	B0			B1			B2			B3			B4		
	$\leqq 4^m,99$	$5^m,00-5^m,99$	$\geqq 5^m,99$	$\leqq 4^m,99$	$5^m,00-5^m,99$	$\geqq 5^m,99$	$\leqq 4^m,99$	$5^m,00-5^m,99$	$\geqq 5^m,99$	$\leqq 4^m,99$	$5^m,00-5^m,99$	$\geqq 5^m,99$	$\leqq 4^m,99$	$5^m,00-5^m,99$	$\geqq 5^m,99$
U''/R	−0,0499	−0,1374	−0,0011	+0,0550	−0,1652	−0,0519	+0,0187	+0,1144	+0,0343	−0,0953	+0,0679		+0,0339	+0,0185	+0,0297
V''/R	+0,2727	+2,3773	+1,6060	+0,4857	+0,8524	+0,5815	+0,6667	+0,0499	+0,4446	+0,9416	+1,1301		+0,9158	+0,9566	+0,9408
W''/R	−0,2273	−0,8511	−0,4972	−0,0091	−0,0790	−0,1761	−0,3043	−0,4971	−0,4164	−0,7735	−0,7141		−0,6424	−0,8596	−0,7506
Ω''/R	+0,3588	+2,529	+1,681	+0,5253	+0,8718	+0,6098	+0,7331	+0,5119	+0,6101	+1,2222	+1,3390		+1,1192	+1,6543	+1,2039
R	95,81	13,56	20,45	65,40	39,44	56,36	46,87	67,16	56,36	28,11	25,66	26,19	30,71	20,75	28,57
M	−3,92	+0,31	−0,56	−3,09	−1,99	−2,76	−2,36	−3,14	−2,76	−1,25	−1,06	−1,10	−1,45	−0,60	−1,29
n	14	23	37	18	8	26	24	16	40	35	102	157	42	62	104

[1] Lund Medd Ser. II, No. 34 (1926).

classes B0—B5 in the New HENRY DRAPER Catalogue. The galactic components of the proper motions are now used.
Thus:

$$u = \mu_l \cos b, \quad v = \mu_b.$$

Then:

$$U = -U'' \sin l + V'' \cos l,$$
$$V = -U'' \sin b \cos l - V'' \sin b \sin l + W'' \cos b,$$
$$W = U'' \cos b \cos l + W'' \cos b \sin l + W'' \sin b,$$

where U'', V'', and W'' are the components of the velocity of a star in the galactic system of coördinates.

Suppose that only stars of the same sub-class are considered and that R as well as M is considered as a constant for all these stars, and take the mean value for all the stars belonging to the same galactic square for which the direction cosines may be considered to have constant values, then:

$$\overline{u'} = -\overline{U''}\,\frac{1}{R} \sin l + \overline{V''}\,\frac{1}{R} \cos l,$$

$$\overline{v'} = -\overline{U''}\,\frac{1}{R} \sin b \cos l - \overline{V''}\,\frac{1}{R} \sin b \sin l + \overline{W''}\,\frac{1}{R} \cos b.$$

The new results for B0 as compared with those found in 1916 make it probable that B0 in Harv Ann (denoted only by B in the earlier Harvard classification) embraces strictly speaking, not only stars of the sub-class B0, but also individual stars of other sub-classes, by which the value of R is considerably reduced. The evolutionary sequence of the B stars seems therefore to be:

B0, B1, B2, B3, B5.

The results of ROBB when he also used the radial velocities as reviewed elsewhere in this chapter are also discussed. As regards the B stars the application of the general method of MALMQUIST will be rather difficult according to CHARLIER. An approximate discussion has led to the adjoined values of R which vary with the apparent magnitudes of the stars.

Spectral class	R		
	$m < 6{,}0$	$6{,}0 \leqq m \leqq 8{,}0$	$m > 8{,}0$
B0	34,5	32,2	29,1
B1	58,2	53,6	49,0
B2	42,1	38,3	35,2
B3	26,8	24,9	23,0
B5	23,0	21,1	19,1

The galactic distribution of the B stars is discussed at length. The catalogue of the individual distances and space coordinates of the 3674 B stars is not published, but is in the possession of the Observatory at Lund.

166. GERASIMOVIČ's Investigations. Practically the same methods and much of the same material have been used as in CHARLIER's investigation, but the proper motion material was more extensive in GERASIMOVIČ's work[1] on account of his inclusion of several sources not used for the derivation of BOSS Preliminary Catalogue. The following means for the absolute magnitudes were found:

Spectral class	m	$\bar{\pi}$	$\overline{M}$	Adopted $\overline{M}$
B0—B2	$4^m{,}5$	0″,0020	$-4^M{,}0$	$-2^M{,}9$
	7 ,0	0 ,0015	−2 ,1	
B3	4 ,5	0′,0049	−2 ,1	−1 ,4
	7 ,0	0 ,0028	−0 ,8	
B5	4 ,5	0 ,0068	−1 ,3	−0 ,7
	7 ,0	0 ,0032	−0 ,5	

[1] V J S 61, p. 219 (1926).

In his moving-cluster work KAPTEYN has found for B0: $\overline{M} = -2^m{,}5$, for B3: $+0^m{,}5$, and for B5: $+1^m{,}6$. GERASIMOVIČ's values are in better agreement with CHARLIER's and it seems reasonable to give greater weight to the method of the latter on account of the doubtfulness of the underlying assumption in KAPTEYN's investigation concerning the systematic motions of the B stars.

167. ROBB's Determination[1]. It has been shown by CHARLIER how the first-order moments of the velocities of the B stars give values for the absolute magnitude M. ROBB started in his memoir from the same assumption as CHARLIER; the following equations are then easily obtained for galactic latitude 0° and 45°:

$$\Sigma (R u_0' - \overline{U}_0)^2 = \Sigma (W_0 - \overline{W}_0)^2,$$
$$\Sigma (R v_0' - \overline{V}_0)^2 = \Sigma (W_0 - \overline{W}_0)^2,$$

where R is the distance at which the star would have an apparent magnitude equal to zero, and U_0, V_0, and W_0 are the linear velocity components in the galactic system (z axis directed towards the northern galactic pole).

If $1/R$ is introduced as unknown, the above equations become:

$$\sum \left(u_0' - \frac{1}{R}\overline{U}_0\right)^2 = \sum (W_0 - \overline{W}_0)^2 \left[\frac{1}{R}\right]^2,$$
$$\sum \left(v_0' - \frac{1}{R}\overline{V}_0\right)^2 = \sum (W_0 - \overline{W}_0)^2 \left[\frac{1}{R}\right]^2.$$

The first equation can be written, if $1/R =$ is put equal to x, and

$$y^2 = \sum \overline{U}_0^2 \left[x - \frac{\sum u_0' \overline{U}_0}{\sum \overline{U}_0^2}\right]^2 + \sum u_0'^2 - \frac{(\sum u_0' \overline{U}_0)^2}{\sum \overline{U}_0^2},$$

$$y^2 = \frac{\sum u_0'^2}{n} - 2x \frac{\sum u_0' U_0}{n} + x^2 \frac{\sum \overline{U}_0^2}{n},$$

$$y^2 = x^2 \frac{\sum (W - \overline{W}_0)^2}{n}.$$

The author solves these equations by plotting the two systems of curves.

The constants are determined from the radial velocities of the B0—B5 stars in the catalogue compiled by GYLLENBERG together with a list from the Yerkes Observatory[2]. Before the latter was used it was examined for its systematic error.

The results are the adjoining:

Spectral class	M from mean velocities	M from average velocities
B0	−1,72	−1,72
B1	−2,83	−2,95
B2	−2,03	−2,28
B3	−1,28	−1,28
B5	−0,84	−0,93

Two solutions were also made on the assumption that $1/R$ is constant for each sub-class. The results were the following.

Spectral subdivision	$\overline{M}$		$\overline{M}$	
B0	$-3^M{,}43$	All latitudes	$-3^M{,}43$	$b = 0°$
B1	−2 ,77		−2 ,91	
B2	−1 ,90		−1 ,87	
B3	−1 ,31		−1 ,37	
B5	−0 ,93		−0 ,97	
B3			−0 ,91	$b = 45°$
B5			−0 ,98	

[1] Lund Medd Ser. II, No. 44 (1926).

[2] Ap J 64, p. 1 (1926).

This order of spectral subdivision is still subject to considerable uncertainty and the B0 stars probably embrace stars of rather varying absolute magnitudes.

168. The Relation between Absolute Magnitude and Proper Motion. The first to attempt to derive the mean parallax from the size of proper motions was H. GYLDÉN[1]. In his first attempt he had only 16 stars of known parallax at his disposal. Later on[2] the discussion of the relation between mean parallax and proper motion was repeated with a more extensive parallax material as basis. From 56 parallaxes he derived the mean parallax $\pi_{m\mu}$ for given magnitude and proper motion. He obtained the formula:

$$\pi_{m\mu} = 0'',204\,(1 + \tfrac{1}{3}\,\mu)\,e^{-0,25\,m - 0,02\,m^2} + 0,070\,\mu\,(1 - e^{-0,04\,m^2\mu})\,.$$

Later on J. C. KAPTEYN[3] deduced his well-known formula:

$$\pi_{m\mu} = a^{m-5,5}\,(b\,\mu)^c\,.$$

H. v. SEELIGER has deduced an analogous relation:

$$\pi_{m\mu} = M\,a^{m-5,5}\,\mu^{(1+5\log a)}\,.$$

These formulae are interpolation-formulae and therefore of limited validity. The same refers to a number of other attemps to find the expression for $\pi_{m\mu}$. G. STRÖMBERG[4] has pointed out that in KAPTEYN's formula $\pi_{m\mu}$ becomes zero for $\mu = 0$, which cannot be correct. He has introduced the modified formula:

$$\bar{\pi} = A\,(\bar{\mu} + c)\,\varepsilon^{\bar{m}}\,,$$

where $\log\varepsilon$ thus is the coefficient of $\bar{m}$. P. J. VAN RHIJN[5] has adopted that view and has derived the formula:

$$\log(\pi - k) = a + b\log\mu + c\log^2\mu + Q\,(\log\mu - 0,60)^2,$$

where a and b are functions of the apparent magnitude, Q of the galactic latitude, and k of the apparent magnitude and galactic latitude, and c a constant.

K. SCHWARZSCHILD[6] was the first to derive the expression for $\pi_{m\mu}$ on the assumption that the distribution of the tangential velocities, not corrected for the Sun's motion, is a normal distribution of the logarithms of the velocities. The mean parallax is then defined by means of:

$$\pi_{m\mu} = \frac{\int\limits_0^\infty d\,r\,r^4\,D(r)\,\varphi\,(i\,r^2)\,\psi\,(\mu\,r)}{\int\limits_0^\infty d\,r\,r^5\,D(r)\,\varphi\,(i\,r^2)\,\psi\,(\mu\,r)}\,.$$

Later on he took into account the influence of a possible correlation between absolute magnitude and velocity.

Of other investigations we only mention that of B. v. HARKANYI[7], which starts from a conclusion of ANDING that if we have the relation:

$$r = f(m,\,\mu)\,,$$

we also have another relation:

$$e = F(M,\,r)\,,$$

---

[1] V J S 12, p. 299 (1877).
[2] A N 136, p. 289 (1894).
[3] Groningen Publ, No. 8 (1902).
[4] Mount Wilson Contr 170 (1919).
[5] Groningen Publ, No. 34 (1923).
[6] A N 190, p. 361 (1912); 198, p. 217 (1914).
[7] A N 223, p. 135 (1924).

where e is the linear velocity and μ the angular proper motion. If T is the tangential velocity expressed in km/sec, the formulae of KAPTEYN:

$$\pi = (0{,}905)^{m-5{,}5} \quad (0{,}116\,\mu)^{1/1{,}11} \quad \text{(B—A stars)},$$
$$\pi = (0{,}905)^{m-5{,}5} \quad (0{,}0262\,\mu)^{1/1{,}54} \quad \text{(G—K stars)},$$
$$\pi = (0{,}905)^{m-5{,}5} \quad (0{,}0387\,\mu)^{1/1{,}405} \quad \text{(All stars)},$$

are transformed by HARKANYI into:

$$\log T = 1{,}105 - 0{,}131 \log \pi + 0{,}0482\, M,$$
$$\log T = 1{,}555 + 0{,}206 \log \pi + 0{,}0668\, M,$$
$$\log T = 1{,}448 + 0{,}100 \log \pi + 0{,}0610\, M.$$

According to v. SEELIGER[1], if the formula of KAPTEYN is written:

$$\pi = a^{m-m_0} (\lambda \mu)^{\alpha},$$

$$1 + 5 \log a = \alpha,$$

and we get:

$$\log T = 1{,}380 + 0{,}022\, M \quad \text{(B—A stars)},$$
$$\log T = 1{,}123 + 0{,}108\, M \quad \text{(G—K stars)},$$
$$\log T = 1{,}237 + 0{,}081\, M \quad \text{(All stars)},$$

whereas ADAMS, STRÖMBERG, and JOY[2] have found for stars of classes F—M:

$$\log T = 1{,}256 + 0{,}0870\, M - 0{,}00303\, M^2.$$

In his lectures in 1917 CHARLIER has given a method of computing from the proper motion of a star the most probable value of its distance, both the size and the direction of the proper motion being taken into consideration. He started from the assumption of a generalized Maxwellian distribution of the velocities of the stars.

This method was further developed by B. FÄNGE[3] on the supposition that the velocity ellipsoid is an ellipsoid of revolution. [Later on MALMQUIST[4] has made a new deduction of the relation between absolute magnitude and proper motion by assuming that the surfaces of equal frequencies of the velocities are ellipsoids of revolution (cf. p. 517).]

If we denote by U', V', W' the velocity components in a system of coordinates coinciding with the axes of the velocity ellipsoid, the velocity distribution may be written:

$$F(U', V', W') = C_0\, e^{-\frac{1}{2}\left[\frac{(U'-U_0)^2}{\sigma_1^2} + \frac{(V'-V_0)^2}{\sigma_2^2} + \frac{(W'-W_0)^2}{\sigma_3^2}\right]}.$$

Further:

$$\sigma_1 = \sigma_2$$

if the velocity ellipsoid is an ellipsoid of revolution.

The sky is divided into 48 equal galactic squares as introduced by CHARLIER. In a system of coordinates that has its z axis directed along the radius vector to the centre of gravity of a square, its x axis in the direction of increasing galactic longitude, and its y axis in the direction of increasing galactic latitude, the velocity ellipsoid takes the form:

$$f = A(U - U_0)^2 + B(V - V_0)^2 + C(W - W_0)^2 + 2D(V - V_0)(W - W_0)$$
$$+ 2E(W - W_0)(U - U_0) + 2F(U - U_0)(V - V_0).$$

[1] München Sitzber d Bayer Akad Math Phys Kl, 1920, p. 126.
[2] Mt Wilson Contr 210 (1921). [3] Lund Medd Ser. II, No. 25 (1921).
[4] Lund Medd Ser. I, No 110 (1925).

The constants A, B, C, D, E, F are functions of the three dispersions $\sigma_1, \sigma_2, \sigma_3$, and of the positions of the vertex and the galactic square. If only the components of velocity perpendicular to the line of sight are considered, we have to integrate the function:

$$F = C_0 e^{-\frac{1}{2}f}.$$

We get:

$$\int_{-\infty}^{+\infty} F\,dW = C_1 e^{-\frac{1}{2}f_1},$$

where f_1 has the form $f_1 = A_1 (U - U_0)^2 + B_1 (V - V_0)^2 + 2 F_1 (U - U_0)(V - V_0)$ If polar coordinates are introduced:

$$U = s\cos\theta$$
$$V = s\sin\theta$$

the frequency function $C_1\,dU\,dV e^{-\frac{1}{2}f_1}$ takes the form:

$$C_1 s\,ds\,d\theta\, e^{-\frac{1}{2}as^2 + bs - \frac{1}{2}c}.$$

The axes of the velocity ellipsoid, A', B', C', are assumed to be:

$$A' = B' = \frac{1}{\alpha^2 \lambda^2}; \quad C' = \frac{1}{\lambda^2}.$$

We may then write:

$$s e^{-\frac{1}{2}as^2 + bs - \frac{1}{2}c} = \frac{2s}{\sqrt{2\pi_e}} e^{-\frac{1}{2}(ks - l)^2} = \varphi(s)$$

where:

$$k = \frac{1}{\lambda K}, \qquad l = \frac{\omega L}{\lambda}.$$

The velocity of the Sun is ω, and K and L are two quantities only dependent on the constant α, the angle θ, and the position of the star with regard to the apex. K and L may be tabulated for different values of θ if the position of the vertex is given.

This has been done by FÄNGE, who has given for each square the values of K and L for every tenth degree according to θ.

Using the proper motion components in galactic longitude and latitude:

$$u = \mu_l \cos B,$$
$$v = \mu_b,$$

and putting:

$$\mu = (u^2 + v^2)^{\frac{1}{2}},$$

expressed in radians per 10^6 years, we have:

$$s = r\mu.$$

Further:

$$r = 10^{0,2(m-M)}.$$

Thus:

$$M = m + 5\log\mu - 5\log s.$$

The most probable value of M for given m and μ is obtained from the most probable value of $5\log s$. Calling this quantity x we get:

$$s = 10^{0,2x} = e^{bx},$$

and:

$$\varphi_1(x)\,dx = \frac{2b e^{2bx}}{\sqrt{2\pi_c}}\, e^{-\frac{1}{2}(k e^{bx}-l)^2}\,dx\,.$$

The most probable value of x is found by taking the logarithm of $\varphi(x)$ and then derivating the expression obtained and putting the result $= 0$.

We then get:

$$2 = (k e^{bx} - l)\, k e^{bx}\,,$$

or:

$$x = 5 \log \frac{1}{2k}\left(l + \sqrt{l^2+8}\right).$$

This formula was derived by FÄNGE and used in his computations. The values of M calculated by the aid of this formula show, however, very great deviations from the spectrographically determined M, in the sense that the bright stars appear too bright and the faint stars too faint. FÄNGE has suggested, as an explanation of this systematic divergence, the existence of a correlation between M and the space velocity. This will undoubtedly explain part of the deviation, but not all of it.

MALMQUIST[1] makes a new deduction of the expression for M, in which he takes into account the distribution of the absolute magnitudes. According to his work which is reviewed elsewhere, we can write:

$$\varphi_m(M) = \frac{1}{\sigma_m\sqrt{2\pi_c}}\, e^{-\frac{(M-\overline{M}_m)^2}{2\sigma_m^2}}\,.$$

With FÄNGE we put:

$$\Phi(s) = \frac{2s}{\sqrt{2\pi_c}}\, e^{-\frac{1}{2}(ks-l)^2}$$

and assume the velocity to be correlated with the absolute magnitude in the following manner:

$$\lambda_M = \lambda_0\, e^{\beta b (M-\overline{M}_m)}\,.$$

In this equation λ_M is the value of the axis of revolution of the velocity ellipsoid for stars of absolute magnitude M. The constant β is dependent on the amount of the correlation assumed, and $b = 0{,}4605$.

The value of λ_0 is found from:

$$\lambda_0 = \overline{\lambda}\, e^{-\beta^2 b^2 \sigma_m^2}\,.$$

The number of stars within the intervals ds and dM is given from:

$$\psi(s, M)\, ds\, dM = C_2 \Phi_M(s)\, \varphi(M)\, ds\, dM\,.$$

This expression can be written:

$$\psi(s, M) = C_3 \frac{2s}{\sqrt{2\pi_c}}\, e^{-\frac{1}{2}(k_M s - l_M)^2 - \frac{(M-M_m)^2}{2\sigma_m^2}}\,.$$

Further:

$$k_M = \frac{1}{\lambda_M K}\,; \qquad l_M = \frac{\omega L}{\lambda_M}\,.$$

The suffix is used on account of the correlation assumed.

[1] Lund Medd Ser. I, No. 110 (1925).

The following new notations are introduced:

$$M - \overline{M}_m = x,$$
$$p = 4{,}8481\,\mu' e^{-b\overline{M}_m},$$
$$s = e^{-bM}\,4{,}8481\,\mu' = e^{-bx}\cdot p,$$

and μ' are the reduced proper motions, expressed in seconds of arc per year.

The number of stars within the interval ds and dM is then:

$$\psi_1(p, x) = C\,p\,e^{-2bx - \frac{e^{-2\beta bx + 2\beta^2 b^2 \sigma_m^2}}{2\lambda^2}\left(e^{-bx}\frac{p}{K} - \omega L\right)^2 - \frac{x^2}{2\sigma_m^2}}.$$

The most probable value of x is found from the equation:

$$0 = -2b - \frac{x}{\sigma_m^2} + \frac{\beta b}{\lambda^2} e^{-2\beta bx + 2\beta^2 b^2 \sigma_m^2}\left(e^{-bx}\frac{p}{K} - \omega L\right)^2 + \frac{pb}{K\lambda^2} e^{-bx - 2\beta bx + 2\beta^2 b^2 \sigma_m^2}\left(e^{-bx}\frac{p}{K} - \omega L\right).$$

For the sake of simplification $p/K\lambda$ is put equal to y. After some reduction, the equation takes the form:

$$\left(y - \frac{(1+2\beta)\,\omega L}{2(1+\beta)\,\lambda} e^{bx}\right)^2 = \frac{\omega^2 L^2 e^{2bx}}{4\lambda^2 (1+\beta)^2} + \frac{x + 2b\,\sigma_m^2}{(1+\beta)\,b\,\sigma_m^2} e^{2(1+\beta)bx - 2\beta^2 b^2 \sigma_m^2}.$$

Considering the most simple case, namely that no correlation between velocity and absolute magnitude exists ($\beta = 0$), we find that when the proper motion tends towards zero, the most probable value of $M - \overline{M}_m$ tends towards a limiting value which is unequivocally determined by the dispersion of the absolute magnitudes for the class of stars considered. This limiting value is equal to $-0{,}921\,\sigma_m^2$.

The influence of the correlation is then discussed, and curves giving x as a function of $\log y$ are constructed for different values of σ_M. It is found that the relation deduced by FÄNGE is valid only if we assume that all values of M are equally probable.

MALMQUIST inquires how great the accuracy is with which the absolute magnitude of a star of a given spectral type and proper motion can be determined. He finds that when the size and direction of the proper motions are considered, the mean error in the absolute magnitude is the same for very small proper motions, but diminishes when the proper motions increase. For very large proper motions the mean error is only $0^{\mathrm{M}},4$.

The method is then applied to the stars classed as G at Mount Wilson and apparently brighter than $6^{\mathrm{m}},0$. It is found that the value $\sigma_m = 0{,}5$ gives the best agreement between the absolute magnitudes according to the Mount Wilson determination and those computed from the value and direction of μ. This result signifies that if the former are free from systematic errors, the dispersion of the absolute magnitude is only $0^{\mathrm{M}},5$ for the G stars, which means that knowledge of the spectral type alone will give M of an individual star with a mean error of only $\pm 0^{\mathrm{M}},5$. The mean error of the determinations in Mt Wilson Contr 199 is estimated to be $\pm 0^{\mathrm{M}},6$. The only way to escape from the paradoxical conclusion that M can be determined more accurately from the spectral class than from the spectrographic absolute magnitude criteria is to assume the existence of a systematic error in the spectrographic magnitudes in such a direction that the scale is too narrow.

169. MALMQUIST's Relations in Stellar Statistics. At the same time as STRÖMBERG, MALMQUIST[1] showed independently the importance of taking into

[1] Lund Medd Ser. I, No. 100 (1922).

consideration the dispersion when computing mean values for the absolute magnitude.

On the basis of the fundamental integral equation in stellar statistics:

$$a(m) = \omega \int_0^\infty r^2 D(r)\, \varphi\left(m - \frac{1}{b} \log r\right) dr,$$

[where $a(m)$ is the frequency of stars of magnitude m, ω is the solid angle embracing the part of the sky considered, and $b = 0{,}4605$] the relative moment about the mean of the nth order is given by the formula:

$$\nu_n(m) = \overline{\left(M - \overline{M}_m\right)^n} = \frac{\omega \int_0^\infty r^2 D(r)\,(M - \overline{M}_m)^n\, \varphi\left(m - \frac{1}{b}\log r\right) dr}{\omega \int_0^\infty r^2 D(r)\, \varphi\left(m - \frac{1}{b}\log r\right) dr}.$$

$\overline{M}_m$ is defined from the relation:

$$\nu_1(m) = 0.$$

The value $\overline{M}_m$ is found to be:

$$\overline{M}_m a(m) = M_0 a(m) - \sigma^2 \frac{d\,a(m)}{dm},$$

or:

$$\overline{M}_m = M_0 - \sigma^2 \frac{d \log a(m)}{dm},$$

where M_0 is given by the equation:

$$\varphi(M) = \frac{1}{\sigma\sqrt{2\pi_c}}\, e^{\frac{-(M-M_0)^2}{2\sigma^2}}.$$

Further the following equation may be derived:

$$\sigma_m^2 = \sigma^2\left(1 + \sigma^2 \frac{d^2 \log a(m)}{dm^2}\right).$$

MALMQUIST has applied this method to the derivation of $\overline{M}_m$ for the A stars[1].

For the moments of the higher orders the following relations are derived:

$$\nu_3(m) = -\sigma^6 \frac{d^3 \log a(m)}{dm^3},$$

$$\nu_4(m) - 3\nu_2^2(m) = \sigma^8 \frac{d^4 \log a(m)}{dm^4},$$

$$\nu_5(m) - 10\nu_2(m)\,\nu_3(m) = -\sigma^{10} \frac{d^5 \log a(m)}{dm^5},$$

$$\nu_6(m) - 10\nu_3^2(m) - 15\nu_2(m)\,\nu_4(m) + 30\nu_2^3(m) = \sigma^{12} \frac{d^6 \log a(m)}{dm^6}.$$

MALMQUIST then makes use of the relations between the moments and the characteristics given by CHARLIER[2]

$$\lfloor 3\, A_3 = \sigma^6 \frac{d^3 \log a(m)}{dm^3},$$

$$\lfloor 4\, A_4 = \sigma^8 \frac{d^4 \log a(m)}{dm^4},$$

$$\lfloor 5\, A_5 = \sigma^{10} \frac{d^5 \log a(m)}{dm^5},$$

$$\lfloor 6\, A_6 = \sigma^{12} \frac{d^6 \log a(m)}{dm^6} + \sigma^{12}\left(\frac{d^3 \log a(m)}{dm^3}\right)^2.$$

[1] Lund Medd Ser. II, No. 22 (1920).

[2] Lund Medd Ser. II, No. 4, p. 7 (1906).

When $a(m) = C e^{\beta m}$ according to SEELIGER's investigations, $\frac{d \log a(m)}{dm} = \beta$ and all higher derivatives vanish. Thus:

$$\overline{M}_m = M_0 - \beta \sigma^2$$

and:

$$\sigma_m^2 = \sigma^2,$$

and the distribution-function is:

$$F_m(M) = \frac{1}{\sigma \sqrt{2\pi_e}} e^{-\frac{(M - M_0 + \beta \sigma^2)^2}{2\sigma^2}}.$$

or a normal curve of type A.

For constant density the following relation is found:

$$\overline{M}_m = M_0 - 1{,}382\, \sigma^2,$$

from which is computed:

σ		$M_0 - \overline{M}_m$
0	....	$0^M{,}00$
1	....	1 ,38
2	....	5 ,35
3	....	12 ,44

If according to CHARLIER and others:

$$a(m) = \frac{N}{\alpha \sqrt{2\pi_e}} e^{-\frac{(m - m_0)^2}{2\alpha^2}}$$

then also the higher derivatives vanish and we have:

$$\overline{M}_m = M_0 + (m - m_0) \frac{\sigma^2}{\alpha^2},$$

and:

$$\sigma_m^2 = \sigma^2 \left(1 - \frac{\sigma^2}{\alpha^2}\right).$$

The two constants M_0 and σ enter in the above relations. It is, therefore, a fundamental problem in stellar statistics to determine the value of these two constants. MALMQUIST[1] investigates therefore the harmonic mean value of the parameter R as defined by CHARLIER in his investigations concerning the B stars[2].

MALMQUIST finds that for stars of a given apparent magnitude we have for any power of the reduced distance R:

$$\overline{R_m^s} = e^{-sbM_0 + \frac{1}{2}s^2 b^2 \sigma^2} \frac{a(m + sb\sigma^2)}{a(m)}$$

where:

$$R = 10^{-0{,}2M} = e^{-bM}.$$

For all stars brighter than a given apparent magnitude we get an analogous relation:

$$\overline{R^s_{-\infty, m}} = e^{-sbM_0 + \frac{1}{2}s^2 b^2 \sigma^2} \frac{A(m + sb\sigma^2)}{A(m)},$$

and for the stars within the magnitude interval k, i. e. fainter than m, but brighter than $m + k$:

$$\overline{R^s_{m, m+k}} = e^{-sbM_0 + \frac{1}{2}s^2 b^2 \sigma^2} \frac{A(m + k + sb\sigma^2) - A(m + sb^2\sigma)}{A(m + k) - A(m)}.$$

These relations are strictly valid regardless of the form of the density function.

[1] Lund Medd Ser. I, No. 100 (1922). [2] Lund Medd Ser. II, No. 14 (1916).

In the special case $s = -1$ we get the following theoretical expression for the mean reduced parallax of all stars brighter than a given m:

$$\overline{\left(\frac{1}{R}\right)}_{-\infty, m} = e^{b M_0 + \frac{1}{2} b^2 \sigma^2} \frac{A(m - b\sigma^2)}{A(m)}.$$

From discussions concerning directly measured parallaxes or proper motions the mean reduced parallax can be determined. We can put:

$$\log A(m) = a + bm + cm^2 + \ldots,$$

and the constants can be determined from a least square solution. Then we get a relation between M_0 and σ. In his paper concerning the A stars MALMQUIST found for the non-galactic Boss stars brighter than $6^m,00$:

$$M_0 = 1{,}03\,\sigma^2 - 1{,}10,$$

and for the galactic stars brighter than $6^m,00$:

$$M_0 = 1{,}47\sigma^2 - 0{,}053\,\sigma^4 - 1{,}22,$$

and for all stars:

$$M_0 = 1{,}31\sigma^2 - 0{,}032\sigma^4 - 1{,}17.$$

If we assume our stellar system to be of limited extent, the coefficient of σ and the value of $\overline{(1/R)}_{-\infty, m}$ must vary with m. Hence if we determine the last quantity for two sufficiently distant values of m, we should get two relations between M_0 and σ, from which the two constants could be determined. The available proper motions do not permit such a determination to be made with any great accuracy.

Then the author discusses the mean value of any power of the distances r of stars selected according to m.

Often we have not determined the value of the mean reduced parallax, but of the mean parallax.

In the first case (constant apparent magnitude) the author finds:

$$\overline{r^s_m} = e^{s b (m - M_0) + \frac{1}{2} s^2 b^2 \sigma^2} \frac{a(m + s b \sigma^2)}{a(m)}.$$

In the second case (all stars brighter than m):

$$A(m)\,\overline{r^s_{-\infty, m}} = e^{-s b M_0 - \frac{1}{2} s^2 b^2 \sigma^2} \sum_{-\infty}^{m + s b \sigma^2} 10^{0{,}2m}.$$

And for all stars brighter than $m + k$, but fainter than m:

$$\{A(m + k) - A(m)\}\,\overline{r^s_{m, m+k}} = e^{-s b M_0 - \frac{1}{2} s^2 b^2 \sigma^2} \int_{m + s b \sigma^2}^{m + k + s b \sigma^2} e^{s b m}\, a(m)\, dm.$$

In the special case of $s = -1$ the formula in the case of all stars brighter than m takes the form:

$$A(m)\,\overline{\left(\frac{1}{r}\right)}_{-\infty, m} = e^{b M_0 - \frac{1}{2} b^2 \sigma^2} \sum_{-\infty}^{m - b\sigma^2} e^{-bm}.$$

Now $e^{-bm} = 10^{-0{,}2m}$. The value of $\sum\limits_{-\infty}^{m} e^{-bm}$ can be computed from the observations for different values of m, and then the constants a, b, c determined from the expression:

$$\log \sum_{-\infty}^{m} e^{-bm} = a + bm + cm^2 + \cdots.$$

Table of e^{-bm}.

Magnitude m	0	1	2	3	4	5	6	7	8	9
0,0	1,000	0,995	0,991	0,986	0,982	0,977	0,973	0,968	0,964	0,959
,1	,955	,951	,946	,942	,938	,933	,929	,925	,920	,916
,2	,912	,908	,904	,899	,895	,891	,887	,883	,879	,875
,3	,871	,867	,863	,859	,855	,851	,847	,843	,839	,836
,4	,832	,828	,824	,820	,817	,813	,809	,805	,802	,798
,5	,794	,791	,787	,783	,780	,776	,773	,769	,766	,762
,6	,759	,755	,752	,748	,745	,741	,738	,735	,731	,728
,7	,724	,721	,718	,714	,711	,708	,705	,701	,698	,695
,8	,692	,689	,685	,682	,679	,676	,673	,670	,667	,664
,9	,661	,658	,655	,652	,649	,646	,643	,640	,637	,634
1,0	,631	,628	,625	,622	,619	,617	,614	,611	,608	,605
,1	,603	,600	,597	,594	,592	,589	,586	,583	,581	,578
,2	,575	,573	,570	,568	,565	,562	,560	,557	,555	,552
,3	,550	,547	,545	,542	,540	,537	,535	,532	,530	,527
,4	,525	,522	,520	,518	,515	,513	,511	,508	,506	,504
,5	,501	,499	,497	,494	,492	,490	,488	,485	,483	,481
,6	,479	,476	,474	,472	,470	,468	,466	,463	,461	,459
,7	,457	,455	,453	,451	,449	,447	,445	,443	,441	,439
,8	,437	,435	,433	,431	,429	,427	,425	,423	,421	,419
,9	,417	,415	,413	,411	,409	,407	,406	,404	,402	,400
2,0	,398	,396	,394	,393	,391	,389	,387	,385	,384	,382
,1	,380	,378	,377	,375	,373	,372	,370	,368	,366	,365
,2	,363	,361	,360	,358	,356	,355	,353	,352	,350	,348
,3	,347	,345	,344	,342	,340	,339	,337	,336	,334	,333
,4	,331	,330	,328	,327	,325	,324	,322	,321	,319	,318
,5	,316	,315	,313	,312	,310	,309	,308	,306	,305	,303
,6	,302	,301	,299	,298	,296	,295	,294	,292	,291	,290
,7	,288	,287	,286	,284	,283	,282	,281	,279	,278	,277
,8	,275	,274	,273	,272	,270	,269	,268	,267	,265	,264
,9	,263	,262	,261	,259	,258	,257	,256	,255	,254	,252
3,0	,251	,250	,249	,248	,247	,245	,244	,243	,242	,241
,1	,240	,239	,238	,237	,236	,234	,233	,232	,231	,230
,2	,229	,228	,227	,226	,225	,224	,223	,222	,221	,220
,3	,219	,218	,217	,216	,215	,214	,213	,212	,211	,210
,4	,209	,208	,207	,206	,205	,204	,203	,202	,201	,200
,5	,200	,199	,198	,197	,196	,195	,194	,193	,192	,191
,6	,191	,190	,189	,188	,187	,186	,185	,185	,184	,183
,7	,182	,181	,180	,179	,179	,178	,177	,176	,175	,175
,8	,174	,173	,172	,171	,171	,170	,169	,168	,167	,167
,9	,166	,165	,164	,164	,163	,162	,161	,161	,160	,159
4,0	,158	,158	,157	,156	,156	,155	,154	,153	,153	,152
,1	,151	,151	,150	,149	,149	,148	,147	,147	,146	,145
,2	,145	,144	,143	,143	,142	,141	,141	,140	,139	,139
,3	,138	,137	,137	,136	,136	,135	,134	,134	,133	,132
,4	,132	,131	,131	,130	,129	,129	,128	,128	,127	,126
,5	,126	,125	,125	,124	,124	,123	,122	,122	,121	,121
,6	,120	,120	,119	,119	,118	,117	,117	,116	,116	,115
,7	,115	,114	,114	,113	,113	,112	,112	,111	,111	,110
,8	,110	,109	,109	,108	,108	,107	,107	,106	,106	,105
,9	,105	,104	,104	,103	,103	,102	,102	,101	,101	,100
5,0	,1000	,0995	,0991	,0986	,0982	,0977	,0973	,0968	,0964	,0959
,1	,0955	,0951	,0946	,0942	,0938	,0933	,0939	,0925	,0920	,0916
,2	,0912	,0908	,0904	,0899	,0895	,0891	,0887	,0883	,0879	,0875
,3	,0871	,0867	,0863	,0859	,0855	,0851	,0847	,0843	,0839	,0836
,4	,0832	,0828	,0824	,0820	,0817	,0813	,0809	,0805	,0802	,0798
,5	,0794	,0791	,0787	,0783	,0780	,0776	,0773	,0769	,0766	,0762
,6	,0759	,0755	,0752	,0748	,0745	,0741	,0738	,0735	,0731	,0728
,7	,0724	,0721	,0718	,0714	,0711	,0708	,0705	,0701	,0698	,0695
,8	,0692	,0689	,0685	,0682	,0679	,0676	,0673	,0670	,0667	,0664
5,9	0,0661	0,0658	0,0655	0,0652	0,0649	0,0646	0,0643	0,0640	0,0637	0,0634

Lastly an inquiry is made concerning the relation between the mean parallax and the mean reduced parallax. For constant m the following is found:

$$\overline{r^s_m}/\overline{R^s_m} = e^{s b m}.$$

For all stars brighter than m:

$$\overline{r^s_{-\infty\, m}}/\overline{R^s_{-\infty,\, m}} = e^{-s^2 b^2 \sigma^2} \frac{\int_{-\infty}^{m + s b \sigma^2} e^{s b m} a(m)\, dm}{A\,(m + s b \sigma^2)};$$

this can be written:

$$\overline{r^s_{-\infty,\, m}}/\overline{R^s_{-\infty,\, m}} = e^{-s^2 b^2 \sigma^2}\, \overline{(e^{s b m})}_{-\infty,\, m + s b \sigma^2}.$$

For all stars brighter than $m + k$, but fainter than m:

$$\overline{r^s_{m,\, m+k}}/\overline{R^s_{m,\, m+k}} = e^{-s^2 b^2 \sigma^2}\, \overline{(e^{s b m})}_{m + s b \sigma^2,\, m + k + s b \sigma^2}.$$

The relation between the mean parallax and the mean reduced parallax for all stars brighter than m is:

$$\frac{\overline{\left(\frac{1}{r}\right)}_{-\infty,\, m}}{\overline{\left(\frac{1}{R}\right)}_{-\infty,\, m}} = e^{-b^2 \sigma^2}\, \overline{(e^{-b m})}_{-\infty,\, m - b \sigma^2}$$

where $\overline{(e^{-bm})}_{-\infty,\, m}$ is the mean value of e^{-bm} for all stars brighter than m.

The value of $e^{-bm} = 10^{-0,2m}$, which is a very important factor, not only in theoretical deductions of the kind reviewed above, but also in work related to computation of absolute magnitude has been tabulated on p. 522.

170. GYLLENBERG's Method. An important method for the derivation of the dispersion in the absolute magnitude of the stars has been given by GYLLENBERG[1]. The author points out the unavoidable selection of the material when results are based upon moving clusters, double stars, or trigonometric parallaxes. The study of the proper motions of the stars will form an excellent way of determining the characteristics of the distribution function of the absolute magnitudes. For the deduction of the distribution function it may be sufficient at present to confine ourselves to the determination of the mean and the dispersion.

The restriction of the method of GYLLENBERG is that an assumption has to be made with regard to the form of the frequency distribution. Generally the following frequency function of M and m has been assumed:

$$\varphi(M) = C e^{-\frac{(M - M_0)^2}{2\sigma_2^2}},$$

$$a(m) = C e^{-\frac{(m - m_0)^2}{2k^2}},$$

where σ_2 and k are the dispersion around the mean values M_0 and m_0, respectively.

The observed proper-motion components of a star are denoted by u and v, and the distance by r. The linear velocity components U and V are then expressed by:

$$U = r u, \qquad V = r V.$$

If U and V are independent of r, the distribution of the linear velocities is determined by the help of the method of moments as used by CHARLIER and

[1] Lund Medd Ser. II, No. 41 (1927).

others. The moments about the origin are defined by means of:

$$\nu'_{ij} = \int_{-\infty}^{+\infty}\int \varphi(u, v)\, u^i v^j\, du\, dv,$$

$$N'_{ij} = \int_{-\infty}^{+\infty}\int \varphi(U, V)\, U^i V^j\, dU\, dV.$$

From the definition of U and V it follows that:

$$u^i v^j = \frac{1}{r^{i+j}} U^i V^j.$$

The following notations are introduced:

$$\nu'_{ij} \text{ the mean value of } u^i v^j = \overline{u^i v^j},$$

$$N'_{ij} \quad ,, \quad ,, \quad ,, \quad ,, \quad U^i V^i = \overline{U^i V^j},$$

$$\vartheta_{ij} \quad ,, \quad ,, \quad ,, \quad ,, \quad \frac{1}{r^{i+j}} = \overline{\frac{1}{r^{i+j}}}.$$

Then:

$$\nu'_{ij} = \vartheta_{ij} N'_{ij}.$$

The fundamental equation of stellar statistics:

$$a(m) = \omega \int_0^{\infty} D(r)\, r^2 \varphi_0\left(m - \frac{1}{b}\log r\right) dr$$

[where ω is the solid angle embracing the part of the sky considered, r the distance, $D(r)$ the density function, and $\varphi_0(m - 1/b \log r)$ expresses the distribution of the absolute magnitudes] can be transformed into:

$$a(m) = \int_{-\infty}^{+\infty} \Delta(y)\, \varphi_0(m + y)\, dy$$

where:

$$\Delta y = \omega b D(e^{-by}) e^{-3by}; \quad r = e^{-by}; \quad b = 0{,}2\,\mathrm{Mod}^{-1}.$$

If the form for $a(m)$ as defined above is introduced, the three following expressions can be substituted for the foregoing equations:

$$a(m) = \frac{C}{k\sqrt{2\pi_c}} e^{-\frac{(m - m_0)^2}{2k^2}},$$

$$\varphi_0(m + y) = \frac{1}{\sigma_2\sqrt{2\pi_c}} e^{-\frac{[y - \overline{(M_0 - m)}]^2}{2\sigma_2^2}},$$

$$\Delta y = \frac{C}{\sqrt{k^2 - \sigma_2^2}\sqrt{2\pi_c}} e^{-\frac{(y - y_0)^2}{2(k^2 - \sigma_2^2)}},$$

$$y_0 = M_0 - m_0.$$

The fundamental equation can be integrated:

$$\vartheta_s = e^{sb[m_1(1-\lambda_1) + M_0\lambda_1 - m\lambda_1] + \frac{s^2}{2} b^2 k^2 \lambda_1 (1 - \lambda_1)},$$

$$\sigma_1^2 + \sigma_2^2 = k^2, \quad \lambda_1 = \sigma_1^2/k^2, \quad m_1 = M_0 - m_0.$$

For $s = 1$ we have:

$$\vartheta_1 = \overline{\frac{1}{r}} = e^{-b[m - \{M_0 - (3 - \varkappa) b \sigma_2^2\} + \frac{1}{2} b^2 \sigma_2^2]}.$$

Further:

$$\vartheta_s = \vartheta_1^s e^{\frac{s^2 - s}{2} b^2 \sigma_2^2 \lambda_1}.$$

The value of ϑ_1 is obtained in the apex solution. The SUN's motion expressed in seconds of arc is $\vartheta_1 S$, where the velocity S in absolute measure is found from the radial velocities.

By another method, independent of the determination of ϑ_1, GYLLENBERG has determined the value of $\vartheta_1 S$. Thus $\overline{M_m}$ is known and from $\overline{M_m}$ the value of M_0 may be computed.

The indirect way of finding the absolute magnitudes from a study of the proper motions will be simplified if the form of the frequency distribution of the linear (absolute) velocities is assumed to be known. According to several investigations by GYLLENBERG and others this distribution is found to be normal or, at least, very nearly normal.

The frequency surface of U and V may be written:

$$\psi(U,V) = \varphi(U,V) + \sum A_{ij} \frac{\partial^{i+j}\varphi(U,V)}{\partial U^i \partial V^j}.$$

The coefficients of A_{ij} are expressed in terms of N_{ij}. In a normal frequency distribution the coefficients A_{40} and A_{04} vanish. If H denotes the excess in linear motion, we have:

$$H_x = (N_{40} - 3N_{20}^2)/8N_{20}^2,$$
$$H_y = (N_{04} - 3N_{02}^2)/8N_{02}^2.$$

If the relation between apparent and linear velocities is written:

$$\nu_{ij} = \vartheta_1^{i+j} e^{\frac{1}{2}[(i+j)^2 - \overline{i+j}] b^2 \sigma_2^2} N_{ij},$$

or

$$\nu_{20} = \vartheta_1^2 e^{b^2\sigma_2^2} N_{20}; \qquad \nu_{40} = \vartheta_1^4 e^{6b^2\sigma_2^2} N_{40},$$

then substituting the linear moments of ν_{20} and ν_{40} in H_x and taking $H_x = 0$ we have:

$$\frac{\nu_{40}}{3\nu_{20}^2} = e^{4b^2\sigma_2^2}, \qquad \text{or} \qquad \sigma_2^2 = 2{,}7143 \log \frac{\nu_{40}}{3\nu_{20}^2}.$$

The excess E in the apparent velocity distribution is defined analogously with H by:

$$E_x = (\nu_{40} - 3\nu_{20}^2)/8\nu_{20}^2$$

and thus:

$$\sigma_2^2 = 2{,}7143 \log\left(\frac{8}{3}E + 1\right).$$

This formula can be applied to both the coordinates independently of the value of ϑ_1. It is also independent of the coefficients in the spatial density law $D(r) = \gamma r^{-\varkappa}$, and thus also valid for $D(r) = \text{const.}$

If the spatial density is constant or proportional to any power of the inverse distance, the excess in the proper motion distribution is a simple function of the dispersion in the absolute magnitude of the stars.

After the introduction of the more general law of density the formula for computing σ_2 is found to be:

$$\sigma_2^2\left(1 - \frac{\sigma_2^2}{\varkappa^2}\right) = 2{,}7143 \log\left(\frac{8}{3}E + 1\right).$$

For the case where the linear velocities are not distributed according to the normal frequency law GYLLENBERG has derived the following formula:

$$\sigma_2^2 = 2{,}7143 \left[\log\left(\frac{8}{3} E_x + 1\right) - \log\left(\frac{8}{3} H_x + 1\right)\right]$$

which is also valid for both the coordinates.

The following table gives the dispersion computed from the first expression for different values of E:

E	σ_2	E	σ_2	E	σ_2
+0,05	0^M,384	+0,80	1^M,160	+1,60	1^M,399
,10	0 ,528	0,90	1 ,201	1,70	1 ,420
,20	0 ,710	1,00	1 ,238	1,80	1 ,439
,30	0 ,832	1,10	1 ,271	1,90	1 ,458
,40	0 ,925	1,20	1 ,301	2,00	1 ,475
,50	0 ,999	1,30	1 ,328	2,50	1 ,550
,60	1 ,061	1,40	1 ,354	+3,00	1 ,609
+0,70	1 ,114	+1,50	1 ,377		

The numerical value of E will have rather a small influence on the corresponding dispersion.

The formulae are applied to the BOSS stars of 5^m. According to CHARLIER $k = 3{,}5$ and $E = 0{,}50$ and thus from the above table $\sigma = 0^M{,}999$. The correction to the value satisfying the general form of the density law is $+0^M{,}049$ and thus the final value is: $\sigma_2 = 1^M{,}048$ (stars $5^m{,}0$ to $6^m{,}0$).

Dividing the material into galactic and non-galactic stars GYLLENBERG finds:

Galactic squares . . . . $\sigma_2 = 1^M{,}092$
Non-galactic squares . . . . $\sigma_2 = 0^M{,}911$

The values are rather low, which indicates that these brighter stars form a rather homogeneous population as regards their absolute magnitudes.

The values of E given in the table of CHARLIER have to be corrected, as has been pointed out by GYLLENBERG, on account of the fact that they are affected by the position of the square in relation to the apex, which clearly appears from the trend in the values of the skewness. The cause of this is the scattering of the stars of fifth magnitude along the line of sight.

Mean displacement of proper motion	Class breadth	Correction ΔE
0",005	0,10	+0,000
0 ,010	0,20	0,004
0 ,015	0,30	0,010
0 ,020	0,40	0,020
0 ,025	0,50	0,033
0 ,030	0,60	0,048
0 ,035	0,70	0,065
0 ,040	0,80	+0,085

GYLLENBERG has derived the correlation between the magnitude of E and the mean displacement in the 48 equal squares into which the sky is divided by CHARLIER. The adjoining table gives the correction which is the increase of the excess due to the influence of the mean displacement.

There is another correction that should be considered. The stars are divided into classes according to their proper motions and each class has a breadth of 0",050. Thus the corrections of SHEPHARD[1] have to be applied to the moments computed. If the true moments are denoted by the index (s) and the observed ones by the index (o), the former are computed from the latter by means of the formulae:

$$\nu_2^{(s)} = \nu_2^{(o)} - \frac{1}{12},$$

$$\nu_4^{(s)} = \nu_4^{(o)} - \frac{1}{2}\,\nu_2^{(o)} - \frac{7}{240}.$$

[1] London Proc Math Soc 29, p. 353 (1898).

Approximately we have:

$$E^{(s)} = E^{(o)} - \frac{1}{16\nu_2^{(o)}}.$$

Using the mean value of ν_2 for all squares GYLLENBERG finds:

$$E^{(s)} = E^{(o)} - 0{,}0367.$$

The final values for the two groups are:

Galactic stars	$\sigma_2 = 1^{\mathrm{M}},146 \pm 0^{\mathrm{M}},043$	$\lambda_1 = 0{,}893$
Non-galactic stars	$\sigma_2 = 0^{\mathrm{M}},938 \pm 0^{\mathrm{M}},053$	$\lambda_1 = 0{,}928$

Lastly GYLLENBERG has determined the value of the excess in the linear motion:

$$H = +0{,}042.$$

This corresponds to a decrease in the dispersion of $0^{\mathrm{M}},057$.

In a subsequent paper[1] GYLLENBERG has published an attempt to make a statistical separation of giant and dwarf stars from a study of the proper motion distribution.

The method is based on the assumption that the observed apparent velocity distribution is composed of two symmetrical, but not normal, frequency distributions due to two groups of stars that are both normally distributed as regards their absolute magnitudes, but have different means.

The following denominations are used:

Number	Giants $N^{(g)}$	Dwarfs $N^{(d)}$
Apparent mean proper motion	$\vartheta_1^{(g)} N_2^{(g)\frac{1}{2}}$	$\vartheta_1^{(d)} N_2^{(d)\frac{1}{2}}$
Dispersion in absolute magnitude	$\sigma_2^{(g)}$	$\sigma_2^{(d)}$

The following system of equations has to be solved:

$$\begin{aligned}
N &= N^{(g)} + N^{(d)},\\
N_{\nu_{|1|}} &= N^{(g)}\,\vartheta_1^{(g)}\,N_{(1)}^{(g)} + N^{(d)}\,\vartheta_1^{(d)}\,N_{(1)}^{(d)},\\
N_{\nu_2} &= N^{(g)}\,e^{b^2\sigma_2^{(g)2}}\,\vartheta_1^{(g)2}\,N_2^{(g)} + N^{(d)}\,e^{b^2\sigma_2^{(d)2}}\,\vartheta_1^{(d)2}\,N_2^{d},\\
N_{\nu_{|3|}} &= N^{(g)}\,e^{3b^2\sigma_2^{(g)2}}\,\vartheta_1^{(g)3}\,N_{(3)}^{(g)} + N^{(d)}\,e^{3b^2\sigma_2^{(d)2}}\,\vartheta_1^{(d)3}\,N_{(3)}^{(d)},\\
N_{\nu_4} &= N^{(g)}\,e^{6b^2\sigma_2^{(g)2}}\,\vartheta_1^{(g)4}\,N_4^{(g)} + N^{(d)}\,e^{6b^2\sigma_2^{(d)2}}\,\vartheta_1^{(d)4}\,N_4^{(d)},\\
N_{\nu_{|5|}} &= N^{(g)}\,e^{10b^2\sigma_2^{(g)2}}\,\vartheta_1^{(g)5}\,N_{(5)}^{(g)} + N^{(d)}\,e^{10b^2\sigma_2^{(d)2}}\,\vartheta_1^{(d)5}\,N_{(5)}^{(d)},\\
N_{\nu_6} &= N^{(g)}\,e^{15b^2\sigma_2^{(g)2}}\,\vartheta_1^{(g)6}\,N_6^{(g)} + N^{(d)}\,e^{15b^2\sigma_2^{(d)2}}\,\vartheta_1^{(d)6}\,N_6^{(d)}.
\end{aligned}$$

The absolute moments of odd orders $\nu_{|s|}$ are defined by means of:

$$\nu_{|s|} = \int_{-\infty}^{+\infty} |\tau|^s \varphi(\tau)\,d\tau,$$

$$s = 1, 3, 5, \ldots$$

where $\varphi(\tau)$ is the distribution of the proper motion components measured at right angles to the parallactic motion.

Owing to the large number of unknown quantities to be derived from the material, which is not very extensive at present it may be profitable to reduce their number. From the investigation of ADAMS and JOY on the spectrographic parallaxes it follows that the dispersion in the absolute magnitudes seems to differ comparatively little in the giant and dwarf groups. GYLLENBERG assumes

[1] Lund Medd Ser. II, No. 41b (1927).

$\sigma_2^{(g)} = \sigma_2^{(d)} = \sigma$. By taking this into account and writing:

$$\frac{N^{(g)}}{N} = n^{(g)}; \qquad \frac{N^{(d)}}{N} = n^{(d)}; \qquad e^{b^2\sigma^2} = \omega$$

the above system of equations is simplified to:

$$1 = n^{(g)} + n^{(d)},$$

$$\left(\frac{\pi_e}{2}\right)^{\frac{1}{2}} \nu_{|1|} = n^{(g)}\,\vartheta_1^{(g)}\,N_2^{(g)\frac{1}{2}} + n^{(d)}\,\vartheta_1^{(d)}\,N_2^{(d)\frac{1}{2}} = a,$$

$$\nu_2 = \omega\,[n^{(g)}\,\vartheta_1^{(g)2}\,N_2^{(g)} + n^{(d)}\,\vartheta_1^{(d)2}\,N_2^{(d)}].$$

If we put:

$$\vartheta_1^{(g)}\,N_2^{(g)\frac{1}{2}} = \mu^{(g)}; \qquad \vartheta_1^{(d)}\,N_2^{(d)\frac{1}{2}} = \mu^{(d)},$$

the equations take the form:

$$1 = n^{(g)} + n^{(d)},$$

$$a = n^{(g)}\,\mu^{(g)} + n^{(d)}\,\mu^{(d)},$$

$$b = \omega\,[n^{(g)}\,\mu^{(g)2} + n^{(d)}\,\mu^{(d)2}],$$

$$c = \omega^3\,[n^{(g)}\,\mu^{(g)3} + n^{(d)}\,\mu^{(d)3}],$$

$$d = \omega^6\,[n^{(g)}\,\mu^{(g)4} + n^{d}\,\mu^{(d)4}].$$

The solution of these gives:

$$n^{(g)} = \frac{a - \mu^{(d)}}{\mu^{(g)} - \mu^{(d)}}.$$

For ω there is the equation:

$$\omega^4 b^3 - 2\omega^3 abc + \omega(c^2 + a^2 d) - bd = 0;$$

when ω is known the auxiliary quantities $y_1 = \mu^{(g)} + \mu^{(d)}$ and $y_2 = \mu^{(g)} \cdot \mu^{(d)}$ are found:

$$y_1 = \frac{b + \omega y_2}{a\omega},$$

$$y_2 = \frac{ac - \omega b^2}{\omega^2 b - \omega^3 a^2} = \frac{ad - \omega^2 bc}{\omega^3 c - \omega^5 ab}.$$

The two roots of the equation:

$$x^2 - y_1 x + y_2 = 0,$$

are equal to $\mu^{(g)}$ and $\mu^{(d)}$.

The method of dissecting the velocity distribution has been applied to the stars of the spectral classes A, F, G, and K. The following table gives a summary of the results obtained. The material used are the proper motions in the Greenwich Catalogue of stars for 1910 (+24° to +32° Decl.).

Spectral class	Apparent magnitude	n	Giants, dispersion in proper motion σ_μ	Percentage of giants	Dwarfs, dispersion in proper motions σ_μ	Percentage of dwarfs	All stars, dispersion in absolute magnitude σ_M
B0—B5	6m,0—8m,5	52	0″,0152	100			[0M,32]
B8—A9	7 ,0—8 ,5	967	0 ,0200	100			0 ,47
F	7 ,0—8 ,5	637	0 ,0129	32,5	0″,0526	67,5	0 ,44
G	7 ,0—8 ,5	480	0 ,0085	48,9	0 ,1037	51,1	0 ,74
K	7 ,0—8 ,5	983	0 ,0204	83,8	0 ,0688	16,2	0 ,73
M	7 ,0—8 ,5	80	0 ,0236	100			0 ,63

These results suggest that the dispersion in the absolute magnitude is probably smaller for the "earlier" spectral classes than for the later. The value for the A stars is in excellent agreement with that found by MALMQUIST ($\pm 0^{M},45$) for the fainter A stars classified according to their colour index.

It will be of considerable interest to use this method for a discrimination of the giants and dwarfs within different apparent magnitudes of stars.

171. A Study of the Adjacent Stars. Such studies have been undertaken in recent years by HAAS[1], LUYTEN[2], and the present writer[3]. Within the space of 10,53 parsecs altogether 207 stars were found, a number of which will certainly be shown to be outside this limit when more accurate parallactic data have been accumulated. On the other hand a considerable number will certainly be included when future measurements of parallaxes are available. We have probably as yet no complete knowledge of spatial distribution even of the stars nearer than 5 parsecs.

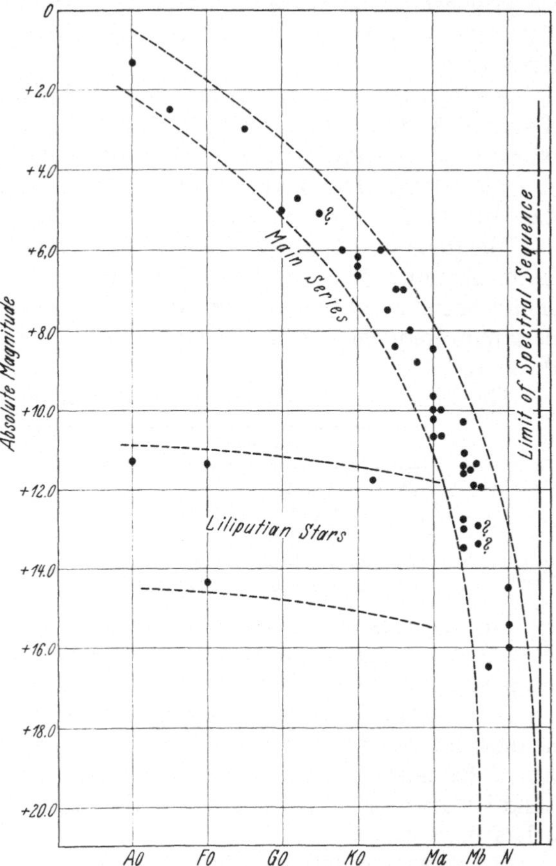

Fig. 137. RUSSELL diagram for stars nearer than 5 parsecs. The diagram is based upon the measures of parallax available at the beginning of 1928. The absence of giants and the presence of four liliputian stars (three white dwarfs and a sub-dwarf of class K2) are interesting features of the diagram.

The following grouping of the stars will show that there is no real star cluster around our Sun:

Distance in parsecs	Absolute magnitude								Sum	Computed number according to KAPTEYN
	0M,0–1M,9	2M,0–3M,9	4M,0–5M,9	6M,0–7M,9	8,M0–9M,9	10M,0–11M,9	12M,0–13M,9	14M,0–15M,9		
,00– 4,64	1	1	2	4	2	9	1	3	23	19
,65– 5,85		1	4	6	3	7			21	19
,86– 6,69			2	3	3	3			11	19
,70– 7,37			7	2	5				14	19
,38– 7,94	2	1	5	3	4	2			17	19
,95– 8,43			1	2	5	1			9	19
,44– 8,88		2	1	4	1	4			12	19
,89– 9,28	1	4	3	2	4				14	19
,29– 9,66		1	4		7				12	19
,67–10,00		1	4		2				7	19
,01–10,53	1		3	6	8	2			20	19
Sum	5	11	36	32	44	28	1	3	160	209

[1] Veröffentl Berlin-Babelsberg, Bd. 3, H. 3 (1923).

[2] Harv Ann 85, p. 73 (1923).

[3] N A T 5, p. 41 (1924).

If there were a real cluster near our Sun the absolute brighter or the more massive stars would concentrate around some distinct centre. The distances are so selected that there should be 19 stars in each spherical shell according to KAPTEYN's density law. The excess in the first sphere suggests that the computed densities are too low.

A star occupies in the mean a space of 16,4 cubic parsecs. That the number of stars is $3 \cdot 10^{10}$ shows that if the density were constant in our stellar system its extent would be 20000 light years in the galactic plane and 4000 in the galactic pole directions. This gives minimum dimensions, as it is well known that the star density decreases in all directions from the Sun.

The frequency of absolute magnitudes shows a remarkable deficiency between the magnitudes $7^M,0$ and $7^M,9$, the very region where the maximum frequency should occur according to the luminosity curve of KAPTEYN. At any rate the maximum frequency of M for neighbouring stars will certainly be situated about 9^M or 10^M.

The high number of visual binaries in our list is a feature of interest.

Number of stars	Binary systems	Percentage
168 (the whole material)	45	27
105 (uncertain parallaxes excluded)	30	29
23 (stars within 4,64 parsecs) . .	7	30
44 (stars within 5,85 parsecs) . .	15	34

The spectroscopic binaries do not seem to be very frequent. This is partly dependent upon the higher frequency of spectroscopic binaries among B and A stars, and partly upon imperfect knowledge of the radial velocities of the adjacent stars.

If the stars are grouped according to spectral class and absolute magnitude the most interesting feature is the comparatively high proportion of white dwarfs. Out of 200 stars at least five are white dwarfs, whereas the number of normal A and F stars is only seven. Thus these liliputian stars may not be such an exception as has been assumed.

The mean absolute magnitudes have been computed from the material and compared with the results from all existing parallax measurements.

Spectral class	$\overline{M}$		Spectral class	$\overline{M}$	
	Nearby stars	Trigonometric parallaxes		Nearby stars	Trigonometric parallaxes
A0	$(1^M,0)$	$1^M,1$	G5	$6^M,6$	$5^M,5$
A2	(2 ,1)	1 ,5	K0	7 ,0	6 ,2
A5	(2 ,8)	2 ,2	K5	8 ,2	7 ,8
F0	(2 ,9)	2 ,8	K8	9 ,0	—
F5	3 ,9	3 ,5	Ma	10 ,1	9 ,8
F0	4 ,7	4 ,2	Mb	11 ,9	11 ,6
G0	5 ,3	4 ,6			

The quantity $H = m + 5 + 5 \log \mu$ is very closely related to M. The coefficient of correlation exceeds 0,8. From a knowledge of the proper motion alone the parallax can be fairly approximated. A combination of spectral class and proper motion determines the parallax within 30 per cent.

The relation between absolute magnitude and radial velocity or absolute magnitude and space velocity is present, but is not very definite.

172. Luminosity Curves for Individual Spectral Classes. P. J. VAN RHIJN has derived the luminosity curve for the separate spectral classes[1]. Already in

[1] Groningen Publ No. 38 (1925).

Mt Wilson Contr No. 147 KAPTEYN had derived curves for the sub-classes of the B stars. The curves in the paper of VAN RHIJN have been derived by means of two independent methods. The first is one initiated by KAPTEYN in Groningen Publ No. 11. Groningen Publ No. 30 contains the number of stars $N(m)$ between determined limits of proper motion and apparent magnitudes. No. 34 of the same Publication contains the mean parallaxes π_m of these stars and the distribution of their true parallaxes. These are distributed in a Gaussian frequency curve where r, the probable deviation, is $0{,}19\,\pi_0$.

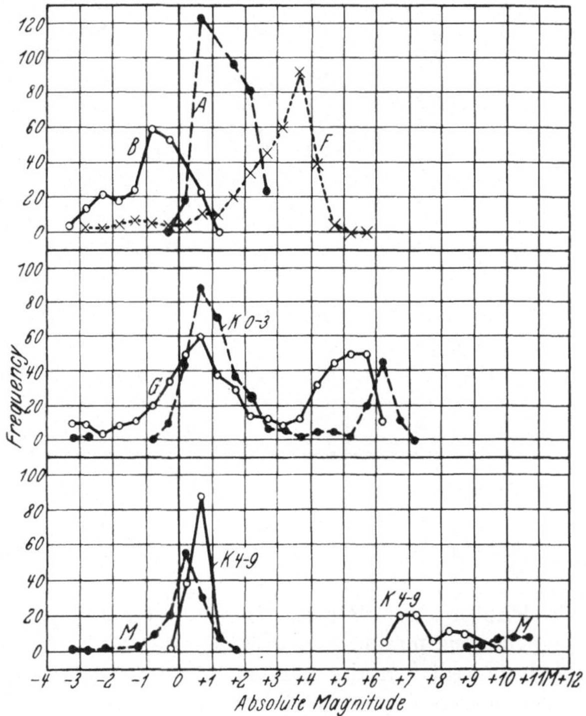

Fig. 138. Frequency curves of absolute magnitude for different spectral classes. The curves are based upon counts made by C. WIRTZ in 1925 and published in MÜLLER-POUILLETS Lehrbuch der Physik. 11. Aufl., Bd. V, 2, Kap. 6. The figure given by WIRTZ has also been used for drawing the above one.

The material used besides the values in Groningen Publ No. 30 consisted of the parallaxes of KAPTEYN for the B stars and the spectrographic parallaxes of EDWARDS, RIMMER, and ADAMS and his collaborators. The luminosity curve was also derived on basis of the trigonometric parallaxes alone. The numbers of stars between determined limits of parallax and apparent magnitude were counted. If the parallaxes were free from observational errors, and if the parallax stars had not been selected on the basis of large proper motion, the numbers counted would represent the number of objects between determined limits of parallax and apparent magnitude actually occurring in the sky. Corrections have to be applied to the numbers counted to allow for the errors of observation in the parallaxes and for the selection. The plenitude of the parallax stars in each group of m and μ has been determined by means of the supposition that the missing stars in each group have the same mean parallax as the observed ones.

The general results are given in the following table:

Log of star number / M	Spectral class						All classes			
	B	A	F	G	K	M	Sum	Method I	Method II	Adopted
$-3^M,44$	3,68	1,37	2,47	2,77	3,30	1,53	3,89	3,27		3,89
$-2\ ,45$	4,28	3,03	2,96	3,00	3,78	3,01	4,47	4,06	4,54	4,47
$-1\ ,46$	4,66	4,11	3,74	3,53	4,48	3,91	5,02	4,78	5,23	5,00
$-0\ ,48$	4,90	5,02	4,28	4,49	5,41	4,50	5,72	5,45	5,72	5,68
$+0\ ,50$	5,20	5,63	5,00	5,21	6,09	4,77	6,33	6,04	6,14	6,24
$+1\ ,50$	5,84	6,09	5,68	5,41	6,32	4,74	6,68	6,52	6,57	6,62
$+2\ ,48$	6,23	6,14	6,39	5,57	5,92	4,50	6,83	6,89	6,89	6,86
$+3\ ,47$	6,07	5,92	6,98	6,30	5,85	4,02	7,16	7,17	7,16	7,16
$+4\ ,45$	5,30		7,05	7,03	6,26		7,38	7,40	7,34	7,36
$+5\ ,43$			6,87	7,47	6,92		7,66	7,56	7,45	7,55
$+6\ ,42$			$<6,62$	6,72	7,30		7,47	7,66	7,40	7,47
$+7\ ,40$			$<6,77$	6,82	7,33	(5,86)	$>7,54$	7,68	7,35	7,54
$+8\ ,39$				$<6,73$	7,40	(6,96)	$>7,53$	7,62	7,63	7,63
$+9\ ,37$						7,36	$>7,36$	7,54	7,62	7,59
$+10\ ,36$						7,51	$>7,51$	7,39	7,47	7,65
$+11\ ,35$						7,85	$>7,85$	7,20		
$+13\ ,3$										8,06

The numbers in the table are the logarithms (augmented by $+10$) of the number of stars per cubic parsec near the Sun.

The absolute magnitudes of the minima and maxima of the luminosity curve are:

Spectral class	Minima	Maxima		
A			$+3^M,3$	
F			$+4\ ,3$	
G	$+1^M,5$	$+1^M,0$		$+5^M,5$
K	$+3\ ,5$	$+1\ ,5$		
M	$+5\ ,3$	$+1\ ,0$		

173. SCHALÉN's Photometric Work. In an extensive paper C. SCHALÉN[1] has derived the photometric magnitudes of 4300 stars and classified the spectra and estimated the luminosity in the classes used by B. LINDBLAD at Upsala.

Four galactic regions have been studied in order to derive distances for the extensive dark nebulae that cover part of the tracts in question. Twenty-four plates were exposed with the Zeiss-Heyde astrograph. The results hitherto published embrace the B and A stars. The apparent photographic magnitudes of the stars have been derived directly from measurements of a certain part of the spectrum. Spectral images were impressed on the plate in such a way that the exposure-ratio between two successive images is 2.

The density or opacity of a spectral region is denoted by 1 for the longest exposure, 2 for the next longest exposure, and so on. When this quantity s increases with one unit the logarithm of the exposure-time t decreases with 0,3. Thus:

$$0{,}3\Delta s = -\Delta \log t.$$

The part of the spectrum chosen for estimates of the opacity is the region $H\gamma - H\delta$. The photographic effective wave length of the white stars is to be found within this interval. As comparison image a spectrum of an A star was used. The opacity of the region of the standard plate was compared with the density in the same region of the star on the plate, and the estimate was put down as 1 if the standard matched the longest exposure. A decimal was added

[1] Stockholm K Vet Akad Handlingar Bd. 6, No. 6 (1928); also Upsala Medd No. 37 and Thesis.

when the opacity of the standard was between that of the longest and next longest exposure and so on. In order to derive the photographic magnitude the density law ought to be known. It was found sufficient to use the SCHWARZSCHILD law $I^q t = \text{const.}$ Introducing the density s, we find:

$$s = C - \tfrac{4}{3} q m .$$

Both the constants C and q can be determined from stars of known magnitude. The standard magnitudes were taken from Harv Ann 50, 54 and 70 and the visual magnitudes were reduced to photographic magnitudes by means of the addition of the mean colour index.

The SCHWARZSCHILD exponent $p = 1/q$ was found to be in the mean 0,79 ±0,029. A correction to the centre, ranging from $0^m,0$ to $0^m,4$, was applied in accordance with PETERSSON's measurements.

From a number of partly overlapping regions on the plates material was furnished for a derivation of the mean error of the apparent magnitudes:

Photographic magnitude	Mean error	n	Photographic magnitude	Mean error	n
$\leqq 7^m,5$	$\pm 0^m,15$	84	$8^m,6-9^m,5$	$\pm 0^m,13$	327
$7^m,6-8^m,5$	$\pm 0\ ,14$	114	$9\ ,6-10\ ,5$	$\pm 0\ ,14$	657

The magnitudes based on the Harvard visual scale have been reduced to the International Scale. For this purpose 600 stars that also occurred in the HENRY DRAPER Extension have been compared. A number of stars in common with SCHALÉN's list gave material for checking the mean error.

A region in Aquila is partly the same as one of E. A. KREIKEN's regions of measured λ_{eff}. No comparison has been made between the effective wave lengths and the spectra.

The methods of classifying spectra and absolute magnitudes are the same as those developed by LINDBLAD[1].

A comparison between the Harvard spectra and the luminosity classes of Upsala has been made by C. SCHALÉN:

Luminosity class acc. to LINDBLAD	B0—B5	B8—A0	A2—A3	A5—F0
	$m < 10^m$			
τ, $\tau-$	242	148	23	8
$\sigma+$ to $\sigma-$	8	385	69	8
ϱ, μ		189	70	8
$\varkappa-$, $\varkappa$		55	140	34
	$m \geqq 10^m$			
τ, $\tau-$	46	21	2	6
$\sigma+$, $\sigma-$	4	61	22	7
ϱ, μ		62	41	10
$\varkappa-$, $\varkappa$		18	26	18

If the luminosity classes and the spectral classes are translated into absolute magnitude by assuming $-2^M,5$, $+0^M,1$, $+1^M,0$, $+2^M,2$, and $-2^M,0$, $+1^M,1$, $+1^M,6$, $+2^M,4$ for the above 8 groups of Upsala classes respectively, the following correlation coefficients have been computed by me:

Material	Coefficients of correlation r	$\varepsilon(r)$
$m < 10$	+0,681	±0,012
$m \geqq 10$	+0,715	±0,026

[1] Upsala designations ciph. 151.

The stars classified as τ, $\tau-$ at Upsala, but as A2—F0 at Harvard should be investigated separately.

The mean error in the classification of luminosity can be estimated as:

Luminosity class	$\varepsilon(M)$	n
τ, $\tau-$	$\pm 0^{M},72$	353
$\sigma+$ to $\sigma-$	$\pm 0\ ,56$	418
ϱ, μ	$\pm 0\ ,72$	266
$\varkappa-$, $\varkappa$	$\pm 0\ ,48$	198

This gives the internal mean error. The external mean error, including the effects of different observers, different instruments and different plates, and different methods of reduction will increase the $\varepsilon(M)$ by at least 30 per cent.

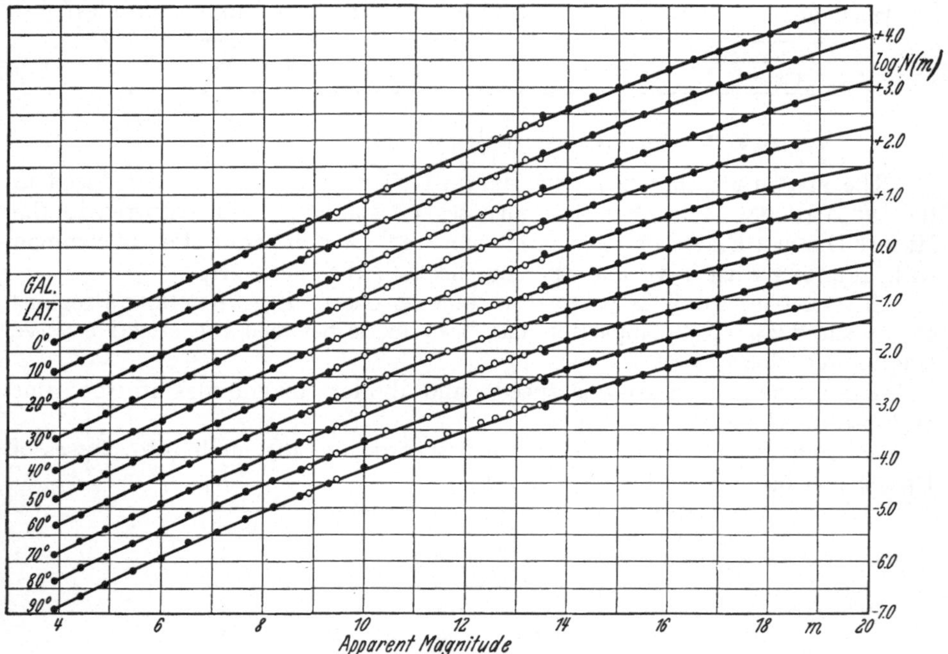

Fig. 139. Curves giving the logarithm of stars per square degree, brighter than magnitude m for different m and each ten degree of galactic latitude according to results by SEARES, VAN RHIJN and MARY C. JOYNER and MYRTLE L. RICHMOND as given in Mount Wilson Contr No. 301 (1925); Ap J 62, p. 320 (1925). The curves represent the mean distribution of stars according to counts in the Mount Wilson Catalogue of Photographic Magnitudes in Selected Areas 1—139 by F. H. SEARES, J. C. KAPTEYN and P. J. VAN RHIJN, assisted by MARY C. JOYNER and MYRTLE L. RICHMOND. The ordinates refer to latitudes 0°. To find $\log N(m)$ for other latitudes add $0{,}05 \times$ latitude to the printed ordinate. This nomogram is very useful when the limiting magnitude of a plate is determined from star counts or when the general course of $N(m)$ will be tested for normality in cases of determination of parallaxes of dark nebulae on basis of star counts.

174. Determinations of the Distance of the Dark Nebulae. PANNEKOEK[1] has given a method for deriving the distance and absorption of a dark nebula, which is assumed to act as an absorbing screen at distance r and which absorbs ε magnitudes of the light of stars at distances greater than r_1.

[1] Proc Kon Akad van Wet Amsterdam 23, No. 5 (1920).

If a star of magnitude m is situated behind the dark nebula, it is observed to be of magnitude $m - \varepsilon$. The number of stars in the dark region is:

$$b_1(m) = \omega \int_0^{r_1} D(r)\,\varphi(M)\,r^2 dr + \omega \int_{r_1}^{\infty} D(r)\,\varphi(M-\varepsilon)\,r^2 dr,$$

where ω is the solid angle embracing the part of the sky considered.

The first integral corresponds to the number of stars seen in front of the nebula and the second integral to the number of stars in or behind the nebula. If we introduce the ratios:

$$\gamma_1 = \frac{\int_0^{r_1} D(r)\,\varphi(M)\,r^2 dr}{\int_0^{\infty} D(r)\,\varphi(M)\,r^2 dr},$$

and:

$$\gamma_2 = \frac{\int_{r_1}^{\infty} D(r)\,\varphi(M-\varepsilon)\,r^2 dr}{\int_0^{\infty} D(r)\,\varphi(M-\varepsilon)\,r^2 dr},$$

the above equation takes the form:

$$b_1(m) = \gamma_1 b(m) + \gamma_2 b(m-\varepsilon).$$

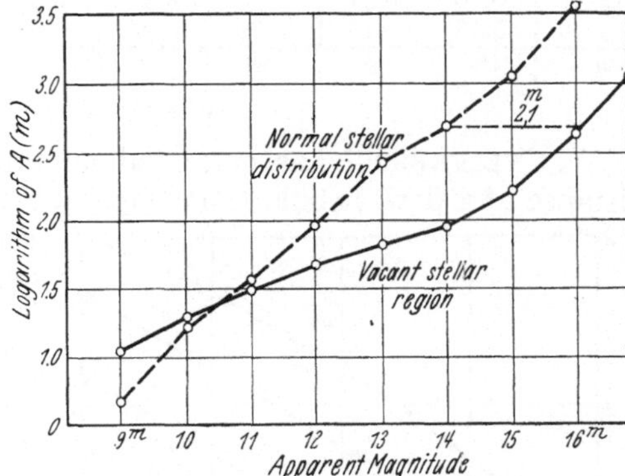

Fig. 140. The influence exercised by the dark nebulae around S Monocerotis on the number of stars $A(m)$ of magnitude m (Max Wolf). The absorption is rather constant within the different magnitude

The number of stars of magnitude m in the dark region is the sum of a proportion γ_1 of the stars of magnitude m and a proportion γ_2 of stars of magnitude $m - \varepsilon$ in a normal region. If analytical expressions for φ and D are introduced, γ_1 and γ_2 can be computed.

Pannekoek[1] has applied this method to the Taurus dark nebula and Schalén[2] has applied it subsequently to five dark nebulae in Cygnus, Cepheus, and Perseus.

His discussion of 4300 stars of the B and A classes led Schalén to derive the following distances of the five objects:

Centre of region		Distance in parsecs	Absorption in magnitudes
α	δ		
20^h25^m	$+35°$	800	$2^m,0$
21 50	$+58$	370	0 ,9
21 45	$+57$	800	1 ,5
$1^h30^m - 1^h50^m$	$+56°$ to $+63°$	900	1 ,0
4 45	$+36$	126	1 ,8

The absorption is thus in the mean $1^M,44 \pm 0^M,53$.

From our point of view the results concerning the luminosity curve of the B and A stars are also of certain interest.

[1] Proc Kon Akad van Wet Amsterdam 23, No. 5 (1920).

[2] Stockholm K Vet Akad Handlingar Bd. 6, No. 6 (1928); also Upsala Medd No. 37 and Thesis.

Luminosity curve of the B and A stars

M	Cygnus		Cepheus		Cassiopeia		Auriga		All regions	
	$r<325$	$r>325$	$r<325$	$r>225$	$r<325$	$r>325$	$r<325$	$r>325$	$r<325$	$r>325$
$-1^{M},0$	0,02	0,03	0,02	0,03	0,00	0,00	0,07	0,03	0,03	0,03
$-0\ ,5$	0,05	0,07	0,02	0,07	0,05	0,05	0,10	0,05	0,05	0,06
$0\ ,0$	0,11	0,15	0,03	0,11	0,14	0,16	0,00	0,09	0,07	0,13
$+0\ ,5$	0,22	0,24	0,22	0,24	0,37	0,25	0,24	0,16	0,26	0,22
$+1\ ,0$	0,23	0,29	0,27	0,36	0,39	0,33	0,36	0,28	0,31	0,32
$+1\ ,5$	0,32	0,32	0,29	0,28	0,26	0,34	0,40	0,43	0,32	0,34
$+2\ ,0$	0,51	0,39	0,46	0,36	0,35	0,45	0,52	0,47	0,46	0,42
$+2\ ,5$	0,40	0,34	0,43	0,34	0,35	0,33	0,26	0,34	0,36	0,34
$+3\ ,0$	0,13	0,15	0,26	0,20	0,09	0,08	0,05	0,14	0,13	0,14
n	164	294	58	109	57	95	42	91	321	589

r is expressed in parsecs.

K. W. GYLLENBERG[1] has presented an interesting method which gives us the distance of a dark nebula from the number of stars in front of it. We have the well-known fundamental equation of stellar astronomy:

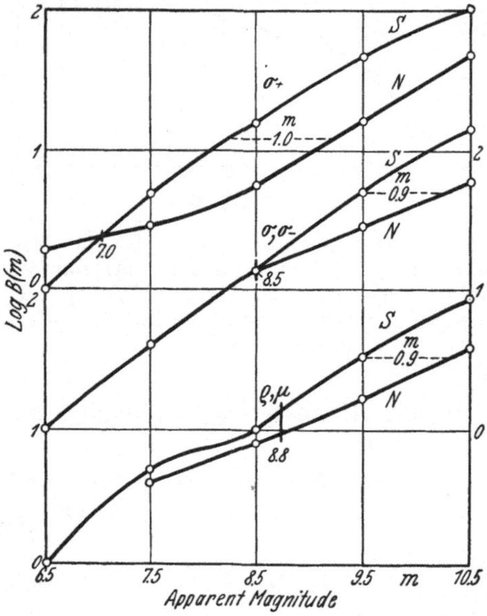

$$a(m) = \omega \int_0^\infty D(r)\, r^2 \varphi(m - 5\log r)\, dr,$$

and

$$5 \log r = \varrho.$$

Fig. 141. The effect exercised by a dark obscuring cloud in the course of $N(m)$ according to SCHALÉN's determinations. The two areas investigated are situated in Cepheus ($\alpha = 22^h$; $\delta = +58°$). The curves S refer to the normal area and N to the area affected by absorption. $B(m)$ is the number of stars brighter than m. The amount of extinction ($1^m \pm$) seems to be fairly constant for the different spectral classes. The curves suggest also the possibility of a differential determination of M for different spectral classes.

The absolute magnitudes are assumed to be normally distributed and further is assumed in accordance with v. SEELIGER[2] $D(r) = C r^{-\varkappa}$ or: $D(e^{b\varrho}) = C e^{-\varkappa b \varrho}$. Further is introduced: $\Delta(\varrho) = D(e^{b\varrho})\, e^{3b\varrho}$, and $b = 0{,}2\ \mathrm{Mod}^{-1}$. The distribution of stars situated within a field where no absorption occurs has then the form:

$$a(m) = C_1 e^{(3-\varkappa)bm}.$$

If the stars are abruptly cut off at the distance R a solution of the equation is found by the aid of the inverse integral of probability or the so-called Err-function of CHARLIER[3]. GYLLENBERG expresses the frequency $a(m)$ in terms of its moments. If $S(m)$ is the total number in front of the nebula he finds finally the simple relation:

$$S(m) = \frac{k_0}{\varepsilon} e^{\varepsilon \varrho_1}$$

where $\varrho_1 = 5 \log R$ and $\varepsilon = (3 - \varkappa)\, b$ and $k_0 = \omega b C$. Further is:

$$\bar{m} = \varrho_1 + M_0 - \frac{1}{\varepsilon},$$

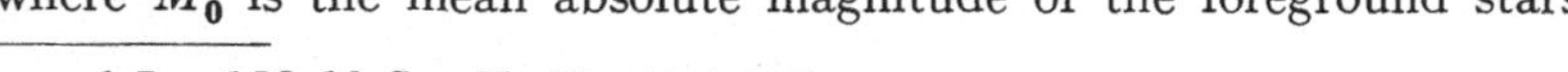

where M_0 is the mean absolute magnitude of the foreground stars.

[1] Lund Medd Ser. II, No. 52 (1929).
[2] München K Bayer Akad Math-Phys Kl Bd. 25, p. 603 (1909).
[3] Lund Medd Ser. II, No. 8 (1912).

Knowledge of M_0 thus gives ϱ_1 when $\bar{m}$ is known.

If the nebula extinguishes μ magnitudes of the stars behind it, the number of these stars per magnitude is reduced by a quantity, q, where:

$$q = C_1 e^{\varepsilon \mu}.$$

A proper value of q can be found only from those m-values where the two curves $\log N(m)$ and $\log N'(m)$ have parallel course. GYLLENBERG obtains ε from the data at hand.

The dispersion in absolute magnitude of the foreground stars is found from:

$$\sigma_M^2 = \sigma_m^2 - \frac{1}{\varepsilon^2}.$$

σ_m being the dispersion in apparent magnitude. From WOLF's[1] star counts near ξ Cygni two parallax values have been found that are in good agreement with determinations according to other methods. In the case of S Monocerotis[2] a much larger parallax is found when other methods are applied. The dispersion in absolute magnitude proves to be in the first case $1^{\mathrm{M}},25$—$1^{\mathrm{M}},34$ and in the second $1^{\mathrm{M}},21$, i. e. surprisingly low.

175. Differential Determinations of Absolute Magnitudes. Investigations of stars in front of a dark nebula or in clusters will give valuable contributions to a knowledge of the difference in absolute magnitude between different spectral classes. From SCHALÉN's data for instance for the mean absolute magnitude of stars of spectral classes B8 and A3 we find:

$$\overline{M}_{\mathrm{B8}} - \overline{M}_{\mathrm{A3}} = -1^{\mathrm{M}},2$$

and from cluster data[3] for stars of spectral classes O and B0:

$$\overline{M}_{\mathrm{O}} - \overline{M}_{\mathrm{B0}} = -1^{\mathrm{M}},44.$$

More accurate determinations will make it necessary to take into account the different distribution of $N(m)$ within different spectral classes.

A determination of the spectra and magnitudes of all stars in front of dark nebulae will also give us an important contribution concerning the spatial proportion or frequencies of different spectral classes.

When a sufficient number of dark nebulae have been investigated it will also be possible to make a determination of the absolute values of the absolute magnitudes for different spectral subgroups. The formulae will be rather long and complicated, but can be worked out along the lines in GYLLENBERG's paper on the distribution of the apparent magnitudes of the foreground stars of dark nebulae.

176. The Distribution of Stars in the DRAPER Catalogue. The contents of the New HENRY DRAPER Catalogue were transformed to a card catalogue at the Observatory of Lund under the supervision of C. V. L. CHARLIER. The statistical results are given by CHARLIER in a very convenient and handy form[4], which also contains 28 charts showing the galactic distribution for different classes of objects. We give the following extract from the paper:

Spectral class	O	B—B5	B8—B9	A	F	G	K	M	N	P	R	Pec.	Con.	S	Sum
All stars . .	170	3675	13111	57850	42988	41999	60620	4397	149	112	84	85	62	3	225305
Stars brighter than $8^{\mathrm{m}},00$.	70	1800	3383	9344	7309	5332	13012	1305	11	0	7	3	0	2	41578

[1] A N 223, p. 89 (1924).
[2] Seeliger Festschr p. 312 (1924).
[3] Obs 47, p. 276 (1924).
[4] Lund Medd Ser. II, No. 36.

Spectral class / m	O	B—B5	B8—B9	A	F	G	K	M	N	P	R	Pec.	Con.	S	Sum
−1				1											1
0		2	1	2	2	2	1	1							11
1		10	2	8	2		6	2							30
2	3	29	2	26	7	7	27	4							105
3	1	51	17	59	40	23	102	10							303
4	5	181	61	231	127	91	265	52							1013
5	14	315	315	804	439	327	941	137	2						3294
6	24	530	934	2403	1583	1226	2855	294	4					1	9854
7	23	682	2051	5810	5109	3656	8815	805	5		7	3		1	26967
8	18	1003	4099	16876	15229	13071	23189	1349	6	2	12	4			74858
9	13	662	3960	20195	15153	16939	19538	938	17		17	3			77435
10	7	175	1453	9697	4781	5943	4371	194	4	2	9				26636
11	9	16	161	1219	330	386	231	35	1		6	2			2396
12				6	4	16	5								31
no magn. given	53	19	55	513	182	312	274	576	110	108	33	73	62	1	2371
Sum	170	3675	13111	57850	42988	41999	60620	4397	149	112	84	85	62	3	225305

With regard to the absolute magnitudes in HDC we can make the following statements.

The B stars have a rather constant absolute magnitude within the different sub-classes. The A stars crowd round the value $+1^{M},2$, and in statistical discussions concerning the distribution of the stars in space a constant value of M can safely be assumed. The same applies to the F stars, which crowd round the value $+3^{M},2$. As regards the G stars the majority are dwarfs of the same absolute magnitude as our Sun, viz. $+4^{M},9$. In the class K there is a considerable dispersion in M, but the majority crowd round $\overline{M} = +0^{M},6$. The main bulk of the 60000 K stars can be taken as giants. In the M class the dwarfs amount to only some 30 or 40, so it can safely be assumed that the M stars in the DRAPER Catalogue are giants with $\overline{M}$ about $0^{M},0$.

The different discussions of the mean absolute magnitudes by CHARLIER[1], LINDBLAD[2], CURTIS[3], SEARES[4], LUNDMARK[5], and ADAMS and JOY[6], and some unpublished investigations of trigonometric parallaxes by LUYTEN have been used by SHAPLEY[7] for the construction of a table giving the distribution of distances of more than 100000 stars. The stars are those included in the New HENRY DRAPER Catalogue and the distances are accurate enough for many statistical purposes. Most of the F stars in the HDC are dwarfs, the A, K, and M stars are mainly giants.

On account of the considerable interest of this table we reproduce it here:

Spectral class	Abs. magn. $\overline{M}$	App. visual magn., distances in parsces, and numbers			
		$6^{m},26$—$7^{m},25$	$7^{m},26$—$8^{m},25$	$8^{m},26$—$9^{m},25$	$6^{m},26$—$9^{m},25$
B8—B9	$0^{M},0 \pm 0^{M},4$	150—340	235—540	375—850	
		1302	2500	4666	8468
A0	0 ,6 ± 0 ,4	110—255	180—405	285—645	
		1400	4184	11016	16600
A2—A3	1 ,2 ± 0 ,4	85—195	135—310	215—490	
		1184	3296	7766	12246

[1] Lund Medd Ser. II, No. 14 (1916) and 34 (1926).

[2] Upsala Medd No. 11 (1926) and 17 (1927). See also ciph. 151—152 in this Chapter.

[3] Publ A S P 34, p. 33 (1922).

[4] Mount Wilson Contr 226 (1922).

[5] Publ A S P 34, p. 147 (1922); 35, p. 211 (1923).

[6] Mt Wilson Contr Nos 89 (1914); 142 (1917); 199 (1921); 244 (1922); 262 (1923); 319 (1926). See also ciph. 140—143 in this Chapter.

[7] Harv Circ 243 (1923).

Spectral class	Abs. magn. $\overline{M}$	App. visual magn., distances in parsces, and numbers			
		$6^m,26$—$7^m,25$	$7^m,25$—$8^m,25$	$8^m,26$—$9^m,25$	$6^m,26$—$9^m,25$
A5	$1^M,7 \pm 0^M,45$	65—160	105—250	165—395	
		280	824	1896	3000
F0	$2\ ,4 \pm 0\ ,4$	50—110	80—180	125—280	
		688	2232	4024	6944
F2	$2\ ,8 \pm 0\ ,4$	40— 95	65—150	105—235	
		512	1320	2768	4600
F5	$3\ ,4 \pm 0\ ,35$	30— 70	50—110	80—175	
		728	2496	5824	9048
K0	$0\ ,6 \pm 0\ ,4$	110—255	180—405	285—645	
		2656	7536	14616	24808
K2	$0\ ,5 \pm 0\ ,3$	125—255	195—405	310—645	
		568	3208	5920	9696
K5	$0\ ,2 \pm 0\ ,3$	140—295	225—470	355—745	
		512	2040	4800	7352
M	$0\ ,0 \pm 0\ ,3$	155—325	245—515	390—815	
		403	998	1366	2767

The tabulated limits of distance for a spectral class and interval of m_{vis} are computed by adding the value of the probable error to the adopted absolute magnitude for the lower limit, and subtracting the p. e. from the adopted M for the upper limit. SHAPLEY is of opinion that nearly 90 per cent of the stars of a given class and magnitude interval fall within the corresponding tabulated limits of distance. For the K0 and K2 stars of 7^m the percentage will probably be a little smaller on account of the increasing proportion of dwarfs.

For the more special groups of stars we refer to the following tabulation of the mean absolute magnitudes or to the figures following.

Mean visual absolute magnitudes of spectral classes of the Main Series.

Group	ADAMS and co-workers	NORMANN LOCKYER	Victoria	LINDBLAD and co-workers	LUNDMARK	HESS	REDMAN	NASSAU	Adopted
Oe					−2,4				− 2,5
Oe5	−2,5				−4,4				− 4,0
B0	−3,1	−3,5			−3,0				− 3,0
B1	−2,4 (*n*)	−2,3 (*n*)			−2,8			−2,3 (Oe5–B2)	− 2,8
	−2,6 (*s*)	−2,9 (*s*)							
B2	−1,5 (*n*)	−1,6 (*n*)							− 2,5
	−2,0 (*s*)	−2,3 (*s*)				−1,0?			
B3	−0,9 (*n*)	−1,1 (*n*)							− 1,8
	−1,5 (*s*)	−1,8 (*s*)							
B5	−0,5 (*n*)	−0,4 (*n*)			−0,5				− 0,6
	−1,1 (*s*)	−1,0 (*s*)							
B8	+0,1 (*n*)	+0,2 (*n*)			+0,8			−1,5 (B5–B9)	+ 0,9
	−0,6 (*s*)	−0,4 (*s*)							
B9	+0,5 (*n*)	+0,3 (*n*)			+0,9				+ 1,0
	−0,2 (*s*)	−0,3 (*s*)							
A0	+0,9 (*n*)	+0,6 (*n*)			+1,0	−1,5	+1,3	+0,3	+ 1,1
	+0,2 (*s*)	−0,1 (*s*)							
A2	+1,7 (*n*)	+1,3 (*n*)			+1,2				+ 1,3
	+0,9 (*s*)	+0,5 (*s*)							
A5		+2,2 (*n*)		+2,3	+1,7		+2,5		+ 1,7
		+1,4 (*s*)							
A8	+2,8	+2,0	+1,7						+ 1,9
dF0	+3,2	+2,8		+2,5	+2,3	+3,0	+2,8	+2,6	+ 2,4
dF2				+2,8	+2,8		+2,6	+2,6	+ 2,6
dF5	+3,5	+3,4		+3,3	+3,3	+3,8	+3,3	+2,9	+ 3,3
dF8				+4,1	+4,4		+3,8	+3,8	+ 4,2
dG0	+4,8	+4,8		+4,2	+4,7	+4,3	+4,2	+4,4	+ 4,6
dG5	+5,2			+4,6	+5,4		+5,1	+5,2	+ 5,2

Group	ADAMS and co-workers	NORMANN LOCKYER	Victoria	LINDBLAD and co-workers	LUNDMARK	HESS	REDMAN	NASSAU	Adopted
dG8				+5,6	+5,9				+ 5,7
dK0	+ 6,1	+6,0		(+4,8)?	+6,3	+ 5,6	+ 5,6	+6,0	+ 6,2
dK2	+ 6,9	+6,8			+7,2			+6,8	+ 7,0
dK5	+ 7,4				+8,0				+ 7,8
dK8	+ 7,8					+ 7,5		+8,0	+ 8,1
dM0	+ 8,3						+ 7,6		+ 8,3
dM1	+ 8,7	+8,9			+9,6 (Ma)				+ 8,7
dM2	+ 9,2								+ 9,2
dM3	+ 9,7					+10,0 (Ma—Md)	+10,3	+10,3 (Ma—Md)	+ 9,7
dM4	+10,2								+10,2
dM5	+11,0				+11,6 (Mb)		+11,8		+11,0
dM6	+12,1								+12,1

Mean absolute magnitudes of special spectral classes.

Groups	ADAMS and co-workers	NORMAN LOCKYER	Victoria	LINDBLAD and co-workers	LUNDMARK	NASSAU	HESS	REDMAN	Adopt[ed]
gF0	−1,8				−1,0			0,5	−0,
gF5	−1,5				+0,5			0,2	0,
gG0	+0,6		−2,5	−0,8 (?)	−0,6	−0,7	+1,8	0,3	−1,
gG5			−0,8	+0,6	+0,2	−0,5		−0,5	−0,
gG8				+0,9	+0,5	−0,5			−0,
gK0	+0,8			+0,9	−0,7	−0,6	0,5	−1,5	+0,
gK2									+0,
gK5	+0,6						0,3	−2,6	−1,
gK8							0,2	−4,3 (Mo)	−3,
gM1									
gM2	+0,3				−0,4 (Ma)				
gM3									
gM4									
gM5	+0,3				−0,2 (Mb)			−2,0	−2,
gM6									
gM7									
gM8									
gM9									
Cepheids					−2,6 (SHAPLEY)				−2,
Novae					−6,0				−6,
Planetary nebulae					(+8[1]) +2				+3
N stars					−1,5 (J.H. MOORE) −1,3 (LUPLAU-JANSSEN and HARRH)				−1,
R stars					0,0				
Globular clusters					−5 to −10			−8,8 SHAPLEY	−8,
Open clusters								−5 to 0 COLLINDER	−3,
Anagalatic objects					−13 to −18			−15 HUBBLE	−16,

For references to the different determinations see the appended source catalogue.

[1] Derived mainly from VAN MAANEN's parallaxes.

177. Relation between Colour and Absolute Magnitude. It is well known that there is a relation between the spectral class and the reduced proper motion and hence also between the absolute magnitude and the colour. The relation between spectral class and mean reduced proper motion is illustrated as follows:

Spectral class	$\bar{\mu}$	n	Spectral class	$\bar{\mu}$	n
O	0″,019	84	G	0″,052	444
B	0 ,024	490	K	0 ,057	1227
A	0 ,046	1647	M	0 ,050	222
F	0 ,077	656			

A more detailed relation is obtained when attention is paid to the giant and dwarf division. Extensive use has been made of the relation between spectral index S and reduced proper motion H; according to the method of LUYTEN and the present writer the correlation surface:

$$M = \psi(H, S),$$

is studied[1].

Using the colour indices of the Göttinger Aktinometrie SCHWARZSCHILD[2] has derived the following relation between the colour index $\bar{C}$ and the mean parallactic motion $\bar{\mu}_\tau$:

$\bar{C}$	$\bar{\mu}_\tau$	n	$\bar{C}$	$\bar{\mu}_\tau$	n
$-0^m,65$	0″,035	64	$+0^m,85$	0″,076	277
$-0\ ,35$	0 ,029	332	$+1\ ,15$	0 ,049	199
$-0\ ,05$	0 ,089	277	$+1\ ,45$	0 0,40	184
$+0\ ,25$	0 ,208	150	$+1\ ,75$	0 ,046	71
$+0\ ,55$	0 ,086	126			

The fact that the majority of the stars in the Göttinger Aktinometrie are giants explains the presence of the maximum for $\bar{C} = +0^m,25$.

B. FÄNGE[3] investigated the dependence of M on the colour classes instead of the spectral classes. The material used consists of the colour estimations in the Göttingen catalogue. The proper motions and radial velocities have been used in the same way as they were by CHARLIER when he was investigating the distribution in space of the B stars. The following table summarizes the results:

$\bar{C}$	K-effect	$\left(\frac{1}{R}\right)$	$\bar{M}$	$\bar{M}'$	n
$< -0^m,25$	+4,26 km/sec	0,40	$-0^M,6$	$-0^M,9$	98
$> -0\ ,25$ $< +0\ ,00$	+3,41	0,55	$+0\ ,1$	$+0\ ,1$	52
$> +0\ ,00$ $< +0\ ,20$	+3,70	0,68	$+0\ ,6$	$+1\ ,5$	38
$> +0\ ,20$ $< +0\ ,65$	+0,47	1,08	$+1\ ,5$	$+1\ ,7$	41
$> +0\ ,65$ $< +1\ ,10$	+3,55	0,60	$+0\ ,2$	$+0\ ,9$	72
$> +1\ ,10$ $< +1\ ,55$	+6,06	0,59	$+0\ ,2$	$+0\ ,2$	52
$> +1\ ,55$	−0,56	0,47	$-0\ ,3$	$-0\ ,2$	34

The $\bar{M}$ have been computed by adopting for the dispersion σ the value $\pm 1^M,2$ for F, G, and K dwarfs, and $\pm 0^M,8$ for G, K, and M giants. The $\bar{M}'$ have been computed on the basis of the common value of the stream velocity.

[1] Publ A S P 35, p. 21 (1923).
[2] Göttinger Aktinometrie Teil B, p. 37 (1912).
[3] Lund Medd Ser. II, No. 49c (1927).

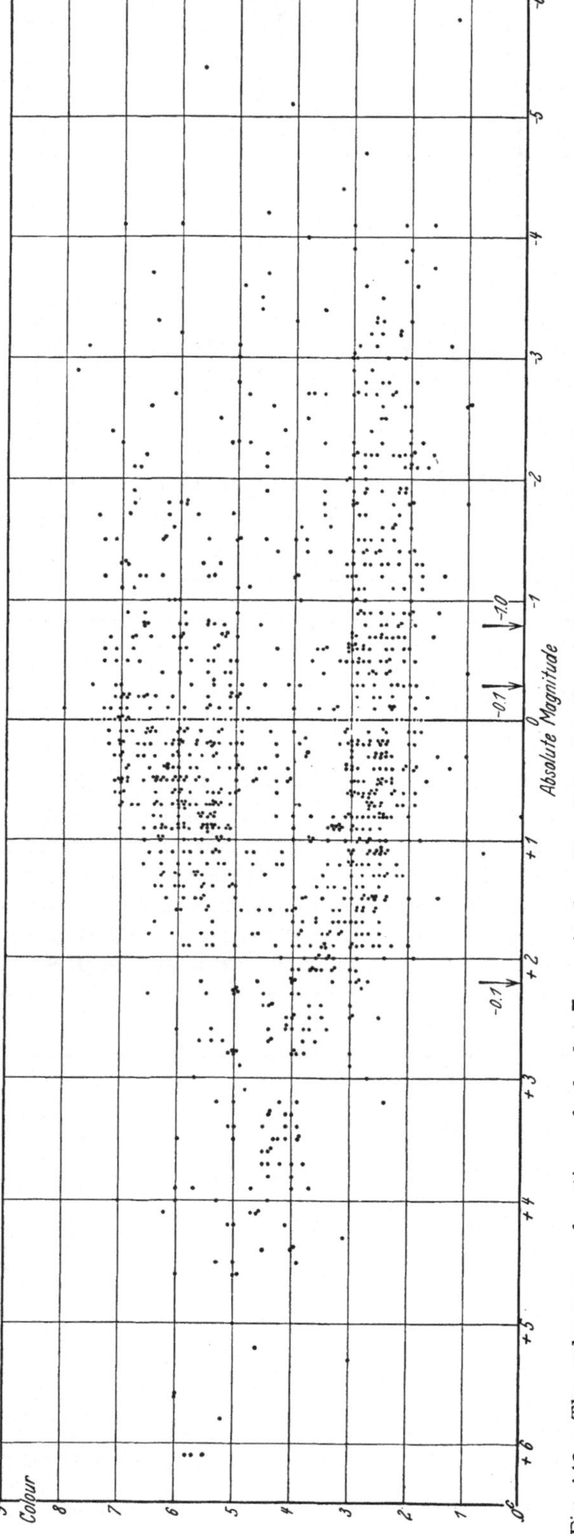

Fig. 142. The colour as a function of absolute magnitude. For this plot the material in the Index Catalogue of KRÜGER [Specola Vaticana IX (1917)] has been used. The colour is rather constant for ordinary giants but varies linearly for dwarfs between $0^{M},5$ to $4^{M},5$.

178. The Magnitude Effect in the Colours and Magnitudes of the Pleiades Stars. R. TRÜMPLER[1] raised the interesting question of the behaviour of colour index and spectrum in the Pleiades cluster. When compared with the spectral types in the DRAPER Catalogue the colour index of the F and G stars was found to be abnormally low. These results appeared clearly in the measurements of HERTZSPRUNG[2] and of Miss PARSONS[3] and were also confirmed by SEARES at Mount Wilson[4].

Provisional values of the Mount Wilson colour indices for giants and dwarfs are as follows:

Giants		Giants		Dwarfs	
Spectral class	Colour index $M=0$	Spectral class	Colour index $M=0$	Colour index	M
B0	$-0^M,32$	F5	$+0^M,62$	$+0^M,62$	$3^M,3$
B5	$-0\ ,17$	G0	$0\ ,86$	$0\ ,72$	$4\ ,4$
A0	$0\ ,00$	G5	$1\ ,15$	$0\ ,83$	$5\ ,2$
A5	$+0\ ,19$	K0	$1\ ,48$	$0\ ,99$	$5\ ,9$
F0	$+0\ ,38$	K5	$1\ ,84$	$1\ ,26$	$7\ ,1$
		Ma	$1\ ,88$	$1\ ,76$	$9\ ,8$
		Mb	$(+1\ ,88)$	$(+2\ ,00)$	$11\ ,0$

Although the difference in colour for giants and dwarfs in the region from G5 to K5 is considerable, it appears that in the interval from F0 to G5, where the abnormal relation makes its appearance in the case of the Pleiades, the influence of luminosity is small, as illustrated in the following table:

Spectral class	PARSONS	n	HERTZSPRUNG	n	SEARES	n	Mt. Wilson colour system	Mt. Wilson Pleiades
B5	$-0^m,15$	6	$-0^m,18$	1	$-0^m,21$	6	$-0^m,17$	$+0^m,01$
B8	$-0\ ,18$	6	$-0\ ,14$	6	$-0\ ,12$	4	$-0\ ,07$	$+0\ ,08$
B9	$-0\ ,07$	7	$-0\ ,11$	6	$-0\ ,03$	6	$-0\ ,03$	$+0\ ,04$
A0	$+0\ ,08$	14	$+0\ ,02$	11	$+0\ ,09$	8	$0\ ,00$	$-0\ ,06$
A2	$+0\ ,15$	5	$+0\ ,13$	15	$+0\ ,14$	4	$+0\ ,07$	$-0\ ,07$
A3	$+0\ ,19$	3	$+0\ ,14$	15	$+0\ ,18$	3	$+0\ ,11$	$-0\ ,04$
A5	$+0\ ,07$	1	$+0\ ,16$	12	$+0\ ,08$	1	$+0\ ,19$	$+0\ ,05$
F0	$+0\ ,31$	1	$+0\ ,19$	10	$+0\ ,14$	1	$+0\ ,38$	$+0\ ,18$
F5	$+0\ ,30$	1	$+0\ ,20$	10	—	—	$+0\ ,62$	$+0\ ,41$
F8	$+0\ ,24$	3	$+0\ ,23$	12	$+0\ ,22$	2	$+0\ ,68$	$+0\ ,45$
G0	$+0\ ,33$	2	$+0\ ,26$	15	$+0\ ,30$	2	$+0\ ,72$	$+0\ ,45$
G5	$+0\ ,47$	1	$+0\ ,37$	26	—	—	$+0\ ,83$	$+0\ ,46$

SEARES and TRÜMPLER mention that a partial explanation may be found in the existence of systematic errors in spectral classification. The stars concerned are faint, ranging from 9^m to 11^m, and the classification, therefore, depends on low dispersion spectrograms, and the conditions are thus favourable for the appearance of an error depending on magnitude.

179. Distances of Star Clouds from the Frequency of λ_{eff}. J. H. PETERSSON[5] has measured a plate taken by LINDBLAD at Upsala by means of the older Zeiss-Heyde astrograph and exposed on two successive nights with two different exposures, one of 45^m and the other of 420^m. The centre of the plate was $\alpha = 19^h\,22^m$; $\delta = +11°,5$, and the region investigated embraces $2° \times 5°,4$. The region investigated consists of two almost equal parts of quite different

[1] Lick Bull 10, p. 110 (1921).

[2] Later on published and communicated to TRÜMPLER in advance of publication.

[3] Ap J 47, p. 42 (1918). [4] Publ A S P 34, p. 56 (1922).

[5] Stockholm Ark Mat Astr Fys 19, No. 3 (1925).

character, one being rich in stars, the other conspicuously poor. The boundary line between them can be fixed with fairly great accuracy.

In order to obtain the apparent magnitudes, the diameters of the images were measured. In addition, eight areas on the plate were compared with a Harvard Standard Region ($\alpha = 19^h\,0^m$; $\delta = +10°,5$), the 36-cm refractor of the observatory being used. Corrections were applied for the distance from the centre of the grating plate according to PETERSSON's determination. This correction reaches a value of $0^m,48$ at a distance from the centre of $3°,0$.

Altogether 1020 stars have been measured for apparent magnitude and effective wave length. The faintest stars are of photographic magnitude of about $13^m,0$. It appears that one can very well distinguish between white and red stars, but scarcely between the successive spectral classes in the Harvard sequence of spectra.

The ratio, bright region/poor region, of the star numbers on the plate is 4,66. The poor region mainly contains the foreground stars of the galactic cloud in this region. The following table shows the difference between the two regions as to the frequency of colours and magnitudes (the numbers in the table are numbers of stars):

λ_{eff} / Apparent magnitude	Poor region							Rich region									
	4150 to 4200	4200 to 4250	4250 to 4300	4300 to 4350	4350 to 4400	4400 to 4450	4450 to 4500	4000 to 4050	4050 to 4100	4100 to 4150	4150 to 4200	4200 to 4250	4250 to 4300	4300 to 4350	4350 to 4400	4400 to 4450	445 to 45
$8^m,5-9^m,5$	2	3				1				1	3	5	5	3	1	1	
9 ,5—10 ,5		10	4	1		1				2	9	14	5	5	3	2	
10 ,5—11 ,5	2	11	12	5	5	3	1	1	1	5	29	47	22	8	8	4	2
11 ,5—12 ,5	4	23	33	21	10	6	3	1	2	14	67	154	109	50	34	8	6

The following summary shows the marked difference as regards the frequency of A stars ($\lambda_{eff} = 4211$) in the two regions:

Mean effective wave length.

Region	$9^m,5-10^m,5$	$10^m,5-11^m,5$	$11^m,5-12^m,5$
Poor region	λ 4225	λ 4275	λ 4275
Cloud region. . . .	λ 4175	λ 4225	λ 4225

The ratio, A stars/all stars, is 0,72 in the interval $10^m,5-11^m,5$, and 0,56 in the interval $11^m,5-12^m,5$. The relative decrease of the A stars for decreasing magnitude is therefore quite distinct.

By the help of the A stars the mean distance of the star cloud is estimated. KAPTEYN's formula is used in the form modified by S. RAAB[1]:

$$\frac{m_1 - \overline{m}}{\sigma} = \frac{m_1 - 5\log r - \overline{M}}{\sigma} + \frac{p(x_1)}{Q(x_1)}.$$

In this formula $\overline{m}$ is the mean value of the magnitudes of all stars considered, brighter than m_1; $\overline{M}$, the mean absolute magnitude of A stars, is taken as $+1^M,4$ and σ is the dispersion in m. The functions $p(x_1)$ and $Q(x_1)$ defined from the equations:

$$p(x) = \frac{e^{-\frac{x^2}{2}}}{\sqrt{2\pi_c}}, \qquad x = \frac{m - 5\log r - \overline{M}}{\sigma},$$

and

$$Q(x) = \int_{-\infty}^{x} p(x)\,dx$$

[1] Lund Medd Ser. II, No. 28, p. 15 (1922).

are tabulated in RAAB's paper. The distance is found to be 5000 light years or $\pi = 0'',00066$.

Later on C. SCHALÉN[1] has classified some 400 stars in the same region with regard to their spectra. Because of the spectra being short, and only the region λ 4100—λ 4400 being in good focus, the classification had to be restricted to the divisions B8, A0, A5, F0, F5, G0, G5, K0, K5, and Ma.

Some 300 stars were in common with PETERSSON's material. The dispersion in the effective wave lengths for the different spectral classes was estimated

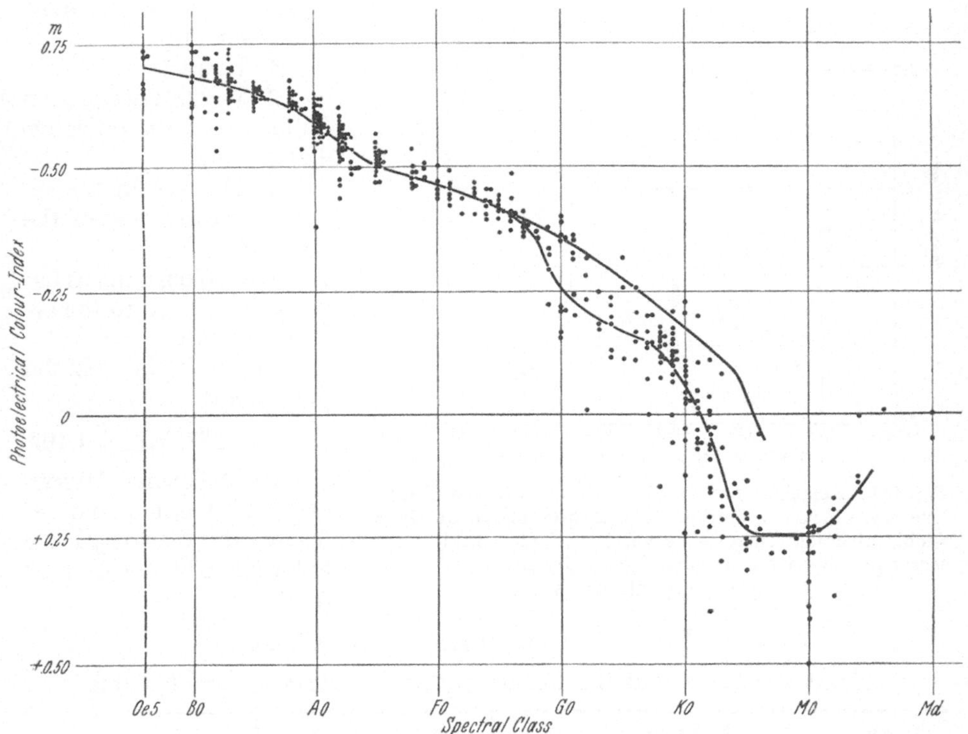

Fig. 143. Photo-electric colour indices as function of the spectral class according to BOTTLINGER's determinations. The two curves correspond to the mean colour indices of giants and dwarfs.

as ±45 AE in general and as ±39 AE for the A stars. The following relation between spectral class and effective wave length was found:

Spectral class	A0	A5	F0	F5	G0	G5	K0	K5	Ma
λ_{eff}	4218	4228	4252	4244	4283	4267	4343	4376	4430

With the aid of the classified A stars a new determination was made of the distance of the cloud. By applying the same formulae and using the same numerical constants the distances were found to be 3000 parsecs. As a by-product it was found that the greater part of the later types, so far classified, were giants.

180. Colour Index from Photo-electric Measurements. The first series of colour indices determined from photo-electric measurements is due to P. GUTHNICK[2], who measured the magnitudes of 67 stars with and without a yellow filter.

[1] Ark Mat Astr Fys 18, No. 36 (1925). [2] A N 210, p. 345 (1920).

Now the difference between spectral classes B and M amounted only to $0^{m},5$, whereas the difference $m_{ph} - m_v$ is four times as large. K. F. BOTTLINGER[1] has enlarged the amplitude of C by using a blue and yellow filter. This method is also favourable for diminishing the difference of intensity between the two light sources.

Two cells of Kalium and one of Calcium-Cadmium were used. The photometer was attached to the 31 cm-refractor of the Observatory of Berlin-Babelsberg. Care was taken to eliminate the systematic errors, of which the most important are:

1. Extinction.
2. Magnitude equation (influence of the objective diaphragm).
3. Changes in the selective sensitiveness of the cell.

The colour indices have been reduced to the system of KING.

The following relation was found:

$$C_{\text{King}} = 1{,}982\, C_{\text{Bottl.}} + 1{,}197.$$

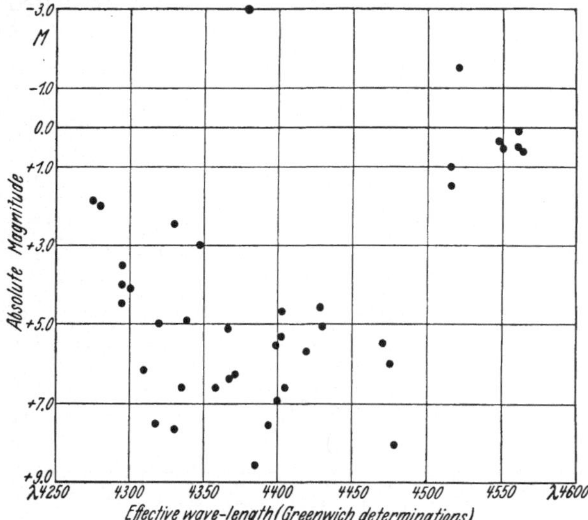

Fig. 144. Relation between absolute magnitude and effective wave length according to measures of λ_{eff} at Greenwich. Although there is no very clear relationship still it seems possible from a given λ_{eff} to discern if two groups of stars are giants or dwarfs.

By following HERTZSPRUNG's method and reducing the C's to c_2/T the following reduction formula resulted:

$$c_2/T = 1{,}20 + 2{,}071\, C_{\text{Bottl.}} - 0{,}088\, C^2_{\text{Bottl.}}.$$

BOTTLINGER has found the following normal values for giant stars:

Spectral class	$\bar{C}$	Spectral class	$\bar{C}$	Spectral class	$\bar{C}$	Spectral class	$\bar{C}$
Oe	−0 ,700	A6	−0 ,495	F9	−0 ,295	K2	+0 ,045
B0	0 ,680	A7	0 ,490	G0	0 ,260	K3	0 ,130
B1	0 ,670	A8	0 ,480	G1	0 ,230	K4	0 ,200
B2	0 ,660	A9	0 ,470	G2	0 ,210	K5	0 ,240
B3	0 ,655	F0	0 ,465	G3	0 ,195	K6	0 ,245
B5	0 ,650	F1	0 ,460	G4	0 ,180	K7	0 ,245
B8	0 ,625	F2	0 ,450	G5	0 ,170	K8	0 ,245
B9	0 ,610	F3	0 ,440	G6	0 ,160	K9	0 ,250
A0	0 ,590	F4	0 ,430	G7	0 ,145	M0	0 ,250
A1	0 ,575	F5	0 ,420	G8	0 ,130	M2	0 ,220
A2	0 ,555	F6	0 ,405	G9	0 ,095	M3	0 ,185
A3	0 ,530	F7	0 ,380	K0	0 ,060	M4	0 ,145
A5	−0 ,500	F8	-0 ,350	K1	−0 ,010	M5	+0 ,120

Attempts have been made to determine two colour indices with a view to eliminate the spectral class and computing M from the equations:

$$C_1 = \varphi(S, M)$$
$$C_2 = \psi(S, M).$$

[1] Veröffentl Berlin-Babelsberg Bd. 3, H. 4. (1923).

In order to obtain the two colour indices the photo-electric magnitudes were used as measured without filter, with yellow filter, and with blue filter. No positive results were found that enabled a determination of M, and in the subsequent work only one colour equivalent was used.

The method of ADAMS and KOHLSCHÜTTER can be formulated in such a way that absolute magnitude and spectral class are determined from two spectral parameters. Since the evolution of a star seems to be dependent only on its mass, the two parameters, mass and temperature, or luminosity and spectral class, will determine the evolution-stage of a star. These two quantities have been called by BOTTLINGER the complete spectral type. C, which also is a spectral argument, ought to give in combination with the spectral class the complete spectral type. The colour-excess ε is defined as the excess of the colour index over the mean value for a certain spectral interval.

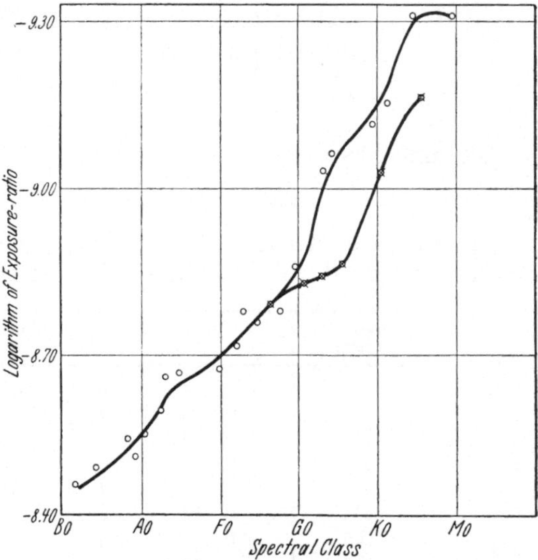

Fig. 145. The colour-absolute magnitude effect in stars of spectral classes later than G0 according to SHAJNS determination[1] (compare ciph. 112).

It was intended to derive the mean secular parallax for different groups of ε and thus obtain a scale that gave the complete spectral type from spectral class and colour index. The material did not prove to be sufficient. Then some calculations were made, and the results are given in the following table:

Groups of objects	$\bar{\varepsilon}$	Absolute magnitude from the proper motions		$\bar{M}$	Mean radial velocity in km/sec	Mean galactic latitude	n
		υ-components	τ-components				
Red supergiants	$+0^m,174$	$-3^M,8$	$-3^M,0$	$-3^M,4$	9	16°	24
c stars	+0 ,003	−3 ,6	−3 ,5	−3 ,6	7	8	17
c and giant stars	+0 ,047	+0 ,1	−0 ,3	−0 ,1	15	38	38
c stars alone	+0 ,045	−3 ,4	−1 ,9	−2 ,6	10	33	11
Giants alone	+0 ,048	+0 ,8	−0 ,4	+0 ,2	17,5	40	27
M stars, c stars	+0 ,175	−6 ,3	−3 ,7	−5 ,0	(11)	10	4
M stars, c stars, giants . .	+0 ,032	+0 ,1	−1 ,3	−0 ,6	(22)	33	15
M stars, giants alone . . .	+0 ,029	+0 ,3	−1 ,2	−0 ,4	(23)	33	13
K and M stars, c stars and giants	−0 ,031	+0 ,2	+0 ,3	+0 ,2	13	30	19
K and M stars, c stars alone	−0 ,039	−4 ,7	−2 ,1	−3 ,4	8	32	9
K and M stars, giants alone	−0 ,024	+2 ,1	+0 ,3	+0 ,7	(19)	29	10

The author justly concludes that a parallax determination from separate colour indices will only be possible in comparatively few cases. He also emphasizes the importance of an accurate spectral classification, and concludes that the colour indices seem to be best adapted to a decision between giants and supergiants.

[1] Pulk Bull X, p. 276 (1925).

181. Colours of Bright Stars. E. S. KING has applied the extra-focal method of SCHWARZSCHILD and determined the photographic and photovisual magnitudes of the brightest stars in the sky. His results together with a number of methodical discussions are given in Harv Ann 59 (1912), 76 (1916), 81 (1923), and 85 (1923) and in Harv Circ 299 (1927). The relation between the visual and photovisual magnitudes is according to KING:

Spectral class	$m_{ph}-m_v$	n	Spectral class	$m_{ph}-m_v$	n
B0—B5	$+0^M,08$	17	F8—K0	$-0^M,10$	6
B8—A3	$+0\ ,09$	13	K2—M	$+0\ ,01$	1
A5—F5	$+0\ ,03$	4			

SEARES[1] has derived the relation:

$$m_{ph} - m_v = -0^M,05 - 0^M,09\,C \quad (m < 5)$$

On the basis of the visual and photovisual magnitudes of bright stars the following standard values of the colour index are derived:

Spectral class	C	n	Spectral class	C	n
B0	$-0^M,24$	6	F0	$+0^M,34$	4
B1	$-0\ ,24$	7	F5	$+0\ ,62$	3
B2	$-0\ ,31$	4	F8	$+0\ ,84$	3
B3	$-0\ ,27$	6	G0	$+0\ ,96$	4
B5	$-0\ ,16$	4	G5	$+1\ ,14$	1
B8	$-0\ ,11$	6	K0	$+1\ ,28$	15
A0	$-0\ ,02$	19	K2	$+1\ ,62$	5
A2	$+0\ ,11$	5	K5	$+1\ ,72$	3
A3	$+0\ ,14$	3	M	$+1\ ,86$	8
A5	$+0\ ,18$	7			

The mean error of the colour indices averages $\pm 0^m,10$. A linear representation of the dependence between spectral index and colour index seems to be permissible. The change $\Delta C/\Delta S$ is $+0^m,437$.

If the colour indices derived on the basis of photovisual magnitudes are grouped according to spectral class and apparent magnitude the following result is obtained:

Spectral class	$-1^m,50$ to $+0^m,50$	$0^m,51$ to $1^m,49$	$1^m,50$ to $2^m,50$	$2^m,51$ to $2^m,99$
B	$-0^m,08$	$-0^m,27$	$-0^m,27$	$-0^m,13$
A—F	$+0\ ,24$	$+0\ ,22$	$+0\ ,23$	$+0\ ,10$
G—M	$+1\ ,11$	$+1\ ,71$	$+1\ ,50$	$+1\ ,29$

It seems that stars of the same spectral class are redder inside the Milky Way than outside. This is probably the effect of some local absorption. The recent measurements of KING confirm his earlier results concerning the existence of a selective absorption, but KING interprets these results as indicating a local absorption.

KING has suggested the use of C for computing M. Plotting 76 stars according to their colour index and absolute magnitude he finds a diagram which resembles the RUSSELL-diagram very closely. The absolute magnitude of giants is sensibly constant, whereas the magnitude of dwarfs can be computed from:

$$C = -0,0986 + 0,1736\,M.$$

[1] Mt Wilson Contr 288; Ap J 61, p. 284 (1925).

182. Spectralphotometry. Determination of Stellar Temperature. The photometric work of which we have given a brief account consists of estimates or measurements of the total intensity of the radiated energy within a certain more or less well defined part of the stellar spectra. The measurements of the intensities within certain parts of the spectrum and the relation of these intensities to the total radiation is the task of spectral photometry. A comparison between the measured distribution in spectra and the energy distribution according to the radiation-laws gives information about the stellar temperatures of the outer layer of the stars. It is not our purpose to deal with the temperature determination which is treated at length by BRILL in Bd. II/1, Kap. 2 of this Handbook, but on account of the importance of the temperature problem for the questions we shall discuss later on we are going to mention briefly the spectral-photometric work.

The first more extensive investigation of the energy distribution I am aware of is the work of J. WILSING and J. SCHEINER at Potsdam[1]. A photometer of the CROVA-type was attached to the 80-cm refractor. Five parts of the visual spectrum were measured viz: λ 4480, λ 4800, λ 5130, λ 5840, and λ 6380. The accuracy of the logarithm of the intensity was found to be the same in the different parts of the spectrum and its numerical value was derived as $\pm 0,024$. The paper gives a detailed account of the determination of the different corrections, such as the correction to the comparison spectrum for different currents through the photometer-lamp, the personal equation of different observers, the absorption of the objective, the focussing, and the atmospheric extinction.

The black-body radiation of fine leaves of platinum was studied, and also the radiation of an electric furnace. The method of measurement when applied to the Sun gave a temperature of 5130° or 5500°, if c_2 was taken as 14200 instead of the earlier accepted value, 14600. A considerable deviation from the black-body radiation was found for the part around λ 5130 (green). The deviation was also found in the lamp of the photometer and later on its cause was located in the selective absorption of a ground glass between the lamp and the Nicols.

At the same time as the Potsdam work was published, CH. NORDMANN[2] communicated his first results concerning the temperatures of the stars. Two parts of the spectrum, one in red and one in blue, were compared by means of the Heterochrom-Photometer constructed by NORDMANN. A linear relation was found between log red/blue and $1/T$ in accordance with the law of PLANCK. The constants were determined by the aid of two electric furnaces and the positive pole of the electric arc. The effective temperature of the Sun is found to be 5990°. The temperatures of 14 other stars are also given. A new reduction of the scale is given in a subsequent paper[3].

The relation between the effective temperatures measured by WILSING and SCHEINER and the colour indices was investigated by J. v. HEPPERGER[4] and v. HARKANYI[5].

In a critical note C. G. ABBOT[6] has warned against deriving temperatures from measurements covering too narrow an interval in spectrum. He feels doubtful if temperatures can really be derived from the Potsdam measurements. But in any case the latter are of high value for our knowledge of the intensity distribution in the visible part of the spectrum.

A photographic determination of the temperatures of 7 stars was published by A. HNATEK[7].

[1] Publ Astrophys Obs Potsdam No. 56 (1909). [2] C R 149, p. 557 (1909).
[3] C R 149, p. 1039 (1909). [4] A N 184, p. 283 (1910). [5] A N 186, p. 161 (1910).
[6] Ap J 32, p. 319 (1910). [7] A N 187, p. 369 (1911).

H. ROSENBERG[1] has applied the photographic method of spectral photometry to a more extensive programme. All northern stars brighter than 3^m-70 in all—have been measured at the Observatory of Göttingen. A detailed comparison between the results and those of NORDMANN showed good agreement between the results of the two methods[2].

The photographic method has also been applied by G. TICHOV[3] at Pulkova, who has used five "selected filters" corresponding to λ 5850—λ 5450, λ 5500—λ 4950, λ 4800—λ 4350, λ 4150—λ 3700, and λ 4100—λ 3500. The intensities of the different spectral regions were determined with the aid of a time-scale and the results were reduced to a mean system. The determinations embrace the 252 Pleiades stars in the list of GAULTIER, for which sufficiently bright spectrograms have been secured with the aid of the Pulkova Astrograph and an objective prism with an angle of 20°.

The determinations of temperatures at Potsdam have been extended by J. WILSING working jointly with J. SCHEINER and W. MÜNCH. In 1919 a new paper was published[4] giving the results for altogether 199 stars (earlier results are included in this paper). The methods are essentially the same as those described in the paper of 1909. The programme includes the stars in the northern sky brighter than 4^m. It is pointed out that identical results cannot be expected in general, when the temperature is determined from measured intensities in certain spectral regions or based on measurements of the total radiation. The fact that the spectral intensities can be represented by means of the radiation equation does not prove that a relation exists between the parameter T in this equation and the temperature of the radiating body. WILSING and SCHEINER have measured intensities in the spectra of the Moon, Jupiter, and Mars and have found that effective temperatures can be computed also for these bodies. This is explained from the fact that the reflected radiation is proportional to the incident radiation, as is found by applying the law of WIEN.

Radiating bodies give effective temperatures from their observed energy-curve provided that this curve can be represented by means of the radiation law; if this is not the case, limiting values can be obtained in several instances. For instance the "gray" radiators have a constant absorption power, and thus it follows according to the law of KIRCHHOFF that the intensities of two parts of the spectrum give equal temperature. The effective temperature thus measured is lower than the actual temperature of "gray" radiating bodies.

The work by LINDBLAD and HERTZSPRUNG concerning the relation between the inverse temperature value c_2/T and the spectral classes of Miss MAURY and ADAMS and JOY, the colours of LAU, and the colour equivalents of KING and of LINDBLAD has been mentioned earlier, especially in ciph. 123 and 131. Very definite relations were obtained which make it possible to reduce spectra or colours to the temperature-value in the Potsdam system.

In 1922 W. W. COBLENTZ[5] published determinations of the temperature of 16 stars based on his observations at Flagstaff Observatory, in which the thermocouple-method, as developed in his paper of 1914, was applied. The spectral energy distribution was determined by means of transmission screens of red and yellow glass, quartz, and water placed in front of a vacuum thermo-

---

[1] Photographische Untersuchung der Intensitätsverteilung in Sternspektren. Abh. d. K Leop Carol Deutsch Akad. d. Naturforscher 101, Nr. 2 (1914).

[2] C R 156, p. 1355 (1913).

[3] Pulk Mitt. 4, p. 35 (1911); Publ Ser. II, p. 17 (1912).

[4] Publ Astrophys Obs Potsdam No. 74 (1919).

[5] Wash Nat Ac Proc 8, p. 49 (1922).

couple. By means of these it was possible to obtain radiations from the regions: λ 3000 to λ 4300, λ 6000 to λ 14000, λ 14000 to λ 41000, λ 41000 to λ 100000. It was found that the maximum radiation intensity for B and A stars lies between 0,3 and 0,4μ, while in the K and M stars it is found between 0,7 and 0,9μ. The black-body temperature thus varies from 3000° to 10000°. Later on a value of 12000° was derived for the B stars. The temperature values of COBLENTZ are certainly very important as they are derived from a very wide spectral interval.

The two principal scales of stellar temperature, that of WILSING and SCHEINER and that of ROSENBERG, have been discussed and reduced by A. BRILL[1]. The conception of spectral colour index is introduced by which is meant the difference between the intensity of a certain wave length and the visual magnitude. The values of C as a function of $1/T$ should form a straight line if the radiation law of PLANCK holds good. It seems that the stars do not radiate as a black body. BRILL introduces further the isophotic wave length (cf. p. 175 of this volume). The relation between these wave lengths and the spectral class is derived from a number of photometric sources.

The measurements of ROSENBERG and of SCHEINER and WILSING have been connected and reduced to an absolute energy curve, the measurements of the former of the spectral intensity of the Sun being used as an auxiliary. The absolute energy curves have been derived for the principal spectral classes. The two series together have been used for computing the energy-values with the temperature and wave length as arguments.

The agreement between ROSENBERG's and the Potsdam measurements is now good. Only in the case of the very hot stars considerable deviations still occur, which, so far, remain unexplained.

The method of using colour filters as suggested by NORDMANN was applied by GALLISSOT[2] to 42 binary stars. If E_R, E_G, E_B are the luminous energies for the red, green, and blue radiations of the primary and e_R, e_G, e_B the corresponding quantities for the companion, the following quantities are defined:

$$M = \log\frac{E_R}{e_R} - \log\frac{E_B}{e_B} = \log\frac{E_R}{E_B} - \log\frac{e_R}{e_B},$$

$$N = \log\frac{E_R}{e_R} - \log\frac{Ee}{e_G} = \log\frac{E_R}{E_G} - \log\frac{e_R}{e_G}.$$

Further a quantity Q is introduced:

$$Q = \alpha \log\frac{E_R}{e_R} + \beta\frac{E_G}{e_G} + \gamma\frac{E_B}{e_B}$$

where α, β, and γ are coefficients immediately obtained from the measurements. The quantity Q is a measure of the light and the colour. This quantity can be determined from the number C in the equation:

$$\log\frac{E_B}{E_R} = KC + H$$

where K and H are constants.

The coloration C is linearly related to the temperature:

$$C = m + \frac{n}{T}$$

where m and n are constants.

A similar expression is found for the colours in OSTHOFF's scale.

---

[1] A N 218 and 219; Berlin-Babelsberg Veröff 5, H. 1 (1924).

[2] Lyon Thèse pour le doctorat (1921).

The temperature scale of GALLISSOT follows:

Colour class	$\overline{T}$	n	Colour according to OSTHOFF	n	Coloration C
W	8360°	5	2°,4	32	2,81
W+	7540	7	2 ,5	65	3,01
GW+	8570	0	2 ,8	155	2,76
GW	6030	9	3 ,3	126	3,73
GW+	6500	2	3 ,8	36	3,49
WG−	5500	2	4 ,1	36	4,05
WG	4390	22	5 ,2	102	4,97
WG+	4500	4	5 ,5	25	4,85
G−	3660	13	6 ,1	85	5,86
G	3770	7	6 ,4	72	5,71
G+	3500	3	6 ,5	27	6,12
RG−	2900	1	6 ,7	20	7,29
RG	2800	1	6 ,9	17	7,53
GR			8 ,8	1	

In his thesis of 1920 LINDBLAD[1] has computed the difference in temperature between giants and dwarfs on the basis of the theory of radiative equilibrium and found in the case of G stars a difference of 6720°−5560° = 1160°, which he compares with the recent value of STORER: 1250°. Later on several determinations have been made of stellar temperatures in the two series of spectra. Some of these, such as HUFNAGEL's[2] and GERASIMOVIČ's[3] are of special value.

Finally we have collected a number of different temperature scales in the following table in order to illustrate the present state of the question of establishment of such scales. No completeness is aimed at.

Different scales of stellar temperature.

Spectral class	WILSING SCHEINER	ROSENBERG	KING	HERTZSPRUNG	SAMPSON	SAHA	HOGG	Spectral class	STORER	HUFNAGEL
	According to BRILL's reductions									
O				10800°		21000°		Oe5	35000°	
B0	12300°	30000°	22000°	9700	25000°	18000		B1p	15000	
B5	11450	18000	15200	10500	16600	14000	16500°	B5	25000	
A0	10250	12000	10900	9700	13100	12000	12000	A0	15000	
A5				7600			8200	A5	11000	
F0	7950	7100	7850	6500	8900	9000		F0	10100	
F5				6000				F5	7800	
G0	5980	6000	5160	5000	6200	7000		dG0	7000	
G5				4550				gG0	5750	
K0	4570	4570	3890	4040	4200			dG5	6400	4800°
K5	4000	3840	3160	3300	3500			gG5	5850	4000
Ma	3550	3580	3000	3160	3400	5000		dK0	6100	4400
Mb				3300				gK0	4500	3700
								dK5	4500	4000
								gK5	4100	(3000)
								dMa		
								gMa	3700	

R. A. SAMPSON[4] has used slit spectrograms for the determinations of T. The measurements were made with a self-registering photoelectric-cell micro-

[1] Upsala Univ Årsskr (1920).

[2] Harv Bull 866 (1929); Harv Circ 339 (1929).

[3] Harv Circ 343 (1929).

[4] M N 83, p. 174 (1923); 85, p. 212 (1925).

photometer. The measured microphotometer tracings were first changed to opacities. An approximate relation between light intensity, expressed in photographic density, and exposure time was assumed, on the basis of which the ratio of the intensities of the radiation from each star and its comparison star was derived. Results have been published for 64 stars. SAMPSON points out that the formula for measuring the intensity of the radiation consists of a pure temperature factor and a qualifying function expressing a feature associated with the stellar atmosphere. Attention should be paid to specifying the pressure factor more precisely and to giving it a clear physical meaning. The cause of variation from uniformity may be explained by such a pressure factor.

H. H. PLASKETT[1] has applied the wedge method to several bright stars. A neutral tinted wedge uniformly illuminated by the star was placed in front of the slit so as to produce a spectrum that was wider in its more intense portions. The measurements determined how much the stellar radiation had to be diminished by the calibrated wedge in order to produce a standard blackening. The comparison spectrum was that of a tungsten lamp photographed on each plate and measured to the same standard blackening.

GREAVES, DAVIDSON, and E. MARTIN[2] have applied the prism-grating method of E. HERTZSPRUNG. A coarse grating with parallel wires is placed in the optical path with its dispersions perpendicular to that of the prism. The central spectrum is then flanked by two spectra with a constant magnitude difference. An accurate photometric scale can then be derived on the basis of SCHWARZSCHILD's method[3]. Results have been published so far for 22 stars.

Numerous investigations of the absorption lines in stellar spectra have been carried out at Harvard Observatory. Miss CECILIA PAYNE and F. HOGG[4] have presented a method for calibrating the intensity scale of objective-prism spectrograms, which makes use of several stars of the same spectral class and known magnitude difference. This method has been applied by F. HOGG[5] to a determination of the effective temperatures of 36 stars in the Pleiades.

N. V. STORER[6] has made an investigation at Lick Observatory of the temperatures of stars of different spectral type and different classes of absolute magnitude (giants and dwarfs). Slit spectrograms of 56 stars were secured with the Crossley reflector. Plates were used that gave a measurable range of spectrum from λ 3800 to λ 7400. On each plate four exposures of different intensities of a standard spectrum were impressed and the Sun served as an ultimate standard. The spectra were studied by means of a MOLL self-registering microphotometer, a simplified method of reduction being applied.

The temperatures so obtained are, both individually and in groups, somewhat higher than those derived by other investigators and a search for sources of systematic errors has been made by STORER. Atmospheric absorption, loss of light at the slit, uneven reflecting power of the mirror, and errors in the process of reduction have been considered. Only the first of these has been found capable of producing temperatures that are too high and the possibility of this seems to be slight. On the other hand the wide range of the wave lengths gives support to the temperature-scale of STORER.

The deviations of each intensity curve from the temperature curve have been computed. STORER has concluded that no colour index that only involves the visual spectrum can distinguish a giant from a dwarf. The suggestion is

---

[1] Publ Dom Astroph Obs Victoria 2, No. 12 (1923).
[2] M N 87, p. 352 (1927).
[3] A N 172, p. 65 (1906).
[4] Harv Circ 301 (1927).
[5] Harv Circ 309 (1927).
[6] Lick Bull 14, p. 41 (1929).

made that the temperatures might be derived from the visual spectrum, spectral regions centred at λ 4700 and λ 6500 being used, and that such determinations together with others in the ultra-violet might distinguish a giant from a dwarf.

The investigations of ionization in the stellar atmospheres have led to a number of determinations of the temperatures of the stars. ADAMS and RUSSELL[1] have also presented a new method of analyzing stellar spectra, which method has been discussed and recommended by the International Astronomical Union. The method is based on the principle that a calibration of ROWLAND's scale of intensities of the solar spectrum leads to a knowledge of the relative numbers of atoms in the solar atmosphere that are active in producing different lines in a certain spectral region. If thermodynamic equilibrium is assumed and if it is also supposed that atoms at different levels are equally effective in producing a line, the investigation can be extended to stellar atmospheres in general. An equation is derived connecting the relative numbers of atoms producing a certain line in different stars with the normal numbers of neutral atoms, the excitation potential, the state of ionization, and the electron pressures and temperatures in the stars.

ADAMS and RUSSELL have found that there appears to be a serious and widespread departure from thermodynamic equilibrium in stellar atmospheres, which causes the relative number of atoms in excited states to be much greater than is theoretically indicated. The behaviour of the hydrogen lines in giants appears to be a particular instance of this phenomenon.

This anomaly does not appear to affect ionization, at least not to any such degree as it does excitation.

The authors have developed an empirical method for deducing the effects of this anomaly, which leads to a new method of determining stellar temperatures. Between the effective temperature T_e and the atmospheric temperature T_a the following relation, which is derived by EDDINGTON, is assumed to exist:

$$T_a = 0{,}88\, T_e .$$

The following table contains the two temperatures T_a and T_e of some well known stars according to ADAMS and RUSSELL.

Object	Spectral class	T_a	T_e	Object	Spectral class	T_a	T_e
α Orionis . .	cM2	2520°	2900°	Sun (zero-point) . .	dG0	5300°	6000°
α Scorpii . .	cM1	2620	3000	α Persei . .	cF5	5940	6700
α Bootis . .	gK0	3380	3850	α Can min. .	cF4	6900	7850
γ Cygni. . .	cF8	4950	5600	α Can maj. .	A2s	8700	9900

The student who requires more information concerning the determination of the temperatures of the stars is advised to consult the preceding chapter of this volume by A. BRILL and Miss CECILIA PAYNE's excellent monograph[2]: "The Atmospheres of the Stars".

183. Bolometric Magnitude. When the temperature is known the visual magnitude of a star can be reduced to bolometric magnitude, m_b, which is the magnitude when the total amount of energy radiated is taken into account. As according to STEFAN this quantity is $\sim T^4$, the temperature gives a means of deriving the total energy. From the definition of magnitudes it follows that, except for a certain zero-point correction, the quantity that reduces the visual magnitude to bolometric magnitude is $-10 \log \frac{c_2}{T}$. It is clear that the tem-

[1] Mt Wilson Rep 1928, p. 141.

[2] Harv Monogr No. 1 (1925).

perature must be very well known before an accurate reduction is possible. We give here a reduction-table derived by BRILL[1].

Spectral class	c_2/T	$m_{vis} - m_{bol}$	Heat index	Spectral class	c_2/T	$m_{vis} - m_{bol}$	Heat index
B0	$0^m,603$	$+2^m,12$	$+0^m,123$	K0	$3^m,140$	$+0^m,461$	$+0^m,659$
B5	0 ,880	+1 ,13	−0 ,068	K5	3 ,727	+0 ,878	+1 ,059
A0	1 ,216	−0 ,068	±0 ,000	Ma	4 ,020	+1 ,128	+1 ,305
A5	1 ,594	−0 ,060	+0 ,041	α Aquilae	1 ,560	−0 ,079	−0 ,003
F0	1 ,816	−0 ,048	+0 ,054	Sun	2 ,158	±0 ,000	+0 ,149
F5	2 ,080	+0 ,026	+0 ,147	3000°	4 ,783	+1 ,847	+2 ,026
G0	2 ,392	+0 ,094	+0 ,256	2500	5 ,740	+2 ,880	+2 ,994
G5	2 ,742	+0 ,232	+0 ,420	2000	7 ,175	+4 ,621	+4 ,628

184. The Absorption of Light in Space. The idea that the intensity of light does not decrease in exact proportion to the inverse square law and that the cause of such a deviation is the existence of some medium in space ("ether") can be traced back to NEWTON. In a letter to FLAMSTEED in 1691 he says that if different wave lengths have different velocities certain colour changes ought to take place immediately before a satellite of Jupiter is totally eclipsed.

The effect of such a dispersion of the light is proportional to distance and ARAGO suggested the use of the variable stars. In 1898 TICHOV[2] suggested observations of the minima of eclipsing stars and in 1908 CH. NORDMANN[3] published results from observations of Algol. By the use of a red, green, and blue filter it was found that the minimum occurred seven minutes earlier for the longer wave lengths than for the shorter ones. These results were substantiated by measurements by TICHOV, but it was also pointed out by LEBEDEW[4] and others that there was no such proportionality in the change as corresponds to the distances of the measured objects. LEBEDEW also suggested that the influence of tides and other phenomena in the atmospheres of eclipsing stars will produce similar changes in the light-curves as a dispersion in space. This suggestion was made very plausible through subsequent investigations by NORDMANN[5] and by BELJAWSKY[6]. In the course of his investigation of globular clusters SHAPLEY[7] measured the light-curves of variable stars in M5 on the basis of "blue" and "yellow"-plates maxima of intensity at λ 4500 and λ 5500. In the mean it was found that the longer wave lengths gave maxima $1,44 \pm 5,32$ minutes earlier than the shorter wave lengths. Thus the two radiations that differ in wave length by about twenty per cent can travel through space for 40000 years without losing more than one or two minutes with respect to each other. It thus seems that the velocity of light is constant within a $^1/_{40\,000\,000}$ part of its value.

The absorption of the light may be of two kinds: 1) an absorption equally affecting all wave lengths and 2) a selective absorption only affecting certain wave lengths. OLBERS[8] assumed an infinite number of suns in space and concluded from the fact that the total light of the luminous bodies in the universe has a comparatively low value that space was not transparent. We have mentioned the results of F. G. W. STRUVE[9] based on HERSCHEL's photometric results as compared with his star gauges. In his extensive work concerning the structure of the stellar system, H. v. SEELIGER[10] has discussed at length the possibility of the existence of a general absorption. It is possible to derive $\varphi(M)$ and a function

[1] Berlin-Babelsberg Veröffentl 5, H. 1 (1924). [2] Mem Soc Spettr Italiani 27, p. 41 (1898).
[3] C R 146, p. 383 (1908). [4] C R 146, p. 1254 (1908). [5] B A 26, p. 5 (1909).
[6] Mitt Pulk III, 31 (1910). [7] Harv Bull 763 (1922). [8] Berl Jahrb 1826, p. 110 (1823).
[9] Etudes d'astronomie stellaire. St Petersburg (1847).
[10] München Abh Akad 25, 3. Abh (1909).

$\Delta\{r/\sqrt{\psi(r)}\}$ of the density $D(r)$ and the absorption function $\psi(r)$ as soon as a sufficiently wide range of $A(m)$ or frequency of apparent magnitudes has been observed. SEELIGER shows that:

$$D(r) = \frac{\psi(r) - \frac{1}{2} r \frac{d\psi(r)}{dr}}{\psi^{\frac{5}{2}}(r)} \Delta\left\{\frac{r}{\sqrt{\psi(r)}}\right\}.$$

D and ψ enter in a somewhat different form into the mean parallax formula and thus it is theoretically possible to find D and ψ. But even our observations nowadays are not extensive and accurate enough to make the solution of the problem possible in practice.

SEELIGER and KAPTEYN[1] have investigated what amount of a general absorption can be assumed without violating the interpretation of the observational data. The assumption is that $D(r)$ cannot increase with increasing distance from the Sun, but is constant through wide regions of space. SEELIGER points out that for appreciable amounts of absorption there will be an enormous concentration of stars in the distant parts of the stellar system. The galactic system seen from a point far outside would look like a ring nebula. The maximal values of the absorption are:

$$\frac{\Delta m}{\Delta r} = 0^{m}{,}00009; \qquad \psi(r) = e^{-0{,}00063\int_0^r D(r)\,dr},$$

$$\frac{\Delta m}{\Delta r} = 0^{m}{,}00006; \qquad \psi(r) = e^{-00006 r}.$$

HALM[2] has shown that the relations between m, $A(m)$, or the number of stars of magnitude m, and π_m can be represented if the following assumptions are made: 1) that $D(r)$ is independent of distance, 2) that the form of $\varphi(M)$ is that of a normal (Gaussian) distribution curve, 3) that there is a general absorption in space proportional to the distance:

$$m = M + 5\log r + \frac{\Delta m}{\Delta r} r.$$

In his first paper HALM found $\frac{\Delta m}{\Delta r} = 0^{m}{,}00305$ and in his second, which was mainly based on photographic data, $0^{m}{,}00206$, when unit of r is one parsec.

It seems that the existence of a general absorption is quite possible. Already the fact that $D(r)$ computed without any assumption concerning absorption shows a general decrease in all directions from the Sun is rather suspicious. We know that our Sun has not any central position in the Galaxy, but is comparatively near its border. It also seems doubtful if the Sun is surrounded by any cluster or local system. It seems quite possible that the theory of constant density and a general absorption will explain quite as much as the classical solutions of the fundamental equation in stellar statistics. But the real question can scarcely be settled before we have extensive star counts in the Sagittarius region down to 18th or 19th magnitude.

It seems that in the regions of the Milky Way the existence of local absorptions mainly in the dark nebulae can be considered to be well established through the work of BARNARD, WOLF, HUBBLE, PANNEKOEK, GYLLENBERG, SCHALÉN, and others. C. SCHALÉN[3] thus finds for five dark nebulae a mean absorption ranging between $0^{m}{,}9$ and $2^{m}{,}0$. But in the galactic regions in general SCHALÉN[4]

---

[1] A J 24, p. 115 (1904) and numerous subsequent papers.
[2] M N 77, p. 243 (1917); 80, p. 162 (1919).
[3] Dissertation Upsala (1928). [4] Unpublished results.

has also found evidence of a general absorption, which he estimates to be of an order of $0^m,0009$ per parsec. To this it may be added that the present writer[1] has found evidence of a very small general absorption of $0^m,00000007$ per parsec in the metagalactic space. This absorption can be explained if the same frequency of meteors is assumed to exist in the metagalactic system as has been observed around the Sun by HOFFMEISTER[2] and others. P. J. VAN RHIJN[3] has found when investigating globular clusters a somewhat higher value for the general absorption.

The selective absorption has been studied in numerous investigations. The method generally employed has been given by KAPTEYN[4] and is based on the equation:

$$C = a + bm + cM + dr,$$

where a, b, c, and d are constants. Of this type are the contributions of KAPTEYN, VAN RHIJN[5], JONES[6], KING[7], and others. Now it is difficult to determine separately the values of c and d. The value of c cannot be taken to equal zero, and ADAMS and KOHLSCHÜTTER[8] and later LINDBLAD[9] have presented good reasons for the opinion that at least part of the absorption in the violet part of the spectrum of the distant stars must be ascribed, not to the selective absorption in space, but to conditions in the stellar atmospheres. In fact the method of LINDBLAD for determining M has its origin in the fact that the colour index can be represented with the expression given above by putting $d = 0$.

When SHAPLEY[10] had determined the colour indices in M13 he pointed out that the existence of negative C values of the same order of magnitude as the mean colour index of the bluest stars in our system gave an upper limit for the coefficient d. Independently of him LINDBLAD and the present writer[11] derived an upper limit for d by comparing the measured effective wave lengths of anagalactic objects and their actually classified spectral classes. The accumulation of data later on has given improved values of d_{max} and as far as the "anagalactic sky" is concerned the selective absorption is so small that even in cases of distances amounting to a hundred million light-years it seems to be negligible.

Determinations of the coefficient of absorption.

Authority	$\Delta m/\Delta r$ (r in parsecs)	Mean error of $\Delta m/\Delta r$	Material	Epoch
TURNER	$0^m,0030$			1908
KAPTEYN	0 ,00031	$\pm 0^m,00006$	Colour indices	1909
E. S. KING	0 ,0019	0 ,0010	,, ,,	1912
E. S. KING	0 ,0023		,, ,,	1914
JONES	0 ,000473	0 ,00035	,, ,,	1914
VAN RHIJN	0 ,00015	0 ,00003	,, ,,	1915
E. S. KING	0 ,00189	0 ,00065	,, ,,	1916
SHAPLEY	0 ,000002		Globular clusters	1918
LUNDMARK and LINDBLAD	0 ,0000002		Anagalactic nebulae	1919
LINDBLAD				
LUNDMARK	0 ,00000004		,, ,,	1925
SHAPLEY	0 ,00000007		,, ,,	1926
SHAPLEY	0 ,000000007		,, ,,	1929
LUNDMARK	0 ,000000003	0 ,000000003	,, ,,	1929
SCHALÉN[12]	0 ,00001		B and A stars	1929

[1] M N 85, p. 865 (1925). [2] Astr Abh Ergh zu den A N 4, No. 5 (1922).
[3] B A N 4, p. 123 (1928). [4] Ap J 40, p. 187 (1914); Mt Wilson Contr 83.
[5] Dissertation Groningen (1915). [6] M N 75, p. 4 (1914).
[7] Harv Ann 76, p. (1916); Nature 95, p. 701 (1916). [8] Ap J 40, p. 385 (1914).
[9] Dissertation Upsala (1920). [10] Wash Nat Ac Proc 2, p. 12 (1916); 3, p. 267 (1917).
[11] Ap J 46, p. 206 (1917); see also LINDBLAD's dissertation Upsala (1920).
[12] This value is valid for the Milky Way regions.

185. Photoelectric Photometry. For an extensive account of the photoelectric photometry we refer to the Chapter 4 by ROSENBERG in Vol II/1 of this Handbook. Here we only give a short sketch of the development. It has been shown by ELSTER and GEITEL[1] that the photoelectric cell can be used for measuring light and gives a very high accuracy. The principle is founded on the fact that when light waves hit a surface a number of electrons will be thrown out (photoelectric effect). The number of electrons is proportional, at least for the alkali metals, to the light intensity.

The first astronomers who applied the method for measuring magnitudes were P. GUTHNICK and R. PRAGER[2]. For an account of the construction of the photometers and for technical details reference should be made to the papers by these authors, and to the thesis of Mrs. EDITH CUMMINGS-TAYLOR[3].

GUTHNICK and PRAGER have measured a number of spectroscopic double stars or ordinary stars, such as β Cephei, α Canum Venaticorum, γ Bootis, γ Pegasi, α and β Arietis, α and β Tauri, α, β, and γ Geminorum and α Leonis. Besides, the planets Mars and Saturn have also been measured.

The many interesting results with regard to the variability of the stars investigated are mentioned in the chapter concerning the variable stars. The accuracy of the method is extremely high. From the first paper by GUTHNICK and PRAGER we quote the following values:

1914	Δm γ Urs$-\zeta$ Drac.	Mean error of one observation	1914	Δm γ Urs$-\zeta$ Drac.	Mean error of one observation
July 2	$+0^m,097$	$\pm 0^m,0115$	Aug. 13	$+0^m,083$	$\pm 0^m,0075$
12	0 ,106	0 ,0082	15	0 ,088	0 ,0090
Aug. 12	0 ,096	0 ,0093	16	0 ,089	0 ,0078

In the second paper by intercomparisons of an artificial light source the authors find the mean error in the method to be $\pm 0^m,0058$. This value is free from atmospherical influence, but this is counterbalanced among other things by the fact that the cells have not got time enough to rest. Altogether it seems possible, in favourable cases, to press down the error within the limits $\pm 0^m,005$.

Among the sources of error the extinction is the one that is most difficult to master. It will be necessary to make new extinction tables by the aid of observations with photoelectric instruments. The establishment of a sequence of stars adapted for photoelectric purposes is one of the urgent needs in future photometry.

When this has been done and measurements have been made at different epochs, it will certainly be possible, within a reasonable time, to determine photometric parallaxes.

Another application of the photoelectric method to astronomical problems was made by Mrs. EDITH CUMMINGS-TAYLOR at the Lick Observatory[3]. In the part of her thesis already published measurements of β Cephei are communicated. Later on measurements have been made at the Lick Observatory by BAKER and STEBBINS.

ROSENBERG[4] has refined the method. According to his investigations it may be possible to reach an accuracy of $0^m,005$ or more. One of the troubles is, of course, that it might be difficult to find stars with a constant light. As

[1] Phys Z 11, p. 212 (1910); 12, p. 609 (1911); 13, p. 739 (1912).
[2] Berlin-Babelsberg Veröffentl 1, part 1 (1914); 2 part 3 (1918).
[3] Lick Bull 11, p. 99 (1924).
[4] Berlin Sitzber 1920, p. 716.

the method increases in accuracy smaller and smaller amplitudes in the light curves will be detected.

STEBBINS and C. M. HUFFER[1] have observed 235 stars in pairs or small groups in order to obtain accurate differences of magnitude. When an extended series showed a mean deviation smaller than $\pm 0^{m},012$ from a constant difference, the stars involved were considered as invariable. The volumes give a number of interesting discoveries with regard to the light variation of variable stars and stars of ellipsoidal figure.

186. The Radiometric Measurements. An attempt has been made to measure stellar radiation by means of thermo-elements and by means of different radiometers. The first attempts were made by HUGGINS[2], who used one or two pairs of elements of bismuth-antimony, in the focus of a refractor, and found positive deflections for some bright stars. STONE[3] applied the thermoelectric method in 1869—70 and obtained results that seem to be reliable. The use of thermo-elements in an evacuated receptacle seems to have increased the sensibility considerably. Among the first to use this method was PFUND[4]. Later on the method was used by W. W. COBLENTZ[5] at the Lick Observatory and results were obtained for more than 100 stars.

The current generated was measured by means of an iron-screened galvanometer. The galvanometer sensitivity was $1,4 \cdot 10^{-10}$ amp for a single swing of four seconds. The thermocouples were made of bismuth-platinum and bismuth-tin alloy. Improvements were invented for maintaining the vacuum by means of metallic calcium.

Measurements were made with the Crossley reflector on the radiation of 112 objects, of which 105 were stars. Quantitative measurements were made on stars down to $5^{m},3$ and high grade qualitative measurements were made on stars down to $6^{m},7$.

Measurements were made on the transmission of radiations from stars by the interposition of an absorption cell of water in the way of the rays.

COBLENTZ has communicated his results expressed in galvanometer deflections. These were translated into magnitudes by K. BURNS[6] in order to facilitate comparison with m_{vis} and m_{phot}. The following table gives the size and distribution of the two new colour indices $C_{rv} = m_{\text{rad}} - m_{\text{vis}}$ and $C_{rp} = m_{\text{rad}} - m_{\text{phot}}$.

Spectral class	Colour index C_{rv}	n	Colour index C_{rp}	n
B —B5	$-0^{M},33 \pm 0^{M},17$	12	$-0^{M},82 \pm 0^{M},19$	8
B8—A5	−0 ,27 0 ,07	27	−0 ,77 0 ,13	14
F —F5	−0 ,6	3	−1 ,3	2
F8—G5	−0 ,99 0 ,13	14	−2 ,02 0 ,15	11
K —K5	−1 ,23 0 ,09	26	−2 ,70 0 ,18	26
Ma	−2 ,20	10	−3 ,70 0 ,21	8
Mb	−2 ,7	2	−4 ,7	1
N	−3 ,2	1		
α Herculis (Mb)	−4 ,2	1	−5 ,8	1

The total intensity of G stars and K stars is twice that of B and A stars, the visual magnitude being the same. This ratio increases with the "lateness" of the type and for N stars it is 15 times that of B and A stars.

[1] Publ Washburn Obs 15, part. I, II (1928).
[2] London R S Proc 17, p. 309 (1868—69).
[3] London R S Proc 18, p. 159 (1869—70).
[4] Allegheny Obs Publ 3, p. 43 (1913).
[5] Lick Bull 8, p. 104 (1915).
[6] Publ A S P 27, p. 110 (1916).

When the stars are grouped according to their spectra it is seen that the values of $m_{rad} - m_{vis}$ agree to within a few tenths of a magnitude in most cases, and also for faint stars.

M was computed for those objects with observed parallax, but as there were only few K. BURNS concludes that the statistics are not of much value.

Among other results it may be mentioned that there is no relation between the spectral energy distribution and the absolute total energy distribution and the absolute total energy of the stars.

187. PETTIT's and NICHOLSON's Measurements of Stellar Radiation. A paper by these authors[1] discusses the measurements of stellar radiation made during the period 1922—27 with vacuum thermocouples attached to the Hooker telescope. Twelve different thermocouples in cells, provided with windows of rock salt, fluorite, quartz, and microscopic cover-glass were employed. The methods of constructing thermocouples described in a paper by PETTIT[2] have been varied since wires of bismuth and of the alloy of this metal with 5 per cent of tin have become available through the work of G. F. TAYLOR[3], who developed the method of drawing these metals in glass tubing, which is later removed with hydrofluoric acid.

The wires are about 0,03 mm in diameter. The operation of producing the thermocouples is carried out under a microscope. The wires are placed on a turn-table, the surface of which consists of a plate of glass resting on a sheet of white paper. A small spot on the glass is moistened with a saturated solution of $ZnCl_2$. The wires are laid on the spot with a speck of solder near the junction and welded together with a nichrome wire loop heated to incandescence with an electric current. The $ZnCl_2$ is washed away with water and alcohol, and the plate glass is warmed over a flame. The receiver is then formed by flattening the globule of metal at the junction with a plate of glass. For the smallest receivers the wires are welded together without solder. Receivers larger than 0,7 mm are made of Cu and soldered to the junctions, although in the case of thermopiles they are attached with white lead.

The receivers are blackened with a thin mixture of lamp-black and platinum-black in alcohol and terpentine. The thermocouple is attached to the solder-covered lead wires by the use of $ZnCl_2$ flux, which is afterwards removed with wet cotton on the point of a stick.

From the electrical properties of the circuit it can be shown that a galvanometer deflexion of 1 mm corresponds to a rise in temperature of $0°,000033$ Celsius in the receiver.

The thermocouple cell is clamped to the double-slide plateholder at the Newtonian focus of the Hooker telescope. A water-cell 1 cm thick can be placed in the convergent beam from the 100-inch mirror. The thermocouple is connected by a cable with a Leeds and Northrap d'ARSONVAL galvanometer of current sensitivity $3 \cdot 10^{-10}$ amp per mm at 8 m scale-distance. The deflections are recorded on a moving photographic plate 6 to 10 m from the galvanometer. The star-image is focussed on the receivers (optical method) and the deflections are produced by shifting the thermocouple with one of the guiding screws so that the image falls alternately on the two junctions at regular intervals of 20 sec, as indicated by a buzzer operated by a motor-driven registering device. There is a definite linear relation between the deflections and the intensity of the incident radiation.

---

[1] Mt Wilson Contr No. 369 (1928); Ap J 68, p. 279 (1928).
[2] Mt Wilson Contr 246; Ap J 56, p. 295 (1922).
[3] Phys Rev 23, p. 655 (1924).

At first 8 deflections are registered and then the water-cell is inserted in the beam from the 100-inch mirror. After refocussing, another set of 8 deflections, produced by the radiation transmitted by the water-cell, is recorded. The deflections are measured on a specially adapted comparator to a hundredth of a mm. The water-cell was cleaned and refilled with distilled water before every second or third night of observation. The authors are now convinced that a plate of crystal alum between quartz plates would be more permanent than the water-cell and give a sharper cut-off at λ 14500.

The measurements are reduced to radiometric magnitudes. m_r is defined by the authors as the visual magnitude of an A0 star which would give the observed galvanometer deflection. The difference:

$$m_v - m_r = C_H$$

gives the heat index, C_H. The following table shows the relation between m_r and the average galvanometer deflection and the heat on the receiver:

Galvanometer deflection	m_r	Heat on the receiver	Galvanometer deflection	m_r	Heat on the receiver
631	−2		1,00	5	0°,00034
250,00	−1	0°,050	0,40	6	0 ,000014
100,00	0	0 ,033	0,16	7	0 ,000006
39,80	1	0 ,020	0,06	8	0 ,000002
15,90	2	0 ,0052	0,02	9	0 ,0000006
6,30	3	0 ,00209	0,01	10	0 ,0000003
2,51	4	0 ,00080			

The magnitudes have to be reduced to the zenith of Mount Wilson. The reduction has the form:

$$\varDelta^1 m_r = -A(\sec Z - 1).$$

The values of A determined from the observations range from $0^m,07$ to $0^m,49$ with a mean value of $0^m,165$. A special determination based on stars of different spectral classes gave a mean of $0^m,16$ and showed little variation with spectral class. The average value of the correction $\varDelta^1 m_r$ was $-0^m,05$ and in only about 6 per cent of the observations it exceeded $-0^m,15$.

During the course of the work it was found that m_r of red stars decreased and that of the blue ones increased with the time elapsed since the silvering of the Newtonian plate, which had been resilvered much less frequently than the 100-inch mirror. This was due to the relatively great decrease in the reflecting power of the mirror in the violet, as the tarnish of the silver increased, and necessitated the determination of a correction $\varDelta^2 m_r$, which reduced the observed m_r to fresh silver. The size S of the daily change in m_r due to the tarnishing of the silver, and the correction $\varDelta^3 m_r$ are given for the different spectral classes.

The quantity: $S\times$ number of days elapsed since silvering is equal to $\varDelta^2 m_r$.

Spectral class	S	$\varDelta^3 m_r$	S'
Bo	$-0^m,000205$	$0^m,000$	$-0^m,000150$
A0	0	0 ,000	−0 ,000112
F0	+0 ,000057	−0 ,004	−0 ,000083
G0	+0 ,000110	−0 ,018	−0 ,000061
K0	+0 ,000156	−0 ,032	−0 ,000049
M0	+0 ,000194	−0 ,046	−0 ,000041
M7e	+0 ,000205	−0 ,056	−0 ,000033
Me min	+0 ,000205	−0 ,067	−0 ,000030

Since there is little energy in stellar radiation with wavelengths exceeding 40000 AE the corrections for measurements made with a thin glass or quartz window on the thermo-couple-cell are small and those for a fluorite window are negligible. The third column in the above table gives the correction $\varDelta^3 m_r$ which reduces m_r determined with cells with glass or quartz windows to values determined with cells with rock-salt windows.

The water-cell absorption, or the fraction of radiation absorbed by the cell, is expressed as $0{,}4 \log \frac{\text{galvanometer defl. before cell is thrown in}}{\text{galvanometer defl. after cell is thrown in}}$. The corrections to zenith are differential in this case and are found to be negligible for the values of Z entering into the set of observations.

The effect of the tarnishing of the reflector on the water-cell absorption was determined in the same way as the similar effect on m_r. The daily rates, S', are entered in the fourth column of the above table.

The size of the deflections given above refers to normal conditions and varies somewhat with the different thermocouples, galvanometers, and scale-distances employed. Sometimes deflections twice as large as those tabulated on p. 561 have been used and sometimes deflections only half as large. The accuracy of the measurements varies especially with the climatic conditions. On quiet mild nights in summer or autumn the galvanometer record shows smooth curves and the probable error of the measurements of such plates is $\pm 0{,}01$ mm or $\pm 0{,}02$ mm. For values smaller than 40 mm the error seems to be nearly independent of the size of the deflection. In winter and spring when the temperature changes rapidly and especially if a cold wind is blowing the lines of the record are seldom smooth. Under extreme conditions the galvanometer may wander over several mm or even cm, thus making accurate determinations impossible. Under good conditions even the small deflections are perfectly definite and their accuracy corresponds to about a tenth of a magnitude.

When the receiver is comparatively large (0,8 mm) observations may be made even with sight 1 on the tenth-scale, providing the star image is not "blowing up". When the receivers are 0,3 to 0,4 mm in diameter a distinct unsteadiness will appear in the records when the visibility is below 3.

Moonlight and sunlight in no way affect the measurements. The thermocouple is compensated against general radiation and since the measurements are differential the stars give the same deflection in a daytime sky as at night. The authors have measured stellar and planetary radiation from objects as near as 11° to the Sun. Except for the generally bad visibility which prevails in a daylight sky, the results are in very good agreement with those obtained at night.

The papers give results for 124 stars together with the values of m_r, the correction to no atmosphere Δm_r, the water-cell absorption WC, the total radiation, heat index C_H, absolute temperatures, apparent diameters, etc. The absolute visual magnitudes were obtained by combining the Harvard m's with parallax values derived by VAN MAANEN from existing data.

By plotting the C_H and WC values against spectral types it is found that the heat index and the water-cell absorption are greater for giants than for dwarfs of the same spectral class, which is in harmony with a similar relation for the ordinary colour index or for other colour equivalents.

The values of C_H and WC were plotted against each other. The rather definite curve splits at $C_H = 2{,}2$ and $\text{WC} = 1{,}2$ into two branches; one, mainly consisting of N stars, extends the original curve rather linearly to 4,8 and 1,8. The other branch is inclined to the direction of the first one at an angle of about 35° and consists of long-period variables and very red stars later than M2. The deviation of these stars indicates that stars of low temperature do not radiate like black bodies. The long-period variables deviate at their minima most widely from a black-body radiator.

Next the temperatures are determined from the heat index. Two factors have to be decided upon that are affected by a comparatively high uncertainty, viz.: the atmospheric absorption and the spectral sensibility-curve of the eye corresponding to the Harvard system of magnitudes. In determining the temperatures from water-cell absorptions the well-determined transmission curve of the water-cell is substituted for the sensibility-curve of the eye. The *m* thus consist of homogeneous observations independent of physiological effects, while the C_H, of course, being anchored on the R H P, are influenced by any magnitude- or colour-errors affecting the Harvard observations.

On account of the small variation in C_H and WC for the classes O, B, and A reliable temperatures cannot be derived for such stars. For stars of very low temperature the results of both methods are affected by the presence of TiO absorption bands, which change the distribution of spectral energy from that of a black body. The results from the C_H are probably influenced more than those from the WC. From the transmission-curves of the atmosphere as given by the Smithsonian observers[1], and the reflection coefficients of silver, the spectral energy curves of black bodies at various temperatures have been derived. The authors have computed the table produced here which shows, for black bodies at different temperatures, the C_H and WC and reduction to no atmosphere, including absorption by the atmosphere and the telescope.

Absolute temperature	Heat index C_H		Water-cell absorption	Reduction to no atmosphere
	λ = 5550	λ = 5290		
24000°	$+0^m,01$	$+0^m,01$	$+0^m,17$	$2^m,84$
20000	0 ,01	0 ,01	0 ,18	2 ,35
16000	0 ,01	0 ,00	0 ,19	1 ,84
14000	0 ,00	0 ,00	0 ,20	1 ,57
12000	0 ,00	0 ,00	0 ,21	1 ,26
10000	0 ,02	0 ,05	0 ,24	1 ,00
8000	0 ,07	0 ,12	0 ,29	0 ,71
7000	0 ,12	0 ,18	0 ,32	0 ,60
6000	0 ,26	0 ,33	0 ,37	0 ,49
5000	0 ,48	0 ,59	0 ,46	0 ,41
4000	0 ,92	1 ,08	0 ,63	0 ,41
3000	1 ,91	2 ,24	1 ,00	0 ,48
2500	2 ,81	3 ,32	1 ,34	0 ,57
2000	4 ,33	5 ,16	1 ,93	0 ,71
1500	+7 ,29	+8 ,58	3 ,03	0 ,96

The visual sensibility curve of the eye depends on the intensity of the source. The maximum shifts from λ 5600 for very bright sources to λ 5100 for very faint sources. If the latter value is adopted as effective wave length for the Harvard magnitudes a system of temperatures results which disagrees with the values derived on the basis of WC quite beyond the reasonable errors of measurement. Although the agreement is better when the curves for brighter sources are used it does not seem proper to use an effective wave length exceeding λ 5550, and the authors have accordingly adopted this value. The value adopted by BRILL, λ 5290, is too small to fit the present data.

The table reproduced below gives the observed heat index and water-cell transmission as found from the mean curves and the temperatures of giants and dwarfs of the various spectral classes.

It is difficult to see why the heat index is not able to give temperature values for stars earlier than G5 consistent with those from the water-cell absorption.

[1] Physical Tables (7th Ed.), p. 418 (1921).

It is likely that both the assumption of black-body radiation and the adopted visual-sensibility curves are not correct. If the stars earlier than G5 have an abnormal visual radiation much greater than that corresponding to the black-body curve, as has been observed in the case of the Sun, the heat index will lead to temperatures higher than those given by the water-cell absorption.

Spectral class	Heat index	Water-cell absorption	Temperature from Heat index $\lambda = 5550$	Temperature from Heat index $\lambda = 5290$	Temperature from Water-cell absorption	Temperature from Colour index
B0	$0^m,05$	$0^m,20$				23000°
B5	0 ,01	0 ,23				15000
A0	0 ,00	0 ,26				11200
A5	0 ,02	0 ,30			7500°	8600
F0	0 ,15	0 ,36	6750°	7300°	6200°	7400
F5	0 ,30	0 ,41	5760	6160	5450	6500
gG0	0 ,47	0 ,50	5000	5450	4700	5500
gG5	0 ,65	0 ,60	4550	4870	4140	4700
gK0	0 ,90	0 ,70	4020	4300	3750	4100
gK5	1 ,57	0 ,93	3240	3480	3130	3300
gM0	1 ,86	1 ,01	3030	3250	2980	3050
gM2	2 ,2	1 ,14	2810	3000	2810	
gM4	3 ,1	1 ,30	2400	2590	2550	
gM6	4 ,2	1 ,46	2050	2200	2390	
gM8	5 ,2	1 ,62	1780	2000	2250	
Me (max)	4 ,4	1 ,5	1990	2160	2350	
Me (min)	8 ,9	2 ,2			1830	
dG0	0 ,32	0 ,42	5700	6100	5350	6000
dG5	0 ,39	0 ,47	5350	5750	4920	5600
dK0	0 ,55	0 ,54	4820	5100	4460	5100
dK5	1 ,10	0 ,76	3720	3980	3550	4400
dM0	1 ,40	0 ,87	3400	3650	3260	3400
dM2	2 ,1	1 ,14	2870	3060	2780	

In order to compute temperatures from the total radiation it is necessary to reduce the m_r to no atmosphere and determine the radiation constant of a star having $m_r = 0$. The thermocouple was calibrated with three different sources, viz: the Sun, a HEFNER lamp, and a 100-watt nitrogen-filled incandescent lamp. The energy of the Sun was reduced by spreading the light reflected from a silvered plane mirror over a circle 1 m in diameter by means of a 1-inch silvered spherical mirror of 2 inches focal length. The following results were found:

Source	Star of zero radiometric magnitude
The Sun	$16,3 \cdot 10^{-12}$ cal cm^{-2} min^{-1}
HEFNER lamp	$17,1 \cdot 10^{-12}$,, ,, ,,
Nitrogen lamp.	$18,4 \cdot 10^{-12}$,, ,, ,,

The radiometric magnitude of the HEFNER lamp at one metre is —20,00 magnitudes and of the international candle $-20^m,11$. The visual magnitude of the metre-candle is $-14^m,18$ and thus the heat index $5^M,82$ corresponding to a temperature of 1900° K which is in perfect agreement with laboratory measurements of the temperature of a yellow gas-flame.

The results obtained by the three methods gave the following equation:

$$E' = 17{,}3 \cdot 10^{-12}\, 2{,}5^{-m_r} \text{ cal cm}^{-2} \text{ min}^{-1},$$

where E' is the energy in calories received per minute by the thermocouple from each exposed square cm of the concave mirror of the telescope. For no atmosphere and a perfect reflecting telescope this equation becomes:

$$\log E = -10{,}7620 - 0{,}4\,(m_r - \Delta m_r),$$

where the reduction Δm_r to no atmosphere includes the absorption by the atmosphere and the silver surfaces of the telescope.

The value of the bolometric magnitude of a star is found from the relation:

$$m_b = m_r - \Delta m_r + (C_H + \Delta m_r)_s,$$

where $(C_H + \Delta m_r)_s$ is the heat index plus the reduction to no atmosphere for the particular star or class of stars for which the radiometric and bolometric scales of magnitudes are made to coincide. Capella A, as chosen by EDDINGTON, is not very suitable because of its uncertain magnitude and parallax. PETTIT and NICHOLSON have taken stars of gG0 as their standard. These stars gave an average zero point for m_b that differed from EDDINGTON's by about $0^{\mathrm{m}},1$.

The above equation then reduces to:

$$m_b = m_r - \Delta m_r + 0{,}9,$$

while the absolute bolometric magnitudes M_b are given by:

$$M_b = m_r - \Delta m_r + 5 \log \pi + 5{,}9 = M_v - C_H - \Delta m_r + 0{,}9.$$

The following tables quote the lists given by PETTIT and NICHOLSON of the ten brightest stars in order of their visual magnitudes, their radiometric magnitudes, and the total radiation reaching the solar system from the star.

The ten brightest stars in order of their visual magnitude.

Name	Spectral class	m_v	m_r	m_b	m_{ph}	m_{phv}	M_b
Sirius	A2s	$-1^{\mathrm{m}},58$	$-1^{\mathrm{m}},27$	$-1^{\mathrm{m}},38$	$-1^{\mathrm{m}},36$	$-1^{\mathrm{m}},40$	+1,4
Canopus	F3	−0 ,86	(−1 ,09)		−0 ,60	−0 ,85	
α Centauri	G6	+0 ,33	(−0 ,08)				
	K4	+1 ,70	(+0 ,70)				
Vega	A1s	+0 ,14	+0 ,10	+0 ,44	+0 ,09	0 ,00	+0,4
Capella	G0	+0 ,21	−0 ,38	+0 ,74 +1 ,24	+1 ,03	+0 ,05	−0,4 −0,2
Arcturus	K0	+0 ,24	−0 ,98	−0 ,46	+1 ,48	−0 ,05	−0,4
Rigel	B8p	+0 ,34	+0 ,23	−0 ,24	+0 ,13	+0 ,21	−6,1
Procyon	F3	+0 ,48	+0 ,22	+0 ,68	+0 ,85	+0 ,35	+3,1
Achernar	B5	+0 ,60	(+0 ,60)		+0 ,26	+0 ,53	
β Centauri	B1	+0 ,86	(+0 ,81)		+0 ,46	0 ,85	

The ten brightest stars in order of their radiometric magnitudes

Name	Spectral class	m_r	m_v
Betelgeuze	M2	$-1^{\mathrm{m}},67$	$+0^{\mathrm{m}},92$
Antares	M1	−1 ,32	+1 ,22
Sirius	A2s	−1 ,27	−1 ,58
Canopus	F3	(−1 ,09)	−0 ,86
γ Crucis	M3	(−1 ,0)	+1 ,61
Arcturus	K0	−0 ,98	+0 ,24
Aldebaran	K5	−0 ,60	+1 ,06
Capella	G0	−0 ,38	+0 ,21
Mira (maximum)	M6e	−0 ,2	+3 ,6
α Centauri	G6	(−0 ,08)	+0 ,33
	K4	(+0 ,70)	+1 ,70

The ten brightest stars in order of their total radiation reaching the solar system

Name	Spectral class	cal cm^{-2} min^{-1} · 10^{12}
Sirius	A2s	145
Betelgeuze	M2	132
Antares	M1	96
β Centauri	B1	(83)
Canopus	F3	(77)
γ Crucis	M3	(69)
Arcturus	K0	64
Achernar	B5	(51)
Rigel	B8p	50
Spica	B3	48

The data within parentheses have been computed from available data.

188. The Extinction. A number of determinations of the constants in the extinction formulae have been undertaken. The first to study the visual extinction was SEIDEL[1]. The observations of MÜLLER in 1879—81, the results of which are to be found in his Photometrie der Gestirne[2], have represented for a long time our best knowledge concerning the amount of the extinction. Among other determinations we may mention those of BEMPORAD[3], GUERRIERI[4], and PADOVA[5], and also MÜLLER's later determination jointly with P. KEMPF[6].

The photographic extinction was first investigated by SCHAEBERLE[7] in 1889—91 (Mt Hamilton and Cayenne). Later on WIRTZ[8] made an investigation

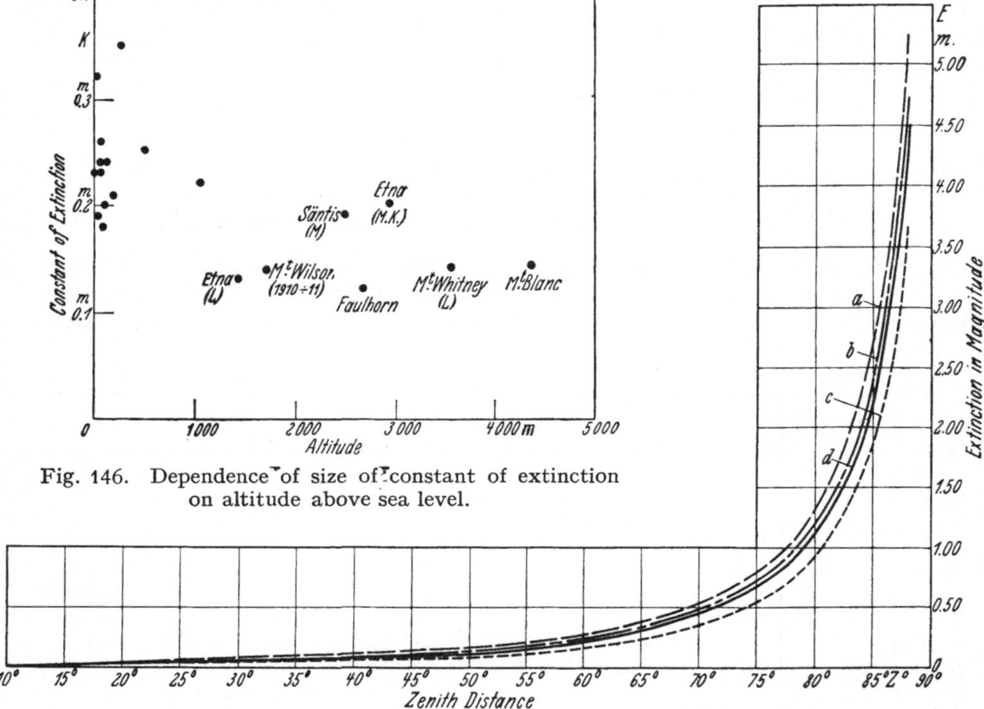

Fig. 146. Dependence of size of constant of extinction on altitude above sea level.

Fig. 147. Curves giving the value of visual extinction for different zenith distances (z) according to VAN DER LINDEN. The curve a corresponds to white stars; b to yellow stars; c to red stars; d is the mean curve.

and gave a table. Recently VAN DER LINDEN[9] has determined the amount of photographic extinction and his results are in good agreement with those of WIRTZ. In the mean VAN DER LINDEN's values are slightly higher. As a practical rule one can say that the amount of the photographic extinction is twice the amount of the visual extinction.

Recently the visual extinction has been investigated independently by G. ZIPLER[10] and by VAN DER LINDEN[11]. Already MÜLLER had found some dependence between the extinction and the colour, but could not reach quite definite

[1] München K Bayer Akad d Wiss, Abh II. Cl. Bd. 6, p. 581 (1850—52).
[2] Photometrie der Gestirne, p. 515 (1897).
[3] Rivista di Astronomia e Scienze affini, p. 580 (1912), p. 388 (1913).
[4] Rend R Acc Napoli (3), T. 19 (1913).
[5] Atti R Instituto Veneto di Science, T. 73, Parte 2a, p. 947 (1914).
[6] Potsdam Publ 8, p. 1 (1891) and 11, p. 213 (1898).
[7] Lick Observ Contr, No. 3 (1892).
[8] A N 154, p. 349 (1901).
[9] Wis- en Natuurkundig Tijdschrift Bd. 1, p. 123.
[10] Berlin-Babelsberg Veröffentl. Bd. 3, H. 2 (1921).
[11] Ann de l'obs Uccle, III Ser., T. 2, Fasc. 1 (1924).

results. ZIPLER as well as VAN DER LINDEN have constructed tables for the three main spectral groups:

B—A, F—G, K—M.

A comparison between the two sources shows good agreement. The values of VAN DER LINDEN are in the mean somewhat larger for spectra of the classes B—G, but for K—M stars the contrary is true. As long as the zenith distances are moderate ($< 50°$), one can consider it certain that the influence of the extinction can be entirely eliminated from photometric measurements, except in case of the most refined methods, e. g. the photoelectric methods where also the difference in extinction in different azimuths may enter and call for improved extinction tables in the future.

189. Concluding Remarks. It is certainly very difficult after an interesting excursion through the wide field of stellar photometry to summarize the most important of the embarras de richesse of impressions which have presented themselves.

The most important of the different sources of errors which affect the photometric estimates and measurements are those of physiological origin. Most of these errors can be overcome, but the one known as the PURKINJE function — the fact that the eye is not able to notice a difference in intensity between two sources before the ratio $\Delta i/i$ has reached a certain minimum value—can never be got over or eliminated. This means that within certain limits varying between $0^m,03$ and $0^m,10$ the eye cannot decide if two light impressions are equal or not. It seems that photometry has not paid enough attention to this very important question; it seems that the capabilities of the complex of photometric methods and devices, which are based on the principle of bringing two sources of light into equality, have been much over-estimated. Other sources of error of a physiological nature, such as the equation of position, colour equation, influencesof light-sources very close to the one to be estimated, effects of background, of the constant irritation of the nerves of the eye, effects of expectation, of imagination, and so forth, cannot be entirely eliminated, but their influence can be minimized to an amount generally smaller than the influence of $\Delta i/i$.

There exist comparatively few investigations concerning the size and behaviour of $\Delta i/i$. In fact, it does not seem even possible to make a correct statement as to which variables enter into the function $\Delta i/i$. Of course, the principal ones are known, such as intensity, colour, time, and angular distance between the light-sources, but even if we restricted our inquiry to the effect of those variables it would be difficult to obtain a satisfactory answer. In many cases nothing is known concerning the peculiarities of the eyesight of a photometric observer. The results themselves, or rather the residuals left when the material is reduced to a certain photometric scale show us the influence of the physiological errors. But in such a case other errors also enter and, besides, it will be extremely complicated, if not entirely impossible, to split up the residuals correctly into the different physiological errors. It seems as if it would be of much value if the eyes of the observers were tested now and then by the aid of the methods of photometry. To mention only one case, I have known of observers who had been at work for years without knowing that their eyes were affected by astigmatism and slight shortsightedness. This might not have affected their observational work, but still it would have been of much advantage if their eyes had been investigated earlier. It may appear a truism to point out that colour-blind observers should not be engaged in any work where the influence of colour is of importance, but still it has happened.

The comparatively small interest regarding a knowledge of the PURKINJE effect is illustrated by the fact that PURKINJE's discovery was related to surfaces. For a long time it was tacitly assumed that the effect also applied to point sources. As far as I know, the PURKINJE effect was not tested in the case of points before 1921, when GALLISSOT made an investigation of the subject in his thesis.

A determination of the size of $\Delta i/i$ should perhaps be possible by comparing results found with the aid of accurate visual photometers and a photoelectric-visual photometer, an apparatus made sensible to the light in as nearly as possible the same way as the eye.

Photoelectric and radiometric photometry are the only methods hitherto known where the physiological effects are eliminated. The method will certainly become of the utmost importance in the future, but it will take much troublesome labour before the scales are established and results reached for, say, the stars brighter than 6^m, but the work will certainly be worth while.

In cases of estimating small differences of light the eye does very well. As soon as it is a question of moderate accuracy in photometry or as soon as the effects to be determined are considerable, e. g. variations in light of high amplitude (novae, long-period variables, etc.), then the step-degree method will be of much value. The considerable gain in time without giving up much of the accuracy attainable with this method should not be forgotten. Besides, it gives a fine opportunity to the amateur to make observations of real value for science. But as soon as it is a question of high accuracy, e. g. of studying variations of small amplitude (Cepheids and other classes of variable stars, eclipsing binaries, rotating non-spherical stars, the derivation of accurate catalogues), then of course actual photometric measurements take place, and then extreme care must be given to the determination of the general accuracy, the relation to fundamental scales, and the influence exercised by the different physiological and instrumental sources of error.

Once very much was hoped for from photographic photometry. As long as no such methods as weighing the amount of reduced silver on a plate or counting the number of grains in a photographic method can be applied, photographic photometry cannot give any higher accuracy than the visual methods. The photographic plate does not register a light impression below a certain limit—also here we have a PURKINJE function—and besides when comparing densities or diameters the eye cannot decide, within a certain limit, if there is equality or inequality between the sources. In the latter case the study of the change of the diameter d of stellar images with logarithmic change of exposure $\Delta d/\Delta \log E$ ($E = It$) or ROSS's quantity astrogamma[1] is a necessary and important task. Now the construction of thermoelectric or photoelectric photometers makes it possible to measure the amount of blackening without any influence of the physiological effects. We can thus nowadays measure the density with a sufficiently high accuracy. But then we have the difficulties connected with the establishment of the relation between the intensity and the photographic density or opacity. This function is very complicated and depends on such factors as chemical properties of the film, properties of the developers, climatic conditions during the exposure, instrumental conditions (aberrations), colour equation, changes in focus during exposure, etc. Still photographic photometry has overcome the effects of most of these factors and minimized their influence. I think it can be stated that the best photographic determinations of magnitude have a (external) mean error of $\pm 0^m,08$ and that it is hardly possible to lower

[1] Ap J 52, p. 86 (1920).

the size of that error. The best visual determinations seem to have an error of the same size. The photographic methods are certainly not more accurate than the visual ones. But the photographic methods have other advantages, such as their being a record of a certain part of the sky at a given moment and the possibility of working out the results in an office, and so on.

Very little seems to be known about the persistency of astronomical negatives. Kept in dry and dark places they will probably be persistent for a very long time, but the matter ought, of course, to be tested. Also varnishing of precious plates is to be recommended.

The conception of stellar magnitude meets us for the first time in PTOLEMAIOS's famous handbook of astronomy, the Almagest. Although a physical element of the stars was thus introduced, there do not seem to have existed, so to say, astrophysical problems among the Greek thinkers. Most of them seem to have believed that the stars were hot and at large distances. In fact, the conception that the "stars are fires", as expressed in HAMLET[1] dates back to the most primitive ideas of cosmogony. In the Northern Mythology as preserved in the Eddas it is expressly stated that the stars are formed in Ginugagap from the sparks emitted from the world of fire, Muspelhem. But in spite of these conceptions the stars were also thought to be dark bodies. HUYGENS's Cosmotheoros of 1698 is the first scientific work in which the stars, for the first time, are compared with the Sun in every respect, and in which the first attempts are made to use the brightness as an indicator of the distance.

No real interest in photometry could arise before astrophysical problems existed. Besides, the Middle Ages suffered from an unhappy faith in the authorities. Although very little observation and inquiry would have sufficed to make it clear that the positions and magnitudes of the stars were not given very accurately in the old sources, scholars copied the authorities and did not observe the sky. It is not lack of activity that characterizes the high and late Middle Ages, it is principally lack of interest in original observational work.

TYCHO BRAHE had his interest directed towards some physical aspects of astronomy, but he believed in astrology, and on that account he was merely standing on mediaeval ground, although as an observer he ranks among the foremost of all times.

The 17th century is characterized by some interest for the astrophysical side of astronomy. Still, it seems that it took a rather long time before the value of the telescope was appreciated. Several leading astronomers during this century declined to use telescopic means.

It is not until the last decades of the 18th century that we meet with a real interest in astrophysics. It is mainly due to Sir WILLIAM HERSCHEL's work that astrophysics in its modern sense has developed. His importance for the development of our knowledge of stellar photometry, of the absolute magnitudes of the stars, and of the distribution of stars in space can scarcely be over-estimated.

Our knowledge of magnitudes and colours did not advance so very rapidly during the first part of the 19th century. Comparatively few contributions to photometry and very few to our knowledge of stellar colours were made. Extensive star catalogues were formed, but no effort was made to obtain decent magnitudes. They were only estimated roughly.

The advances within the said branches of astronomy during the second part of the last century are therefore so much the more wonderful. Most of what we now know about magnitudes and colours is due to that period. The general

---

[1] Act 2, Scene 2.

aim was to improve the accuracy of the methods, to increase the data, and to extend the results so as to obtain a conception of the construction of the universe that was based not mainly on speculations, but on observational data.

If we turn for a moment to the first three decades of our century, it seems that there is not quite the same interest in the further accumulation of data as existed a generation ago. Many astronomers may have the feeling that the field of stellar photometry is nearly exhausted, others are engaged in problems in astrophysics of more actual interest than those connected with photometric researches. The knowledge we now have of the absolute magnitudes and their distribution is mainly founded on work carried out during the last two decades of the 19th century. It is not astonishing if the interest in stellar photometry had to be kept waiting during the pioneer-time taken up by the absolute magnitudes of the stars.

The results of stellar photometry are certainly imposing. Our knowledge of the stars visible to the unaided eye is very extensive. For some 3000 stars we have visual observations extending over sixty years. For some 1000 of the brightest stars the oldest observations date back to antiquity and these observations, although rough and inaccurate in individual cases, fix an upper limit of the general change, $\Delta m/\Delta t$, at $0^m,0001$. The stars brighter than $7^m,5$, numbering some 40000, are accurately known as regards their visual magnitudes, thanks to the splendid work done at Harvard and at Potsdam. If we extend the limit to $9^m,0$, the remaining 120000 stars are fairly well known through the estimates in the Astronomische Gesellschaft catalogues, and the estimates in BD, CPD, and CD. When the Cordoba Durchmusterung is finished we shall possess a fair knowledge of the magnitudes of more than a million stars completely covering the interval down to 10^m and also including scores of fainter stars. Besides, the diameter-measurements or estimates in CdC give a possibility of obtaining at least rough magnitudes of stars down to $12^m,5$. When finished the CdC will extend our knowledge of the magnitudes of the stars to about three millions in number.

With regard to the photographic magnitudes it is rather remarkable that more has not been done in the field of the brighter stars. We have accurate magnitudes down to $8^m,0$ for stars in the northern zone 0°—20° (Göttinger Aktinometrie), we have also accurate magnitudes north of 73° to the pole (PARKHURST, Greenwich and Pulkova), we have also many valuable special catalogues reviewed in this chapter. But it is singular that general knowledge of the photographic magnitudes of stars brighter than $7^m,5$ is lacking. The very brightest stars have been measured at Harvard by KING. But in general the photographic magnitude of an intermediate star (brighter than $7^m,5$) cannot be found at present. If we pass over to photovisual magnitudes our knowledge is still more scanty.

For stars fainter than 10^m we have the Selected Areas of Groningen-Harvard and the extension of the Selected Areas performed at Mount Wilson by SEARES, KAPTEYN and VAN RHIJN[1], which together give accurate (in the latter case very accurate) magnitudes of 200000 stars well distributed over the sky, which are thus also of value as secondary standards.

The work concerning the primary standards is in a very satisfactory state as far as the Northern Sky is concerned. The two scales of the North Polar Sequence are now established down to 21^m and probably without appreciable systematic errors down to 18^m. Besides, there is or there will very soon be a possibility of

[1] Mount Wilson Catalogue of Photographic Magnitudes in Selected Areas 1—139 by F. H. SEARES, J. C. KAPTEYN and P. J. VAN RHIJN assisted by MARY C. JOYNER and MYRTLE L. RICHMOND. Carnegie Inst Publ 402. Washington 1930.

using the Pleiades, the Praesepe, and other areas as standards. And again, there are the Harvard secondary standards, which form an appreciable number. At the Cape work has been started on the establishment of a South Polar Sequence and final results can be expected after a few years.

However wonderful the general results won in photometric work then may be, it seems that several times there has been too little understanding of the importance of cooperation so that not enough consideration has been paid to the fact that many regions are well observed, and scarcely need any new observations, whereas other regions are very imperfectly known as regards the magnitudes. In the Pleiades region at least 20 different determinations exist for a number of stars, but in other very important cluster regions our knowledge concerning the magnitudes and colours is very imperfect indeed. It is to be hoped that the International Astronomical Union will become of much importance as an organizer of photometric and colour work. Very much has already been done by this body concerning the standardization of the fundamental photometric scales. It seems also to be an important task to reduce the observations of stars brighter than $7^{m},5$ to a common (absolute) scale, and to undertake the determination of the photographic magnitudes of these stars, to mention only two of the many desiderata of stellar photometry.

Another feature that seems to be significant of part of the photometric work is the interest in the construction of new photometers or in the derivation of new photometric methods. Of course, we need new methods and improved instrumental means. But sometimes the interest of astronomers seems to have been directed more towards methodical or instrumental questions than towards the accumulation of new data. Many constructions of photometers have not been successful; sometimes the instruments have not even been tested with regard to their general appropriateness for the designed purpose. It would take up too much space to give review in this textbook of the many methods and instruments that have not led to any results of value. I am not underestimating the importance of the technical side of photometric problems and I understand that many devices have to be failures during the course of the development of a certain branch of science. This is the course of scientific evolution. Still, I cannot help feeling that a general word of advice with regard to photometry should be: "Please, observe, observe, and observe, but do not invent new methods and instruments!" These pia desiderata can be expressed so much the more freely as both in the case of visual and photographic methods the errors seem to be practically minimized with the application of the best methods. It seems to be a safe statement that new instruments and methods cannot sensibly increase the accuracy attainable. In order to obtain much higher accuracy we have to turn to other methods, e.g. photoelectric photometry and radiometric measurements.

In the long run it is the observational data that count. We do not know anything of the methods and instruments used by PTOLEMAIOS and we probably never shall. This is regrettable, but still the observations are of high value already, and their value increases very much with increasing time.

It might seem that the coming time as far as visual and photographic photometry is concerned will be a dull, uninteresting time. More data have to be collected, there is not much room for the invention of new methods and instrumental devices, and immediate results of general interest can scarcely be expected. It is true that the accumulation of more data has to be made in order to increase our general knowledge of the magnitudes of the stars. But it would be premature to conclude that few results of immediate interest could be expected from such an accumulation. The problems of photometry are by no means exhausted, even if the highest attainable accuracy as regards m_v and m_{ph} seems to have been reached. It is enough to refer to the photometric work at Harvard, which was inaugurated through the enthusiasm of one man, E. C. PICKERING. An excursion into the Harvard Annals, that rich mine of photometric information, will show the student what a large number of problems have presented themselves during the course of the Harvard work.

There are numerous special problems within stellar photometry of much interest that can be solved by small or moderate instrumental means. We have very few determinations of the magnitudes of the components of double or multiple stars. The importance of observations of m_A and m_B or of the difference $m_A - m_B$ for the solution of a number of problems, such as the frequency of absolute magnitudes, the dispersion in absolute magnitudes, the frequency of spectral classes, etc., does not seem to have been wholly recognized. Many of the open clusters are awaiting photometric measurements. The magnitudes of the nuclei of planetary nebulae and the magnitudes of stars related to galactic nebulae (so-called nebulous stars) are very imperfectly known. What do we know about variations of very small periods in the stars? Much evidence speaks in favour of sudden changes in the starlight, but the evidence is uncertain and many more observations are needed. HERTZSPRUNG found what may be termed a blink-nova on plates of one night, which showed a variation of nearly one magnitude of the same kind as in the novae that had taken place during a few hours. It is quite possible that such changes of a periodical and non-periodical nature occur comparatively often. Many of the stars with very large proper motions or well determined parallaxes or accurate radial velocities are imperfectly known as regards their magnitudes. The same applies to stars showing peculiar spectra or unusual colours. Do the white dwarfs vary in brightness? Are the giant stars the ones that change their light and is the light of the dwarf stars of sensible constancy? The facts point in that direction, but many more observational data are needed before the question can be settled.

Here only a few photometric problems have been mentioned that can all be solved with moderate and many with small instrumental means. The determination of the total magnitudes of stellar agglomerations, such as small open clusters, globular clusters, and anagalactic nebulae could also be added.

When we pass to the giant telescopes, the number of available problems increases, of course. It is unnecessary to mention them here. I should only like to call attention to two difficult problems, as the objects are situated on the lower end of the photometric scale, namely the determination of the magnitudes of resolved stars in the anagalactic nebulae, and the determination of the total magnitudes of the clusters in other galaxies (heterogalactic clusters).

The determination of the absolute magnitude is a comparatively young branch of astronomy. It depends on the photometric scales, so that any error or pecularity in the latter is carried over into the scale of absolute magnitudes. Besides, the determination of M is based on knowledge of the parallax. The determination of that quantity is made according to direct methods or indirect ones; of the indirect methods the spectrographic ones are the most important, being based on the determination of line intensities or intensities of a certain portion of the spectrum, and on the application of empirical relations between these intensities and M, computed from trigonometric parallaxes or from moving cluster parallaxes derived by the aid of certain statistical methods. The so-called spectroscopic parallaxes are a by-product in the determination of absolute magnitudes with the aid of spectrographic methods.

Our knowledge concerning the absolute magnitudes of the stars has advanced wonderfully from 1914 up till now. The inauguration of the spectral method of ADAMS and KOHLSCHÜTTER for determining M marks the beginning of a new era in astrophysics. As much of the material available had been secured for other purposes, such as the determination of radial velocities, it is clear that a few drawbacks could not be avoided. The line intensities thus depend on the development of the plates and such a fact will cause difficulties when reduction-curves are being derived. But new data are steadily being accumulated from the point of view of the determination of the absolute magnitudes.

It seems that too little attention has sometimes been given to methodological questions relating to the spectrographic determinations of absolute magnitude. In some papers very little information is given about the reduction-curves, the differences between different plates and different observers, different pairs of lines, and so on. Very often the number of plates is not given, nor samples of the reduction-curves, nor data concerning the quality of the plates, nor the results of estimates of line intensities as compared with photometrical measurements, etc.

All determinations of absolute magnitudes are provisional as far as the reduction to the magnitude scales is concerned. More extensive parallax data will make possible a more accurate reduction to the magnitude scale. It thus seems to be a good plan to publish the estimates of the intensities themselves and the data concerning the reduction-curves. Then it will be possible to standardize the scale when more parallax measurements have been performed. It might also be worth while re-examining part of the oldest material for which few absolute-magnitude criteria were applied with regard to the criteria discovered later on.

It also seems of importance that all spectrograms already existing should be used for a derivation of absolute magnitudes. There is urgent need that the extensive collections of high dispersion spectrograms of southern stars that are in possession of two of the leading observatories should be immediately examined for absolute magnitudes.

The investigation of double stars for spectrographic parallaxes is of much importance. The physical pairs furnish the valuable criterion $\Delta m = \Delta M$, which can be applied in various ways. It will thus be possible to derive the function $f(M$, spectral class) by a process of integration, or the mass-luminosity relation, or the dispersion in M. Sometimes the criterion can also be used with advantage for the decision if a double star is a physical or optical system. Other problems of much interest are, e. g., determinations of spectrographic parallaxes of members of star clusters. Here we have the criterion $\bar{\pi} = \underline{\pi}$, which can also be applied to stars in the Magellanic Clouds, i. e. the two nearest of the anagalactic nebulae. A search for stars nearer than, say, 10 parsecs is also an important task, because a thorough knowledge of the stars within a certain element of space is invaluable when questions relating to the luminosity law, the spectral-luminosity law, or the mass-luminosity relation, etc. are being dealt with, etc. Investigations of stars nearer than 10 parsecs can be substituted by investigations of the stars seen in front of a certain (adjacent) dark nebula. The fact that many of these objects are comparatively near will give statistical populations although of not too many members still appropiate for a detailed investigation with regard to the values of M of different spectral classes.

(To be continued.)

Corrections and Addenda to Chap. 4.

p. 246. After the table insert:

In the above table the columns headed I, II, ..., VI give the values of the symbols derived from each of HERSCHEL's six catalogues, whereas the corresponding columns headed n give the numbers of stars in each catalogue which PICKERING used for his comparisons. The column headed All gives the values deduced from all six catalogues taken together, and the last column the adopted mean.

p. 306. After line 35 from above insert:

Further the time corresponding to the optimal intensity and required to produce the desired actinic effect is denoted by t_0. Likewise a quantity τ is introduced defined by

$$\tau = t/t_0.$$

p. 309. Line 7 from above read:

The numerical value of C is 3,51. From this formula the values of $\log\tau$ corresponding to those values of t which correspond to given values of D, are computed. Then using table IX of the mentioned Cape Zone Catalogue, which table gives the relation between $m - m_0$ and $\log\tau$, the magnitudes m corresponding to given values of D are found.

p. 309. After the table insert:

It should be remarked that the negative values of the diameter belong to the faintest stars. These stars giving images of sensibly equal dimensions but showing marked contrast in the degree of intensity, an estimate of the intensity of their images was recorded. Five different degrees of density were recognised and they were indicated by the numbers $-1, -2, \ldots, -5$; -1 indicates that the star image only just failed to attain the full density of the images of the brighter stars on the plate and -5 refers to the faintest stars whose images were distinctly visible.